Vibrational Excitations in Multilayer Nanostructures

Properties and manifestations

Stepan I Beril

Department of Fundamental Physics, Electronics and Communication Systems, T G Shevchenko Pridnestrovian State University, Tiraspol, Republic of Moldova

Vladimir M Fomin

Leibniz Institut für Festkörper und Werkstoffforschung Dresden, Dresden, Germany

and

State University of Moldova, Chişinău, Republic of Moldova

Alexander S Starchuk

Department of Fundamental Physics, Electronics and Communication Systems, T G Shevchenko Pridnestrovian State University, Tiraspol, Republic of Moldova

IOP Publishing, Bristol, UK

ISBN 978-0-7503-6164-4 (ebook)
ISBN 978-0-7503-6162-0 (print)
ISBN 978-0-7503-6165-1 (myPrint)
ISBN 978-0-7503-6163-7 (mobi)

DOI 10.1088/978-0-7503-6164-4

Version: 20241201

IOP ebooks

British Library Cataloguing-in-Publication Data: A catalogue record for this book is available from the British Library.

Published by IOP Publishing, wholly owned by The Institute of Physics, London

IOP Publishing, No.2 The Distillery, Glassfields, Avon Street, Bristol, BS2 0GR, UK

US Office: IOP Publishing, Inc., 190 North Independence Mall West, Suite 601, Philadelphia, PA 19106, USA

Vibrational Excitations in Multilayer Nanostructures

Properties and manifestations

Online at: https://doi.org/10.1088/978-0-7503-6164-4

In blessed memory of the outstanding scientist teacher and friend

Professor Evgeny Petrovich Pokatilov,

whose ideas became the conceptual basis of the present book.

Contents

Preface

The most incomprehensible thing about the Universe is that it is comprehensible.
—Albert Einstein

The book is dedicated to theoretical studies of the role of fast polarization (plasmons of valence and free electrons) and slow polarization (polar optical phonons) in the formation of polaronic, exciton, bipolaronic, and polariton states in multilayer structures of various geometries: planar, cylindrical, spherical, and in superlattices.

The research is based on the theory of potential and collective states developed for these purposes, taking into account and without taking into account the delay, which made it possible to obtain spectra of vibrational excitations in these structures. A rigorous derivation of the exact Hamiltonians of the electron–phonon interaction in such structures is presented, specifically taking into account the influence of boundaries, material and geometric parameters of layers, and types of polarization. This made it possible to correctly describe the interaction of free charge carriers with all types of polarization and generalize the Pekar–Froelich theory to the case of polarons in arbitrary multilayer structures; describe the state of the plasmon polaron (a generalization of Toyozawa theory) by constructing a quantum theory of potential and image forces; and generalize the theories of Haken and Mayer of polar excitons in infinite polar crystals in the case of contact between two polar crystals and multilayer polar structures. Based on the exact Hamiltonian of the electron–phonon interaction for heterogeneous structures, the theory of large-radius bipolarons was developed, and the possibility of bipolaronic high-temperature superconductivity in structures of the Ginzburg sandwich type was investigated: monolayer FeSe/$SrTiO_3$;FeSe/SrTi; $SrTiO_3$/monolayer FeSe/$SrTiO_3$. The application of the exact Hamiltonian of the electron–phonon interaction has proved fruitful for the analysis of other relevant phenomena, such as thermo-autoelectronic emission. In particular, it was shown that in the Fowler–Nordheim and Richardson–Schottky formulas in super-strong electric fields, the output operation becomes dependent on the electric field strength.

One of the most significant manifestations of the potential theory in multilayer planar systems and the theory of collective states in such systems based on it is the theoretical prediction followed by experimental detection of the phenomenon of propagation in superlattices of vibrational excitations of a new type, called in the works of the authors surface spatially extended (P3) polaritons (spatially extended interface polaritons SEIP). This served as the basis for the development of new directions in solid-state theory: the theory of spatially extended optical vibrational excitations and the theory of their interaction with charge carriers in superlattices. The study of phenomena caused by the existence of P3 polaritons contributed to the expansion of fundamental and applied tasks.

This short list of the results given in the book, which are based on exact solutions of the Poisson equation for arbitrary multilayer structures, indicates the important and decisive role of the crystal surface and the geometric boundaries of the layers of multilayer structures, generating a variety of new properties of these materials.

The present book is intended for specialists in solid-state physics, micro-nanoelectronics, and optoelectronics, as well as for graduate students and undergraduates.

Stepan I Beril, Vladimir M Fomin, and Alexander S Starchuk

Acknowledgements

In the course of our work, discussions of specific problems and theories included in the monograph with colleagues of physics departments and scientific laboratories of the T G Shevchenko Pridnestrovian State University, the State University of Moldova, the Institute of Applied Physics of the Academy of Sciences of the Republic of Moldova as well as presentations at seminars at the Faculty of Physics of the Lomonosov Moscow State University, at the Ioffe Institute of Physics and Technology and the Lebedev Physical Institute of the Russian Academy of Sciences, at the Institute of Theoretical Physics and the Institute of Semiconductors of the Academy of Sciences of Ukraine, at the Martin Luther University of Halle-Wittenberg and the Max Planck Institute for Solid State Research (Stuttgart) proved fruitful and important.

V M Fomin is deeply grateful to the late Prof. Manuel Cardona, Prof. Lutz Wendler, Prof. Friedhelm Bechstedt, Prof. Carlos Trallero-Giner, and Dr. Sergei N. Klimin for valuable discussions on the subject of this book.

We are grateful to the researchers, graduate students, and students of the T G Shevchenko Pridnestrovian University, who creatively listened and commented on the special courses written based on individual sections of this book.

We also express our deep gratitude to Olga Abahina, coordinator of the Center for English Language and American Culture, and Elena Polshakova, Chief specialist of the General Department of T G Shevchenko Pridnestrovian State University, for their fruitful participation in the preparation of the manuscript of this book for publication. The authors express their heartfelt gratitude to Ms. Phoebe Hooper and Ms. Emily Tapp for their valuable help in preparing this book for publication.

Author biographies

Stepan I Beril

Stepan I Beril received PhD and Doctor of Science degrees from the Institute of Applied Physics Academy of Moldova in 1979 and 1991, respectively. He is currently President of T G Shevchenko Pridnestrovian State University, Head of Department of Fundamental Physics, Electronics and Communication Systems, and Head the scientific laboratory 'Polaron'. He has authored and co-authored more than 300 scientific articles and five monographs. Doctor Beril has been a member of Russian Academy of Natural Science since 1998. He received a diploma of a Scientific Discovery of the *Phenomenon of the Propagation of Spatially-Extended Interface Phonon Polaritons in Composite Superlattices* (Academy of Natural Sciences of Russia, 1999, together with E P Pokatilov and V M Fomin) and Medal 'Academician P L Kapitsa' (Academy of Natural Sciences of Russia, 2000). His current research interests include the physics of surface of solid state and physics of multilayer structures of low dimension.

Vladimir M Fomin

Vladimir M Fomin, Research Professor at the Leibniz Institute for Solid State and Materials Research (IFW) Dresden and Professor at the State Univeresity of Moldova, Chişinău. PhD in theoretical physics (Chişinău, State University of Moldova, 1978). Dr habil. in physical and mathematical sciences (Academy of Sciences of Moldova, 1990). University Professor in Theoretical Physics (State University of Moldova, 1995). Member of APS, German Physical Society, European Physical Society, Physical Society of the Republic of Moldova, IEEE, 'Superconducting Nanodevices and Quantum Materials for Coherent Manipulation' COST Action (European Cooperation in Science and Technology), Mediterranean Institute of Fundamental Physics. Scientific Sectional Editor of the Encyclopedia of Condensed Matter Physics, 2nd edn (Oxford: Elsevier, 2024). State Prize of Moldova (1987). Research Fellow of the Alexander von Humboldt Foundation (Martin-Luther-University of Halle-Wittenberg, 1993–94). Diploma of a Scientific Discovery of the *Phenomenon of the Propagation of Spatially-Extended Interface Phonon Polaritons in Composite Superlattices* (Academy of Natural Sciences of Russia, 1999, together with E P Pokatilov and S I Beril). Medal 'Academician P L Kapitsa' (Academy of Natural Sciences of Russia, 2000). Honorary Member of the Academy of Sciences of Moldova (2007). Outstanding Reviewer of APS (2023). Medal 'Dimitrie Cantemir' (Academy of Sciences of Moldova, 2023). Research interests in nanophysics: topological effects in quantum rings, topology- and geometry-induced properties of 3D superconductor micro- and nanoarchitectures and patterned superconductors, topological states of light and spin-orbit coupling in optical microcavities, optical properties of quantum dots, superconducting properties of metallic nanograins, thermoelectric properties of semiconductor nanostructures, quantum transport in sub-0.1 μm semiconductor devices, spin-dependent phenomena in semiconductor micro- and nanoparticles, phonons, vibrational excitations and polaronic effects in nanostructures, polarons in transition metal dichalcogenides, propulsion mechanisms of catalytic tubular micromotors, theory of self-propelled micromotors for cleaning polluted water. 6 monographs, including '*Self-rolled Micro- and Nanoarchitectures: Effects of Topology and Geometry* (De Gruyter, 2021); *Physics of Quantum Rings* (Editor) (Springer, 2014; 2nd edn, Springer International Publishing, 2018), 3 textbooks, 14 review papers, 10 patents and 222 scientific articles. 5984 citations, h-index: 38 (Google Scholar as of 24 October 2024).

Alexander S Starchuk

Alexander S Starchuk defended his PhD thesis for the title of Candidate of Physico-Mathematical Sciences in the specialty 'Physics of Semiconductors' in 2006 at the Faculty of Physics of Lomonosov Moscow State University (Russia) and currently he is an associate professor at the Department of Fundamental Physics, Electronics and Communication Systems of the T G Shevchenko Pridnestrovian State University. Starchuk is the author and co-author of about 30 scientific articles, two textbooks, and one monograph. His research interests lie in the field of physics of multilayer structures and nanophysics.

Introduction

The idea of enriching the properties and expanding the functionality of electronic devices by using compositions of known semiconductor materials in the mid-twentieth century turned out to be extremely fruitful [1]. The wide range of various solid-state electronics devices created on this basis (diodes, transistors, thyristors, etc.) played a revolutionary role in the development of instrumentation. That provided a powerful acceleration to semiconductor physics and the physics of contact phenomena [2–4].

The progressing the manufacturing technology of composite semiconductor structures made it possible to obtain first thin films on substrates (single quantum wells), and subsequently thin-film multilayer structures and superlattices [5–7].

A wide range of semiconductor materials and the geometries of the system (the number of layers, their thicknesses, and their location in multilayer structures) made it possible to manufacture structures unique in their properties with characteristics unattainable for a homogeneous substance. Thus, Esaki and Tsu [8] theoretically justified the possible use of superlattices as a material with negative differential resistance. Soon after that Zh I Alferov *et al* [9] manufactured superlattices based on GaP_xAs_{1-x} with a split conduction band.

The development of technology for obtaining perfect thin-film structures and superlattices [10, 11] led to a rapid increase in scientific research in this field and the accumulation of information about the properties of multilayer structures and superlattices. The idea of a deep qualitative change in the electronic properties of a stratified (e.g., by using a powerful ultrasonic wave) system (artificial superlattice) was expressed in 1962 by L V Keldysh [12]. In composite multilayer structures and superlattices, changing from layer to layer of (i) elastic constants leads to the 'fragmentation' of acoustic modes [13, 14], and (ii) dielectric permittivity and optical phonon frequencies causes the 'capture' of optical modes [15, 16].

Due to the presence of interface boundaries, vibrations of a new type occur as surface vibrations [17] in multilayer structures, surface spatially extended vibrations [18, 19] in infinite superlattices, and surface spatially decreasing vibrations [20, 21] in semibounded superlattices.

Due to the spatial variability of the composition, the additional potential in multilayer structures and superlattices leads to a radical restructuring of the energy spectrum of charge carriers, especially significant for energies close to the edges of the former bands. Levels of size quantization of electrons and holes are formed in the layers, splitting into minibands when decreasing the thickness of the layers and increasing their number.

The quantum structure of the electronic spectrum gives the voltage characteristics in multilayer structures and superlattices an oscillating character (resonant tunneling [22, 23]), which made multilayer structures and superlattices important materials for the manufacturing of amplifiers and generators of electromagnetic energy controlled by low-voltage sources.

An outstanding breakthrough was a discovery of the effect of stabilizing the Hall resistance of a two-dimensional electron gas at discrete magnetic field values determined by a constant fine structure and the speed of light (Quantum Hall Effect) [24]. Due to the spatial separation of carriers in multilayer structures and superlattices, concentration relaxation processes acquire a peculiar character; in particular, nonequilibrium states may be frozen for a long time. The 'fragmentation' of the energy spectrum of charges into levels of size quantization and minibands leads to oscillatory dependencies of the absorption and reflection coefficients on the frequency of light.

The position and shape of the pass bands are easily controlled by the geometry of multilayer structures and superlattices, which allowed for the manufacturing of strip filters, mirrors, and antireflection coatings. Due to the high values of the conversion coefficient of light from the visible to the IR range in Raman scattering, the possibility of creating IR lasers based on this effect (Raman lasers) was unveiled [25]. The replacement of a single-layer structure with a multilayer one in the manufacturing of lasers (multiheterolasers) led to a significant improvement in their characteristics: a decrease in the beam divergence, a decrease in the threshold currents, and an improvement in the temperature dependence [26–28].

The was a particular interest in exciton effects in multilayer structures of various geometries and planar superlattices. It turned out that with a suitable choice of geometry and structure parameters, the binding energy of the Wannier–Mott exciton, weakly bound in a massive sample, can be increased by many times, and the exciton spectra can be observed at room temperatures. The 'compression' of the exciton in the layer increases the probability of its recombination and increases the intensity of the radiation bands [29–31]. Biexciton states [32] and other exciton complexes [33] are also more easily observed in thin layers and quantum wells. However, the intensity of impurity transitions in multilayer structures and superlattices is lower than that of exciton transitions due to a strong change in the symmetries of the ground and excited states. A strong electric field in a multilayer structure does not destroy the exciton, since the mutual separation of charges (electron and hole) is limited by the thickness of the layer.

Another interesting phenomenon is important for understanding the nature of high-temperature superconductivity (HTSC) in multilayer structures formed by alternating thin polar and nonpolar semiconductor layers. It is the change in the Coulomb interaction of the charge carriers of the same sign from repulsion to attraction and the occurrence of bipolaron states with high binding energy [34–37].

The above overview allows us to get an idea of the richness and high efficiency of controlling the physical properties of multilayer structures and superlattices. Therefore, it is not surprising that these structures first became the focus of fundamental and applied scientific research, and subsequently the basis for creating new element base for solid-state nanoelectronics. In the reviews by A P Silin [38] and A Ya Shik [39], the problems of the structure of band as well as optical and acoustic properties of superlattices are well represented, and the transport phenomena are described. It is noteworthy that thin-film layers, multilayer structures, and superlattices are examples of objects of a new branch of condensed matter physics—

mesoscopic physics [40, 41]. Their common feature is the fragmentation of a physical system into regions of intermediate scales, which, on the one hand, are much larger than atomic ones, in which a macroscopic description can still be used, and, on the other hand, small enough (less than the free path of the carriers) that the effects of size quantization are manifested. Examples of objects of mesoscopic physics are also filamentous structures fabricated of quantum wells (so-called quantum wires) [42], which are quasi-one-dimensional (1D) structures, and Watanabe's 'superatoms' [43]. Monographs by Y Imri [44] and M Cardona and P Yu [45] as well as collective publications [46] edited by L Challis and [47] edited by J-P Leburton amply represent studies of 'mesoscopic' systems (with sizes from a few to hundreds of nanometers), in which quantum effects play an essential role.

Despite the extensive literature relating to multilayer structures of various geometries and superlattices, several important problems related to vibrational excitations and the electron–phonon interaction [37] in multilayer structures and superlattices need to be consistently considered and described. In some monographic publications [48–50, 55], this deficiency is to some extent compensated. The books by M Strosio and M Dutta [50, 51] set out the theories necessary for understanding the properties of nanostructures, studying phonons and effects with their participation. The book by A I Gusev [48] outlines the current state of research on nanocrystalline materials and analyzes model representations explaining the structural features of substances in the nanocrystalline state. The book by V V Pogosov [49] represents the research of surface and size effects and low-dimensional systems.

The authors of the present monograph have set themselves the task of filling in the insufficiency of existing theories for describing the electrical and optical properties of multilayer systems in what concern the following issues. Firstly, the analysis of the vibrational spectra of multilayer systems, the identification of a new type of oscillations in infinite and limited superlattices, the derivation of potential energies of self-action and the electron-vibrational interactions, and derivation of a Hamiltonian of the Pekar-Fröhlich type for an arbitrary 2-layered structure. Secondly, the analysis of electronic states in quantum wells, taking into account the change in the profile of their walls by the effect of self-action, the polaron renormalization of the effective masses of electrons and holes in spatially limited systems of various types, the conclusions based on the quasiclassical limit of quantum mechanics, the potential energy of the image forces, the establishment and quantitative description of various size effects, and quantum mechanical problems concerning the states of charge carriers, excitons, biexcitons, and bipolarons in quantum wells and in structures of various geometries, for which the exact Hamiltonian was derived by some of the present authors. This Hamiltonian is also necessary for solving optical and transport problems, such as Raman scattering of light in semi-infinite superlattices, scattering of electrons by polar optical phonons in structures with quantum wells, and infrared absorption of light by free charge carriers during inelastic scattering by optical phonons.

It should be emphasized that the theory based on the continuum approach proved to be heuristically valuable not only in the eighties of the twentieth century, but also for qualitatively new systems studied in the present monograph: multilayer structures of different geometries with the manifestation of quantum size effects. The validity and criteria for the applicability of the continuum theory were determined based on comparing the results of the theory with the corresponding results of experimental studies and obtaining quasiclassical limits from the quantum mechanical description of the systems under study.

In chapter 1, a theory of potential is constructed for planar multilayer systems with an arbitrary number of polar layers by solving the electrostatic Poisson equation with the appropriate boundary conditions. Potentials are found, in general, due to the following sources of fields: charges on the interface of the structure, charges in the bulk of layers, polarization of matter, and external stress sources. On their basis, exact Hamiltonians were derived describing the interaction of charges in these systems and necessary for the study of charge transfer processes in the multilayer systems under consideration, including in quantum structures such as 'Ginzburg sandwiches', in which high-temperature superconductivity was experimentally detected [52]. Methods for calculating potentials for various relevant cases have been developed.

In chapter 2, a theory of potential in multilayer structures of cylindrical and spherical geometries is constructed based on solving a similar problem.

In chapter 3, a set of equations for polarization oscillations ('slow' and 'fast') is generally formulated and solved without taking into account retardation in multilayer structures and superlattices, normal bulk and surface modes are found. The Hamiltonian of the electron-polarization interaction in multilayer structures and superlattices is explicitly obtained. In particular, in superlattices it describes the interaction of an electron with bulk ('confined' and 'free') phonons, as well as the surface spatially extended modes predicted by some of the authors of the present monograph and subsequently experimentally discovered. The influence of boundary conditions in layers on the spectrum of bulk vibrations is analyzed. The Hamiltonians of the interaction of an electron with optical vibrations in multilayer cylindrical and spherical systems are explicitly obtained.

In chapter 4, the wave equations describing the propagation of polaritons in multilayer structures and superlattices are solved. A consistent classification of polariton waves is performed according to the spatial distribution of electric and magnetic field strengths in the layer, as well as by changes of these fields over the period of the structure. The dispersion laws for surface spatially extended and surface spatially decreasing polaritons are derived, and the optical properties of multilayer structures and superlattices (reflection and transmission of light) are analyzed.

In chapter 5, the theory of various types of polarons in planar systems is developed: surface, near-surface, film, external, levitating, electronic and others, and criteria for the existence of surface polaron states are investigated. The quantum theory of the image potential is presented and the dielectric function of a quantum dielectric is derived based on the polaron theory for planar multilayer systems. The potential energy of self-action is found and the polaron states in multilayer cylindrical and spherical structures are investigated. Polaron states in homogeneous electric and magnetic fields in planar, cylindrical, and spherical nanostructures are studied.

The specific features of the electron–phonon interaction in superlattices are studied, of which the most important is the occurrence of new elementary excitations surface spatially extended optical vibrations in a composite superlattice consisting of alternating layers of two types. The dispersion law for interacting surface modes is established. The Hamiltonian of the electron–phonon interaction is found and the polaron states in a composite superlattice, in nanoscale cylindrical and spherical structures are studied, and the polaron states in homogeneous electric and magnetic fields are analyzed. The binding energy and the effective mass of a magnetopolaron in a quantum wire embedded in a dielectric medium are obtained.

In chapter 6, investigation of electron and exciton states in multilayer homopolar systems is carried out based on a development of the theory of Wannier–Mott exciton states in composite superlattices. The effective Hamiltonian of the electron–hole interaction at the contact of two polar crystals is derived and the polaronic exciton states within the Haken ($R_{ex} > R_p$) and Mayer ($R_{ex} < R_p$) models are investigated, where R_{ex} and R_p are the radius of the exciton and the electron (hole = polaron), respectively. The developed theory of polaronic excitons is used to interpret the results of experimental studies of Wannier–Mott excitons in thin films PbI_2 and CdTe.

The biexciton states and exciton complexes on the surface of a polar crystal are considered. The effect of a strong magnetic field on the energy of a polaronic exciton in a quantum well is investigated. Vanier–Mott exciton states in polar semiconductor quantum wires are theoretically analyzed, and the experimental results are compared with the theory

In chapter 7, large-radius bipolaron states in multilayer polar planar and cylindrical structures are investigated. The Hamiltonian of the electron–phonon interaction is derived for two cases: (i) in a quantum layer (δ-layer), separating semi-infinite polar crystals, and (ii) in quantum layers (δ-layer), spatially separated by a polar layer (interlayer bipolaron states) [34].

The Hamiltonian of the electron–phonon interaction in a quantum wire in a polar medium is derived, based on which the effective potential energy of the electron–phonon interaction and the binding energy of a bipolaron in a quantum wire are found.

A study of bipolaron states in inhomogeneous systems is carried out for the above cases. It is established that significantly more favorable conditions for the formation of bipolarons arise in the composite structures under consideration due to the possibility of an independent choice of polarization parameters of neighboring media, which are responsible for the formation of bipolarons.

The possibility of the occurrence of high-temperature bipolaron superconductivity in two structures of the ‘Ginzburg sandwich’-type is investigated: FeSe (monolayer)/$SrTiO_3$, $SrTiO_3$/FeSe (monolayer)/$SrTiO_3$. The critical temperature (T_c) in the studied multilayer structures is estimated.

In chapter 8, results of applying the theory of the electron–phonon interaction in arbitrary multilayer structures to the study of several transport effects in structures with quantum wells are presented: electron scattering on surface optical phonons in inversion layers of MDS (Metal–Dielectric–Semiconductor) structures: scattering of charge carriers in polaron semiconductor quantum wells with participation of surface and bulk optical phonons, and cyclotron–phonon absorption of electromagnetic radiation in thin semiconductor polar films in a quantizing magnetic field.

A study of emission processes from metal–dielectric film–vacuum structures is carried out. Based on it, a generalization of the theory of thermo-autoelectronic emission in MDS structures is carried out in case of manifestation of the quantum nature of image forces under conditions close to atomic electric fields.

In chapter 9, the theory of Raman scattering of light by phonons in multilayer structures with an arbitrary number of layers and in superlattices is developed based on the modulation transmission method presented in chapter 1 in the theory of potential in multilayer systems of planar geometry. In the previous chapters, it is shown that in multilayer structures, there occurs a radical restructuring of the energy spectrum of phonons and other vibrational excitations, compared with those spectra in massive crystals of the same materials, of which individual layers constitute. In particular, 'confined' bulk and spatially extended surface optical phonons arise in superlattices [18, 19, 21], the existence of which is discovered experimentally and interpreted theoretically [53–56], through the Raman light spectra.

References

[1] Shockley V 1953 *Theory of Electronic Semiconductors* (Moscow: IL) p 714

[2] Zi M 1973 *Physics of Semiconductor Devices* (Moscow: Energiya) p 665

[3] Mills A and Feucht D 1975 *Heterojunctions and Metal-Semiconductor Transitions* (Moscow: Mir)

[4] Semiconductors in Science and Technology 1957 (Moscow: Publishing House of the CCCP Academy of Sciences) *T. 1* **2** 470 658 p

[5] Ando T, Fowler A and Stern F 1985 *Electronic Properties of Two–Dimensional Systems* (Moscow: Mir) p 415

[6] Foreign Electronic Technology 1981 *Electronics* (Moscow: Central Research Institute) p 65

[7] Esaki L 1986 A bird's-eye view on the evolution of semiconductor superlattices and quantum wells *IEEE J. Quantum Electron.* **22** 1611–24

[8] Esaki L and Tsu R 1970 Superlattice and negative differential conductivity in semiconductors *IBM J. Res. Dev.* **14** 61–5

[9] Alferov Zh I, Zhilyaev Yu V and Shmartsev Yu V 1971 Splitting of the conduction band in a 'superlattice' based on GaP_xAs_{1-x} *FTP* **5** 196–8

[10] Gaponov S V, Luskin B M and Salashchenko N N 1979 On the possibility of obtaining superlattice structures by laser sputtering *Lett. ZhTF* **5** C 516–21

[11] Razeghi M and Duchemin J P 1984 MOCVD growth for heterostructures and two-dimensional electronic systems springer ser *Sol. State Sol.* **53** 100–14

[12] Keldysh L V 1962 On the effect of ultrasound on the electronic spectrum of a crystal *FTT* **4** 2265–7

[13] Rytov S M 1956 Acoustic properties of a finely layered medium *Acoust. J.* **2** 71–83

[14] Merlin R, Colvard C, Klein M V *et al* 1980 Raman scattering in superlattices: anisotropy of polar phonons *Appl. Phys. Lett.* **36** 43–5

[15] Barker A S, Mers J L and Gossard A C 1978 Study of zone-folding effects on phonons in alternating monolayers of GaAs–AlAs *Phys. Rev.* **B 17** 3171–9

[16] Colvard C, Gant T A, Klein M V *et al* 1985 Folded acoustic and quantized optic phonons in GaAs/AlAs superlattices *Phys. Rev. B.* **31** 2080–91

[17] Fuchs R and Kliewer K L 1965 Optical modes of vibration in an ionic crystal slab *Phys. Rev.* **140** A2076–88
[18] Pokatilov E P and Beril S I 1983 Electron–phonon interaction in periodic two-layer structures *Phys. Stat. Sol.* B **118** 567–73
[19] Pokatilov E and Beril S I 1982 Spatially extended optical interface modes in a two-layer periodic structure *Phys. Stat. Sol.* B **110** K75–78
[20] Camley R E and Mills D L 1984 Collective excitations of semi-infinite superlattice structures: surface plasmons, bulk plasmons, and the electron-energy-loss spectrum *Phys. Rev.* B **29** 1695–706
[21] Fomin V M and Pokatilov E P 1986 Optical properties of multi-layer structures. I. Polaritons *Phys. Stat. Sol.* B **136** 187–99
[22] Aleksanyan A G, Belenov E M and Companets I N 1982 Investigation of metal-barrier-metal-barrier-metal structures *ZHETF* **83** 1389–97
[23] Chang L, Esaki L and Tsu R 1974 Resonant tunneling in semiconductor double barriers *Appl. Phys. Lett.* **24** 593–5
[24] Klitzing K, Dorda G and Pepper M 1980 New method for high-accuracy determination of the fine-structure constant based on quantized hall resistance *Phys. Rev. Lett.* **45** 494–7
[25] Tsang W T 1979 Low-current-threshold and high-lasing uniformity GaAs–$Al_xGa_{1-x}As$ double-heterostructure lasers grown by molecular beam epitaxy *Appl. Phys. Lett.* **34** 473–5
[26] Holonyak N, Kolbas R M, Laidig W D *et al* 1978 Low-threshold continuous laser operation (300–337 K) of multilayer MO-CVD $Al_xGa_{1-x}As$ – GaAs quantum-well heterostructures *Appl. Phys. Lett.* **33** 737–9
[27] Miller R C, Dingle R, Gossard A C *et al* 1976 Laser oscillation with optically pumped very thin GaAs – $Al_xGa_{1-x}As$ multilayer structures and conventional double heterostructures *J. Appl. Phys.* **47** 4509–17
[28] Tsang W T, Weisbuch C, Miller R C and Dingle R 1979 Current injection GaAs – $Al_xGa_{1-x}As$ multi-quantum-well heterostructure lasers prepared by molecular beam epitaxy *Appl. Phys. Lett.* **35** 673–5
[29] Calman E V, Dorow C J, Fogler M M, Butov L V, Hu S, Mishchenko A and Geim A K 2016 Control of excitons in multi-layer van der Waals heterostructures *Appl. Phys. Lett.* **108** 101901
[30] Malic E, Perea-Causin R, Rosati R, Erkensten D and Brem S 2023 Exciton transport in atomically thin semiconductors *Nat. Commun.* **14** 3430
[31] Lee H, Kim Y B, Ryu J W, Kim S, Bae J, Koo Y, Jang D and Park K-D 2023 Recent progress of exciton transport in two-dimensional semiconductors *Nano Converg.* **10** 57
[32] Glutsch S 2004 *Excitons in Low-Dimensional Semiconductors: Theory, Numerical Methods, Applications* (Berlin: Springer) p 298
[33] Molas M R 2023 Excitons and phonons in two-dimensional materials: from fundamental to applications *Nanomaterials* **13** 3047–9
[34] Pokatilov E P, Beril S I, Fomin V M and Ryabukhin G J 1992 Polaron pairing in multi-layer structures. Part 1. Bipolaron states in multi-layer structures with quantum wells *Phys. Stat. Sol.* B **169** 429–41
Pokatilov E P, Beril S I, Fomin V M, Yu. Riabukhin G and Gorjachkovskii E R 1992 Polaron pairing in multi-layer structures. Part 2. Interlayer bipolaron states in structures with quantum wells *Phys. Stat. Sol.* B **171** 437–45

[35] Beril S and Starchuk A 2023 On the bipolaronic mechanism of high-temperature superconductivity in 'ginzburg sandwiches' FeSe-$SrTiO_3$; $SrTiO_3$-FeSe-$SrTiO_3$ *Am. J. Phys. Appl.* **11** 8–20

[36] Sadovsky M V 2016 High-temperature superconductivity in monolayers FeSe *Uspekhi Fiz. Nauk* **186** 1035–57

[37] Alexandrov A S and Devreese J T 2009 *Advances in Polaron Physics. Springer Series in Solid State Science* (Berlin: Springer) p 165

[38] Silin A P 1985 Semiconductor SR *UFN* **147** 485–521

[39] Shik A Y 1974 Superlattices periodic semiconductor structures *FTP* **8** C 1841–64

[40] Andryushin E A and Bykov A A 1988 From superlattices to superatoms *UFN* **154** 123–32

[41] Schwarzschild B 1986 Currents in normal-metal rings exhibit Aharonov–Bohm effect *Phys. Today* **39** 17–20

[42] Petroff P M, Gossard A C, Logan R A and Wiegman W 1982 Toward quantum well wires: fabrication and optical properties *Appl. Phys. Lett.* **41** 635–8

[43] Watanabe H and Inoshita T 1986 Electronic structure and various types the realization of the superatom optoelectron *Dev. Technol.* **1** 33–8

[44] Imri Y 2002 *Introduction to Mesoscopic Physics* (Moscow: Fizmatlit) p 304 Translated from the English edition
Imry Y 2002 *Introduction to Mesoscopic Physics* (Oxford: University Press)

[45] Yu P Y and Cardona M 2010 *Fundamentals of Semiconductors: Physics and Materials Properties* (Berlin: Springer) p 775

[46] Challis L (ed) 2003 *Electron–Phonon Interaction in Low-Dimensional Structures* (Oxford: Oxford University Press)

[47] 1993 *Phonons in Semiconductor Nanostructures* Series E: Applied Sciences vol 236 ed J-P Leburton, J Pascual and C S Torres (Dordrecht: Springer)

[48] Gusev A I 2005 *Nanomaterials, Nanostructures, Nanotechnology* (Moscow: Fizmatlit) p 410

[49] Pogosov V V 2006 Introduction to the physics of charge and dimensional effects *Surface, Clusters, Low-dimensional Systems* (Moscow: Fizmatlit) p 328

[50] Strosio M and Dutta M 2006 *Phonons in Nanostructures* (Moscow: Fizmatlit) p 319 translated from the English edition
Stroscio M S and Dutta M (ed) (Cambridge: Cambridge University Press) p 2001

[51] Ridley B K 2017 *Hybrid Phonons in Nanostructures* (Oxford: Oxford University Press) p 192

[52] Lee J J *et al* 2014 Interfacial mode coupling as the origin of the enhancement of in FeSe films on *Nature* **515** 245–8

[53] Klein M V 1986 Phonons in semiconductor superlattices *IEEE J. Quantum Electron.* **QE–22** 1760–70

[54] Sood A K, Menendez J, Cardona M and Ploog A K 1985 Interface vibrational modes in GaAs–AlGaAs superlattices *Phys. Rev. Lett.* **54** 2115–8

[55] Miller R C, Dingle R, Gossard A C *et al* 1976 Laser oscillation with optically pumped very thin GaAs–$Al_xGa_{1-x}As$ multilayer structures and conventional double heterostructures *J. Appl. Phys.* **47** 4509–17

[56] Trallero-Giner C, Pérez-Alvarez R and García-Moliner F 1998 *Long Wave Polar Modes in Semiconductor Heterostructures* (London: Pergamon Elsevier Science) p 164

IOP Publishing

Vibrational Excitations in Multilayer Nanostructures
Properties and manifestations
Stepan I Beril, Vladimir M Fomin and Alexander S Starchuk

Chapter 1

Potentials in multilayer planar systems

1.1 Introduction

In recent decades, thanks to the achievements of modern nanotechnology, applied fields related to the use of low-dimensional systems — multilayer structures of various geometries — planar structures with quantum wells, including superlattices, quantum wires, and quantum dots — have been intensively developing. Such systems have unique physical properties (mechanical, electrical, optical, etc), which are fundamentally different from those of massive structures made of the same materials. This is due to the significant role of quantum effects in the case when the dimensions of the structural elements of the system become comparable to the de Broglie wavelength of charge carriers [1]. These distinctive properties resulted in the broad use of low-dimensional systems in science and technology, which stimulated an exponential growth of theoretical and experimental work devoted to the study of such structures.

To explain the charge transfer processes in low-dimensional multilayer systems, including materials with high-temperature superconductivity, it is necessary to construct theories based on the exact solutions of Poisson equation for multilayer systems of various geometries with an arbitrary number of layers (the theory) of potential and the exact Hamiltonians describing the interaction of charges in these systems.

This first chapter is devoted to the construction of the theory of potential in multilayer systems of planar geometries. Methods of calculating the potential for various actual cases have been developed.

1.2 Equations of the spatial distribution of potentials in planar multilayer systems

Consider the spatial distribution of the electrostatic potential in a multilayer system (figure 1.1). We take into account the following factors that determine the potentials and

doi:10.1088/978-0-7503-6164-4ch1

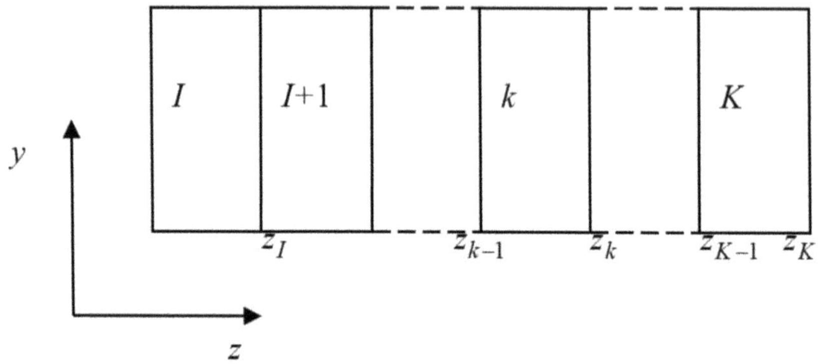

Figure 1.1. Scheme of the multilayer system.

fields in the system: external voltage sources, surface charges located at the interfaces between layers with density the $\sigma_k(\mathbf{y})$, $\mathbf{y} = x\mathbf{e}_1 + y\mathbf{e}_2$, bulk charges with density $\rho_k(\mathbf{r})$ in kth layer, and polarization, which is subdivided in two types—fast (inertial) and slow (inertial) in relation to the characteristic frequencies of external fields.

The spatial change of the electric field in the kth layer is described by a set of equations [2]:

$$\text{rot}\, \mathbf{E}_k(\mathbf{r}) = 0, \ \ \text{div}\, \mathbf{D}_k(\mathbf{r}) = \rho_k(\mathbf{r}), \tag{1.2.1}$$

where the electric induction vector can be represented as

$$\mathbf{D}_k(\mathbf{r}) = \varepsilon_0 \mathbf{E}_k(\mathbf{r}) + \varepsilon_0 \overleftrightarrow{\chi}_k \mathbf{E}_k(\mathbf{r}) + \mathbf{P}_k(\mathbf{r}). \tag{1.2.2}$$

Here ε_0 is the electric constant; $\overleftrightarrow{\chi}_k$ is the dielectric permittivity tensor; and $\mathbf{P}_k(\mathbf{r})$ is the inertial polarization vector. The fast polarization follows the field without inertia: $\mathbf{P}_{\text{fast}}(\mathbf{r}) = \varepsilon_0 \overleftrightarrow{\chi}_k \mathbf{E}_k(\mathbf{r})$.

The first of the equation (1.2.1) is satisfied by a potential field

$$\mathbf{E}_k(\mathbf{r}) = -\text{grad}\, V_k(\mathbf{r}), \tag{1.2.3}$$

and the potential is described by the equation

$$\text{div}\, [\overleftrightarrow{\varepsilon}_k \text{grad}\, V_k(\mathbf{r})] = -\varepsilon_0^{-1}[\rho_k(\mathbf{r}) - \text{div}\, \mathbf{P}_k(\mathbf{r})], \tag{1.2.4}$$

where the high-frequency permittivity tensor is introduced

$$\overleftrightarrow{\varepsilon}_k = \overleftrightarrow{I} + \overleftrightarrow{\chi}_k. \tag{1.2.5}$$

With sufficiently large transverse dimensions of the system, its properties can be considered homogeneous in the XOY planes. In real structures, the concept of 'partition boundary' is conditional. More precisely, we should talk about a transition layer [3] consisting of two, three, or more atomic planes, in which the crystal lattice of one layer of the structure changes into the crystal lattice of a neighboring layer. In the continuum theory, when calculating potentials, the existence of a transition layer can be taken into account by entering a transition function $S(z)$ for a contact between the k-th and $(k+1)$-th layers, which changes from 0 to 1 within the transition layer: $\varepsilon(z) = \varepsilon_k + (\varepsilon_{k+1} - \varepsilon_k)S(z_k - z)$. In what follows, we assume that $S(z_n - z) = \theta(z_n - z)$, where

$$\theta(z) = \begin{cases} 0 & (z < 0), \\ 1 & (z \geqslant 0), \end{cases}$$

thus limiting ourselves to the assumption that the length of the transition region is smaller than other effective lengths.

By performing a Fourier transform with respect to coordinates $\mathbf{y} = x\mathbf{e}_1 + y\mathbf{e}_2$:

$$V_k(\mathbf{r}) = \iint d^2\eta e^{i\boldsymbol{\eta}\mathbf{y}} V_k(\boldsymbol{\eta}, z), \tag{1.2.6}$$

from the partial differential equation (1.2.4), we obtain an ordinary differential equation:

$$\left(\frac{\partial}{\partial z}\mathbf{e}_3 + i\boldsymbol{\eta}\right)\mathbf{E}_k\left(\frac{\partial}{\partial z}\mathbf{e}_3 + i\boldsymbol{\eta}V_k(\eta, z)\right) = -\varepsilon_0^{-1}\tilde{\rho}_k(\eta, z), \tag{1.2.7a}$$

$$\tilde{\rho}_k(\boldsymbol{\eta}, z) \equiv \rho_k(\boldsymbol{\eta}, z) - \frac{\partial}{\partial z}P_k^{||}(\boldsymbol{\eta}, z) - i\boldsymbol{\eta}\mathbf{P}_k^{\perp}(\boldsymbol{\eta}, z), \tag{1.2.7b}$$

where $\boldsymbol{\eta} = \eta_1\mathbf{e}_1 + \eta_2\mathbf{e}_2$ is a two-dimensional spatial frequency vector; and $P_k^{||}(\boldsymbol{\eta}, z) = \mathbf{P}_k(\boldsymbol{\eta}, z)\mathbf{e}_3$ and $P_k^{\perp} = \mathbf{P}_k(\boldsymbol{\eta}, z) - P_k^{||}\mathbf{e}_3$ are, respectively, longitudinal and two-dimensional transverse (relative to the axis Oz) components of the vector $\mathbf{P}_k(\boldsymbol{\eta}, z)$.

Assuming that the tensors $\overleftrightarrow{\varepsilon_k}$ are uniaxial:

$$\varepsilon_k^{ij} = \delta^{ij}[\varepsilon_k^{\perp}(\delta^{i1} + \delta^{i2}) + \varepsilon_k^{||}\delta^{i3}],$$

we simplify (1.2.7a):

$$\left(\frac{\partial^2}{\partial z^2} - \epsilon_k\eta^2\right)V_k(\boldsymbol{\eta}, z) = -(\varepsilon_0\varepsilon_k^{||})^{-1}\tilde{\rho}_k(\boldsymbol{\eta}, z), \tag{1.2.7c}$$

where the notations are used:

$$\varepsilon_k^{11} = \varepsilon_k^{22} = \varepsilon_k^{\perp};\ \varepsilon_k^{33} = \varepsilon_k^{||};\ \varepsilon_k^{\perp}/\varepsilon_k^{||} = \epsilon_k. \tag{1.2.7d}$$

In this case, a solution of the equation (1.2.7c) for the kth layer, satisfying the condition of continuity of potential

$$V_{k-1}(\boldsymbol{\eta}, z)\big|_{z=z_{k-1}} = V_k(\boldsymbol{\eta}, z)\big|_{z=z_{k-1}} \equiv V(\boldsymbol{\eta}, z_{k-1}), \tag{1.2.8}$$

can be represented as

$$\begin{aligned} V_k(\boldsymbol{\eta}, z) = F_k(z - z_{k-1})V(\boldsymbol{\eta}, z_k) + F_k(z_k - z)V(\boldsymbol{\eta}, z_{k-1}) + \\ + \int_{z_{k-1}}^{z_k} dz' G_k(z, z')\varepsilon_0^{-1}\tilde{\rho}(\boldsymbol{\eta}, z'),\ z_{k-1} \leqslant z \leqslant z_k, \end{aligned} \tag{1.2.9a}$$

where

$$F_k = \frac{\sinh(\sqrt{\epsilon_k}\eta z)}{\sinh \zeta_k}; \ \varepsilon_k = \sqrt{\varepsilon_k^{\perp}\varepsilon_k^{\parallel}} = \sqrt{\epsilon_k}\varepsilon_k; \ \zeta_k = \sqrt{\epsilon_k}\eta l_k. \tag{1.2.9b}$$

The first two terms in the right part of the equation (1.2.9a) represent a partial solution of the homogeneous equation corresponding to (1.2.7c) that satisfies the boundary conditions (1.2.8), and the last one is a partial solution of (1.2.7b). Using general methods for solving a second-order inhomogeneous differential equation, we find the kernel of the integral in (1.2.9a):

$$\begin{aligned} G_k(z, z') = \frac{\sinh \zeta_k}{\eta \varepsilon_k}\{&\theta(z' - z)F_k(z - z_{k-1}) \\ &+ \theta(z - z')F_k(z_k - z)F_k(z' - z_{k-1}\}, \end{aligned} \tag{1.2.10}$$

where $\theta(z)$ is the Heaviside step function.

Taking into account the Fourier transform (1.2.6), we obtain the components of the electric field:

transverse

$$\mathbf{E}_k^{\perp}(\boldsymbol{\eta}, z) = -i\boldsymbol{\eta} V_k(\boldsymbol{\eta}, z) \tag{1.2.11a}$$

and longitudinal

$$E_k^{\parallel}(\boldsymbol{\eta}, z) = -\frac{\partial}{\partial z} V_k(\boldsymbol{\eta}, z) \tag{1.2.11b}$$

in the kth layer.

1.3 Modulation transmission matrix

The electric induction vectors (1.2.2) satisfy the boundary conditions

$$D_{k+1}^{\parallel}(\mathbf{y}, z) - D_k^{\parallel}(\mathbf{y}, z) = \varepsilon_0^{-1}\sigma_k(\mathbf{y}), \ k = I, \ldots, K - 1. \tag{1.3.1}$$

Substituting the definitions $\mathbf{D}_k$ according to (1.2.2) in the equation (1.3.1), by taking the Fourier transform (1.2.6) and expressing the field components according to (1.2.11) in terms of potentials, we obtain a set of equations for the boundary values of potentials:

$$\begin{aligned} &-\mu_k(\eta)V(\boldsymbol{\eta}, z_{k-1}) + \nu_k(\eta)V(\boldsymbol{\eta}, z_k) \\ &-\mu_{k+1}(\eta)V(\boldsymbol{\eta}, z_{k+1}) = \varepsilon_0^{-1}\tilde{\sigma}_k(\boldsymbol{\eta}), \end{aligned} \tag{1.3.2}$$

where in the right parts

$$\begin{aligned} &\tilde{\sigma}_k(\boldsymbol{\eta}) \equiv \sigma_k(\boldsymbol{\eta}) - P_{k+1}^{\parallel}(\boldsymbol{\eta}, z_k) + P_k^{\parallel}(\boldsymbol{\eta}, z_k) \\ &+ \int_{z_k}^{z_{k+1}} dz' F_{k+1}(z_{k+1} - z')\tilde{\rho}_{k+1}(\boldsymbol{\eta}, z') + \int_{z_{k-1}}^{z_k} dz' F_k(z' - z_{k-1})\tilde{\rho}_k(\boldsymbol{\eta}, z'), \end{aligned} \tag{1.3.3a}$$

and the coefficients in the left parts are

$$\mu_k(\eta) \equiv \frac{\varepsilon_k \eta}{\sinh \zeta_k}; \; \nu_k(\eta) \equiv (\varepsilon_k \coth \zeta_k + \varepsilon_{k+1} \coth \zeta_{k+1})\eta. \tag{1.3.3b}$$

The set (1.3.2) consists of $K - 1$ equations and allows us to express potentials at all interfaces in terms of the values of potentials at the externalexternal surfaces $V(\eta, z_{I-1})$, $V(\eta, z_K)$ and distributions of the internal field sources $\tilde{\sigma}_k(\eta)$. In systems with a finite number of layers, the potentials at the externalexternal surfaces are set, for example, by the external voltage sources. In systems with an infinite number of layers, the actual value of the potentials at the external surfaces does not matter for finding the potential distribution.

Transferring terms from $V(\eta, z_{I-1})$ and $V(\eta, z_K)$ in the right parts of the equations (1.3.2), we bring this set to the form ($k = I, \ldots, K - 1$)

$$\sum_{n=I}^{K-1} b_{kn} V(\eta, z_n) = \varepsilon_0^{-1} \sigma'_k(\eta), \tag{1.3.4}$$

where the symmetric matrix of coefficients $\|b_{ij}\|$ contains non-zero elements only along the main diagonal ($b_{jj} = \nu_j$) and for pairs of indexes that differ by one ($b_{j,j+1} = b_{j+1,j} = -\mu_{j+1}$), all other elements are equal to zero. The free members of the set (1.3.2) are

$$\begin{cases} \sigma'_I(\eta) = \tilde{\sigma}_I(\eta) + \varepsilon_0 \mu_I(\eta) v\,(\eta, z_{I-1}); \\ \sigma'_k(\eta) = \tilde{\sigma}_k(\eta),\; k = I + 1, \ldots, K - 2; \\ \sigma'_{K-1}(\eta) = \tilde{\sigma}_{K-1}(\eta) + \varepsilon_0 \mu_K(\eta) V(\eta, z_K). \end{cases} \tag{1.3.5}$$

The solution of the set (1.3.4) has the form

$$V(\eta, z_k) = \sum_{n=I}^{K-1} D_{nk}(I\,|K) \varepsilon_0^{-1} \sigma'_k(\eta), \tag{1.3.6}$$

where the coefficients in the right side are elements of the inverse matrix $\|b_{ij}\|$.

Additive according to field sources form $\sigma'_k(\eta)$ (1.3.5) according to (1.3.6) leads to the same additive structure of boundary potentials:

$$V(\eta, z_n) = V^e(\eta, z_n) + V^s(\eta, z_n) + V^b(\eta, z_n) + V^p(\eta, z_n), \tag{1.3.7a}$$

where

$$V^e(\eta, z_n) = D_{nI} \mu_I(\eta) V(\eta, z_n) + D_{n,K-1} \mu_K(\eta) V(\eta, z_K) \tag{1.3.7b}$$

is the contribution due to the external voltage sources, which is determined by the values of potentials at the external surfaces of the system.

Note that due to the considered dependency $V(\eta, z_{I,K})$ on η, their spatial distribution on surfaces xOy is not uniform;

$$V^s(\eta, z_n) = \sum_{k=I}^{K-1} D_{nk} \varepsilon_0^{-1} \sigma_k(\eta) \tag{1.3.7c}$$

is the contribution due to the interface charges;

$$V^b(\eta, z_n) = \sum_{k=I}^{K} \int_{z_{k-1}}^{z_k} dz' G_{nk}(z') \varepsilon_0^{-1} \rho_k(\eta, z') \tag{1.3.7d}$$

is the contribution due to the volume charges;

$$V^p(\eta, z_n) = \sum_{k=I}^{K} \int_{z_{k-1}}^{z_k} dz' \varepsilon_0^{-1} \left[\frac{\partial G_{nk}(z')}{\partial z'} P_k^{||}(\eta, z') - i\eta G_{nk}(z') P_k^{\perp}(\eta, z') \right] \tag{1.3.7e}$$

is the contribution due to the polarization. The used notation is

$$G_{nk}(z') = D_{n,\,k-1} F_k(z_k - z')\theta(k > I) + D_{nk} F_k(z' - z_{k-1})\theta(K > k); \tag{1.3.8a}$$

$$\theta(k > I) = 1,\ k > 1;\ \theta(k > I) = 0,\ k \leqslant 1. \tag{1.3.8b}$$

Hence, the potential (1.2.9a) in the nth layer has an additive structure:

$$V_n(\eta, z) = V_n^e(\eta, z) + V_n^s(\eta, z) + V_n^b(\eta, z) + V_n^p(\eta, z), \tag{1.3.9a}$$

where, according to the previously listed terms

$$V_n^e(\eta, z) = F_n(z - z_{n-1}) V^e(\eta, z_n) + F_n(z_n - z) V^e(\eta, z_{n-1}); \tag{1.3.9b}$$

$$V_n^s(\eta, z) = \sum_{k=I}^{K-1} \left[D_{nk} F(z - z_{n-1})\theta(K > n) + D_{n-1,\,k} F(z_n - z)\theta(n > I) \right] \varepsilon_0^{-1} \sigma_k(\eta); \tag{1.3.9c}$$

$$V_n^b(\eta, z) = \sum_{k=I}^{K} \int_{z_{k-1}}^{z_k} dz' g_{nk}(z, z') \varepsilon_0^{-1} \rho_k(\eta, z'), \tag{1.3.9d}$$

with the Green's function

$$\begin{aligned} g_{nk}(z, z') = G_{nk}(z') F_n(z - z_{k-1})\theta(K > n) + G_{n-1,\,k}(z') F_n(z_n - z)\theta(n > I) \\ + G_k(z, z')\delta_{kn}; \end{aligned} \tag{1.3.9e}$$

$$V_n^p(\eta, z) = \sum_{k=I}^{K} \int_{z_{k-1}}^{z_k} dz' \varepsilon_0^{-1} \left[\frac{\partial g_{nk}(z, z')}{\partial z'} P_k^{||}(\eta, z') - i\eta g_{nk}(z, z') \mathbf{P}_k^{\perp}(\eta, z) \right]. \tag{1.3.9f}$$

In particular, if the entire space is filled with a single material $n = 1$, from the equations (1.3.9d), (1.3.9e), and (1.2.10), the expression for the Green's function follows:

$$G_{11}(z, z') = (2\eta\varepsilon_1)^{-1} e^{-\sqrt{\varepsilon_1}\eta|z-z'|}. \tag{1.3.9g}$$

As follows from (1.3.7c), the coefficient $D_{nk} = D_{nk}(I|K)$ has the sense of the modulation transmission function from the charge at the surface z_k to the potential at the surface z_n:

$$D_{nk}(I\,|K) = \varepsilon_0 \frac{\delta V(\eta, z_n)}{\delta \sigma_k(\eta)}. \tag{1.3.10}$$

Therefore, the set of coefficients $\|D_{nk}\|$ forms a spatial frequency-dependent modulation transmission matrix for the system. In accordance with the equation (1.3.9b–e), the modulation transmission function from the field sources in the layer k to potentials at the surface and in the bulk of the layer k are completely determined by the modulation transmission matrix elements. According to the rules of linear algebra [4], the elements of the modulation transmission matrix are related to the elements of the original matrix $\|b_{ij}\|$ by

$$D_{nk}(I|K) = \frac{B_{nk}(I|K)}{B(I|K)}, \tag{1.3.11a}$$

where

$$B(I|K) \equiv det\|b_{nk}\| \tag{1.3.11b}$$

is the determinant, and $B_{nk}(I|K)$ is the algebraic complement of an element b_{nk} of the original matrix, which is at $k \leqslant n$ can be represented as a product of determinants of diagonal blocks:

$$B_{kn}(I|K) = B(I|k)\left(\prod_{j=k+1}^{k} \mu_j\right)B(n+1|K), \tag{1.3.12a}$$

where the denotation (1.3.11b) is used twice. For $k = 1$, $B(I|I) = 1$ in the expression (1.3.12a), and similarly $B(K|K) = 1$ for $n = K - 1$. If $k = n$, $\prod_{j=n+1}^{n} \mu_j = 1$ should be considered in the expression (1.3.12a). The matrix $\|b_{ij}\|$ is symmetric, and the matrix of algebraic extensions $\|B_{kn}\|$ has the same property. Therefore, when $k \geqslant n$

$$B_{kn}(I|K) = B_{nk}(I|K) = B(I|n)\left(\prod_{j=n+1}^{k} \mu_j\right)B(k+1|K) \tag{1.3.12b}$$

is obtained from (1.3.12a) by replacement $k \leftrightarrow n$. It follows from equations (1.3.12a) and (1.3.12b) that the algebraic complements $B_{nk}(I|K)$ are generally expressed in terms of products of determinants of matrices structurally similar to $\|b_{ij}\|$, but of a lower dimension, and some diagonal elements of the original matrices. Using the expansion of the determinant $B(I|K)$ with respect to the kth line

$$B(I|K) = \sum_{n=I}^{K-1} b_{kn}B_{kn}(I|K), \tag{1.3.13}$$

we obtain recurrence relations for determinants of different dimensions. When $k = K - 1$, from the equation (1.3.13) with account for (1.3.12b) an important recurrence relation is found:

$$B(I|K) = \nu_{K-1}B(I|K-1) - \mu_{K-1}^2 B(I|K-2). \tag{1.3.14}$$

Because for a two-layer system ($K = I + 1$)

$$B(I|I+1) = \nu_1, \tag{1.3.15}$$

then for three-layer ($K = I + 2$) we find from (1.3.14):

$$B(I|I+2) = \nu_{I+1}\nu_I - \mu_{I+1}^2. \tag{1.3.16}$$

Continuing to increase by one and using the already known determinants, we obtain the determinants for any system with a finite number of layers in the same way using equation (1.3.14). The described recurrent procedure is easily algorithmized for numerical calculation of elements of the modulation transmission matrix for finite multilayer systems.

1.4 Modulation transmission matrix for three-layer and five-layer systems

For a three-layer system (1|2|3), we find the determinant according to (1.3.16):

$$B(1|3) = \eta^2[\varepsilon_2^2 + (\varepsilon_1 \coth\zeta_1 + \varepsilon_3 \coth\zeta_3)\varepsilon_2 \coth\zeta_2 + \varepsilon_1 \coth\zeta_1\varepsilon_3 \coth\zeta_3, \tag{1.4.1}$$

and according to the equations (1.3.12a), (1.3.12b), and (1.3.15)—algebraic complements:

$$B_{11}(1|3) = \nu_2 = (\varepsilon_2 \coth\zeta_2 + \varepsilon_3 \coth\zeta_3)\eta; \tag{1.4.2a}$$

$$B_{12}(1|3) = \mu_2 = \frac{\varepsilon_2}{\sinh\zeta_2}\eta = B_{21}(1|3); \tag{1.4.2b}$$

$$B_{22}(1|3) = \nu_1 = (\varepsilon_1 \coth\zeta_1 + \varepsilon_2 \coth\zeta_2)\eta; \tag{1.4.2c}$$

Therefore, the modulation transmission matrix (1.3.11a) has the form

$$\|D_{nk}(1|3)\| = \frac{1}{B(1|3)}\begin{Vmatrix} B_{11}(1|3) & B_{12}(1|3) \\ B_{21}(1|3) & B_{22}(1|3) \end{Vmatrix} \tag{1.4.3}$$

with the determinant $B(1|3)$ and the matrix elements $B_{kn}(1|3)$, expressed by equations (1.4.1) and (1.4.2). In the case of semi-infinite external layers ($l_1 \to 1$; $l_3 \to 1$), the replacements $\coth\zeta_1 \to 1$; $\coth\zeta_3 \to 1$ should be performed in the results (1.4.1) and (1.4.2).

For a five-layer system (0|1|2|3|4) with the external semi-infinite layers ($l_0 \to \infty$, $l_4 \to \infty$), using the recurrent relation (1.3.14), we get the determinant

$$\begin{aligned} B(0|4) = {} & \{\varepsilon_2\coth\zeta_2[\varepsilon_1(\varepsilon_0\coth\zeta_1 + \varepsilon_1)(\varepsilon_3\coth\zeta_3 + \varepsilon_4) \\ & + \varepsilon_3(\varepsilon_0 + \varepsilon_1\coth\zeta_1)(\varepsilon_3 + \varepsilon_4\coth\zeta_3)] \\ & + \varepsilon_2^2(\varepsilon_0 + \varepsilon_1\coth\zeta_1)(\varepsilon_4 + \varepsilon_3\coth\zeta_3) + \varepsilon_1\varepsilon_3(\varepsilon_0\coth\zeta_1 + \varepsilon_1) \\ & (\varepsilon_3 + \varepsilon_4\coth\zeta_3)\}\eta^4, \end{aligned} \tag{1.4.4}$$

The algebraic complements calculated using the equations (1.3.12a) and (1.3.12b) are as follows:

$$B_{11}(0|4) = (\varepsilon_0 + \varepsilon_1 \coth \zeta_1)[\varepsilon_3(\varepsilon_3 + \varepsilon_4 \coth \zeta_3) + \varepsilon_2 \coth \zeta_2(\varepsilon_3 \coth \zeta_3 + \varepsilon_4)]\eta^3 \quad (1.4.5a)$$

$$B_{12}(0|4) = B_{21}(0|4) = (\varepsilon_0 + \varepsilon_1 \coth \zeta_1)\frac{\varepsilon_2}{\sinh \zeta_2}(\varepsilon_4 + \varepsilon_3 \coth \zeta_3)\eta^3; \quad (1.4.5b)$$

$$B_{22}(0|4) = (\varepsilon_3 \coth \zeta_3 + \varepsilon_4)[\varepsilon_1(\varepsilon_1 + \varepsilon_0 \coth \zeta_1) + \varepsilon_2 \coth \zeta_2(\varepsilon_0 + \varepsilon_1 \coth \zeta_1)]\eta^3. \quad (1.4.5c)$$

The elements $D_{11}(0|4)$, $D_{12}(0|4) = D_{21}(0|4)$, $D_{22}(0|4)$ of the the modulation transmission matrices can be written explicitly using the definition (1.3.11a) and the equations (1.4.4) and (1.4.5). In the limit ($l_1 \to 0$, $l_3 \to 0$) they turn into matrix elements (1.4.1), (1.4.2) for a three-layer system having the structure (1|2|3) with the external semi-infinite layers. In the limit ($l_1 \to 0$, $l_3 \to 0$) the results (1.4.5) turn into the matrix elements of the system (0|2|4) with the external semi-infinite layers. Finally, in the limit of pairwise identical substances that are made of layers 0 and 1, as well as 3 and 4, the elements of the modulation transmission matrix lose their dependence on l_1 and l_3 and turn into the results for the three-layer system (1|2|3) with the external semi-infinite layers.

1.5 Modulation transmission matrix for periodic systems

Thanks to advances in molecular vacuum epitaxy technology [1, 5–7], it was possible to create solid state electronic systems that consist of alternating layers of different materials, called superlattices in the case of a large number of periods. In the present section we consider a periodic system of two alternating layers of the type *a* and *b* (figure 1.2). First, we derive the elements of the modulation transmission matrix for that system. According to the equations (1.3.12a) and (1.3.12b), it is necessary to calculate the determinants $B(I|K)$ matrix blocks $\|b_{ij}\|$. We introduce the dimensionless quantities:

$$\gamma(I|k) \equiv \frac{B(I|k)}{\nu_{k-1}B(I|k-1)}, \quad k = I + 2, \ldots, K, \quad (1.5.1)$$

related to each other, as follows from the expression (1.3.14), by recurrent relations:

$$\gamma(I|k) = 1 - \frac{\xi_{k-1}}{\gamma(I|k-1)}, \quad k = I + 3, \ldots, K, \quad (1.5.2)$$

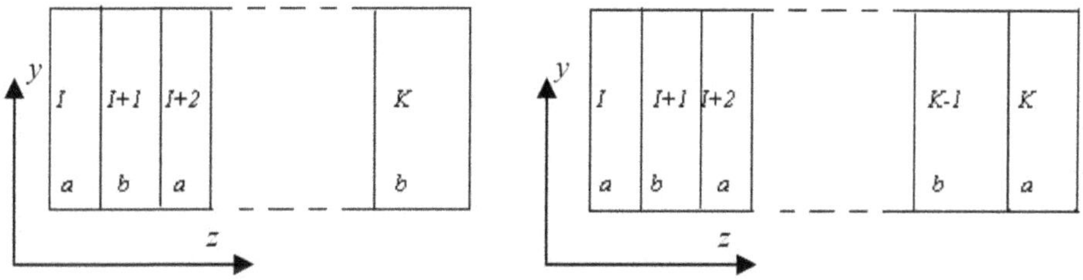

Figure 1.2. Variants of periodic systems with different (left) and identical (right) external layers.

where the dimensionless parameters are used

$$\xi_{k-1} \equiv \frac{\mu_{k-1}^2}{\nu_{k-1}\nu_{k-2}}. \tag{1.5.3}$$

According to (1.5.1), the determinant

$$B(I|K) = \prod_{k=I+2}^{K} \gamma(I|K) \prod_{j=I}^{k-1} \nu_j. \tag{1.5.4}$$

Because $B(I|I) = 1$, in accordance with (1.3.15) $\gamma(I|I+2) = 1 - \xi_{I+1}$, and the recurrence ratio (1.5.2) gives the finitechain fractions:

$$\gamma(I|k) = 1 - \cfrac{\xi_{k-1}}{1 - \cfrac{\vdots}{1 - \cfrac{\xi_{I+2}}{1-\xi_{I+1}}}}. \tag{1.5.5}$$

A method of converting them to normal fractions for a periodic structure that consists of layers a, b is described in [8]:

$$\gamma(I|I+2n) = \frac{\sum_{j=0}^{n}\alpha^{2j} + \lambda a \sum_{j=0}^{n}\alpha^{2j}}{A\sum_{j=0}^{n-1}\alpha^{2j}}; \tag{1.5.6a}$$

$$\gamma(I|I+2n+1) = \frac{\sum_{j=0}^{n}\alpha^{2j}}{\sum_{j=0}^{n}\alpha^{2j} + \lambda_b\sum_{j=0}^{n-1}\alpha^{2j}}, \tag{1.5.6b}$$

where

$$\lambda_a = A\xi_a, \ \lambda_b = A\xi_b; \tag{1.5.7a}$$

$$A = \frac{1}{\sqrt{\xi_a\xi_b}}(\psi + \sqrt{\psi^2 - 1}); \ \ \psi = \frac{1 - \xi_a - \xi_b}{2\sqrt{\xi_a\xi_b}} > 1; \tag{1.5.7b}$$

$$\alpha \equiv \sqrt{\lambda_a\lambda_b} = \psi + \sqrt{\psi^2 - 1} > 1. \tag{1.5.7c}$$

Using equation (1.5.4), we find the determinant of the structure with different

$$B(I|I+2N-1) = \frac{\nu^{2N-1}(\alpha^{2N} - 1)}{A^{N-1}(\alpha^{2N} - 1)} \tag{1.5.8a}$$

and similar

$$B(I|I+2N) = \frac{\nu^{2N-1}[\alpha^{2N}\lambda_a(1+\lambda_b) - (1+\lambda_a)]}{A^N(\alpha^2 - 1)}. \tag{1.5.8b}$$

external layers. From (1.5.8a) and (1.5.8b), for systems consisting of two and three layers, we obtain:

$$B(I|I+1) = \nu, \;\; B(I|I+2) = \nu^2 - \mu_b^2. \tag{1.5.8c}$$

These results match with the known ones (1.3.15) and (1.3.16), in which the identity of layers with indexes of the same parity must be taken into account.

For a multilayer system with large enough number of pairs of layers N, so that $\alpha^{2N} >> 1$ in the equarion (1.5.7c), the determinant takes the power form:

$$B(I|I+2N-1) = \frac{\alpha^2\nu}{\alpha^2-1}\left(\frac{\alpha^2\nu^2}{A}\right)^{N-1}; \tag{1.5.9a}$$

$$B(I|I+2N) = \frac{\alpha^2\nu^2\lambda_a}{(\alpha^2-1)(1+\lambda_a)}\left(\frac{\alpha^2\nu^2}{A}\right)^{N-1}. \tag{1.5.9b}$$

Calculations of the elements of the modulation transmission matrix will be performed using the equations (1.3.11a), (1.3.11b), (1.3.12a), and (1.3.12b), selecting for certainty a system with different external layers $I = 2i+1, k = I + 2N - 1$. For this system, four cases should be distinguished according to different parity of the matrix indices:

$$D_{2J,\,2L-1} = \begin{cases} \dfrac{1+\lambda_b}{\mu_b(\alpha^2-1)}\alpha^{L-J}, & L \leqslant J; \\ \dfrac{1+\lambda_a}{\mu_a(\alpha^2-1)}\alpha^{J-L+1}, & L > J; \end{cases} \tag{1.5.10a}$$

$$D_{2J,\,2L} = D_{2J-1,\,2L-1} = \frac{\nu\alpha^{1-|J-L|}}{\mu_a\mu_b(\alpha^2-1)}; \tag{1.5.10b}$$

$$D_{2J-1,\,2L} = D_{2L,\,2J-1}. \tag{1.5.10c}$$

For a system with the same external layers I and $k = I + 2N$, in the limit of large N the same elements of the modulation transmission matrix as (1.5.10a)–(1.5.10c) are obtained. The received elements of the modulation transmission matrix do not depend on the total number of layer pairs N in the system, since this number is large, but depend on the differences $L - J$, determining the relative position of the layers, which satisfies the conditions of translational symmetry:

$$D_{nk} = D_{n+2J,\,k+2J}. \tag{1.5.11}$$

We consider further the limit of extremely thick layers of the type a, when $l_a \to \infty$, $\mu_a \to 0$, $\xi_a \to 0$, but μ_b, ν, ξ_b remain final, $A \to (1-\xi_b)/(\xi_a\xi_b)$, $\lambda_a \to (1-\xi_b)/\xi_b$, $\lambda_b \to (1-\xi_b)/\xi_a$. Therefore, only those elements of the modulation transmission matrix, in which the multiplier proportional to $\alpha \to \infty$ is contained in the zero degree take the finite values:

$$D_{2L,\,2L} = D_{2L-1,\,2L-1} = \frac{\nu}{\nu^2-\mu_b^2}; \quad D_{2L,\,2L-1} = D_{2L-1,\,2L} = \frac{\mu_b}{\nu^2-\mu_b^2},$$

where $\nu = (\varepsilon_a + \varepsilon_b \coth \xi_b)\eta$, what coincides with the results (1.4.1), (1.4.2) for a three-layer system, in which the second layer is of b-type, and the semi-infinite layers 1 and 3 are of the same type a. All other elements of the modulation transmission matrix vanish. Therefore, in the limit under consideration, a charge at any surface causes potential modulation only at two surfaces: at the same surface and on the surface divided from it by a layer of finite thickness.

The elements of the modulation transmission matrix completely determine Green's functions of various field sources used for calculating potentials. In what follows, potentials due to bulk charges and polarization will be obtained.

1.6 The potential of bulk charges in multilayer systems

Further, we substitute the explicit matrix elements of the modulation transmission matrix into the equation (1.3.8a). Using the latter to write down the Green's function $G_{nk}(z, z')$, we obtain [according to the equation (1.3.9d) in the integral form] the potential in the layer n, caused by an arbitrary distribution of charges in the system. However, due to the involved nature of this equation we limit ourselves to the special case when the potential is defined in one of the inner layers ($I < n < K$) of a multilayer system containing bulk charges:

$$V_n^b(\eta, z) = \int_n^{n-1} dz' G_{nn}(z, z')\varepsilon_0^{-1}\rho_n(\eta, z'), \tag{1.6.1}$$

where according to the equation (1.3.8a)

$$\begin{aligned} G_{nn}(z, z') = -\frac{1}{2\eta\varepsilon_n}\Bigg\{ & e^{-\sqrt{(\epsilon_n)}\eta|z-z'|} + \frac{1}{\sinh \zeta_n}\Big([e^{-\zeta_n} + \frac{\eta\varepsilon_n}{\sinh \zeta_n}(2D_{n,\, n-1}\cosh \zeta_n - D_{nn} \\ & - D_{n-1,\, n-1})] + \left[\frac{\eta\varepsilon_n}{\sinh \zeta_n}((D_{n,\, n} + D_{n-1,\, n-1})\cosh \zeta_n - 2D_{n,\, n-1}) - 1\right] \\ & \times \cosh[\sqrt{\epsilon_n}\eta(z + z' - z_n - z_{n-1})] + \eta\varepsilon_n(D_{nn} - D_{n-1,\, n-1}) \\ & \times \sinh[\sqrt{\epsilon_n}\zeta_n\eta(z + z' - z_n - z_{n-1})]\Big)\Bigg\}. \end{aligned} \tag{1.6.2}$$

If the thickness of the layer l_n is small enough, so that the condition $\zeta_n \ll 1$ is met, the function (1.6.2) takes the form

$$\begin{aligned} & G_{nn}(z, z') \cong \widetilde{G}_{nn}(\eta) \\ & = \frac{1}{2\eta\varepsilon_n}\frac{\cosh \zeta_n - 1}{\sinh \zeta_n}\left[1 + \frac{\eta\varepsilon_n}{sh\zeta_n}(2D_{n,\, n-1} + D_{n,\, n} + D_{n-1,\, n-1})\right]\zeta_n \ll 1. \end{aligned} \tag{1.6.3}$$

For a three-layer system, using the previously found modulation transmission matrix (1.4.3) and taking into account (1.4.1) and (1.4.2), we obtain

$$\begin{aligned} G_{22}(z, z') = \frac{1}{2\eta\varepsilon_2}\{ & e^{-\sqrt{\epsilon_2}\eta|z-z'|} + \frac{1}{\widetilde{B}(1\mid 3)\sinh \zeta_2}[e^{-\zeta_2}(\varepsilon_2 - \varepsilon_1 \coth \zeta_1)] \\ & \times (\varepsilon_2 - \varepsilon_3 \coth \zeta_3)\cosh[\sqrt{\epsilon_2}\eta(z' - z)] \\ & + \varepsilon_2(\varepsilon_1 \coth \zeta_1 - \varepsilon_3 \coth \zeta_3)\sinh[\sqrt{\epsilon_2}\eta(z + z' - z_2 - z_1)]\}, \end{aligned} \tag{1.6.4a}$$

where

$$\widetilde{B}(1|3) \equiv \frac{B(1|3)}{\eta^2} \tag{1.6.4b}$$
$$= \varepsilon_2^2 + \varepsilon_2 \coth \zeta_2 (\varepsilon_1 \coth \zeta_1 + \varepsilon_3 \coth \zeta_3) + \varepsilon_1 \coth \zeta_1 \varepsilon_3 \coth \zeta_3.$$

This modulation transmission function is a generalization of the results [9–12] in what concerns (i) anisotropy in all three layers and (ii) accounting for the finite thickness of the layers 1 and 3. The function (1.6.4a) for the considered three-layer system is:

$$\overline{G}_{22}(\eta) = \frac{1}{2\eta\varepsilon_2 \widetilde{B}(1|3)} \times \left[\varepsilon_2^2 \coth \frac{\zeta_2}{2} + \varepsilon_2(\varepsilon_1 \coth \zeta_1 + \varepsilon_3 \coth \zeta_3) + \varepsilon_1 \coth \frac{\zeta_1}{2} \coth \frac{\zeta_3}{2} \tanh \frac{\zeta_2}{2}\right]\Bigg|_{\zeta_2 \ll 1}, \tag{1.6.5a}$$

where in the case when $\zeta_1, \zeta_3 \gg 1$:

$$\overline{G}_{22}(\eta) \cong \frac{1}{\eta}[(\varepsilon_1 + \varepsilon_3) + F\eta]^{-1}; \; F = (\varepsilon_2^2 + \varepsilon_1\varepsilon_3)\frac{l_2}{\varepsilon_2^{||}}; \tag{1.6.5b}$$

and in the case when $\zeta_1, \zeta_2 \ll 1$:

$$\overline{G}_{22}(\eta) \cong \frac{1}{D}(\eta^2 + d^{-2})^{-1}; \quad D^{-1} = \frac{1}{g}\prod_{j=1,3}\left(\frac{l_j}{\varepsilon_j^{||}} + \frac{l_2}{2\varepsilon_2^{||}}\right); \tag{1.6.5c}$$

$$d^{-2} = \frac{1}{g}\sum_{j=1}^{3}\frac{l_j}{\varepsilon_j^{||}}; \quad g = \frac{l_1}{\varepsilon_1^{||}} l_2 \varepsilon_2^{\perp} \frac{l_3}{\varepsilon_3^{||}} + \frac{1}{3}\sum_{j=1}^{3}\varepsilon_j l_j^2 \sum_{k\neq j}\frac{l_k}{\varepsilon_k^{||}}. \tag{1.6.5d}$$

Based on the equations (1.6.1) and (1.2.6), representing the electric charge density as

$$\rho(\mathbf{r}' - \mathbf{r}_e) = e\delta(\mathbf{y}' - \mathbf{y}_e)\delta(z' - z_e), \tag{1.6.6}$$

we get the potential of interaction between an electron and a hole from (1.6.5b):

$$U(\boldsymbol{\rho}) = -\frac{e^2}{4\varepsilon_0 F}\left\{H_0\left[\frac{\rho(\varepsilon_1 + \varepsilon_3)}{F}\right] - N_0\left[\frac{\rho(\varepsilon_1 + \varepsilon_3)}{F}\right]\right\} \tag{1.6.7}$$

and from (1.6.5c):

$$U(\boldsymbol{\rho}) = -\frac{e^2}{2\pi\varepsilon_0 D}K_0\left(\frac{\rho}{D}\right), \; \boldsymbol{\rho} = \mathbf{y} - \mathbf{y}_e, \tag{1.6.8}$$

These results serve as the basis for calculating the electronic energy spectrum in the corresponding limiting cases.

An explicit form of the Green's function $G_{22}(z, z')$ for the five-layer system is defined by the equations (4.3a) from [11]. Here we will confine ourselves to the limiting case $\zeta_2 < 1$:

$$\begin{aligned}\overline{G}_{22}(\eta) = \; & \frac{1}{2\eta\varepsilon_2\widetilde{B}(0\,|\,4)}\Bigg\{\varepsilon_2^2\coth\frac{\zeta_2}{2}(\varepsilon_0 + \varepsilon_1\coth\zeta_1)(\varepsilon_4 + \varepsilon_3\coth\zeta_3) \\ & +\varepsilon_2\Big[\varepsilon_1(\varepsilon_1 + \varepsilon_0\coth\zeta_1)(\varepsilon_4 + \varepsilon_3\coth\zeta_3) \\ & \quad + \varepsilon_3(\varepsilon_0 + \varepsilon_1\coth\zeta_1)(\varepsilon_3 + \varepsilon_4\coth\zeta_3)\Big] \\ & +\varepsilon_1\varepsilon_3\left(\varepsilon_1 + \varepsilon_0\coth\zeta_1\right)\left(\varepsilon_3 + \varepsilon_4\coth\zeta_3\right)\coth\frac{\zeta_2}{2}\Bigg\}; \end{aligned} \tag{1.6.9a}$$

$$\widetilde{B}(0|4) = \frac{B(0|4)}{\eta^4},$$

from this equation in the limit $\zeta_1,\zeta_3 \gg 1$, the expression follows:

$$\overline{G}_{22}(\eta) \cong \frac{1}{\eta}\left[(\varepsilon_1 + \varepsilon_3) + (\varepsilon_2^2 + \varepsilon_1\varepsilon_3)\frac{l_2}{\varepsilon_2^{\parallel}}\eta\right]^{-1}, \tag{1.6.9b}$$

which is independent of the properties of layers 0 and 4 and coinciding with the result (1.6.5b) for a three-layer system (1|2|3). In the opposite limit $\zeta_1,\zeta_3 \ll 1$, the expression results

$$\overline{G}_{22}(\eta) \cong \frac{1}{\eta}\left\{(\varepsilon_0 + \varepsilon_4) + \left[\sum_{j=1}^{3}(\varepsilon_j^2 + \varepsilon_0\varepsilon_4)\frac{l_j}{\varepsilon_j^{\parallel}}\right]\eta\right\}^{-1}, \tag{1.6.9c}$$

transferring when $l_1 = l_3 = 0$ to a result (1.6.5b) for a three-layer system (0|2|4). Finally, in the case of pairwise identical substances of layers, the limiting function (1.6.9a) loses its dependence on l_1 and l_3 and takes the previously found form (1.6.5a) for a system with semi-infinite external layers. In particular, the expression (1.6.9c) in this case is reduced to the form

$$\overline{G}_{22}(\eta) \cong \frac{1}{\eta}\left\{(\varepsilon_0 + \varepsilon_4)(1 + \zeta_1 + \zeta_3) + (\varepsilon_2^2 + \varepsilon_0\varepsilon_4)\frac{l_2}{\varepsilon_2^{\parallel}}\eta\right\}^{-1},$$

where $1 + \zeta_1 + \zeta_3 \cong 1$ in view of the assumption of smallness of ζ_1 and ζ_3, coincides with the result (1.6.5b) for a three-layer system (0|2|4).

1.7 Potential of bulk charges in periodic systems

For the periodic multilayer system considered in section 1.5, in which the volume charges are only present in the layers $n = 2I$ of type b, substituting the elements of the modulation transmission matrices (1.5.10a), (1.5.10b), (1.5.10c), into (1.6.2), we find:

$$G_{nn}(z, z') = \frac{1}{2\eta\varepsilon_b}\left\{e^{-\sqrt{\epsilon_b}\eta|z-z'|} + \frac{1}{\sqrt{\psi^2-1}\,\sinh\zeta_b}\left[(e^{-\zeta_b}\sqrt{\psi^2-1} + \frac{\cosh\zeta_b}{\alpha}\right.\right.$$
$$- \cosh\zeta_a)\cosh[\sqrt{\epsilon_b}\eta\,|z - z'|] \tag{1.7.1}$$
$$\left.\left. + \frac{\sinh\zeta_a\sinh\zeta_b}{2\varepsilon_a\varepsilon_b}(\varepsilon_b^2 - \varepsilon_a^2)\cosh[\sqrt{\epsilon_b}\eta(z + z' - z_n - z_{n-1})]\right]\right\}.$$

In the limiting case when the layers of types a and b consist of the same material

$$\psi = \cosh(\zeta_a + \zeta_b),\quad \sqrt{\psi^2-1} = \sinh(\zeta_a + \zeta_b),\quad \alpha = e^{\zeta_a + \zeta_b},$$

the coefficients for hyperbolic cosines vanish and the function (1.7.1) becomes the Green's function for an infinite space filled with the medium b.

In the limit of thick layers a ($\zeta_a \gg 1$)

$$\psi \cong \sqrt{\psi^2-1} \cong (2\varepsilon_a\varepsilon_b)^{-1}[\varepsilon_a^2 + \varepsilon_b^2 + 2\varepsilon_a\varepsilon_b\coth\zeta_b]\sinh\zeta_a\sinh\zeta_b \gg 1,$$

the function (1.7.1) takes the form

$$\begin{aligned} G_{nn}(z, z') = \frac{1}{2\eta\varepsilon_b}\Big\{ & \mathrm{e}^{-\sqrt{\epsilon_b}\eta|z-z'|} \\ & + [\sinh\zeta_b(\varepsilon_a^2 + \varepsilon_b^2 + 2\varepsilon_a\varepsilon_b\coth\zeta_b)]^{-1} \\ & \Big[\mathrm{e}^{-\zeta_b}(\varepsilon_a - \varepsilon_b)^2\cosh[\sqrt{\epsilon_b}\eta|z - z'|] \\ & + (\varepsilon_b^2 - \varepsilon_a^2)\cosh[\sqrt{\epsilon_b}\eta(z + z' - z_n - z_{n-1})]\Big]\Big\} \end{aligned} \tag{1.7.2a}$$

and coincides with the result (1.6.4a) for a symmetric system ($a\,|b|\,a$) with infinite external layers.

The limiting form of the function (1.6.3) for the considered periodic system is as follows:

$$\begin{aligned} \overline{G}_{nn}(\eta) = & \frac{1}{2\eta\varepsilon_b}\tanh\frac{\zeta_b}{2} \\ & \times\left.\left\{1 + \frac{1}{\sqrt{\psi^2-1}\,\alpha}\left[1 + \frac{\alpha\sinh\zeta_a}{\varepsilon_a}\left(\varepsilon_b\coth\frac{\zeta_b}{2} + \varepsilon_a\coth\zeta_a\right)\right]\right\}\right|_{\zeta_b \ll 1}. \end{aligned} \tag{1.7.2b}$$

When both types of layers are thin enough,

$$\psi \cong 1 + \frac{s^2\eta^2}{2},\quad s^2 = (\varepsilon_a^{\perp} l_a + \varepsilon_b^{\perp} l_b)\left(\frac{l_a}{\varepsilon_a^{\parallel}} + \frac{l_b}{\varepsilon_b^{\parallel}}\right),$$

keeping one largest term in the numerator and two terms in the denominator, we get a symmetric with respect to the indices a and b expression for the Green's function:

$$\overline{G}_{nn}(\eta) = \frac{\frac{l_a}{\varepsilon_a^{\parallel}} + \frac{l_b}{\varepsilon_b^{\parallel}}}{2\eta s(1 + s\eta)}. \tag{1.7.2c}$$

Using this result and equations (1.2.6), (1.6.1), (1.6.6), we find the potential of interaction between an electron and a hole:

$$U(\rho) = -\frac{e^2\left(\frac{l_a}{\varepsilon_a^{\parallel}} + \frac{l_b}{\varepsilon_b^{\parallel}}\right)}{2\varepsilon_0 s^2}\left\{H_0\left[\frac{\rho}{s}\right] - N_0\left[\frac{\rho}{s}\right]\right\}, \tag{1.7.2d}$$

It will be used below for calculating the exciton energy spectraum in the periodic systems.

1.8 Polarization potential in multilayer systems

Potential in the nth layer $V_n^P(\eta, z)$, due to polarization in a multilayer system, is found above in section 1.3 in the form (1.3.9f) along with (1.3.9c) and (1.3.8b). For the purpose of numerical integration over z', we decompose the polarization vector in the kth layer into a Fourier series:

$$\mathbf{P}_k(\eta, z') = \frac{2\pi}{l_k}\sum_{q_k} e^{iq_k(z' - z_{k-1})}\mathbf{P}_k(\mathbf{Q}_k); \;\; \mathbf{P}_k(-\mathbf{Q}_k) = \mathbf{P}_k^*(\mathbf{Q}_k), \tag{1.8.1a}$$

where

$$q_k = \frac{2\pi\ n_k}{l_k}, \;\; n_k = 0, \;\; \pm 1, \;\; \pm 2, \;\; \pm 3, \;\; \ldots; \mathbf{Q}_k = \{\eta, q_k\}. \tag{1.8.1b}$$

Here $\mathbf{Q}_k$ is a vector with components: η in plane xOy, and q_k—along the axis z.

Substituting (1.8.1a) in (1.3.9f), performing numerical integration over z' and introducing the notation

$$\hat{\mathbf{Q}}_k = \frac{\mathbf{Q}_k}{T_k}; \;\; \hat{\mathbf{R}}_k = \frac{\mathbf{R}_k}{T_k} \equiv T_k^{-1} = \left\{\frac{q_k\eta}{\sqrt{\epsilon_k}\eta}; -\sqrt{\epsilon_k}\eta\right\}; \; T_k^2 = \epsilon_k\eta^2 + q_k^2, \tag{1.8.1c}$$

we bring the polarization potential in the nth layer to the form

$$V_n^P(\eta, z) = \varepsilon_0^{-1}\sum_{k=I}^{K}\frac{2\pi}{l_k}\sum_{q_k}\mathbf{F}_{nk}(z, \mathbf{Q}_k)\mathbf{P}_k(\mathbf{Q}_k), \tag{1.8.2a}$$

where

$$\mathbf{F}_{nk}(z, \mathbf{Q}_k) = -\frac{i\hat{\mathbf{Q}}_k}{\varepsilon_n^{\parallel}T_n}e^{iq_n(z - z_{n-1})}\delta_{n,\,k} + \frac{\sqrt{\epsilon_n}}{2T_k}$$

$$\left\{[K_{1n,\,1k}'i\hat{\mathbf{Q}}_k + K_{1n,\,2k}'\hat{\mathbf{R}}_n]\frac{\cosh w_n}{\sinh\left(\frac{\zeta_n}{2}\right)} + [K_{2n,\,1k}'i\hat{\mathbf{Q}}_k + K_{2n,\,2k}'\hat{\mathbf{R}}_k]\frac{\sinh w_n}{\sinh\left(\frac{\zeta_n}{2}\right)}\right\} \tag{1.8.2b}$$

with the denotation $w_n = \sqrt{\epsilon_n}\eta[z - (z_n + z_{n-1})/2]$ and the coefficients

$$K'_{1n,\,1k} = \left(\frac{2}{\varepsilon_k}\delta_{n,\,k} - \eta \tanh\frac{\zeta_n}{2}A_{nk}(++++)\right)\tanh\frac{\zeta_n}{2};$$

$$K'_{1n,\,2k} = \eta \tanh\frac{\zeta_n}{2}A_{nk}(+\pm\,-); \tag{1.8.2c}$$

$$K'_{2n,\,1k} = -\eta \tanh\frac{\zeta_k}{2}A_{nk}(\pm+\,-);\; K'_{2n,\,2k} = \eta A_{nk}(+\;\;-\mp);$$

where in the matrices

$$\begin{aligned} &A_{nk}(i_1 i_2 i_3 i_4) \\ &= [i_1 D_{n,\,k-1}\theta(K-n-1/2) + i_2 D_{n-1,\,k-1}\theta(n-I-1/2)]\theta(k-I-1/2) \\ &+ [i_3 D_{n,\,k}\theta(K-n-1/2) + i_4 D_{n-1,\,k}\theta(n-I-1/2)]\theta(k-K-1/2), \end{aligned} \tag{1.8.2d}$$

$i_m = "\pm"$ $(m = 1, 2, 3, 4)$ are the sign coefficients. Using the fact that the modulation transmission matrix (1.3.11a) is symmetric, the following property can be derived from (1.8.2d):

$$A_{kn}(i_1 i_2 i_3 i_4) = A_{nk}(i_4 i_2 i_3 i_1), \tag{1.8.2e}$$

from which the symmetry of the coefficients follows

$$K'_{jk,ln} = K'_{ln,jk}. \tag{1.8.2f}$$

According to the definition (1.8.1c)

$$\mathbf{R}_k \overleftrightarrow{\varepsilon}_k \mathbf{Q}_k = 0, \tag{1.8.3a}$$

from which we get orthogonality in the isotropic case:

$$\mathbf{RQ} = 0. \tag{1.8.3b}$$

For the scalar products of the input symbols

$$P_{1k}(\mathbf{Q}_k) \equiv i\hat{\mathbf{Q}}_k \mathbf{P}_k(\mathbf{Q}_k); \quad P_{2k}(\mathbf{Q}_k) \equiv \hat{\mathbf{R}}_k \mathbf{P}_k(\mathbf{Q}_k), \tag{1.8.4}$$

and then for the summs $(l = 1, 2)$

$$\tilde{P}_{lk}(\boldsymbol{\eta}, 0) = \sum_{q_k} S_{0q_k} P_{lk}(\mathbf{Q}_k), \tag{1.8.5a}$$

$$S_{0q_k} = C_k \frac{\sqrt{\epsilon_k}\eta}{T_k}; \quad C_k = \left[\frac{2}{\zeta_k}\tanh\left(\frac{\zeta_k}{2}\right)\right]^{1/2}. \tag{1.8.5b}$$

The constant C_k is chosen so as to satisfy the condition $\sum_{q_k} S^2_{0q_k} = 1$. After substitution of (1.8.5a) into (1.8.2a), the potential can be represented in the form

$$\varepsilon_0 V_n^P(\eta, z) = -\frac{2\pi}{l_n}\sum\nolimits_{q_n} \frac{1}{\varepsilon_n^{\|} T_n} e^{iq_n(z-z_{n-1})} P_{1n}(\mathbf{Q}_n) + \frac{1}{\eta}\sum\nolimits_{k=I}^{K} \frac{\pi}{C_k l_k} \frac{1}{\sinh \frac{\zeta_n}{2}}$$
$$\left\{\left[\left(\frac{2}{\varepsilon_n} \tanh \frac{\zeta_n}{2}\delta_{n,\,k} + K_{1n,\,1k}\right)\widetilde{P}_{1k}(\eta,\,0) + K_{1n,\,2k}\widetilde{P}_{2k}(\eta,\,0)\right]\cosh w_n + \right. \tag{1.8.6}$$
$$\left. + \left[\sum\nolimits_{l=1}^{2} K_{2n,\,lk}\widetilde{P}_{lk}(\eta,\,0)\right]\sinh w_n\right\},$$

where

$$K_{1n,\,1k} = K'_{1n,\,1k} - \frac{2}{\varepsilon_k}\tanh\frac{\zeta_n}{2}\delta_{n,\,k};\;\; K_{\mathrm{l}n,\,jk} = K'_{\mathrm{l}n,\,1k} - \frac{2}{\varepsilon_k}\tanh\frac{\zeta_n}{2}\delta_{n,\,k} \tag{1.8.7}$$

for all l, j except $l = j = 1$.

Equation (1.8.5a) can be considered as a special case (when $q'_k = 0$) of the unitary transformation:

$$\tilde{P}_{lk}(\mathbf{Q}'_k) = \sum_{q_k} S_{q'_k q_k} P_{lk}(\mathbf{Q}'_k). \tag{1.8.8}$$

Here

$$S_{q'_k q_k} = S_{q_k q'_k},$$

for $q'_k \neq 0$, $q_k \neq 0$;

$$S_{q'_k q_k} = \delta_{q'_k q_k} - \frac{C_k^2}{1 - C_k}\frac{\epsilon_n \eta^2}{T'_k T_k},\; (T'^2_k = \epsilon_k \eta^2 + q'^2_k). \tag{1.8.9}$$

It is easy to see that this matrix satisfies the unitarity condition

$$\sum_{q_k} S_{q'_k q_k} S_{q_k q''_k} = \delta_{q'_k q''_k}; \tag{1.8.10}$$

therefore

$$P_{jn}(\mathbf{Q}_n) = \sum_{q'_n} S_{q_n q'_n} \tilde{P}_{jn}(\mathbf{Q}_n). \tag{1.8.11}$$

Substituting the result of the transformation (1.8.11) into the first term of (1.8.6), we get the expression for the potential in terms of $\tilde{P}_{lk}(\mathbf{Q}_k)$:

$$\varepsilon_0 V_n^P(\eta, z) = -\frac{2\pi}{\varepsilon_n^{\|} l_n}\sum\nolimits_{q_n \neq 0} \frac{1}{T_n}\left[e^{iq_n(z-z_{n-1})} + \frac{1}{1 - C_n}\left(C_n - \frac{\cosh w_n}{\cosh \frac{\zeta_n}{2}}\right)\right]\widetilde{P}_{\mathrm{l}n}(\mathbf{Q}_n)$$
$$+ \frac{1}{\eta}\sum\nolimits_{k=I}^{K} \frac{\pi}{C_k l_k}\frac{1}{\sinh\frac{\zeta_n}{2}}\left\{\left[\sum\nolimits_{l=1}^{2} K_{1n,\,lk}\widetilde{P}_{lk}(\eta,\,0)\right]\cosh w_n \right. \tag{1.8.12}$$
$$\left. + \left[\sum\nolimits_{l=1}^{2} K_{2n,\,lk}\widetilde{P}_{lk}(\eta,\,0)\right]\sin\mathrm{h}w_n\right\}.$$

Note the main features of the potential described by the equation (1.8.12): the first term of the equation gives a contribution of the bulk polarization, the second one, with multipliers $\tilde{P}_{lk}(\eta, 0)$, that of the surface polarization. Thus, in (1.8.12), the bulk and surface contributions are separated. Of the three components of the polarization vector, the bulk contribution contains only the longitudinal part $\tilde{P}_{1n}(\mathbf{Q}_n)$, parallel to the $\mathbf{Q}_n$ component, while the surface contribution contains two parts: $\tilde{P}_{1n}(\eta, 0)$ and $\tilde{P}_{2n}(\eta, 0)$ along $\mathbf{Q}_n$ and $\mathbf{R}_{n-}$, respectively. The third component of the polarization is perpendicular to the plane, in which the vectors $\mathbf{Q}_n$ and $\mathbf{R}_n$ lie, is not present at all in (1.8.12).

That's why during the derivation of the potential $V_n^P(\eta, z)$ no special assumptions were made about the physical nature of the polarization, and the type of the coordinate dependence of the function $\mathbf{P}_k(\mathbf{r})$ was not limited. Thus, the formula (1.8.12) remains valid in cases where $\mathbf{P}_k(\mathbf{r})$ is a sum of various contributions, for example, due to ions and plasma of free charge carriers, generally speaking, inhomogeneously distributed within the layers.

Using the modulation transmission matrix (1.4.3) and the denotations (1.4.1) and (1.4.2), we provide here the coefficients for the three-layer structure (1|2|3) using the equations from [13]:

$$\|K_{jk,\,lk}\| = -\frac{\overline{\varepsilon}_2 + \overline{\varepsilon}_k}{\widetilde{B}(1|3)} \left\| \begin{matrix} \tanh^2 \frac{\zeta_k}{2} & (-1)^{\frac{(k-1)}{2}} \tanh \frac{\zeta_k}{2} \\ (-1)^{\frac{(k-1)}{2}} \mathrm{th} \frac{\zeta_2}{2} & 1 \end{matrix} \right\|; \qquad (1.8.13a)$$

$$\| K_{j2,l2} \| = \frac{\tanh \frac{\zeta_2}{2}}{\widetilde{B}(1 \mid 3)}$$

$$\left\| \begin{matrix} -\left[2\varepsilon_2 + (\overline{\varepsilon}_1 + \overline{\varepsilon}_3)\tanh \frac{\zeta_2}{2}\right] & \overline{\varepsilon}_3 - \overline{\varepsilon}_1 \\ \overline{\varepsilon}_3 - \overline{\varepsilon}_1 & -\left[2\varepsilon_2 + (\overline{\varepsilon}_1 + \overline{\varepsilon}_3)\coth \frac{\zeta_2}{2}\right] \end{matrix} \right\|; \qquad (1.8.13b)$$

$$\| K_{jk,l2} \| = \frac{1}{\widetilde{B}(1 \mid 3)}$$

$$\left\| \begin{matrix} -\left[\varepsilon_2 + \overline{\varepsilon}_{\overline{k}} \tanh \frac{\zeta_2}{2}\right] & (-1)^{\frac{k-1}{2}} \left[\varepsilon_2 \tanh \frac{\zeta_2}{2} + \overline{\varepsilon}_{\overline{k}}\right] \tanh \frac{l_{12}}{2} \\ (-1)^{\frac{k-1}{2}} \left[\varepsilon_2 + \overline{\varepsilon}_{\overline{k}} \tanh \frac{\zeta_2}{2}\right] & \varepsilon_2 \tanh \frac{\zeta_2}{2} + \overline{\varepsilon}_{\overline{k}} \end{matrix} \right\|; \qquad (1.8.13c)$$

$$\|K_{j3,\,l1}\| = \frac{\varepsilon_2}{\widetilde{B}(1\mid 3)\mathrm{sh}\zeta_2}\left\|\begin{matrix} -\tanh\frac{\zeta_1}{2}\mathrm{th}\frac{\zeta_3}{2} & -\tanh\frac{\zeta_3}{2} \\ \tanh\frac{\zeta_1}{2} & 1 \end{matrix}\right\|; \tag{1.8.13d}$$

where $k = 1, 3$; $\bar{k} = 4 - k$; $\bar{\varepsilon}_n = \varepsilon_n \coth\zeta_n$ and $\tilde{B}(1|3) = \dfrac{B(1|3)}{\eta^2} = \varepsilon_2^2 + \bar{\varepsilon}_2(\bar{\varepsilon}_1 + \bar{\varepsilon}_3) + \bar{\varepsilon}_1\bar{\varepsilon}_3$.

The remaining coefficients for the three-layer structure are obtained from the above-mentioned ones, if we apply the symmetry properties (1.8.2e) and (1.8.2f). In particular, for symmetric three-layer systems, when $\overleftrightarrow{\varepsilon_1} = \overleftrightarrow{\varepsilon_3}$ and $l_1 = l_3$, from (1.4.1) a multiplicative representation follows:

$$\tilde{B}(1|3) = \varepsilon_2^{(1)}\varepsilon_2^{(2)}, \tag{1.8.14a}$$

where

$$\varepsilon_2^{(1)} = \varepsilon_2 + \varepsilon_1 \coth\zeta_1 \coth(\zeta_2/2), \tag{1.8.14b}$$

$$\varepsilon_2^{(2)} = \varepsilon_2 + \varepsilon_1 \coth\zeta_1 \coth(\zeta_2/2), \tag{1.8.14c}$$

and the coefficients take the form:

$$\|K_{j2,l2}\| = -2\tanh\frac{\zeta_2}{2}\left\|\begin{matrix} [\varepsilon_2^{(1)}]^{-1} & 0 \\ 0 & [\varepsilon_2^{(2)}]^{-1} \end{matrix}\right\|; \tag{1.8.15a}$$

$$K_{12,12} = \frac{2\varepsilon_1 \coth\zeta_1}{\varepsilon_2\varepsilon_2^{(1)}}; \tag{1.8.15b}$$

$$\|K_{j2,l2}\| = \left\|\begin{matrix} -\dfrac{\tanh\frac{\zeta_1}{2}}{\varepsilon_2^{(1)}} & \dfrac{\tanh\frac{\zeta_1}{2}\tanh\frac{\zeta_2}{2}}{\varepsilon_2^{(2)}} \\ -\dfrac{1}{\varepsilon_2^{(1)}} & \dfrac{\tanh\frac{\zeta_2}{2}}{\varepsilon_2^{(2)}} \end{matrix}\right\|; \tag{1.8.15c}$$

$$\|K_{j3,\,l2}\| = \left\|\begin{matrix} -\dfrac{\tanh\frac{\zeta_1}{2}}{\varepsilon_2^{(1)}} & -\dfrac{\tanh\frac{\zeta_1}{2}\tanh\frac{\zeta_2}{2}}{\varepsilon_2^{(2)}} \\ \dfrac{1}{\varepsilon_2^{(1)}} & \dfrac{\tanh\frac{\zeta_2}{2}}{\varepsilon_2^{(2)}} \end{matrix}\right\|. \tag{1.8.15d}$$

If the polarization exists only in the middle layer, the resulting potential in the same middle layer includes the bulk and surface parts:

$$\varepsilon_0 V_2^P(\mathbf{\eta}, z) = -\frac{2\pi}{\varepsilon_2^{\parallel} l_2} \sum_{q_2} \frac{1}{T_2} e^{iq_2(z-z_1)} P_{12}(\mathbf{Q}_2)$$

$$+ \frac{2\pi}{\eta C_2 l_2} \left[\frac{\varepsilon_1 \coth\zeta_1}{\varepsilon_2 \varepsilon_2^{(1)}} \widetilde{P}_{12}(\mathbf{\eta}, 0) \frac{\cosh \omega_2}{\sinh \frac{\zeta_2}{2}} - \frac{1}{\varepsilon_2^{(2)}} \widetilde{P}_{22}(\mathbf{\eta}, 0) \frac{\sinh \omega_2}{\cosh \frac{\zeta_2}{2}} \right]. \quad (1.8.16a)$$

In the external layers, the potential is due to only the surface part of the polarization:

$$\varepsilon_0 V_1^P(\mathbf{\eta}, z) = -\frac{2\pi}{\eta C_2 l_2} \left[\frac{1}{\varepsilon_2^{(1)}} \widetilde{P}_{12}(\mathbf{\eta}, 0) - \frac{1}{\varepsilon_2^{(2)}} \tanh \frac{\zeta_2}{2} \widetilde{P}_{22}(\mathbf{\eta}, 0) \right] \frac{\sinh\left[\sqrt{\epsilon_1}\, \eta (z - z_0)\right]}{\sinh \zeta_1}; \quad (1.8.16b)$$

$$\varepsilon_0 V_3^P(\mathbf{\eta}, z) = -\frac{2\pi}{\eta C_2 l_2} \left[\frac{1}{\varepsilon_2^{(1)}} \widetilde{P}_{12}(\mathbf{\eta}, 0) + \frac{1}{\varepsilon_2^{(2)}} \tanh \frac{\zeta_2}{2} \widetilde{P}_{22}(\mathbf{\eta}, 0) \right] \frac{\sinh\left[\sqrt{\epsilon_3}\, \eta (z_3 - z)\right]}{\mathrm{sh}\zeta_3}. \quad (1.8.16c)$$

This potential is a generalization of the result (15) from [14] in what concerns (i) taking into account the anisotropy of both the inner and external layers of the system, (ii) taking into account the final thickness of layers 1 and 3, and (iii) describing the properties of the external layers by means of a high-frequency permittivity that is not generally equal to one.

For an isotropic semiconductor layer in a vacuum, from equations (1.8.16a) to (1.8.16c), the result (15) from [14] follows. Similarly, calculating the matrices (1.8.2e) and the coefficients (1.8.2c), (1.8.7) using the modulation transmission matrix of section 1.4, we determine the potential due to polarization in a five-layer system.

Note that the form of the Fourier expansion (1.8.1a) fixes only the equality of the boundary values of the polarization $\mathbf{P}_k(\mathbf{\eta}, z_k) = \mathbf{P}_k(\mathbf{\eta}, z_{k-1})$, but it does not define those values themselves. The latter can be naturally achieved by specifying the other forms of Fourier expansion as discussed below.

In the first case, corresponding to the zeros of polarization at the boundaries, we substitute the sinusoidal expansion in (1.3.9f)

$$\mathbf{P}_k(\mathbf{\eta}, z) = \frac{2\pi}{l_k} \sum_{q_k} \mathbf{P}(\mathbf{Q}_k) \sin\left[q_k(z - z_{k-1})\right], \quad (1.8.17a)$$

where

$$q_k = \frac{\pi N_k}{l_k}, \quad N_k = 0, 1, 2, \ldots, \quad \mathbf{Q}_k = (\mathbf{\eta}, q_k), \quad (1.8.17b)$$

and we find the polarization potential in the form (1.8.2a). It is convenient to introduce here a representation

$$V_n^P(\boldsymbol{\eta}, z) = V_n^{P(b)}(\eta, z) + V_n^{P(s)}(\boldsymbol{\eta}, z), \tag{1.8.18a}$$

where in terms

$$V_n^{P(b,\, s)}(\boldsymbol{\eta}, z) = \varepsilon_0^{-1}\sum_{k=I}^{K}\frac{2\pi}{l_k}\sum_{q_k}\mathbf{F}_{nk}^{(b,\, s)}(z, \mathbf{Q}_k)\mathbf{P}_k(\mathbf{Q}_k) \tag{1.8.18b}$$

a vector function

$$\begin{aligned}\mathbf{F}_{nk}^{(b)}(z, \mathbf{Q}_k) = &-\frac{1}{\varepsilon_n^{\|}T_n^2}\left\{q_n\mathbf{e}_3\left[\cos q_n(z - z_{n-1}) - \frac{\cosh \omega_n}{\cosh \frac{\zeta_n}{2}}\delta_{N_n,\, 2m}\right.\right.\\ &\left.\left.+\frac{\sinh \omega_n}{\sinh \frac{\zeta_n}{2}}\delta_{N_n,\, 2m+1}\right] + i\eta \sin\left[q_n(z - z_{n-1})\right]\right\}\delta_{n,\, k}\end{aligned} \tag{1.8.19a}$$

describes the potential created in a certain layer by the polarization from the same layer ($k = n$). A vector function

$$\begin{aligned}\mathbf{F}_{nk}^{(s)}(z, \mathbf{Q}_k) = &\frac{1}{2}[\mathbf{F}_{nk}(\mathbf{Q}_k)\theta(k > n) + \mathbf{F}_{n-1,\, k}(\mathbf{Q}_k)\theta(n > I)]\frac{\cosh \omega_n}{\cosh \frac{\zeta_n}{2}}\\ &+\frac{1}{2}[\mathbf{F}_{nk}(\mathbf{Q}_k)\theta(k > n) - \mathbf{F}_{n-1,\, k}(\mathbf{Q}_k)\theta(n > I)]\frac{\sinh \omega_n}{\sinh \frac{\zeta_n}{2}},\end{aligned} \tag{1.8.19b}$$

expressed in terms of the elements of the modulation transmission matrix of section 1.3

$$\begin{aligned}\mathbf{F}_{nk}(\mathbf{Q}_k) = &\frac{1}{\mathbf{C}_k^1}\mathbf{S}_{0q_k}^1 \tanh\frac{\zeta_k}{2}[D_{n,\, k-1}\theta(k > I) + D_{nk}\theta(k > K)]\\ &+\frac{1}{\mathbf{C}_k^2}\mathbf{S}_{0q_k}^2[D_{n,\, k-1}\theta(k > I) - D_{nk}\theta(k > K)],\end{aligned} \tag{1.8.19c}$$

describes the potential created in this layer by the polarization from all layers of the system. Considering the vectors

$$\mathbf{S}_{0q_k}^1 = C_k^1\frac{q_k}{T_k^2}\left[\sqrt{\theta_k}\,\eta\mathbf{e}_3\delta_{N_k,\, 2m} + i\boldsymbol{\eta}\coth\frac{\zeta_k}{2}\delta_{N_k,\, 2m+1}\right], \tag{1.8.20a}$$

$$\mathbf{S}_{0q_k}^2 = -C_k^2\frac{q_k}{T_k^2}\left[i\boldsymbol{\eta}\delta_{N_k,\, 2m} + \sqrt{\theta_k}\,\eta\mathbf{e}_3\coth\frac{\zeta_k}{2}\delta_{N_k,\, 2m+1}\right] \tag{1.8.20b}$$

as matrix rows of unitary projectors satisfying the conditions [cf (1.8.10)]

$$\sum_{q_k}{}' \mathbf{S}^{l_1}_{q_k' q_k} \mathbf{S}^{l_2*}_{q_k'' q_k} = \delta_{l_1 l_2} \delta_{q_k' q_k''}, \tag{1.8.21}$$

we find (when $\in_k = 1$) that $C_k^1 = C_k^2 = [\frac{4}{\zeta_k} \tanh \frac{\zeta_k}{2}]^{1/2}$, and further, in accordance with the rule for denoting the polarization amplitudes [cf (1.8.8)]:

$$\tilde{P}_{ln}(\boldsymbol{\eta}, q_k') = \sum_{q_k} \mathbf{S}^l_{q'_k q_k} \mathbf{P}_k(\mathbf{Q}_k); \; l = 1, 2; \tag{1.8.22}$$

$$\begin{aligned} V_n^{P(s)}(\boldsymbol{\eta}, z) &= \varepsilon_0^{-1} \sum\nolimits_{k=I}^{K} \frac{2\pi}{\eta l_k} \frac{1}{\sinh \frac{\zeta_n}{2}} \frac{1}{C_k^1} \\ &\left\{ \left[\sum\nolimits_{l=1}^{2} \kappa_{1n, l_k} \widetilde{P}_{lk}(\boldsymbol{\eta}, 0) \right] \cosh \omega_n + \left[\sum\nolimits_{l=1}^{2} \kappa_{2n, l_k} \widetilde{P}_{lk}(\boldsymbol{\eta}, 0) \right] \cosh \omega_n \right\} \end{aligned} \tag{1.8.23}$$

with the coefficients $K_{jn,\, lk}$, exactly the same as those previously found in the equation (1.8.7).

Finally, by completely reconstructing the matrices of unitary projectors according to their strings (1.8.20a) and (1.8.20b), we find the contribution to the potential from the polarization sources:

$$\begin{aligned} V_n^{P(b)}(\boldsymbol{\eta}, z) &= \varepsilon_0^{-1} \frac{2\pi}{l_n} \frac{C_n^1 \eta}{\varepsilon_n} \sum_{q_n \neq 0} \left\{ \left[\frac{q_n}{T_n^2} \left(\cos\left[q_n(z - z_{n-1}) \right] - \frac{\cosh \omega_n}{\sinh \frac{\zeta_n}{2}} \right) \delta_{N_n, 2m} \right. \right. \\ &\left. + \frac{\eta}{T_n^2} \sin\left[q_n(z - z_{n-1}) \right] \delta_{N_n, 2m+1} \right] \widetilde{P}_{1n}(\mathbf{Q}_n) \\ &+ \left[\frac{q_n}{T_n^2} \left(\cos\left[q_n(z - z_{n-1}) \right] + \frac{\cosh \omega_n}{\sinh \frac{\zeta_n}{2}} \right) \right. \\ &\left. \left. \delta_{N_n, 2m+1} + \frac{\eta}{T_n^2} \sin\left[q_n(z - z_{n-1}) \right] \delta_{N_n, 2m} \right] \widetilde{P}_{2n}(\mathbf{Q}_n) \right\}. \end{aligned} \tag{1.8.24}$$

In the second case, corresponding to the extrema of polarization at the boundaries, using the cosinusoidal decomposition

$$\begin{aligned} \mathbf{P}_k(\boldsymbol{\eta}, z) &+ \frac{2\pi}{l_k} \left\{ \frac{1}{2} \mathbf{P}_k(\boldsymbol{\eta}, 0) + \sum_{q_k \neq 0} \mathbf{P}_k(\mathbf{Q}_k) \cos\left[q_k(z - z_{k-1}) \right] \right\} \\ &\equiv \frac{2\pi}{l_k} \sum_{q_k}{}' \mathbf{P}_k(\mathbf{Q}_k) \cos\left[q_k(z - z_{k-1}) \right] \end{aligned} \tag{1.8.25}$$

to perform integration in (1.3.9f), we get the potential again in the form of (1.8.18a) and (1.8.18b), where now

$$\mathbf{F}_{nk}^{(b)}(z, \mathbf{Q}_k) = \frac{1}{\varepsilon_n^{\|} T_n^2}\Big\{q_n \mathbf{e}_3 \sin\left[q_n(z - z_{n-1})\right]$$
$$- i\eta\left[\cos q_n(z - z_{n-1}) - \frac{\cosh \omega_n}{\cosh \frac{\varsigma_n}{2}}\delta_{N_n,\,2m} + \frac{\sinh \omega_n}{\sinh \frac{\varsigma_n}{2}}\delta_{N_n,\,2m+1}\right]\Big\}\delta_{n,\,k}, \tag{1.8.26}$$

Moreover, the equations (1.8.19b) and (1.8.19c) remain valid with the inclusion of new vectors in them:

$$\begin{cases} \mathbf{S}_{0q_k}^1 = C_k^1 \dfrac{\sqrt{\epsilon_k}\eta}{T_k^2}\left[i\eta\delta_{N_k,\,2m} + \sqrt{\epsilon_k}\eta\mathbf{e}_3\coth\dfrac{\zeta_k}{2}\delta_{N_k,\,2m+1}\right]; \\ \mathbf{S}_{0q_k}^2 = -C_k^2 \dfrac{\sqrt{\epsilon_k}\eta}{T_k^2}\left[\sqrt{\epsilon_k}\eta\mathbf{e}_3\delta_{N_k,\,2m} + i\eta\coth\dfrac{\zeta_n}{2}\delta_{N_k,\,2m+1}\right]. \end{cases} \tag{1.8.27a,b}$$

Repeating the above procedure, we find the contribution to the potential from surface polarization in the previously obtained form (1.8.23) and the contribution from the bulk polarization, which has a structure similar to (1.8.24) and is not provided here for the sake of saving space.

In the third case, the following combination of sinusoidal and cosinusoidal decompositions is used:

$$\mathbf{P}_k(\eta, z) = \frac{2\pi}{l_k}\sum_{q_k}{}'\Big\{\mathbf{P}_k^{\perp}(\mathbf{Q}_k)\sin\left[q_k(z - z_{k-1})\right] + \mathbf{e}_3 P_k^{\|}(\mathbf{Q}_k)\cos\left[q_k(z - z_{k-1})\right]\Big\}. \tag{1.8.28}$$

In this case, additional boundary conditions are naturally described when the polarization component perpendicular to the stratification axis has zeros at the boundaries, while the component parallel to the stratification axis has extrema at the boundaries. Inserting (1.8.28) into (1.3.9f) gives the potential in the standard form (1.8.18a) and (1.8.18b) with a relatively simple vector function

$$\mathbf{F}_{nk}^{(b)}(z, \mathbf{Q}_k) = \frac{i\eta + q_n\mathbf{e}_3}{\varepsilon_n^{\|} T_n^2}\sin\left[q_n(z - z_{n-1})\right]\delta_{n,\,k}, \tag{1.8.29}$$

and the equations (1.8.19b) and (1.8.19c) now contain the vectors

$$\begin{cases} \mathbf{S}_{0q_k}^1 = C_k^1 \dfrac{i\eta q_k + \epsilon_k\eta^2\mathbf{e}_3}{T_k^2}\coth\dfrac{\zeta_k}{2}\delta_{N_k,\,2m+1}; \\ \mathbf{S}_{0q_k}^2 = -C_k^2 \dfrac{i\eta q_k + \epsilon_k\eta^2\mathbf{e}_3}{T_k^2}\delta_{N_k,\,2m+1}. \end{cases} \tag{1.8.30a,b}$$

In the same way as above, we find the contribution to the potential from the surface polarization again in the form of (1.8.23). The matrix of unitary projectors for (1.8.22) is restored in this case (where $\epsilon_k = 1$) in an exceptionally simple form:

$$\mathbf{S}^1_{q'_k q_k} = \frac{-i\eta\varepsilon_k + q_k\mathbf{e}_3}{Q_k}\delta_{q'_k q_k}, \quad q'_k \neq 0, \tag{1.8.31}$$

which allows us to get a contribution from the bulk polarization:

$$V_n^{P(b)}(\eta, z) = \varepsilon_0^{-1}\frac{2\pi}{\varepsilon_n^{||}l_n}\sum_{q_n \neq 0}\frac{1}{Q_n}\sin\left[q_n(z - z_{n-1})\right]\tilde{P}_{1n}(\mathbf{Q}_n). \tag{1.8.32}$$

In the fourth case, a different combination of sinusoidal and cosinusioidal expansions is used:

$$\mathbf{P}_k(\eta, z) = \frac{2\pi}{l_k}\sum_{q_k}{}'\left\{\mathbf{P}_k^{\perp}(\mathbf{Q}_k)\cos\left[q_k(z - z_{k-1})\right] + \mathbf{e}_3P_k^{||}(\mathbf{Q}_k)\sin\left[q_k(z - z_{k-1})\right]\right\}, \tag{1.8.33a}$$

which naturally describes additional boundary conditions when the polarization component parallel to the stratification axis has zeros at the boundary:

$$P_k^{||}(z_k) = P_k^{||}(z_{k-1}) = 0, \tag{1.8.33b}$$

and the component perpendicular to the stratification axis has its extrema. Note that the polarization confinement condition is direct and was used in [15] and also in [16], where physical justifications and bibliography are provided. In this case, from (1.3.9f) a potential follows in the form (1.8.18a), (1.8.18b) with a vector function

$$\mathbf{F}_{nk}^{(b)}(z, \mathbf{Q}_k) = -\frac{i\eta + q_n\mathbf{e}_3}{\varepsilon_n T_n^2}$$
$$\left\{\cos\left[q_n(z - z_{n-1})\right] - \frac{\cosh\omega_n}{\cosh\frac{\zeta_n}{2}}\delta_{N_n,2m} + \frac{\sinh\omega_n}{\sinh\frac{\zeta_n}{2}}\delta_{N_n,2m+1}\right\}\delta_{n,k}, \tag{1.8.34}$$

Moreover, in the equations (1.8.19b) and (1.8.19c), the vectors are

$$\begin{cases} \mathbf{S}^1_{0q_k} = C_k^1\sqrt{\epsilon_k}\,\eta\dfrac{i\eta + q_k\varepsilon_3}{T_k^2}\delta_{N_k,\,2m+1}; \\ \mathbf{S}^2_{0q_k} = -C_k^2\sqrt{\epsilon_k}\,\eta\dfrac{i\eta + q_k\varepsilon_3}{T_k^2}\coth\dfrac{\zeta_n}{2}\delta_{N_k,\,2m+1}. \end{cases} \tag{1.8.35a,b}$$

The above procedure provides a contribution to the potential from the surface polarization in a universal form (1.8.23) and, after finding the direct matrices of unitary operators by their rows (1.8.35a) and (1.8.35b), the contribution from the bulk polarization (at $\epsilon_k = 1$):

$$V_n^{P(b)}(\eta, z) = \varepsilon_0^{-1}\frac{2\pi}{\varepsilon_n^{||}l_n}\sum\nolimits_{q_n \neq 0}\frac{1}{Q_n}$$
$$\left\{\cos\left[q_n(z - z_{n-1})\right] - \frac{\cosh\omega_n}{\cosh\frac{\zeta_n}{2}}\delta_{N_n,2m} + \frac{\sinh\omega_n}{\sinh\frac{\zeta_n}{2}}\delta_{N_n,2m+1}\right\}\widetilde{P}_{1n}(\mathbf{Q}_n) \tag{1.8.36}$$

As the calculation shows, the electric field corresponding to the potentials (1.8.32) and (1.8.36) has the only component $E_{1n}^{P(b)}(\mathbf{Q}_n)$, from which it follows that the dynamics of the polarization component $\tilde{P}_{1n}(\mathbf{Q}_n)$ is characterized by the frequency $\omega_{n,\,LO}$ of the longitudinal vibrations. The other component of polarization $\tilde{P}_{2n}(\mathbf{Q}_n)$ in the latter cases does not contribute to the potential at all.

Thus, as a result of using alternative Fourier expansions of polarization, various forms of the bulk contribution to the potential and, consequently, the corresponding Hamiltonians of the interaction of electrons with bulk vibrations are obtained. Identification of the coordinate dependence of the potential $V_n^{P(b)}(\boldsymbol{\eta}, z)$ in multilayer structures and superlattices is possible, for instance, by the spectra of Raman scattering of light. Since there is not always detailed information about the spectral composition and properties of individual modes of vibrational excitations, and because discrepancies in the parameters of integral effects (polar and excitonic) when using different types of bulk contributions to the potential, as the calculation shows, are as small as a few percent, then a choice of one of them is made according to the features of the problem, for example, the system's geometry.

1.9 Polarization potential in periodic systems

Here we use the following numbering: a for odd layers ($n = 2j - 1$) and b for even layers ($n = 2j$). Due to the translational symmetry, polarization vectors in the same type of layers differ only by a phase multiplier that depends on the difference in the layer numbers. Therefore, the Fourier images of polarization in layers of the type a and b can be represented as series of the following types:

$$\mathbf{P}_n(\mathbf{Q}_n) = \sum_{\kappa} e^{in\kappa L/2}\mathbf{P}_f(\kappa, \mathbf{Q}_f), \tag{1.9.1}$$

where for $n = 2j - 1$ $f = a$, for $n = 2j$ $f = b$, $L = l_a - l_b$ is the period. The wave vector $\boldsymbol{\kappa}$ changes within an area of length $2\pi/L$: $-\pi/L \leqslant \kappa \leqslant \pi/L$. As a result, the potential (1.8.12) takes the form

$$\begin{aligned}
&\varepsilon_0 V_n^P(\boldsymbol{\eta}, z) = \sum_{\kappa} e^{i\frac{n}{2}\kappa L} \\
&\left\{-\frac{2\pi}{\varepsilon^{\parallel}_f l_f}\sum_{q_f\neq 0}\frac{1}{T_f}\left[e^{iq_f(z-z_{n-1})} + \frac{1}{1-C_f}\left(C_f - \frac{\cosh\omega_n}{\cosh\frac{\zeta_f}{2}}\right)\right]\widetilde{P}_{1f}\left(\kappa\widetilde{Q}_f\right)\right. \\
&+\frac{1}{\eta}\sum_{C=a,b}\frac{\pi}{C_c l_c}\frac{1}{\sinh\left(\frac{\zeta_f}{2}\right)} \\
&\left.\times\left(\left[\sum_{l=1}^{2}K_{1f,lc}(\kappa)\widetilde{P}_{lc}(\kappa, \boldsymbol{\eta}, 0)\right]\cosh\omega_n + \left(\left[\sum_{l=1}^{2}K_{2f,lc}(\kappa)\widetilde{P}_{lc}(\kappa, \boldsymbol{\eta}, 0)\right]\sinh\omega_n\right)\right)\right\}
\end{aligned} \tag{1.9.2}$$

.

The coefficients in (1.9.2) are the sums:

$$\begin{Vmatrix} K_{ja,\,la}(\kappa) & K_{ja,\,lb}(\kappa) \\ K_{jb,\,la}(\kappa) & K_{jb,\,lb}(\kappa) \end{Vmatrix} = \sum_{L=-\infty}^{\infty} \begin{Vmatrix} K_{j,\,2I-1,\,2L-1}; & e^{\frac{i\kappa L}{2}} K_{j,\,2L-1;l,\,2L} \\ e^{-\frac{i\kappa L}{2}} K_{j,\,2I;l2L-1}; & K_{j,\,2I;l,\,2L} \end{Vmatrix} e^{i(L-J)\kappa L}. \quad (1.9.3)$$

Substituting into the right part (1.9.3) the coefficients $K_{jn,\,lk}$ according to the equation (1.8.7) with matrices (1.5.10a), (1.5.10b), and (1.5.10c) and performing summation over L, we find the coefficients $K_{jf,\,lc}$ in a symmetrical form:

$$K_{1f,\,1c}(\kappa) = -\eta \tanh\frac{l_{1f}}{2} \tanh\frac{l_{1c}}{2} A_{fc}(\kappa,\ ++++); \quad (1.9.4a)$$

$$K_{1f,\,2c}(\kappa) = \eta \tanh\frac{l_{1f}}{2} A_{fc}(\kappa,\ +\pm-); \quad (1.9.4b)$$

$$K_{2f,\,1c}(\kappa) = -\eta \tanh\frac{l_{1c}}{2} A_{fc}(\kappa,\ \pm\pm); \quad (1.9.4c)$$

$$K_{2f,\,2c}(\kappa) = \eta A_{fc}(\kappa,\ \pm\mp). \quad (1.9.4d)$$

The obtained coefficients are related to matrices

$$\|A_{fc}(\kappa,\ +\pm\pm+)\| = \frac{2}{\Delta(\kappa)} \begin{Vmatrix} (\pm\nu + \mu_a + \mu_b \cos(\kappa L)) & (\nu \pm \mu_a \pm \mu_b)\cos\left(\frac{\kappa L}{2}\right) \\ (\nu \pm \mu_a \pm \mu_b)\cos\left(\frac{\kappa L}{2}\right) & \pm\nu + \mu_a \cos(\kappa L) + \mu_b \end{Vmatrix}; \quad (1.9.5a)$$

$$\|A_{fc}(\kappa,\ +\pm\mp-)\| = \frac{2i}{\Delta(\kappa)} \begin{Vmatrix} \mu_b \sin(\kappa L)) & (\nu \pm \mu_a \mp \mu_b)\sin\left(\frac{\kappa L}{2}\right) \\ (\nu \mp \mu_a \pm \mu_b)\sin\left(\frac{\kappa L}{2}\right) & \mu_a \sin(\kappa L) \end{Vmatrix}, \quad (1.9.5b)$$

where $\Delta(\kappa) \equiv \nu^2 - \mu_a^2 - \mu_b^2 - 2\mu_a\mu_b \cos(\kappa L)$, and the symbols ν, μ_a, μ_b are defined by the expressions (1.3.3b).

1.10 Conclusion

For planar multilayer systems of an arbitrary number of polar layers, by solving the electrostatic Poisson equation with the appropriate boundary conditions, potentials are found due to the following sources of fields: charges at the interfaces of the structure, charges in the bulk of layers, polarization of matter, and external sources of stresses.

References

[1] Alferov Z I, Zhilyaev Y V and Shmartsev Y V 1971 Splitting the conduction band of a 'superlattice' based on GaP_xAs_{1-x} *FTP* **5** 196–8

[2] Landau L D and Lifshits E M 2005 *Theoretical Physics in 10 Volumes. Volume 8. Electrodynamics of Continuous Media* (Moscow: Fizmatlit) p 656

[3] Ando M, Fowler A and Stern F 1985 *Electronic Properties of Two-Dimensional Systems* (Moscow: Mir) 415 p

[4] Smirnov V I 1974 *Course of Higher Mathematics* (Moscow: Nauka) p 480

[5] Gaponov S V, Luskin B M and Salanenko N N 1979 On the possibility of using a structure with strict clarity of the laser spraying method *Lett. LJ* **5** 516–21

[6] Esaki L 1986 A bird's-eye view of the evolution of semiconductor superlattices and quantum wells *IEEE J. Quantum Electron* **22** 1611–24

[7] Esaki L and Tsu R 1970 Superlattice and negative differential conductivity in semiconductors *IBM J. Res. Dev.* **14** 61–5

[8] Kaneko S, Nasu H, Ikegami T, Matsuoka J and Kamiya K 1992 Influence of production conditions on the properties of thin SiO_2 glass films doped with CdSe microcrystals obtained by radio frequency sputtering *Jpn. Phys.* **31** 2206

[9] Wendler L 1985 Electron–phonon interaction in dielectric two-layer systems. The effect of electronic polarizability *Phys. Status Solidi* B **129** 513–30

[10] Muradian A A and Wright G B 1966 Observation of the interaction of plasmons with longitudinal optical phonons in GaAs *Phys. Rev. Lett.* **16** 999–1001

[11] Pokatilov E P and Fomin V M 1984 *Interaction of an Electron with Polarization Optical Vibrations in Periodic Structures* No. 508M-05.85 (Chisinau: MoldNIINTI) p 14

[12] Fasol G, Tanaka M, Sasaki H and Horikoshi Y 1988 The roughness of the interface and the dispersion of limited LO-phonons in GaAs quantum wells/AlAs *Phys. Rev.* B **38** 6056–65

[13] Hai G K, Peeters F M and Devreese J T 1993 Electron-optical–phonon interaction in $GaAs/Al_xGa_{1-x}As$ quantum wells due to interface, layer and half-space modes *Phys. Rev.* B **48** 4666

[14] Bryksin V V and Firsov Y 1971 A. interaction of an electron with surface phonons in an ion crystal plate *FTT* **13** 496–503

[15] Pokatilov E P, Beril S I and Fomin V M 1988 Image potentials and image forces in the theory of polarons *Phys. Status Solidi* B **147** 163–72

[16] Kash K, Van der Haag B P, Matoni D D, Gozdz A S, Flores L T, Harbison J P and Sturge M D 1991 Observation of quantum confinement using deformation gradients *Phys. Rev. Lett.* **67** 1326–9

Chapter 2

Potentials in multilayer cylindrical and spherical systems

2.1 Introduction

Mesoscopic systems are attracting attention because of the observation of new physical effects that can be used in electronic and optoelectronic devices. In recent decades, planar multilayer structures and superlattices have been intensively studied due to advances in molecular beam epitaxy (MBE) technology [1–6].

The specific features of all low-dimensional structures are as follows: (1) the confinement of vibrational excitations, as well as the movements of charge carriers; (2) the presence of interface vibrational modes. The spectra of optical polar oscillations in the approximation of a dielectric medium for a layer in vacuum were studied in [7–10], The separation of bulk-like and surface vibrations was performed. The Hamiltonian of the electron–phonon interaction has been considered for single- and multielectron systems in the same approximation in a number of papers [7, 11–16]. Volume-like and interface modes have been experimentally observed in superlattices [17]. Reviews of experimental and theoretical studies of the effects of phonon confinement were presented in [10, 18].

New advances in technological methods such as MBE, in combination with etching and microwave plasma technology [19], magnetron and cathode sputtering [20], and electron beam lithography [21], made it possible to create quasi-one-dimensional (quantum wires) and quasi-two-dimensional (quantum dots) structures, which stimulated the growth of interest in theoretical [22–24] and experimental [25–27] studies.

The spectra of optical polar oscillations in spherical points and cylindrical wires were first theoretically investigated in [28]. Quantum wires with an oval cross-section were considered in [22, 29], where it was shown that the potential of interface polar oscillations for the case of an elliptical cross-section can be described analytically. The calculation of the ground state energy and the effective mass of the polaron in a

doi:10.1088/978-0-7503-6164-4ch2

quantum wire with a rectangular cross-section was performed by the Feynman variational method in [30], where only the electron–phonon interaction with bulk phonon modes was taken into account. The confinement of bulk phonons and interface phonons in rectangular quantum wires were studied in [31–33]. The separation of bulk and interface vibrations is a difficult problem that always exists when deriving the Hamiltonian of the electron–phonon interaction. This difficulty was circumvented by the introduction of a parabolic potential to limit the movement of the electron, so that interaction with interface phonons could be neglected. The polaron parameters for rectangular quantum wires with a parabolic potential were calculated in [31, 34]. The relationship between energy and momentum for polarons in a rectangular quantum wire was studied in [35], taking into account the interaction of an electron with bulk phonon modes. The scattering coefficients due to electron–phonon interaction with bulk LO-phonons in quantum wires were calculated in [36].

It should be noted that in all the above-mentioned works, the exact separation of bulk and interface vibrational excitations was not performed, despite the fact that without this procedure, the exact shape of the vibrational modes and the Hamiltonian of the electron–phonon interaction cannot be obtained. We determine the bulk and interface vibrational modes and the Hamiltonian of the electron–phonon interaction based on the theory of electrostatic potential in multilayer cylindtrical and spherical structures.

2.2 Spatial distribution equations potentials in multilayer cylindrical systems

Consider a cylindrical multilayer structure (figure 2.1).

The equations for the electrostatic field in the first layer have the form:

$$\begin{cases} \operatorname{rot} \mathbf{E}_n = 0, \\ \operatorname{div} \mathbf{D}_n = \rho_n, \end{cases} \tag{2.2.1}$$

where $\mathbf{E}_n$, $\mathbf{D}_n$ are, accordingly, the intensity and induction of the electrostatic field in the nth layer, $n = I, I+1, \ldots, K-1, K$;

$$\mathbf{D}_n = \varepsilon_0 \overleftrightarrow{\varepsilon}_n \mathbf{E}_n + \mathbf{P}_n, \tag{2.2.2}$$

$\overleftrightarrow{\varepsilon}_n$, $\mathbf{P}_n$ are the high-frequency permittivity tensor and the inertia-free polarization vector in the nth layer correspondingly.

Using the first equation of the set (2.2.1), the strength of the electrostatic field in each layer can be expressed as

$$\mathbf{E}_n(\mathbf{r}) = -\operatorname{grad} V_n(\mathbf{r}), \tag{2.2.3}$$

where $V_n(\mathbf{r})$ is the potential is in the same layer.

From equation (2.2.2) we get

$$\operatorname{div} \mathbf{D}_n = \operatorname{div} (\varepsilon_0 \overleftrightarrow{\varepsilon}_n \mathbf{E}_n) + \operatorname{div} \mathbf{P}_n, \tag{2.2.4}$$

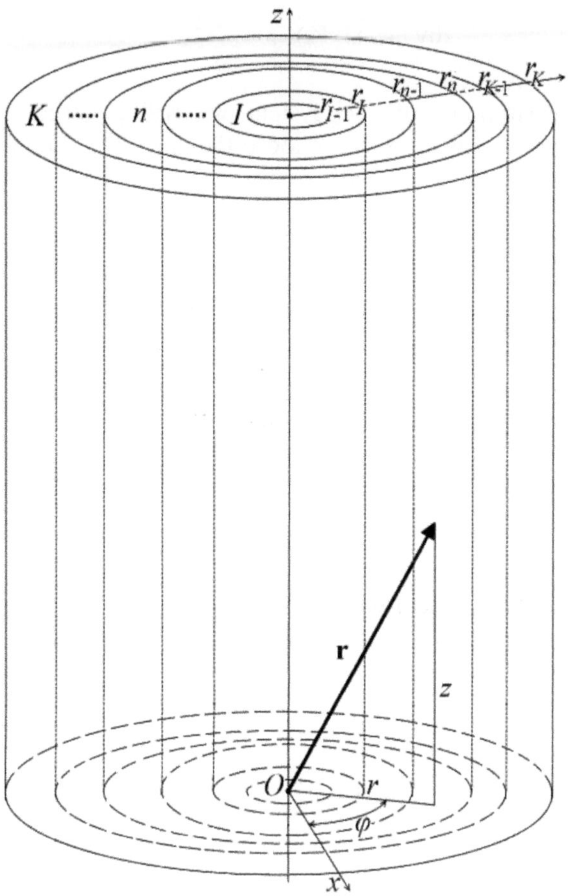

Figure 2.1. Cylindrical multilayer system.

then the second equation of the set (2.2.1) is written as

$$\mathrm{div}\,(\varepsilon_0 \overset{\leftrightarrow}{\varepsilon}_n \mathbf{E}_n) + \mathrm{div}\,\mathbf{P}_n = \rho_n,$$

or

$$\mathrm{div}\,(\varepsilon_0 \overset{\leftrightarrow}{\varepsilon}_n \mathbf{E}_n) = \rho_n - \mathrm{div}\,\mathbf{P}_n. \tag{2.2.5}$$

Denote:

$$\tilde{\rho}_n = \rho_n - \mathrm{div}\,\mathbf{P}_n, \tag{2.2.6}$$

then (2.2.5) taking into account (2.2.6) will take the form:

$$\mathrm{div}\,(\varepsilon_0 \overset{\leftrightarrow}{\varepsilon}_n \mathbf{E}_n) = \tilde{\rho}_n. \tag{2.2.7}$$

Substituting (2.2.3) in (2.2.7), we get the equation for the potential in each layer in vector form:

$$\operatorname{div}(\overleftrightarrow{\varepsilon}_n \nabla V_n) = -\frac{1}{\varepsilon_0}\tilde{\rho}_n. \tag{2.2.8}$$

Based on the symmetry of the system we will proceed to the cylindrical coordinates (r,φ, z), and as an axis z choose the axis of the axial symmetry of the structure (see figure 2.1). Then within the nth layer the following inequality is satisfied:

$$r_{n-1} < r < r_n. \tag{2.2.9}$$

We use expressions for the gradient of the scalar field and the divergence of the vector field in cylindrical coordinates:

$$\nabla V(r, \varphi, z) = \left(\mathbf{e}_r\frac{\partial}{\partial r} + \mathbf{e}_\varphi\frac{1}{r}\frac{\partial}{\partial\varphi} + \mathbf{e}_z\frac{\partial}{\partial z}\right)V(r, \varphi, z), \tag{2.2.10}$$

$$\operatorname{div}\mathbf{u} = \frac{1}{r}\frac{\partial}{\partial r}(ru_r) + \frac{1}{r}\frac{\partial u_\varphi}{\partial\varphi} + \frac{\partial u_z}{\partial z}, \tag{2.2.11}$$

where $(\mathbf{e}_r, \mathbf{e}_\varphi, \mathbf{e}_z)$ are cylindrical basic orts.

We assume that due to the axial symmetry of the physical properties of the structure under consideration, the permittivity tensor in cylindrical coordinates will take a diagonal form:

$$\overleftrightarrow{\varepsilon}_n = \begin{pmatrix} \varepsilon_n^r & 0 & 0 \\ 0 & \varepsilon_n^\varphi & 0 \\ 0 & 0 & \varepsilon_n^z \end{pmatrix}. \tag{2.2.12}$$

Then, substituting (2.2.10)–(2.2.12) in (2.2.8), taking into account the homogeneity of the layers, we get the equation for the potential in cylindrical coordinates:

$$\frac{\varepsilon_n^r}{r}\frac{\partial}{\partial r}\left(r\frac{\partial V_n}{\partial r}\right) + \frac{\varepsilon_n^\varphi}{r^2}\frac{\partial^2 V_n}{\partial\varphi^2} + \varepsilon_n^z\frac{\partial^2 V_n}{\partial z^2} = -\frac{1}{\varepsilon_0}\tilde{\rho}_n. \tag{2.2.13}$$

We decompose the potential and density of bulk charges in each layer into a Fourier series by and into the Fourier integral by z:

$$V_n(r, \varphi, z) = \sum_{m=-\infty}^{\infty}\int_{-\infty}^{\infty} d\eta V_n(r|m, \eta)\ e^{im\varphi+i\eta z}, \tag{2.2.14}$$

$$\tilde{\rho}_n(r, z, \varphi) = \sum_{m=-\infty}^{\infty}\int_{-\infty}^{\infty} d\eta\ \tilde{\rho}_n(r|\eta, m)\ e^{i\eta z+im\varphi}, \tag{2.2.15}$$

where $V_n(r|m, \eta)$ is the Fourier transform of the image potential in the layer:

$$V_n(r|m, \eta) = \frac{1}{4\pi^2}\int_{-\infty}^{\infty} dz \int_0^{2\pi} d\varphi\ V_n(r, \varphi, z)\ e^{-im\varphi-i\eta z}; \tag{2.2.16}$$

$\tilde{\rho}_n(r|m, \eta)$ is the Fourier image of the density of bulk charges:

$$\tilde{\rho}_n(r|m, \eta) = \frac{1}{4\pi^2}\int_{-\infty}^{\infty} dz \int_0^{2\pi} d\varphi\ \tilde{\rho}_n(r, \varphi, z)\ e^{-i\eta z - im\varphi}. \qquad (2.2.17)$$

Then the equation for the Fourier image of the potential will take the form:

$$\frac{\varepsilon_n^r}{r}\frac{\partial}{\partial r}\left\{r\frac{\partial}{\partial r}(V_n r|m, \eta)\right\} - \left(\frac{\varepsilon_n^\varphi m^2}{r^2} + \varepsilon_n^z\eta^2\right)V_n(r|m, \eta) = -\varepsilon_0^{-1}\tilde{\rho}_n(r|m, \eta),$$

or

$$\left\{\frac{\partial^2}{\partial r^2} + \frac{1}{r}\frac{\partial}{\partial r} - \left(\frac{\varepsilon_n^\varphi m^2}{\varepsilon_n^r r^2} + \frac{\varepsilon_n^z}{\varepsilon_n^r}\eta^2\right)\right\}V_n(r|m, \eta) = -\frac{1}{\varepsilon_0\varepsilon_n^r}\tilde{\rho}_n(r|m, \eta). \qquad (2.2.18a)$$

Given that the equation (2.2.16) includes derivatives only for r, next, we will consider it as an equation in ordinary derivatives, and the variables φ и z—as parameters. Then (2.2.16) can be represented as

$$\left\{r^2\frac{d^2}{dr^2} + r\frac{d}{dr} - \left(\frac{\varepsilon_n^\phi}{\varepsilon_n^r}m^2 + \frac{\varepsilon_n^z}{\varepsilon_n^r}\eta^2 r^2\right)\right\}V_n(r|m, \eta) = -\frac{r^2}{\varepsilon_0\varepsilon_n^r}\tilde{\rho}_n(r|m, \eta) \qquad (2.2.18b)$$

Denote

$$\varepsilon_n = \frac{\varepsilon_n^z}{\varepsilon_n^r}; \qquad (2.2.19)$$

$$\eta_n = \sqrt{\varepsilon_n}\ |\eta|; \qquad (2.2.20)$$

$$\nu_n \equiv \sqrt{\frac{\varepsilon_n^\varphi}{\varepsilon_n^r}}\,|m|; \qquad (2.2.21)$$

$$\xi = \eta_n r. \qquad (2.2.22)$$

Then

$$r^2\frac{d^2}{dr^2} + r\frac{d}{dr} - \left(\frac{\varepsilon_n^\varphi}{\varepsilon_n^r}m^2 + \frac{\varepsilon_n^z}{\varepsilon_n^r}\eta^2 r^2\right) = \xi^2\frac{d^2}{d\xi^2} + \xi\frac{d}{d\xi} - (\nu_n^2 + \xi^2); \qquad (2.2.23)$$

$$\frac{r^2}{\varepsilon_0\varepsilon_n^r} = \frac{\xi^2}{\eta_n^2\varepsilon_0\varepsilon_n^r} = \frac{\xi^2}{\frac{\varepsilon_n^z}{\varepsilon_n^r}\eta^2\varepsilon_0\varepsilon_n^r} = \frac{\xi^2}{\varepsilon_0\varepsilon_n^z\eta^2}. \qquad (2.2.24)$$

Substituting (2.2.23) and (2.2.24) in (2.2.16), we get:

$$\left\{\xi^2\frac{d^2}{d\xi^2} + \xi\frac{d}{d\xi} - (\nu_n^2 + \xi^2)\right\}\ V_n(\xi\ |\ m, \eta) = -\frac{\xi^2}{\varepsilon_0\varepsilon_n^z\eta^2}\tilde{\rho}_n(\xi\ |\ m, \eta). \qquad (2.2.25)$$

Equation (2.2.25) is a linear inhomogeneous second-order differential equation. The corresponding homogeneous equation is a modified Bessel equation of order ν_n. The fundamental set of solutions of the latter consists of modified Bessel functions $I_{\nu_n}(\xi)$ and $K_{\nu_n}(\xi)$. Therefore, the solution of equation (2.2.25) will be found in the form [35]

$$V_n(r|m, \eta) = a_n(r)I_{\nu_n}(\eta_n r) + b_n(r)K_{\nu_n}(\eta_n r). \tag{2.2.26}$$

Then

$$\frac{d}{dr}V_n(r|\eta, m) = \frac{da_n(r)}{dr}I_{\nu_n}(\eta_n r) + \frac{db_n(r)}{dr}K_{\nu_n}(\eta_n r) + a_n(r)\frac{dI_{\nu_n}(\eta_n r)}{dr} + b_n(r)\frac{dK_{\nu_n}(\eta_n r)}{dr} \tag{2.2.27}$$

$$\begin{aligned}\frac{d^2}{dr^2}V_n(r|\eta, m) = {} & \frac{d^2a_n(r)}{dr^2}I_{\nu_n}(\eta_n r) + \frac{d^2b_n(r)}{dr^2}K_{\nu_n}(\eta_n r) + a_n(r)\frac{d^2I_{\nu_n}(\eta_n r)}{dr^2} \\ & + b_n(r)\frac{d^2K_{\nu_n}(\eta_n r)}{dr^2} + 2\frac{da_n(r)}{dr}\frac{dI_{\nu_n}(\eta_n r)}{dr} + 2\frac{db_n(r)}{dr}\frac{dK_{\nu_n}(\eta_n r)}{dr}.\end{aligned} \tag{2.2.28}$$

Entering functions:

$$S_n(r, r'|m, \eta) = I_{\nu_n}(\eta_n r)K_{\nu_n}(\eta_n r') - I_{\nu_n}(\eta_n r')K_{\nu_n}(\eta_n r); \tag{2.2.29}$$

$$s_n(m, \eta) = S_n(r_n, r_{n-1}|m, \eta); \tag{2.2.30}$$

$$F_n(r, r'|m, \eta) = \frac{S_n(r, r'|m, \eta)}{s_n(m, \eta)}. \tag{2.2.31}$$

Then

$$\begin{aligned}V_n(r|m, \eta) = {} & V(r_n|m, \eta)F_n(r, r_{n-1}|m, \eta) + V(r_{n-1}|m, \eta)F_n(r_n, r|m, \eta) \\ & + \frac{1}{\varepsilon_0}\int_{r_{n-1}}^{r_n} G_n(r, r'|m, \eta)\tilde{\rho}_n(r'|m, \eta)\, r'dr',\end{aligned} \tag{2.2.32}$$

where

$$\begin{aligned}G_n(r, r'|m, \eta) & = \frac{1}{\varepsilon_n^r s_n}\{\theta(r > r')S_n(r_n, r|m, \eta)S_n(r', r_{n-1}|m, \eta) \\ & \quad + \theta(r < r')S_n(r_n, r'|m, \eta)S_n(r, r_{n-1}|m, \eta)\} \\ & = \frac{s_n}{\varepsilon_n^r}\{\theta(r > r')F_n(r_n, r|m, \eta)F_n(r', r_{n-1}|m, \eta) \\ & \quad + \theta(r < r')F_n(r_n, r'|m, \eta)F_n(r, r_{n-1}|m, \eta)\}\end{aligned} \tag{2.2.33}$$

is Green's function for the Fourier image of a potential in the nth layer.

To find the boundary values of the Fourier image of the potential, we use the boundary conditions for the normal component of the electric field induction:

$$D_{n+1,\,\perp} - D_{n\,\perp} = \sigma_n, \tag{2.2.34}$$

where σ_n is density of surface third-party charges.

Let us substitute a material equation for this equality

$$\mathbf{D}_n = \varepsilon_0 \overleftrightarrow{\varepsilon}\, \mathbf{E}_n + \mathbf{P}_n, \tag{2.2.35}$$

we get

$$\varepsilon_0[\, (\overleftrightarrow{\varepsilon}_{n+1}\mathbf{E}_{n+1})_\perp - (\overleftrightarrow{\varepsilon}_n\mathbf{E}_n)_\perp]_{r=r_n} = \sigma_n - P_{n+1,\,\perp} + P_{n\,\perp}, \tag{2.2.36}$$

or, given the cylindrical symmetry of the system,

$$\varepsilon_0[\varepsilon^r_{n+1}E^r_{n+1} - \varepsilon^r_n E^r_n]_{r=r_n} = \sigma_n - (P^r_{n+1} - P^r_n)_{r=r_n}. \tag{2.2.37}$$

Next we use the relationship between the strength and potential of the electrostatic field:

$$\mathbf{E} = -gradV, \tag{2.2.38}$$

in particular for the radial component of tension:

$$E^r_n = -\frac{\partial V_n(r)}{\partial r}. \tag{2.2.39}$$

Note that equations (2.2.37) and (2.2.39) refer to both the originals and images of the corresponding functions.

Substitute (2.2.32) in (2.2.39), and after some transformations we get:

$$-\mu_n V(r_{n-1}|m,\eta) + \tilde{\upsilon}_n V(r_n|m,\eta) - \mu_{n+1}V(r_{n+1}|m,\eta) = \varepsilon_0^{-1} r_n \tilde{\sigma}_n(m,\eta);$$

$$n = I, \ldots, K-1, \tag{2.2.40}$$

where

$$\mu_n = \frac{\varepsilon^r_n}{s_n(m,\eta)}; \tag{2.2.41}$$

$$\begin{aligned}\tilde{\sigma}_n(m,\eta) &= \sigma_n m, \eta + P^r_n(r_n|m,\eta) - P^r_{n+1}(r_n|m,\eta)\\ &+\frac{1}{r_n}\int_{r_n}^{r_{n+1}} F_{n+1}(r_{n+1}, r'|m,\eta)\tilde{p}_{n+1}(r'|m,\eta)\, r'dr'\\ &+\frac{1}{r_n}\int_{r_{n-1}}^{r_n} F_n(r', r_{n-1}|m,\eta)\tilde{p}_n(r'|m,\eta)r'dr',\end{aligned} \tag{2.2.42}$$

$$\tilde{\upsilon}_n = r_n(\varepsilon^r_n \eta_n C^+_n + \varepsilon^r_{n+1}\eta_{n+1}C_{n+1}), \tag{2.2.43}$$

$$C^+_n = \frac{1}{s_n}[I'_{\nu_n}(\eta_n r_n)K_{\nu_n}(\eta_n r_{n-1}) - I_{\nu_n}(\eta_n r_{n-1})K'_{\nu_n}(\eta_n r_n)], \tag{2.2.44}$$

$$C_n = \frac{1}{s_n}[I'_{\nu_n}(\eta_n r_{n-1})K_{\nu_n}(\eta_n r_n) - I_{\nu_n}(\eta_n r_n)K'_{\nu_n}(\eta_n r_{n-1})]; \tag{2.2.45}$$

$$\tilde{\rho}_n(r|m,\eta) = \rho_n(r|m,\eta) - \frac{1}{r}\frac{d}{dr}\left[rP_n^r(r|m,\eta) \right] - \left(i\eta\, P_n^z(r|m,\eta) + i\frac{m}{r}P_n^\phi(r|m,\eta)\right). \quad (2.2.46)$$

Denote

$$\sigma'_n(m,\eta) = \tilde{\sigma}_n(m,\eta) + \delta_{n,I}\frac{\varepsilon_0\varepsilon_I^r}{r_I s_I}V(r_{I-1}|m,\eta) + \delta_{n,K-1}\frac{\varepsilon_0\varepsilon_K^r}{r_{K-1}s_K}V(r_K|m,\eta), \quad (2.2.47)$$

then the set (2.2.40) can be represented as follows:

$$\sum_{k=I}^{K-1} b_{nk}\, V(r_k|m,\eta) = \frac{r_n}{\varepsilon_0}\sigma'_n(m,\eta); \quad n = I, \ldots, K-1, \quad (2.2.48)$$

where

$$\|b_{nk}\| = \begin{vmatrix} \tilde{\nu}_I & -\mu_{I+1} & 0 & 0 & 0 & 0 & 0 & 0 & \ldots \\ -\mu_{I+1} & \tilde{\nu}_{I+1} & -\mu_{I+2} & 0 & 0 & 0 & 0 & 0 & \ldots \\ \ldots & \ldots & \ldots & \ldots & \ldots & \ldots & \ldots & \ldots & \ldots \\ \ldots & 0 & -\mu_{n-1} & \tilde{\nu}_{n-1} & -\mu_n & 0 & 0 & 0 & \ldots \\ \ldots & 0 & 0 & -\mu_n & \tilde{\nu}_n & -\mu_{n+1} & 0 & 0 & \ldots \\ \ldots & 0 & 0 & 0 & -\mu_{n+1} & \tilde{\nu}_{n+1} & -\mu_{n+2} & 0 & \ldots \\ \ldots & \ldots & \ldots & \ldots & \ldots & \ldots & \ldots & \ldots & \ldots \\ \ldots & 0 & 0 & 0 & 0 & 0 & -\mu_{K-2} & \tilde{\nu}_{K-2} & -\mu_{K-1} \\ \ldots & 0 & 0 & 0 & 0 & 0 & 0 & -\mu_{K-1} & \tilde{\nu}_{K-1} \end{vmatrix}. \quad (2.2.49)$$

The solution of this set is written as

$$V(r_n|m,\eta) = \frac{1}{\varepsilon_0}\sum_{k=I}^{K-1} D_{nk}(I|K) r_k \sigma'_n(m,\eta), \quad (n = I, \ldots, K-1), \quad (2.2.50)$$

where

$$\|D_{nk}\| = \|b_{nk}\|^{-1}. \quad (2.2.51)$$

Denote:

$$B(I|K) = \det \|b_{nk}\|; \quad (2.2.52)$$

$$B_{kn}(I|K) = (-1)^{k+n}\mathrm{minor}(b_{kn}), \quad (2.2.53)$$

where $minor(b_{kn})$ is the minor of the element b_{kn}.

Then:

$$B_{kn}(I|K) = B(I|k)B(n+1|K)\prod_{j=k+1}^{n}\mu_j; \quad (k \leqslant n); \quad (2.2.54)$$

$$B_{kn} = B_{nk}. \tag{2.2.55}$$

Hence:

$$D_{nk}(I|K) = \frac{B(I|K)B(n+1|K)}{B(I|K)} \prod_{j=k+1}^{n} \mu_j; \quad (k \leqslant n); \tag{2.2.56}$$

$$D_{kn} = D_{nk}; \tag{2.2.57}$$

$$D_{nk}(I|K) = \frac{B(I|n)B(n+1|K)}{B(I|K)}. \tag{2.2.58}$$

Here:

$$B(I|K) = \tilde{\nu}_{K-1}B(I|K-1) - \mu_{K-1}^2 B(I|K-2). \tag{2.2.59}$$

2.3 Classification of potential contributions by sources in multilayer cylindrical systems

From the equations (2.2.50)–(2.2.58) it follows that the Fourier image of the potential created at the boundary of the nth and $(n+1)$-th layers $(n = I, \ldots, K-1)$ can be written, as in the case of a planar system, as the sum of contributions from various sources:

$$V(r_n|m, \eta) = V^e(r_n|m, \eta) + V^s(r_n|m, \eta) + V^b(r_n|m, \eta) + V^p(r_n|m, \eta). \tag{2.3.1}$$

where

$$\begin{aligned} V^e(r_n|m, \eta) &= \frac{1}{\varepsilon_0}\sum_{k=I}^{K-1} D_{nk}(I|K)r_k \left\{\delta_{k,\,I}\frac{\varepsilon_0\varepsilon_I^r}{r_I s_I}V(r_{I-1}|m, \eta) + \delta_{k,\,K-1}\frac{\varepsilon_0\varepsilon_K^r}{r_{K-1}s_K}V(r_K|m, \eta)\right\} \\ &= D_{nI}\frac{\varepsilon_I^r}{s_I}V(r_{I-1}|m, \eta) + D_{n,\,K-1}\frac{\varepsilon_K^r}{s_K}(r_K|m, \eta) \end{aligned} \tag{2.3.2}$$

is the potential created by the external sources of the fields;

$$V^s(r_n|m, \eta) = \varepsilon_0^{-1}\sum_{k=I}^{K-1} D_{nk} r_k \sigma_k(m, \eta) \tag{2.3.3}$$

is the potential created by surface charges located at the interface of layers;

$$V^b(r_n|m,\eta)=\frac{1}{\varepsilon_0}\sum_{k=I}^{K-1}D_{nk}r_k\left\{\frac{1}{r_k}\int_{r_{k-1}}^{r_k}F_k(r',r_{k-1}|m,\eta)\rho_k(r'|m,\eta)r'dr'\right.$$
$$\left.+\frac{1}{r_k}\int_{r_k}^{r_{k+1}}F_{k+1}(r_{k+1},r'|m,\eta)\rho_{k+1}(r'|m,\eta)r'dr'\right\}$$
$$=\varepsilon_0^{-1}\sum_{k=I}^{K-1}D_{nk}\int_{r_{k-1}}^{r_k}F_k(r',r_{k-1})\rho_k(r'|\eta,m)r'dr'$$
$$+\varepsilon_0^{-1}\sum_{k=I+1}^{K}D_{n,k-1}\int_{r_{k-1}}^{r_k}F_k(r_k,r'|m,\eta)\rho_k(r'|m,\eta)r'dr' \tag{2.3.4}$$
$$=\varepsilon_0^{-1}\sum_{k=I}^{K}\int_{r_{k-1}}^{r_k}\{\theta(k<K)D_{nk}F_k(r',r_{k-1}|m,\eta)+\theta(k>I)D_{n<k-1}F_k(r_k,r'|m,\eta)\}$$
$$\rho_k(r'|m,\eta)r'dr'$$
$$=\varepsilon_0^{-1}\sum_{k=I}^{K}\int_{r_{k-1}}^{r_k}G_{nk}(r'|m,\eta)\rho_k(r'|m,\eta)r'dr',\quad (n=I,\bar{K}-1),$$

is the potential created by the volume charges inside the layers, where

$$G_{nk}(r|m,\eta)\equiv\theta(k<K)D_{nk}F_k(r,r_{k-1}|m,\eta)+\theta(k>I)D_{n,k-1}F_k(r_k,r|m,\eta); \tag{2.3.5}$$

$$V^P(r_n|m,\eta)=\varepsilon_0^{-1}\sum_{k=I}^{K}\left\{\frac{\partial G_{nk}(r'|m,\eta)}{\partial r'}P_k^r(r'|m,\eta)\right.$$
$$\left.-iG_{nk}(r'|m,\eta)\left[\eta P_k^z(r'|m,\eta)+\frac{m}{r'}P_k^\varphi(r'|m,\eta)\right]\right\}r'dr',\quad (n=I,\bar{K}-1). \tag{2.3.6}$$

is the potential created by the slow polarization.

From equations (2.2.6), (2.2.11), and (2.2.17) we get:

$$V_n(r|m,\eta)=V(r_n|m,\eta)F_n(r,r_{n-1}|m,\eta)+V(r_{n-1}|m,\eta)F_n(r_n,r|m,\eta)$$
$$+\varepsilon_0^{-1}\int_{r_{n-1}}^{r_n}G_n(r,r'|m,\eta)$$
$$\left[\rho_n(r'|m,\eta)-\frac{1}{r'}\frac{\partial}{\partial r'}[r'P_n^r(r'|m,\eta)]-\frac{im}{r'}P_n^\phi(r'|m,\eta)\right. \tag{2.3.7}$$
$$\left.-i\eta P_n^z(r'|m,\eta)\right]r'dr'.$$

Using the ratio $G_n(r,r_{n-1}|m,\eta)=G_n(r,r_n|m,\eta)=0$, in the right part (2.3.7), we can perform integration by parts:

$$V_n(r|m,\eta)=V(r_n|m,\eta)F_n(r,r_{n-1}|m,\eta)+V(r_{n-1}|m,\eta)F_n(r_n,r|m,\eta)$$
$$+\varepsilon_0^{-1}\int_{r_{n-1}}^{r_n}\left\{G_n(r,r'|m,\eta)\left[\rho_n(r'|m,\eta)-\frac{im}{r'}P_n^\varphi(r'|m,\eta)-i\eta\,P_n^z(r'|m,\eta)\right]\right. \tag{2.3.8}$$
$$\left.+\frac{\partial G_n(r,r'|m,\eta)}{\partial r'}P_n^z(r'|m,\eta)r'dr'\right\}.$$

Then get:

$$\begin{aligned} &V_n^e(r|m,\eta) \\ &= \left[D_{nI}\frac{\varepsilon_I^r}{s_I(m,\eta)}V(r,r_{I-1}|m,\eta) + D_{n,K-1}\frac{\varepsilon_K^r}{s_K(m,\eta)}V(r_K|m,\eta)\right] \\ &\theta(n<K)F_n(r,r_{n-1}|m,\eta) \\ &+\left[D_{n-1,I}\frac{\varepsilon_I^r}{s_I(m,\eta)}V(r_{I-1}|m,\eta) + D_{n-1,K-1}\frac{\varepsilon_K^r}{s_K(m,\eta)}V(r_K|m,\eta)\right] \\ &\theta(n>I)F_n(r_n,r|m,\eta); \end{aligned} \tag{2.3.9}$$

$$\begin{aligned} V_n^s(r|m,\eta) &= \varepsilon_0^{-1}\sum_{k=I}^{K-1} D_{nk}r_k\sigma_k F_n(r,r_{n-1}|m,\eta)\theta(n<K) \\ &+\varepsilon_0^{-1}\sum_{k=I}^{K-1} D_{n-1,k}r_k\sigma_k F_n(r_n,r|m,\eta)\theta(n>I) \\ &=\varepsilon_0^{-1}\sum_{k=I}^{K-1}\left[D_{nk}\theta(n<K)F_n(r,r_{n-1}|m,\eta)\right. \\ &\left.+D_{n-1,k}\theta(n>I)F_n(r_n,r|m,\eta)\right]\; r_k\sigma_k; \end{aligned} \tag{2.3.10}$$

$$\begin{aligned} V_n^b(r|m,\eta) &= \varepsilon_0^{-1}\sum_{k=I}^{K} G_{nk}(r'|m,\eta)\rho_k(r'|m,\eta)r'dr'\theta(n<K)F_n(r,r_{n-1}|m,\eta) \\ &+\varepsilon_0^{-1}\sum_{k=I}^{K}\int_{r_{k-1}}^{r_k} G_{nk}(r'|m,\eta)\rho_k(r'|m,\eta)r'dr'F_n(r_n,r)\theta(n>I) \\ &+\varepsilon_0^{-1}\int_{r_{n-1}}^{r_n} G_n(r,r'|m,\eta)\rho_n(r'|m,\eta)r'dr' \\ &=\varepsilon_0^{-1}\sum_{k=I}^{K}\int_{r_{k-1}}^{r_k} G_{nk}(r,r'|m,\eta)\rho_k(r'|m,\eta)r'dr'; \end{aligned} \tag{2.3.11}$$

$$\begin{aligned} V_n^p(r|m,\eta) &= \varepsilon_0^{-1}\sum_{k=I}^{K}\int_{r_{k-1}}^{r_k}\left\{\frac{\partial G_{nk}(r,r'|m,\eta)}{\partial r'}P_k^r(r'|m,\eta)\right. \\ &\left.-iG_{nk}(r,r'|m,\eta)\left[\frac{m}{r'}P_k^z(r'|m,\eta)+\eta P_k^z(r'|m,\eta)\right]\right\}r'dr', \end{aligned} \tag{2.3.12}$$

where

$$\begin{aligned} G_{nk}(r,r'|m,\eta) &= \theta(n<K)G_{nk}(r'|m,\eta)F_n(r,r_{n-1}|m,\eta) \\ &+\theta(n>I)G_{n-1,k}(r'|m,\eta)F_n(r_n,r|m,\eta)+\delta_{nk}G_k(r,r'|m,\eta). \end{aligned} \tag{2.3.13}$$

2.4 Equations of the spatial distribution potentials in multilayered spherical systems

Consider a spherical multilayer structure (figure 2.2).

The boundaries of the nth layer are defined by the equations $R_{n-1} < r < R_n$ ($n = 1, \ldots, K$). Boundary values $V(R_0, \theta, \varphi)$ and $V(R_K, \theta, \varphi)$ of the potential on the external borders are set. Here (r, θ, φ) are the spherical coordinates. You need to find a value $V_n(\mathbf{r})$ potential at any point.

Let's write down the usual electrostatic equations:

$$\begin{cases} \text{rot}\, \mathbf{E}_n = 0, \\ \text{div}\, \mathbf{D}_n = \rho_n, \\ \mathbf{D}_n = \varepsilon_0 \overleftrightarrow{\varepsilon}_n \mathbf{E}_n + \mathbf{P}_n, \end{cases} \tag{2.4.1}$$

where ρ_n is the third-party charge density and $\mathbf{P}_n$ is the vector of inertial polarization.

Further, using the relationship between the strength and potential of the electrostatic field $\mathbf{E}_n = -\text{grad}\, V_n$ and substituting this ratio in the system (2.4.1), we get the equation for the potential in the nth layer:

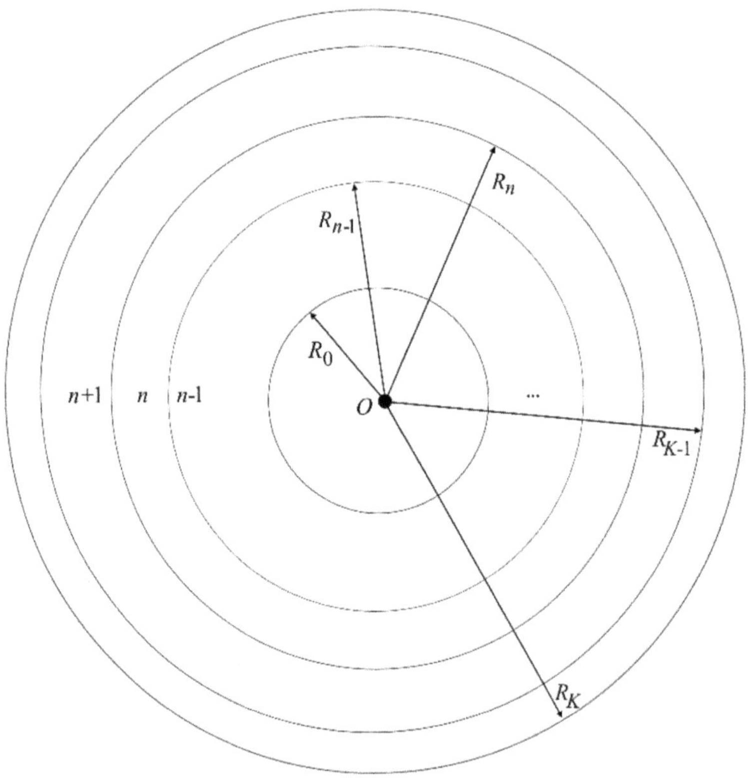

Figure 2.2. Spherical multilayer structure.

$$\operatorname{div}\left(-\varepsilon_0 \overset{\leftrightarrow}{\varepsilon}_n \operatorname{grad} V_n + \mathbf{P}_n\right) = \rho_n;$$

$$-\varepsilon_0 \operatorname{div}\left(\overset{\leftrightarrow}{\varepsilon}_n \operatorname{grad} V_n\right) + \operatorname{div} \mathbf{P}_n = \rho_n;$$

$$-\varepsilon_0 \operatorname{div}\left(\overset{\leftrightarrow}{\varepsilon}_n \operatorname{grad} V_n\right) = \tilde{\rho}_n,$$

where the following notation is entered:

$$\tilde{\rho}_n \equiv \rho_n - \operatorname{div} \mathbf{P}_n. \tag{2.4.2}$$

Then

$$\operatorname{div}\left(\overset{\leftrightarrow}{\varepsilon}_n \operatorname{grad} V_n\right) = -\varepsilon_0^{-1} \tilde{\rho}_n. \tag{2.4.3}$$

Let the permittivity tensor of each layer have a diagonal form in a spherical coordinate system:

$$\overset{\leftrightarrow}{\varepsilon}_n = \begin{pmatrix} \varepsilon_n^{||} & 0 & 0 \\ 0 & \varepsilon_n^{\perp} & 0 \\ 0 & 0 & \varepsilon_n^{\perp} \end{pmatrix}, \tag{2.4.4}$$

where $\varepsilon_n^{||}$ and $\varepsilon_n^{\perp}$ are the permittivity of the nth layer in the radial and perpendicular directions, respectively.

Then, writing equation (2.4.3) in spherical coordinates and substituting the condition (2.4.4), we get the desired equation for the potential in the nth layer:

$$\begin{gathered} \frac{\varepsilon_n^{||}}{r^2} \frac{\partial}{\partial r}\left[r^2 \frac{\partial V_n(\mathbf{r})}{\partial r}\right] + \frac{\varepsilon_n^{\perp}}{r^2 \sin\theta} \frac{\partial}{\partial \theta}\left[\sin\theta \frac{\partial V_n(\mathbf{r})}{\partial \theta}\right] \\ + \frac{\varepsilon_n^{\perp}}{r^2 \sin^2\theta} \frac{\partial^2 V_n(\mathbf{r})}{\partial \phi^2} = -\varepsilon_0^{-1} \tilde{\rho}_n(\mathbf{r}) \end{gathered} \tag{2.4.5}$$

Decompose the scalar coordinate functions in a series of spherical functions $Y_{lm}(\theta, \varphi)$ by equations of the form

$$f_n(r, \theta, \varphi) = \sum_{l=0}^{\infty} \sum_{m=-l}^{l} f_n(r; l, m) Y_{lm}(\theta, \varphi), \tag{2.4.6a}$$

where

$$f_n(r; l, m) = \frac{1}{4\pi} \int_0^{2\pi} d\varphi \int_0^{\pi} d\theta \sin\theta \, f_n(r, \theta, \varphi) Y_{lm}^*(\theta, \varphi). \tag{2.4.6b}$$

Using a decomposition of the form (2.4.6a) for the potential in each layer and considering the identity that the functions satisfy $Y_{lm}(\theta, \varphi)$:

$$\Delta_\Omega Y_{lm}(\theta, \phi) \equiv \frac{1}{\sin\theta} \frac{\partial}{\partial \theta}\left[\sin\theta \frac{\partial Y_{lm}(\theta, \phi)}{\partial \theta}\right] + \frac{1}{\sin^2\theta} \frac{\partial Y_{lm}(\theta, \phi)}{\partial \phi^2} = -l(l+1) Y_{lm}(\theta, \phi)$$

where Δ_Ω is the angular part of the Laplace operator, and for spherical potential harmonics in the nth layer, we obtain the following equation:

$$\frac{\varepsilon_n^{||}}{r^2}\frac{d}{dr}\left[r^2\frac{dV_n(r;\, l,\, m)}{dr}\right] - \frac{\varepsilon_n^{\perp} l(l+1)}{r^2}V_n(r;\, l,\, m) = -\varepsilon_0^{-1}\tilde{\rho}_n(r;\, l,\, m). \qquad (2.4.7)$$

(it is written as an equation in ordinary derivatives, since l and m play the role of parameters).

We introduce the following notation:

$$\epsilon_n \equiv \frac{\varepsilon_n^{\perp}}{\varepsilon_n^{||}}. \qquad (2.4.8)$$

Then equation (2.4.7) will take the form

$$\left\{\frac{d^2}{dr^2} + \frac{2}{r}\frac{d}{dr} - \frac{\epsilon_n l(l+1)}{r^2}\right\}V_n(r;\, l,\, m) = -(\varepsilon_0\varepsilon_n^{||})^{-1}\tilde{\rho}_n(r;\, l,\, m). \qquad (2.4.9)$$

Let's first solve the corresponding homogeneous equation:

$$\left\{\frac{d^2}{dr^2} + \frac{2}{r}\frac{d}{dr} - \frac{\epsilon_n l(l+1)}{r^2}\right\}V_n(r;\, l,\, m) = 0. \qquad (2.4.10)$$

Let's write this equation in the following form:

$$f'' + \frac{2f'}{r} - \frac{Af}{r^2} = 0. \qquad (2.4.11)$$

To solve it we use substitution $f = r^\alpha$. Then for α we get the characteristic equation

$$\alpha(\alpha - 1) + 2\alpha - A = 0,$$

or

$$\alpha^2 + \alpha - A = 0. \qquad (2.4.12)$$

Find its roots:

$$\alpha_{1,\,2} = -\frac{1}{2} \pm \sqrt{\frac{1}{4} + A} = -\frac{1}{2} \pm \sqrt{\epsilon_n l(l+1) + \frac{1}{4}}. \qquad (2.4.13)$$

Denote:

$$l_n = \sqrt{\epsilon_n l(l+1) + \frac{1}{4}} - \frac{1}{2}. \qquad (2.4.14)$$

Then the general solution of equation (2.4.11) can be represented as

$$f = C_1 r^{l_n} + \frac{C_2}{r^{l_n+1}}. \qquad (2.4.15)$$

Entering functions

$$S_n(r, r') \equiv \frac{r^{l_n}}{(r')^{l_n+1}} - \frac{(r')^{l_n}}{r^{l_n+1}} \tag{2.4.16}$$

and coefficients

$$s_n \equiv S_n(R_n, R_{n-1}) \tag{2.4.17}$$

(note that the functions $S_n(r, r')$ and coefficients s_n depend on l, but not on m), also functions

$$F_n(r, r') \equiv \frac{S_n(r, r')}{s_n}; \tag{2.4.18a}$$

$$F_{n1}(r) \equiv F_n(r, R_{n-1}) \equiv \frac{S_n(r, R_{n-1})}{s_n}; \tag{2.4.18b}$$

$$F_{n2}(r) \equiv F_n(R_n, r) \equiv \frac{S_n(R_n, r)}{s_n}, \tag{2.4.18c}$$

satisfying the homogeneous equation (2.4.10) with boundary conditions

$$F_{n1}(R_n) = 1, \tag{2.4.19a}$$

$$F_{n1}(R_{n-1}) = 0, \tag{2.4.19b}$$

$$F_{n2}(R_n) = 0, \tag{2.4.19c}$$

$$F_{n2}(R_{n-1}) = 1. \tag{2.4.19d}$$

We look for the solution (2.4.9) in the form

$$\begin{aligned} V_n(r; l, m) = {} & V(R_n; l, m)F_{n1}(r) + V(R_{n-1}; l, m)F_{n2}(r) \\ & + \varepsilon_0^{-1} \int_{R_{n-1}}^{R_n} g_n(r, r')\tilde{\rho}_n(r'; l, m) r'^2 dr'. \end{aligned} \tag{2.4.20}$$

Then to find the function $g_n(r, r')$ we obtain an integro-differential equation

$$\begin{aligned} & \left\{ \frac{d^2}{dr^2} + \frac{2}{r}\frac{d}{dr} - \frac{\epsilon_n l(l+1)}{r^2} \right\} \varepsilon_0^{-1} \int_{R_{n-1}}^{R_n} g_n(r, r')\tilde{\rho}_n(r'; l, m) r'^2 dr' \\ & = \varepsilon_0^{-1} (\varepsilon_n^{||})^{-1} \tilde{\rho}_n(r; l, m) \end{aligned} \tag{2.4.21}$$

from which

$$\left\{ \frac{d^2}{dr^2} + \frac{2}{r}\frac{d}{dr} - \frac{\epsilon_n l(l+1)}{r^2} \right\} g_n(r, r') = -\frac{1}{\varepsilon_n^{||} r^2} \delta(r - r'). \tag{2.4.22}$$

Boundary condition:

$$g_n(R_n, r') = 0, \tag{2.4.23a}$$

$$g_n(R_{n-1}, r') = 0. \tag{2.4.23b}$$

We look for the solution of equation (2.4.22) in the form

$$g_n(r, r') = A(r')[\theta(r > r')F_{n1}(r')F_{n2}(r) + \theta(r < r')F_{n1}(r)F_{n2}(r')]. \tag{2.4.24}$$

Then:

$$\left\{\frac{d^2}{dr^2} + \frac{2}{r}\frac{d}{dr} - \frac{\epsilon_n l(l+1)}{r^2}\right\}g_n(r, r') = A(r')\delta(r - r')[F_{n1}(r')F'_{n2}(r) - F'_{n1}(r)F_{n2}(r')] =$$

$$\left\langle F'_{n1}(r) = \frac{1}{s_n}\frac{d}{dr}S_n(r, R_{n-1}) = \frac{1}{s_n}\frac{d}{dr}\left(\frac{r^{l_n}}{R_{n-1}^{l_n+1}} - \frac{R_{n-1}^{l_n}}{r^{l_n+1}}\right) = \frac{1}{s_n}\left(l_n\frac{r^{l_n-1}}{R_{n-1}^{l_n+1}} + (l_n+1)\frac{R_{n-1}^{l_n}}{r^{l_n+2}}\right);\right.$$

$$\left. F'_{n2}(r) = \frac{1}{s_n}\frac{d}{dr}S_n(R_n, r) = \frac{1}{s_n}\frac{d}{dr}\left(\frac{R_n^{l_n}}{r^{l_n+1}} - \frac{r^{l_n}}{R_n^{l_n+1}}\right) = \frac{1}{s_n}\left(-(l_n+1)\frac{R_n^{l_n}}{r^{l_n+2}} - l_n\frac{r^{l_n-1}}{R_n^{l_n+1}}\right)\right\rangle$$

$$= -\frac{A(r')\delta(r - r')}{s_n^2}\left\{\left(\frac{r^{l_n}}{R_{n-1}^{l_n+1}} - \frac{R_{n-1}^{l_n}}{r^{l_n+1}}\right)\left[(l_n+1)\frac{R_n^{l_n}}{r^{l_n+2}} + l_n\frac{r^{l_n-1}}{R_n^{l_n+1}}\right]\right.$$

$$\left. + \left(\frac{R_n^{l_n}}{r^{l_n+1}} - \frac{r^{l_n}}{R_n^{l_n+1}}\right)\left[l_n\frac{r^{l_n-1}}{R_{n-1}^{l_n+1}} + (l_n+1)\frac{R_{n-1}^{l_n}}{r^{l_n+2}}\right]\right\}$$

$$= -\frac{A(r')\delta(r - r')}{s_n^2}\left\{(l_n+1)\frac{R_n^{l_n}}{R_{n-1}^{l_n+1}r^2} - (l_n+1)\frac{R_{n-1}^{l_n}{}_{n-1}^{l_n}R_n^{l_n}}{r^{2l_n+3}} - l_n\frac{R_{n-1}^{l_n}}{R_n^{l_n+1}r^2} + l_n\frac{r^{2l_n-1}}{R_{n-1}^{l_n+1}R_n^{l_n+1}}\right.$$

$$\left. + (l_n+1)\frac{R_{n-1}^{l_n}R_n^{l_n}}{r^{2l_n+3}} + l_n\frac{R_n^{l_n}}{R_{n-1}^{l_n+1}r^2} - (l_n+1)\frac{R_{n-1}^{l_n}}{R_n^{l_n+1}r^2} - l_n\frac{r^{2l_n-1}}{R_{n-1}^{l_n+1}R_n^{l_n+1}}\right\}$$

$$= -\frac{A(r')\delta(r - r')}{s_n^2}\left\{(2l_n+1)\frac{R_n^{l_n}}{R_{n-1}^{l_n+1}r^2} - (2l_n+1)\frac{R_{n-1}^{l_n}}{R_n^{l_n+1}r^2}\right\} = -\frac{A(r')\delta(r - r')(2l_n+1)s_n}{s_n^2 r^2};$$

$$\left\{\frac{d^2}{dr^2} + \frac{2}{r}\frac{d}{dr} - \frac{\epsilon_n l(l+1)}{r^2}\right\}g_n(r, r') = -\frac{A(r')\delta(r - r')(2l_n+1)}{s_n r^2}. \tag{2.4.26a}$$

Comparing the right parts of equations (2.4.22) and (2.4.26a), we get

$$A(r') = \frac{s_n}{(2l_n+1)\varepsilon_n^{||}}. \tag{2.4.26b}$$

Then

$$g_n(r, r') = \frac{s_n}{(2l_n+1)\varepsilon_n^{||}}[\theta(r > r')F_{n1}(r')F_{n2}(r) + \theta(r < r')F_{n1}(r)F_{n2}(r')]. \tag{2.4.26c}$$

Write down the boundary conditions for the normal components of the electric displacement vector $D_n^r(\mathbf{r})$:

$$D_{n+1}^r(R_n, \theta, \varphi) - D_n^r(R_n, \theta, \varphi) = \sigma_n(\theta, \varphi). \tag{2.4.27}$$

Then for spherical harmonics of the vector $\mathbf{D}$ we get

$$D_{n+1}^r(R_n; l, m) - D_n^r(R_n; l, m) = \sigma_n(l, m). \tag{2.4.28}$$

For the electric field strength, this equation gives

$$\begin{aligned}&\varepsilon_0\varepsilon_{n+1}^{||}E_{n+1}^r(R_n; l, m) - \varepsilon_0\varepsilon_n^{||}E_n^r(R_n; l, m) + P_{n+1}^r(R_n; l, m)\\&\quad - P_n^r(R_n; l, m)\\&= \sigma_n(l, m).\end{aligned} \tag{2.4.29}$$

By recording the relationship between the strength and potential of the electric field $\mathbf{E}_n = -gradV_n$, from (2.4.29) we obtain a boundary condition for spherical harmonics of the potential:

$$\begin{aligned}&-\varepsilon_0\varepsilon_{n+1}^{||}\frac{dV_{n+1}(r; l, m)}{dr}\Big|_{r=R_n} + \varepsilon_0\varepsilon_n^{||}\frac{dV_n(r; l, m)}{dr}\Big|_{r=R_n} =\\&= \sigma_n(l, m) - P_{n+1}^r(R_n; l, m) + P_n^r(R_n; l, m).\end{aligned} \tag{2.4.30}$$

Substitute (2.4.20) in (2.4.30):

$$\begin{aligned}&\varepsilon_0\varepsilon_n^{||}\Big\{V(R_n; l, m)F'_{n1}(r)\Big|_{r=R_n} + V(R_{n-1}; l, m)F'_{n2}(r)\Big|_{r=R_n}\\&\qquad +\varepsilon_0^{-1}\int_{R_{n-1}}^{R_n}\frac{\partial g_n(r, r')}{\partial r}\Big|_{r=R_n}\tilde{\rho}_n(r')r'^2dr'\Big\}\\&-\varepsilon_0\varepsilon_{n+1}^{||}\Big\{V(R_{n+1}; l, m)F'_{n+1,1}(r)\Big|_{r=R_n} + V(R_n; l, m)F'_{n+1,2}(r)\Big|_{r=R_n}\\&\qquad +\varepsilon_0^{-1}\int_{R_n}^{R_{n+1}}\frac{\partial g_{n+1}(r, r')}{\partial r}\Big|_{r=R_n}\tilde{\rho}_{n+1}(r')r'^2dr'\Big\}\\&\qquad =\sigma_n(l, m) + P_n^r(R_n; l, m) - P_{n+1}^r(R_n; l, m).\end{aligned} \tag{2.4.31}$$

Find the derivatives:

$$F'_{n1}(r) = \frac{1}{s_n}\frac{\partial S_n(r, R_{n-1})}{\partial r} = \frac{1}{s_n}\left\{l_n\frac{r^{l_n-1}}{R_{n-1}^{l_n+1}} + (l_n+1)\frac{R_{n-1}^{l_n}}{r^{l_n+2}}\right\}; \tag{2.4.32a}$$

$$F'_{n2}(r) = \frac{1}{s_n}\frac{\partial S_n(R_n, r)}{\partial r} = -\frac{1}{s_n}\left\{l_n\frac{r^{l_n-1}}{R_n^{l_n+1}} + (l_n+1)\frac{R_n^{l_n}}{r^{l_n+2}}\right\}. \tag{2.4.32b}$$

Denote:

$$\xi(l, x) = (l+1)x^l + lx^{-l-1};\ \xi(l_n, x) \equiv \xi_n(x), \tag{2.4.33}$$

Then:

$$\begin{cases} F'_{n1}(r) = \dfrac{1}{s_n r^2}\xi\left(l_n, \dfrac{R_{n-1}}{r}\right); \\ F'_{n2}(r) = -\dfrac{1}{s_n r^2}\xi\left(l_n, \dfrac{R_n}{r}\right); \end{cases} \Rightarrow \begin{cases} F'_{n1}(r) = \dfrac{1}{s_n r^2}\xi_n\left(\dfrac{R_{n-1}}{r}\right); \\ F'_{n2}(r) = -\dfrac{1}{s_n r^2}\xi_n\left(\dfrac{R_n}{r}\right). \end{cases} \tag{2.4.34}$$

We also calculate the derivative of the Green's function:

$$\begin{aligned} \frac{\partial g_n(r, r')}{\partial r} &= \frac{s_n}{(2l_n + 1)\varepsilon_n^{||}} \\ &\left[-\theta(r > r')F_{n1}(r')\frac{1}{s_n r^2}\xi_n\left(\frac{R_n}{r}\right) + \theta(r' > r)F_{n2}(r')\frac{1}{s_n r^2}\xi_n\left(\frac{R_{n-1}}{r}\right)\right] \\ &= \frac{1}{(2l_n + 1)\varepsilon_n^{||} r^2}\left[\theta(r < r')\xi_n\left(\frac{R_{n-1}}{r}\right)F_{n2}(r') - \theta(r > r')\xi_n\left(\frac{R_n}{r}\right)F_{n1}(r')\right]; \\ \left.\frac{\partial g_n(r, r')}{\partial r}\right|_{r=R_n} &= -\frac{1}{(2l_n + 1)\varepsilon_n^{||}R_n^2}F_{n1}(r')(2l_n + 1) = -\frac{F_{n1}(r')}{\varepsilon_n^{||}R_n^2}; \end{aligned} \tag{2.4.35a}$$

$$\left.\frac{\partial g_{n+1}(r, r')}{\partial r}\right|_{r=R_n} = \frac{1}{(2l_{n+1} + 1)\varepsilon_{n+1}^{||}R_n^2}F_{n+1,\,2}(r')(2l_{n+1} + 1) = \frac{F_{n+1,\,2}(r')}{\varepsilon_{n+1}^{||}R_n^2}. \tag{2.4.35b}$$

Substituting (2.4.34), (2.4.35a), and (2.4.35b) in (2.4.31), we get:

$$\begin{aligned} &\varepsilon_0\varepsilon_n^{||}\left\{V(R_n; l, m)\frac{1}{s_n R_n^2}\xi_n\left(\frac{R_{n-1}}{R_n}\right) + V(R_{n-1}; l, m)\left(-\frac{2l_n + 1}{s_n R_n^2}\right)\right. \\ &\qquad\left. + \varepsilon_0^{-1}\int_{R_{n-1}}^{R_n}\left[-\frac{F_{n1}(r')}{\varepsilon_{n+1}^{||}R_n^2}\right]\tilde{\rho}_n(r'; l, m)r'^2 dr'\right\} \\ &-\varepsilon_0\varepsilon_{n+1}^{||}\left\{V(R_{n+1}; l, m)\frac{1}{s_{n+1}R_n^2}(2l_{n+1} + 1) - V(R_n; l, m)\frac{1}{s_{n+1}R_n^2}\xi_{n+1}\left(\frac{R_{n+1}}{R_n}\right)\right. \\ &\qquad\left. + \varepsilon_0^{-1}\int_{R_n}^{R_{n+1}}\frac{F_{n+1,2}(r')}{\varepsilon_{n+1}^{||}R_n^2}\tilde{\rho}_{n+1}(r'; l, m)r'^2 dr'\right\} \\ &= \sigma_n + P_n^r(R_n; l, m) - P_{n+1}^r(R_n; l, m). \end{aligned} \tag{2.4.36}$$

Denote:

$$\mu_n \equiv \frac{\varepsilon_n^{||}(2l_n + 1)}{s_n}; \quad \nu_n \equiv \frac{\varepsilon_n^{||}}{s_n}\xi_n\left(\frac{R_{n-1}}{R_n}\right) + \frac{\varepsilon_{n+1}^{||}}{s_{n+1}}\xi_{n+1}\left(\frac{R_{n+1}}{R_n}\right). \tag{2.4.37}$$

Then (2.4.36) takes the look of

$$\frac{\varepsilon_0}{R_n^2}\nu_n V(R_n;\, l,\, m) - \frac{\varepsilon_0}{R_n^2}\mu_n V(R_{n-1};\, l,\, m) - \frac{\varepsilon_0}{R_n^2}\mu_{n+1} V(R_{n+1};\, l,\, m)$$
$$= \sigma_n(l,\, m) + P_n^r(R_n;\, l,\, m) - P_{n+1}^r(R_n;\, l,\, m) \tag{2.4.38}$$
$$+ \frac{1}{R_n^2}\int_{R_{n-1}}^{R_n} F_{n1}(r)\tilde{\rho}_n(r;\, l,\, m) r^2 dr + \frac{1}{R_n^2}\int_{R_n}^{R_{n+1}} F_{n+1,\,2}(r)\tilde{\rho}_{n+1}(r;\, l,\, m) r^2 dr.$$

Denote:

$$\begin{aligned}\tilde{\sigma}_n(l,\, m) &= \sigma_n(l,\, m) + P_n^r(R_n;\, l,\, m) - P_{n+1}^r(R_n;\, l,\, m)\\ &+ \frac{1}{R_n^2}\int_{R_{n-1}}^{R_n} F_{n1}(r)\tilde{\rho}_n(r;\, l,\, m) r^2 dr\\ &+ \frac{1}{R_n^2}\int_{R_n}^{R_{n+1}} F_{n+1,2}(r)\tilde{\rho}_{n+1}(r;\, l,\, m) r^2 dr.\end{aligned} \tag{2.4.39}$$

As a result we get the following set of equations for finding the boundary values of the potential:

$$-\mu_n V(R_{n-1};\, l,\, m) + \nu_n V(R_n;\, l,\, m) - \mu_{n+1} V(R_{n+1};\, l,\, m) = \varepsilon_0^{-1} R_n^2 \tilde{\sigma}_n(l,\, m) \tag{2.4.40}$$

We introduce an additional symbol:

$$\sigma'_n(l,\, m) \equiv \tilde{\sigma}_n(l,\, m) + \delta_{n,\,1}\varepsilon_0 R_n^{-2}\mu_n V(R_0;\, l,\, m) + \delta_{n,\,K-1}\varepsilon_0 R_n^{-2}\mu_{n+1} V(R_K;\, l,\, m) \tag{2.4.41}$$

$$b_{nn} = \nu_n;\; b_{n,\,n-1} = b_{n-1,\,n} = -\mu_n. \tag{2.4.42}$$

Then the set of equations (2.4.40) will take the form

$$\sum_{k=1}^{K-1} b_{nk} V(R_K;\, l,\, m) = \varepsilon_0^{-1} R_n^2 \sigma'_n(l,\, m). \tag{2.4.43}$$

Its solution can also be represented in matrix form:

$$V(R_n;\, l,\, m) = \varepsilon_0^{-1} \sum_{k=1}^{K-1} D_{nk} R_k^2 \sigma'_k(l,\, m), \tag{2.4.44}$$

where

$$\|D_{nk}\| = \|b_{nk}\|^{-1} \tag{2.4.45}$$

is the the modulation transmission matrix.

2.5 Classification of potential contributions in multilayer spherical systems by source

The potential boundary values can be represented as a superposition of contributions from multiple sources:

$$
\begin{aligned}
V(R_n;\,l,\,m) &= \varepsilon_0^{-1}\sum_{k=1}^{K-1} D_{nk}R_k^2\{\delta_{k,\,1}\varepsilon_0 R_k^{-2}\mu_k V(R_0;\,l,\,m)\\
&+ \delta_{k,\,K-1}\varepsilon_0 R_k^{-2}\mu_{k+1}V(R_K;\,l,\,m) + \sigma_k(l,\,m) + P_k^r(R_k;\,l,\,m) - P_{k+1}^r(R_k;\,l,\,m)\\
&+ \frac{1}{R_k^2}\int_{R_{k-1}}^{R_k} F_{k1}(r)\tilde{\rho}_k(r;\,l,\,m)r^2dr + \frac{1}{R_k^2}\int_{R_k}^{R_{k+1}} F_{k+1,\,2}(r)\tilde{\rho}_{k+1}(r;\,l,\,m)r^2dr\}\\
&= V^e(R_n;\,l,\,m) + V^s(R_n;\,l,\,m) + V^b(R_n;\,l,\,m) + V^p(R_n;\,l,\,m),
\end{aligned}
\tag{2.5.1}
$$

where:

(a) $V^e(R_n;\,l,\,m)$ is the contribution from the external boundary of:

$$V^e(R_n;\,l,\,m) = D_{n1}\mu_1 V(R_0;\,l,\,m) + D_{n,\,K-1}\mu_K V(R_K;\,l,\,m); \tag{2.5.2}$$

(b) $V^s(R_{n;};\,l,\,m)$ is contribution from surface third-party charges:

$$V^s(R_{n;};\,l,\,m) = \varepsilon_0^{-1}\sum_{k=1}^{K-1} D_{nk}R_k^2\sigma_k(l,\,m); \tag{2.5.3}$$

(c) $V^b(R_n;\,l,\,m)$ is the contribution from bulk charges:

$$
\begin{aligned}
V^b(R_n;\,l,\,m) &= \varepsilon_0^{-1}\sum_{k=1}^{K-1} D_{nk}\int_{R_{k-1}}^{R_k} F_{k1}(r)\rho_k(r;\,l,\,m)r^2dr\\
&+\varepsilon_0^{-1}\sum_{k=2}^{K} D_{n,k-1}\int_{R_{k-1}}^{R_k} F_{k2}(r)\rho_k(r;\,l,\,m)r^2dr\\
&= \varepsilon_0^{-1}\sum_{k=1}^{K}\int_{R_{k-1}}^{R_k}[\theta(k<K)D_{nk}F_{k1}(r) + \theta(k>1)D_{n,k-1}F_{k2}(r)]\rho_k(r;\,l,\,m)r^2dr.
\end{aligned}
\tag{2.5.4}
$$

Redefining matrix elements D_{nk}:

$$D_{nk}\theta(k<K)\theta(n<K)\theta(k\geqslant 1)\theta(n\geqslant 1) \to D_{nk}, \tag{2.5.5}$$

so

$$D_{nk} \equiv 0,\ \text{if}\ n,\,k \geqslant K\ \text{or}\ n,\,k<1.$$

Let's enter the function

$$\gamma_{nk}(r) \equiv D_{nk}F_{k1}(r) + D_{n,\,k-1}F_{k2}(r). \tag{2.5.6}$$

Then the formula (2.5.4) can be represented as

$$V^b(R_n;\,l,\,m) = \varepsilon_0^{-1}\sum_{k=1}^{K}\int_{R_{k-1}}^{R_k}\gamma_{nk}(r)\rho_k(r;\,l,\,m)r^2dr; \tag{2.5.7}$$

d) $V^p(R_n;\,l,\,m)$ is the contribution from the slow polarization:

$$
\begin{aligned}
V^P(R_n;\,l,\,m) &= \varepsilon_0^{-1}\sum_{k=1}^{K-1} D_{nk}R_k^2[P_k^r(R_k;\,l,\,m) - P_{k+1}^r(R_k;\,l,\,m)]\\
&+ \varepsilon_0^{-1}\sum_{k=1}^{K}\int_{R_{k-1}}^{R_k}\gamma_{nk}(r)\rho_k^P(r;\,l,\,m)r^2dr
\end{aligned}
\tag{2.5.8}
$$

where $(r; l, m)$ are spherical harmonics of the volume density of bound charges

$$\rho_k^P(\mathbf{r}) = -\mathrm{div}\mathbf{P}_k(\mathbf{r}). \tag{2.5.9}$$

Decompose $\mathbf{P}_k(\mathbf{r})$ by ball vectors:

$$\mathbf{P}_k(r, \theta, \varphi) = \sum_{l=0}^{\infty}\sum_{m=-l}^{l}\sum_{\lambda=-1}^{1} P_k^{(\lambda)}(r; l, m)\mathbf{Y}_{lm}^{(\lambda)}(\theta, \varphi), \tag{2.5.10}$$

where

$$\mathbf{Y}_{lm}^{(-1)}(\theta, \varphi) = \mathbf{n}\Upsilon_{lm}(\theta, \varphi), \ \mathbf{n} = \frac{\mathbf{r}}{r} = (\sin\theta\cos\varphi, \ \sin\theta\sin\varphi, \ \cos\theta); \tag{2.5.11}$$

$$\mathbf{Y}_{lm}^{(1)}(\theta, \varphi) = \frac{1}{\sqrt{l(l+1)}}\nabla_\Omega \Upsilon_{lm}(\theta, \varphi) = \frac{r}{\sqrt{l(l+1)}}\nabla \Upsilon_{lm}(\theta, \varphi); \tag{2.5.12a}$$

$$\nabla_\Omega \Upsilon_{lm}(\theta, \varphi) = \mathbf{e}_r\frac{\partial}{\partial\theta} + \frac{\mathbf{e}_\varphi}{\sin\varphi}\frac{\partial}{\partial\varphi}; \tag{2.5.12b}$$

$$\mathbf{Y}_{lm}^{(0)}(\theta, \varphi) = [\mathbf{n}\times\mathbf{Y}_{lm}^{(1)}(\theta, \varphi)] = \frac{1}{\sqrt{l(l+1)}}[\mathbf{r}\times\nabla \Upsilon_{lm}(\theta, \varphi)]. \tag{2.5.13}$$

Using the decomposition (2.5.10)–(2.5.13), calculate the divergence of the vector $\mathbf{P}_k$:

$$\begin{aligned}\mathrm{div}\mathbf{P}_k(r, \theta, \phi) = \sum_{l=0}^{\infty}\sum_{m=-l}^{l}\sum_{\lambda=-1}^{1}\{&P_k^{(\lambda)}(r; l, m)div\ \mathbf{Y}_{lm}^{(\lambda)}(\theta, \phi)\\ +(\nabla P_k^{(\lambda)}(r; l, m), \mathbf{Y}_{lm}^{(\lambda)}(\theta, \phi)\}\end{aligned} \tag{2.5.14}$$

We have

$$\nabla P_k^{(\lambda)}(r; l, m) = \mathbf{n}\frac{dP_k^{(\lambda)}(r; l, m)}{dr}, \tag{2.5.15}$$

whence

$$(\nabla P_k^{(\lambda)}(r; l, m), \mathbf{Y}_{lm}^{(\lambda)}(r; \theta, \varphi)) = \frac{dP_k^{(\lambda)}(r)}{dr}\delta_{\lambda,-1}\Upsilon_{lm}(\theta, \varphi). \tag{2.5.16}$$

Then we get:

$$\mathrm{div}\mathbf{Y}_{lm}^{(-1)}(\theta, \phi) = \mathrm{div}(\mathbf{n}\Upsilon_{lm}(\theta, \phi)) = \Upsilon_{lm}(\theta, \phi)\ \mathrm{div}\,\mathbf{n} + (\mathbf{n}, \nabla \Upsilon_{lm}(\theta, \phi)) \tag{2.5.17}$$

Using the expression for the divergence of a vector field in a rectangular Cartesian coordinate system, we get:

$$\mathrm{div}\,\mathbf{n} = \frac{\partial n_x}{\partial x} + \frac{\partial n_y}{\partial y} + \frac{\partial n_z}{\partial z} = \frac{\partial}{\partial x}\left(\frac{x}{r}\right) + \frac{\partial}{\partial y}\left(\frac{y}{r}\right) + \frac{\partial}{\partial z}\left(\frac{z}{r}\right); \tag{2.5.18}$$

$$\frac{\partial}{\partial x}\left(\frac{x}{r}\right)=\frac{1}{r}+x\frac{\partial}{\partial x}\left(\frac{1}{r}\right)=\frac{1}{r}-\frac{x}{r^2}\frac{\partial r}{\partial x}=\frac{1}{r}-\frac{x}{r^2}\frac{\partial\sqrt{x^2+y^2+z^2}}{\partial x}$$
$$=\frac{1}{r}-\frac{x}{r^2}\frac{2x}{2r}=\frac{1}{r}-\frac{x^2}{r^3} \tag{2.5.19a}$$

Similarly:

$$\frac{\partial}{\partial y}\left(\frac{y}{r}\right)=\frac{1}{r}-\frac{y^2}{r^3};\ \frac{\partial}{\partial z}\left(\frac{z}{r}\right)=\frac{1}{r}-\frac{z^2}{r^3}. \tag{2.5.19b}$$

Substitute (2.5.19a) and (2.5.19b) in (2.5.18):

$$\text{div}\,\mathbf{n}=\frac{3}{r}-\frac{x^2+y^2+z^2}{r^3}=\frac{3}{r}-\frac{r^2}{r^3}=\frac{2}{r}. \tag{2.5.20}$$

Then, taking into account (2.5.20), (2.5.17) will take the form

$$\text{div}\mathbf{Y}_{lm}^{(-1)}(\theta,\varphi)=\frac{2}{r}\Upsilon_{lm}(\theta,\varphi)+\frac{\partial}{\partial r}\Upsilon_{lm}(\theta,\varphi)=\frac{2}{r}\Upsilon_{lm}(\theta,\varphi). \tag{2.5.21}$$

Then from (2.5.12a) and (2.5.13), respectively, we get:

$$\text{div}\mathbf{Y}_{lm}^{(1)}(\theta,\phi)=\text{div}\left(\frac{r}{\sqrt{l(l+1)}}\,\nabla\,\Upsilon_{lm}(\theta,\phi)\right)$$
$$=\frac{1}{\sqrt{l(l+1)}}\{(\nabla\,r,\nabla\,\Upsilon_{lm}(\theta,\phi))+r\Delta\Upsilon_{lm}(\theta,\phi)\} \tag{2.5.22}$$
$$=\frac{r}{\sqrt{l(l+1)}}\left(-\frac{l(l+1)}{r^2}\right)\Upsilon_{lm}(\theta,\phi)=-\frac{\sqrt{l(l+1)}}{r}Y_{lm}(\theta,\phi);$$

$$\text{div}\mathbf{Y}_{lm}^{(0)}(\theta,\phi)=(\nabla\,,[\mathbf{n}\times\mathbf{Y}_{lm}^{(1)}])=\frac{1}{\sqrt{l(l+1)}}(\nabla\,,[\mathbf{r}\times\nabla\,\Upsilon_{lm}(\theta,\phi)])$$
$$=\frac{1}{\sqrt{l(l+1)}}\{([\nabla\times\mathbf{r}],\nabla\,\Upsilon_{lm}(\theta,\phi))-(\mathbf{r},[\nabla\times\nabla\,\Upsilon_{lm}(\theta,\phi)]\}=0. \tag{2.5.23}$$

In that way,

$$\text{div}\mathbf{P}_k(r,\theta,\phi)$$
$$=\sum_{l=0}^{\infty}\sum_{m=-l}^{l}\sum_{\lambda=-1}^{1}\left\{P_k^{(\lambda)}(r;l,m)div\ \mathbf{Y}_{lm}^{(\lambda)}(\theta,\phi)+\delta_{\lambda,-1}\frac{dP_k^{(-1)}(r;l,m)}{dr}\Upsilon_{lm}(\theta,\phi)\}\right\}$$
$$=\sum_{l=0}^{\infty}\sum_{m=-l}^{l}\left\{P_k^{(-1)}(r;l,m)\frac{2}{r}\Upsilon_{lm}(\theta,\phi)-\frac{\sqrt{l(l+1)}}{r}\Upsilon_{lm}(\theta,\phi)P_k^{(1)}(r;l,m)\right.$$
$$\left.+\frac{dP_k^{(-1)}(r;l,m)}{dr}\Upsilon_{lm}(\theta,\phi)\right\} \tag{2.5.24}$$
$$=\sum_{l=0}^{\infty}\sum_{m=-l}^{l}\left\{\left(\frac{d}{dr}+\frac{2}{r}\right)P_k^{(-1)}(r;l,m)-\frac{\sqrt{l(l+1)}}{r}P_k^{(1)}(r;l,m)\right\}\Upsilon_{lm}(\theta,\phi).$$

We decompose the density of bound charges due to slow polarization into spherical functions:

$$\rho_k^P(\mathbf{r}) = \sum_{l=0}^{\infty}\sum_{m=-l}^{l} \rho_k^P(r;\, l,\, m)\Upsilon_{lm}(\theta,\, \varphi). \tag{2.5.25}$$

Then, comparing the right parts of the decompositions (2.5.9) and (2.5.24), we get

$$\rho_k^P(r;\, l,\, m) = -\left(\frac{d}{dr} + \frac{2}{r}\right)P_k^{(-1)}(r;\, l,\, m) + \frac{\sqrt{l(l+1)}}{r}P_k^{(1)}(r;\, l,\, m). \tag{2.5.26}$$

Thus, the contribution from polarization can be represented as

$$V^P(R_n;\, l,\, m) = \varepsilon_0^{-1}\sum_{k=1}^{K-1} D_{nk}R_k^2[P_k^r(R_k;\, l,\, m) - P_{k+1}^r(R_k;\, l,\, m)] + \varepsilon_0^{-1}\sum_{k=1}^{K}\int_{R_{k-1}}^{R_k}\gamma_{nk}(r)$$

$$\times\left\{-\frac{1}{r^2}\frac{d}{dr}(r^2P_k^{(-1)}(r;\, l,\, m) + \frac{\sqrt{l(l+1)}}{r}P_k^{(1)}(r;\, l,\, m)\right\}r^2dr$$

$$=\varepsilon_0^{-1}\sum_{k=1}^{K}\int_{R_{k-1}}^{R_k}\left\{\frac{d\gamma_{nk}(r)}{dr}P_k^{(-1)}(r;\, l,\, m) + \frac{\sqrt{l(l+1)}}{r}\gamma_{nk}(r)P_k^{(1)}(r;\, l,\, m)\right\}r^2dr;$$

$$V^P(R_n;\, l,\, m) = \varepsilon_0^{-1}\sum_{k=1}^{K}\int_{R_{k-1}}^{R_k}\frac{d\gamma_{nk}(r)}{dr}P_k^{(-1)}(r;\, l,\, m) + \frac{\sqrt{l(l+1)}}{r}\gamma_{nk}(r)P_k^{(1)}(r;\, l,\, m)\Big\}r^2dr. \tag{2.5.27}$$

The form will be used more often

$$\begin{aligned} V^P(R_n;\, l,\, m) &= \varepsilon_0^{-1}\sum_{k=1}^{K-1} D_{nk}\Big\{R_k^2\big[P_k^r(R_k;\, l,\, m) - P_{k+1}^r(R_k;\, l,\, m)\big] \\ &+ \int_{R_{k-1}}^{R_k} F_{k1}(r)\rho_k^P(r;\, l,\, m)r^2dr + \int_{R_k}^{R_{k+1}} F_{k+1,\,2}(r)\rho_{k+1}^P(r;\, l,\, m)r^2dr\Big\}, \end{aligned} \tag{2.5.28}$$

or

$$\begin{aligned} V^P(R_n;\, l,\, m) &= \varepsilon_0^{-1}\sum_{k=1}^{K}\Big\{\int_{R_{k-1}}^{R_k}\gamma_{nk}(r)\rho_k^P(r;\, l,\, m)r^2dr \\ &+ D_{nk}R_k^2(R_k;\, l,\, m) - D_{n,k-1}R_{k-1}^2P_k^r(R_{k-1};\, l,\, m)\Big\} \end{aligned} \tag{2.5.29}$$

Using the derived equations, we represent the potential inside an arbitrary layer, as well as the boundary potentials, as a superposition of contributions from different types of sources:

$$V_n(r;\, l,\, m) = V^e(r;\, l,\, m) + V^s(r;\, l,\, m) + V^b(r;\, l,\, m) + V^p(r;\, l,\, m), \tag{2.5.30}$$

where $V^e(r;\, l,\, m)$ is the contribution from the external boundary of:

$$V_n^e(r;\, l,\, m) = V^e(R_n;\, l,\, m)F_{n1}(r) + V^e(R_{n-1};\, l,\, m)F_{n2}(r); \tag{2.5.31}$$

$V^s(r; l, m)$ is the contribution from surface third-party charges:

$$V^s(r; l, m) = F_{n1}(r)\varepsilon_0^{-1}\sum_{k=1}^{K-1} D_{nk}R_k^2\sigma_k(l, m) + F_{n2}(r)\varepsilon_0^{-1}\sum_{k=1}^{K-1} D_{n-1,\,k}R_k^2\sigma_k(l, m)$$

$$= \sum_{k=1}^{K-1}\varepsilon_0^{-1}[D_{nk}F_{n1}(r) + D_{n-1,\,k}F_{n2}(r)]R_k^2\sigma_k(l, m); \tag{2.5.32}$$

$$V_n^s(r; l, m) = \varepsilon_0^{-1}\sum_{k=1}^{K-1}\gamma_{kn}(r)R_k^2\sigma_k(l, m);$$

$V^b(r; l, m)$ is the contribution from bulk charges:

$$V_n^b(r; l, m) = \varepsilon_0^{-1}\sum_{k=1}^{K}\int_{R_{k-1}}^{R_k}\{\gamma_{nk}(r')F_{n1}(r) + \gamma_{n-1,\,k}(r')F_{n2}(r) + \delta_{nk}\gamma_n(r, r')\}\rho_k(r'; l, m)r'^2dr', \tag{2.5.33}$$

or

$$V_n^b(r; l, m) = \varepsilon_0^{-1}\sum_{k=1}^{K}\int_{R_{k-1}}^{R_k} G_{nk}(r, r')\rho_n(r'; l, m)r'^2dr', \tag{2.5.34}$$

where

$$\begin{aligned} G_{nk}(r, r') &= F_{n1}(r)\gamma_{nk}(r') + F_{n2}(r)\gamma_{n-1,k}(r') + \delta_{nk}\gamma_n(r, r') \\ &= F_{n1}(r)[D_{nk}F_{k1}(r') + D_{n,k-1}F_{k2}(r')] + F_{n2}(r) \\ &[DF_{k1}n - 1, k(r') + D_{n-1,k-1}F_{k2}(r')] \\ &+ \delta_{nk}\gamma_n(r, r'). \end{aligned} \tag{2.5.35}$$

Let's enter for each element D_{nk} transmission matrix modulation a new matrix size 2×2 by the formula

$$\left\|\hat{D}_{nk}^{\alpha\beta}\right\| = \begin{pmatrix} D_{nk} & D_{n,\,k-1} \\ D_{n-1,\,k} & D_{n-1,\,k-1} \end{pmatrix}. \tag{2.5.36}$$

Note that this matrix is symmetric with respect to double index substitution:

$$D_{nk}^{\alpha\beta} = D_{kn}^{\beta\alpha}. \tag{2.5.37}$$

Then the Green's matrix function (2.5.35) with consideration for (2.5.36) can be represented as

$$G_{nk}(r, r') = \sum_{\alpha=1}^{2}\sum_{\beta=1}^{2} D_{nk}^{\alpha\beta}F_{n\alpha}(r)F_{k\beta}(r') + \delta_{nk}\gamma_n(r, r'); \tag{2.5.38}$$

$V_n^p(r; l, m)$ is the contribution from the inertial polarization:

$$V_n^P(r;\,l,\,m) = V^P(R_n;\,l,\,m)F_{n1}(r) + V^P(R_{n-1};\,l,\,m)F_{n2}(r) + \varepsilon_0^{-1}\int_{R_{n-1}}^{R_n} \gamma_n(r,\,r')\rho_n^P(r';\,l,\,m)r'^2dr' \tag{2.5.39}$$

whence

$$V_n^P(r;\,l,\,m) = \varepsilon_0^{-1}\sum_{k=1}^{K}\int_{R_{k-1}}^{R_k}\left\{\frac{\partial G_{nk}(r,\,r')}{\partial r'}P_k^{(-1)}(r';\,l,\,m) + \frac{\sqrt{l(l+1)}}{r'}G_{nk}(r,\,r')P_k^{(1)}(r';\,l,\,m)\right\}r'^2dr' \tag{2.5.40}$$

2.6 Conclusion

For multilayer cylindrical and spherical systems, based on the solution of the Poisson equation with boundary conditions, potentials are found due to the following sources of fields: charges on the interface surfaces of a cylindrical structure, charges in the volumes of cylindrical layers, polarization of matter, and external stress sources.

References

[1] Alferov Z I, Zhilyaev Y V and Shmartsev Y V 1971 Splitting the conduction band of a 'superlattice' based on GaP_xAs_{1-x} *FTP* **5** 196–8

[2] Wendler L 1985 Electron–phonon interaction in dielectric two-layer systems. The effect of electronic polarizability *Phys. Status Solidi* B **129** 513–30

[3] Pokatilov E P, Beril S I and Fomin V M 1988 Image potentials and image forces in the theory of polarons *Phys. Status Solidi* B **147** 163–72

[4] Degani M H 1989 The ratio of energy and momentum for polarons in wires with quantum wells *Phys. Rev.* B **40** 11937–9

[5] Knipp P A and Reinecke T L 1992 Interface phonons of quantum wires *Phys. Rev.* B **45** 9091–102

[6] Puppin R and Engleman R 1970 Optical phonons of small crystals *Rep. Prog. Phys.* **33** 149–97

[7] Esaki L 1986 A bird's-eye view of the evolution of semiconductor superlattices and quantum wells *IEEE J. Quantum Electron* **22** 1611–24

[8] Likari J J and Evrard R 1977 Electron–phonon interaction in a dielectric plate: the effect of electron polarizability *Phys. Rev.* B **15** 2254–64

[9] Fuchs R and Kliewer K L 1965 Optical modes of vibrations in an ion–crystal plate *Phys. Rev.* **140** A2076–88

[10] Cardona M 1989 Folded, closed, boundary, surface and lamellar vibrational modes in semiconductor superlattices *Superlattices Microstruct* **5** 27–42

[11] Esaki L and Tsu R 1970 Superlattice and negative differential conductivity in semiconductors *IBM J. Res. Dev.* **14** 61–5

[12] Keldysh L V 1962 On the effect of ultrasound on the electronic spectrum of a crystal *Solid State Phys.* **4** 2265–7

[13] Komas F, Trallero-Giner S and Riera R 1988 The effect of local phonon retention on the properties of polarons in semiconductor heterostructures with a quantum well *Phys. In* **154** 17–26

[14] Lee U S, Gu S V, Au-Eng T S and Eng Yu. Y 1992 Electron–phonon interaction in a quantum wire with a parabolic potential *Phys. Lett.* A **166** 377–82

[15] Sood A K, Menendez J, Cardona M and Ploog A K 1985 Vibrational modes at the interface of phases in GaAs–AlGaAs superlattices *Phys. Rev. Lett.* **54** 2115–8

[16] Klein M V 1986 Phonons in semiconductor superlattices *IEEE J. Quantum Electron.* **22** 1760–70

[17] Sibert J, Petroff P M, Dolan J G, Pirton S J, Gossard A S and English J H 1986 Optically detectable carrier retention in single and zero dimensions in wires and boxes with GaAs quantum wells *Appl. Phys. Lett.* **49** 1275–8

[18] Fomin V M and Pokatilov E P 1985 Phonons and electron–phonon interaction in multilayer systems *Phys. Status Solidi* B **132** 69–82

[19] Brus L E 1984 Electron–electron and electron–hole interactions in small semiconductor crystallites: dependence on the size of the lowest excited electronic state *J. Chem. Phys.* **80** 4403–9

[20] Vasilopoulos P, Warmenbol P, Peeters F M and Devreese J T 1989 Magnetic phonon resonances in quasi-one-dimensional wires *Phys. Rev.* B **40** 1810–6

[21] Pokatilov E P, Beril S I, Fomin V M and Pogorilko G A 1984 Vanier–Mott exciton states in two-layer periodic structures *Phys. Status Solidi* B **130** 278–88

[22] Ando M, Fowler A and Stern F 1985 *Electronic Properties of Two-Dimensional Systems* (Moscow: Mir) p 415

[23] Smirnov V I 1974 *Course of Higher Mathematics* (Moscow: Nauka) p 480

[24] Gaponov S V, Luskin B M and Salanenko N N 1979 On the possibility of using a structure with strict clarity of the laser spraying method *Lett. LJ* **5** 516–21

[25] Bryksin V V and Firsov Y 1971 A interaction of an electron with surface phonons in an ion crystal plate *FTT* **13** 496–503

[26] Zhu K D and Gu S V 1992 The intrinsic energy of polarons due to the retention of phonons in quantum boxes and wires *J. Phys.: Condens. Matter* **4** 1291–9

[27] Strossio M 1989 Interaction between longitudinal optical-phonon modes of a rectangular quantum wire and charge carriers of a one-dimensional electron gas *Phys. Rev.* B **40** 6428–31

[28] Nash K J 1992 Electron–phonon interactions and lattice dynamics of optical phonons in semiconductor heterostructures *Phys. Rev.* B **46** 7723–44

[29] Degani M H and Ippolito O 1988 Interaction of optical phonons with electrons in GaAs quantum wires *Solid State Commun.* **65** 1185–7

[30] Fasol G, Tanaka M, Sasaki H and Horikoshi Y 1988 The roughness of the interface and the dispersion of limited LO-phonons in GaAs quantum wells/AlAs *Phys. Rev.* B **38** 6056–65

[31] Iimura D, Shimomura S, Nagata K, Dan S, Aoyagi Y and Naomba S 1989 New fabrication technique of quantum wire structures with dimensions precisely controlled by the CBE method *Jpn. Phys.* **28** L1083

[32] Pokatilov E P, Fomin V M and Beril S I 1990 *Vibrational Excitations, Polarons and Excitons in Multilayer Systems and Superlattices* (Chisinau: Stiinza) p 280

[33] Silin A P 1985 Semiconductor SR *UFN* **147** 485–521

[34] Asada M, Miyamoto Y and Suematsu Y 1985 Theoretical gain factor of wire lasers on quantum wells *Jpn. Phys.* **24** L95–7

[35] Hai G K, Peeters F M and Devreese J T 1993 Electron-optical–phonon interaction in $GaAs/Al_xGa_{1-x}As$ quantum wells due to interface, layer and half-space modes *Phys. Rev.* B **48** 4666

[36] Pokatilov E P and Fomin V M 1984 *Interaction of an Electron with Polarization Optical Vibrations in Periodic Structures* No. 508M-05.85 (Chisinau: MoldNIINTI) p 14

Chapter 3

Collective states in multilayer planar systems (excluding retardation). Hamiltonian of the electron-polarization interactions

3.1 Introduction

In this chapter, a system of equations for polarization oscillations (slow and fast) is written down and solved in a general form without taking into account the retardation in multilayer planar structures of general appearance, and superlattices, normal bulk, and interface modes are found. The Hamiltonian of the electron-polarization interaction in multilayer planar structures and superlattices is obtained explicitly, describing, in particular, in superlattices, the interaction of an electron with bulk ('captive' and 'free') phonons, as well as with the interface spatially extended modes predicted by the authors of the monograph and subsequently discovered experimentally. The influence of boundary conditions in layers on the spectrum of bulk oscillations is analyzed. The Hamiltonians of the interaction of an electron with optical vibrations in multilayer cylindrical and spherical systems are obtained explicitly.

3.2 Equations of evolution of slow polarizations

The formulas derived in chapters 1 and 2 for potentials due to polarization in multilayer systems serve as the basis for obtaining the frequency spectrum and finding the amplitudes of normal vibrations. Both of these problems are solved by composing evolution equations for polarization and reducing these equations to a normal (single-oscillator) form. This approach can be called a method of equations of motion. For a simple system (ion crystal plate in vacuum) it was developed in [1]. Subsequently, in [2–4], this method was generalized to multi-layer structures, the specificity of which is to confuse the interface vibrations of different layers. In many crystals, along with ion polarization, a significant role is played by the polarization

doi:10.1088/978-0-7503-6164-4ch3

of the charge carrier plasma, whose vibrations are mixed with the vibrations of ions, leading to the appearance of a new type of vibrations—phonon-plasma. In this chapter, the theory is developed with both types of vibrations in mind.

The main result of this theory is the derivation of Hamiltonians of electron–polarization interactions for multilayer structures, taking into account two types of polarization, while explicitly obtaining constants of electron-polarization interactions. These results overlap and summarize all the studies known by the author from the literature on this topic, devoted, as a rule, to solving individual problems (the general approach is also later in the works [5, 6]).

One of the well-known approaches to the problem of polarization oscillations is the method of the dielectric function, which consists of the fact that in formulas (3.2.1) and (3.2.2) and in boundary conditions for the induction vector, the material equation is used:

$$\mathbf{P}_n = \varepsilon_0(\varepsilon_n(\omega) - \varepsilon_n)\mathbf{E},$$

where $\varepsilon_n = \varepsilon_n(\omega)|_{\omega\to\infty}$, by which polarization is excluded from them. Due to this reduction of the problem, secular equations for frequencies are easily derived, but the possibility of obtaining amplitudes of normal vibrations is completely lost, without which it is impossible to consistently derive constants of electron-polarization interactions. In one of the most general works performed by the method of the dielectric function [7], the spectral part of the problem for various multilayer structures was studied quite fully, but due to these difficulties, the Hamiltonians of electron-polarization interactions were not obtained.

Energy of interaction of polarization with the electric field created by it $\mathbf{E}^p$

$$U_{\text{int}} = -\frac{1}{2}\sum_{n=I}^{K}\int d^2y \int_{z_{n-1}}^{z_n} dz\, \mathbf{P}_n(\mathbf{r})\mathbf{E}_n^p(\mathbf{r}). \qquad (3.2.1)$$

After switching to a Fourier representation with a quasi-discrete set of wave vector values $\boldsymbol{\eta}$, corresponding to the final area of the system $L_x L_y$ in the XOY plane, takes the form

$$U_{int} = -\frac{1}{2}\sum_{n=I}^{K}\frac{(2\pi)^6}{V_n}\sum_{\mathbf{Q}_n}\mathbf{P}_n^*(\mathbf{Q}_n)\mathbf{E}_n^p(\mathbf{Q}_n), \qquad (3.2.2)$$

where $V_n = l_n L_x L_y$ is the volume of the n-th layer. Based on the formula (1.8.12) the Fourier image of the potential is calculated.

$$V_n^p(\mathbf{Q}_n) = \frac{1}{2\pi}\int_{z_{n-1}}^{z_n} dz e^{-iq_n(z-z_{n-1})} V_n^p(\boldsymbol{\eta}, z) =$$

$$= -\frac{1}{\varepsilon_0\varepsilon_n^{||}T_n}\sum_{q'_n\neq 0} S_{q_n,\, q'_n}\tilde{P}_{1n}(\mathbf{Q}'_n) + \frac{1}{\varepsilon_0 C_n\eta\, T_n}S_{q_n,\,0}\sum_{k=I}^{K}\frac{1}{C_k l_k}\times \qquad (3.2.3)$$

$$\times\left\{\left[\sum_{l=1}^{2}K_{1n,\, lk}\tilde{P}_{lk}(\boldsymbol{\eta}, 0)\right] + \left[\sum_{l=1}^{2}K_{2n,\, lk}\tilde{P}_{lk}(\boldsymbol{\eta}, 0)\right]\frac{iq_n}{\sqrt{\epsilon_n\eta}}\right\},$$

using which and formulas (1.2.11a) and (1.2.11b) the Fourier image of the electric field intensity created in the nth layer by the inertial polarization of all layers of the system is found:

$$\varepsilon_0 \mathbf{E}_n^p(\mathbf{Q}_n) = \frac{i\hat{\mathbf{Q}}}{\varepsilon_{\mathrm{T}}^{||}} \sum_{q'_n \neq 0} S_{q_n, q'_n} \tilde{P}_{1n}(\mathbf{Q}'_n) + \frac{S_{q_n, 0}}{C_n \eta} \sum_{k=I}^{K} \frac{1}{C_k l_k} \times$$
$$\times \left\{ \left[\sum_{l=1}^{2} K_{1n, lk} \tilde{P}_{ln}(\eta, 0) \right] (-i\hat{\mathbf{Q}}_n) + \left[\sum_{l=1}^{2} K_{2n, lk} \tilde{P}_{lk}(\eta, 0) \right] \left(\hat{\mathbf{R}}_n + \frac{C_n}{S_{q_n, 0}} e_3 \right), \right. \tag{3.2.4}$$

required according to (3.2.2) for calculating the interaction enegy

$$U_{\mathrm{int}} = \frac{1}{2} \sum_{n=I}^{K} \frac{(2\pi)^6}{V_n \varepsilon_0}$$
$$\sum_{\eta} \left\{ \frac{1}{\varepsilon_n^{||}} \sum_{q_n \neq 0} |\tilde{P}_{1n}(\mathbf{Q}_n)|^2 - \frac{1}{C_n \eta} \sum_{k=I}^{k} \frac{1}{C_k l_k} \sum_{j, l=1}^{2} \tilde{P}_{jn}^*(\eta, 0) K_{jn, lk} \tilde{P}_{lk}(\eta, 0) \right\} \tag{3.2.5}$$

As noted in [8] after the formula (1.8.12), the polarization in the expressions for the potential (1.8.12), electric field strength (3.2.4), and interaction energy (3.2.5) can represent the sum of contributions from different sources, and each of these contributions can, generally speaking, have a multimode structure with different independent frequencies [8]:

$$\mathbf{P}_n = \sum_x \mathbf{P}_{nx}, \tag{3.2.6}$$

where x is a certain type of polarization and a certain mode.

In the future, we will limit ourselves to the single-mode approximation and consider in detail two types of inertial polarization caused by: (a) mutual displacements of positive and negative lattice ions (optical vibrations of the lattice); and (b) displacement of charge carriers relative to charged lattice nodes (optical vibrations of a single-component carrier plasma). The additivity of full polarization for mechanisms a and b is proved in [9]. If necessary, the results can be extended to any number of summands in the sum (3.2.6). The polarization of each of these types obeys the evolution equation (in the isotropic approximation):

$$\ddot{\mathbf{P}}_{nx}(\mathbf{Q}_n) = -\sum_{j=1}^{3} \Omega_{jnx}^2(\mathbf{Q}_n) P_{jnx}(\mathbf{Q}_n) e_j + \omega_{nx}^2 \varepsilon_n \varepsilon_0 \mathbf{E}_n(\mathbf{Q}_n). \tag{3.2.7}$$

Thus, for the polarization of the lattice ($x = 0$—optical phonons)

$$\Omega_{jn0}^2(\mathbf{Q}_n) \equiv \Omega_n^2;\; \omega_{n0}^2 \equiv \omega_{n, l0}^2 - \omega_n^2 = \omega_n^2 \frac{(\varepsilon_{n0} - \varepsilon_n)}{\varepsilon_n}, \tag{3.2.8a}$$

where ε_n and ε_{n0} are high-frequency and static permittivity, if the Einstein law of dispersion is accepted. In the stationary field from (3.2.7) follows

$$\mathbf{P}_{n0}(\mathbf{Q}_n) = \varepsilon_0(\varepsilon_{n0} - \varepsilon_n) \mathbf{E}_n(\mathbf{Q}_n) \tag{3.2.8b}$$

In a variable frequency field ω from (3.2.8a) we find

$$\mathbf{P}_n(\mathbf{Q}_n) = \omega_n^2 \frac{\varepsilon_0(\varepsilon_{n0} - \varepsilon_n)}{(\omega_n^2 - \omega^2)} \mathbf{E}_n(\mathbf{Q}_n), \tag{3.2.8c}$$

which is equivalent to describing the lattice polarization by a dielectric function

$$\varepsilon_n(\omega) = \varepsilon_n + \frac{\omega_n^2(\varepsilon_{n0} - \varepsilon_n)}{\omega_n^2 - \omega^2}. \tag{3.2.8d}$$

For polarization caused by the displacement of charge carriers ($x = p$—plasmons), according to [10] in the plasma wave model:

$$\Omega_{jnp}^2(\mathbf{Q}_n) = \frac{\gamma_{jk} Q_n^2}{m_n N_n}; \quad \omega_{np}^2 \equiv \frac{e^2 N_n}{m_n \varepsilon_0 \varepsilon_n}. \tag{3.2.9a}$$

In the variable frequency field ω, the plasma polarization from (3.2.7) in disregard of the elasticity of the electron gas has the form

$$\mathbf{P}_{np}(\mathbf{Q}_n) = -\frac{\omega_{np}^2 \varepsilon_n \varepsilon_0}{\omega^2} \mathbf{E}_n(\mathbf{Q}_n), \tag{3.2.9b}$$

where the dielectric function of the plasma comes from

$$\varepsilon_n(\omega) = 1 - \frac{\omega_p^2}{\omega^2}; \quad \omega_p^2 = \frac{e^2 N_n}{m_n \varepsilon_0}, \tag{3.2.9c}$$

where m_n is the effective mass; N_n is the concentration of charge carriers in the nth layer; and γ_{1n} is the compression module of an ideal electron gas, the shear modules $\gamma_{2n} = \gamma_{3n} = 0$. Given a more general model, for example, plasma of valence electrons (see section 15 of the book [8]), shear deformations must also be taken into account.

Methods for introducing plasma (collective) variables to describe the motion of a charge carrier system, as well as the restrictions imposed by such a description, have been discussed in a large number of papers (see, e.g., [11]).

Renormalize the polarization amplitude:

$$\mathbf{P}_{nx}(\mathbf{Q}_n) = \omega_{nx} \sqrt{V_n \varepsilon_0 \varepsilon_n} (2\pi)^{-3} \mathbf{W}_{nx}(\mathbf{Q}_n) \tag{3.2.10}$$

and, having made the transition from W_{jnx} to $\tilde{W}_{jnx}$ using the unitary matrix (1.8.9) by a formula similar to (1.8.8), we obtain as kinetic

$$T = \frac{1}{2} \sum_{n=I}^{K} \sum_{x} \sum_{\mathbf{Q}_n} \sum_{j=1}^{3} \left| \dot{\tilde{W}}_{jnx}(\mathbf{Q}_n) \right|^2, \tag{3.2.11}$$

so the potential energy of quasi-elastic forces is

$$U_{qe} = \frac{1}{2} \sum_{n=I}^{K} \sum_{x} \sum_{\mathbf{Q}_n} \sum_{j=1}^{3} \Omega_{jnx}^2(\mathbf{Q}_n) \left| \tilde{W}_{jnx}(\mathbf{Q}_n) \right|^2 \tag{3.2.12}$$

in standard forms, and the potential energy of interaction in the form of is

$$U_{\text{int}} = \frac{1}{2}\sum_{n=I}^{K}\sum_{x,y}\sum_{\eta}\left\{\sum_{q_n\neq 0}\omega_{nx}\omega_{ny}\tilde{W}^*_{1nx}(\mathbf{Q}_n)\tilde{W}_{1ny}(\mathbf{Q}_n) - \sum_{k=I}^{K}\sum_{j,l=1}^{2} M_{jkx,\ lky}\tilde{W}^*_{jnx}(\eta,0)\tilde{W}_{lky}(\eta,0)\right\} \quad (3.2.13)$$

with coefficients

$$M_{jkx,\ lky} = \frac{1}{2}\frac{\omega_{nx}\sqrt{\varepsilon_n}}{\sqrt{\tanh\frac{\varsigma_n}{2}}}\frac{\omega_{ky}\sqrt{\varepsilon_k}}{\sqrt{\tanh\frac{\varsigma_k}{2}}}K_{jn,\ lk}. \quad (3.2.14)$$

Adding up (3.2.12) and (3.2.13), we get the total potential energy of polarization

$$\begin{aligned} U = \frac{1}{2}\sum_{n=I}^{K}\sum_{x,y}\sum_{\eta}\Bigg\{&\sum_{q_n\neq 0}\sum_{j=1}^{3}\left[\Omega^2_{jnx}(\mathbf{Q}_n)\delta_{x,\ y} + \omega_{nx}\omega_{ny}\delta_{j,\ 1}\right]\times \\ &\times \tilde{W}^*_{jnx}(\mathbf{Q}_n)\tilde{W}_{jny}(\mathbf{Q}_n) + \Omega^2_{3nx}(\eta,0)\delta_{x,\ y}\,|\tilde{W}_{3nx}(\eta,0)|^2 + \\ &+\sum_{k=I}^{K}\sum_{j,l=1}^{2}\left[\Omega^2_{jnx}(\mathbf{Q}_n)\delta_{j,\ l}\delta_{n,\ k}\delta_{x,\ y} - M_{jnx,\ lky}\right]\tilde{W}^*_{jnx}(\eta,0)\tilde{W}_{lky}(\eta,0)\Bigg\}. \end{aligned} \quad (3.2.15)$$

Adding the kinetic energy (3.2.15) to (3.2.11), we get the total energy of polarization

$$H = T + U. \quad (3.2.16)$$

According to (3.2.15) the interface and bulk modes of all polarizations j and types x completely separated from each other, and as a result, the system of equations of motion for amplitudes

$$\ddot{\tilde{W}}_{jnx}(\mathbf{Q}_n) \equiv -\omega^2\tilde{W}_{jnx}(\mathbf{Q}_n) = -\frac{\partial H}{\partial\tilde{W}^*_{jnx}(\mathbf{Q}_n)} \quad (3.2.17)$$

decomposes into mutually independent systems for bulk modes of polarization oscillations

$$\sum_{y}\left\{\left[\omega^2 - \Omega^2_{jnx}(\mathbf{Q}_n)\right]\delta_{x,\ y} - \omega_{nx}\omega_{ny}\delta_{j1}\right\}\tilde{W}_{jny}(\mathbf{Q}_n) = 0 \quad (3.2.18a)$$

(with free parameters η, $q_n \neq 0, j, n, x$) and amplitudes of interface modes of polarization oscillations

$$\sum_{k=I}^{K}\sum_{l=1}^{2}\sum_{y}\left\{\left[\omega^2 - \Omega^2_{jnx}(\eta,0)\right]\delta_{j,\ l}\delta_{n,\ k}\delta_{x,\ y} + M_{jnx,\ lky}\right\}\tilde{W}_{lky}(\eta,0) = 0 \quad (3.2.18b)$$

(with free parameters η, $j = 1, 2, n, x$)

$$[\omega^2 - \Omega^2_{3nx}(\eta,0)]\tilde{W}_{3nx}(\eta,0) = 0 \quad (3.2.18c)$$

(for each η, n, x). Note that the problem of finding normal variables that diagonalize a quadratic form (3.2.15) is equivalent to the problem of solving systems of equations (3.2.18a)–(3.2.18c).

3.3 Normal bulk fluctuations

As follows from (3.2.18a), transverse bulk fluctuations of different types are independent:

$$\left[\omega^2 - \Omega_{jnx}^2(\mathbf{Q}_n)\right]\tilde{W}_{jnx}(\mathbf{Q}_n) = 0 \tag{3.3.1a}$$

(for each η; $q_n \neq 0$; $j = 2, 3$; n, x). Thus, the transverse ($j = 2, 3$) bulk fluctuations (optical $x = 0$, plasma $x = p$) $\tilde{W}_{jnx}(\mathbf{Q}_n)$ with natural frequencies $\Omega_{jnx}(\mathbf{Q}_n)$ are normal. According to (1.8.12) and (3.2.3), they do not create microscopic potentials (and hence electric fields, as can be seen directly from (3.2.4)), so they are not active in forming the interaction of an electron with an inertial polarization field.

Longitudinal bulk fluctuations of different types ($x = 0$, p) mix in each of the layers separately:

$$\sum_y \left\{ [\omega^2 - \Omega_{1nx}^2(\mathbf{Q}_n)]\, \delta_{x,\,y} - \omega_{nx}\omega_{ny} \right\} \tilde{W}_{1ny}(\mathbf{Q}_n) = 0. \tag{3.3.1b}$$

The frequency spectrum of normal longitudinal bulk vibrations is determined from the secular equation for this system:

$$\left| [(\omega^2 - \Omega_{1nx}^2(\mathbf{Q}_n))]\, \delta_{x,\,y} - \omega_{nx}\omega_{ny} \right| = 0. \tag{3.3.1c}$$

If type fluctuations are mixed $x = 0$ and $x = p$, then from (3.3.1b) taking into account the definitions (1.9), we obtain a system of two equations:

$$\begin{cases} (\omega^2 - \omega_{n,\,LO})\tilde{W}_{1n0}(Q_n) - \omega_{n0}\omega_{np}\tilde{W}_{1np}(Q_n) = 0; \\ -\,\omega_{n0}\omega_{np}\tilde{W}_{1n0}(Q_n) + \left[\omega^2 - \Omega_{1np}^2(Q_n) - \omega_{np}^2\right]\tilde{W}_{1np}(Q_n) = 0. \end{cases} \tag{3.3.1d}$$

The secular equation (3.3.1) for this system is:

$$\omega^4 - \left[\omega_{n,\,LO}^2 + \Omega_{1np}^2(\mathbf{Q}_n) + \omega_{np}^2\right]\omega^2 + \omega_{n,\,LO}^2\left[\Omega_{1np}^2 + \omega_{np}^2\right] - \omega_{n0}^2\omega_{np}^2 = 0$$

it has two solutions:

$$\begin{aligned} \omega_{(1,\,2)n}^2(\mathbf{Q}_n) &= \frac{1}{2}\Big(\left[\omega_{n,\,LO}^2 + \Omega_{1np}^2(\mathbf{Q}_n) + \omega_{np}^2\right] \pm \\ &\pm \left\{ \left[\omega_{n,\,LO}^2 - \Omega_{1np}^2(\mathbf{Q}_n) - \omega_{np}^2\right]^2 + 4\omega_{n0}^2\omega_{np}^2 \right\}^{1/2} \Big), \end{aligned} \tag{3.3.1e}$$

which are natural frequencies of mixed bulk modes. Recall that the neglect of mixing ($\omega_{n0}\omega_{np} \to 0$) (3.3.1e) defines the phonon

$$\lim_{\omega_{n0}\omega_{np}\to 0} \omega_{1n}^2(\mathbf{Q}_n) = \omega_{n,\,LO}^2 \tag{3.3.2a}$$

and plasma dispersing

$$\lim_{\omega_{n0}\omega_{np}\to 0} \omega_{2n}^2(\mathbf{Q}_n) = \Omega_{1np}^2(\mathbf{Q}_n) + \omega_{np}^2 \tag{3.3.2b}$$

branches of bulk polarization oscillations. The criterion of weak mixing follows directly from the comparison of terms in curly brackets (3.3.1e):

$$2\omega_{n0}\omega_{np} \ll \left|\omega_{n,\,LO}^2 - \Omega_{1np}^2(\mathbf{Q}_n) - \omega_{np}^2\right|. \tag{3.3.2c}$$

Let us consider some limiting cases of the formula (3.3.1e). According to (3.2.9), the frequency dispersion of plasma modes is determined by the elasticity coefficient of the ideal gas $\gamma = V\,|(\partial P/\partial V)|$, which in the case of statistical degeneration, $\sim N_n^{5/3}$, a $\Omega_{1np}^2 \sim N_n^{2/3} Q_n^2$, whereas $\omega_{np}^2 \sim N_n$. Therefore, at sufficiently high concentrations of Nn, the plasma frequency becomes greater than the frequency of elastic vibrations. For example, according to an estimate from [10, p 93] at the concentration of carriers $N_n \sim 10^{23}\,\mathrm{cm}^{-3}$ $\Omega_{1np} \sim 5 \cdot 10^{13}\mathrm{c}^{-1}$, and $\omega_{np} \sim 2{\cdot}10^{16}\,\mathrm{c}^{-1}$, that is, three orders of magnitude higher. In the case of a non-degenerate state of the electron gas $\gamma/N_n = P/N_n = T$ does not depend on concentration.

If you neglect the elastic properties of the electron gas ($\Omega_{1n}(\mathbf{Q}_n) = 0$), then from (2.1e) we get

$$\omega_{1,2}^2 = \frac{1}{2}\left\{ \omega_{np}^2 + \omega_{n,LO}^2 \pm \left[(\omega_{np}^2 + \omega_{n,LO}^2)^2 - 4\omega_{n,LO}^2\omega_{np}^2\right]^{1/2}\right\}. \tag{3.3.2d}$$

The formula (3.3.2d) for mixed bulk plasmon-phonon modes was obtained in [12] from the condition of vanishing of the dielectric function

$$\varepsilon_T(0,\,\omega) = \varepsilon - \frac{\varepsilon_0 - \varepsilon}{1 - \frac{\omega^2}{\omega_{TO}^2}} - \frac{e^2 N}{m\varepsilon_0\varepsilon}\frac{1}{\omega^2}.$$

It was also used in [13] to interpret the experiment.

In the limit of low concentrations and, consequently, small ω_{np}, preserving in (3.3.2d) only ω_{np}^2, we get

$$\omega_1^2 \approx \omega_{n,LO}^2 + \frac{\omega_{n,LO}^2 - \omega_{n,TO}^2}{\omega_{n,LO}^2 - \Omega_{1n}^2(\mathbf{Q}_n)}\omega_{np}^2; \quad \omega_2^2 \approx \Omega_{1n}^2(\mathbf{Q}_n) + \frac{\omega_{n,TO}^2 - \Omega_{1n}^2(\mathbf{Q}_n)}{\omega_{n,LO}^2 - \Omega_{1n}^2(\mathbf{Q}_n)}\omega_{np}^2 \tag{3.3.2e}$$

which are respectively phonon-like and plasmon-like branches of vibrations, the latter (due to inequality $\omega_{n,TO}, \omega_{n,LO} > \Omega_{1n}$) shielded by longitudinal optical vibrations of the grid:

$$\omega_2^2 \approx \frac{\varepsilon_n}{\varepsilon_{n0}}\omega_{np}^2 = \frac{e^2 N_n}{\varepsilon_0\varepsilon_{n0}m_n}. \tag{3.3.2f}$$

In the opposite limit of high concentrations decomposition by ω_{np}^2 in (3.3.2d) gives

$$\omega_{1,2}^2 \approx \frac{1}{2}\left\{ \omega_{n,LO}^2 + \Omega_{1n}^2(\mathbf{Q}_n) + \omega_{np}^2 \pm \right. \\ \left. \pm \left[\omega_{np}^2 + \Omega_{1n}^2(\mathbf{Q}_n) + \omega_{n,LO}^2 - 2\omega_{n,TO}^2\right] \right\}, \tag{3.3.2g}$$

where we get modes from

$$\omega_1^2 \approx \omega_{np}^2 + \omega_{n,LO}^2 + \Omega_{1n}^2(\mathbf{Q}_n) - \omega_{n,TO}^2; \;\; \omega_2^2 \approx \omega_{n,TO}^2 \tag{3.3.2h}$$

respectively close to the plasmon and shielded by plasma vibrations longitudinal phonon mode. The theoretical work [9] was followed by experimental work [12], which reported the detection of mixed modes and some results of their research.

Find the amplitudes of normal modes that oscillate with frequencies $\omega_{(1,\,2)n}$ (3.3.1e). According to equations (3.3.1d), the ratio of the amplitudes of lattice and plasma oscillations with the same frequencies

$$\frac{\tilde{W}_{1np}(\mathbf{Q}_n, \omega_{1n})}{\tilde{W}_{1n0}(\mathbf{Q}_n, \omega_{1n})} = -\frac{\tilde{W}_{1n0}(\mathbf{Q}_n, \omega_{2n})}{\tilde{W}_{1np}(\mathbf{Q}_n, \omega_{2n})} \equiv \phi_n \\ = \omega_{n0}\omega_{np} \times \left(\frac{1}{2}\left[\omega_{n,LO}^2 - \Omega_{1np}^2(\mathbf{Q}_n) - \omega_{np}^2\right] \right. \\ \left. + \left\{ \left[\omega_{n,LO}^2 - \Omega_{1np}^2(\mathbf{Q}_n) - \omega_{np}^2\right]^2 + 4\omega_{n0}\omega_{np}^2 \right\}^{1/2}\right)^{-1} \tag{3.3.3}$$

For all frequencies, they are expressed using a single common parameter φ_n. Substituting new amplitudes in the first of the equations (3.3.1d)

$$\tilde{W}_{1n0}(\mathbf{Q}_n, \omega_{1n}) = \frac{Z_{1n}(\mathbf{Q}_n, \omega_{1n})}{\sqrt{1+\phi_n^2}}, \;\; \tilde{W}_{1np}(\mathbf{Q}_n, \omega_{1n}) = \frac{\phi_n Z_{1n}(\mathbf{Q}_n, \omega_{1n})}{\sqrt{1+\phi_n^2}}, \tag{3.3.4}$$

in which the coefficient is determined from the condition of preserving the type of kinetic energy of the oscillation with frequency ω_{1n}

$$\frac{1}{2}\sum_{x=0,p} \dot{\tilde{W}}_{1nx}^2 = \frac{1}{2}\dot{Z}_{1n}^2, \tag{3.3.5}$$

We get the equation for the normal mode:

$$(\omega^2 - \omega_{1n}^2)Z_{1n} = 0. \tag{3.3.6}$$

In the same way, we find the connection of the previous amplitudes with the amplitude of the normal mode, which has a second frequency:

$$\tilde{W}_{n0}(\mathbf{Q}_n, \omega_{2n}) = -\frac{\phi_n Z_{2n}(\mathbf{Q}_n, \omega_{2n})}{\sqrt{1+\phi_n^2}}; \;\; \tilde{W}_{np}(\mathbf{Q}_n, \omega_{2n}) = \frac{Z_{2n}(\mathbf{Q}_n, \omega_{2n})}{\sqrt{1+\phi_n^2}}. \tag{3.3.7}$$

3.4 Normal interface vibrations

It follows directly from (3.2.18c) that the transverse interface vibrations $j = 3$ whose polarization vector lies in the XOY plane are normal with eigenfrequencies $\Omega_{3nx}(\eta, 0)$. These oscillations are not active in the formation of the interaction of the electron with the field of inertial polarization, since their amplitude is absent in the formulas for potentials, for example, in (3.2.3).

The frequency spectrum of normal interface vibrations is determined from the secular equation for the system (3.2.18b):

$$\left|\left[\omega^2 - \Omega^2_{jnx}(\eta, 0)\right] \delta_{j,l}\delta_{n,k}\delta_{x,y} + M_{jnx,lky}\right| = 0, \tag{3.4.1}$$

having when mixing two types of oscillation $2N_0$ numbers of solutions

$$\Omega_s^2 \quad (s = 1, \dots, 2N_0), \quad (N_0 = 2(K - I + 1)).$$

Given the property of coefficients (32a) [14], which leads to the form of linear combinations of interface vibration amplitudes (32b) [14] for each type of polarization $j = 1, 2$, we perform unitary transformations of interface vibration amplitudes in the external layers:

$$W'_{1nx}(\eta) = \left(\tanh^2\frac{\zeta_n}{2} + 1\right)^{-\frac{1}{2}}\left[\widetilde{W}_{1nx}(\eta, 0) - (-1)^{\frac{n-I}{K-I}}\tanh\frac{\zeta_n}{2}\widetilde{W}_{2nx}(\eta, 0)\right]; \tag{3.4.2a}$$

$$W'_{2nx}(\eta) = \left(\tanh^2\frac{\zeta_n}{2} + 1\right)^{-1/2}\left[(-1)^{\frac{n-I}{K-I}}\tanh\frac{\zeta_n}{2}\widetilde{W}_{1nx}(\eta, 0) + \widetilde{W}_{2nx}(\eta, 0)\right], \tag{3.4.2b}$$

and in the inner layers we will leave them unchanged:

$$W'_{jnx}(\eta) = \tilde{W}_{jnx}(\eta, 0). \tag{3.4.2c}$$

The new amplitudes follow a system of equations:

$$\sum_{k=I}^{K}\sum_{l=1}^{2}\sum_{y}\left\{\left[\omega^2 - \Omega^2_{jnx}(\eta, 0)\right]\delta_{j,l}\delta_{n,k}\delta_{x,y} + M'_{jnx,lky}\right\} W'_{lky}(\eta) = 0. \tag{3.4.3}$$

Coefficients are entered here:
for $n, k = I + 1, \dots, K - 1$

$$M'_{jnx,lky} = M_{jnx,lky}; \tag{3.4.4a}$$

for $n = I + 1, \dots, K - 1, \quad k = I, K$

$$M'_{jnx,lky} = \delta_{l,2}\left(1 + \tanh^2\frac{\zeta_k}{2}\right)^{1/2} M_{jnx,2ky}; \tag{3.4.4a}$$

for $n = I, K, \quad k = I + 1, \dots, K - 1$

$$M'_{jnx,\, lky} = \delta_{j,\,2}\left(1 + \tanh^2\frac{\zeta_n}{2}\right)^{1/2} M_{2nx,\, lky}; \tag{3.4.4c}$$

for $n,\, k = I,\, K$

$$M'_{2nx,\, 2ky} = \left(1 + \tanh^2\frac{\zeta_n}{2}\right)^{\frac{1}{2}}\left(1 + \tanh^2\frac{\zeta_k}{2}\right)^{\frac{1}{2}} M_{2nx,\, 2ky}$$

$$-\delta_{n,\,k}\delta_{x,\,y}\mathrm{th}^2\frac{\zeta_n}{2}\left(1 + \tanh^2\frac{\xi_n}{2}\right)^{-1}\left[\Omega^2_{1nx}(\eta,\, 0) - \Omega^2_{2nx}(\eta,\, 0)\right]. \tag{3.4.4d}$$

$$M'_{1nx,\, 2ky} = M'_{2nx,\, lky} = -\delta_{n,\,k}\delta_{x,\,y}(-1)^{\frac{n-I}{K-I}}\tanh\frac{\zeta_n}{2}\left(1 + \tanh^2\frac{\zeta_n}{2}\right)^{-1}$$

$$\times\left[\Omega_{1nx}(\eta,\, 0) - \Omega_{2nx}(\eta,\, 0)\right]; \tag{3.4.4e}$$

$$M'_{1nx,1ky} = -\delta_{n,\,k}\delta_{x,\,y}\tanh^2\frac{\zeta_n}{2}\left(1 + \tanh^2\frac{\zeta_n}{2}\right)^{-1}\left[\Omega^2_{2nx}(\eta,\, 0) - \Omega^2_{1nx}(\eta,\, 0)\right], \tag{3.4.4f}$$

which obviously form a symmetric matrix.

Further compaction is achieved by the introduction of a single index $(j,\, n,\, x) \to r;\ (l,\, k,\, y) \to r'$ for any particular rule. If two types of oscillation are mixed, these rules can be set as follows:

$x = 0$	$x = p$		
$n = I + 1,\, \ldots,\, K - 1$	$j = 1,\, 2$	$r = 2(n - I - 1) + j$	
$r = N_0 - 2 + 2(n - I - 1) + j$			
$n = I$	$j = 2$	$r = N_0 - 3$	$r = 2N_0 - 5$
$n = K$	$j = 2$	$r = N_0 - 2$	$r = 2N_0 - 4$
$n = I$	$j = 1$	$r = 2N_0 - 3$	$r = 2N_0 - 1$
$n = K$	$j = 1$	$r = 2N_0 - 2$	$r = 2N_0$.

Then the system of equations (3.2.18b) takes the form

$$\sum_{r'=1}^{2N_0} N_{rr'}(\omega)W'_{r'}(\eta) = 0, \quad r = 1,\, \ldots,\, 2N_0. \tag{3.4.5a}$$

with a symmetrical matrix of order $2N_0$:

$$\|N_{rr'}(\omega)\| = \|[\omega^2 - \Omega^2_r(\eta,\, 0)]\,\delta_{r,\,r'} + M'_{r,\,r'}\|. \tag{3.4.5b}$$

The spectrum of normal interface vibrations $\Omega_s(\eta)$ $(s = I,\, \ldots,\, 2N_0)$ is defined by solutions of the secular equation for the system of equations (3.4.5a):

$$|N_{rr'}(\omega)| = 0;\ r;\, r' = 1,\, \ldots,\, 2N_0. \tag{3.4.5c}$$

In case if, $\Omega^2_{2nx}(\eta, 0) = \Omega^2_{1nx}(\eta, 0)$, then from the system (3.4.5a) four equations are split off:

$$[\omega^2 - \Omega_r^2(\eta, 0)]W_r'(\eta) = 0, \ r = 2N_0 - 3, \ldots, 2N_0, \tag{3.4.6a}$$

describing normal fluctuations $Z_s(\eta) = W_s'(\eta)$ with natural frequencies $\Omega_s(\eta) = \Omega_{r=s}(\eta, 0)$, where $s = 2N_0 - -3, \ldots, 2N_0$. These oscillations in the external layers are inactive in the formation of the interaction of the electron with the field of inertial polarization, since their amplitudes in accordance with the definitions (3.4.2a) and (3.4.2b) and the ratios (32b) in [14] are not included in the potential (1.8.12) and in the further formulas for potentials following from (1.8.12).

The rest of the range of normal variation $\Omega_s(\eta)$, $s = I, \ldots, 2N_0 - 4$ is determined from the condition of existence of non-trivial solutions of the system of equations

$$\sum_{r'=1}^{2N_0-4} N_{rr'}(\omega)W_{r'}'(\eta) = 0; \ r = 1, \ldots, 2N_0 - 4 \tag{3.4.6b}$$

with a symmetric matrix of the form (3.4.5b), but of a lower order $(2N_0 - 4)$. The specified condition is formulated as a secular equation, similarly to (3.4.5c):

$$|N_{rr'}(\omega)| = 0, \ r, r' = 1, \ldots, 2N_0 - 4, \tag{3.4.6c}$$

from which, in particular, it follows that the number of active interface vibrations in a system with two types of polarization is equal to $4(K - I + 1) - 4$. This rule is true when all the layers in the system are polar. In other cases, the number of active vibrations depends not only on the number of polar layers, but also on their location in the system. When the external layers are polar, the above rule remains valid. If one of the external layers is non-polar, then subtract 2 from the fourfold number of polar layers. If both external layers are non-polar, then the number of active vibrations is maximum and equal to the fourfold number of polar layers. A special case of mixing interface optical vibrations with two-dimensional plasmon modes is considered in [12].

For a system with one type of polarization, all the results are divided into two. Using these examples, it is easy to consider all other options. Substituting in (3.4.6b) the roots Ω_s, we obtain a system of homogeneous equations for amplitudes $W_r'(\eta, \Omega_s)$, with which all amplitudes can be expressed as one; for example, $W_s'(\eta, \Omega_s)$. The summand with this amplitude in all equations of the system is moved to the right:

$$\sum_{r' \neq s} N_{rr'}(\Omega_s)W_{r'}'(\eta, \Omega_s) = -N_{rs}(\Omega_s)W_s'(\eta, \Omega_s),$$

$$r = 1, \ldots, 2N_0 - 4, \ r \neq s, \tag{3.4.7}$$

where does it come from

$$W_r'(\eta, \Omega_s) = [\delta_{r,s} - (1 - \delta_{r,s})F_{r,s}]\,(1 + \sum_{r' \neq s} |F_{r',s}|^2)^{-1/2} Z_s(\eta, \Omega_s), \tag{3.4.8a}$$

where the matrix elements are

$$F_{rs} = \sum_{r' \neq s} N_{rr'}^{-1}(\Omega_s)N_{r's}(\Omega_s), \tag{3.4.8b}$$

$N_{rr'}^{-1}(\Omega_s)$—elements of the matrix, inverse

$$\|N_{rr'}(\Omega_s)\|,\quad r \neq s;\ r' \neq s, \tag{3.4.8c}$$

and $Z_s(\eta, \Omega_s)$ is the amplitude of a normal oscillation with a natural frequency Ω_s, normalized from the condition of preserving the type of kinetic energy in the same way (3.3.5).

3.5 Interface vibrations in a three-layer system

As follows from the formulas in section 3.4, the number of active normal oscillations even in a system of three layers with two types of polarization is 8. The eighth order of the secular equation excludes the possibility of its analytical solution. Further simplifications are needed to obtain analytical solutions. For example, you can limit yourself to considering only one type of polarization. This immediately reduces the order of the secular equation to 4. (Note that in the analytical solution in does not matter which type of polarization is taken into account, since all coefficients for the transition to normal amplitudes and frequencies are expressed in the same way—in terms of permittivity and thickness. You can specify the type of polarization and, accordingly, the type of dielectric function after extracting the roots.) But this is still not enough. Therefore, we will limit ourselves to considering a three-layer system in which only layers 1 and 2 are polar. We write down a system of three equations for this case, following the general formula (3.4.6b) and denoting the amplitude of the active oscillation in the external layer $\widetilde{W}_{11x} = Z_1$, $W'_{12x} = W'_{12}$, $W_{22x} = W'_{22}$:

$$\begin{cases} (\omega^2 - \omega_{10}^2)Z_1 - R\left(\varepsilon_2 + \overline{\varepsilon}_3 \tanh\frac{\zeta_2}{2}\right)W'_{12} + R\left(\varepsilon_2 \tanh\frac{\zeta_2}{2} + \overline{\varepsilon}_3\right)W'_{22} = 0; \\ -R\left(\varepsilon_2 + \overline{\varepsilon}_3 \tanh\frac{\zeta_2}{2}\right)Z_1 + (\omega^2 - \omega_{21}^2)W'_{12} + R'(\overline{\varepsilon}_3 - \varepsilon_1)W'_{22} = 0; \\ R\left(\varepsilon_2 \tanh\frac{\zeta_2}{2} + \overline{\varepsilon}_3\right)Z_1 + R'(\overline{\varepsilon}_3 - \varepsilon_1)W'_{12} + (\omega^2 - \omega_{22}^2)W'_{22} = 0, \end{cases} \tag{3.5.1}$$

where notation is used

$$\begin{cases} \omega_{10}^2 = \dfrac{\omega_1^2}{\widetilde{B}(1\mid 3)}[\varepsilon_2^2 + \overline{\varepsilon}_2(\overline{\varepsilon}_{10} + \overline{\varepsilon}_3) + \overline{\varepsilon}_{10}\overline{\varepsilon}_3]; \\ \omega_{21}^2 = \dfrac{\omega_2^2}{\widetilde{B}(1\mid 3)}\left[\varepsilon_{20}\varepsilon_2 + \dfrac{1}{2}\left(\varepsilon_{20}\tanh\dfrac{\zeta_2}{2} + \varepsilon_2\coth\dfrac{\zeta_2}{2}\right)(\overline{\varepsilon}_1 + \overline{\varepsilon}_3) + \overline{\varepsilon}_1\overline{\varepsilon}_3\right]; \\ \omega_{22}^2 = \dfrac{\omega_2^2}{\widetilde{B}(1\mid 3)}\left[\varepsilon_{20}\varepsilon_2 + \dfrac{1}{2}\left(\varepsilon_{20}\mathrm{cth}\dfrac{\zeta_2}{2} + \varepsilon_2\mathrm{th}\dfrac{\zeta_2}{2}\right)(\overline{\varepsilon}_1 + \overline{\varepsilon}_3) + \overline{\varepsilon}_1\overline{\varepsilon}_3\right]; \\ R = \omega_1\omega_2\sqrt{(\overline{\varepsilon}_{10} - \overline{\varepsilon}_1)(\varepsilon_{20} - \varepsilon_2)}\Big/\left(\widetilde{B}(1\mid 3)\sqrt{2\tanh\dfrac{\zeta_2}{2}}\right); \\ R' = \omega_2^2(\varepsilon_{20} - \varepsilon_2)/2\widetilde{B}(1\mid 3); \\ \widetilde{B}(1\mid 3) = \varepsilon_2^2 + \overline{\varepsilon}_2(\overline{\varepsilon}_1 + \overline{\varepsilon}_3) + \overline{\varepsilon}_1\overline{\varepsilon}_3;\ \overline{\varepsilon}_n = \varepsilon_n\coth\zeta_n;\ \overline{\varepsilon}_{n0} = \varepsilon_{n0}\coth\zeta_n. \end{cases} \tag{3.5.2a}$$

Here it is explicitly assumed that fluctuations of the type are considered $x = 0$ (polar optical phonons). Going to $x = p$ you can use formal notation in finite formulas for frequencies and coefficients of normal amplitudes

$$\omega_n^2 = 0,\ \varepsilon_n = 1,\ \omega_n^2(\varepsilon_{n0} - \varepsilon_n) = \omega_p^2, \tag{3.5.2b}$$

Transferring

$$\varepsilon_n(\omega) = \varepsilon_n + \frac{(\varepsilon_{n0} - \varepsilon_n)\omega_n^2}{\omega_n^2 - \omega^2} \tag{3.5.2c}$$

is the dielectric function of a polar crystal in:

$$\varepsilon_\perp(\omega) = \frac{\omega^2 - \omega_p^2}{\omega^2} \tag{3.5.2d}$$

To reduce parallel calculations, this method of obtaining final formulas for $x = p$ will be used in the future.

The frequencies of normal vibrations are determined from the secular equation for the system (3.5.1), which is given as:

$$\omega^6 - a_1\omega^4 + a_2\omega^2 - a_3 = 0 \tag{3.5.3}$$

with coefficients

$$\begin{cases} a_1 = \dfrac{1}{\tilde{B}(1\,|3)}\{\omega_1^2[\varepsilon_2^2 + \bar{\varepsilon}_2(\bar{\varepsilon}_{10} + \bar{\varepsilon}_3)] + \\ + \omega_2^2[2\varepsilon_{20}\varepsilon_2 + (\bar{\varepsilon}_{20} + \bar{\varepsilon}_2)(\bar{\varepsilon}_1 + \bar{\varepsilon}_3) + 2\bar{\varepsilon}_1\bar{\varepsilon}_3]\}\ ; \\ a_2 = \dfrac{\omega_2^2}{\tilde{B}(1\,|3)}\{\omega_2^2[\varepsilon_{20}^2 + \bar{\varepsilon}_{20}(\bar{\varepsilon}_1 + \bar{\varepsilon}_3) + \bar{\varepsilon}_1\bar{\varepsilon}_3] + \\ + \omega_1^2[2\varepsilon_{20}\varepsilon_2 + (\bar{\varepsilon}_{20} + \bar{\varepsilon}_2)(\bar{\varepsilon}_{10} + \bar{\varepsilon}_3) + 2\bar{\varepsilon}_{10}\bar{\varepsilon}_3]\}\ ; \\ a_3 = \dfrac{\omega_1^2\omega_2^4}{\tilde{B}(1\,|3)}\{\varepsilon_{20}^2 + \bar{\varepsilon}_{20}(\bar{\varepsilon}_{10} + \bar{\varepsilon}_3) + \bar{\varepsilon}_{10}\bar{\varepsilon}_3\}, \end{cases} \tag{3.5.4}$$

according to known formulas for solving cubic equations [15].

In particular, if the substance of the first layer is non-polar ($\varepsilon_{10} = \varepsilon_1$, $\omega_1 = 0$), then the two frequencies of normal vibrations are equal:

$$\Omega_{1,2}^2 = \frac{\omega_2^2}{\tilde{B}(1\,|3)}\left[\varepsilon_{20}\varepsilon_2 + \frac{\bar{\varepsilon}_{20} + \bar{\varepsilon}_2}{2}(\bar{\varepsilon}_1 + \bar{\varepsilon}_3) \mp \frac{\bar{\varepsilon}_{20} - \bar{\varepsilon}_2}{2}\overline{S} + \bar{\varepsilon}_1\bar{\varepsilon}_3\right], \tag{3.5.5}$$

where $\overline{S} = [(\bar{\varepsilon}_1 + \bar{\varepsilon}_3)^2 - 4\bar{\varepsilon}_1\bar{\varepsilon}_3 \tanh^2 \zeta_2]^{1/2}$. The system (3.5.1) is reduced to two equations, of which when setting the frequencies $\Omega_{1,2}^2$ the amplitudes of previous vibrations is expressed in terms of the amplitudes of normal ones using the formulas:

$$W'_{12}(\Omega_1) = \frac{z_1}{\sqrt{1 + l^2}};\ W'_{22}(\Omega_1) = \frac{lz_1}{\sqrt{1 + l^2}}; \tag{3.5.6a}$$

$$W'_{12}(\Omega_2) = -\frac{lz_2}{\sqrt{1+l^2}};\ W'_{12}(\Omega_2) = \frac{z_2}{\sqrt{1+l^2}} \tag{3.5.6b}$$

(similar to (3.3.3) and (3.3.4)). Counting layers 1 and 3 as semi-infinite with $\varepsilon_1 = \varepsilon_3 = 1$ (vacuum), we obtain the frequencies of interface phonon vibrations derived in [1]:

$$\Omega_1^2 = \omega_2^2 \frac{\varepsilon_{20} + \coth\frac{\zeta_2}{2}}{\varepsilon_2 + \coth\frac{\zeta_2}{2}};\ \Omega_2^2 = \omega_2^2 \frac{\varepsilon_{20} + \tanh\frac{\zeta_2}{2}}{\varepsilon_2 + \tanh\frac{\zeta_2}{2}}. \tag{3.5.7}$$

Limit of the thick middle layer:

$$\tilde{B}(1\,|3) = (\varepsilon_2 + \bar{\varepsilon}_1)(\varepsilon_2 + \bar{\varepsilon}_3);\ \bar{S} = \bar{\varepsilon}_1 - \bar{\varepsilon}_3 \tag{3.5.8}$$

gives

$$\Omega_1^2 = \omega_2^2 \frac{\varepsilon_{20} + \bar{\varepsilon}_1}{\varepsilon_2 + \bar{\varepsilon}_1};\ \Omega_2^2 = \omega_2^2 \frac{\varepsilon_{20} + \bar{\varepsilon}_3}{\varepsilon_2 + \bar{\varepsilon}_3} \tag{3.5.9}$$

—doubly degenerate vibrations on the surfaces of the partition (1 |2) and (2 |3) media whose dispersion is due to the finite thickness of layers 1 and 3 adjacent to layer 2. in the limiting case of infinite thicknesses of layers 1 and 3, these formulas for frequencies are easily obtained from boundary conditions written using dielectric functions:

$$\varepsilon_2(\omega) = -\varepsilon_1;\ \varepsilon_2(\omega) = -\varepsilon_3. \tag{3.5.10}$$

To obtain (3.5.9) and (3.5.10) by this method, it is necessary to pre-calculate the distribution of potentials in the layers.

Using the formula (3.5.5), it is also possible to obtain plasma frequencies of an electron gas in an inversion layer at the boundary of a semiconductor 1 with a dielectric 3, interpreting this system as a three-layer system with a thin average polar layer 2 and non-polar layers 1 and 3 of finite thickness. From (3.5.5) in this approximation, ignoring the elastic properties of the electron gas and using (3.5.2b), at the last stage of calculations we find

$$\tilde{B}(1\,|3) \approx \varepsilon_2^2 + \frac{\varepsilon_2}{\zeta_2}(\bar{\varepsilon}_1 + \bar{\varepsilon}_3) + \bar{\varepsilon}_1\bar{\varepsilon}_3;\ \bar{S} = \bar{\varepsilon}_1 + \bar{\varepsilon}_3. \tag{3.5.11}$$

Substituding (3.5.7) in (3.5.3), we get:

$$\Omega_1^2 \cong \frac{\varepsilon_2 \omega_{2p}^2 \eta}{\varepsilon_1 \coth\zeta_1 + \varepsilon_3 \coth\zeta_3} = \frac{e^2 N_s \eta}{m\varepsilon_0(\varepsilon_1 \coth\zeta_1 + \varepsilon_2 \coth\zeta_2)}; \tag{3.5.12}$$

$$\Omega_2^2 = \omega_{2p}^2 \tag{3.5.13}$$

which is the plasma frequency of a two-dimensional electron gas, given, in particular, in [14].

It is interesting to note that the dependency Ω_1 in the limit under consideration ($l_2 \to 0$) from η is very powerful: Ω_1 is proportional to $\sqrt{\eta}$ (especially in the field of small η), this is unusual for optical branches of vibrations, for which the dispersion-free approximation is usually performed well.

In the other limit of the thick polar layer 2:

$$\tilde{B}(1\,|3) \cong (\varepsilon_2 + \bar{\varepsilon}_1)(\varepsilon_2 + \bar{\varepsilon}_3); \;\; \overline{S} = \bar{\varepsilon}_1 - \bar{\varepsilon}_3;$$

$$\Omega_1^2 \cong \frac{\varepsilon_2 \omega_{2p}^2}{\varepsilon_2 + \bar{\varepsilon}_1}; \;\; \Omega_2^2 \cong \frac{\varepsilon_2 \omega_{2p}^2}{\varepsilon_2 + \bar{\varepsilon}_3}. \tag{3.5.14}$$

It is obvious that Ω_1 is the frequency of the interface oscillation related to the surface separating layers 1 and 2, and Ω_2—layers 2 and 3. Both of them differ from the plasma frequency of bulk vibrations Ω_{2p} by the amount and presence of dispersion due to the finite thickness of layers 1 and 3 (see formulas (3.5.2)). Within the l_1, $l_3 \to \infty$ interface plasma vibrations become non-dispersive.

3.6 Interface vibrations in periodic systems

Equations of polarization evolution $\mathbf{P}_{f,\,x}(\kappa, \mathbf{Q}_\perp)$ in a periodic system of two types of layers, a and b, have the same appearance as (3.2.7) if $n = f$. The formulas for kinetic energy are written accordingly:

$$\Gamma = \frac{1}{2} \sum_{f=a,b} \sum_{x} \sum_{\mathbf{Q}_f, \kappa} \sum_{j=1}^{3} \left| \dot{\tilde{W}}_{jfx}(\kappa, \mathbf{Q}_f) \right|^2, \tag{3.6.1}$$

potential energy of quasi-elastic forces

$$U_{qe} = \frac{1}{2} \sum_{f=a,b} \sum_{x} \sum_{\mathbf{Q}_f, \kappa} \sum_{j=1}^{3} \Omega_{jfx}^2(\mathbf{Q}_f) \left| \tilde{W}_{jfx}(\kappa, \mathbf{Q}_f) \right|^2 \tag{3.6.2}$$

and the energy of interaction

$$\begin{aligned} U_{\text{int}} = \frac{1}{2} \sum_{f=a,\,b} \sum_{x,\,y} \sum_{\boldsymbol{\eta},\,\kappa} \Bigg\{ & \sum_{q_n \neq 0} \omega_{fx} \omega_{fy} \widetilde{W}_{lfx}^{*}(\kappa, \mathbf{Q}_f) \widetilde{W}_{lfy}(\kappa, \mathbf{Q}_f) \\ & - \sum_{c=a,\,b} \sum_{j,\,l=1,\,2} M_{jfx,\,lcy} \widetilde{W}_{jfx}^{*}(\kappa, \boldsymbol{\eta}, 0) \widetilde{W}_{lcy}(\kappa, \boldsymbol{\eta}, 0) \Bigg\} \end{aligned} \tag{3.6.3}$$

with coefficients

$$M_{jfx,\,lcy} = \frac{1}{2} \cdot \frac{\omega_{fx}\sqrt{\varepsilon_f}}{\sqrt{\tanh \frac{\zeta_f}{2}}} \cdot \frac{\omega_{cy}\sqrt{\varepsilon_c}}{\sqrt{\tanh \frac{\zeta_c}{2}}} \cdot K_{jf,\,lc}(\kappa). \tag{3.6.4}$$

Adding up (3.6.1), (3.6.2), and (3.6.3), we get the total polarization energy:

$$H = \Gamma + U_q + U_{\text{int}} = \Gamma + U = \frac{1}{2}\sum_{f=a,b}\sum_{x}\sum_{\mathbf{Q}_f,\kappa}\sum_{j=1}^{3}\left|\dot{\tilde{W}}_{jfx}(\kappa, \mathbf{Q}_f)\right|^2$$

$$+\frac{1}{2}\sum_{f=a,b}\sum_{x,y}\sum_{\kappa,\boldsymbol{\eta}}\left\{\sum_{q_f\neq 0}\sum_{j=1}^{3}\left[\Omega^2_{jfx}(\mathbf{Q}_f)\delta_{x,\,y} + \omega_{fx}\omega_{fy}\delta_{j1}\right]\right.$$

$$\times\tilde{W}^*_{jfx}(\kappa, \mathbf{Q}_f)\tilde{W}_{jfy}(\kappa, \mathbf{Q}_f) + \Omega^2_{3fx}(\boldsymbol{\eta}, 0)\delta_{x,\,y}\left|\tilde{W}_{3fx}(\kappa, \boldsymbol{\eta}, 0)\right|^2$$

$$+\sum_{c=a,b}\sum_{j,l=1}^{2}\left[\Omega^2_{jfx}(\boldsymbol{\eta}, 0) + \delta_{j,\,l}\delta_{f,\,c}\delta_{x,\,y} - M_{jfx,\,lcy}(\kappa)\right]\times$$

$$\left.\tilde{W}^*_{jfx}(\kappa, \boldsymbol{\eta}, 0)\tilde{W}_{lcy}(\kappa, \boldsymbol{\eta}, 0)\right\}. \tag{3.6.5}$$

From the total energy (3.6.5) follows the equation for amplitudes:

$$\ddot{\tilde{W}}_{jfx}(\kappa, \mathbf{Q}_f) \equiv -\omega^2\tilde{W}_{jfx}(\kappa, \mathbf{Q}_f) = -\frac{\partial H}{\partial W^*_{jfx}(\kappa, \mathbf{Q}_f)}, \tag{3.6.6}$$

which, as well as for the periodic system, form mutually independent subsystems for amplitudes of bulk vibrations

$$\sum_{y}\left\{[\omega^2 - \Omega^2_{jfx}(\mathbf{Q}_f)]\,\delta_{x,\,y} - \omega_{fx}\omega_{fy}\delta_{j,\,1}\right\}\tilde{W}_{jfy}(\mathbf{Q}_f) = 0 \tag{3.6.7}$$

(with free parameters $\boldsymbol{\eta}$, $q_{||} \neq 0$, κ, j, f, x); amplitudes of interface vibrations with $j = 1, 2$

$$\sum_{c=a,b}\sum_{l=1}^{2}\sum_{y}\left\{[\omega^2 - \Omega^2_{jfx}(\boldsymbol{\eta}, 0)]\delta_{j,\,l}\delta_{f,\,c}\delta_{x,\,y} + M_{jfx,\,lcy}(\kappa)\right\}\tilde{W}_{lcy}(\kappa, \boldsymbol{\eta}, 0) = 0 \tag{3.6.8}$$

(with free parameters $\boldsymbol{\eta}$, $\kappa, j = 1, 2, f, x$); amplitudes of interface vibrations with polarization $j = 3$

$$[\omega^2 - \Omega^2_{3fx}(\boldsymbol{\eta}, 0)]\,\tilde{W}_{3fx}(\kappa, \boldsymbol{\eta}, 0) = 0 \tag{3.6.9}$$

(with free parameters $\boldsymbol{\eta}$, κ, f, x).

According to (3.6.7) transverse bulk fluctuations $j = 2, 3$ are normal

$$[\omega^2 - \Omega^2_{jfx}(\mathbf{Q}_f)]\,\tilde{W}_{jfx}(\mathbf{Q}_n) = 0;\ j = 2, 3 \tag{3.6.10}$$

for each type of polarization $x = 0, p$, with the seed frequencies $\Omega_{jfx}(\mathbf{Q}_f)$, defined in (3.2.8a) and (3.2.9a). Their amplitudes are absent in the interaction energy (3.2.2b), and, consequently, they are not active in the formation of electric fields and potentials.

Longitudinal bulk fluctuations $j = 1$ are obtained mixed by the types of polarizations $x = 0$, p, but separated in sublattices of layers a and b:

$$\sum_{y}\left\{ [\omega^2 - \Omega_{1fx}^2(\mathbf{Q}_f)]\delta_{x,\,y} - \omega_{fx}\omega_{fy} \right\} \tilde{W}_{1fy}(\mathbf{Q}_f) = 0. \tag{3.6.11}$$

From (3.6.11) we obtain two secular equations with coefficients depending on the parameters of the layers a and b, whose solutions coincide with (3.3.1b). In disregard of mixing $\omega_{ax}\omega_{by} = 0$, just as for a non-periodic structure, the frequencies of normal modes pass into a bulk longitudinal phonon $\omega_{f,\,L0}$ and bulk plasma ω_{fp} dispersing frequencies. In normal interface vibrations according to (3.6.8) all states are mixed $j = 1, 2$; $f = a, b$; $x = 0, p$, which leads to a system of eight equations:

$$\sum_{r'} N_{rr'}(\kappa, \omega)\tilde{W}_{r'}(\kappa, \eta, 0) = 0; \tag{3.6.12a}$$

$$(j, f, x) \to r; \;\; (l, c, y) \to r'$$

and the secular equation:

$$|N_{rr'}(\kappa, \omega)| \equiv \left| (\omega^2 - \Omega_{jfx}^2(\eta, 0))\delta_{x,\,y}\delta_{f,\,c}\delta_{j,\,l} + M_{jfx,\,lcy}(\kappa) \right| = 0 \tag{3.6.12b}$$

eighth degree relative ω^2. In general, it allows only a numerical solution.

As an example, let's consider special cases of either phonon or plasma polarization. The secular equation now of the fourth order can be represented in the form

$$\prod_{\beta=1}^{2} (\omega^4 - R_\beta^2\omega^2 + S_\beta) = 0 \tag{3.6.13}$$

with coefficients ($\beta = 1, 2$), recorded for $x = 0$:

$$R_\beta = \omega_a^2\frac{\varepsilon_{a0} + \phi_\beta\varepsilon_b}{\varepsilon_a + \phi_\beta\varepsilon_b} + \omega_b^2\frac{\varepsilon_a + \phi_\beta\varepsilon_{b0}}{\varepsilon_a + \phi_\beta\varepsilon_b}; \qquad S_\beta = \omega_a^2\omega_b^2\frac{\varepsilon_{a0} + \phi_\beta\varepsilon_{b0}}{\varepsilon_a + \phi_\beta\varepsilon_b} \tag{3.6.14a}$$

in which notation

$$\phi_1 = \frac{1}{\phi_2} = \Gamma + (\Gamma^2 - 1)^{\frac{1}{2}}; \; \Gamma = \frac{\cosh\zeta_a ch\zeta_b - \cos(\kappa L)}{\sinh\zeta_a \sinh\zeta_b} \tag{3.6.14b}$$

is used. The solution of the secular equation (3.6.13) gives four frequencies defined by the formulas:

$$\Omega_{1,\,4}^2(\kappa, \eta) = \frac{R_1}{2} \pm \sqrt{\left(\frac{R_1}{2}\right)^2 - S_1}; \quad \Omega_{2,\,3}^2(\kappa, \eta) = \frac{R_2}{2} \mp \sqrt{\left(\frac{R_2}{2}\right)^2 - S_2} \tag{3.6.14c}$$

for normal interface optical vibrations that relate to all surfaces of the layer system, and not to the surfaces of one of the layers, which are called interface spatially extended normal vibrations due to this circumstance. A spatially extended fashion to form a zone in κ, the width of which is determined by the thickness of the layers l_a and l_b.

If $l_f \to \infty$, then according to the definition Γ (3.6.14b) frequency dependence on κ disappears, and zones degenerate into levels. Examples of dependencies $\Omega_s(\kappa, \eta)$ are shown in figures 3.1 and 3.2 (see works [4, 16, 17], in which this law of dispersion was obtained, as well as in [7].

Consider plasma oscillations for the case when the electron gas is present only in one of the layers b. Then $\omega_{ap}^2 = 0$ and $S_\beta = 0$. From (3.6.13), using the rules of transition to plasmon parameters, according to (3.5.2b) we obtain a formula for the frequency of spatially extended plasma oscillations

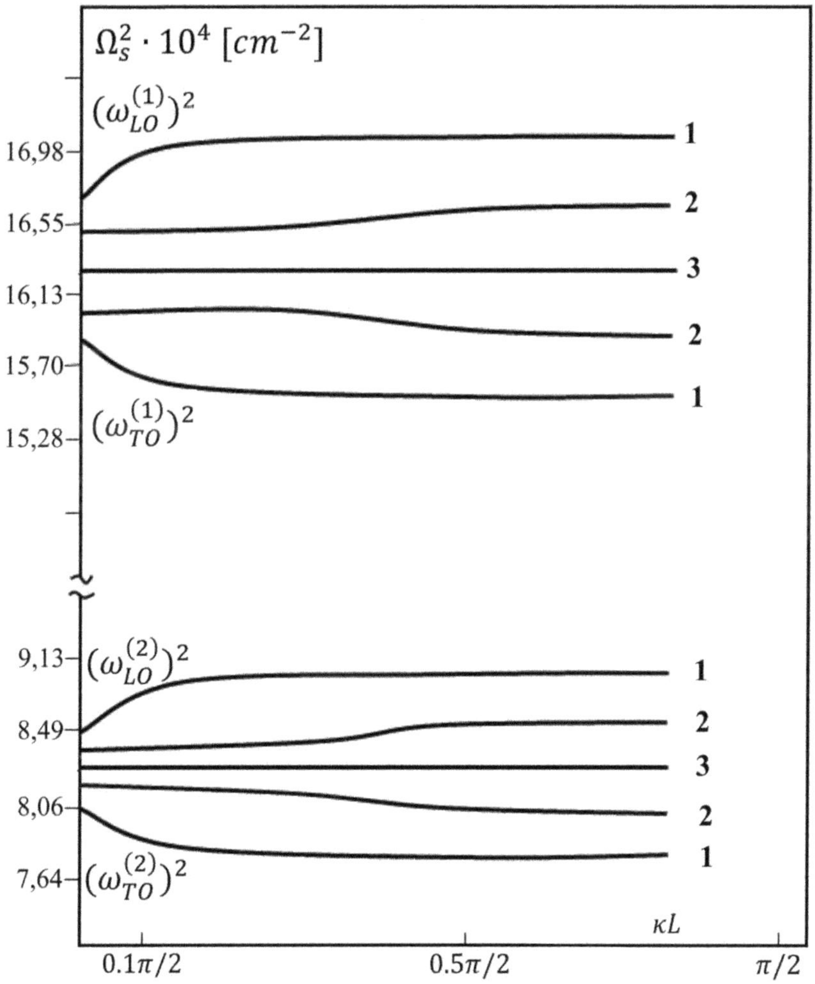

Figure 3.1. Dependence Ω_s^2 on κL for a two-layer periodic structure with an elementary cell containing layers AlSb and GaAs: 1—$\eta l_a = 0.1$; $\eta l_b = 0.2$; 2—$\eta l_a = 1$; $\eta l_b = 2$; 3—$\eta l_a = 10$; $\eta l_b = 20$.

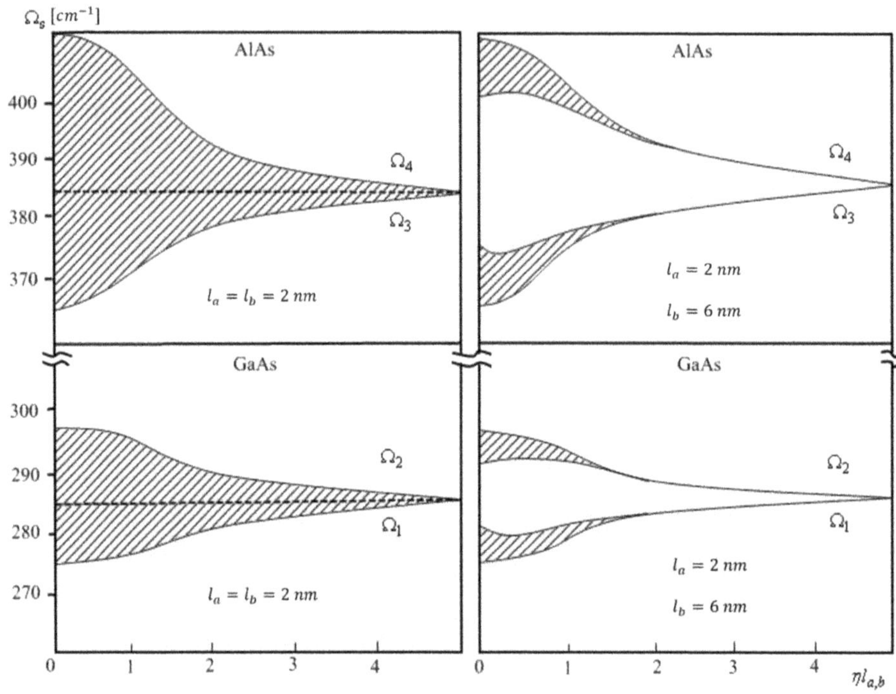

Figure 3.2. Dependence Ω_s on the structure AlAs/GaAs [19]. Solid curves correspond $\kappa L = \pi$, dashed ones —$\kappa L = 0$.

$$\omega_\beta^2 = \frac{\varepsilon_b \omega_{bp}^2 \phi_\beta}{\varepsilon_a^2 + \phi_\beta \varepsilon_b} = \varepsilon_b \omega_{bp}^2 \frac{\varepsilon_a(\Gamma \pm \sqrt{\Gamma^2 - 1}) + \varepsilon_b}{2\varepsilon_a \varepsilon_b \Gamma + \varepsilon_a^2 + \varepsilon_b^2}, \tag{3.6.15}$$

coinciding with formula (7) from [18].

For a separate plate, we get the frequencies by taking $\zeta_a \to \infty$; $\varepsilon_a = 1$; $\varepsilon_b = 1$:

$$\Omega_{1,2}^2 = \frac{\omega_p^2}{2}(1 \mp e^{-\zeta_b}).$$

At small $\zeta_b < 1$:

$$\Omega_{1,2}^2 = \frac{\omega_p^2}{2}\left(1 \mp \left(1 + \frac{\zeta_b \sinh \zeta_a}{\cos \kappa L - \cosh \zeta_a}\right)\right). \tag{3.6.16b}$$

One of the frequencies defined by the last formula, corresponding to the sign «minus»:

$$\Omega_1^2 = \frac{\omega_p^2}{2} \cdot \frac{\zeta_b \sinh \zeta_a}{\cosh \zeta_a - \cos \kappa L},$$

describes a spatially extended plasma mode with strong dispersion Ω_1^2, proportional η. Dependency graphs $\Omega_\beta(\kappa, \eta)$ for periodic systems with plasma vibrations are given in [19, 20].

From the equations (3.6.8), the amplitudes of normal vibrations are found in the way described in section 3.4. The result has a form similar to that described by the formulas (3.4.8a) and (3.4.8b), with the only difference that the indices r and r' take eight values:

$$\begin{aligned} &r = 1(f = a, j = 1, x = 0);\ r = 2(f = a, j = 2, x = 0);\\ &r = 3(f = b, j = 1, x = 0);\ r = 4(f = и, j = 2, x = 0);\\ &r = 5(f = a, j = 1, x = p);\ r = 6(f = a, j = 2, x = p);\\ &r = 7(f = b, j = 1, x = p);\ r = 8(f = b, j = 2, x = p). \end{aligned} \tag{3.6.17}$$

In normal variables, the energy of elastic vibrations of a periodic system has the usual diagonal form, the explicit form of which will be written out further.

3.7 Hamiltonian of the interaction of an electron with a field of slow polarization (slow polarization)

The total polarization energy (3.2.16) of a multilayer system decays into the sum of the bulk H_B and interface H_S parts:

$$H = H_B + H_S, \tag{3.7.1a}$$

where is the energy of bulk:

$$\begin{aligned} H_B = \frac{1}{2}\sum_{\eta}\sum_{n=I}^{K}\sum_{q_n\neq 0}\Bigg\{\sum_{x}\sum_{j=2}^{3}\left[\left|\dot{\tilde{W}}_{jnx}(\mathbf{Q}_n)\right|^2 + \Omega^2_{jnx}(\mathbf{Q}_n)\left|\tilde{W}_{jnx}(\mathbf{Q}_n)\right|^2\right]\\ + \sum_{s=1}^{2}[|\dot{z}_{sn}(\mathbf{Q}_n)|^2 + \omega^2_{sn}(\mathbf{Q}_n)|z_{sn}(\mathbf{Q}_n)|^2]\Bigg\} \end{aligned} \tag{3.7.1b}$$

and interface vibrations:

$$\begin{aligned} H_S = \frac{1}{2}\sum_{\eta}\Bigg\{\sum_{n=I}^{K}\sum_{x}\left[\left|\dot{\tilde{W}}_{3nx}(\eta, 0)\right|^2 + \Omega^2_{3nx}(\eta, 0)|\tilde{W}_{3nx}(\eta, 0)|^2\right]\\ + \sum_{s=1}^{2N_0}[|\dot{z}_s(\eta)|^2 + \Omega^2_s(\eta)|z_s(\eta)|^2]\Bigg\} \end{aligned} \tag{3.7.1c}$$

Standard transition to the representation of secondary quantization by formulas:

$$\tilde{W}_{jnx}(\mathbf{Q}_n) = \sqrt{\frac{\hbar}{2\omega_{jnx}}}\left[\hat{b}^{+}_{jnx}(-\mathbf{Q}_n) + \hat{b}_{jnx}(\mathbf{Q}_n)\right]; \tag{3.7.2a}$$

$$\dot{\tilde{W}}_{jnx}(\mathbf{Q}_n) = i\sqrt{\frac{\hbar\omega_{jnx}}{2}}\left[\hat{b}^{+}_{jnx}(-\mathbf{Q}_n) + \hat{b}_{jnx}(\mathbf{Q}_n)\right] \tag{3.7.2b}$$

for unmixing vibrations (which have retained the original numbering j and x) and

$$\hat{z}_s(\eta) = \sqrt{\frac{\hbar}{2\Omega_s}}\left[\hat{b}_s^+(-\eta, 0) + \hat{b}_s(\eta, 0)\right]; \tag{3.7.3a}$$

$$\dot{\hat{z}}(\eta) = i\sqrt{\frac{\hbar\Omega_s}{2}}\left[\hat{b}_s^+(-\eta, 0) + \hat{b}_s(\eta, 0)\right] \tag{3.7.3b}$$

for mixing ones with $\hat{b}_s^+(-\eta, 0)$, $\hat{b}_s(\eta, 0)$—by Bose operators, respectively, the generation and destruction of quanta of normal oscillations gives the Hamiltonian of the field of inertial polarization:

$$\hat{H} = \hat{H}_B + \hat{H}_S, \tag{3.7.4a}$$

where

$$\begin{aligned}\hat{H}_B = \sum_{\eta}\sum_{n=I}^{K}\sum_{q_n\neq 0}\Bigg\{&\sum_{x}\sum_{j=2}^{3}\hbar\Omega_{jnx}(\mathbf{Q}_n)\left[\hat{b}_{jnx}^+(\mathbf{Q}_n)\hat{b}_{jnx}(\mathbf{Q}_n) + \frac{1}{2}\right]\\ &+\sum_{s=1}^{2}\hbar\omega_{sn}(\mathbf{Q}_n)\left[\hat{b}_{sn}^+(\mathbf{Q}_n)\hat{b}_{sn}(\mathbf{Q}_n) + \frac{1}{2}\right]\Bigg\};\end{aligned} \tag{3.7.4b}$$

$$\begin{aligned}\hat{H}_S = \sum_{\eta}\Bigg\{&\sum_{n=I}^{K}\sum_{x}\hbar\Omega_{3nx}(\eta, 0)\left[\hat{b}_{3nx}^+(\eta, 0)\hat{b}_{3nx}(\eta, 0) + \frac{1}{2}\right]\\ &+\sum_{s=1}^{2N_0}\hbar\Omega_s(\eta)\left[\hat{b}_s^+(\eta, 0)\hat{b}_s(\eta, 0) + \frac{1}{2}\right]\Bigg\}.\end{aligned} \tag{3.7.4c}$$

The potential energy of interaction of charge carriers with the polarization field in a multilayer structure is

$$H_{n-P} = \int d^2y\sum_{n=I}^{K}\int_{z_{n-1}}^{z_n} dz\ \rho_n(\mathbf{r})\ V_n^P(\mathbf{r});\ r^2 = y^2 + z^2 \tag{3.7.5}$$

In the proposition that a point charge located at a point interacts with the polarization nth layer $\rho_{n'}(\mathbf{r}') = -e\delta_{n'n}\delta(\mathbf{r}' - \mathbf{r})$, the potential energy of the interaction has the form

$$H_{n-P} = -eV_n^P(\mathbf{r}), \tag{3.7.6}$$

which, in accordance with the structure of the potential of the polarization field (3.2.3), takes the form

$$H_{n-P} = H_{n-B}(\mathbf{r}) + H_{n-S}(\mathbf{r}). \tag{3.7.7}$$

After renormalization (3.2.10) of amplitudes, the potential energy of the electron interaction with bulk fluctuations of polarization

$$H_{n-B}(\mathbf{r}) = \frac{e}{\sqrt{\varepsilon_0}}\sum_{\boldsymbol{\eta}} \mathrm{e}^{i\boldsymbol{\eta}\mathbf{v}} \frac{1}{\sqrt{V_n\varepsilon_n}} \sum_{q_n\neq 0} \frac{1}{\mathbf{Q}_n}\left[\mathrm{e}^{iq_n(z-z_{n-1})} + \frac{1}{1-C_n}\left(C_n - \frac{\cosh w_n}{\cosh\frac{\zeta_n}{2}}\right)\right]$$

$$\times\sum_x \omega_{nx}\tilde{W}_{1nx}(\mathbf{Q}_n) \tag{3.7.8}$$

and with interface polarization fluctuations

$$H_{n-S}(\mathbf{r}) = -\frac{e}{\sqrt{\varepsilon_0}}\sum_{\boldsymbol{\eta}} \mathrm{e}^{i\boldsymbol{\eta}\mathbf{v}} \frac{1}{2\sqrt{\eta L_x L_y}\,\sinh\frac{\zeta_n}{2}}$$

$$\times\sum_{k=1}^{K}\sum_{l=1}^{2}\sum_y [L_{1n,\,lky}\cosh w_n + L_{2n,\,lky}\sinh w_n]\quad \widetilde{W}_{lky}(\boldsymbol{\eta}, 0), \tag{3.7.9}$$

where coefficients are entered:

$$L_{jn,\,lky} = \frac{1}{\sqrt{2}}\frac{\omega_{ky}\sqrt{\varepsilon_k}}{\sqrt{\tanh\frac{\zeta_k}{2}}}K_{jn,\,lk}. \tag{3.7.10}$$

As a result of the unitary transformation of amplitudes (3.4.2a)–(3.4.2c) from (3.7.9) follows:

$$H_{n-S}(\mathbf{r}) = -\frac{e}{\sqrt{\varepsilon_0}}\sum_{\boldsymbol{\eta}} \mathrm{e}^{i\boldsymbol{\eta}\mathbf{v}} \frac{1}{2\sqrt{\eta L_x L_y}\,\sinh\frac{\zeta_n}{2}}$$

$$\times\sum_{k=I}^{K}\sum_{l=1}^{2}\sum_y \left[L'_{1n,\,lky}\cosh w_n + L'_{2n,\,lky}\sinh w_n\right]\quad W'_{lky}(\boldsymbol{\eta}) \tag{3.7.11}$$

with coefficients: for $k = I + 1, \ldots, K - 1$

$$L'_{jn,lky} = L_{jn,lky}, \tag{3.7.12a}$$

for $k = I, K$

$$L'_{jn,lky} = \delta_{l,2}\left(1 + \tanh^2\frac{\zeta_k}{2}\right)^{\frac{1}{2}} L_{jn,2ky}. \tag{3.7.12b}$$

Using the unified indexes entered above, we obtain from (3.7.11):

$$H_{n-S}(\mathbf{r}) = -\frac{e}{\sqrt{\varepsilon_0}}\sum_{\boldsymbol{\eta}} \mathrm{e}^{i\boldsymbol{\eta}\mathbf{v}} \frac{1}{2\sqrt{\eta L_x L_y}\,\sinh\frac{\zeta_n}{2}}$$

$$\times\sum_{r'=1}^{2N_0-4}\left(L'_{1n,r'}\cosh w_n + L'_{2n,r'}\sinh w_n\right)\quad W'_{r'}(\boldsymbol{\eta}). \tag{3.7.13}$$

The transition to normal amplitudes (3.3.4) and (3.3.7) provides for interaction with bulk fluctuations:

$$H_{n-B}(\mathbf{r}) = \frac{e}{\sqrt{\varepsilon_0}} \sum_{\boldsymbol{\eta}} \mathrm{e}^{i\boldsymbol{\eta}\mathbf{v}} \frac{1}{\sqrt{V_n \varepsilon_n}} \sum_{q_n \neq 0} \frac{1}{Q_n} \left[\mathrm{e}^{iq_n(z-z_{n-1})} + \frac{1}{1-C_n}\left(C_n - \frac{\cosh w_n}{\cosh \frac{\varsigma_n}{2}} \right) \right]$$

$$\times \left[\frac{\omega_{n0} + \omega_{np} l_n}{\sqrt{1+l_n^2}} z_{1n}(\mathbf{Q}_n) + \frac{\omega_{np} - \omega_{n0} l_n}{\sqrt{1+l_n^2}} z_{2n}(\mathbf{Q}_n) \right] \tag{3.7.14}$$

and from (3.7.13):

$$H_{n-S}(\mathbf{r}) = -\frac{e}{\sqrt{\varepsilon_0}} \sum_{\boldsymbol{\eta}} \mathrm{e}^{i\boldsymbol{\eta}\mathbf{v}} \frac{1}{2\sqrt{\eta L_x L_y} \sinh \frac{\varsigma_n}{2}}$$

$$K^{\pm}_{n,\,n',\,s}(\Omega) = A_2(\Omega)\left\{ 1 - \exp\left(-\frac{\hbar(\Omega)}{k_0 T} \right) \right\}\left(-\frac{E_0}{k_0 T} \right) 2 \sinh\left(\frac{\hbar\omega_c}{2k_0 T} \right)$$

$$\times [\delta_{r',\,s} - (1-\delta_{r',\,s}) F_{r',\,s}]\left(1 + \sum\nolimits_{r'' \neq s} |F_{r''s}|^2 \right)^{-\frac{1}{2}} \cdot z_s(\boldsymbol{\eta}). \tag{3.7.15}$$

Finally, moving to the representation of secondary quantization by the above formulas, we come to the desired Hamiltonian of the interaction of an electron with polarization:

$$\hat{H}_{n-P} = \hat{H}_{n-B}(\mathbf{r}) + \hat{H}_{n-S}(\mathbf{r}), \tag{3.7.16a}$$

where

$$H_{n-B}(\mathbf{r}) = \frac{e}{\sqrt{\varepsilon_0}} \sum_{\boldsymbol{\eta}} \mathrm{e}^{i\boldsymbol{\eta}\mathbf{v}} \sqrt{\frac{\hbar}{2V_n \varepsilon_n}} \sum_{q_n \neq 0} \frac{1}{Q_n} \left[\mathrm{e}^{iq_n(z-z_{n-1})} + \frac{1}{1-C_n}\left(C_n - \frac{\cosh w_n}{\cosh \frac{\varsigma_n}{2}} \right) \right]$$

$$\times \left[\frac{\omega_{n0} + \omega_{np} l_n}{\sqrt{1+l_n^2}} \cdot \frac{1}{\sqrt{\omega_{1n}(\mathbf{Q}_n)}} \left[\hat{b}^{+}_{1n}(-\mathbf{Q}_n) + \hat{b}_{1n}(\mathbf{Q}_n) \right] \right.$$

$$\left. + \frac{\omega_{n0} - \omega_{np} l_n}{\sqrt{1+l_n^2}} \cdot \frac{1}{\sqrt{\omega_{2n}(\mathbf{Q}_n)}} \left[\hat{b}^{+}_{2n}(-\mathbf{Q}_n) + \hat{b}_{2n}(\mathbf{Q}_n) \right] \right]; \tag{3.7.16b}$$

$$\hat{H}_{n-S}(\mathbf{r}) = -\frac{e}{\sqrt{\varepsilon_0}} \sum_{\boldsymbol{\eta}} \mathrm{e}^{i\boldsymbol{\eta}\mathbf{v}} \frac{\sqrt{\hbar}}{2\sqrt{2\eta L_x L_y} \sinh \frac{\varsigma_n}{2}}$$

$$\times \sum_{s=1}^{2N_0-4} \frac{1}{\sqrt{\Omega_s(\eta)}} \sum_{r'=1}^{2N_0-4} \left(L'_{1n,\,r'} \cosh w_n + L'_{2n,\,r'} \sinh w_n\right) \times [\delta_{r',\,s} - (1 - \delta_{r',\,s})F_{r',\,s}]$$

$$\times \left(1 + \sum\nolimits_{r''\neq s} |F_{r''s}|^2\right)^{-\frac{1}{2}} \left[\hat{b}_s^{+}(-\eta,\,0) + \hat{b}_s(\eta,\,0)\right]. \tag{3.7.16c}$$

After the introduction of normal oscillation amplitudes, the energy of bulk

$$H_B = \frac{1}{2}\sum_{\vec{\eta}}\sum_{\kappa}\sum_{f=a,b}\sum_{q_f\neq 0}\left\{\sum_{x}\sum_{j=2}^{3}\left[\left|\dot{\tilde{W}}_{jnx}(\kappa,\,\mathbf{Q}_f)\right|^2 + \Omega_{jnx}^2(\mathbf{Q}_f)\left|\tilde{W}_{jfx}(\kappa,\,\mathbf{Q}_f)\right|^2\right]\right.$$

$$\left. + \sum_{s=1}^{2}\left[\left|\dot{z}_{sf}(\kappa,\,\mathbf{Q}_f)\right|^2 + \omega_{sf}^2(\mathbf{Q}_f)\left|z_{sf}(\kappa,\,\mathbf{Q}_f)\right|^2\right]\right\} \tag{3.7.17a}$$

and interface vibrations of the periodic system

$$H_S = \frac{1}{2}\sum_{\eta}\sum_{\kappa}\left\{\sum_{f=a,b}\sum_{x}\left[\left|\dot{\tilde{W}}_{3fx}(\kappa,\,\eta,\,0)\right|^2 + \Omega_{3fx}^2(\eta,\,0)\left|\tilde{W}_{3fx}(\kappa,\,\eta,\,0)\right|^2\right]\right.$$

$$\left. + \sum_{s=1}^{8}\left[\left|\dot{z}_s(\kappa,\,\eta,\,0)\right|^2 + \Omega_s^2(\kappa,\,\eta)|z_s(\kappa,\,\eta,\,0)|^2\right]\right\}. \tag{3.7.17b}$$

The standard transition to providing secondary quantization (3.7.2a)–(3.7.3b) gives

$$\hat{H}_B = \sum_{\eta}\sum_{\kappa}\sum_{f=a,b}\sum_{q_f\neq 0}\left\{\sum_{x}\sum_{j=2}^{3}\hbar\Omega_{jfx}(\mathbf{Q}_f)\left[\hat{b}_{jfx}^{+}(\kappa,\,\mathbf{Q}_f)\hat{b}_{jfx}(\kappa,\,\mathbf{Q}_f) + \frac{1}{2}\right]\right.$$

$$\left. + \sum_{s=1}^{2}\hbar\omega_{sf}(\mathbf{Q}_f)\left[\hat{b}_{jfx}^{+}(\kappa,\,\mathbf{Q}_f)\hat{b}_{jfx}(\kappa,\,\mathbf{Q}_f) + \frac{1}{2}\right]\right\}; \tag{3.7.18a}$$

$$\hat{H}_S = \sum_{\eta}\sum_{\kappa}\left[\sum_{f=a,b}\sum_{x}\hbar\Omega_{3fx}(\eta,\,0)\left[\hat{b}_{3fx}^{+}(\kappa,\,\eta,\,0)\hat{b}_{3fx}(\kappa,\,\eta,\,0) + \frac{1}{2}\right]\right.$$

$$\left. + \sum_{s=1}^{8}\hbar\Omega_s(\kappa,\,\eta)\left[\hat{b}_s^{+}(\kappa,\,\eta,\,0)\hat{b}_s(\kappa,\,\eta,\,0) + \frac{1}{2}\right]\right]. \tag{3.7.18b}$$

We express the potential energy of the interaction of an electron with polarization (3.7.8) and (3.7.9) through the found amplitudes of normal vibrations, and then, passing to the provision of secondary quantization, we find the Hamiltonian of the specified interaction with the following terms:

$$\hat{H}_{n-B}(\mathbf{r}) = \frac{e}{\sqrt{\varepsilon_0}} \sum_{\eta} e^{i\eta y} \frac{1}{\sqrt{N}}$$

$$\times \sum_{\kappa} e^{i\frac{n}{2}\kappa L} \sqrt{\frac{\hbar}{2V_f \varepsilon_f}} \sum_{q_f \neq 0} \frac{1}{Q_f} \left[e^{iqf(z-z_{n-1})} + \frac{1}{1-C_f} \left(C_f - \frac{\cosh w_n}{\cosh \frac{\zeta_f}{2}} \right) \right]$$

$$\times \left\{ \frac{\omega_{f0} + \omega_{fp} \ell_f}{\sqrt{i + \ell_f^2}} \frac{1}{\sqrt{\omega_{1f}(\mathbf{Q}_f)}} \left[\hat{b}_{1f}^{+}(-\kappa, -\mathbf{Q}_f) + \hat{b}_{1f}(\kappa, \mathbf{Q}_f) \right] \right.$$

$$\left. + \frac{\omega_{fp} - \omega_{f0} \ell_f}{\sqrt{1 + \ell_f^2}} \frac{1}{\sqrt{\omega_{2f}(\mathbf{Q}_f)}} \left[\hat{b}_{2f}^{+}(-\kappa, -\mathbf{Q}_f) + \hat{b}_{2f}(\kappa, \mathbf{Q}_f) \right] \right\}; \qquad (3.7.19a)$$

$$\hat{H}_{n-S}(\mathbf{r}) = -\frac{e}{\sqrt{\varepsilon_0}} \sum_{\eta} e^{i\eta y} \frac{1}{\sqrt{N}} \sum_{\kappa} e^{i\frac{n}{2}\kappa L} \frac{\sqrt{\hbar}}{2\sqrt{2\eta L_x L_y} \sinh \frac{\zeta_f}{2}}$$

$$\times \sum_{s=1}^{8} \frac{1}{\sqrt{\Omega_s(\kappa, \eta)}} \sum_{r'=1}^{8} [L_{1f,\, r'}(\kappa) \cosh w_n + L_{2f,\, r'}(\kappa) \sinh w_n]$$

$$\times \left(1 + \sum_{r'' \neq s} |F_{r'',\, s}''|^2\right)^{-1/2} \left[\hat{b}_s^{+}(-\kappa, -\eta, 0) + \hat{b}_s(\kappa, \eta, 0) \right]. \qquad (3.7.19b)$$

Coefficients are entered here

$$L_{jf,\, l_{cy}}(\kappa) = \frac{1}{\sqrt{2}} \frac{\omega_{cy} \sqrt{\varepsilon_c}}{\sqrt{\tanh \frac{\zeta_c}{2}}} K_{jf,\, lcy}(\kappa),$$

which use the notations (3.1.8.4a)–(3.1.8.4d).

3.8 Interaction Hamiltonian of an electron with fast polarization (plasma of valence electrons)

In Chapter 1, deriving the equations for potentials in multilayer structures, we attributed the plasma of valence electrons to a fast subsystem, the state of which follows the external field without inertia, and on this basis excluded the polarization vector of valence electrons from the definition of the induction vector, writing the latter in the form (3.2.2). In essence, this means using a purely classical description of the shielding effect created by the valence electron plasma. Since the plasma frequencies of the valence electron plasma are of the order $10^{15} \ldots 10^{16}$ s^{-1}, then this approximation for many problems should be considered reasonable. For example, when considering the movement of an electron interacting with a plasma, we can assume that the polaron state caused by this interaction (in the terminology of Toyozawa [21]—an electronic polaron) practically does not depend on the nature of the movement (whether it is moving through the volume of the crystal or orbiting in a small impurity center, etc) or changes in the state of the electron (scattering, transitions between impurity centers, etc) and manifests itself only in renormalization of the mass and shielding by means of the interaction potential with other charges. The situation, however, changes significantly in spatially limited systems. When the electron polaron approaches the border, the non-precision of the polaron state appears, which significantly depends on the distance to the border. Therefore, a quantum mechanical description of the system using the electron-plasmon interaction Hamiltonian is necessary. We will briefly trace the main stages of derivation of such a Hamiltonian.

The plasma state of valence electrons is described by the electrostatic equations (3.2.1). Material management in this case is as follows:

$$\mathbf{D}_n = \varepsilon_0 \mathbf{E}_n + \mathbf{P}_n \tag{3.8.1}$$

Here $\mathbf{P}_n$ is polarization of the valence electron system.

Assuming that the bulk charges are absent, the equation for the potential in n-th surrounding in accordance with (3.2.1) and (3.8.1) we obtain as:

$$\text{div grad} V_n(\mathbf{r}) = -\varepsilon_0{}^{-1} \text{div} \mathbf{P}_n, \tag{3.8.2}$$

where $V_n(\mathbf{r})$ is potential in n-th surrownding, $\mathbf{E}_n = -\text{grad} V_n(\mathbf{r})$. Assuming the elastic compression and shear modules of valence electrons are equal, we write down a system of equations for the evolution of polarization:

$$\ddot{\mathbf{P}}_n = -\Delta^2 \mathbf{P}_n + \omega_{np}^2 \varepsilon_0 \mathbf{E}_n^p \text{ где } n = 1, \ldots, K, \tag{3.8.3}$$

where Δ is frequency of the return elastic force; ω_{np} is the plasma frequency of the valence electrons; and $\omega_{np} = e^2 N/(m_n \varepsilon_0)$; $\mathbf{E}_n^p$ is the intensity of the electric field induced by polarization. The method for solving equations (3.8.2) and (3.8.3) is described in sections 3.2 and 3.3. It consists of calculating the potential $V_n(\mathbf{r})$ from equation (3.8.2), finding the field $\mathbf{E}_n^p$, solving the system of equations (3.8.3), and obtaining normal frequencies and normal amplitudes. In sections 3.2 and 3.3, these operations are performed with respect to equations (3.2.4) and (3.2.7), which pass into equations (3.8.2) and (3.8.3) when $\varepsilon_n = 1$, $\omega_{TO}^2 = \Delta^2$, $\omega_{TO}^2(\varepsilon_{n0} - \varepsilon_n) = \omega_p^2$. Making the same substitutions in the Hamiltonians (3.7.16a) and (3.7.16b), we obtain the Hamiltonian of the interaction of an electron with a plasma of valence electrons.

3.9 Conclusion

A system of equations for polarization oscillations is solved without taking into account the delay in multilayer planar structures of a general type in superlattices. At the same time, the plasma of valence electrons was assigned to a fast subsystem. Since the plasma frequencies of valence electrons lie in the range of $10^{15} - 10^{16}$ s^{-1}, its state inertially follows the external field. In essence, this means using a purely classical description of the shielding effect created by the plasma of valence electrons. Thus, this approximation should be considered reasonable for many tasks. Normal volumetric and interface modes are found. The exact Hamiltonian of the electron-polarization interaction in multilayer planar structures and superlattices, including superlattices, for which we theoretically predicted the appearance of interface spatially extended optical modes, has been explicitly derived. The Hamiltonian of the electron-phonon interaction was obtained separately for this relevant case. The Hamiltonians of the interaction of an electron with optical vibrations in multilayer cylindrical and spherical systems are derived explicitly.

References

[1] Bryksin B B and Firsov Yu A 1971 Interaction of an electron with interface phonons in an ion crystal plate *FTT* **13** 496–503

[2] Pokatilov E P and Fomin V M 1984 *Theory of Potential in Multilayer Systems* No. 500M-05.85 (Chisinau: MoldNIINTI) p 19

[3] Pokatilov E P and Fomin V M 1984 *Interaction of an Electron with Polarization Optical Vibrations in Periodic Structures* No. 508M-05.85 (Chisinau: MoldNIINTI) p 14

[4] Fomin V M and Pokatilov E R 1985 *Phys. Status Solidi* B **132** 69–82

[5] Wendler L and Pechsted R 1986 Dynamical screening, collective excitations, and electron-phonon interaction in heterostructures and semiconductor quantum wells. General theory *Phys. Status Solidi* B **138** 197–217

[6] Wendler L and Haupt E 1987 Electron-phonon interaction in semiconductor superlattices *Phys. Status Solidi* B **143** 487–510

[7] Camley R E and Mills D Z 1984 Collective excitations of semi-infinite superlattice structures: surface plasmons, bulk plasmons and the electron-energy-loss spectrum *Phys. Rev.* B **29** 1695–706

[8] Beril S I, Fomin V M and Starchuk A S 2020 *Theory of Polarons, Excitons, Bipolarons and Kinetic Effects in Multilayer Structures of Various Geometries and Superlattices* (Tiraspol: Pridnestrovian University Publishing House) p 696

[9] Varga B B 1965 Coupling of plasmons to polar phonons in degenerate semiconductors *Phys. Rev.* **137** A1896–902

[10] Davydov A S 1972 *Theory of a Solid Body* (Moscow: Nauka) p 640

[11] Pines D 1965 *Elementary Excitations in Solids* (Moscow: Mir) p 382

[12] Mooradian A and Wright G B 1966 Observation of the interaction of plasmons with longitudinal optical phonons in GaAs *Phys. Rev. Lett.* **16** 999–1001

[13] Lapeyere G and Anderson J R 1975 Evidence for a surface–states exciton on GaAs (110) *Phys. Rev.* B **35** 117–19

[14] Ando T, Fowler A and Stern F 1985 *Electronic Properties of Two-Dimensional Systems* (Moscow: Mir) p 415

[15] Smirnov V I 1974 *Course of Higher Mathematics* (Moscow: Nauka) p 480

[16] Pokatilov E P and Beril S I 1982 Spatially extended optical modes in two-layer periodical structures *Phys. Status Solidi* B **110** K75–8

[17] Pokatilov E P and Beril S I 1983 Electron-phonon interaction in periodic two-layer structures *Phys. Status Solidi* B **118** 567–73

[18] Bloss W L 1983 Plasmon modes of a superlattice–classical vs quantum limits *Solid State Commun.* **48** 927–31

[19] Bloss W L 1982 Optic and acoustic plasmon modes of a semiconductor superlattice *Solid State Commun.* **44** 363–7

[20] Bloss W L and Brody E M 1982 *Solid State Commun.* **43** 523–8

[21] Toyozawa Y 1954 Theory of electronic polaron and ionization of trapped electron by excited *Prog. Theor. Phys.* **12** 421–42

Chapter 4

Vibrational spectra and optical properties of multilayer planar systems (taking into account retardation)

4.1 Introduction

In this chapter, the wave equations describing the propagation of polaritons in multilayer planar structures and superlattices are solved. For the first time, a sequential classification of polariton waves was performed by the nature of the spatial distribution of electric and magnetic field strengths in the layer, as well as by the change of these fields during the period of the structure. The laws of dispersion of interface spatially extended and interface spatially decreasing polaritons are obtained, and the optical properties of multilayer structures and superlattices (reflection and passage of light) are described.

4.2 Solution of wave equations. First classification of polaritons

Polarizing vibrations generate not only potential (electrostatic), but also vortex (electromagnetic) fields. The potential component of the field can be taken into account in the electrostatic approximation. In the previous chapter, we considered the electrically active oscillations in multilayer structures that cause the electron-phonon interaction. This chapter will take into account the electromagnetic fields generated by polarization fluctuations with multilayer structures. Normal vibrations of the system formed as a result of mixing polarizing and electromagnetic vibrations are called polaritons.

To study polaritons in multilayer systems it is necessary to solve the system of Maxwell's equations for each of the layers:

$$\mathrm{rot}\mathbf{H}_n = \frac{\partial \mathbf{D}_n}{\partial t}; \quad \mathrm{rot}\mathbf{E}_n = -\frac{\partial \mathbf{B}_n}{\partial t}; \tag{4.2.1a}$$

doi:10.1088/978-0-7503-6164-4ch4

$$\text{div}\,\mathbf{B}_n = 0; \quad \text{div}\,\mathbf{D}_n = 0, \tag{4.2.1b}$$

by subjecting solutions to boundary conditions.

In this chapter, we will consider the classification of polaritons in multilayer structures, for which it is sufficient to use an approach that can be considered as a generalization of the method of dielectric function:

$$\mathbf{D}_n = \varepsilon_0 \mathbf{E}_n + \mathbf{P}_n = \varepsilon_0 \overset{\leftrightarrow}{\varepsilon} \mathbf{E}_n; \quad \mathbf{B}_n = \mu_0 \mathbf{H}_n + \mathbf{M}_n = \mu_0 \overset{\leftrightarrow}{\boldsymbol{\mu}}_n \mathbf{H}_n \tag{4.2.2}$$

where μ_0 is the magnetic constant; $\mathbf{M}_n$ is the magnetization vector; and $\overset{\leftrightarrow}{\boldsymbol{\mu}}_n$ is the tensor of magnetic permeability (designation $\boldsymbol{\mu}_n$ adopted to distinguish magnetic permeability from coefficients μ_n from the potential modulation transfer matrix).

Substituting in (4.2.1a) and (4.2.1b) the definitions of (4.2.2) and excluding the magnetic field from (4.2.1a), after the Fourier transform in time we find

$$\text{rot}\,\overset{\leftrightarrow}{\boldsymbol{\mu}}_n^{-1}(\omega)\text{rot}\,\mathbf{E}_n(\mathbf{r}) = \omega^2 c^{-2} \overset{\leftrightarrow}{\varepsilon}_n(\omega)\mathbf{E}_n(\mathbf{r}); \tag{4.2.3a}$$

$$\text{div}\,\overset{\leftrightarrow}{\varepsilon}_n(\omega)\mathbf{E}_n = 0. \tag{4.2.3b}$$

The magnetic field strength is expressed in terms of the electric field strength by the ratio following from the second equation of the system (4.2.1a):

$$\mathbf{H}_n(\mathbf{r}) = (i\omega\mu_0)^{-1} \overset{\leftrightarrow}{\boldsymbol{\mu}}_n^{-1}(\omega)\text{rot}\mathbf{E}_n(\mathbf{r}). \tag{4.2.3c}$$

To simplify the problem of classifying vibrations, consider isotropic media or crystals with cubic symmetry, for which the electric and magnetic permeability tensors become scalars. Just as in the previous chapters, the layers are considered homogeneous and sufficiently extended in the *XOY* plane, so electric and magnetic fields can be represented as Fourier decompositions:

$$\mathbf{E}_n(\mathbf{r}) = \int d^2\eta e^{i\boldsymbol{\eta}\mathbf{y}} \mathbf{E}_n(\boldsymbol{\eta}, z); \tag{4.2.4a}$$

$$\mathbf{E}_n(\boldsymbol{\eta}, z) = \frac{1}{(2\pi)^2} \int d^2 y e^{-i\boldsymbol{\eta}\mathbf{y}} \mathbf{E}_n(\mathbf{y}, z), \tag{4.2.4b}$$

then the wave equation (4.2.3a) takes the form:

$$\frac{\partial^2 \mathbf{E}_n(\boldsymbol{\eta}, z)}{\partial z^2} = \kappa_n^2(\omega)\mathbf{E}_n(\boldsymbol{\eta}, z), \tag{4.2.5}$$

where the frequency function is entered:

$$\kappa_n^2(\omega) \equiv \eta^2 - \omega^2 c^{-2} \varepsilon_n(\omega) \boldsymbol{\mu}_n(\omega). \tag{4.2.6}$$

Denote that $\varepsilon_n(\omega)$ and $\boldsymbol{\mu}_n(\omega)$ correspond to full electrical polarization and magnetization. In a substance with single-mode polarizations of types $x = 0, p$, for example, $\varepsilon(\omega)$, it has the form given in section 3.2 (formulas (3.2.8d), (3.2.9c)) and section 3.3.

From equation (4.2.3b) follows the relation

$$\varepsilon_n(\omega)\left[i\boldsymbol{\eta}\mathbf{E}_n^{\perp}(\boldsymbol{\eta}, z) + \frac{\partial E_n^{||}(\boldsymbol{\eta}, z)}{\partial z}\right] = 0 \tag{4.2.7}$$

between components of the electric field that are perpendicular to and parallel to the axis Oz:

$$\mathbf{E}_n(\boldsymbol{\eta}, z) = \mathbf{E}_n^{\perp}(\boldsymbol{\eta}, z) + E_n^{||}(\boldsymbol{\eta}, z)\mathbf{e}_3; \tag{4.2.8a}$$

$$\mathbf{E}_n^{\perp}(\boldsymbol{\eta}, z) = \mathbf{E}_n^{x}(\boldsymbol{\eta}, z)\mathbf{e}_1 + \mathbf{E}_n^{y}(\boldsymbol{\eta}, z)\mathbf{e}_2. \tag{4.2.8b}$$

In conclusion, from (4.2.3c) we find the Fourier image of the magnetic field:

$$\begin{aligned}\mathbf{H}_n(\boldsymbol{\eta}, z) = \left[i\omega\mu_0\boldsymbol{\mu}_n(\omega)\right]^{-1}\Bigg\{\left[i\eta_y E_n^{||}(\vec{\boldsymbol{\eta}}, z) - \frac{\partial \mathbf{E}_n^{y}(\boldsymbol{\eta}, z)}{\partial z}\right]\mathbf{e}_1 \\ - \left[i\eta_x E_n^{||}(\boldsymbol{\eta}, z) - \frac{\partial \mathbf{E}_n^{x}(\boldsymbol{\eta}, z)}{\partial z}\right]\mathbf{e}_2 + i\left[\eta_x \mathbf{E}_n^{y}(\boldsymbol{\eta}, z) - \eta_y \mathbf{E}_n^{x}(\boldsymbol{\eta}, z)\right]\mathbf{e}_3\Bigg\}.\end{aligned} \tag{4.2.9}$$

The solution of the wave equation (4.2.5) satisfying the condition of continuity of electric field components perpendicular to the axis Oz, similar to the electrostatic (3.2.7c), obtained at $\tilde{p} = 0$ is:

$$\mathbf{E}_n^{\perp}(\boldsymbol{\eta}, z) = \frac{1}{\sinh \zeta_n}\{\sinh[\kappa(\omega)(z - z_{n-1})]E^{\perp}(\boldsymbol{\eta}, z_n) + \sinh[\kappa(\omega)(z_n - z)]E^{\perp}(\boldsymbol{\eta}, z_{n-1})\}, \tag{4.2.10}$$

where $\zeta_n \equiv \kappa_n(\omega)l_n$. If $\mathrm{Im}[\kappa^2(\omega)]$ is equal to or close to zero, it is convenient to enter a parameter

$$I_n \equiv \mathrm{sign}\,\mathrm{Re}[\kappa_n^2(\omega)]. \tag{4.2.11}$$

Then it follows from (4.2.10) that the electric field in the nth layer depends on the coordinate z through hyperbolic functions (in particular, it decreases exponentially when moving away from the boundary of a semi-infinite medium $l_n \to \infty$) in the area where $I_n = 1$, and via trigonometric functions (giving waveguide polariton waves) in the region $I_n = -1$. These areas of the (η, ω) space are separated from each other by curves defined from the condition $\kappa_n(\omega) = 0$, which gives the dispersion equation for bulk polaritons in the nth layer. For example, for $\varepsilon_n(\omega)$ that defines the lattice polarization (3.2.8c), the dispersion equation ($\boldsymbol{\mu}_n \equiv 1$)

$$\varepsilon_n\omega^4 - \omega^2(\varepsilon_{n0}\omega_n^2 + \eta^2c^2) + \eta^2c^2\omega_n^2 = 0$$

gives two functions:

$$\Omega_{1,2}^2 = \frac{1}{2\varepsilon_n}\left[\varepsilon_{n0}\omega_n^2 + c^2\eta \pm \sqrt{(\varepsilon_{n0}\omega_n^2 + c^2\eta^2)^2 - 4\omega_n^2\varepsilon_n c^2\eta^2}\right]. \tag{4.2.12a}$$

With the small η decisions Ω_1^2, Ω_2^2 take a look of

$$\Omega_1^2 = \omega_n^2 \frac{\varepsilon_{n0}}{\varepsilon_n} + \frac{c^2\eta^2}{\varepsilon_n}; \tag{4.2.12b}$$

$$\Omega_2^2 = \frac{c^2\eta^2}{\varepsilon_n}. \tag{4.2.12c}$$

With the largeη

$$\Omega_1^2 = \frac{c^2\eta^2}{\varepsilon_n}; \tag{4.2.12d}$$

$$\Omega_2^2 = \omega_n^2. \tag{4.2.12e}$$

Based on (4.2.12b) and (4.2.12d) frequency Ω_1 when $\eta = 0$ has the meaning of $\omega_n(\varepsilon_{n0}/\varepsilon_n)^{1/2}$, coinciding with the frequency of bulk longitudinal polarization oscillations of the lattice, then increases at large η $\Omega_1 = c\eta/\varepsilon_n$ coincide with the frequency of an electromagnetic wave propagating in a medium with a dielectric constant ε_n. And Ω_2 with the small η corresponds to the frequency of an electromagnetic wave propagating in a medium with permittivity ε_{n0}, and at large η is equal to the frequency of transverse optical vibrations of the lattice, and the frequency branches Ω_1 and Ω_2 devided by the interval $(\Omega_1 |_{\eta=0} - \Omega_2 |_{\eta=\infty}) = \omega_n^2(\varepsilon_{n0}/\varepsilon_n - 1)$. In the area where straight lines intersect ω_n and $c\eta/\varepsilon_{n0}$, representing the laws of dispersion of transverse optical phonons and photons in a medium with permittivity ε_{n0}, in polariton energy, electromagnetic and mechanical components are mixed in significant proportions.

In the common case, when $\varepsilon_n \neq 0$ in all substances, for p-polarized polaritons

$$\mathbf{E}_n^{\perp}(\boldsymbol{\eta}, z);\ \mathbf{E}_n^{\perp}(\boldsymbol{\eta}, z) = \frac{\boldsymbol{\eta}}{\eta}\mathbf{E}_n^{\perp}(\boldsymbol{\eta}, z); \quad E_n^{||}(\boldsymbol{\eta}, z) \neq 0, \tag{4.2.13}$$

the ratio (4.2.7) is performed when

$$\frac{\partial E_n^{||}(\boldsymbol{\eta}, z)}{\partial z} = -i\eta\, \mathbf{E}_n^{\perp}(\boldsymbol{\eta}, z). \tag{4.2.14a}$$

From (4.2.14a) and the wave equation (4.2.5) for the longitudinal component of p-polarization

$$\mathbf{E}_n^{||}(\boldsymbol{\eta}, z) = -\frac{i\eta}{\kappa_n^2(\omega)} \frac{\partial \mathbf{E}_n^{\perp}(\boldsymbol{\eta}, z)}{\partial z}. \tag{4.2.14b}$$

From (4.2.9) we find that the magnetic field is perpendicular to the Oz axis and $\boldsymbol{\eta}$:

$$\mathbf{H}_n^{\perp}(\boldsymbol{\eta}, z) = \frac{\eta_y \mathbf{e}_1 - \eta_x \mathbf{e}_2}{\eta}\mathbf{H}_n^{\perp}(\boldsymbol{\eta}, z); \quad H_n^{||}(\boldsymbol{\eta}, z) = 0, \tag{4.2.14c}$$

where the squared value is

$$\mathbf{H}_n^{\perp}(\eta, z) = \frac{\omega\varepsilon_0\varepsilon_n(\omega)}{i\kappa_n^2(\omega)}\frac{\partial \mathbf{E}_n^{\perp}(\eta, z)}{\partial z}. \tag{4.2.15}$$

determines the spatial distribution of the energy of p-polarized polariton waves in the considered region nth layer.

To find the boundary values of a field component $\mathbf{E}_n^{\perp}$ we use the condition of continuity of normal components $\mathbf{D}_n(\eta, z) = \varepsilon_0\varepsilon_n(\omega)\mathbf{E}_n(\eta, z)$:

$$\varepsilon_n(\omega)E_n^{||}(\eta, z_n - 0) = \varepsilon_{n+1}(\omega)E_{n+1}^{||}(\eta, z_n + 0). \tag{4.2.16a}$$

Using the relation (4.2.14b) and performing the differentiation on z in (4.2.10), we get (the numbering of layers according to figure 4.1 is convenient to start with $I - 1$ and finish with the $K + 1$, so that, $n = I - 1, I, \ldots, K$):

$$-\mu_n\mathbf{E}_n^{\perp}(\eta, z_{n-1}) + \nu_n\mathbf{E}_n^{\perp}(\eta, z_n) - \mu_{n+1}\mathbf{E}_n^{\perp}(\eta, z_{n+1}) = 0, \tag{4.2.16b}$$

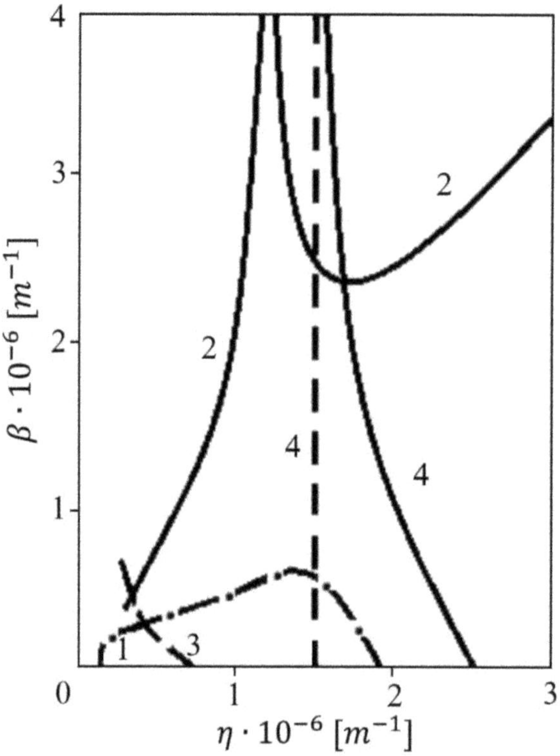

Figure 4.1. Dispersion curves of the coefficient of decrease for some branches of spatially decreasing p-polaritons, calculated by numerical solution of the equation (4.3.18), with frequencies Ω_S: 1—4994 · 10^{13}; 2—5452 · 10^{13}; 3—6671 · $10^{13}\mathrm{s}^{-1}$ at $\eta = 5 \cdot 10^5$ m^{-1}; 1—5000 · 10^{13}; 2—5458 · 10^{13} s^{-1} at $\eta = 1 \cdot 10^6$ m^{-1}; 3—5463 · 10^{13}; 4—5014 · 10^{13} s^{-1} at $\eta = 2 \cdot 10^6$ m^{-1}.

where the following notation is entered:

$$\mu_n = \frac{\tilde{\mu}_n}{sh\zeta_n}; \quad \tilde{\mu}_n = \tilde{\mu}_n^p \equiv \frac{\varepsilon_n(\omega)}{\kappa_n(\omega)}; \quad \nu_n = \hat{\mu}_n + \hat{\mu}_{n+1}; \quad \hat{\mu}_n = \mu_n ch\zeta_n. \tag{4.2.17}$$

As already noted in section 4.2, in real structures at the interface of layers there are transition regions, the parameters of which (and, consequently, the coefficients (4.2.17)), generally speaking, change both in the plane and along the axis. As a result, there is a violation of the coherence of the radiation propagating in the structure. To calculate the observed values, additional averaging operations are required, taking into account the statistics of the spread of transition layer parameters. In this chapter, all numerical calculations are performed for an ideal transition layer that is a plane.

The not-equal to zero component of the field s-polarized polariton is

$$\mathbf{E}_n^{\perp}(\boldsymbol{\eta}, z) = \frac{\eta_y \mathbf{e}_1 - \eta_x \mathbf{e}_2}{\eta} E_n^{\perp}(\boldsymbol{\eta}, z); \quad E_n^{||}(\boldsymbol{\eta}, z) = 0$$

orthogonal $\boldsymbol{\eta}$, which is why (4.2.7) is satisfied automatically. The square of the magnitude $E_n^{\perp}(\boldsymbol{\eta}, z)$ it characterizes the spatial distribution of the energy of an *s*-polarized polariton wave in the considered layer. It follows from (4.2.9) that the magnetic field has components

$$\mathbf{H}_n^{\perp}(\boldsymbol{\eta}, z) = \frac{\boldsymbol{\eta}}{\eta} H_n^{\perp}(\boldsymbol{\eta}, z); \quad H_n^{||}(\boldsymbol{\eta}, z) \neq 0, \tag{4.2.18a}$$

where

$$\mathbf{H}_n^{\perp}(\boldsymbol{\eta}, z) = \left[i\omega\mu_0\boldsymbol{\mu}_n(\omega)\right]^{-1} \frac{\partial \mathbf{E}_n^{\perp}(\boldsymbol{\eta}, z)}{\partial z}; \tag{4.2.18b}$$

$$H_n^{||}(\boldsymbol{\eta}, z) = -\left[\omega\mu_0\boldsymbol{\mu}_n(\omega)\right]^{-1} \eta \mathbf{E}_n^{\perp}(\boldsymbol{\eta}, z). \tag{4.2.18c}$$

Boundary conditions for continuity of tangential magnetic field components

$$H_n^{\perp}(\boldsymbol{\eta}, z_n - 0) = H_{n+1}^{\perp}(\boldsymbol{\eta}, z_n + 0) \tag{4.2.19}$$

on the basis of (4.2.18b) and (4.2.10) can be presented in a form similar to (4.2.16b), with a new designation:

$$\tilde{\boldsymbol{\mu}}_n = \tilde{\boldsymbol{\mu}}_n^{(S)} \equiv \frac{\kappa_n(\omega)}{\boldsymbol{\mu}_n(\omega)}. \tag{4.2.20}$$

In general, polarizing waves in the multilayer structures can be classified by a set of parameters $\{I_n\}_{n=I-1}^{K+1}$. In a special case, when under certain conditions $\varepsilon_n(\omega) = 0$ in one or more substances, the ratio (4.2.7) is satisfied in these substances regardless of the properties of the electric field intensity vector. This condition corresponds to the excitation of longitudinal vibrations in the nth surrounding. This case is considered in detail in [1], which gives a classification of polaritons. The conclusion is that from the layers in which the special case is realized, the electric field does not penetrate into those for which the general case is realized.

So, polariton waves in the multilayer structures can be classified by the values of a set of parameters $\{I_m\}$, where m numbers of surroundings with $\varepsilon_m(\omega) \neq 0$, and distribution of electric and magnetic fields in substances with $\varepsilon_n(\omega) = 0$. This is the first classification of polaritons in a multilayer structure.

4.3 The ratio of the dispersion of polaritons. The second classification of polaritons

The system of algebraic equations (4.2.16b) can be solved by the method described in chapter 1. According to the formula (3.3.7b) values of the electric field on internal interlayer surfaces $z_n(n = I, \ldots, K - I)$ and structures $(I\,|\ldots|\,K)$ are connected:

$$E^{\perp}(\eta, z_n) = D_{nI}(I\,|K)\mu_I E^{\perp}(\eta, z_{I-1}) + D_{n,\,K-1}(I\,|K)\mu_K E^{\perp}(\eta, z_K) \qquad (4.3.1)$$

with field values $E^{\perp}(\eta, z_{I-1})$, $E^{\perp}(\eta, z_K)$. Elements of a symmetric matrix $\|D_{n,\,K}(I\,|K)\|$ are expressed in terms of coefficients of the system of equations (4.2.16b) by the ratios (3.3.11a) and (3.3.12a), (3.3.12b). The two remaining equations of the set (4.2.16a) with $n = I - 1$, K have the form:

$$[\nu_{I-1} - \mu_I^2 D_{I,\,I}(I\,|K)]E^{\perp}(\eta, z_{I-1}) - \mu_I \mu_K D_{I,\,K-1}(I\,|K)E^{\perp}(\eta, z_K) = \mu_{I-1} E^{\perp}(\eta, z_{I-2}); \quad (4.3.2a)$$

$$-\mu_I \mu_K D_{K-1,\,I}(I\,|K)E^{\perp}(\eta, z_{I-1})$$

$$+\left[\nu_K - \mu_K^2 D_{K-1,\,K-1}(I\,|K)\right]E^{\perp}(\eta, z_K) = \mu_{K+1} E^{\perp}(\eta, z_{K+1}). \qquad (4.3.2b)$$

Using the sets (4.3.2a) and (4.3.2b), you can find $E^{\perp}(\eta, z_{I-1})$, $E^{\perp}(\eta, z_K)$, knowing the values of the fields on the external surfaces of the system $(I - 1|\ \ldots\ |K + 1)$.

In order to find normally polarized waves inside the structure, it is necessary to put $E^{\perp}(\eta, z_{I-2}) = 0$ and $E^{\perp}(\eta, z_{K+1}) = 0$ for external surfaces. Then the condition for the existence of non-trivial solutions of homogeneous systems (4.3.2a) and (4.3.2b)

$$[\nu_{I-1} - \mu_I^2 D_{II}(I\,|K)]\,[\nu_K - \mu_K^2 D_{K-1,\,K-1}(I\,|K)] - \mu_I^2 \mu_K^2 D_{I,\,K-1}^2(I\,|K) = 0 \quad (4.3.3)$$

gives dispersion relations for localized polariton waves.

Roots Ω_S of the transcendental equation (4.3.3) give the eigenfrequencies of localized polariton waves. For each Ω_S any of the equations of homogeneous sets (4.3.2a) and (4.3.2b) allows us to express $E^{\perp}(\eta, z_K)$ through $E^{\perp}(\eta, z_{I-1})$. As a result the values of electric fields for all internal interlayer surfaces become proportional $E^{\perp}(\eta, z_{I-1})$ and the spatial distribution of fields is determined by the formulas (4.2.8a), (4.2.8b), (4.2.10), (4.2.14b), and (4.2.15).

Now the method developed above will be applied to describe localized polariton waves in a limited periodic system that consists of sequentially arranged layers of matter c, N pairs of alternating layers of the type a and b, and layer of substance type f.

Using for the periodic part $(I \mid \ldots \mid K)$ systems of the matrix $\|D_{n,\,K}(I\,|K)\|$, obtained in section 3.5, in accordance with (4.3.1) we find the electric field on all internal surfaces:

$$E^{\perp}(\eta,\, z_{I+2J-1}) = \frac{1}{\alpha^{2N}-1}\Big\{\alpha^{J}[\alpha^{2(N-J)}-1]E^{\perp}(\eta,\, z_{I-1}) + \alpha^{N-J}(\alpha^{2J}-1)E^{\perp}(\eta,\, z_K)\Big\};\quad J=0,\,1,\,\ldots,\,N; \tag{4.3.4a}$$

$$E^{\perp}(\eta,\, z_{I+2J}) = \frac{\alpha^{2}-1}{\nu(\alpha^{2N}-1)}\Big\{\alpha^{J}\Big[\frac{\alpha^{2(N-J-1)}-1}{\alpha^{2}-1}\alpha(\mu_b+\alpha\mu_a)+\mu_a\Big] \times E^{\perp}(\eta,\, z_{I-1}) + \alpha^{N-J-1}\Big[\frac{\alpha^{2J}-1}{\alpha^{2}-1}\alpha(\mu_a+\alpha\mu_b)+\mu_b\Big]E^{\perp}(\eta,\, z_K)\Big\}; \tag{4.3.4b}$$

$$J=0,\,1,\,\ldots,\,N-1,$$

where

$$\nu = \hat{\mu}_a + \hat{\mu}_b;\quad \nu = \hat{\mu}_a + \hat{\mu}_b \ \text{ and } \ \alpha = \psi + \sqrt{\psi^{2}-1};$$

$$\psi = \frac{\tilde{\mu}_a^{2}+\tilde{\mu}_b^{2}}{2\mu_a\mu_b} + \cosh\zeta_a\cosh\zeta_b. \tag{4.3.4c}$$

Given the power-law nature of the dependencies of the right parts (4.3.4a) and (4.3.4b) on the indices of surfaces, it can be shown that

$$\frac{E^{\perp}(\eta,\, z_{n+2}) + E^{\perp}(\eta,\, z_{n-2})}{E^{\perp}(\eta,\, z_n)} = \alpha + \frac{1}{\alpha}, \tag{4.3.5}$$

where $n = I+1,\,\ldots,\,K-2$. This means that the sum of the values α and the inverse of it determines the sum of the ratios of the field components that differ in two periods $L = l_a + l_b$, to the field component on the middle identical surface between them. In particular, in the case of an unbounded periodic system in which translational symmetry takes place:

$$\frac{E^{\perp}(\eta,\, z_{n+2})}{E^{\perp}(\eta,\, z_n)} = e^{iq_{||}L}, \tag{4.3.6}$$

from (4.3.5) and definitions (4.3.4c) the dispersion ratio of spatially extended polaritons is obtained:

$$\tilde{\mu}_a^{2} + \tilde{\mu}_b^{2} + 2\tilde{\mu}_a\tilde{\mu}_b(\cosh\zeta_a ch\zeta_b - \cos q_{||}L) = 0. \tag{4.3.7}$$

In accordance with (4.3.6), the wave excitations described by the dispersion relation (4.3.7) are non-attenuating: $|E^{\perp}(\eta,\, z_{n+2})| = |E^{\perp}(\eta,\, z_n)|$. It follows from (4.2.16b) and (4.3.6) that the ratio of the values of electric fields on the two nearest neighboring surfaces will satisfy the equality

$$\frac{E^{\perp}(\eta, z_{n+1})}{E^{\perp}(\eta, z_n)} = \frac{\nu}{\mu_n e^{-iq_{||}L} + \mu_{n+1}}, \tag{4.3.8}$$

which together with (4.2.10) defines the spatial distribution of the field within the layers.

The dispersion relations (9) and (30) for electromagnetic waves in a multilayer system obtained in [2] are determined from (4.3.7) when $\eta = n\omega/c$, $q_{||} = 0$ and $\eta = 0$, $q_{||} = n\omega/c$, respectively, with the refractive index p, whereas the spatial distribution of the fields follows from (4.2.8a), (4.2.8b), (4.2.10), (4.2.14b), (4.2.15), (4.3.6), and (4.3.8).

For p-polarized spatially extended polaritons, the statement (4.2.20) in (4.3.7) gives the dispersion relations studied numerically in [3]. The dispersion relations for s-polarized spatially extended polaritons are obtained after substituting (4.2.20) in (4.3.7).

For a structure with a longitudinal number of pairs of layers N, the system of equations (4.3.2a) and (4.3.2b) acquires the form

$$\begin{aligned} &R_c E^{\perp}(\eta, z_{I-1}) - \frac{\mu_a \mu_b}{\nu} \frac{\alpha^1 - 1}{\alpha^{2N} - 1} \alpha^{N-1} E^{\perp}(\eta, z_K) = \mu_c E^{\perp}(\eta, z_{I-2}) \\ &- \frac{\mu_a \mu_b}{\nu} \frac{\alpha^2 - 1}{\alpha^{2N} - 1} \alpha^{N-1} E^{\perp}(\eta, z_{I-1}) + R_f E^{\perp}(\eta, z_K) = \mu_f E^{\perp}(\eta, z_{K+1}), \end{aligned} \tag{4.3.9}$$

in which the following symbols are entered:

$$R_c = \frac{\mu_a \mu_b}{\nu} \left[\frac{1}{\alpha_c} - \frac{\alpha^{2N} - \alpha^2}{\alpha(\alpha^{2N} - 1)} \right]; \tag{4.3.10a}$$

$$\frac{1}{\alpha_c} = \frac{\hat{\mu}_c \nu + \tilde{\mu}_a^2}{\mu_a \mu_b} + \cosh \zeta_a \cosh \zeta_b; \tag{4.3.10b}$$

R_f, $1/\alpha_f$ differ from R_c, $1/\alpha_c$ replacements $c \to f$, $a \leftrightarrow b$. The condition for the existence of non-trivial solutions of a homogeneous system of equations (4.3.9) gives a dispersion relation for localized polariton waves in the system under consideration

$$\left(\frac{\mu_a \mu_b}{\nu} \right)^2 \left[\frac{1}{\alpha_c \alpha_f} - \left(\frac{1}{\alpha_c} + \frac{1}{\alpha_f} \right) \frac{\alpha^{2N} - \alpha^2}{\alpha(\alpha^{2N} - 1)} + \frac{\alpha^{2N} - \alpha^4}{\alpha^2(\alpha^{2N} - 1)} \right] = 0, \tag{4.3.11}$$

which is a concrete example of the general dispersion relation (4.3.3). For each eigenfrequency of a localized polariton Ω_S, which is the solution of the transcendental equation (4.3.11), any of the equations of the homogeneous system (4.3.9) allows us to express $E^{\perp}(\eta, z_K)$ through $E^{\perp}(\eta, z_{I-1})$. Further by means of (4.3.4a) and (4.3.4b) the values of the electric field on the internal surfaces are determined, and the formulas' (4.2.8a), (4.2.8b), (4.2.10), (4.2.14b), (4.2.15), (4.3.6), and (4.3.8) spatial distributions are found.

In particular, for a two-layer structure ($N = 1$) when considering non-magnetic media ($\mu_a(\omega) = \mu_b(\omega) \equiv 1$), bordering on vacuum ($\varepsilon_c(\omega) = \varepsilon_f(\omega) \equiv 1$; $l_{c,f} \rightarrow \infty$) in cases of p-polarization and s-polarization from (4.3.11) follow the laws of dispersion of surface polaritons studied in [4] (formula (9)) and [5] (formula (16)), whereas (4.2.10) together with (4.3.4a), (4.3.4b) and (4.3.9) lead to the spatial distribution of the electric field (formulas (5) from [4] and (11) from [5]).

If the structure consists of a sufficiently large number of periods $N \gg 1$ and the condition is met

$$|\alpha| > 1, \tag{4.3.12}$$

then $\alpha^{2N} - 1 \approx \alpha^{2N}$ and for final values J expressions (4.3.4a) and (4.3.4b) have the form

$$E^{\perp}(\eta, z_{I+2J-1}) \approx \alpha^{-J} E^{\perp}(\eta, z_{I-1}); \tag{4.3.13a}$$

$$E^{\perp}(\eta, z_{I+2J}) \approx \alpha^{-J-1} \frac{(\mu_b + \alpha\mu_a)}{\nu} E^{\perp}(\eta, z_{I-1}). \tag{4.3.13b}$$

Hence, it follows in force $|\alpha|^{-1} < 1$ that the field amplitude decreases when moving away from the boundary surface z_{I-1} on the structure period. Note that when $|\alpha| < I$ $(1 - \alpha^{2N} \approx 1)$ expressions (4.3.4a) and (4.3.4b) lead to relations of the form (4.3.13a) and (4.3.13b) with substitution $\alpha \rightarrow \alpha^{-1}$, and, accordingly, the value of $|\alpha|^{-1}$ takes on the meaning of a decrement of decrease. Equation $|\alpha| = 1$ determines the dispersion of spatially extended waves, as established above.

For end users $(N - J)$ formulas similar to (4.3.3a) and (4.3.3b) describe the decreasing field in the direction from the second bounding surface on the structure period. In the case under consideration, the dispersion relation of spatially decreasing polaritons follows from (4.3.11):

$$\left(\frac{1}{\alpha_c} - \frac{1}{\alpha}\right)\left(\frac{1}{\alpha_f} - \frac{1}{\alpha}\right) = 0, \tag{4.3.14}$$

which explicitly reflects the mutual independence of wave excitations decreasing in amplitude from one or another limiting surface of the periodic part of the system. Substitution of (4.3.10b) and definitions (4.3.4c) in the dispersion relation for spatially decreasing polaritons

$$\frac{1}{\alpha_c} - \frac{1}{\alpha} = 0 \tag{4.3.15a}$$

allows you to get the variance ratio in the form

$$\hat{\mu}_a(\tilde{\mu}_b^2 - \hat{\mu}_c^2) + \hat{\mu}_b(\tilde{\mu}_a^2 - \hat{\mu}_c^2) + \hat{\mu}_c(\tilde{\mu}_b^2 - \tilde{\mu}_a^2) = 0. \tag{4.3.15b}$$

The solution of this transcendental equation gives the eigenfrequencies of spatially decreasing polaritons Ω_s. Then using one of the formulas (4.3.4c) and (4.3.10b) or

$$\frac{1}{\alpha} = -\frac{\mu_b(\tilde{\mu}_a^2 - \hat{\mu}_c^2)}{\mu_a(\tilde{\mu}_b^2 - \hat{\mu}_c^2)} \tag{4.3.16}$$

where the descending decrement is defined as $|\alpha|$ and the coefficient of decrease as

$$\beta = L^{-1} \ln |\alpha|. \tag{4.3.17}$$

where a spatial distribution of fields is given by the formulas (4.2.8a), (4.2.8b), (4.2.10), (4.2.14a), (4.2.14b), (4.2.15), (4.3.6), and (4.3.8) with boundary values (4.3.4a) and (4.3.4b).

In particular, for *p*-polarized spatially decreasing polaritons, substituting the notation (4.2.17) in (4.3.15b) gives the dispersion relation

$$\hat{\mu}_a^{(p)}((\tilde{\mu}_b^{(p)})^2 - (\hat{\mu}_c^{(p)})^2) + \hat{\mu}_b^{(p)}((\tilde{\mu}_a^{(p)})^2 - (\hat{\mu}_c^{(p)})^2) + \hat{\mu}_c^{(p)}((\tilde{\mu}_b^{(p)})^2 - (\tilde{\mu}_a^{(p)})^2) = 0, \tag{4.3.18}$$

where $\hat{\mu}_n^{(p)} = \tilde{\mu}_n^{(p)} \coth \zeta_n$. Ignoring the lag effect and for $l_c \to \infty$ from (4.3.18) we obtain the dispersion relation for interface phonon waves considered in [6].

The dispersion relations for *s*-polarized spatially decreasing polaritons can be obtained after substituting the notation (4.2.20) in (4.3.15b):

$$\hat{\mu}_a^{(s)}((\tilde{\mu}_b^{(s)})^2 - (\hat{\mu}_c^{(s)})^2) + \hat{\mu}_b^{(s)}((\tilde{\mu}_a^{(s)})^2 - (\hat{\mu}_c^{(s)})^2) + \hat{\mu}_c^{(s)}((\tilde{\mu}_b^{(s)})^2 - (\tilde{\mu}_a^{(s)})^2) = 0, \tag{4.3.19}$$

where $\hat{\mu}_n^{(s)} = \tilde{\mu}_n^{(s)} \coth \zeta_n$.

To sum up, we note that polaritons in a periodic multilayer structure can be divided into two classes with respect to their change in the structure period—spatially extended and spatially decreasing.

4.4 Properties of spatially decreasing numbers polaritons

Spatially extended polaritons were discussed in detail in the previous section. Here we will focus on the study of new object-spatially decreasing polaritons. To bring the calculation results closer to reality, we use numerical values of parameters of a well-known periodic structure consisting of layers GaAs (*a*) and $Ga_{1-x}Al_x$ As (*b*), bordering on vacuum. For these substances, the magnetic permeability $\mu_a = 1$ and the permittivity has the form [7]:

$$\varepsilon_a(\omega) = \varepsilon_a \frac{\omega_{La}^2 - \omega^2}{\omega_{Ta}^2 - \omega^2}; \quad \varepsilon_b(\omega) = \varepsilon_b \prod_{j=1}^{2} \frac{\omega_{Lbj}^2 - \omega^2}{\omega_{Tbj}^2 - \omega^2}.$$

Values of high-frequency dielectric permittivity ε_n, longitudinal and cross-sections ω_{Tn} phonon frequencies according to [7, 8] when $x = 0.3$ the following: $\varepsilon_a = 10.9$; $\omega_{La} = 5.50 \cdot 10^{13}\ \text{s}^{-1}$; $\omega_{Ta} = 5.05 \cdot 10^{13}\ \text{s}^{-1}$; $\varepsilon_b = 10.1$; $\omega'_{Lb1} = 7.04 \cdot 10^{13}\ \text{s}^{-1}$; $\omega'_{Tb1} = 6.71 \cdot 10^{13}\ \text{s}^{-1}$; $\omega_{Lb2} = 5.46 \cdot 10^{13}\ \text{s}^{-1}$; $\omega_{Tb2} = 5.01 \cdot 10^{13}\ \text{s}^{-1}$.

With the help of numerical solutions of the transcendental equations (4.3.18) and (4.3.19) with respect to ω for various fixed values of a two-dimensional wave vector η eigenfrequencies Ω_s are obtained of the spatially decreasing polaritons, after which the decreasing decrements $|\alpha|$ and the coefficients of decreasing β are calculated.

For the structure c $l_a = 1.2\ \mu m$, $l_b = 0.3\ \mu m$ is shown, what frequency *p*-polarized generally speaking, there are more polaritons than frequencies *s*-polarized. Polariton branches of all types are characterized by weak dispersion. Figures 4.1 and 4.2 show the dispersion curves of the coefficient of decrease for some branches of the spatially decreasing *p*- and *s*-polarized polaritons, respectively. Polaritons of the type $(I_a, I_b) = (1, 1)$ correspond to solid lines, $(-1, 1)$—dashed lines, and $(-1, -1)$—dash-dotted lines.

It follows from the graphs that polaritons of the type $(-1, -1)$ attenuate, generally speaking, weaker than other types of polaritons. As the wave vector increases, polariton branches of type $(-1, 1)$ become branches of type $(1, 1)$. $l_a = 1.2\ \mu m$ polariton attenuation coefficient c $\Omega_s = 5.46 \cdot 10^{13}\ s^{-1}$ experiences a gap associated with a change in the sign that is positive when small η and negative for large η. Figure 4.3 shows the spatial distributions of electric and magnetic fields calculated using (4.2.10), (4.2.15), and (4.3.13a), (4.3.13b) for the first period of the

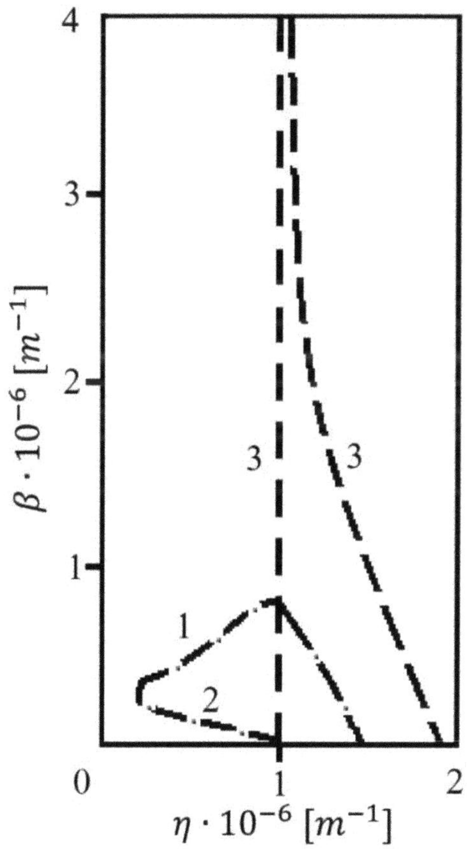

Figure 4.2. Dispersion curves of the coefficient of decrease for some branches of spatially decreasing *s*-polaritons, calculated by numerical solution of the equation (4.3.19), with frequency Ω_S: 1—4988 · 10^{13}; 2—5035 · 10^{13} s^{-1} at $\eta = 5 \cdot 10^5$ m^{-1}. 1—5007 · 10^{13}; 2—5036 · 10^{13} s^{-1} at $\eta = 1 \cdot 10^6$ m^{-1}. 1—5008 · 10^{13}; 3—5013 · 10^{13} s^{-1} at $\eta = 1.2 \cdot 10^6$ m^{-1}.

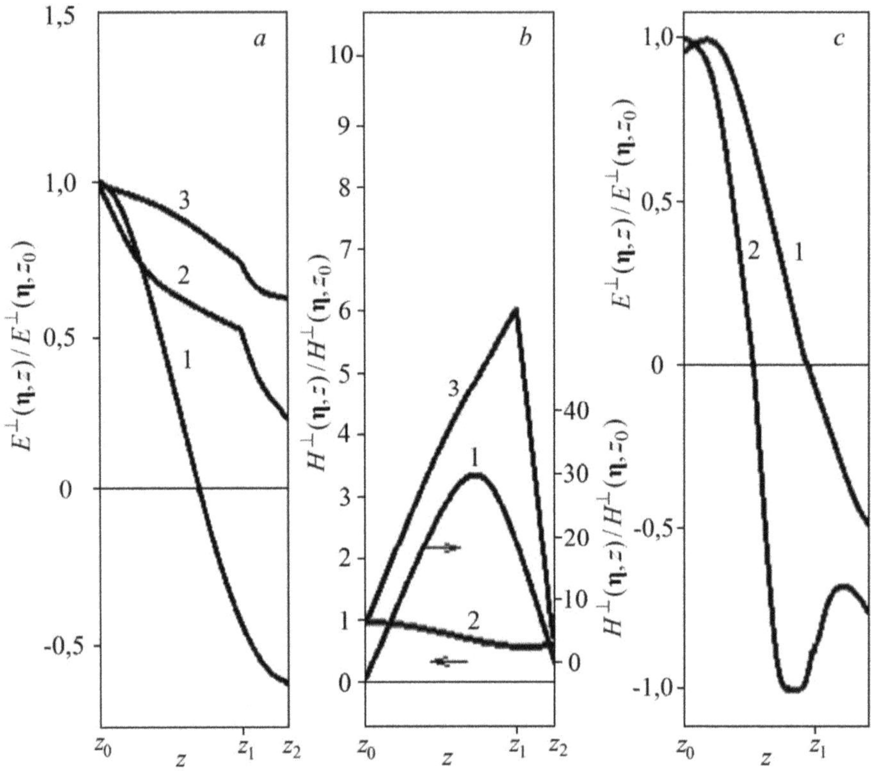

Figure 4.3. Spatial distribution of electric (a, b) and magnetic (b) fields for *p*-polarized (a, b) and *s*-polarized (b) spatially decreasing polaritons at $\eta = 5 \cdot 10^5 \ \text{m}^{-1}$.

structure. On their basis, you can make a general conclusion that the branch difference in spatially decreasing polaritons is characterized by significantly different properties spatial distributions of fields, and, consequently, energy.

Thus, the established properties of spatially decreasing polaritons in semi-bounded periodic systems open up additional possibilities for controlling the size of the regions where the energy of the electromagnetic field is concentrated by selecting its frequency, as well as by changing the parameters of the superlattice.

4.5 Solution of the wave equation for external layers. Optical characteristics of multilayer structures

In contrast to the other paragraphs, where the proper polariton waves were considered, the intensity of which decreases exponentially in the external layers, so usually called non-radiation, this and further paragraphs of the chapter explore the polariton waves that pass through the structure and reflect from it and are therefore called radiation.

Consider an electromagnetic wave in a multilayer system $(I - 1 \mid \ldots \mid K + 1)$. Perpendicular axis Oz, the component of the electric field strength, is given by the formula (4.2.10), which we will rewrite, introducing more convenient notation in this case:

$$[k''_n(\omega)]^2 = \omega^2 c^2 \varepsilon_n(\omega)\mu_n(\omega) - \eta^2 \equiv -\kappa_n^2(\omega); \tag{4.5.1a}$$

$$\lambda_n = k''_n(\omega) l_n \equiv i\zeta_n; \tag{4.5.1b}$$

$$E_n^{\perp}(\eta,\ z) = \frac{1}{\sin\lambda_n}\{\sin[k''_n(\omega)(z - z_{n-1})]E^{\perp}(\eta,\ z_n) + \sin[k''_n(\omega)(z_n - z)]E^{\perp}(\eta,\ z_{n-1})\}. \tag{4.5.1c}$$

Let us assume that an electromagnetic wave generated by an external source propagates in an external environment that forms a (I–1)-first layer of the system, in the direction from the border z_{I-2} and z_{I-1}. The component of the field that is perpendicular to the Oz axis is denoted by $\mathbf{E}_i^{\perp}(\eta)$. In the same layer in the opposite direction from the border z_{I-1} to the border z_{I-2}, a reflected electromagnetic wave propagates, the transverse component of the electric field which we denote $\mathbf{E}_r^{\perp}(\eta)$.

Converting the sinuses to (4.5.1c) using Euler's formula and writing the result as a superposition of two waves, one of which moves towards the surface z_{I-1}, and the other from it, we get

$$E_{I-1}^{\perp}(\eta,\ z) = e^{ik''_{I-1}(\omega)(z-z_{I-1})}E_r^{\perp}(\eta) + e^{-ik''_{I-1}(\omega)(z-z_{I-1})}E_r^{\perp}(\eta), \tag{4.5.2a}$$

where the amplitudes of the incident and reflected waves are

$$E_i^{\perp}(\eta,\ z) = \frac{i}{2\sin\lambda_{I-1}}[E^{\perp}(\eta,\ z_{I-2}) - e^{i\lambda_{I-1}}E^{\perp}(\eta,\ z_{I-1})]; \tag{4.5.2b}$$

$$E_r^{\perp}(\eta) = \frac{1}{2i\sin\lambda_{I-1}}[E^{\perp}(\eta,\ z_{I-2}) - e^{-i\lambda_{I-1}}E^{\perp}(\eta,\ z_{I-1})]. \tag{4.5.2c}$$

The electromagnetic field in the other external layer $K + 1$ is determined only by the past wave propagating from z_K and z_{K+1}, the amplitude of which we denote $\mathbf{E}_t^{\perp}(\eta)$:

$$E_{K+1}^{\perp}(\eta) = e^{ik''_{K+1}(z-z_K)}E_t^{\perp}(\eta), \tag{4.5.3a}$$

where

$$E_t^{\perp}(\eta) = \frac{1}{2i\sin\lambda_{K+1}}[E^{\perp}(\eta,\ z_{K+1}) - e^{-i\lambda_{K+1}}E^{\perp}(\eta,\ z_K)], \tag{4.5.3b}$$

and the condition for the absence of a counter wave is satisfied when

$$E^{\perp}(\eta,\ z_{K+1}) = e^{i\lambda_{K+1}}E^{\perp}(\eta,\ z_K). \tag{4.5.3c}$$

This ratio allows you to provide the amplitude of the passing wave in a simple form:

$$E_t^{\perp}(\eta) = E^{\perp}(\eta,\ z_K). \tag{4.5.3d}$$

Component $E''_n(\eta, z)$, the electric field and magnetic field, can be expressed in terms of (4.2.3c) and (4.2.5) by the formulas (4.2.14a), (4.2.15), (4.2.18b), and (4.2.18c). In particular, for a *p*-polarized electromagnetic field, the component of the field perpendicular to the *Oz* axis in the $(I-1)$th layer has a form similar to (4.5.2a):

$$H^{\perp}_{I-1}(\eta,\ z) = e^{ik''_{I-1}(\omega)(z-z_{I-1})}H^{\perp}_i(\eta) + e^{-ik''_{I-1}(\omega)(z-z_{I-1})}H^{\perp}_r(\eta), \qquad (4.5.4a)$$

where

$$H^{\perp}_i(\eta) = -\frac{\omega\varepsilon_0\varepsilon_{I-1}(\omega)}{k''_{I-1}(\omega)}\,E^{\perp}_i(\eta); \qquad (4.5.4b)$$

$$H^{\perp}_r(\eta) = -\frac{\omega\varepsilon_0\varepsilon_{I-1}(\omega)}{k''_{I-1}(\omega)}\,E^{\perp}_i(\eta). \qquad (4.5.4c)$$

In the same way (4.5.3a) in the $(K+1)$th layer the same component of the magnetic field has the form

$$H^{\perp}_{K+1}(\eta,\ z) = e^{ik''_{K+1}(\omega)(z-z_K)}H^{\perp}_t(\eta), \qquad (4.5.5a)$$

where

$$H^{\perp}_t(\eta) = -\frac{\omega\varepsilon_0\varepsilon_{K+1}(\omega)}{k''_{K+1}(\omega)}\,E^{\perp}_t(\eta). \qquad (4.5.5b)$$

As follows from the form of formulas (4.5.2a) and (4.5.3a), for the value $k''_K(\omega)$ it makes sense to parallel the Oz axis of the components of the wave vector of the electromagnetic wave, while η according to the original definitions (4.2.4a) and (4.2.4b) represent a two-dimensional wave vector of an electromagnetic wave in the *XOY* plane. Therefore, taking into account the definition (4.2.1a), the square of the module of the three-dimensional wave vector is:

$$\mathbf{k}_n(\omega) = \eta + k''_n(\omega)\mathbf{e}_3$$

$$k_n^2(\omega) \equiv [k''_n(\omega)]^2 + \eta^2 = \omega^2 c^{-2}\varepsilon_n(\omega)\mu_n(\omega). \qquad (4.5.6)$$

Introducing the angle $\theta_n = (\mathbf{k}_n(\omega)\hat{;}\ \mathbf{e}_3)$ characterizing the direction of propagation of an electromagnetic wave in the *n*th layer, you can write the components of the wave vector as

$$k''_n(\omega) = k_n(\omega)\cos\theta_n;\quad \eta = k_n(\omega)\sin\theta_n. \qquad (4.5.7)$$

It should be noted that for an electromagnetic wave in a multilayer structure, the two-dimensional wave vector η is uniform in all layers of the system, so the value of η does not have an index *n*.

Using three equations: (4.3.1) for internal surfaces z_{I-1} and z_K and (4.5.3d) any three of the four boundary values $E^{\perp}(\eta, z_n)$ can be expressed through the fourth, such as $E^{\perp}(\eta, z_{I-2})$. Substituting them in the right parts of (4.5.2b), (4.5.2c), and (4.5.3d), we get the electric fields of the incident

$$E_i^{\perp}(\eta) = \frac{\gamma^+}{2i \sin \lambda_{I-1} R_K} E^{\perp}(\eta, z_{I-2}), \tag{4.5.8a}$$

reflected

$$E_r^{\perp}(\eta) = \frac{\gamma^-}{2i \sin \lambda_{I-1} R_K} E^{\perp}(\eta, z_{I-2}), \tag{4.5.8b}$$

and passing electromagnetic waves

$$E_t^{\perp}(\eta) = \frac{1}{R_K} D_{K,\, I-1}(I-1|K+1)\mu_{I-1} E^{\perp}(\eta, z_{I-2}), \tag{4.5.8c}$$

fn8where all symbols are used

$$R_K = 1 - D_{KK}(I-1|K+1)\mu_{K+1} e^{i\lambda_{K+1}}; \tag{4.5.9a}$$

$$R_{I-1}^{\pm} = 1 - D_{I-1,\, I-1}(I-1|K+1)\mu_{I-1} e^{\pm i\lambda_{I-1}}; \tag{4.5.9b}$$

$$\gamma^{\pm} = R_{I-1}^{\pm} R_K - D_{I-1,\, K}^2 (I-1|K+1)\mu_{K+1}\mu_{I-1} e^{i(\lambda_{K+1} \pm \lambda_{I-1})}. \tag{4.5.9c}$$

In case of *s*-, a polarized electromagnetic wave for which $E''_n = 0$, from the ratio of the amplitudes of reflected and incident electromagnetic waves, we directly obtain the amplitude of the reflection of a polarized electromagnetic wave, for which

$$r \equiv \frac{E_r^{\perp}(\eta)}{E_i^{\perp}(\eta)} = -\frac{\gamma^-}{\gamma^+} \tag{4.5.10a}$$

and the transmission amplitude

$$t \equiv \frac{E_t^{\perp}(\eta)}{E_i^{\perp}(\eta)} = -\frac{2i \sin \lambda_{I-1}}{\gamma^+} D_{K,\, I-1}(I-1|K+1)\mu_{I-1}. \tag{4.5.10b}$$

In case of *p*- for a polarized electromagnetic wave, the direction vector of the magnetic field lies in the *XOY* plane. Therefore, it is convenient to express the amplitudes of reflection and transmission through it. Using (4.5.4b), (4.5.4c), and (4.5.5b), we come to the formulas

$$r \equiv \frac{H_r^{\perp}(\eta)}{H_i^{\perp}(\eta)} = \frac{\gamma^-}{\gamma^+}; \tag{4.5.11a}$$

$$t \equiv \frac{H_t^{\perp}(\eta)}{H_i^{\perp}(\eta)} =$$

$$= -\frac{\varepsilon_{K+1}(\omega) k''_{I-1}(\omega) 2i \sin \lambda_{I-1}}{k''_{K+1}(\omega) \varepsilon''_{I-1}(\omega) \gamma^+} D_{K,\, I-1}(I-1|K+1)\mu_{I-1} \tag{4.5.11b}$$

.

The squares of the modules of the received amplitudes allow us to obtain the reflection coefficient

$$R \equiv |\, r\,|^2 \tag{4.5.12a}$$

and the transmission rate

$$T \equiv |\, t\,|^2 \tag{4.5.12b}$$

for multilayer systems. If there are absorbing substances in the multilayer structures the absorption coefficient is

$$A \equiv 1 - R - T. \tag{4.5.12c}$$

4.6 Optical characteristics of periodic systems

We now apply the approach developed above to calculate the optical characteristics of the periodic system described in section 4.3. The solutions of the system of algebraic equations (4.3.9) can be written as

$$E^{\perp}(\eta,\; z_{I-1}) = \frac{\mu_c R_f}{\Delta} E^{\perp}(\eta,\; z_{I-2}) + \frac{\mu_f \mu_a \mu_b}{\Delta \nu} \frac{\alpha^2 - 1}{\alpha^{2N} - 1} \alpha^{N-1} E^{\perp}(\eta,\; z_{K+1}); \tag{4.6.1a}$$

$$E^{\perp}(\eta,\; z_K) = \frac{\mu_f R_c}{\Delta} E^{\perp}(\eta,\; z_{K+1}) + \frac{\mu_c \mu_a \mu_b}{\Delta \nu} \frac{\alpha^2 - 1}{\alpha^{2N} - 1} \alpha^{N-1} E^{\perp}(\eta,\; z_{I-2}), \tag{4.6.1b}$$

where the determinant of the mentioned set of equations is

$$\Delta = R_c R_f - \left(\frac{\mu_a \mu_b}{\nu} \frac{\alpha^2 - 1}{\alpha^{2N} - 1} \alpha^{N-1} \right)^2 \tag{4.6.1c}$$

and must not be zero. Substituting (4.6.1b) in (4.5.3c), you can get the amplitude of the electric field on the external surface $E^{\perp}(\eta,\; z_{K+1})$, expressed in terms of amplitude $E^{\perp}(\eta,\; z_{I-2})$, and obtain in accordance with (4.5.3d) the value of the field of the passing electromagnetic wave:

$$E_t^{\perp}(\eta) = \frac{\mu_c \mu_a \mu_b}{\nu(\Delta - \mu_f R_c e^{i\lambda_f})} \frac{\alpha^2 - 1}{\alpha^{2N} - 1} \alpha^{N-1} E^{\perp}(\eta,\; z_{I-2}). \tag{4.6.2}$$

Using (4.6.1a), meaning $E^{\perp}(\eta,\; z_{I-1})\; E^{\perp}(\eta,\; z_{I-2})$, the electric fields of the incident and reflected electromagnetic waves are obtained as

$$E_i^{\perp}(\eta) = -\frac{S^+}{2i \sin \lambda_c(\Delta - \mu_f R_c e^{i\lambda_f})} E^{\perp}(\eta,\; z_{I-2}); \tag{4.6.3a}$$

and can be expressed using

$$E_r^{\perp}(\eta) = \frac{S^-}{2i \sin \lambda_c(\Delta - \mu_f R_c e^{i\lambda_f})} E^{\perp}(\eta,\; z_{I-2}), \tag{4.6.3b}$$

where the following notation is entered:

$$S^{\pm} = \Delta - \mu_c R_f e^{\pm i\lambda_c} - \mu_f R_c e^{i\lambda_f} + \mu_c \mu_f e^{i(\lambda_f \pm \lambda_c)}. \tag{4.6.4}$$

Hence, for a periodic structure with an arbitrary number N pairs of layers of reflection and passage amplitudes have the form

$$r = -\frac{S^-}{S^+}; \tag{4.6.5a}$$

$$t = -\frac{2i\,\overline{\mu}_c \mu_a \mu_b(\alpha^2 - 1)}{\nu\, S^+(\alpha^{2N} - 1)}\alpha^{N-1} \tag{4.6.5b}$$

for the s-polarized electromagnetic wave in accordance with the general formulas (4.5.10a) and (4.5.10b) and

$$r = \frac{S^-}{S^+}; \tag{4.6.6a}$$

$$t = \frac{\varepsilon_f(\omega)k''_c}{k''_f(\omega)\varepsilon_c(\omega)}\frac{2i\,\overline{\mu}_c \mu_a \mu_b}{\nu\, S^+}\frac{(\alpha^2 - 1)}{(\alpha^{2N-1} - 1)}\alpha^{N-1} \tag{4.6.6b}$$

—for the p-polarized electromagnetic wave in accordance with (4.5.11a) and (4.5.11b), where

$$\overline{\mu}_n = \mu_n \sin \lambda_n. \tag{4.6.6c}$$

Substituting the determinant (4.6.1c) in (4.3.10a) and (4.3.10b), we get the formula

$$S^{\pm} = \tilde{R}_c^{\mp}\,\tilde{R}_f^{-} - \left(\frac{\mu_a \mu_b}{\nu}\frac{(\alpha^2 - 1)}{(\alpha^{2N} - 1)}\alpha^{N-1}\right)^2, \tag{4.6.7a}$$

where the following notation is entered:

$$\tilde{R}_c^{\mp} = \frac{\mu_a \mu_b}{\nu}\left[\frac{1}{\tilde{\alpha}_c^{\pm}} - \frac{\alpha^{2N} - \alpha^2}{\alpha(\alpha^{2N} - 1)}\right]; \tag{4.6.7b}$$

$$\frac{1}{\tilde{\alpha}_c^{\mp}} = \frac{1}{\mu_a \mu_b}(\mp i\,\overline{\mu}_c \nu - \overline{\mu}_a^2) + \cos\lambda_a \cos\lambda_b, \tag{4.6.7c}$$

with the fact that $\tilde{R}_f^-$ and $1/\tilde{\alpha}_f^-$ differs from $\tilde{R}_c^-$ and $1/\tilde{\alpha}_c^-$ replacement $c \to f$, $a \leftrightarrow b$. Since according to (4.2.9) and (4.2.20) for s-polarized waves

$$\overline{\mu}_n = \overline{\mu}_n^{(s)} \equiv -\frac{k''_n(\omega)}{\mu_n(\omega)} \tag{4.6.8a}$$

And for the p-polarized

$$\overline{\mu}_n = \overline{\mu}_n^{(p)} \equiv -\frac{\varepsilon''_n(\omega)}{k''_n(\omega)}, \tag{4.6.8b}$$

it follows directly from (4.6.7c) that the reflection and transmission amplitudes (4.6.5a), (4.6.5b), (4.6.6a), and (4.6.6b) do not depend on the thickness of the external layers. Using (4.6.3a) and (4.6.3b) we find for the case of s-polarization:

$$\psi = \cos\lambda_a \cos\lambda_b - \frac{1}{2}\left(\frac{\overline{\mu}_a^{(s)}}{\overline{\mu}_b^{(s)}} + \frac{\overline{\mu}_b^{(s)}}{\overline{\mu}_a^{(s)}}\right)\sin\lambda_a \sin\lambda_b. \tag{4.6.9a}$$

And for the case of p-polarization:

$$\psi = \cos\lambda_a \cos\lambda_b - \frac{1}{2}\left(\frac{\overline{\mu}_a^{(p)}}{\overline{\mu}_b^{(p)}} + \frac{\overline{\mu}_b^{(p)}}{\overline{\mu}_a^{(p)}}\right)\sin\lambda_a \sin\lambda_b. \tag{4.6.9b}$$

The amplitudes of reflection and transmission (4.6.5a), (4.6.5b) and (4.6.6a), (4.6.6b) together with the definitions (4.6.6c)–(4.6.7c) obtained in the above approach were equivalent to the results of the work [9, 10]. However, in the latter (see formulas (49), (50), (88), and (90) from [11]) the main formulas are expressed in terms of Chebyshev polynomials. It is interesting to note that when formulas for optical coefficients of periodic structures were widely used in theory and numerical calculations, the fact that they can be expressed in terms of elementary functions remained unnoticed. The transition from formulas with Chebyshev polynomials to formulas expressed in terms of elementary functions is described in detail in [12] on the optical properties of multilayer structures, following which the material is described in sections 4.5–4.7 of the current chapter.

Let's consider some special cases of general formulas. For a two-layer structure ($N = 1$) (4.6.7a) simplifies:

$$S^{\pm} = \left(\frac{\mu_a \mu_b}{\nu}\right)^2 \left(\frac{1}{\tilde{\alpha}_c^{\mp}\tilde{\alpha}_f^{-}} - 1\right). \tag{4.6.10}$$

In the extreme case $l_b \to 0$ of the a-type single layer system with the finite thickness the formulas (4.6.7a) and (4.6.10) lead to the results of (7), (17) [13]. If, in addition, $l_a \to 0$, then

$$S^{\pm} = -i(\pm\overline{\mu}_c + \overline{\mu}_f)\frac{\mu_a \mu_b}{\nu}, \tag{4.6.11a}$$

which means that

$$-\frac{S^{-}}{S^{+}} = \frac{\overline{\mu}_c - \overline{\mu}_f}{\overline{\mu}_c + \overline{\mu}_f}; \tag{4.6.11b}$$

$$-\frac{2i\,\overline{\mu}_c \overline{\mu}_a \overline{\mu}_b}{v\,S^{+}} = \frac{2\overline{\mu}_c}{\overline{\mu}_c + \overline{\mu}_f}. \tag{4.6.11c}$$

Substituting here the values μ_n according to definitions (4.2.17), we find from expressions (4.5.10a), (4.5.10b), (4.5.11a), and (4.5.11b) the Fresnel formulas (86.4) and (86.6) from [14].

For a significantly larger number of periods in the structure $N \gg 1$ with $|\alpha| > 1$, when $\alpha^{2N} - 1 \approx \alpha^{2N}$, keeping members of the highest order by α, we get from (4.6.7a), (4.6.7b)

$$S^{\pm} \approx \tilde{R}_c^{\pm}\tilde{R}_f^{-}; \ \tilde{R}_c^{\pm} \approx \frac{\mu_a \mu_b}{\nu}\left(\frac{1}{\tilde{\alpha}_c^{\mp}} - \frac{1}{\alpha}\right). \tag{4.6.12}$$

As a result, the reflection amplitudes take the form

$$r = \mp \frac{(1/\tilde{\alpha}_c^{+}) - (1/\alpha)}{(1/\tilde{\alpha}_c^{-}) - (1/\alpha)}, \tag{4.6.13}$$

in which the upper sign refers to s-polarized and the lower sign refers to p-polarized by electromagnetic waves, while the transmission amplitudes in both cases are small:$t \sim \alpha^{-N-1}$. With $|\alpha| < 1$, when $1 - \alpha^{2N} \approx 1$, leaving the members of the lowest order by, we find from (4.6.7b) and (4.6.7c) an expression of the form (4.6.12) and (4.6.13), in which it is necessary to make a substitution $\alpha^{-1} \to \alpha$, and we get that the transmission amplitudes $t \sim \alpha^{N-1}$ are also small. Finally, the condition corresponds to a continuous wave in a multilayer system.

When the dielectric and magnetic permeability of both layers, a- and b-types, and also the functions $k_c(\omega)$ and α are real, then in accordance with (4.6.7c), (4.6.8a), and (4.6.8b) it can be shown,

$$\frac{1}{\alpha_c^{-}} = \left(\frac{1}{\tilde{\alpha}_c^{+}}\right)^{*} \tag{4.6.14}$$

and the reflection coefficient (4.5.12a) (according to (4.6.13)) is equal to one. It can take other values if the above conditions are not met. In particular, it can be seen from (4.6.13) that the reflection amplitude vanishes when the equality is satisfied:

$$\frac{1}{\tilde{\alpha}_c^{*}} - \frac{1}{\alpha} = 0. \tag{4.6.15}$$

The comparison of (4.6.7c) and (4.3.10b) shows that (4.6.15) is formally equivalent to the dispersion relation (4.3.15a) for polariton waves that are spatially attenuated from the boundary between the periodic part of the structure and the semi-infinite medium c-type, received for the area where $\mathrm{Re}[\chi_c^2(\omega)] > 0$. This equivalence can be interpreted as a manifestation of the resonance of proper polariton waves in a multilayer structure with an incident electromagnetic wave. It clearly follows that the features of the optical characteristics spectra are a source of information about the spectrum of elementary excitations in a multilayer system.

4.7 Numerical calculations of optical characteristics

As in the previous chapter, let us turn to the periodic structure GaAs (a-type) —$Ga_{1-x}Al_xAs$ (b-type), bordering on vacuum. For these substances, the magnetic permeability can be considered equal to one, while the dielectric permittivity with attenuation is taken into account:

$$\varepsilon_a(\omega) = \varepsilon_a \frac{\omega_{La}^2 - \omega^2 - i\omega\omega_{Ta}\gamma_a}{\omega_{Ta}^2 - \omega^2 - i\omega\omega_{Ta}\gamma_a}; \tag{4.7.1a}$$

$$\varepsilon_b(\omega) = \varepsilon_b + \sum_{j=1}^{2} \frac{s_j\omega_{Ta_j}^2}{\omega_{Tb_j}^2 - \omega^2 - i\omega\omega_{Tb_j}\gamma_{b_j}} \tag{4.7.1b}$$

with the designation

$$s_1 = \varepsilon_b \frac{(\omega_{Lb_1}^2 - \omega_{Tb_1}^2)(\omega_{Tb_1}^2 - \omega_{Lb_2}^2)}{\omega_{Tb_1}^2(\omega_{Tb_1}^2 - \omega_{Tb_2}^2)}; \tag{4.7.1c}$$

$$s_2 = \varepsilon_b \frac{(\omega_{Lb_2}^2 - \omega_{Tb_2}^2)(\omega_{Lb_1}^2 - \omega_{Tb_2}^2)}{\omega_{Tb_2}^2(\omega_{Tb_1}^2 - \omega_{Tb_2}^2)}. \tag{4.7.1d}$$

Values of all parameters except γ_a, in the formulas (4.7.1a)–(4.7.1d) for $x = 0.3$ given in section 4.4 of the current chapter. Factors of attenuation γ_a, γ_{b_j} taken equal to 0.01.

For systems with $l_a = 1.2$ μm and $l_b = 0.3$ μm and a different number of periods N, the amplitudes of reflection and transmission are calculated numerically using the formulas (4.6.5a) and (4.6.5b), after which the transmission coefficients are found by the formulas (4.5.12a) and (4.5.12b). Figure 4.4 shows the spectra of reflection and transmission coefficients for the case of a normal incidence of an electromagnetic wave on structures with a different number of periods. In the case under consideration, these spectra do not depend on the type of polarization. It can be seen that in the vicinity of the frequency of the transverse optical phonon, the reflection coefficient reaches a maximum value that practically does not depend on the number of periods N. In the same frequency range, the transmission coefficient becomes very small. As the number of periods increases, the transmission coefficient decreases, while the plateau expands. In the low-frequency region, the coefficients of both reflection and transmission have a rich spectral structure determined by the values of N. At a fixed frequency ω, the reflection and transmission coefficients depend significantly on the angle θ_c of incidence for both types of polarization. For example, if $\omega = 6.5 \cdot 10^{13}$ s^{-1} and $N = 5$, for $\theta_c = 0^{\text{x000B0;}}$, 30° and $60^{\text{x000B0;}}$ the transmission coefficients of an s-polarized electromagnetic wave are 0.145, 0.122, and 0.080, respectively. Thus, multilayer structures are good filters of electromagnetic radiation. Filtration characteristics are easily controlled by both the number of periods and the geometric and material parameters of the constituent layers.

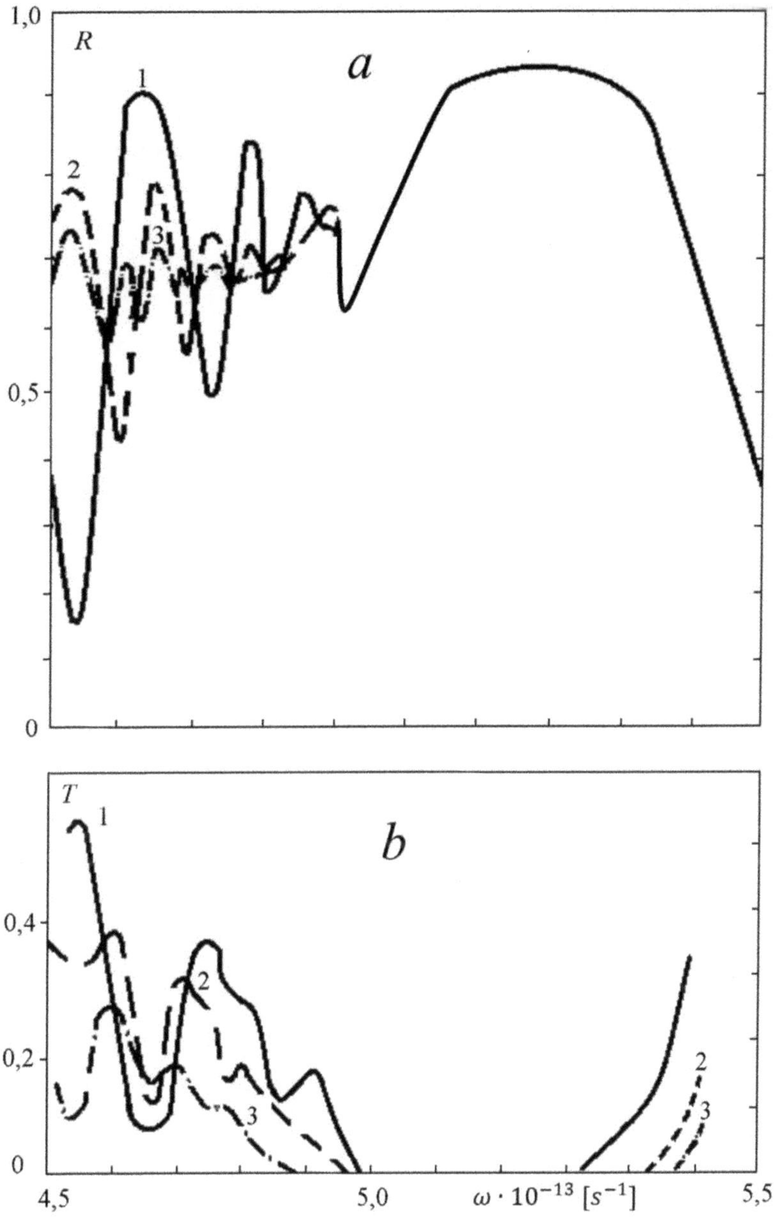

Figure 4.4. Spectra of reflection coefficients (a) and transmission coefficients (b) in the case of a normal electromagnetic wave incident on the structure with the number of periods $N = 5$ (1), 10 (2), 20 (3).

4.8 Conclusion

Wave equations for the propagation of polaritons in multilayer planar structures and superlattices were solved. For the first time, a consistent classification of polariton waves was performed by the nature of the spatial distribution of electric and magnetic field strengths in the layer and by the change of these fields over the period of the structure. The laws of dispersion of interface spatially extended and interface spatially decreasing polaritons were obtained. The optical properties of planar multilayer structures and superlattices (reflection and passage of light) were described and numerical calculations of optical characteristics performed.

References

[1] Fomin V M and Pokatilov E P 1986 Optical properties of multi-layer structures. I. Polaritons *Phys. Status Solidi* B **136** 187–99

[2] Rytov S M 1955 *JETF* **29** 605–16

[3] Pokatilov E P and Fomin V M 1984 Tez. dokl. VI Vsesoyuz. conf. on the nonresonant interaction of radiation with matter *Vilnius* 423–84

[4] Wendler L 1984 Phonon-polaritons in bilayer systems. p-polarized guided wave phonon-polaritons and guided wave surface phonon-polaritons *Phys. Status Solidi* B **123** 469–77

[5] Wendler L 1985 Phonon-polaritons in bilayer systems s-polarized guided wave phonon-polaritons *Phys. Status Solidi* B **128** 425–37

[6] Camley R E and Mills D Z 1984 Collective excitations of semi-infinite superlattice structures: surface plasmons, bulk plasmons and the electron-energy-loss spectrum *Phys. Rev.* B **29** 1695–706

[7] Ilegems M and Pearson G L 1970 Infrared reflection spectra of $Ga_{1-x}Al_xAs$ mixed crystals *Phys. Rev.* B **1** 1576–82

[8] Keyoi X and Panish M 1981 *Lasers on Heterostructures* (Moscow: Mir) p 306

[9] Abelès F 1950 Recherches sur la propagation des ondes électromagnétiques sinusoïdales dans les milieux stratifiés. Première partie *Ann. Phys.* **12** 596–640

[10] Abelès F 1950 Recherches sur la propagation des ondes électromagnétiques sinusoïdales dans les milieux stratifiés. Deuxième partie *Ann. Phys.* **12** 706–82

[11] Born M and Wolf E 1973 *Fundamentals of Optics. Part 1, Section 6* (Moscow: Nauka) p 856

[12] Fomin V M and Pokatilov E P 1986 Optical properties of multi-layer structures. II. Reflection and transmission *Phys. Status Solidi* B **136** 593–602

[13] Fuchs R, Kliewer K Z and Pardey W J 1966 Optical properties of an ionic crystal slab *Phys. Rev.* **150** 589–96

[14] Landau L D and Livshits E M 1982 *Electrodynamics of Continuous Media* (Moscow: Nauka) p 620

[15] Beril S I, Fomin V M and Starchuk A S 2020 *Theory of Polarons, Excitons, Bipolarons and Kinetic Effects in Multilayer Structures of Various Geometries and Superlattices* (Tiraspol: Pridnestrovian University Publishing House) p 696

[16] Bryksin B B and Firsov Yu A 1971 Interaction of an electron with interface phonons in an ion crystal plate *FTT* **13** 496–503

[17] Pokatilov E P and Fomin V M 1984 *Theory of Potential in Multilayer Systems* No. 500M-05.85 (Chisinau: MoldNIINTI) p 19

[18] Pokatilov E P and Fomin V M 1984 *Interaction of an Electron with Polarization Optical Vibrations in Periodic Structures* No. 508M-05.85 (Chisinau: MoldNIINTI) p 14

[19] Fomin V M and Pokatilov E R 1985 *Phys. Status Solidi* B **132** P 69–82

[20] Wendler L and Pechsted R 1986 Dynamical screening, collective excitations, and electron-phonon interaction in heterostructures and semiconductor quantum wells. General theory *Phys. Status Solidi* B **138** 197–217

[21] Wendler L and Haupt E 1987 Electron-phonon interaction in semiconductor superlattices *Phys. Status Solidi* **B 143** 487–510
[22] Varga B B 1965 Coupling of plasmons to polar phonons in degenerate semiconductors *Phys. Rev.* **137** A1896–902
[23] Davydov A S 1972 *Theory of a Solid Body* (Moscow: Nauka) p 640
[24] Pines D 1965 *Elementary Excitations in Solids* (Moscow: Mir) p 382
[25] Mooradian A and Wright G B 1966 Observation of the interaction of plasmons with longitudinal optical phonons in GaAs *Phys. Rev. Lett.* **16** 999–1001
[26] Lapeyere G and Anderson J R 1975 Evidence for a surface–states exciton on GaAs (110) *Phys. Rev.* **B 35** 117–9
[27] Reshina N I, Herbstein Y M and Mirlin D I *FT* 1972**14** 1280–2
[28] Ando T, Fowler A and Stern F 1985 *Electronic Properties of Two-Dimensional Systems* (Moscow: Mir) p 415
[29] Smirnov V I 1974 *Course of Higher Mathematics* (Moscow: Nauka) p 480
[30] Pokatilov E P and Beril S I 1982 Spatially extended optical modes in two-layer periodical structures *Phys. Status Solidi* **B 110** K75–8
[31] Pokatilov E P and Beril S I 1983 Electron-phonon interaction in periodic two-layer structures *Phys. Status Solidi* **B 118** 567–73
[32] Bloss W L 1983 Plasmon modes of a superlattice—classical vs quantum limits *Solid State Commun.* **48** 927–31
[33] Sood A K, Menendez J, Cardona M and Ploog A K 1985 Interface vibrational modes in GaAs–AlGaAs superlattices *Phys. Rev. Lett.* **54** 2115–8
[34] Bloss W L 1982 Optic and acoustic plasmon modes of a semiconductor superlattice *Solid State Commun.* **44** 363–7
[35] Bloss W L and Brody E M 1982 Collective modes of a superlattice - plasmons, LO phonon-plasmons, and magnetoplasmons *Solid State Commun.* **43** 523–8
[36] Toyozawa Y 1954 Theory of electronic polaron and ionization of trapped electron by excited *Prog. Theor. Phys.* **12** 421–42

IOP Publishing

Vibrational Excitations in Multilayer Nanostructures
Properties and manifestations
Stepan I Beril, Vladimir M Fomin and Alexander S Starchuk

Chapter 5

Theory of surface polaronic states

5.1 Introduction to the theory

The idea of the possibility of 'autolocalization' of an electron in an ideal crystal in the field of deformation caused by the field of the electron itself was expressed by Landau [1] in 1933. In 1951, Pekar developed it for ionic crystals into a quantitative theory, which is based on taking into account the dielectric polarization of the crystal by the electric field of the conduction electron. The resulting local polarization is inertial because it is caused by the displacement of ions. This polarization cannot follow a fast-moving electron and therefore forms a potential well for the electron (about 0.5–1.0 eV), in which there are discrete energy levels of the electron. Being in a local state at one of these levels, an electron can maintain local polarization with its field. Such states of a crystal with a polarizing potential well in which an electron is localized were called polarons by Pekar [2]. The polaron moves in the crystal like a particle with an electron charge and with some effective mass.

Further theoretical and experimental studies of polaronic states in dielectrics and semiconductors with an ionic crystal lattice have led to a new understanding of the polaronic effect as fundamental, playing an important role in electrical, photoelectric, optical, and other phenomena. The Polaron effect plays a special role in the phenomenon of superconductivity, and this was again manifested in the phenomenon of high-temperature superconductivity discovered in 1986.

The existence of a surface in polar crystals introduces a change in their vibrational spectrum. In addition to the usual bulk polarization oscillations, surface optical vibrations (SV-1 type) appear, attenuating exponentially from the boundary deep into the crystal on a scale significantly larger than the lattice constant a [3], during which macroscopic polarization occurs, with which an electron from the conduction band of the crystal can interact. (SV-1 are divided into two groups: radiation and nonradiative ones. Radiation fluctuations are characterized by the fact that they can directly interact with light; nonradiative—they do not interact with light without the participation of a third body.) The interaction of the conduction electron with surface optical vibrations under certain conditions can lead to the localization of the electron in the near-surface domain and the formation of surface polaronic states. Their specific difference from bulk polaron states is that the polaron is localized in the direction perpendicular to the surface (hereinafter referred to as the z-axis).

The interaction of an electron with surface optical phonons was considered in the works [4–9], of which the works of Evans and Mills can be distinguished [5, 6]. For the first time, an attempt was made to establish criteria for the formation of surface polaronic states at the contact of a polar crystal with a vacuum. However, since a continuous polaron was considered a seed quasiparticle, an incorrect conclusion was made that surface polaron states would occur even at arbitrarily small constants of the electron–phonon interaction. If we consider the interaction of an electron with the surface and bulk optical vibrations of the same order, then, as will be shown below, the interaction with the surface will have the character of attraction only if certain criteria are met. The correct criteria can be established only if the interaction of the charge carrier with inertial (plasmons of valence electrons) and inertial (bulk and surface optical phonons) polarizations is correctly taken into account.

The theory that allows us to establish criteria for the existence of surface polaronic states at the contact of two polar crystals and in various special cases was developed in the works of Pokatilov, Beril [10–17] and summarized in the monograph [18]. Before proceeding to the description of surface polaronic states, it is necessary to make the following remark. The crystal surface can be considered as a defect that strongly perturbs the periodic potential of the crystal lattice in a direction perpendicular to the surface and leads to the formation of a potential well. Taking into account the electron–polarization interaction in such a one-dimensional potential well turns this problem into a fundamentally new quantum mechanical problem, which is available only to approximate research methods since it boils down to the need to coordinate the movements of two subsystems—the particle and polarization. The latter has an infinite number of degrees of freedom. The specifics of the problem in this case, in comparison with an infinite crystal, is that the 'isolation' of a free surface electron does not always make sense, because the motion of an electron is determined not only by the electron–polarization interaction but also by the influence of the potential of the defect field. There are two possible limiting cases here.

1. Localization caused by all interactions is weak; the electron movement is almost free; the oscillation frequency of the localized electron is less than the frequency of polarization oscillations (shallow local level):

$$\frac{E_{\text{loc}}}{\hbar} < \omega_{\text{polariz}}; \tag{5.1.1a}$$

$$\bar{z} > R_p, \tag{5.1.1b}$$

where $\bar{z}$ is the average distance of the polaron from the surface and R_p is the polaron radius. Inequalities (5.1.1a) and (5.1.1b) can occur at any strength of the electron–polarization coupling, in particular, the small value of the E_{loc}

with a strong electron–polarization interaction may be a consequence, for example, of the repulsion of an electron from the interface by an external field. The solution of the problem splits into two stages: the 'isolation' of a quasi-particle–polaron localized in the vicinity of an arbitrary point z, and the study of the motion of the polaron in a field of forces determined by potential fields and electron–polarization interaction.

2. Localization caused by all interactions is strong. The state of polarization of the medium is determined not by the current z coordinate, but by the state of the electron. In this approximation, the solution of the problem also contains two stages: the derivation of the adiabatic potential of the slow subsystem (polarization oscillations), the role of which is played by the total energy of the fast subsystem, and the solution of the equation of motion for the slow subsystem.

In chapter 5, the theory of polarons in planar systems is developed: surface, near-surface, film, external, levitating, electronic, and others, for the description of which exact Hamiltonians of the electron–polarization interaction are derived and criteria for the existence of surface polaronic states are investigated. The quantum theory of image potential and strength is presented and the dielectric function of a quantum dielectric is derived based on the polaronic theory for planar systems. The potential energy of self-action is found and the polaronic states in multilayer cylindrical and spherical structures are considered. Polaronic states in homogeneous electric and magnetic fields in planar, cylindrical, and spherical nanostructures are studied.

The specific features of the electron–phonon interaction in superlattices have been studied, of which the most important is the occurrence of new elementary excitations—surface spatially extended optical vibrations in a composite superlattice consisting of alternating polar layers of two types. The law of dispersion of interacting surface modes is established. The Hamiltonian of the electron–phonon interaction is found and the polaron states in a composite superlattice, in nanoscale cylindrical and spherical structures are studied.

The energy and effective mass of a magnetopolaron in a quantum wire in a dielectric medium are obtained.

5.2 Hamiltonian of the electron–phonon interaction in a three-layer structure

To describe the interaction of a charge carrier with surface optical vibrations, the use of electrodynamic methods is difficult or even impossible. This applies to a wide range of tasks presented in this book, which uses a quantum mechanical approach: scattering of band electrons on surface phonons, polaron pairing on surface optical phonons, dynamic shielding of electron–hole interaction, and polaron exciton. To solve these and other problems, the Hamiltonian of the electron–phonon interaction is necessary. Section 5.1 describes a general procedure for deriving the Hamiltonian of the electron–phonon interaction for spatially inhomogeneous systems such as section contacts: polar crystal vacuum; polar–nonpolar crystals $(1x|2)$; polar–polar

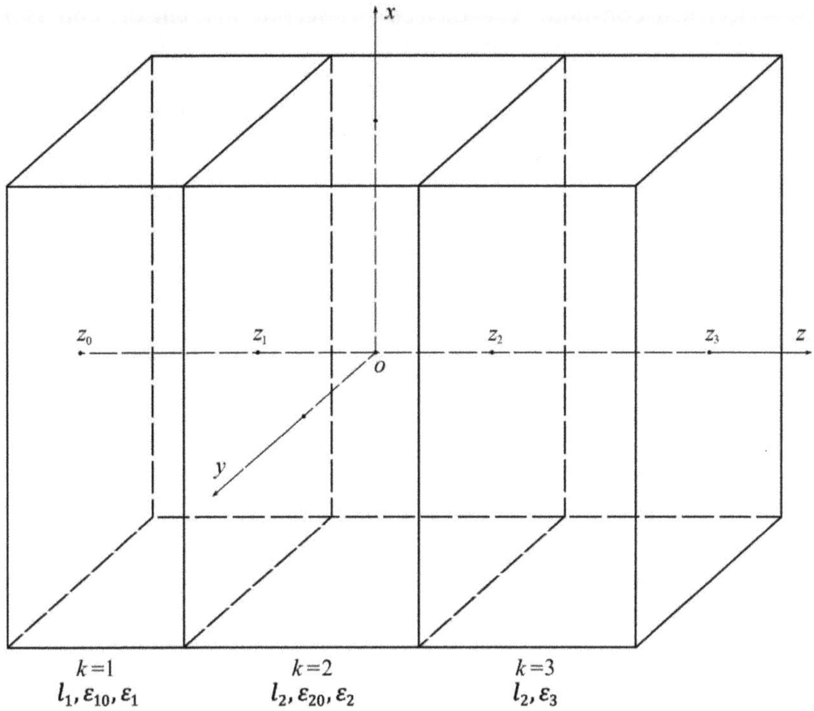

Figure 5.1. Three-layer structure.

crystals ($1x|2x$); the polar layer in nonpolar medium plates ($1|2x|3$), where the numbers 1, 2, 3 number the crystalline plates; and x—the polar environment. The most common of these is the electron–polarization Hamiltonian of the system ($1x|2x|3$), which is shown in figure 5.1 and is discussed below.

5.2.1 Contact: polar–polar–homeopolar crystals

The interaction of an electron with surface phonons is performed without taking into account the delay in the electric field. A generalization for the case of delay is given in the monograph [18].

To derive the Hamiltonian of the electron–phonon interaction with optical vibrations in a three-layer system (figure 5.1), it is necessary first of all to find the electric field that arises during lattice vibrations. Bryksin and Firsov [19] showed that for this purpose it is possible to use Maxwell's electrostatic equations:

$$\text{rot } \mathbf{E}_k(\mathbf{r}) = 0; \text{ div } \mathbf{D}_k(\mathbf{r}) = 0 \tag{5.2.1}$$

(k—layer number), where, in accordance with the division of polarization into inertia-free and inertial, the electric induction vector can be represented as

$$\mathbf{D}_k(\mathbf{r}) = \varepsilon_0 \mathbf{E}_k(\mathbf{r}) + \varepsilon_0 \overleftrightarrow{\chi}_k \mathbf{E}_k(\mathbf{r}) + \mathbf{P}_k(\mathbf{r}); \tag{5.2.2}$$

where ε_0—electrical constant, $\overleftrightarrow{\chi}_k$—dielectric susceptibility tensor, and $\mathbf{P}_k(\mathbf{r})$—the vector of inertial polarization. Fast polarization follows the field inertially and is completely determined by it:

$$\mathbf{P}_{\text{fast}}(\mathbf{r}) = \varepsilon_0 \overleftrightarrow{\chi}_k \mathbf{E}_k(\mathbf{r}). \tag{5.2.3}$$

The first of the equations (5.2.1) is satisfied by a potential field

$$\mathbf{E}_k(\mathbf{r}) = -\text{grad } V_k(\mathbf{r}), \tag{5.2.4}$$

and the potential is described by the equation

$$\text{div } \overleftrightarrow{\varepsilon}_k \text{ grad } V_k(\mathbf{r}) = \varepsilon_0^{-1} \text{ div } \mathbf{P}_k(\mathbf{r}), \tag{5.2.5}$$

where the high-frequency dielectric constant tensor is introduced:

$$\overleftrightarrow{\varepsilon}_k = \overleftrightarrow{1} + \varepsilon_0^{-1} \overleftrightarrow{\chi}_k. \tag{5.2.6}$$

With sufficiently large transverse dimensions of the system, its properties can be considered homogeneous in the plane xOy. In real structures, the concept of a 'dividing line' is conditional. More precisely, we should talk about the transition layer [20], consisting of two or three or more atomic planes in which the crystal lattice of one layer 'transitions' into the crystal lattice of another neighboring layer of the structure. In the continuum theory, when calculating potentials, the existence of a transition layer can be taken into account by introducing a transition function $S(z)$ for the contact kth and $(k+1)$th layers:

$$\varepsilon(z) = \varepsilon_k + (\varepsilon_{k+1} - \varepsilon_k) S(z), \tag{5.2.7}$$

which varies from zero to one within the transition layer. In the following, we assume that

$$S(z_n - z) = \theta(z_n - z), \tag{5.2.8}$$

where

$$\theta(z) = \begin{cases} 0, & z < 0; \\ 1, & z > 0, \end{cases} \tag{5.2.9}$$

thus limiting ourselves to the assumption that the length of the transition domain is less than the effective lengths.

By performing the Fourier transform on the coordinates $\boldsymbol{\rho} = x\mathbf{e}_1 + y\mathbf{e}_2$:

$$V_k(\mathbf{r}) = \iint d^2\eta e^{i\boldsymbol{\eta}\boldsymbol{\rho}} V_k(\boldsymbol{\eta}, z), \; k = 1, 2, 3, \tag{5.2.10}$$

from the partial differential equation we obtain an ordinary differential equation:

$$\left(\frac{\partial}{\partial z}\mathbf{e}_3 + i\boldsymbol{\eta}\right) \overleftrightarrow{\varepsilon}_k \left(\frac{\partial}{\partial z}\mathbf{e}_3 + i\boldsymbol{\eta}\right) V_k(\boldsymbol{\eta}, z) = -\varepsilon_0^{-1} \tilde{\rho}_k(\boldsymbol{\eta}, z), \tag{5.2.11}$$

where

$$\tilde{\rho}_k(\boldsymbol{\eta}, z) = -\frac{\partial}{\partial z}P_k^{||}(\boldsymbol{\eta}, z) - i\boldsymbol{\eta}\mathbf{P}_k^{\perp}(\boldsymbol{\eta}, z); \tag{5.2.12}$$

$\boldsymbol{\eta} = \eta_1\mathbf{e}_1 + \eta_2\mathbf{e}_2$—two-dimensional wave vector; and $P_k^{||}(\boldsymbol{\eta}, z) = \mathbf{P}_k(\boldsymbol{\eta}, z)\mathbf{e}_3$; $\mathbf{P}_k^{\perp}(\boldsymbol{\eta}, z) = \mathbf{P}_k(\boldsymbol{\eta}, z) - P_k^{||}(\boldsymbol{\eta}, z)\mathbf{e}_3$, respectively, the longitudinal and two-dimensional transverse components of the vector relative to the z-axis $\mathbf{P}_k(\boldsymbol{\eta}, z)$. Assuming that the $\overleftrightarrow{\varepsilon}_k$ tensors are uniaxial:

$$\varepsilon_k^{ij} = \delta^{ij}\left[\varepsilon_k^{\perp}(\delta^{i1} + \delta^{i2}) + \varepsilon_k^{||}\delta^{i3}\right], \tag{5.2.13}$$

simplify (5.2.11):

$$\left(\frac{\partial^2}{\partial z^2} - \epsilon_k\eta^2\right)V_k(\boldsymbol{\eta}, z) = -\left(\varepsilon_0\varepsilon_0^{||}\right)^{-1}\tilde{\rho}_k(\boldsymbol{\eta}, z), \tag{5.2.14}$$

where are the designations introduced

$$\varepsilon_k^{11} = \varepsilon_k^{22} = \varepsilon_k^{\perp},\ \varepsilon_k^{33} = \varepsilon_k^{||},\ \varepsilon_k^{\perp}/\varepsilon_k^{||} = \epsilon_k. \tag{5.2.15}$$

The solution of equation (5.2.14) for the kth layer satisfying the condition of continuity of potential is

$$V_{k-1}(\boldsymbol{\eta}, z)|_{z=z_{k-1}} = V_k(\boldsymbol{\eta}, z)|_{z=z_{k-1}} \equiv V(\boldsymbol{\eta}, z_{k-1}), \tag{5.2.16}$$

and can be represented as

$$\begin{aligned} V_k(\boldsymbol{\eta}, z) = {} & F_k(z - z_{k-1})V(\boldsymbol{\eta}, z_k) \\ & + F_k(z_k - z)V(\boldsymbol{\eta}, z_{k-1}) + \int_{z_{k-1}}^{z_k} dz'\gamma_k(z, z')\tilde{\rho}_k(\boldsymbol{\eta}, z), \end{aligned} \tag{5.2.17}$$

with symbols

$$F_k(z) = \frac{\sinh(\sqrt{\epsilon_k}\eta z)}{\sinh\zeta_k};\ \varepsilon_k = \sqrt{\varepsilon_k^{\perp}\varepsilon_k^{||}} = \sqrt{\epsilon_k}\varepsilon_k^{||};\ \zeta_k = \sqrt{\epsilon_k}\,\eta l_k. \tag{5.2.18}$$

The first two terms in the right part (5.2.17) represent the general solution of the homogeneous equation corresponding to (5.2.14) satisfying the boundary conditions (5.2.16) with $\tilde{\rho}_k(\boldsymbol{\eta}, z) = 0$ and the latter is a particular solution (5.2.14). Using general methods for solving an inhomogeneous second-order equation, we find the kernel of the integral in (5.2.17):

$$\begin{aligned} \gamma_k(z, z') = \frac{1}{2\eta\varepsilon_k}\Bigg\{ & \mathrm{e}^{-\sqrt{\epsilon_k}\eta|z-z'|} + \frac{\mathrm{e}^{-\zeta_k}}{\sinh\zeta_k}\mathrm{co} \\ & \sinh[\sqrt{\epsilon_k}\eta(z - z')] - \\ - \frac{1}{\sinh\zeta_k}\cosh[\sqrt{\epsilon_k}\eta(z' + z - z_k - z_{k-1})]\Bigg\}, \end{aligned} \tag{5.2.19}$$

which can be reduced to the form

$$\gamma_k(z, z') = \frac{\sinh \zeta_k}{\eta \varepsilon_k} \{\theta(z' - z)F_k(z - z_{k-1})F_k(z_k - z') + \theta(z - z')F_k(z_k - z)F_k(z' - z_{k-1})\}. \tag{5.2.20}$$

where $\theta(z)$—step function.

Taking into account the transformation (5.2.10), we obtain the components of the electric field:

a) the longitudinal

$$E_k^{\|}(\eta, z) = -i\eta V_k(\eta, z); \tag{5.2.21}$$

b) transverse

$$E_k^{\perp}(\eta, z) = -\frac{\partial}{\partial z} V_k(\eta, z). \tag{5.2.22}$$

The electrostatic induction vectors (5.2.2) satisfy the boundary conditions

$$D_{k+1}^{\|}(\mathbf{y}, z) - D_k^{\|}(\mathbf{y}, z) = \varepsilon_0^{-1}\sigma_k(\mathbf{y}), \ k = 1, 2. \tag{5.2.23}$$

Substituting the definition of D_k, in (5.2.23) according to (5.2.2), performing Fourier transforms of the form (5.2.10) and expressing the components of the field according to (5.2.21)–(5.2.22) in terms of potentials, we obtain a system of equations for boundary potentials:

$$-\mu_k(\eta)V(\eta, z_{k-1}) + \nu_k(\eta)V(\eta, z_{k+1}) = \varepsilon_0^{-1}\tilde{\sigma}_k(\eta), \tag{5.2.24}$$

where

$$\tilde{\sigma}_k(\eta) = -P_{k+1}^{\|}(\eta, z_k) + P_k^{\|} + \int_{z_k}^{z_{k+1}} dz' F_{k+1}(z_{k+1} - z')\tilde{\rho}_{k+1}(\eta, z') + \int_{z_{k-1}}^{z_k} dz' F_k(z' - z_{k-1})\tilde{\rho}_k(\eta, z'), \tag{5.2.25}$$

and the coefficients $\mu_k(\eta)$ and $\nu_k(\eta)$ have the form

$$\mu_k(\eta) \equiv \frac{\varepsilon_k \eta}{\sinh \zeta_k}; \ \nu_k(\eta) \equiv (\varepsilon_k \coth \zeta_k + \varepsilon_{k+1} \coth \zeta_{k+1})\eta. \tag{5.2.26}$$

Considering that layers 1 and 3 are semi-infinite (l_1, $l_3 \to \infty$), we obtain a system of four equations for finding potentials on surfaces $z_{1,2}$ and the distribution of internal sources of the field $\tilde{\sigma}_k$. The formulas obtained for the potentials due to polarization serve as the basis for obtaining the frequency spectrum and finding the amplitudes of normal oscillations. Both problems are solved by composing the evolution equation for polarization and bringing these equations to normal form. For a simple system (a polar crystal plate in vacuum), this method was developed in [19]. In [18, 20–22] this method is generalized to multilayer structures, the specificity of which lies in the 'mixing' of surface vibrations of different layers.

Next, we omit the calculations (they are given in [20–22] and monograph [18]) and give only a comment on the content of these calculations.

After calculating the potentials of electric fields, we proceed to the quantum mechanical formulation of the problem. To do this, we find the potential energy of lattice vibrations using the equations of motion in the optical branch for this purpose. Then we record the total energy, and find the normal vibrations and their frequencies. Finally, the Hamiltonian is written in quantum mechanical form. To do this, operators of secondary quantization of the phonon field are introduced based on the well-known correspondence between classical and quantum mechanics.

We also emphasize some specific differences of the considered three-layer system with two polar layers when the electron is in the $k = 2$ layer. In this case, it interacts with the fields of three active surface modes and longitudinal bulk optical vibrations.

The Hamiltonian of EPI for an electron in the $k = 2$ layer can be represented as:

$$\hat{H}_{\text{e-ph}} = \hat{H}^{S}_{\text{e-ph}} + \hat{H}^{V}_{\text{e-ph}}, \tag{5.2.27}$$

where $\hat{H}^{S}_{\text{e-ph}}$—the Hamiltonian of interaction with surface optical phonons:

$$\hat{H}^{S}_{\text{e-ph}} = \sum_{s,\boldsymbol{\eta}} C(\eta) e^{i\boldsymbol{\eta}\boldsymbol{\rho}_e} L_s(\boldsymbol{\eta}, \Omega_s, z_e)[\hat{b}^{+}_{s}(-\boldsymbol{\eta}, 0) + \hat{b}_s(\boldsymbol{\eta}, 0)]; \tag{5.2.28}$$

$$C(\eta) = \frac{e\sqrt{\hbar}}{4\sqrt{\varepsilon_0 \eta L_x L_y}}; \tag{5.2.29a}$$

$$L_s(\boldsymbol{\eta}, \Omega_s, z_e) = \frac{L_{sc}\cosh W_2 + L_{ss}\sinh W_2}{\sinh W_2}; \quad W_2 = \sqrt{\epsilon_2}\,\eta\left(z_e - \frac{l_2}{2}\right). \tag{5.2.29b}$$

The s index numbers the surface modes. The frequencies of surface vibrations Ω_1, Ω_2, Ω_3 are found from the equation

$$\Omega^6 - \bar{a}_1\Omega^4 + \bar{a}_2\Omega^2 - \bar{a}_3 = 0; \tag{5.2.30}$$

$\hat{H}^{V}_{\text{e-ph}}$—the Hamiltonian of interaction with bulk optical vibrations:

$$\hat{H}^{V}_{\text{e-ph}} = \sum_{\mathbf{q}(q_z \neq 0)} C_q e^{i\mathbf{q}_\perp \boldsymbol{\rho}_e} g_{nV}(q_z, q_\perp, l, z_e)\left[\hat{b}^{+}_{-\mathbf{q}} + \hat{b}_{\mathbf{q}}\right]; \tag{5.2.31a}$$

$$|C_q|^2 = \frac{1}{L_x L_y l} \cdot \frac{4\pi\alpha_V(\hbar\omega_0)^2}{\beta_V q^2}; \tag{5.2.31b}$$

$$\alpha_V = \frac{e^2}{4\pi\varepsilon_0\hbar}\left(\frac{1}{\varepsilon_2} - \frac{1}{\varepsilon_{20}}\right)\left(\frac{m_e^*}{2\hbar\omega_0}\right)^{\frac{1}{2}};\ \beta_V = \left(\frac{2m_e^*\omega_0}{\hbar}\right)^{\frac{1}{2}}; \tag{5.2.31c}$$

$$C = \left[\left(\frac{2}{q_\perp}\right)\mathrm{th}\left(\frac{q_\perp l}{2}\right)\right]^{\frac{1}{2}};\ q^2 = q_\perp^2 + q_z^2;\ q_z = \frac{2\pi N}{l};\ N = 0, \pm 1, \ldots; \tag{5.2.31d}$$

$$g_{nV}(q_z, q_\perp, l, z) = e^{iq_z z} + \frac{1}{1 - C}\left(C - \frac{\cosh q_\perp\left(\frac{l}{2} - z\right)}{\cosh\frac{q_\perp l}{2}}\right). \tag{5.2.31e}$$

The last term in the right part (5.2.31e) describes the effect of attenuation of the interaction of an electron at the boundary of a polar plate with bulk optical modes. It follows from (5.2.31e) that $g_{nv}(z = 0, l) = 0$. In [7], based on the formalism of the dielectric function, another expression for $\hat{H}^V_{\text{e-ph}}$:

$$\hat{H}^V_{\text{e-ph}} = \sum_{\mathbf{q}_\perp}\left\{C_V e^{i\mathbf{q}_\perp \mathbf{\rho}_e}\left[\sum_{m=1,3,5,\ldots}^{\frac{N}{2}} \frac{\cos\left[\frac{m\pi}{2l}z\right]}{\left[q_\perp^2 + \left(\frac{m\pi}{2l}\right)^2\right]^{\frac{1}{2}}}\hat{b}^+_{m^+}(\mathbf{q}_\perp) + \sum_{m=2,4,6,\ldots}^{\frac{N}{2}} \frac{\sin\left[\frac{m\pi}{2l}z\right]}{\left[q_\perp^2 + \left(\frac{m\pi}{2l}\right)^2\right]^{\frac{1}{2}}}\hat{b}_m(-\mathbf{q}_\perp)\right]\right\}, \tag{5.2.32}$$

which was used to describe polaronic effects in semi-infinite crystals, at the contact of two polar crystals and a plate of a polar crystal of finite thickness, in a quantum well. As will be shown below, the contribution to the energy of the polaron calculated on the Hamiltonian (5.2.32) is greater than the corresponding contribution on the Hamiltonian (5.2.31a–e) by about 1%–3%. At the same time, in the problems of scattering of charge carriers in thin films, Raman scattering of light, etc, the difference can be very significant. In the polaronic problem for a bounded crystal, the use of the Hamiltonian (5.2.32) is of considerable convenience.

The Hamiltonian (5.2.27)–(5.2.31) contains all known special cases of Hamiltonians that are used in solving the polaron problem in dimensionally limited systems of the following types: a semi-infinite polar crystal, a contact of polar and nonpolar crystals, a contact of two polar crystals, and a plate of a polar crystal of finite thickness in symmetrical and asymmetric plates. Here we present those of them that were used to discuss the problem of surface polariton in the works of Pokatilov, Beril [10–17].

5.2.2 Contact 'polar–nonpolar crystals' (and in a special case–contact of a polar crystal with a vacuum)

The Hamiltonian of EPI for this particular case is obtained from the general expression (5.2.27)–(5.2.31) if we put in it:

$$l_2 \to \infty, \;\; \varepsilon_{20} = \varepsilon_2 \equiv \varepsilon \text{ (in the vacuum } \varepsilon = 1); \tag{5.2.33}$$

(a) for an electron in a polar crystal ($z > 0$)

$$\hat{H}^{S}_{\text{e-ph}} = \sum_{\boldsymbol{\eta}} C_{\eta} e^{i\boldsymbol{\eta}\boldsymbol{\rho}_e} e^{-\eta z_e} \left[\hat{b}^{+}_{-\boldsymbol{\eta}} + \hat{b}_{\boldsymbol{\eta}} \right]; \tag{5.2.34}$$

$$\hat{H}^{V}_{\text{e-ph}} = \sum_{\mathbf{q}(q_z \neq 0)} C_q e^{i\mathbf{q}_\perp \boldsymbol{\rho}_e} [e^{iq_z z} - e^{-q_\perp z}] \left[\hat{b}^{+}_{-\boldsymbol{\eta}} + \hat{b}_{\boldsymbol{\eta}} \right]. \tag{5.2.35}$$

Here

$$C_\eta = \frac{e\omega_1}{\varepsilon_1 + \varepsilon_2} \sqrt{\frac{\hbar(\varepsilon_{10} - \varepsilon_1)}{2\varepsilon_0 L_x L_y \eta \Omega_1}}. \tag{5.2.36a}$$

$$C_q = \frac{e}{q} \sqrt{\frac{\hbar\omega_{10}(\varepsilon_{10} - \varepsilon_1)}{2\varepsilon_0 \varepsilon_{10} L_x L_y L_z}}. \tag{5.2.36b}$$

The polaron constants of interactions with surface and bulk optical phonons are respectively equal:

$$\alpha_S = \frac{e^2(\varepsilon_{10} - \varepsilon_1)}{4\pi\varepsilon_0(\varepsilon_{10} + \varepsilon_2)(\varepsilon_1 + \varepsilon_2)\hbar\Omega_1 R_S}; \; R_S = \left(\frac{\hbar}{2m^* \Omega_1} \right)^{\frac{1}{2}}; \tag{5.2.37}$$

$$\alpha_V = \frac{e^2(\varepsilon_{10} - \varepsilon_1)}{8\pi\varepsilon_0 \varepsilon_{10} \varepsilon_1 \hbar\omega_1 R_V}; \; R_V = \left(\frac{\hbar}{2m^* \omega_{10}} \right)^{\frac{1}{2}}; \tag{5.2.38}$$

$$\Omega_1 = \omega_1 \left(\frac{\varepsilon_{10} + \varepsilon_2}{\varepsilon_1 + \varepsilon_2} \right)^{\frac{1}{2}}; \tag{5.2.39}$$

(b) for an electron in a polar crystal ($z<0$). In this case, the electron interacts only with surface optical vibrations (external polaron). The Hamiltonian EFB differs from (5.2.34) only by replacing $x \to -z$.

5.2.3 Contact of two polar crystals

$$l_2 \to \infty \ (\varepsilon_3 = 1) \tag{5.2.40}$$

(the electron is in the medium $k = 2$, the origin of the coordinate system is placed on the boundary of the layers). In this case, the electron interacts with a bulk longitudinal and two surface optical modes. $\hat{H}^V_{\text{e-ph}}$ remains the same as in the previous case (see (5.2.35)). $\hat{H}^S_{\text{e-ph}}$ is obtained from the general expression (5.2.28)–(5.2.30) in the limit (5.2.40):

$$\hat{H}^S_{\text{e-ph}} = \sum_{s,\boldsymbol{\eta}} V_s(\eta) e^{i\boldsymbol{\eta}\boldsymbol{\rho}_e} e^{-\eta z_e} \left[\hat{b}^+_{s,\,-\boldsymbol{\eta}} + \hat{b}_{s,\,\boldsymbol{\eta}} \right]; \tag{5.2.41}$$

s is the number of the surface mode.

$$V_1(\eta) = -\frac{e\omega_1\sqrt{\hbar}}{\sqrt{2\varepsilon_0 \eta L_x L_y}} \cdot \frac{(-a_1 + a_2 F)}{(1 + F^2)^{\frac{1}{2}}(\varepsilon_1 + \varepsilon_2)\Omega_1^{\frac{1}{2}}}; \tag{5.2.42}$$

$$V_2(\eta) = -\frac{e\omega_1\sqrt{\hbar}}{\sqrt{2\varepsilon_0 \eta L_x L_y}} \cdot \frac{(a_1 A + a_2)}{(1 + F^2)^{\frac{1}{2}}(\varepsilon_1 + \varepsilon_2)\Omega_2^{\frac{1}{2}}}. \tag{5.2.43}$$

(Formulas (5.2.41)–(5.2.43) are written in the isotropic limit.)

The polaron constants for surface modes have the form

$$\alpha_{S_1} = \frac{e^2(a_1 - a_2 F)^2}{4\pi\varepsilon_0 \hbar\Omega_1 R_1 (\varepsilon_1 + \varepsilon_2)^2 \theta_1^2 (1 + F^2)}; \tag{5.2.44}$$

$$\alpha_{S_2} = \frac{e^2(a_1 F + a_2)^2}{4\pi\varepsilon_0 \hbar\Omega_2 R_2 (\varepsilon_1 + \varepsilon_2)^2 \theta_2^2 (1 + F^2)}, \tag{5.2.45}$$

where

$$a_1 = (\varepsilon_{10} - \varepsilon_1)^{\frac{1}{2}}; \ a_2 = \nu(\varepsilon_{20} - \varepsilon_2)^{\frac{1}{2}}; \ R_{1,\,2} = \left(\frac{\hbar}{2m^*\Omega_{1,\,2}} \right)^{\frac{1}{2}}; \tag{5.2.46}$$

$$\nu^2 = \frac{\omega_2^2}{\omega_1^2}; \ \theta_{10}^2 = \frac{\varepsilon_{10} + \varepsilon_2}{\varepsilon_1 + \varepsilon_2}; \ \theta_{20}^2 = \nu^2 \frac{\varepsilon_{20} + \varepsilon_1}{\varepsilon_1 + \varepsilon_2}; \tag{5.2.47}$$

$$F \equiv -F_{21}(\Omega_1) = \frac{\nu\omega_1^2(\varepsilon_{10} - \varepsilon_1)^{\frac{1}{2}}(\varepsilon_{20} - \varepsilon_2)^{\frac{1}{2}}}{(\varepsilon_1 + \varepsilon_2)\left(\Omega_1^2 - \Omega_{20}^2\right)}, \tag{5.2.48}$$

and the frequencies of surface vibrations are found from the equation

$$\left(\Omega^2 - \Omega_{10}^2\right)\left(\Omega^2 - \Omega_{20}^2\right) - \omega_1^2\omega_2^2(\varepsilon_{10} - \varepsilon_1)^{\frac{1}{2}}(\varepsilon_{20} - \varepsilon_2)^{\frac{1}{2}}(\varepsilon_1 + \varepsilon_2)^{-2} = 0, \tag{5.2.49}$$

arising from equation (5.2.30) and having two solutions:

$$\Omega_{1,2}^2 = \frac{\omega_1^2}{2}\left[\theta_{10}^2 + \theta_{20}^2\right] \pm \frac{1}{2}\left\{\left[\theta_{10}^2 - \theta_{20}^2\right]^2 + \frac{4\nu^2(\varepsilon_{10} - \varepsilon_1)(\varepsilon_{20} - \varepsilon_2)}{(\varepsilon_1 + \varepsilon_1)^2}\right\}^{\frac{1}{2}} \equiv \omega_1^2\theta_{1,2}; \quad (5.2.50)$$

here

$$\Omega_{10}^2 = \omega_1^2\frac{(\varepsilon_{10} + \varepsilon_2)}{\varepsilon_1 + \varepsilon_2}; \quad \Omega_{10}^2 = \omega_1^2\frac{(\varepsilon_{10} + \varepsilon_2)}{\varepsilon_1 + \varepsilon_2}. \quad (5.2.51)$$

5.2.4 A plate of a polar crystal of finite thickness in asymmetric plates

In this case, we have $\varepsilon_{10} = \varepsilon_1;\ \varepsilon$. The electron in the $k = 2$ layer will interact with two surface optical modes. $\hat{H}^S_{\text{e-ph}}$, in this case, has the form

$$\hat{H}^S_{\text{e-ph}} = \hat{H}^{S_1}_{\text{e-ph}} + \hat{H}^{S_2}_{\text{e-ph}}, \quad (5.2.52)$$

where

$$\hat{H}^{S_{1,2}}_{\text{e-ph}} = \sum_{s,\eta}\left\{C_{S_1}\cosh\left[\eta\left(z_e - \frac{l}{2}\right)\right] + C_{S_2}\sinh\left[\eta\left(z_e - \frac{l}{2}\right)\right]\right\}e^{i\eta\rho_e}\left[\hat{b}^+_{s,-\eta} + \hat{b}_{s,\eta}\right]; \quad (5.2.53)$$

with interaction constants

$$|C_{S_1}(\eta)|^2 = \frac{1}{L_xL_y}\cdot\frac{2\pi\alpha_{S_1}(\hbar\Omega_{S_1})^2}{\beta_{S_1}\eta}F_{1,2}^2(\eta); \quad (5.2.54)$$

$$|C_{S_2}(\eta)|^2 = \frac{1}{L_xL_y}\cdot\frac{2\pi\alpha_{S_2}(\hbar\Omega_{S_2})^2}{\beta_{S_2}\eta}F_{3,4}^2(\eta); \quad (5.2.55)$$

and

$$\alpha_{S_{1,2}} = \frac{e^2}{4\pi\varepsilon_0\hbar}\left(\frac{1}{\tilde{\varepsilon}_2} - \frac{1}{\tilde{\varepsilon}_{20}^{1,2}}\right)\left(\frac{m_e^*}{2\hbar\Omega_{S_{1,2}}}\right)^{\frac{1}{2}}; \quad \beta_{S_{1,2}} = R^{-1}_{S_{1,2}}. \quad (5.2.56)$$

A special case of this system is a three-layer symmetric system in which the $k = 2$ layer is polar. The Hamiltonian of EPI in this system can be obtained from (5.2.52)–(5.2.55) if we put $\varepsilon_1 = \varepsilon_3 \equiv \varepsilon$:

$$\hat{H}^{S_1}_{\text{e-ph}} = \sum_{\eta}C_{S_1}(\eta)e^{i\eta\rho_e}\frac{\cosh\left[\eta\left(z_e - \frac{l}{2}\right)\right]}{\cosh\frac{\eta l}{2}}\left[\hat{b}^+_{S_1,-\eta} + \hat{b}_{S_1,\eta}\right]; \quad (5.2.57)$$

$$\hat{H}^{S_2}_{\text{e-ph}} = \sum_{\eta}C_{S_2}(\eta)e^{i\eta\rho_e}\frac{\sinh\left[\eta\left(z_e - \frac{l}{2}\right)\right]}{\cosh\frac{\eta l}{2}}\left[\hat{b}^+_{S_2,-\eta} + \hat{b}_{S_2,\eta}\right], \quad (5.2.58)$$

where the squares of the modules of the interaction constants with surface optical phonons have the form

$$\left|C_{S_{1,2}}(\eta)\right|^2 = \frac{1}{L_x L_y} \cdot \frac{2\pi\alpha_{S_{1,2}}(\hbar\Omega_{S_{1,2}})^2}{\beta_{S_{1,2}}\eta \tanh\left(\frac{\eta l}{2}\right)}; \tag{5.2.59}$$

$$\alpha_{S_{1,2}} = \frac{e^2}{4\pi\varepsilon_0\hbar}\left\{\left[\varepsilon_2 + \varepsilon\begin{pmatrix}\tanh\dfrac{\eta l}{2}\\ \coth\dfrac{\eta l}{2}\end{pmatrix}\right]^{-1} - \left[\varepsilon_{20} + \varepsilon\begin{pmatrix}\tanh\dfrac{\eta l}{2}\\ \coth\dfrac{\eta l}{2}\end{pmatrix}\right]^{-1}\right\} R_{S_{1,2}}; \tag{5.2.60}$$

$$\Omega^2_{S_{1,2}} = \omega_2^2\left[\varepsilon_{20} + \varepsilon\begin{pmatrix}\tanh\dfrac{\eta l}{2}\\ \coth\dfrac{\eta l}{2}\end{pmatrix}\right]\left[\varepsilon_2 + \varepsilon\begin{pmatrix}\tanh\dfrac{\eta l}{2}\\ \coth\dfrac{\eta l}{2}\end{pmatrix}\right]^{-1}. \tag{5.2.61}$$

5.2.5 Hamiltonian of EPI in the case when the layer $k = 1$ is polar, and the layers $k = 2$ and $k = 3$ are nonpolar (the electron is in the layer $k = 3$)

This Hamiltonian is used in the levitating polaron problem and is obtained from the general expressions (5.2.27)–(5.2.31) if we put $\varepsilon_{20} = \varepsilon_2$. Being in the domain $k = 3$, the electron interacts with only one surface optical mode:

$$\hat{H}^S_{e-ph} = \sum_{\eta} C_S(\eta) e^{i\eta\rho_e} e^{-\eta(z_e - l_2)}\left[\hat{b}^+_{s,-\eta} + \hat{b}_{s,\eta}\right];\; z \geqslant l_2; \tag{5.2.62}$$

here

$$|C_S(\eta)|^2 = \frac{1}{L_x L_y} \cdot \frac{4\pi\alpha_S(\hbar\Omega_S)\overline{F}^2}{\beta_S\eta}; \tag{5.2.63}$$

$$\begin{aligned}\alpha_S = \frac{e^2 R_S^{-1}}{8\pi\varepsilon_0\hbar\Omega_S}\Bigg\{&\frac{1}{\varepsilon_2^2 + \varepsilon_2(\varepsilon_1 + \varepsilon_3)\coth\eta l_2 + \varepsilon_1\varepsilon_3}\\ &- \frac{1}{\varepsilon_2^2 + \varepsilon_2(\varepsilon_{10} + \varepsilon_3)\coth\eta l_2 + \varepsilon_{10}\varepsilon_3}\Bigg\};\end{aligned} \tag{5.2.64}$$

$$\begin{aligned}\Omega_S^2 &= \omega_1^2\left[\frac{\varepsilon_2^2 + \varepsilon_2(\varepsilon_{10} + \varepsilon_3)\coth\eta l_2 + \varepsilon_{10}\varepsilon_3}{\varepsilon_2^2 + \varepsilon_2(\varepsilon_1 + \varepsilon_3)\coth\eta l_2 + \varepsilon_1\varepsilon_3}\right];\\ \overline{F}^2 &= \frac{\varepsilon_2^2}{(\varepsilon_3 + \varepsilon_3\coth\eta l_2)\sinh^2\eta l_2}.\end{aligned} \tag{5.2.65}$$

If the media are $k = 2$ and $k = 3$, respectively, a liquid helium film ($\varepsilon_2 \approx 1$) and ($\varepsilon_3 = 1$), then from (5.2.62)–(5.2.65) we get

$$\hat{H}^S_{\text{e-ph}} = \sum_{\boldsymbol{\eta}} C_S(\eta) e^{i\boldsymbol{\eta}\boldsymbol{\rho}_e} e^{-\eta z_e} \left[\hat{b}^+_{S,-\boldsymbol{\eta}} + \hat{b}_{S,\boldsymbol{\eta}} \right], \tag{5.2.66}$$

where

$$|C_S(\eta)|^2 = \frac{1}{L_x L_y} \cdot \frac{4\pi\alpha_S(\hbar\Omega_S)^2}{\beta_S \eta}; \tag{5.2.67}$$

$$\alpha_S = \frac{e^2}{8\pi\varepsilon_0\hbar\Omega_S R_S}\left(\frac{1}{\varepsilon_1 + 1} - \frac{1}{\varepsilon_{10} + 1}\right); \; \Omega_S^2 = \omega_1^2\left(\frac{\varepsilon_{10} + 1}{\varepsilon_1 + 1}\right). \tag{5.2.68}$$

It follows from (5.2.66)–(5.2.68) that the helium layer has a very weak effect on the interaction of the electron with the surface optical phonons of the polar substrate, and $\hat{H}^S_{\text{e-ph}}$ has the same appearance as for the polar crystal–vacuum contact.

5.2.6 Metal–dielectric–semiconductor structure

Finally, let's look at another relevant system: A metal–dielectric–semiconductor (MDS) is a structure in which the dielectric layer is polar. $\hat{H}^S_{\text{e-ph}}$, for an electron located in a semiconductor, can be obtained from the expressions (5.2.52)–(5.2.56) by putting $\varepsilon_1 \to \infty$:

$$\hat{H}^S_{\text{e-ph}} = \sum_{\boldsymbol{\eta}} C_S(\eta) e^{i\boldsymbol{\eta}\boldsymbol{\rho}_e} e^{-\eta(z_e - l)} \left[\hat{b}^+_{S,-\boldsymbol{\eta}} + \hat{b}_{S,\boldsymbol{\eta}} \right], \tag{5.2.69}$$

where

$$|C_S(\eta)|^2 = \frac{1}{L_x L_y} \cdot \frac{4\pi\alpha_S(\hbar\Omega_S)^2\left(1 + \tanh^2\frac{\eta l_2}{2}\right)}{2\beta_S \eta}; \tag{5.2.70}$$

$$\alpha_S = \frac{e^2}{8\pi\varepsilon_0\hbar\Omega_S R_S}\left(\frac{1}{\varepsilon_3 + \varepsilon_2 \tanh\frac{\eta l_2}{2}} - \frac{1}{\varepsilon_3 + \varepsilon_{20}\tanh\frac{\eta l_2}{2}}\right); \tag{5.2.71}$$

$$\Omega_S^2 = \omega_2^2\left(\varepsilon_{20} + \varepsilon_3 \coth\frac{\eta l_2}{2}\right)\left(\varepsilon_2 + \varepsilon_3 \coth\frac{\eta l_2}{2}\right)^{-1}. \tag{5.2.72}$$

In this particular case, as can be seen from (5.2.69)–(5.2.72), the metal substrate completely shields the interaction with one of the surface modes, and the interaction with the second surface mode weakens (is shielded by the metal plasma) as it decreases l_2: $\lim\limits_{\eta l_2 \to 0} \alpha_S = 0$.

All the above Hamiltonians of the electron–phonon interaction satisfy the limiting transitions—(a) to an infinite crystal—the Pekar–Fröhlich Hamiltonian; (b) the contact of a semi-infinite crystal with a vacuum—the Hamiltonian of Sack, Evans, Mills; (c) the contact of a polar crystal with a nonpolar one—the Hamiltonian of Beril, Pokatilov, Fomin; (d) the plate of a polar crystal in vacuum—Hamiltonian of Bryksin, Firsova; (e) the plate of a polar crystal in the plates of nonpolar crystals—the Beril and Pokatilov Hamiltonian—and allow us to construct a closed theory of surface polaronic states in these spatially inhomogeneous systems.

As already noted, the Hamiltonian of the electron–polarization interaction in the systems under consideration can be obtained on the basis of another approach—the method of the dielectric function, which consists of the fact that in equations (5.2.1) and (5.2.2) and in boundary conditions, the material equation is used for the induction vector

$$\mathbf{P}_n = \varepsilon_0(\varepsilon_n(\omega) - \varepsilon_n)\mathbf{E}, \tag{5.2.73}$$

where $\varepsilon_n = \varepsilon_n(\omega)|_{\omega\to\infty}$, by which polarization is excluded from them. Due to this reduction of the problem, secular equations for frequencies are easily derived, but the possibility of obtaining the amplitudes of normal oscillations is completely lost, without which it is impossible to consistently derive the constants of electron–polarization interactions. In one of the most general works performed by the dielectric function method [23], the spectral part of the problem for various multilayer systems was studied quite fully; however, due to these difficulties, the Hamiltonians of electron–polarization interactions were not obtained.

In the work [13], the method of limit ratios for the derivation of EPI constants in the dielectric function method was developed. However, as shown in the subsequent work [17], it gives correct results only when the structure contains no more than one polar layer. Otherwise, the interaction of the surface modes of different layers leads to other frequency dependencies of the electron–phonon interaction constants.

5.3 Hamiltonian of the interaction of an electron with a plasma of valence electrons

When deriving the equation for the potential of a spatially inhomogeneous system, we attributed the plasma of valence electrons to a fast subsystem, the state of which follows the external field inertially, and on this basis excluded the polarization vector of valence electrons from the definition of the induction vector, writing the latter in the form (5.2.2). In essence, this means using a purely classical description of the shielding effect created by the plasma of valence electrons. Since the plasma frequencies of valence electrons are of the order of $10^{15} - 10^{16}\, c^{-1}$, then this approximation should be considered reasonable for many tasks. For example, when considering the motion of an electron interacting with a plasma, we can assume that the polaron state caused by this interaction (in Toyozawa's terminology [24]—'electron polaron') practically does not depend on the nature of the motion (whether it is movement through the bulk of the crystal or orbiting in a shallow impurity center, etc) or changes in the state of the electron (scattering, transitions between impurity centers, etc), and manifests itself only in the renormalization of the mass and shielding by means of high-frequency dielectric permittivity of the interaction potential with other charges. The situation, however, changes significantly in spatially heterogeneous systems. When an electron polaron approaches the boundary, the 'inaccuracy' of the polaron state manifests itself, which significantly depends on the distance to the boundary. Therefore, a quantum mechanical description of the system using the Hamiltonian of the electron–phonon interaction is necessary. Let us briefly trace the main stages of the derivation of such a Hamiltonian.

The state of the plasma of valence electrons is described by electrostatic equations (5.2.1). The material equation in the case under consideration is as follows:

$$\mathbf{D}_n = \varepsilon_0 \mathbf{E}_n + \mathbf{P}_n; \tag{5.3.1}$$

where $\mathbf{P}_n$—polarization of the valence electron system.

Considering the bulk charges to be absent, the equation for the potential in the nth medium according to (5.2.1) and (5.2.5) is obtained as

$$\text{div grad } V_n(\mathbf{r}) = -\varepsilon_0^{-1} \text{div } \mathbf{P}_n, \tag{5.3.2}$$

where $V_n(\mathbf{r})$—potential in the nth environment; $\mathbf{E}_n = -\text{grad } V_n(\mathbf{r})$.

Assuming the elastic compression and shear modulus of valence electrons to be equal, we write down the system of the polarization evolution equation:

$$\ddot{\mathbf{P}}_n = -\Delta^2 \mathbf{P}_n + \omega_{np}^2 \varepsilon_0 \mathbf{E}_n^P, \quad n = 1,\ 2,\ 3; \tag{5.3.3}$$

where Δ—frequency of the returning elastic force, $\omega_{np} = e^2 N/(m_n^* \varepsilon_0)$—plasma frequency of valence electrons, and $\mathbf{E}_n^P$—the intensity of the electric field induced by polarization. The method of solving equations (5.3.2), (5.3.3) is described in [18]. It consists of calculating the potential $V_n(\mathbf{r})$ from equation (5.3.2), finding the field $\mathbf{E}_n^P$, solving the set of equations (5.3.3), and obtaining normal frequencies and normal amplitudes. This problem is mathematically identical to the problem of inertial polarization, if at the same time substitutions are performed:

$$\varepsilon_k \Rightarrow 1; \ \omega_{TO}^2 \Rightarrow \Delta^2; \ \omega_{TO}^2(\varepsilon_{k0} - \varepsilon_k) \Rightarrow \omega_p^2. \tag{5.3.4}$$

Thus, if in Hamiltonians $\hat{H}_{\text{e-ph}}^S$ and $\hat{H}_{\text{e-ph}}^V$ to replace (5.3.4), we obtain the Hamiltonians of the electron–plasmon interaction (on valence electrons), which allow us to correctly solve the problem of isolating a seed quasi-particle—an electron polaron—when discussing the problem of an optical polaron in spatially inhomogeneous systems.

5.4 Theory of image potential and strength. The dielectric function of a quantum dielectric

When using the quantum mechanical method of describing the interaction of an electron with the polarization of a medium, an electron turns into a polaron—a quasi-particle having finite dimensions determined by its radius [24–26, 124]. If the radius of the polaron's orbit in external fields is greater than the radius of the polaron itself (while the step of quantization of the energy of the state in the external field, for example, the impurity field, is much less than the corresponding energy of the internal state of the polaron), the internal structure of the quasi-particle-polaron practically does not manifest itself during its movement in the bulk. When the polaron collides with the interface. the situation is changing significantly. The polaron is deformed (the radii in different layers differ), a strong disturbance of the internal state leads to quantum transitions with absorption or emission of a quantum of the polarization field. Consequently, when the polaron approaches the interface of the media by a distance of the order of its radius, the classical description completely loses its force. For a consistent continuous description of a large-radius polaron, it is necessary to use a medium model in which the discreteness is taken into account by using a dispersing dielectric constant. Let's consider the interaction of an electron with plasma oscillations of valence electrons (Toyozawa electron polaron [24–26]). We use the dielectric function $\varepsilon(q, \omega)$ in the form given in [25, 26]:

$$\varepsilon(q, \omega) = 1 + \frac{\omega_p^2}{\Delta^2(q) - \omega^2 + i\omega\,\Gamma}, \tag{5.4.1}$$

where ω_p

$$\omega_p^2 = \frac{e^2 N_n}{m_n \varepsilon_0 \varepsilon_n} \tag{5.4.2}$$

is the plasma frequency of valence electrons.

The physical meaning of the remaining parameters follows from the general properties of the dielectric function $\varepsilon(q, \omega)$. At the limit $\Gamma \to 0$ the imaginary part $\varepsilon(q, \omega)$, proportional to the absorption coefficient, has the form $\varepsilon'' \sim [\Delta(q) - \omega]^{-1}$, which means that $\Delta(q)$—the resonant frequency of the radiation absorbed by the substance, the minimum value of which determines the width of the band gap $E_g = \hbar\Delta(0)$. The natural oscillation frequency is determined from the condition $\varepsilon(q, \omega) = 0$, which is when $\Gamma = 0$ gives

$$\omega_{pV}^2 = \omega_p^2 + \Delta^2(q). \tag{5.4.3}$$

In the low-frequency limit ($\omega \to 0$) the formula (5.4.3) follows:

$$\varepsilon(q, 0) = \varepsilon(q) = 1 + \frac{\omega_p^2}{\Delta^2(q)}, \tag{5.4.4}$$

that is, the static dielectric constant $\varepsilon(q)$. Excluding $\Delta^2(q)$ from (5.4.3) and ω_p^2 from (5.4.1) (for $\Gamma = 0$), we find:

$$\omega_{pV}^2 = \frac{\omega_p^2 \varepsilon(q)}{\varepsilon(q) - 1}; \tag{5.4.5}$$

$$\varepsilon(q, \omega) = 1 + \frac{(\varepsilon(q) - 1)}{\left(1 - \frac{\omega^2}{\Delta^2(q)}\right)}. \quad (5.4.6)$$

The frequency of longitudinal plasma oscillations of valence electrons ω_{pV} is greater than ω_p, because in addition to the electric return force in the crystal, the electrons are affected by the elastic coupling force with the lattice. It follows from the sum rule for $\varepsilon(q, \omega)$ [25, 26] that the dependence Δ(q) must be chosen in the form

$$\Delta(q) = \Delta(0) + \frac{\hbar^2 q^2}{2m_0}. \quad (5.4.7)$$

It follows that at high frequencies $\omega \gg \Delta(0)$ the energy spectrum of the system acquires a quasi-partial form: $\hbar\omega \sim \hbar^2 q^2/(2m_0)$ where m_0—the mass of an electron in a vacuum.

The spectrum of surface polaritons at the contact of a substance with a vacuum ($n = 2$) is determined from the condition

$$\varepsilon_1(q, \omega) = -1, \quad (5.4.8)$$

which, for $q = 0$ and $\Gamma = 0$, together with (5.4.1) gives:

$$\Omega_{1p}^2 = \frac{\varepsilon_1 + 1}{2(\varepsilon_1 - 1)}\omega_p^2 = \frac{\varepsilon_1 + 1}{2}\Delta(0), \quad (5.4.9)$$

where $\varepsilon_1(0) \equiv \varepsilon_1$. (Putting in the formula for the frequency of surface optical vibrations at the contact of a polar crystal with a nonpolar one

$$\Omega_1^2 = \frac{\omega_1^2(\varepsilon_{10} + 1)}{\varepsilon_1 + 1}; \ \varepsilon_1 = \varepsilon_2 = 1; \ \varepsilon_{10} \to \varepsilon_1,$$

we are convinced of the analogy of the parameters $\omega_{TO} \equiv \omega_1$ and Δ, which was mentioned in the previous paragraph.)

5.4.1 Electronic polaron at the contact of two media

The Hamiltonian of the electron–plasmon interaction for the considered case of a homeopolar crystal with a vacuum is obtained from the Hamiltonian (5.2.40)–(5.2.51) if substitutions are made in it (5.3.4). As a result, we get

$$\hat{H}_{n-S} = \sum_{\eta} V_S(\eta) e^{i\eta\rho_e} g_{nS}(\eta, z)\left[\hat{b}^{+}_{s,-\eta} + \hat{b}_{s,\eta}\right], \quad (5.4.10)$$

where

$$V_S(\eta) = \frac{e}{\sqrt{\eta}}\sqrt{\frac{\hbar\Omega_{1p}(\varepsilon_1 - 1)}{\varepsilon_0(\varepsilon_1 + 1)L_x L_y}}; \ g_{nS}(\eta, z) = e^{\pm\eta z}; \ n = 1, \ 2; \quad (5.4.11)$$

whereΩ_{1p}—the surface plasma frequency at the interface of the medium ($n = 1$) with vacuum ($n = 2$). The '+' sign in the exponent refers to $n = 1$, and the '−' sign refers to $n = 2$. The origin of the coordinates is placed at the interface. (This Hamiltonian was first obtained in the work of Beril, Pokatilov [11].)

The constant of the electron–plasmon interaction has the form

$$\alpha_{pS} = \frac{e^2(\varepsilon_1 - 1)}{8\pi\varepsilon_0(\varepsilon_1 + 1)\hbar\Omega_{1p}R_{1pS}}; \quad R_{1pS} = \left(\frac{\hbar}{2m^*\Omega_{1pS}}\right)^{\frac{1}{2}}. \tag{5.4.12}$$

Let the electron be in a vacuum ($n = 2$). Then it will interact only with surface plasma vibrations of valence electrons. Let's calculate the potential energy of the electron. Since the α_{pS} constant is small in the actual domain (on the order of 0.01–0.1), perturbation theory can be used to calculate the energy. The frequency of the electron along the z-axis due to polaronic effects is much lower than the plasma frequencies of valence electrons (on the order of $10^{15} - 10^{16}\ \mathrm{s}^{-1}$); therefore, the z-coordinate of the electron is considered fixed, and the movement of the electron along the surface plane is described by a plane wave due to translational symmetry. In this approximation, according to the standard theory of second-order perturbations of the electron–plasmon interaction, the potential energy of interaction with the surface mode is in the form [11]

$$U_{pS}(z, \mathbf{P}_\perp) = -\sum_\eta \frac{|V_S(\eta)|^2\, e^{-2\eta z}}{\hbar\Omega_{1p} + \frac{\hbar^2\eta^2}{2m^*} - \frac{\hbar\eta\mathbf{P}_\perp}{m^*}}, \tag{5.4.13}$$

where $\mathbf{P}_\perp$—the component of the electron momentum in the plane of the surface. The η—dependent terms in the denominator on the right side (5.4.13)—describe the recoil effect experienced by an electron during scattering on a plasmon. This is one of the factors that determine the convergence of the sum (5.4.13) at large η. The second one is related to the dependence of ε on η. Moving in (5.4.13) from summation to integration by η, we obtain:

$$U_{pS}(z) = -\int_0^\infty \frac{2\alpha_{pS}\hbar\Omega_{1p}R_{pS}e^{-2\eta z}d\eta}{1 + R_{pS}^2\eta^2}. \tag{5.4.14}$$

To establish the main mechanism for ensuring convergence at $z \to 0$, the integral in (5.4.14) can be calculated in two ways: taking into account the return, but neglecting the variance (i.e., considering Ω_{1p}, α_{pS}, R_{pS} constant), and neglecting the return (i.e., discarding $\eta^2 R_{pS}^2$, but taking into account the variance $\varepsilon(\eta)$) according to the formulas (5.4.4) and (5.4.6).

As a result of comparing the integration results, the criteria for circumcision according to the impact effect are obtained:

$$\frac{2^{\frac{3}{2}}m^*}{m_0(\varepsilon_1 + 1)^{\frac{1}{2}}} < 1, \ n = 1; \tag{5.4.15a}$$

$$\frac{2^{\frac{3}{2}}}{(\varepsilon_1 + 1)^{\frac{1}{2}}} < 1, \; n = 2; \tag{5.4.15b}$$

according to which recoil plays a major role for small m^* and large ε_1. Since our main interest is related to the electron–polarization interaction, we will consider the criteria (5.4.15a, b) fulfilled and take into account only the recoil effect. In this approximation, we have a dispersion-free case in which the potential energy is determined by the formula

$$\begin{aligned} U_{pS}(z) &= -2\alpha_{pS}\hbar\Omega_{1p}R_{pS}\int_0^\infty \frac{e^{-2\eta z}d\eta}{1 + R_{pS}^2\eta^2} \\ &= -2\alpha_{pS}\hbar\Omega_{1p}\left\{\mathrm{ci}\left(\frac{2z}{R_{pS}}\right)\sin\left(\frac{2z}{R_{pS}}\right) - \mathrm{si}\left(\frac{2z}{R_{pS}}\right)\cos\left(\frac{2z}{R_{pS}}\right)\right\} \\ &\equiv -2\alpha_{pS}\hbar\Omega_{1p}S\left(\frac{2z}{R_{pS}}\right), \end{aligned} \tag{5.4.16}$$

where si (x), ci (x)—the integral sine and cosine, respectively.

In the limit of large z ($z \gg R_{pS}$) from (5.4.16) we find:

$$U_{pS}(z) = -\frac{e^2(\varepsilon_1 - 1)}{16\pi\varepsilon_0(\varepsilon_1 + 1)z} \equiv -\frac{ee^*}{8\pi\varepsilon_0 z}. \tag{5.4.17a}$$

It follows that at large z the potential energy of an electron is described by the interaction with the image charge $e^*=e(\varepsilon_1 - 1)/(2(\varepsilon_1 + 1))$. In the general case of arbitrary z, this interpretation becomes incorrect; it more accurately describes the meaning of the formula (5.4.16), the definition we have adopted [27, 28] as the energy of self-action.

At low z in accordance with (5.4.16)

$$U_{pS}(0) = -\frac{\pi}{2}\alpha_{pS}\hbar\Omega_{1p}. \tag{5.4.17b}$$

For typical values of semiconductor parameters $\alpha_{pS} = 0.01 - 0.1$; $\hbar\Omega_{1p} \sim 1 - 20$ eV the depth of the potential pit $U_{pS} \sim 0.1 - 2$ eV. Note that according to equation (5.4.16), the potential energy $U_{pS}(z)$ of the interaction of an electron with a surface mode does not depend on which side of the boundary it is located on.

In the medium $n = 1$, the electron also interacts with the bulk plasmon mode; the Hamiltonian of this interaction is obtained from the general formulas (5.2.40)–(5.2.48) by replacing (5.3.4):

$$\hat{H}_{n-V} = \sum_{\mathbf{Q}(\eta, q)} V_{1V}(\mathbf{Q})q_{1V}(\eta, q, z)e^{i\eta\rho}e^{-\eta z_e}\left[\hat{b}_{-\mathbf{Q}}^{+} + \hat{b}_{\mathbf{Q}}\right]; \tag{5.4.18}$$

$$V_{1V}(\mathbf{Q}) = \frac{e}{Q}\sqrt{\frac{(\varepsilon_1 - 1)\hbar\omega_{pV}}{2\varepsilon_0\varepsilon_1 L_x L_y L_z}};\ g_{1V} = 1 - e^{\eta z - iqz}, \tag{5.4.19}$$

where ω_{pV} is the frequency of longitudinal plasma oscillations in the medium $n = 1$ (5.4.5) without taking into account the dispersion: $\varepsilon_n(q) \Rightarrow \varepsilon_n(0) \equiv \varepsilon_n$. Therefore,

$$\alpha_{pV} = \frac{e^2(\varepsilon_1 - 1)}{8\pi\varepsilon_0\hbar\omega_{pV}R_{pV}\varepsilon_1};\ R_{pV} = \left(\frac{\hbar}{2m^*\omega_{pV}}\right)^{\frac{1}{2}}. \tag{5.4.20}$$

Using the perturbation theory in the second order, we obtain the potential energy of the interaction:

$$U_{pS}(z, \mathbf{P}_\perp) = -\sum_{\mathbf{Q}(\eta, q)} \frac{|V_{1V}(Q)|^2 |e^{iqz} - e^{\eta z}|^2}{\hbar\omega_{pV} + \frac{\hbar^2 Q^2}{2m^*}}. \tag{5.4.21}$$

Moving from the sum to the integral and performing the integration, we find the bulk part $U_{pV}(z)$:

$$\begin{aligned} U_{pV}(z) = &- \alpha_{pV}\hbar\omega_{pV}\left\{2 - \frac{R_{pV}}{2z} - \left(1 - \frac{R_{pV}}{z}\right)e^{-\frac{z}{R_{pV}}} - \left(\frac{z}{R_{pV}}\right)Ei\left(-\frac{z}{R_{pV}}\right)\right. \\ &\left. - \frac{\pi}{2}\left[H_1\left(\frac{2z}{R_{pV}}\right) - N_1\left(\frac{2z}{R_{pV}}\right)\right]\right\} \equiv -\alpha_{pV}\hbar\omega_{pV}B\left(\frac{z}{R_{pV}}\right), \end{aligned} \tag{5.4.22}$$

where $Ei(x)$ is an integral exponential function and $H_1(x)$, $N_1(x)$ are the Struve and Neumann functions, respectively.

At the interface ($z = 0$), the formula (5.4.22) gives $U_{pV}(z = 0) = 0$, and at the depth of the crystal ($z \gg R_{pV}$)

$$U_{pV}(z) \approx -\alpha_{pV}\hbar\omega_{pV} + \frac{e^2(\varepsilon_1 - 1)}{16\pi\varepsilon_0\varepsilon_1 |z|}, \tag{5.4.23}$$

with $z \to \infty$

$$U_{pV} = -\alpha\hbar\omega_{pV}. \tag{5.4.24}$$

Adding up $U_{pV}(z)$ and $U_{pS}(z)$, we get the total energy of an electron in a substance:

$$U_p(z) = U_{pV}(z) + U_{pS}(z). \tag{5.4.25}$$

$$U_p(z) \approx -\alpha_{pV}\hbar\omega_{pV} + \frac{e^2(\varepsilon_1 - 1)}{16\pi\varepsilon_0\varepsilon_1(\varepsilon_1 + 1)|z|}, \tag{5.4.26}$$

which is described by the formula for the potential energy of the image forces of an electron in a substance.

At low z

$$U_p(0) = U_{pS}(0) = -\frac{\pi}{2}\alpha_{pS}\hbar\Omega_{1p}; \quad U_{pV}(0) = 0. \tag{5.4.27}$$

Note that dimensionless energies $\overline{U}_{pV} \equiv U_{pV}/(\hbar\omega_{pV})$ and $U_{pS}/(\hbar\Omega_{1p})$ are universal functions of dimensionless coordinates $\bar{z} = z/R_{pV}$ and $\bar{z}_S = z/R_{pS}$.

Figure 5.2 shows the graphs: 1—$\overline{U}_{pV}(\bar{z})$; 2—$\overline{U}_{pS}(\bar{z})$ according to the formula (5.4.16); 3—$\overline{U}_{pV}(\bar{z}) + \overline{U}_{pS}(\bar{z})$; 4—$\overline{U}_{pS}(\bar{z})$–potential energy in a vacuum. Curves 5 and 6 are the potential energies of the image in the medium and vacuum, respectively. In vacuum, the energy value $U_{pS}(|z| \to \infty)$ is taken as the zero-reference level; in matter, it is shifted by $\alpha_{pV}\hbar\omega_{pV}$ below the bottom of the conduction band, according to (5.4.24). The characteristic form of the function $U_p(z)$ is determined by the ratio of potential energies taken on the surface and in the bulk of the substance:

$$\frac{U_p(|z| \to 0)}{U_p(|z| \to \infty)} = \frac{\pi}{\sqrt{2}} \cdot \left(\frac{\varepsilon_1}{2(\varepsilon_1 + 1)}\right)^{\frac{3}{4}}. \tag{5.4.28}$$

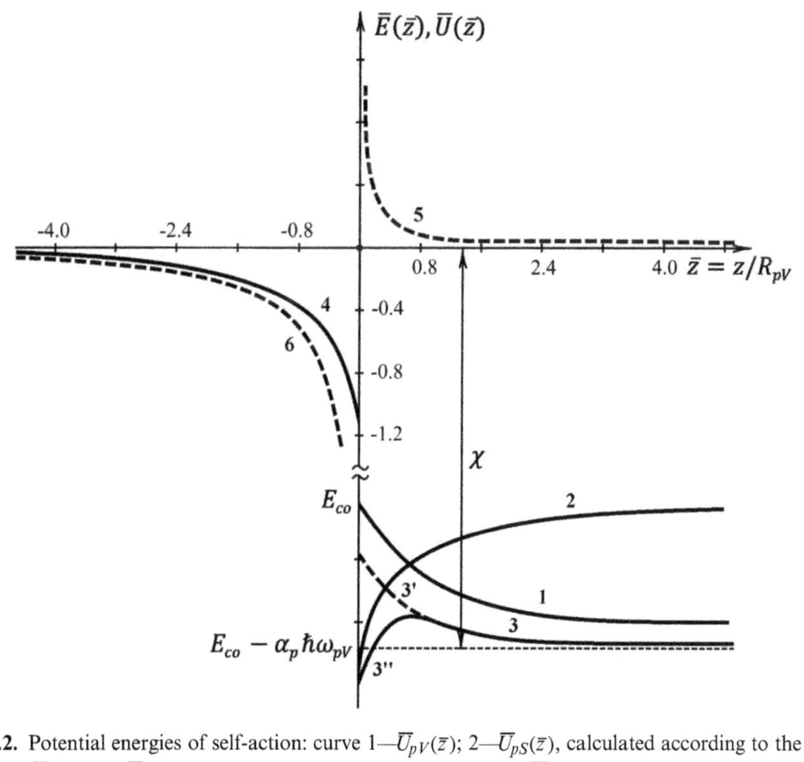

Figure 5.2. Potential energies of self-action: curve 1—$\overline{U}_{pV}(\bar{z})$; 2—$\overline{U}_{pS}(\bar{z})$, calculated according to the formula (5.4.16); 3—$\overline{U}_{pV}(\bar{z}) + \overline{U}_{pS}(\bar{z})$; 4 is a graph of the potential energy of $\overline{U}_{pS}(\bar{z})$ in a vacuum. Curves 5 and 6 are the potential energies of the image in the medium and vacuum, respectively.

At $\varepsilon_1 \leqslant 2.2$, the bottom of the conduction band forms a potential barrier gradually increasing towards the boundary (curve 3′). At $\varepsilon_1 > 2.2$, the behavior of the $U_p(z)$ becomes nonmonotonic (curve 3″). Before reaching its maximum at $z = 0$, the potential energy passes through the maximum in the bulk of matter. Note that noticeable deviations of the values of the potential energy $U_p(z)$ from the potential energy of the image take place in a fairly wide area of $0 \leqslant z \leqslant 5R_{pV}$.

Let's clarify the course of $U_p(z)$ near the surface. Decomposing $U_p(z)$ (5.4.25), taken in dimensionless form $|\bar{z}| < 1$, we find:

$$\bar{U}_p(\bar{z}) = -\frac{3}{2}\bar{z} + \bar{z}\ln(\gamma\bar{z}) - A\left[\frac{\pi}{2} + 2b\bar{z}\ln(2b\gamma\bar{z}) - 2b\bar{z}\right], \tag{5.4.29}$$

where

$$A = \frac{1}{\sqrt{2}}\left(\frac{\varepsilon_1}{\varepsilon_1 + 1}\right)^{\frac{3}{4}};\ b = \left(\frac{\varepsilon_1 + 1}{2\varepsilon_1}\right)^{\frac{1}{4}};\ \ln\gamma \equiv C \approx 0.58, \tag{5.4.30}$$

and for an electron in a vacuum

$$\bar{U}_p(\bar{z}) = -A\left[\frac{\pi}{2} + 2b\bar{z}\ln(2b\gamma\,\bar{z}) - 2b\bar{z}\right]. \tag{5.4.31}$$

Thus, instead of the classical limit for small $\bar{z}$ ($\bar{U}_p(\bar{z}) \sim \bar{z}^{-1}$)

(a) $\bar{z} \ll 1$:

$$\bar{U}_p(\bar{z}) = -\frac{\pi}{2}A + \bar{z}\ln(\gamma\bar{z}) - 2bA\bar{z}\ln(2b\gamma\,\bar{z}) \text{– in the crystal;} \tag{5.4.32}$$

$$\bar{U}_p(\bar{z}) = -\frac{\pi}{2}A - 2bA\bar{z}\ln(2b\gamma\,\bar{z}) \text{– in a vacuum;} \tag{5.4.33}$$

(b) $\bar{z} \sim 1$:

$$\bar{U}_p(\bar{z}) = -\frac{\pi}{2}A - \frac{3}{2}\bar{z} + 2bA\bar{z} \text{ – in the crystal;} \tag{5.4.34}$$

$$\bar{U}_p(\bar{z}) = -\frac{\pi}{2}A + 2bA\bar{z} \text{ – in a vacuum;} \tag{5.4.35}$$

Thus, it is shown that using the polaron approach to calculate the potential energy of a charge due to its interaction with the polarization of matter allows us to obtain a formula for potential energy without divergence at $|z| \to 0$ and naturally take into account the polaron shift $-\alpha\hbar\omega_{pV}$ for the conduction band in the depth of matter.

Using the potential energy of self-action, it is possible to calculate the corresponding force:

$$F_{SA}(z) = -\frac{\partial U_p(z)}{\partial z}. \tag{5.4.36a}$$

Substituting expressions (5.4.16) and (5.4.22), we obtain for the force of self-action:

$$F_{SA}^{V}(z) = \alpha_{pV}\hbar\omega_{pV}\left\{\frac{R_{pV}}{2z^2} - \frac{1}{z}\left(1 + \frac{R_{pV}}{z}\right)e^{-\frac{z}{R_{pV}}} - \frac{1}{R_{pV}}Ei\left(-\frac{z}{R_{pV}}\right)\right. \\ \left. - \frac{\pi}{R_{pV}}\left[H_0\left(\frac{2z}{R_{pV}}\right) - N_0\left(\frac{2z}{R_{pV}}\right)\right] + \frac{\pi}{2z}\left[H_1\left(\frac{2z}{R_{pV}}\right) - N_1\left(\frac{2z}{R_{pV}}\right)\right]\right\}; \tag{5.4.36b}$$

$$F_{SA}^{S}(z) = \frac{e^2}{R_{pS}}\left(\frac{\varepsilon_1 - 1}{\varepsilon_1 + 1}\right)\left[\mathrm{ci}\left(\frac{2z}{R_{pS}}\right)\cos\left(\frac{2z}{R_{pS}}\right) + \mathrm{si}\left(\frac{2z}{R_{pS}}\right)\sin\left(\frac{2z}{R_{pS}}\right)\right]. \tag{5.4.36c}$$

Plots of the functions $F_{SA}(z)$ are shown in figure 5.3. The dimensionless force $\overline{F}(\overline{z}) = F_{SA}(z)/F_0$ is deposited on the vertical axis, where $F_0 = e^2(\varepsilon_1 - 1)/(2R_{pV}^2\varepsilon_1)$, along the horizontal axis, is the dimensionless z coordinate. Curves 1—$\overline{F}_V(\overline{z})$ and

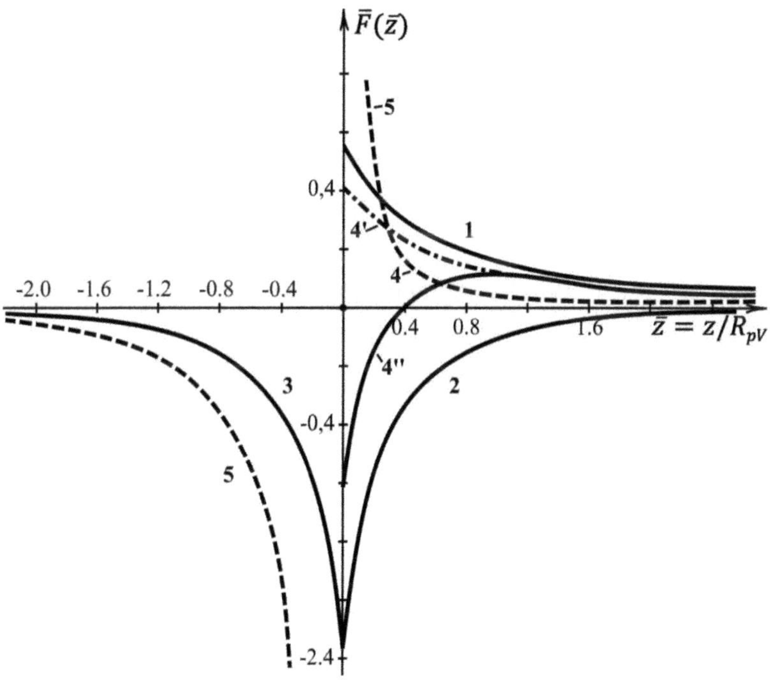

Figure 5.3. Plots of the dependence of the self-action force on the coordinate. The dimensionless force $F(z)$ is deposited on the vertical axis=$\overline{F}(\overline{z}) = F_{SA}(z)/F_0$,where $F_0 = e^2(\varepsilon_1 - 1)/(2R_{pV}^2\varepsilon_1)$, along the horizontal axis is the dimensionless z-coordinate. Curves 1—$\overline{F}_V(\overline{z})$ and 2—$\overline{F}_S(\overline{z})$ correspond to the interaction forces of an electron with bulk and surface modes, respectively; 3 and 4 are their sum in vacuum and in matter, respectively. Curves 4′ and 4″ correspond to potentials 3′ and 3″in figure 5.2. Image forces 5—in vacuum; $F_2(z) = -\varepsilon_1/(8\pi\varepsilon_0(z/R_{pV}^2))$; 6 in matter $\overline{F}_1 = 1/(8\pi\varepsilon_0(z/R_{pV}^2)(\varepsilon_1 + 1))$.

2—$\overline{F}_S(\overline{z})$ correspond to the interaction forces of an electron with bulk and surface modes, respectively; 3 and 4 are their sum in vacuum and in matter, respectively. Curves 4′ and 4″ correspond to potentials 3′ and 3″ in figure 5.2. Image forces 5—in vacuum; $F_2(z) = -\varepsilon_1/(8\pi\varepsilon_0(z/R_{pV}^2))$. The 'step' of force $F_2(z)$ at the boundary of matter with vacuum can be smoothed taking into account the 'penetration' of an electron into a neighboring medium.

5.4.2 Dielectric function in the theory of polarons

Let's move on to the problem of the dielectric function. As is known, in electrostatics, the potential energy $U(z)$ of a charge placed near the interface of two media is described by the expression

$$U(z) = \frac{e^2}{4z} \cdot \frac{(\varepsilon_1 - \varepsilon_2)}{\varepsilon_1(\varepsilon_1 + \varepsilon_2)}. \tag{5.4.37}$$

On the surface ($z = 0$) $U(z)$ has a singularity. One of the obvious reasons for the singularity is a sharp jump in the dielectric constant at the interface. If, as already noted, we introduce a transition function $S(z_n - z)$, smoothly varying from 0 to 1 at lengths of the order of several lattice constants, then we can obtain a smooth, singularity-free change in the potential energy of the charge during its transition through the domain separating substances with different permittivity. In fact, however, the transition domain in many cases extends only two to four lattice constants and cannot be described by models of the continuum theory. The elimination of the singularity from the potential energy is in principle possible if the dispersion of the dielectric constant is taken into account. The main disadvantage of this area of work is [29–33]. It is the use of model dielectric functions, the strict quantum mechanical derivation of which is not yet available. It is also significant that with this approach, the charge interacting with the medium is stationary (i.e., associated with a classical particle ($M \to \infty$)).

Let's find the variance due to the recoil effect. Let us select $\varepsilon(\eta)$ in such a way that from the electrostatic formulas [29, 30, 34] for potential energy it is possible to arrive at potential energy (5.4.25). The potential energy of an electron in a vacuum interacting with a dielectric is represented by the general electrostatic formula (3) from [30]. Comparing it with $U_{pS}(z)$ we derive the polaron function $a(\eta, 0)$ for expression (8) from [30]:

$$a(\eta, 0) = \frac{1}{\eta} - \frac{4\mu_S}{\eta\left(1 + 2\mu_S + R_S^2\eta^2\right)}; \tag{5.4.38}$$

here

$$\mu_S = \frac{\varepsilon - 1}{2(\varepsilon + 1)}. \tag{5.4.39}$$

Equating it to the dielectric function $a(\eta, 0)$ (formula (10) from [30]), we obtain the equation

$$\frac{1}{\eta} - \frac{4\mu_S}{\eta\left(1 + 2\mu_S + R_S^2\eta^2\right)} = \frac{1}{\pi}\int_0^\infty \frac{dq_z}{(q^2 + \eta^2)\varepsilon(q, z)}.$$

Solving this equation with respect to $\varepsilon^{-1}(\eta, q)$, we find

$$\varepsilon^{-1}(\eta, q) = 1 - \frac{4\mu_S(1 + 2\mu_S)(\eta^2 + q^2)}{(1 + 2\mu_S)^2\eta^2 + \left[(1 + 2\mu_S) + R_S^2\eta^2\right]^2\eta^2}. \qquad (5.4.40)$$

In the limit of a point particle ($R_{pS} \to 0$) we obtain a dispersion-free dielectric function $\varepsilon^{-1}(\eta, q) = \varepsilon_1^{-1}$; within the limits of $\eta \to 0$ and $\eta \to \infty$ we obtain $\varepsilon^{-1}(\eta, q)|_{\eta\to 0} = \varepsilon_1^{-1}$; $\varepsilon^{-1}(\eta, q)|_{\eta\to\infty} = 1$. The anisotropy of the function (5.4.40) reflects the geometry of the problem; it can be seen from the structure of the right part that the variance is significant at $\eta \sim (1 + 2\mu_S)/R_{pS} \sim R_{pS}^{-1}$ (i.e., it is determined by the size of the polaron).

Note that the continuum approach developed in [24–26] and used in this paragraph in the theory of the electron polaron does not claim accuracy in $z \sim R_{pS}$ if R_{pS} is less than the lattice constant. In this case, to solve the problems considered, it is necessary to use the theory of small-radius polarons.

In conclusion of this section, we note that the theory of self-action allows us to calculate the parameter λ^{-1}, which plays the role of the shielding length of the bound (valence) electrons of the Coulomb field of a free electron and is included in the dielectric functions proposed in [29–31, 33]:

$$\varepsilon(\eta) = 1 + (\varepsilon - 1)/[1 + \eta^2/\lambda^2(\varepsilon - 1)]; \qquad (5.4.41a)$$

where ε—the dielectric constant in a homogeneous electric field a $\eta \to 0$, and λ^{-1} is a parameter that is not determined within the framework of the theory [10, 29, 30] and plays the role of the shielding length by bound (valence) electrons.

For large transmitted pulses, $\varepsilon(\eta) = 1 + \lambda^2/\eta^2$, which formally coincides with the Thomas–Fermi approximation for conduction electrons in a metal. Taking into account quantum mechanical effects leads to another interpolation formula:

$$\varepsilon(\eta) = 1 + \frac{(\varepsilon - 1)}{\left[1 + \frac{\eta^2}{\lambda^2}(\varepsilon - 1)\right]\left(1 + \frac{3\eta^2}{4\eta_F^2}\right)}, \qquad (5.4.41b)$$

which, at $\eta \to \infty$, leads to the correct formula:

$$\varepsilon(\eta) = 1 + \frac{\xi}{3}\left(\frac{2\eta_F}{\eta}\right)^4; \ \xi = (\pi a_0\eta_F)^{-1}. \qquad (5.4.41c)$$

Comparing the results of the theory developed in this section for $U_{SA}(z)$ with the results of the theory [29, 31, 35], we obtain:

$$\lambda^{-1} = \frac{2(\varepsilon_1 + 1)}{\pi\sqrt{\varepsilon_1 - 1}}R_{pS}\left\{\sqrt{\varepsilon_1} + (\varepsilon_1 - 1)\mathrm{arctg}\sqrt{\varepsilon_1} - \frac{\pi}{2}(\varepsilon_1 - 1)\right\}. \qquad (5.4.41d)$$

The estimate λ^{-1} for Si, which has $\hbar\omega_{pV} = 17$ eV, $\varepsilon_1 = 11, 7$, $m_{\|}^* = 0, 97m_0$, $m_{\perp}^* = 0, 19m_0$, according to the formula (5.4.41) gives $\lambda_1^{-1} \sim R_{pS}$ (i.e., it is equal in order of magnitude to the electron polaron radius).

Let us supplement the dielectric function $\varepsilon(q, \omega)$, defined by the formula (5.4.1), with the contribution from interaction with the polar optical vibrations of the lattice, taken in the oscillatory approximation [36, 37]:

$$\varepsilon_T(\eta, \omega) = \varepsilon(\eta, \omega) + \frac{a}{\omega_{TO}^2 - \omega^2}. \tag{5.4.42}$$

The formula (5.4.42) takes into account that the frequency of radiation absorbed by the lattice vibrations $\omega = \omega_{TO}$ coincides with the frequency of transverse optical vibrations. Taking into account that the frequency of electronic transitions, at which the first term has resonance, is two to three orders of magnitude higher than the frequency of phonon transitions, at which the second term has resonance, it is possible to replace the function $\varepsilon(\eta, \omega)|_{\eta,\, \omega\to\infty} = \varepsilon_1$ where ε_1 is the high-frequency dielectric constant. Using the definition of static $\varepsilon_T(\omega)|_{\omega\to 0} = \varepsilon_{10}$ and equation (5.4.42), we find

$$\omega_{LO}^2 \equiv \omega_0^2 = \omega_{TO}^2 \cdot \frac{\varepsilon_{10}}{\varepsilon_1} \tag{5.4.43}$$

(formula (5.4.43) was obtained by Lidden–Sachs–Teller [20]). And from the boundary conditions for the polarization oscillation field $\varepsilon_T(\omega) = -1$ calculate the frequency of surface optical phonons for the polar crystal–vacuum contact:

$$\omega_{SO}^2 = \omega_{TO}^2\left(\frac{\varepsilon_{10} + 1}{\varepsilon_1 + 1}\right). \tag{5.4.44}$$

Using the Hamiltonian (5.2.10)–(5.2.12), we calculate, in the same way as it was done to calculate $U_{pS}(z)$ and $U_{pV}(z)$, the contributions of $U_{phS}(z)$ and $U_{phV}(z)$ to the total potential energy. Then

$$U_T(z) = \begin{cases} U_{pS}(z) + U_{pV}(z) + U_{phS}(z) + U_{phV}(z), & z \geqslant 0; \\ U_{pS}(z) + U_{phS}(z), & z < 0. \end{cases} \tag{5.4.45}$$

In an area close to the surface

$$\frac{U_{pS}(0)}{U_{phS}(0)} = \frac{(\varepsilon_1 - 1)\varepsilon_{10}}{\varepsilon_{10} - \varepsilon_1}\sqrt{\frac{\Omega_{1p}}{\omega_{phS}}} \tag{5.4.46}$$

the plasmonic contribution is usually greater than the phonon contribution. The reverse situation corresponds to only very small values of the difference $\varepsilon_1 - 1 \ll 10^{-1}$.

For large z ($z \gg R_{phS}$) we get

$$U_{phS}(z) = -\frac{e^2}{2z}\left(\frac{1}{\varepsilon_1 + 1} - \frac{1}{\varepsilon_{10} + 1}\right); \tag{5.4.47}$$

$$U_{pS}(z) = -\frac{e^2}{4z} \cdot \frac{\varepsilon_1 - 1}{\varepsilon_1 + 1}, \tag{5.4.48}$$

so

$$U_T(z) = U_{pS}(z) + U_{phS}(z) = -\frac{e^2}{4z} \cdot \frac{\varepsilon_{10} - 1}{\varepsilon_{10} + 1}. \tag{5.4.49}$$

From the comparison (5.4.49) and (5.4.17) it follows that the total potential energy of an electron in a vacuum contains only the contribution from interaction with surface optical phonons.

The potential energy inside the crystal at $\bar{z} \gg R_S,\ R_{pS}$ is equal to

$$\widetilde{U}_T(z) = U_T(z) - \alpha_{phV}\hbar\omega_0 - \frac{e^2}{16\pi\varepsilon_0 z} \cdot \frac{\varepsilon_1 - 1}{\varepsilon_1(\varepsilon_1 + 1)} + \frac{e^2}{16\pi\varepsilon_0 z} \cdot \frac{\varepsilon_{10} - 1}{\varepsilon_{10}(\varepsilon_{10} + 1)}. \tag{5.4.50}$$

The sum of the last two terms in (5.4.50) is always positive if

$$\varepsilon_{10} < 1 + \sqrt{2}, \tag{5.4.51}$$

and it can be positive or negative for $\varepsilon_{10} > 1 + \sqrt{2}$. Therefore, the phonon contribution to potential energy can both decrease and increase the repulsive force.

The phonon contribution to the dielectric function $\varepsilon(\eta, q)$ is similar to the plasmon contribution, with the replacement of μ_S by

$$\widetilde{\mu}_S = \frac{1}{\varepsilon_1 + 1} - \frac{1}{\varepsilon_{10} + 1}. \tag{5.4.52}$$

In conclusion, we note that both contributions to $U_T(z)$ are significant only for slow electrons: $\omega_e < \omega_0 < \omega_V$. If the inequality holds

$$\omega_0 \ll \omega_e < \omega_V, \tag{5.4.53}$$

the phonon contribution (the polaron effect) becomes insignificant. In the case of large distances $\omega_1 \sim v$ where v is the electron velocity, the condition for frequencies is better satisfied.

5.4.3 Localized states of the charge carrier in the field of the quantum potential of the image forces. A new type of surface states at the crystal–vacuum contact

Finally, it is of interest to discuss the question of the bound states of an electron in the field of self-action potential when $\varepsilon_1 > 2.4$ (i.e., a potential well appears in the area of contact with vacuum, separated by a barrier from the level of the bulk polaron $-\alpha_{1pV}\hbar\omega_{pV}$. It was shown in [11] that in a potential well formed by surface and bulk plasma vibrations of valence electrons, there are bound electron states that can be found by the variational method. The variational energy of the ground state is determined by the expression

$$E(\beta) = \frac{\hbar^2 \beta^2}{2m_e^*} + \int_0^\infty |\psi(z)|^2 U_p(z) dz, \tag{5.4.54}$$

where

$$\psi(z) = 2\beta^{\frac{3}{2}} z \, e^{-\beta z}, \tag{5.4.55}$$

is a trial wave function with a variational parameter β, and U_p is defined by the expression (5.4.25). Minimizing by β (5.4.54), we obtain:

$$E_0 = -\alpha_{pV}^2 \hbar\omega_{pV} \frac{(f - 0,\, 5)^2}{4(1 + 0,\, 5\alpha_{pV})}; \tag{5.4.56}$$

$$\beta^{-1} = R_{pV} \frac{2(1 + 0,\, 5\alpha_{pV})}{\alpha_{pV}(f - 0,\, 5)}. \tag{5.4.57}$$

Estimates for InSb–vacuum; HgTe–vacuum contacts are given, respectively: $E_0 = -0.15$ and -0.17 eV.

Thus, in addition to localization on Tamm states, the formation of surface states having a polaronic nature with a binding energy of the order of 0.1 eV can take place. Since the effect of self-action of the charge is fundamental in the sense of its mandatory presence in all cases when the charge carrier is in a polarizing medium, the observed values of the parameters characterizing the surface states of the charge carriers (effective mass, band gap, output operation, etc) can be largely determined by the electronic polaron effect.

5.4.4 Dynamic potential of the image at the contact of two media

The self-action potentials obtained in sections 5.4.1–5.4.3 describe the interaction of a resting charge with self-induced inertial and inertial polarizations at the contact of two media. If the charge carrier moves relative to the interface, then the times $\tau^{-1} \sim \omega_{ph,\, pl}$ and $\tau_2 = z/v_{\|}$ should be compared, where $v_{\|}$ is the velocity of the charge carrier along the z-axis. Since $\omega_{pl} \sim 10^{16} - 10^{15}\ \mathrm{s}^{-1}$, $\omega_{ph} \sim 10^{13}$ s, and $\tau_2 \sim 10^{-13} - 10^{-14}\ \mathrm{s}^{-1}$, then with respect to the electron, the polarization of valence electrons can be considered as static, and the crystal lattice dynamically. In this case, the self-action potential becomes a function of the velocity of the charge carrier ($\rho = v_{\perp} t$, $z = v_{\|} t$) and is called the dynamic self-action potential ($U_{SA}(z)$).

To calculate $U_{SA}[z(v)]$ we will use the phenomenological method developed in [8]. Consider the contact of a polar crystal with a nonpolar one when an electron moves in a nonpolar crystal. Let's average the Hamiltonians (5.2.42) and (5.2.44) on the wave function

$$\psi(z) = \exp\left(-\frac{1}{2} |I(t)|^2\right) \exp\left\{-iI^*(t)\hat{b}^+_{S,\, -\eta}\right\} |\, 0\rangle, \tag{5.4.58}$$

where

$$I(t) = \frac{V_{S,\eta}}{\hbar} \int_{-\infty}^{t} dt' e^{-i\Omega t'} e^{-\eta t'} e^{i\eta\rho}, \qquad (5.4.59)$$

found from the solution of the Schrödinger time equation for the system under consideration.

The phonon part of the dynamic potential of self-action has the form:

$$U_{pS}[z(v)] = -\sum_{s,\eta} \hbar\Omega_s \, |I(\eta, v, t)|^2. \qquad (5.4.60)$$

Substituting (5.2.42), (5.2.44), and (5.4.59) into (5.4.60) and performing integration over η and t, as well as assuming $v_\perp = 0$, we obtain

$$U_{pS}(z, v) = -\frac{e^2(\varepsilon_{20} - \varepsilon_2)}{(\varepsilon_{20} + \varepsilon_1)(\varepsilon_2 + \varepsilon_1)} \cdot \frac{\Omega}{v_z}\left[\mathrm{ci}\left(\xi\frac{z}{R_V}\right)\sin\left(\xi\frac{z}{R_V}\right) - \mathrm{si}\left(\xi\frac{z}{R_V}\right)\cos\left(\xi\frac{z}{R_V}\right)\right], \qquad (5.4.61)$$

where

$$\xi = \frac{2\Omega_s R_V}{v_z}. \qquad (5.4.62)$$

Figure 5.4 shows graphs of phonon contributions to the potential energy of self-action: $U_S^{\text{кл}}(z)$ is the classical potential, $U_S^{\text{кв}}(z)$, $U_S(v, z)$ for the Ge – CdS contact.

5.5 Surface polaronic states of weak, intermediate, and strong electron–phonon coupling (general approach)

For the study of surface polaronic states in various approaches, when the localization due to all interactions in the surface domain is weak or strong, the same Hamiltonian is used, with consideration of which we begin the discussion of this problem.

We derive the basic formulas for both approximations of weak and strong localization without specifying the type of electron–polarization interaction. Let's write down a Hamiltonian of the general form for the contact of two media:

$$\hat{H} = \frac{\hat{P}_\parallel^2}{2m^*} + \frac{\hat{P}_\perp^2}{2m^*} + \hat{H}_{ph}^V + \hat{H}_{ph}^S + \hat{H}_{e-ph}^V + \hat{H}_{e-ph}^S + U_{SA}(z) + U^0. \qquad (5.5.1)$$

Here, the first two terms are the kinetic energy of the electron; m^* is the effective mass in the periodic potential field; and $\hat{H}_{\mathrm{ph}}^V$, $\hat{H}_{\mathrm{ph}}^S$ are the operators of bulk and surface phonons, respectively:

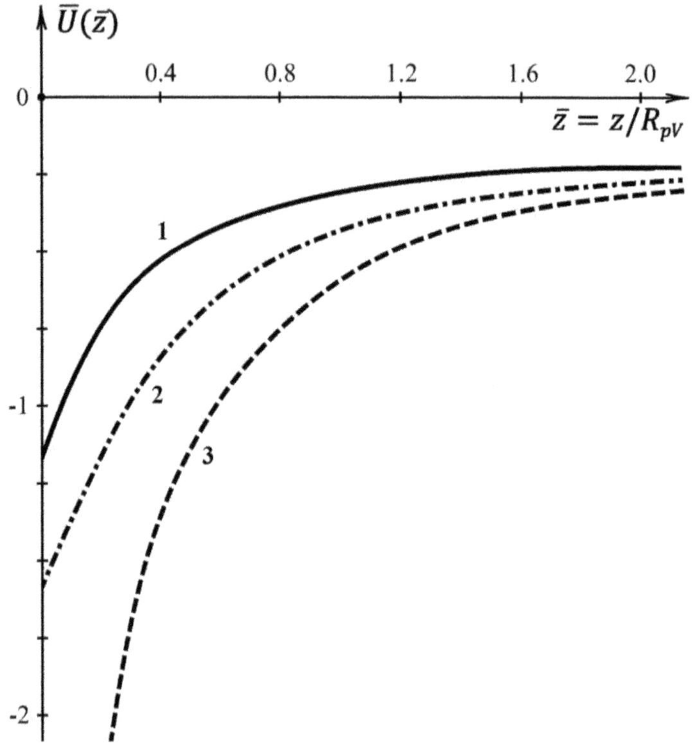

Figure 5.4. Plots of phonon contributions to the potential energy of self-action for the Ge – CdS contact. Curve 1 is the quantum potential $U_{qu}(z, v)$; curve 2 is the quantum potential at $v = 0$; and curve 3 is the classical potential $U_S^{cl}(z)$. The following parameter values were used in the calculation: $\varepsilon_1 = 16$, $m^*=0.19m_0$, $\varepsilon_{20} = 8.42$, $\varepsilon_2 = 5.27$, $\omega_0 = 5.7 \cdot 10^{13}\ \mathrm{s}^{-1}$, $\Omega_s = 4.83 \cdot 10^{13}\ \mathrm{s}^{-1}$.

$$\hat{H}_{\mathrm{ph}}^{V} = \sum_{\mathbf{Q}} \hbar\omega_0\left(\hat{b}_{\mathbf{Q}}^{+}\hat{b}_{\mathbf{Q}} + \frac{1}{2}\right); \tag{5.5.2}$$

$$\hat{H}_{\mathrm{ph}}^{S} = \sum_{s,\eta} \hbar\Omega_s\left(\hat{b}_{s,\,\eta}^{+}\hat{b}_{s,\,\eta} + \frac{1}{2}\right); \tag{5.5.3}$$

$$\hat{H}_{\mathrm{e-ph}}^{V} = \sum_{\mathbf{Q}} V_V(\mathbf{Q})\mathrm{e}^{i\mathbf{q}_\perp\boldsymbol{\rho}_e} g_{nV}(q_\perp, q_z, z_e)\left[\hat{b}_{-\mathbf{Q}}^{+} + \hat{b}_{\mathbf{Q}}\right]; \tag{5.5.4}$$

$$\hat{H}_{\mathrm{e-ph}}^{S} = \sum_{s,\eta} V_s(\eta)\mathrm{e}^{i\boldsymbol{\eta}\boldsymbol{\rho}_e} g_{ns}(\boldsymbol{\eta}, z_e)\left[\hat{b}_s^{+}(-\boldsymbol{\eta}, 0) + \hat{b}_s(\boldsymbol{\eta}, 0)\right], \tag{5.5.5}$$

where g_{ns} and g_{nV} are factors of attenuation of interaction with surface vibrations as they move away from the boundary, and with bulk vibrations as they approach the boundary. As follows from section 5.1, the simple form of the dependence of g_{ns} and g_{nV} on z takes place only for the contact of two semi-infinite crystals; $U_{SA}(z)$ is the potential energy of the self-action of the charge at the contact of two media, which was discussed in detail in section 5.3 (in the general quantum mechanical representation and in the classical limit); and U^0 is the height of the potential barrier at the contact, which is further assumed to be infinite.

5.5.1 Weak localization and weak connection with fluctuations. Perturbation theory

According to the criterion inequalities (5.2.1) and (5.1.1b), it can be assumed that in the case of weak coupling with the surface, a polaron is formed, which performs slow oscillations at the surface. We will describe various variants of the electron–polarization interaction.

For a fixed z $U_{SA}(z) = \text{const}$. The Hamiltonian of the zero approximation

$$\hat{H}_0 = \frac{\hat{P}_{\|}^2}{2m^*} + \frac{\hat{P}_{\perp}^2}{2m^*} + \hat{H}_{ph}^V + \hat{H}_{ph}^S. \tag{5.5.6}$$

The eigenfunction of this Hamiltonian

$$\psi_p(r, 0) = \frac{1}{\sqrt{V}} \exp\left(\frac{ipr}{\hbar}\right) | \, 0\rangle, \tag{5.5.7}$$

where $| \, 0\rangle$ is the wave function of polarization oscillations. Matrix elements of interaction operators:

$$\begin{aligned} &\left\langle \psi_{\mathbf{p}'}(\mathbf{r}, 1, 0),\left|, , V_V(Q) g_{nV}(\mathbf{q}_{\perp}, q_z, z_e) \hat{b}^{+}_{-\mathbf{Q}}, ,\right|, , \psi_{\mathbf{p}}(\mathbf{r}, 0, 0) \right\rangle = \\ &= \left\langle \psi_{\mathbf{p}}(\mathbf{r}, 0, 0),\left|, , V_V^*(Q) g_{nV}^*(\mathbf{q}_{\perp}, q_z, z_e) \hat{b}_{\mathbf{Q}}, ,\right|, , \psi_{\mathbf{p}'}(\mathbf{r}, 1, 0) \right\rangle = \\ &= V_V(Q) g_{nV}(\mathbf{q}_{\perp}, q_z, z_e) \delta\left[\frac{\mathbf{p}'_{\perp} - \mathbf{p}_{\perp}}{\hbar} + \mathbf{Q}\right]; \end{aligned} \tag{5.5.8a}$$

$$\begin{aligned} &\left\langle \psi_{\mathbf{p}}(\mathbf{r}, 0, 1),\left|, , V_S(\eta) g_{nS}(\boldsymbol{\eta}, z_e) \hat{b}^{+}_{S, -\boldsymbol{\eta}}, ,\right|, , \psi_{\mathbf{p}'}(\mathbf{r}, 0, 0) \right\rangle = \\ &= V_S(\eta) g_{nS}(\boldsymbol{\eta}, z_e) \delta\left[\frac{\mathbf{p}'_{\perp} - \mathbf{p}_{\perp}}{\hbar} + \boldsymbol{\eta}\right]. \end{aligned} \tag{5.5.8b}$$

When calculating matrix elements in the factors g_{nV}, g_{nS} the z-coordinate is considered fixed in accordance with the accepted approximations (a similar approach was used in [5]).

The energy of the interaction of an electron with polarization oscillations in the vicinity of z is described in the second order of the perturbation theory. For bulk fluctuations

$$\begin{aligned} U_{nV}(z, \mathbf{P}) &= -\sum_{\mathbf{Q}} \frac{|V_{nV}(Q)|^2 \left|g_{nV}(\mathbf{q}_{\perp}, q_z, z_e)\right|^2}{\hbar\omega_0 + \frac{\hbar^2 Q^2}{2m^*} - \frac{\hbar \mathbf{QP}}{m^*}} \approx \\ &\approx U_{nV}(z) + a_{nV}(z)\frac{P^2}{2m^*} + b_{nV}(z)\left(\frac{P^2}{2m^*}\right)^2 + \ldots, \end{aligned} \tag{5.5.9a}$$

where

$$U_{nV}(z) = -\sum_{\mathbf{Q}} \frac{|V_{nV}(Q)|^2 |g_{nV}(\mathbf{q}_\perp, q_z, z_e)|^2}{\hbar\omega_0 + \frac{\hbar^2 Q^2}{2m*}}; \tag{5.5.9b}$$

$$a_{nV}(z) = -\frac{2\hbar^2}{P^2 m*} \sum_{\mathbf{Q}} \frac{|V_{nV}(Q)|^2 |g_{nV}(\mathbf{q}_\perp, q_z, z_e)|^2 (\mathbf{QP})^2}{\left(\hbar\omega_0 + \frac{\hbar^2 Q^2}{2m*}\right)^3}. \tag{5.5.9c}$$

The way to get $b_{nV}(z)$ is also obvious. Adding the calculated interaction energy to the zero-approximation energy and considering $\mathbf{P}$ as the polaron momentum operator, we obtain the bulk part of the polaron Hamiltonian:

$$\hat{H}_{nV}(z, \mathbf{P}) = \frac{P^2}{2m_p^*} + U_{nV}(z), \tag{5.5.9d}$$

where

$$\left(m_p^*\right)^{-1} = (m^*)^{-1}(1 + a_{nV}). \tag{5.5.9e}$$

Similar calculations for the interaction of an electron with surface vibrations give:

$$U_{nS}(z, \mathbf{P}_\perp) = -\sum_{s,\boldsymbol{\eta}} \frac{|V_S(\eta)|^2 |g_{nS}(\boldsymbol{\eta}, z)|^2}{\hbar\Omega_s + \frac{\hbar^2\eta^2}{2m*} - \frac{\hbar\boldsymbol{\eta}\mathbf{P}_\perp}{m*}}; \tag{5.5.10a}$$

$$U_{nS}(z, \mathbf{P}_\perp) \approx U_{nS}(z) + a_S(z)\frac{P_\perp^2}{2m^*} + b_S(z)\left(\frac{P_\perp^2}{2m^*}\right)^2, \tag{5.5.10b}$$

where

$$U_{nS}(z) = -\sum_{s,\boldsymbol{\eta}} \frac{|V_S(\eta)|^2 |g_{nS}(\boldsymbol{\eta}, z)|^2}{\hbar\Omega_s + \frac{\hbar^2\eta^2}{2m*}}; \tag{5.5.10c}$$

$$a_S(z) = -\frac{2\hbar^2}{P_\perp^2 m*} \sum_{s,\boldsymbol{\eta}} \frac{|V_s(\eta)|^2 |g_{nS}(\boldsymbol{\eta}, z)|^2 (\boldsymbol{\eta}\mathbf{P}_\perp)^2}{\left(\hbar\Omega_s + \frac{\hbar^2\eta^2}{2m*}\right)^3}. \tag{5.5.10d}$$

So, the operator of the total energy of the polaron

$$\hat{H}_T = \frac{\hat{P}_\parallel^2}{2m_{\parallel p}^*} + \frac{\hat{P}_\perp^2}{2m_{\perp p}^*} + U_{nV}(z) + U_{nS}(z) + U_{SA}(z) + U^0; \tag{5.5.11}$$

here

$$\left(m_{\parallel p}^*\right)^{-1}(z) = (m^*)^{-1}(1 + a_V(z)); \left(m_{\perp p}^*\right)^{-1}(z) = (m^*)^{-1}(1 + a_V(z) + a_S(z)). \tag{5.5.12}$$

In general, the choice of the method of averaging dependencies on z in $m^*_{\|p}$ and $m^*_{\perp p}$ is determined by the peculiarities of specific processes of interaction of the polaron with radiation, impurities, etc. Taking into account the accepted approximation of the weak electron–polarization coupling, it is possible to neglect the additions of a_V and a_S to unity and perform averaging over this function when calculating the eigenfunction $\varphi(z)$ of the operator (5.5.11):

$$\left(m^*_{\|p}\right)^{-1} = (m^*)^{-1}\int_0^\infty |\varphi(z)|^2(1 + a_V(z))dz; \qquad (5.5.13a)$$

$$\left(m^*_{\perp p}\right)^{-1} = (m^*)^{-1}\int_0^\infty |\varphi(z)|^2(1 + a_V(z) + a_S(z))dz. \qquad (5.5.13b)$$

In the future, we will consider $m^*_{\|p}$, $m^*_{\perp p}$ constant values obtained using one or another averaging procedure.

Let's consider an example when the potential energies $U_{nV}(z)$ and $U_{nS}(z)$, having an electron–polarization interaction as their source, can be obtained explicitly. For squares of matrix elements, standard designations containing polaron parameters are usually used: $\alpha_{V,\,S}$ constants of the electron–polarization interaction; $R_{V,\,S}$—radii of polaron states; $U_{nV}(z)$ and $U_{nS}(z)$ have the same form as in the case of interaction with plasma polarization oscillations of valence electrons (formulas (5.4.13) and (5.4.16) for $U_{pS}(z)$ and (5.4.18) and (5.4.22) for $U_{pV}(z)$) (with replacement of R_{pS}, R_{pV} by $R_{S,\,V}$ and $\alpha_{pS,\,V}$—on $\alpha_{S,\,V}$. The behavior of the functions $U_S(z)$ and $U_V(z)$ for large and small values of the arguments z/R_S and z/R_V is described by the asymptotic formulas (5.4.17a), (5.4.17b), (5.4.23), and (5.4.27). In particular, it follows from (5.4.17b) and (5.4.27) that on the surface (contact), the interaction with bulk optical vibrations vanishes (attenuation effect), and with surface vibrations it has a maximum value. The limit of large values of the argument $z/R_{S,\,V}$ $U_t(z) = U_S(z) + U_V(z)$ has the form of a one-dimensional potential and, in the case of attraction to the surface, determines the spectrum of bound surface states.

5.5.2 Intermediate communication. The Lee, Low, and Pines method

The method of exclusion of oscillatory variables, proposed back in 1953, has been widely used [38, 39]. It was used not only in single-electron (polaron) tasks, but also in exciton, etc. It combines exact transformations of the Hamiltonian with a variational approach. This technique turned out to be very flexible and implemented in a large number of variants. A variant has also been found in which the limits of strong and weak bonds are combined in a continuous way [40]. The Lee, Low, and Pines method is based on two unitary transformations—the exclusion of electronic variables from the Hamiltonian of the electron–phonon interaction and the polaron shift of the amplitudes of the lattice oscillators.

First, let's consider the option corresponding to the weak localization approximation. Due to the translational symmetry in the XOY plane of the Hamiltonian of the electron–polarization interaction of the component of the total momentum of the system

$$\mathbf{P}_{\perp} = \mathbf{p}_{\perp} + \sum_{s,\eta} \hbar\eta \hat{b}^{+}_{s,\eta} \hat{b}_{s,\eta} + \sum_{\mathbf{Q}} \hbar\mathbf{Q} \hat{b}^{+}_{\mathbf{Q}} \hat{b}_{\mathbf{Q}} \tag{5.5.14}$$

commutes with the Hamiltonian and is an integral of motion. By choosing a unitary transformation in the form

$$\hat{S}_1 = \exp\left\{ -i\rho \left[\sum_{s,\eta} \hbar\eta \hat{b}^{+}_{s,\eta} \hat{b}_{s,\eta} + \sum_{\mathbf{Q}} \hbar\mathbf{Q} \hat{b}^{+}_{\mathbf{Q}} \hat{b}_{\mathbf{Q}} \right] \right\}, \tag{5.5.15}$$

ensuring the exclusion of the coordinate ρ from the electron–polarization interaction, we have

$$\hat{S}_1^{-1} \hat{b}_{s,\eta} \hat{S}_1 = \hat{b}_{s,\eta} e^{-i\eta\rho}\ ;\ \hat{S}_1^{-1} \hat{b}^{+}_{s,\eta} \hat{S}_1 = \hat{b}^{+}_{s,\eta} e^{i\eta\rho}\ ; \tag{5.5.16a}$$

$$\hat{S}_1^{-1} \hat{b}_{\mathbf{Q}} \hat{S}_1 = \hat{b}_{\mathbf{Q}} e^{-i\mathbf{Q}\rho};\ \hat{S}_1^{-1} \hat{b}^{+}_{\mathbf{Q}} \hat{S}_1 = \hat{b}^{+}_{\mathbf{Q}} e^{i\mathbf{Q}\rho}. \tag{5.5.16b}$$

In this case, the conversion of the electronic part of the pulse

$$\hat{S}_1^{-1} \mathbf{p}_{\perp} \hat{S}_1 = \mathbf{P}_{\perp} - \sum_{s,\eta} \hbar\eta \hat{b}^{+}_{s,\eta} \hat{b}_{s,\eta} - \sum_{\mathbf{Q}} \hbar\mathbf{Q} \hat{b}^{+}_{\mathbf{Q}} \hat{b}_{\mathbf{Q}} \tag{5.5.17}$$

is such that

$$\hat{S}_1^{-1} \mathbf{P}_{\perp} \hat{S}_1 = \mathbf{p}_{\perp}. \tag{5.5.18}$$

The integral of motion in the transformed system becomes the component $\mathbf{p}_{\perp}$ of the electronic pulse, which can now be considered a C-number. Transform the Hamiltonian (5.5.1):

$$\begin{aligned} \hat{H}_1 = \hat{S}_1^{-1} \hat{H} \hat{S}_1 = {} & \frac{1}{2m^*} \left[\mathbf{P}_{\perp} - \sum_{s,\eta} \hbar\eta \hat{b}^{+}_{s,\eta} \hat{b}_{s,\eta} - \sum_{\mathbf{Q}} \hbar\mathbf{Q} \hat{b}^{+}_{\mathbf{Q}} \hat{b}_{\mathbf{Q}} \right]^2 + \frac{P_{\parallel}^2}{2m^*} \\ & + \sum_{\mathbf{Q}} \hbar\omega_{n0} \hat{b}^{+}_{\mathbf{Q}} \hat{b}_{\mathbf{Q}} + \sum_{\mathbf{Q}} V_V(Q) g_{nV}(\eta, \mathbf{q}_{\perp}, z) \left[\hat{b}^{+}_{-\mathbf{Q}} + \hat{b}_{\mathbf{Q}} \right] \\ & + \sum_{s,\eta} \hbar\Omega_s \hat{b}^{+}_{s,\eta} \hat{b}_{s,\eta} + \sum_{s,\eta} V_s(\eta) g_{ns}(\eta, z) \left[\hat{b}^{+}_{s,-\eta} + \hat{b}_{s,\eta} \right]. \end{aligned} \tag{5.5.19}$$

Now we will perform the second unitary transformation of the oscillator amplitude displacement and average the result on the wave functions of the phonon vacuum:

$$\left\langle 0\left|\hat{S}_2^{-1}\hat{H}_1\hat{S}_2\right|0\right\rangle = \frac{1}{2m^*}\left[\mathbf{P}_\perp - \sum_{s,\eta}\hbar\eta\left|f_{\eta,s}\right|^2 - \sum_{\mathbf{Q}}\hbar\mathbf{q}_\perp\left|f_Q\right|^2\right]^2$$
$$+\frac{1}{2m^*}\left(P_{\|} - \sum_{\mathbf{Q}}\hbar q_z\left|f_Q\right|^2\right)^2 + \sum_{\mathbf{Q}}\left|f_Q\right|^2\left(\hbar\omega_0 + \frac{\hbar^2 q_\perp^2}{2m^*}\right)$$
$$+\sum_{s,\eta}\left|f_{\eta,s}\right|^2\left(\hbar\Omega_s + \frac{\hbar^2\eta^2}{2m^*}\right) + \sum_{\mathbf{Q}}\left|f_Q\right|^2\frac{\hbar^2 q_z^2}{2m^*} \tag{5.5.20}$$
$$+ U_{SA}(z) + \sum_{\mathbf{Q}}\left[V_V(Q)g_{nV}f_Q^* + V_V^*(Q)g_{nV}^*f_Q\right]$$
$$+\sum_{s,\eta}V_s(\eta)g_{ns}\left[V_s(\eta)g_{ns}f_{\eta,s}^* + V_s^*(\eta)g_{ns}^*f_{\eta,s}\right].$$

$$\hat{S}_2 = \exp\left\{\sum_{s,\eta}\left[\hat{b}_{s,\eta}^+ f_{s,\eta} - \hat{b}_{s,\eta}f_{s,\eta}^*\right] + \sum_{\mathbf{Q}}\left[\hat{b}_{\mathbf{Q}}^+ f_Q e^{iq_z z} - \hat{b}_{\mathbf{Q}}f_Q^* e^{-iq_z z}\right]\right\}. \tag{5.5.21}$$

Using the unitary transformation $\hat{S}_2(z)$, which depends on z, means adopting an approximation according to which the criterion of oscillator displacement is determined by the z-coordinate (approximation of slow motion along the z-axis).

Following [5, 41], we introduce the parameters:

$$\xi_\perp\mathbf{P}_\perp \equiv \sum_{s,\eta}\hbar\eta\left|f_{\eta,s}\right|^2 + \sum_{\mathbf{Q}}\hbar\mathbf{q}_\perp\left|f_Q\right|^2; \tag{5.5.22a}$$

$$\xi_{\|}P_z = \sum_{\mathbf{Q}}\hbar q_z\left|f_Q\right|^2, \tag{5.5.22b}$$

which we substitute in (5.5.20) and minimize the resulting expression by $f_{\eta,s}$, f_Q, which gives the relations:

$$f_{\eta,s} = -\frac{V_s(\eta)g_s(\eta, z)}{\hbar\Omega_s + \frac{\hbar^2\eta^2}{2m^*} - (1-\xi_\perp)\frac{\hbar\eta\mathbf{P}_\perp}{m^*}}; \tag{5.5.23a}$$

$$f_Q = -\frac{V_V(Q)g_V(\mathbf{q}_\perp, q_z, z)}{\hbar\omega_0 + \frac{\hbar^2 q_\perp^2}{2m^*} + \frac{\hbar^2 q_z^2}{2m^*} + \xi_r\frac{\hbar^2 q_z P_z}{m^*} - (1-\xi_\perp)\frac{\hbar\mathbf{q}_\perp\mathbf{P}_\perp}{m^*}}, \tag{5.5.23b}$$

by which the variational parameters are excluded from the formula for minimized energy. After decomposing it in terms of $\mathbf{P}_\perp$ and P_z while preserving the squares of the momentum components, using the definition (5.5.17), we arrive at exactly the same expressions for potential energy and effective mass (5.5.12)–(5.5.13), which were obtained in clause 5.4.1 by the method of perturbation theory.

5.5.3 The Lee, Low, and Pines method with weak coupling elements

Within the framework of the Lee, Low, and Pines method, we can consider the case of rapid movement along the z-axis. In this case, the displacements of the oscillators will be determined not by the z-coordinate, but by the quantum state of the electron. The elimination of the electron coordinates x and y is performed using the same unitary operator $\hat{S}_1$, as in clause 5.4.2. But now, even before the operation of shifting the oscillator amplitudes, the transformed Hamiltonian $\hat{H}_1 = \hat{S}_1^{-1}\hat{H}\hat{S}_1$ is averaged on the wave functions $\psi_\kappa(z)$, where k is a set of quantum numbers of the state the electron:

$$\hat{H}_{1\kappa} \equiv \left\langle \psi_\kappa \middle| \hat{H}_1 \middle| \psi_\kappa \right\rangle = \frac{1}{2m^*}\left\langle \psi_\kappa \mid \mathbf{P}_\perp - \sum_{s,\eta}\hbar\eta\hat{b}^+_{s,\eta}\hat{b}_{s,\eta} - \sum_{\mathbf{Q}}\hbar\mathbf{q}_\perp\hat{b}^+_{\mathbf{Q}}\hat{b}_{\mathbf{Q}} \mid \psi_\kappa \right\rangle$$
$$+\frac{1}{2m^*}\left\langle \psi_\kappa \mid P_z^2 \mid \psi_\kappa \right\rangle + \sum_{s,\eta}\hbar\Omega_s\hat{b}^+_{s,\eta}\hat{b}_{s,\eta} + \sum_{\mathbf{Q}}\hbar\omega_0\hat{b}^+_{\mathbf{Q}}\hat{b}_{\mathbf{Q}}$$
$$+\sum_{s,\eta}V_s(\eta)\overline{g}_{ns,\kappa}\left[\hat{b}^+_{s,-\eta} + \hat{b}_{s,\eta}\right]$$
$$+\sum_{\mathbf{Q}}V_V(Q)\overline{g}_{nV,\kappa}\left[\hat{b}^+_{-\mathbf{Q}} + \hat{b}_{\mathbf{Q}}\right] + \left\langle \psi_\kappa \mid U_{SA}(z) \mid \psi_\kappa \right\rangle. \tag{5.5.24}$$

where

$$\overline{g}_{ns,\kappa} = \int g_{ns}(\eta, s)|\psi_\kappa|^2 dz;\ \overline{g}_{nV} = \int g_{nV}(\mathbf{q}_\perp, q_z, z)|\psi_\kappa|^2 dz. \tag{5.5.25}$$

In the second unitary transformation

$$\hat{S}_2(\eta) = \exp\left\{-\sum_{s,\eta}\left(\hat{b}^+_{s,\eta}f_{\kappa,s} - \hat{b}_{s,\eta}f^*_{\kappa,s}\right)\right\}; \tag{5.5.26}$$

$$\hat{S}_2(Q) = \exp\left\{-\sum_{\mathbf{Q}}\left(\hat{b}^+_{\mathbf{Q}}f_{\kappa,Q} - \hat{b}_{\mathbf{Q}}f^*_{\kappa,Q}\right)\right\}. \tag{5.5.27}$$

The offsets of the oscillator amplitudes are determined by the state of the electron. After converting $\hat{S}_2$, averaging on a phonon vacuum and performing other operations described in clause 5.4.2, the above ground state energy (at $\mathbf{P}_\perp = 0$) is obtained as

$$E_\kappa = \frac{1}{2m^*}\left\langle \psi_\kappa \middle| P_z^2 \middle| \psi_\kappa \right\rangle$$
$$+\left\langle \psi_\kappa \middle| U_{SA}(z) \middle| \psi_\kappa \right\rangle - \sum_{s,\eta}\frac{|V_s(\eta)|^2\left|\overline{g}_{ns,\kappa}\right|^2}{\hbar\Omega_s + \frac{\hbar^2\eta^2}{2m^*}} - \sum_{\mathbf{Q}}\frac{|V_V(Q)|^2\left|\overline{g}_{nV,\kappa}\right|^2}{\hbar\omega_0 + \frac{\hbar^2Q^2}{2m^*}}. \tag{5.5.28}$$

The comparison of the results obtained in clauses 5.4.2 and 5.4.3 will be carried out further in the levitating polaron problem. Note that the considered case of fast electron oscillations along the z-axis can be realized due to an electric field perpendicular to the surface, even with relatively small values of the electron–phonon interaction constant.

5.5.4 The Lee, Low, and Pines method with strong coupling elements

In the variant where there is a strong coupling in the XOY plane and a weak coupling in the z-direction (with the surface), instead of the first unitary transformation $\hat{S}_1$, it is necessary to average the Hamiltonian on the wave function $\psi(x, y)$, describing the localization in the XOY plane. Field operators from the averaged Hamiltonian are eliminated by the methods described in clauses 5.4.2 and 5.4.3.

5.5.5 Strong communication and strong localization

The Hamiltonian is averaged on the wave function $\psi(x, y, z)$, after which the shift transformation $\hat{S}_2$ is performed with variational parameters depending on the quantum numbers of the wave function. In [41–43], a generalized variational procedure was developed that combines the limits of weak and strong bonds for motion in the XOY plane. To this end, the authors introduce the variational parameter A into the first unitary transformation:

$$\hat{S}_1 = \exp\left\{-iA_1\sum_{s,\eta}(\eta\rho)\hat{b}^+_{s,\eta}\hat{b}_{s,\eta} - iA_2\sum_{\mathbf{Q}}\left(\mathbf{Q}\rho\right)\hat{b}^+_{\mathbf{Q}}\hat{b}_{\mathbf{Q}}\right\}, \tag{5.5.29}$$

according to which it is necessary to minimize the variational energy obtained after all transformations. Obtaining the values $A_{1,2} \approx 1$ indicates a weak connection, and the values $A_{1,2} \ll 1$ in accordance with the above, means a strong connection. However, the variational energy in the case of an arbitrary coupling has a very cumbersome appearance. Therefore, they are often limited to considering the limiting cases $A_{1,2} = 1$ and $A_{1,2} = 0$ [42, 43].

5.6 Polaron at the contact of a polar crystal with a nonpolar one and its phase diagram

5.6.1 Criteria for the localization of a polaron at the contact of a polar crystal with a nonpolar one

Most of the work on flat-surface and near-surface polarons is devoted to the study of polaronic states at the contact of two media: polar and nonpolar crystals. The first works [4–6, 8, 9] appeared back in 1970–73. In general theoretical terms, the problem of the polaron on contact is of interest as another real example of a well-known problem in the interaction of a Fermi particle with a bosonic field, which allows us to develop and test field-theoretical methods.

A significant role in drawing attention to the problem was played by its specificity —multiparametricity and lower symmetry than in the bulk case. The values of the parameters ε_1, ε_{10}, ε_2 depend on which state is realized—localized at the boundary surface or bulk. When formulating boundary conditions for the electron wave function, the relative positions of the edges of the conduction bands and valence bands in neighboring layers play an essential role. In the sense of implementing a two-dimensional system, the limit of the surface polaron is of interest—a 'flat' polaron, the Hamiltonian of which depends only on two coordinates of the electron in the *XOY* plane. The Hamiltonian of a 'flat' polaron can be derived by assuming that the electron is closed in a narrow near-surface layer with a thickness of one or two permanent lattices [44] (the Tamm surface band). The z-coordinate describing the fast oscillations of an electron in a narrow and deep potential well can be excluded in the adiabatic approximation, after which only x and y remain of the electronic variables in the Hamiltonian. From an applied point of view, surface polaronic states are of great interest due to the important role they play in systems with a developed surface (heterostructures, inversion layers, superlattices), which have found wide application in solid-state electronics. The key point in solving polaronic problems is the derivation of the Hamiltonian of the electron–polarization interaction, which was described in section 5.1.

Many authors use the dielectric function method [5], the advantage of which is simplicity, but the final shape of the matrix element of the electron–phonon interaction is determined from comparison with known limiting cases (e.g., with the potential energy of the image forces). The methodology developed on the basis of the approach is more consistent [45, 46]. It was used to obtain the Hamiltonian for an electron in a polar plate [7]. The method proposed in [19] turned out to be fruitful, in which the exact Hamiltonian of the electron–phonon interaction for the polar plate was found for the first time. Based on the development and generalization of this method, the general Hamiltonians given in section 5.1 are derived.

The basic properties of the contact of a polar crystal with a nonpolar one are in many ways similar to the previously considered nonpolar crystal–vacuum contact. The qualitative difference is that $R_{0S,\,V} \equiv R_{SV} \gg R_{pS,\,V}$, a $\omega_1 \ll \omega_{1p}$. Consequently, phonon and quantum transitions occur at lower energies than plasma ones. Therefore, the electron polaron can be considered a fairly stable seed quasiparticle, the potential energy of interaction of which with the plasma of valence electrons of both media in this case is described quite accurately by $U_{SA}(z)$. The Hamiltonians of the interaction of an electron with surface and bulk phonons have the form (5.2.42)–(5.2.43) with the interaction constants (5.2.44)–(5.2.48). The problem of determining potential energy is reduced to a known dispersion-free one. The final results are related to the formulas by substitutions: $\varepsilon_{10} \to \varepsilon_1$, $\varepsilon_1 \to 1$, $\Omega_{10} \to \Omega_{1p}$, $\omega_{10} \to \omega_{1p}$.

The potential energy of the interaction of an electron in a second (nonpolar) medium (an external polaron with a surface dispersion-free mode) is determined by the formula (5.5.10c). The potential energy at the interface due to the interaction with phonons is equal to

$$U_S(0) = -\frac{\pi}{2}\alpha_S \hbar\Omega_{10}. \tag{5.6.1}$$

Potential energy ratios of the nonpolar crystal–vacuum and polar crystal–vacuum systems

$$\frac{U_S(0)}{U_{pS}(0)} = \frac{(\varepsilon_{10} - \varepsilon_1)}{(\varepsilon_1 - 1)(\varepsilon_{10} + 1)}\left(\frac{\Omega_{10}}{\Omega_{1p}}\right)^{\frac{1}{2}} \tag{5.6.2}$$

in the general case, it is much less than one, because $\Omega_{10} \ll \Omega_{1p}$, i.e. the plasma potential well is much deeper than the phonon one.

The reverse situation occurs when ε_1 differs very little from one. The potential energy of interaction with the bulk vibrations of the lattice is given by the formula (5.5.9b). In the limit $z/R_{S,\,V} \gg 1$ for $U_S(z)$ and $U_V(z)$ we obtain

$$U_S(z) = -\frac{e^2}{8\pi\varepsilon_0}\cdot\frac{(\varepsilon_{10} - \varepsilon_1)}{(\varepsilon_{10} + \varepsilon_2)(\varepsilon_1 + \varepsilon_2)|z|}; \tag{5.6.3}$$

$$U_V(z) = -\alpha\hbar\omega_0 + \frac{e^2}{16\pi\varepsilon_0}\cdot\frac{(\varepsilon_{10} - \varepsilon_1)}{\varepsilon_{10}\varepsilon_1\,|z|}. \tag{5.6.4}$$

Adding $U_S(z)$ and $U_V(z)$, we get the phonon

$$U(z) = \frac{e^2}{16\pi\varepsilon_0}\left[\frac{\varepsilon_2 - \varepsilon_1}{\varepsilon_1(\varepsilon_1 + \varepsilon_2)} - \frac{\varepsilon_{10} - \varepsilon_2}{\varepsilon_{10}(\varepsilon_{10} + \varepsilon_2)|z|}\right]\frac{1}{|z|} \tag{5.6.5}$$

and plasma

$$U_p(z) = \frac{e^2(\varepsilon_1 - \varepsilon_2)}{16\pi\varepsilon_0\varepsilon_1(\varepsilon_1 + \varepsilon_2)|\,z\,|} \tag{5.6.6}$$

of the potential energy in the medium ($n = 1$), and we obtain a well-known potential energy of the image forces, taking into account both types of polarizations:

$$U_T(z) = U(z) + U_p(z) = \frac{e^2(\varepsilon_{10} - \varepsilon_2)}{16\pi\varepsilon_0\varepsilon_{10}(\varepsilon_{10} + \varepsilon_2)|z|}. \tag{5.6.7}$$

Using the variational method with the Hamiltonian obtained in the weak coupling approximation (5.5.11), we calculate the energy of an electron localized at the polar–nonpolar crystal contact. To simplify calculations, consider the contact of two media, for which a dispersion-free approximation is performed. Large z $U_T(z)$ gives a Coulomb well, so it is convenient to choose a variational function in a form similar to the exact solution of the Schrödinger equation of the ground state with a one-dimensional Coulomb potential:

$$\psi(z) = 2\beta^{3/2}z\mathrm{e}^{-\beta z}. \tag{5.6.8}$$

The variational energy of the ground state has the form

$$E(\beta) = \frac{\hbar^2\beta^2}{2m_{||}} + \frac{e^2}{16\pi\varepsilon_0}\frac{(\varepsilon_1 - \varepsilon_2)\beta}{\varepsilon_1(\varepsilon_1 + \varepsilon_2)} + \langle\psi|U_S(z) + U_V(z)|\psi\rangle. \tag{5.6.9}$$

In the second term, the potential energy is determined by the formulas (5.5.10c) and (5.5.9b):

$$\langle\psi|U_S(z)|\psi\rangle = -\alpha_S\hbar\Omega_{10}J_1(2\beta R_S); \tag{5.6.10}$$

$$\langle\psi|U_V(z)|\psi\rangle = -\frac{2}{\pi}\hbar\omega_{10}J_2(2\beta R_V); \tag{5.6.11}$$

We introduce a dimensionless variational parameter

$$\xi = 2\beta R_S, \tag{5.6.12}$$

then

$$E(\xi) = \hbar\Omega_{10}\left\{\frac{1}{4}\xi^2 + \alpha_S[\xi a - J_1(\xi) - bpJ_2(p\xi)]\right\}, \tag{5.6.13}$$

where

$$p = \frac{R_V}{R_S} = \left[\frac{\varepsilon_1(\varepsilon_{10}+\varepsilon_2)}{\varepsilon_{10}(\varepsilon_1+\varepsilon_2)}\right]^{\frac{1}{4}}. \tag{5.6.14}$$

In the approximation of the isotropic masses of an electron

$$J_1(\xi) = \int_0^\infty dt(1+\xi^2t^2)^{-1}(1+t)^{-6}; \tag{5.6.15a}$$

$$J_2(p,\xi) = \int_0^\infty\int_0^\infty ydydx(x^2+y^2)^{-1}(1+y^2)^{-1}\left\{\left(1+\frac{x^2}{p^{2\xi^2}}\right)^{-3}+\left(1+\frac{y}{\xi p}\right)^{-6}\right.$$
$$\left.-2\left(1+\frac{y}{\xi p}\right)^3\left(1-\frac{2x^2}{\xi^2p^2}\right)\left(1+\frac{x^2}{\xi^2p^2}\right)^{-3}\right\}. \tag{5.6.15b}$$

Expanding (5.6.14a), (5.6.14b) by ξ (the assumption of smallness ξ is consistent with the approximation $z/R_S \gg 1$) and preserving the terms of the first order of smallness, we obtain

$$E(\beta) = \frac{\hbar^2\beta^2}{2m_\parallel} - \frac{\beta}{2}\left[\alpha_S\hbar\Omega_{10}R_S + \alpha_V\hbar\omega_{10}R_V - \frac{e^2}{8\pi\varepsilon_0}\frac{(\varepsilon_1-\varepsilon_2)}{\varepsilon_1(\varepsilon_1+\varepsilon_2)}\right]. \tag{5.6.16}$$

The condition for the existence of a minimum in β implies a sufficient criterion for the existence of a surface-bound state of an electron in a polar medium ($n = 1$):

$$\alpha_S\hbar\Omega_{10}R_S + \alpha_V\hbar\omega_{10}R_V - \frac{e^2}{8\pi\varepsilon_0}\frac{(\varepsilon_1-\varepsilon_2)}{\varepsilon_1(\varepsilon_1+\varepsilon_2)} > 0, \tag{5.6.17}$$

from where, after substituting the values of the polaron parameters, we get a simple inequality:

$$\varepsilon_2 > \varepsilon_{10}, \tag{5.6.18}$$

clarifying the results of the work [42–44, 47]. Minimizing the β function (5.6.16), we obtain

$$\beta_{\min} = \alpha_S R_S^{-1} f_1(\varepsilon_1, \varepsilon_{10}, \varepsilon_2), \tag{5.6.19}$$

where

$$f_1(\varepsilon_1, \varepsilon_{10}, \varepsilon_2) = \frac{(\varepsilon_2 - \varepsilon_{10})(\varepsilon_1 + \varepsilon_2)}{8\varepsilon_{10}(\varepsilon_{10} - \varepsilon_1)}.$$

Excluding $\beta_{\min}$, we write out the energy of the surface polaron taking into account higher degrees of α_S, than the first one:

$$E_0(\beta_{\min}) = -\alpha_V \hbar\omega_{10} - \alpha_S^2 \hbar\Omega_{10} f_1^2 - \frac{1}{2} f_1^2 \alpha_S^4 \hbar\Omega_{10}(1 + 2\ln(\alpha_S f_1)). \tag{5.6.20}$$

This result justifies the accepted approximation of the weak coupling of the polaron with the surface at small $\alpha_{S,\,V}$, since the intrinsic energy of the polaron is of the order $\sim\alpha_V$, and the binding energy is $\sim\alpha_S^2$.

In the energy of the external polaron (in a nonpolar medium: $n = 2$), there is no interaction with bulk phonons $\langle\psi|U_V(z)|\psi\rangle$. The β-linear result from (5.6.16) has the form

$$E(\beta) = \frac{\hbar^2\beta^2}{2m_{\parallel}} - \frac{1}{2}\alpha_S \hbar\Omega_{10} R_S \beta + \frac{e^2}{16\pi\varepsilon_0} \cdot \frac{(\varepsilon_2 - \varepsilon_1)\beta}{\varepsilon_2(\varepsilon_2 + \varepsilon_1)}, \tag{5.6.21}$$

from which the conditions for the existence of a minimum of energy follow:

$$\alpha_S \hbar\Omega_{10} R_S \beta - \frac{e^2}{8\pi\varepsilon_0} \cdot \frac{(\varepsilon_2 - \varepsilon_1)}{\varepsilon_2(\varepsilon_2 + \varepsilon_1)} > 0 \tag{5.6.22}$$

and a sufficient criterion for the localization of an electron at the outer surface: $\varepsilon_2 > \varepsilon_{10}$ in the form of an inequality of the opposite meaning compared to (5.6.18). From (5.6.21) follows the energy of the external polaron:

$$E_0 = -\alpha_S^2 \hbar\Omega_{10} f_2^2(\varepsilon_1, \varepsilon_{10}, \varepsilon_2); \tag{5.6.23a}$$

$$f_2(\varepsilon_1, \varepsilon_{10}, \varepsilon_2) = \frac{(\varepsilon_{10} - \varepsilon_2)(\varepsilon_1 + \varepsilon_2)}{8\varepsilon_2(\varepsilon_{10} - \varepsilon_1)}. \tag{5.6.23b}$$

There is no bulk polaron effect in the external environment, so the formula (5.6.22) gives a net binding energy that is proportional to α_S^2.

In both cases of surface states, it was assumed that the penetration of an electron into another medium was prevented by an infinitely high potential barrier at the interface.

5.6.2 Phase transitions of the surface polaron

The problem of phase transition for a continuous Fröhlich polaron, which consists of the existence of a certain critical value (α_c) of the electron–phonon interaction constant, in which the polaron 'self-captures' and becomes localized, has been discussed in many theoretical papers. In [48], the results of various approaches are analyzed in detail and it is shown that for massive crystals, the phase transitions of a polaron from a free, nonlocalized state to a localized one are rather features of approximations made in various theories than an internal property of the Fröhlich Hamiltonian.

A surface polaron at the contact of two media has one specific difference from a polaron in a massive polar crystal: its formation significantly depends on the parameters of the contacting medium, and this introduces into the problem of phase transition an additional parameter to the Fröhlich constant α characterizing the strength of the electron–surface bond. At the contact of two crystals, such a parameter is the self-action potential constant, which in the classical limit ($z \gg R_{S,\,V}$) has the form

$$\beta = \frac{e^2(\varepsilon_1 - \varepsilon_2)}{4\varepsilon_1(\varepsilon_1 + \varepsilon_2)} \tag{5.6.24}$$

and it can be both positive and negative.

For the first time, the problem of the phase transition of a surface polaron was considered in [40, 48–55], in which it was shown that there is a value of the electron–phonon interaction constant α_c, at which the state of the surface polaron changes from almost free to autolocalized (local phase transition). However, the contact of a polar crystal with a vacuum considered in [55] is a special case of a more general situation: the contact of two crystals, since in this case the parameter β varies from values $\beta < 0$ to values $\beta > 0$. In addition, the interaction with bulk optical vibrations, which creates an alternative—a surface or bulk polaron—was not taken into account.

In this section, we will discuss this problem for the contact of a polar crystal with a nonpolar one when an electron is in a polar crystal.

The Hamiltonian of the system has the form (5.5.1)–(5.5.5), where we take the potential energy of self-action in the classical limit (formula (5.4.37)) and represent it as

$$U_{SA}(z) = \frac{\beta}{z}. \tag{5.6.25}$$

Following the method of Lee, Low, and Pines, in which the first unitary transformation is chosen in the form (5.5.29), we obtain for the variational energy of the ground state ($\mathbf{P}_\perp = 0$) an expression similar to (5.5.28):

$$E_0 = \left\langle \psi(z) \left| \frac{P_z^2}{2m^*} + U_{SA}(z) \right| \psi(z) \right\rangle - \sum_{s,\eta} \frac{|\widetilde{V}_s(\eta)|^2}{\hbar\Omega_s + A_1 \frac{\hbar^2\eta^2}{2m^*}} - \sum_{Q} \frac{|\widetilde{V}_V(Q)|^2}{\hbar\omega_0 + A_2 \frac{\hbar^2 Q^2}{2m^*}}. \tag{5.6.26}$$

where

$$\widetilde{V}_S(\eta) = V_S(\eta)\langle\psi(\rho, z)|e^{-\eta z}e^{-i(1-A_1)\eta\rho}|\psi(\rho, z)\rangle; \tag{5.6.27a}$$

$$\widetilde{V}_V(\eta) = V_V(\eta)\langle\psi(\rho, z)|(e^{iq_z z} - e^{-q_\perp z})e^{-i(1-A_2)q_\perp\rho}|\psi(\rho. z)\rangle. \tag{5.6.27b}$$

The test wave function of the electronic subsystem is selected as

$$\psi(\rho, z) = \left(\frac{\lambda^3\nu^2}{4\pi}\right)^{\frac{1}{2}} z\, e^{-\frac{\lambda}{2}z}\, e^{-\frac{\nu}{2}\rho}, \tag{5.6.28}$$

where λ and ν are variational parameters.

After averaging on the wave function (5.6.5), we obtain from the expression (5.6.3)

$$\frac{E_0}{\hbar\Omega_s} = \frac{\lambda^2}{8} + \frac{\nu^2}{8} + \frac{\beta\lambda}{2} - \frac{\alpha_S}{\sqrt{2}}J_1(\lambda, \nu, A_1) - \frac{\sqrt{2}}{\pi}\alpha_V\nu^6\lambda^6\sqrt{\frac{\omega_0}{\Omega_s}}J_2(\lambda, \nu, A_2), \tag{5.6.29}$$

where

$$J_1(\lambda, \nu, A_1) = \int_0^\infty \frac{\lambda^6\nu^6 dx}{(x+\lambda)^6(\nu^2 + (1-A_1)^2x^2)^3}; \tag{5.6.30a}$$

$$\begin{aligned} J_2(\lambda, \nu, A_2) = \int_0^\infty \frac{x^2(3\lambda^2 - x^2)^2}{(\lambda^2 + x^2)^6}\Bigg\{ & \frac{2\ln[x(1-A_2)/\nu]}{[x^2(1-A_2)^2 - \nu^2]^3} - \frac{2\ln[y(1-A_2)/\nu]}{[y^2(1-A_2)^2 - \nu^2]^3} \\ & - \frac{1}{\nu^2[x^2(1-A_2)^2 - \nu^2]^2} + \frac{1}{\nu^2[y^2(1-A_2)^2 - \nu^2]^2} \\ & + \frac{(y^2 - x^2)(1-A_2)^2}{2\nu^4[x^2(1-A_2)^2 - \nu^2][y^2(1-A_2)^2 - \nu^2]}\Bigg\} dx; \quad y = \sqrt{x^2 + A_2^{-2}}. \end{aligned} \tag{5.6.30b}$$

It is of interest to calculate the average number of surface (N_S) and bulk (N_V) optical phonons in a phonon cloud:

$$N_S = \sum_{s,\eta}\hbar\Omega_s\,|f_s(\eta)|^2; \quad N_V = \sum_{\mathbf{Q}}\hbar\omega_0\,|f_V(\mathbf{Q})|^2, \tag{5.6.31}$$

where the displacement amplitudes $f_s(\eta)$ and $f_V(\eta)$ have the form

$$f_s(\eta) = V_S(\eta)\langle\psi(\rho, z)|e^{-\eta z}e^{-i(1-A_1)\eta\rho}|\psi(\rho. z)\rangle\left[\hbar\Omega_s + A_1^2\frac{\hbar^2\eta^2}{2m^*}\right]^{-1}; \tag{5.6.32a}$$

$$f_V(\mathbf{Q}) = V_V(Q)\langle\psi(\rho, z)|(e^{iq_z z} - e^{-q_\perp z})e^{-i(1-A_2)q_\perp\rho}|\psi(\rho. z)\rangle\left[\hbar\omega_0 + A_2^2\frac{\hbar^2Q^2}{2m^*}\right]^{-1}. \tag{5.6.32b}$$

Substituting (5.6.11a) and (5.6.11b) into (5.6.10) and performing transformations, we obtain

$$N_S = \frac{\alpha_S}{\sqrt{2}} \int_0^\infty \frac{dx}{(1 - \frac{1}{2}A_1^2 x^2)} \cdot \frac{\lambda^6}{(x+\lambda)^6} \cdot \frac{\nu^6}{[\nu^2 + (1-A_1)^2 x^2]^3}; \tag{5.6.33a}$$

$$N_V = \frac{\sqrt{2}}{\pi}\left(\frac{\omega_0}{\Omega_s}\right)^{\frac{1}{2}} \alpha_V \int_0^\infty t dt \int_0^\infty \frac{dx}{(t^2 + x^2)} \cdot \frac{x^2[3\lambda^2 - x^2]^2}{(\lambda^2 + x^2)^6} \times \frac{\lambda^6 \nu^6}{[\nu^2 + (1-A_2)^2 t^2]^3 \left[1 + A_2^2 \frac{\Omega_S}{\omega_0}(x^2 + t^2)\right]^2}. \tag{5.6.33b}$$

The ground state energy of the surface polaron can be obtained from (5.6.6)–(5.6.7) by minimizing the parameters λ, ν, A_1, and A_2. Figures 5.5 and 5.6 show the results of numerical calculations of E_0, $N = N_S + N_V$ and the values $\langle z \rangle = 2/\lambda$;

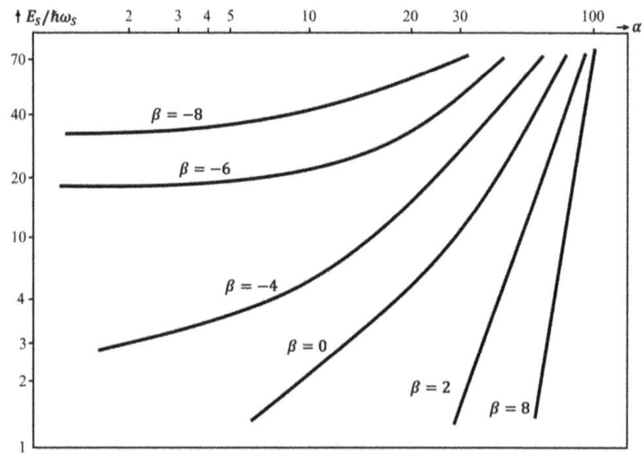

Figure 5.5. The results of numerical calculation of the dependence of the value $E_S/(\hbar\omega_S)$ on the constant of the electron–phonon interaction for different values of the parameter β.

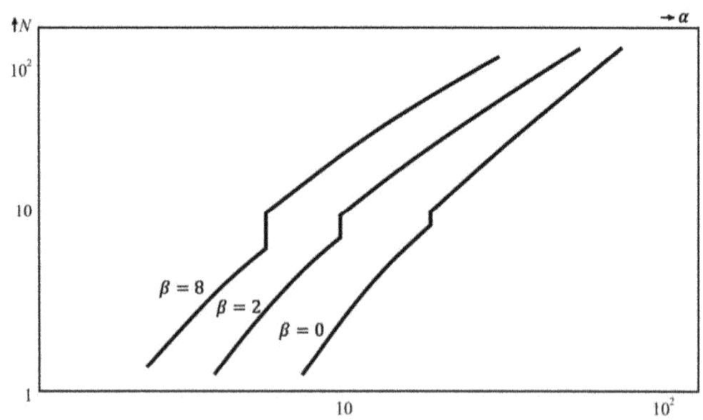

Figure 5.6. The results of numerical calculation of the dependence of the average number of phonons on the constant of the electron–phonon interaction for different values of the parameter β.

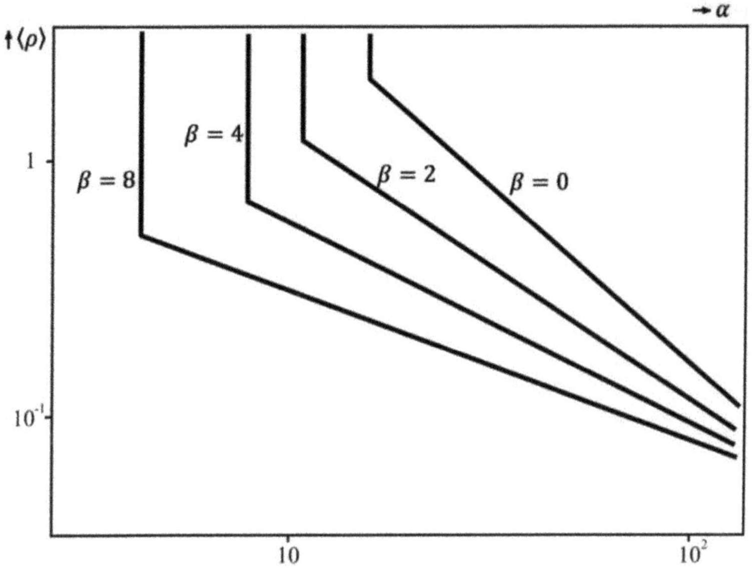

Figure 5.7. The results of numerical calculation of the dependence of the parameter $\langle\rho\rangle$ on the constant of the electron–phonon interaction for different values of the parameter β.

$\langle\rho\rangle = 2/\nu$, characterizing the spatial localization of the electron perpendicular to and along the crystal interface for a number of values of the electron–phonon interaction constant α_V and the parameter β. For each of the considered values of β, there are two branches, the first of which corresponds to small values of α_V and has the same form as in the Lee, Low, and Pines approximation. The second branch, intersecting with the first one at point α_c, corresponds to the approximation of a strong connection.

Thus, the state of the surface polaron at the point $\alpha = \alpha_c$ changes from almost free to autolocalized. Moreover, the growth of β (increased localization in the contact area) shifts the critical point to a domain of smaller α. A characteristic feature of this phase transition is manifested in the dependence of the average number of phonons and parameters characterizing spatial localization along and perpendicular to the surface on the constant of the electron–phonon interaction α, as shown in figures 5.7 and 5.8.

5.7 Surface polaronic states at the contact of two polar crystals

The problem of electron–polarization interaction for the contact of two polar media is significantly more complex than the similar problem for the contact of a polar medium with a nonpolar one. None of the published papers on this topic (see, e.g. [44, 47, 56]) solved it correctly, because an electrostatic approximation was used to obtain the constant of the electron–phonon interaction. The reason for the complication is the mixing of the surface polarization modes of the crystals.

The effective polaron Hamiltonian has the form

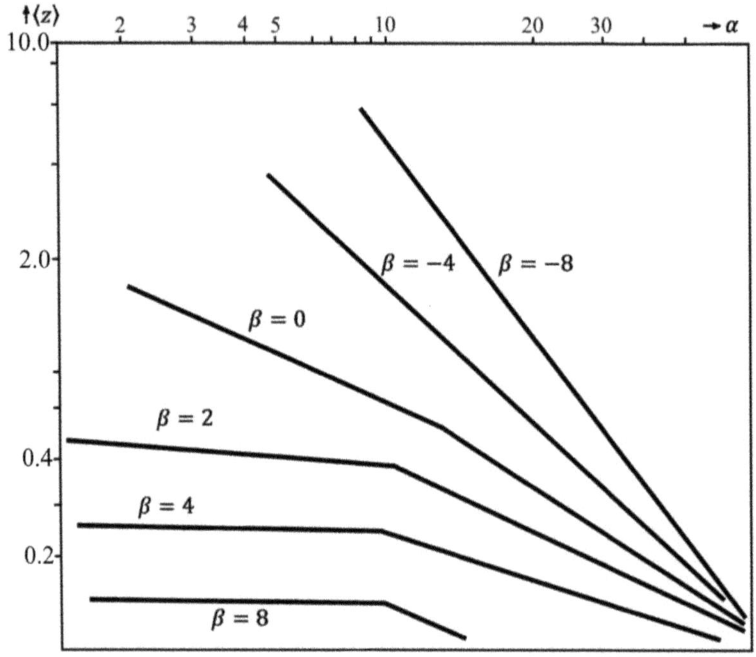

Figure 5.8. The results of numerical calculation of the dependence of the parameter $\langle z\rangle$ on the constant of the electron–phonon interaction for different values of the parameter β.

$$\hat{H} = \frac{\hat{P}_{||}^2}{2m_{||}^*} + \frac{\hat{P}_{\perp}^2}{2m_{\perp}^*} + U_{nV}(z) + U_{nS}(z) + U_{SA}(z), \tag{5.7.1}$$

where

$$U_{nV}(z) = -\sum_{\mathbf{Q}} \frac{|V_{nV}(Q)|^2 |g_{nV}(\mathbf{Q}, z)|^2}{\hbar\omega_{n0} + \frac{\hbar^2 Q^2}{2m^*}}; \tag{5.7.2a}$$

$$U_{nS}(z) = -\sum_{s,\boldsymbol{\eta}} \frac{|V_S(\eta)|^2 |g_{nS}(\boldsymbol{\eta}, z)|^2}{\hbar\Omega_s + \frac{\hbar^2\eta^2}{2m^*}}; \tag{5.7.2b}$$

$$\left(m_{p||}^*\right)^{-1} = (m^*)^{-1}(1 + a_{nV}); \ \left(m_{p\perp}^*\right)^{-1} = (m^*)^{-1}(1 + a_{nV} + a_{nS}); \tag{5.7.3}$$

$$a_{nV} = -\frac{2\hbar^2}{m^* P^2} \sum_{\mathbf{Q}} \frac{|V_{nV}(Q)|^2 |g_{nV}(\mathbf{Q}, z)|^2 (\mathbf{QP})^2}{\left(\hbar\omega_{n0} + \frac{\hbar^2 Q^2}{2m^*}\right)^3}; \tag{5.7.4a}$$

$$a_{nS} = -\frac{2\hbar^2}{m^* P_{\perp}^2} \sum_{s,\boldsymbol{\eta}} \frac{|V_S(\eta)|^2 |g_{nS}(\boldsymbol{\eta}, z)|^2 (\boldsymbol{\eta}\mathbf{P}_{\perp})}{\left(\hbar\Omega_s + \frac{\hbar^2\eta^2}{2m^*}\right)^3}; \tag{5.7.4b}$$

obtained using (5.2.34), (5.2.42)–(5.2.52) and (5.5.10a)–(5.5.10d). By collecting these terms, we obtain the total potential energy of the interaction of an electron with inertial and inertialess polarizations at the contact of two polar media:

$$
\begin{aligned}
U_T(z) \;=\; &-\frac{e^2\omega_1^2}{16\pi\varepsilon_0\left(1+F^2\right)\left(\varepsilon_1+\varepsilon_2\right)^2} \\
&\left\{\frac{\left(a_1-a_2F\right)^2}{R_{1S}\Omega_1^2}S\left(\frac{|z|}{R_{1S}}\right)+\frac{\left(a_1F+a_2\right)^2}{R_{2S}\Omega_2^2}S\left(\frac{|z|}{R_{2S}}\right)\right\} \\
&-\alpha_{nV}\hbar\omega_{n0}B\left(\frac{|z|}{R_{nV}}\right)+e^2\left(\varepsilon_n-\varepsilon_{\bar{n}}\right);\ \bar{n}=3-n.
\end{aligned}
\tag{5.7.5}
$$

At large z ($z \gg R_{nS,\,V}$) the quantum mechanical potential energy (5.7.5) coincides with the classical one:

$$U_i(z) \equiv U_{SA}(z) = \frac{e^2(\varepsilon_{n0}-\varepsilon_{\bar{n}0})}{16\pi\varepsilon_0\varepsilon_{n0}(\varepsilon_{n0}+\varepsilon_{\bar{n}0})|\,z\,|}. \tag{5.7.6}$$

For small z ($z \ll R_{nS,\,V}$) their dependences on z are significantly different (figures 5.9 and 5.10), which may affect the results of calculations of the energy of polarons localized at the surface.

At large distances from the interface ($|z|/R_{nV} \gg 1$) the total potential energy of the electron (5.7.5) has the form

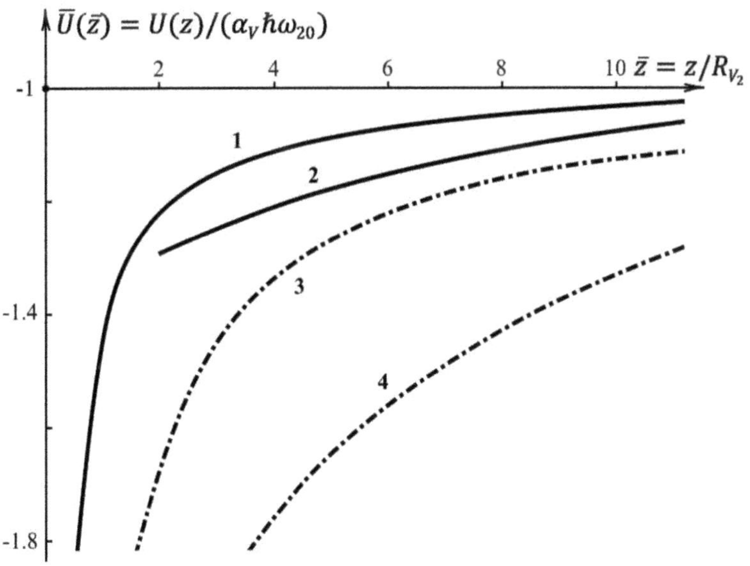

Figure 5.9. Coordinate dependence of the potential energy of the interaction of an electron with its induced polarization at the contact of two polar media: curve 1—$\bar{U}_i(\bar{z})$ at the CdS – TlBr; contact; curve 2—$\bar{U}_i(\bar{z})$ at the GaAs – TlCl; contact; and curve 3—$\bar{U}_T(\bar{z})$ at the GaAs – TlCl.

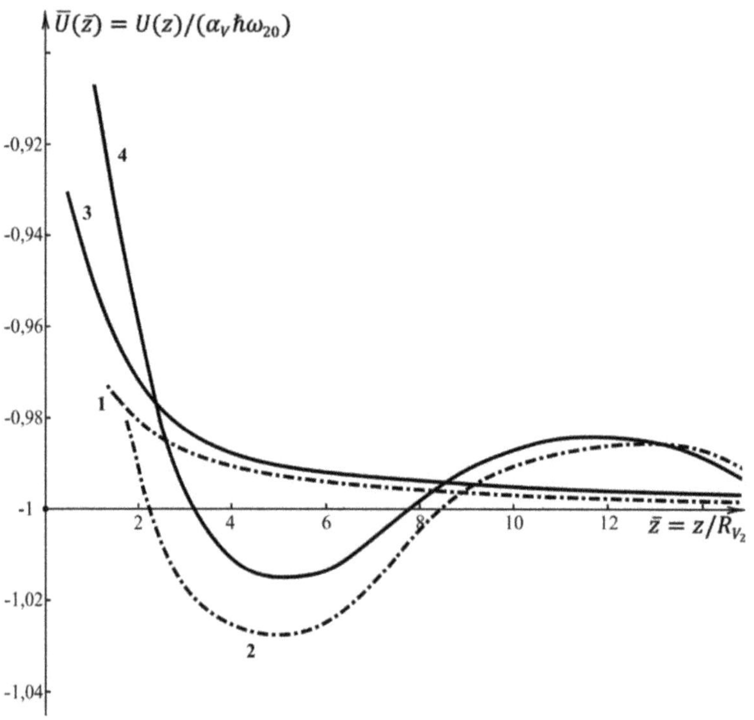

Figure 5.10. Coordinate dependence of the potential energy of the interaction of an electron with its induced polarization at the contact of two polar media: curve 1—$\overline{U}_i(\bar{z})$ at the TlCl – GaAs contact; curve 2—$\overline{U}_T(\bar{z})$ at the TlCl – GaAs contact; and curve 3—$\overline{U}_i(\bar{z})$ at the TlBr – CdTe; curve 4—$\overline{U}_T(\bar{z})$ on the TlBr – CdTe contact.

$$U_T(z) = -\alpha_{nV}\hbar\omega_{n0} + \alpha_{nV}\hbar\omega_{n0}\frac{R_{nV}}{2\,|z|} - \frac{\alpha_1\hbar\Omega_1 R_{1S} + \alpha_2\hbar\Omega_2 R_{2S}}{2\,|z|} + \frac{e^2(\varepsilon_n - \varepsilon_{\bar{n}})}{16\pi\varepsilon_0\varepsilon_n(\varepsilon_n + \varepsilon_{\bar{n}})|\,z\,|}. \tag{5.7.7}$$

Taking into account the definitions α_{nV}, α_{S_1}, α_{S_2} (5.2.38a), (5.2.45), (5.2.46), as well as the formula for frequencies $\Omega_{1,\,2}^2$ (5.1.52), after some transformations we obtain

$$U_T(z) = -\alpha_{nV}\hbar\omega_{n0} + \alpha_{nV}\hbar\omega_{n0}\frac{R_{nV}}{2\,|z|} - \frac{\alpha_1\hbar\Omega_1 R_{1S} + \alpha_2\hbar\Omega_2 R_{2S}}{2\,|z|} + \frac{e^2(\varepsilon_n - \varepsilon_{\bar{n}})}{16\pi\varepsilon_0\varepsilon_n(\varepsilon_n + \varepsilon_{\bar{n}})|\,z\,|}. \tag{5.7.8}$$

Expression (5.7.8) represents the classical limit (potential energy of the image forces), taking into account both types of polarization (electronic and ionic) as a consequence of the general quantum mechanical formulas. In this limit, the parameter $\nu = \omega_2/\omega_1$ has fallen out of the formula for potential energy, since the polarization created by the classical charge does not depend on frequencies and is completely determined by the permittivity ε_{n0}, $\varepsilon_{\bar{n}0}$.

The Schrödinger equation with potential energy (5.7.5) is not solved exactly. Therefore, to calculate the energy of the ground state of the polaron, we use a variational approach with a variational function (5.6.8). The diagonal matrix element of the Hamiltonian (5.7.1) on the wave function (5.6.8), which gives the energy of the ground state of the polaron associated with $E_0(\beta)$, is calculated accurately and expressed in terms of elementary functions:

$$\begin{aligned} E_0(\beta) = &- \alpha_{nV}\hbar\omega_{n0} + \frac{\hbar^2\beta^2}{2m^*} + \frac{e^2(\varepsilon_{\bar{n}0} - \varepsilon_n)\beta}{16\pi\varepsilon_0\varepsilon_n(\varepsilon_{\bar{n}0} + \varepsilon_n)\varepsilon_{\bar{n}0}} \\ &- \alpha_{S_1}\hbar\Omega_1 J_1(\beta R_{S_1}) - \alpha_{S_2}\hbar\Omega_2 J_2(\beta R_{S_2}) - \frac{2\alpha_{nV}\hbar\omega_{n0}}{\pi} J_3(\beta R_V); \end{aligned} \tag{5.7.9}$$

where

$$J_{j=1,2}(\beta R_{S_j}) = \int_0^\infty dx(1 + x^2)^{-1}\left(1 + \frac{x}{\beta R_{S_j}}\right)^{-3}; \tag{5.7.10}$$

$$\begin{aligned} J_3(\beta R_V) = &\int_0^\infty \int_0^\infty \frac{y dx dy}{(x^2 + y^2)(1 + x^2 + y^2)} \\ &\left\{ \frac{1}{\left(1 + \frac{y}{\beta R_V}\right)^3} - \frac{1}{\left(1 + \frac{y}{\beta R_V} + \frac{x^2}{2\beta R_V(y + 2\beta R_V)}\right)^3} \right\}. \end{aligned} \tag{5.7.11}$$

The integrals $J_{j=1,2}$ and J_3 after calculation have the form

$$J_j = \frac{1}{2(a_j^2 + 1)}\left\{ a_j + \frac{4a_j}{a_j^2 + 1} + \frac{\pi(1 - 3a_j^2)}{(1 + a_j^2)^2} + \frac{a_j(3 - a_j^2)\ln a_j^2}{(1 + a_j^2)^3} \right\}, \tag{5.7.12a}$$

where

$$a_j = (\beta R_{S_j})^{-1}; \tag{5.7.12b}$$

$$\begin{aligned} J_3 = 2 &+ \frac{b}{2} + \left(\frac{2b}{1 + 2b}\right)^2 \cdot \frac{b(2b - 3)}{(1 + 2b)} - \frac{22b^2 + 15b + 3}{(2b + 1)^3} - \frac{3}{2b}\ln(1 + 2b) \\ &- \frac{(2b^2 + 1)(3b^2 + 2)}{2(b^2 + 1)^2} - b^4\left\{ \frac{2b^2 - 1}{2(b^2 + 1)^{5/2}} + \frac{1}{(b^2 + 1)^{3/2}} \right\}\ln A \\ &+ \frac{3b^4 \ln B}{2(b^2 + 1)^3} + \frac{b^3(b^2 - 2)}{2(b^2 + 1)^{5/2}}, \end{aligned} \tag{5.7.13}$$

where

$$b = \beta R_V;\; A = b(\sqrt{b^2 + 1} - 1)^{-1};\; B = \sqrt{b^2 + 1} - b. \tag{5.7.14}$$

Table 5.1. Results of calculation of the binding energy of the surface polaron.

No.	Contact of polar connections	α_{V1}	$\alpha_{V1}\hbar\omega_{10}$ (meV)	β_m (10^8 cm^{-1})	β_m^{-1} (Å)	R_V (Å)	$\frac{E_m}{\alpha_{V1}\hbar\omega_{10}}$	$\frac{E_0}{\alpha_{V1}\hbar\omega_{10}}$
1	GaAs/PbTe	0.07	2.57	0.25	400	40.0	0.115	0.104
2	GaAs/TlCl	0.07	2.57	0.12	833	40.0	0.037	0.032
3	GaAs/TlBr	0.07	2.57	0.12	833	40.0	0.039	0.029
4	CdTe/PbTe	0.36	7.84	0.52	192	42.8	0.141	0.047
5	CdTe/GaAs	0.36	7.84	0.16	625	42.8	0.016	0,001
6	CdS/PbTe	0.64	24.60	1.35	74.1	23.7	0.157	0.063
7	CdS/TlBr	0.64	24.60	0.70	143	23.7	0.056	0,029
8	CdS/CdTe	0.64	24.60	0.40	250	23.7	0.014	0.001
9	CdS/GaAs	0.64	24.60	0.42	238	23.7	0.016	0.003
10	TlBr/PbTe	2.05	29.36	1.00	100	38.5	0.063	0.003
11	TlCl/PbTe	2.60	55.83	2.20	45.5	22.0	0.070	0.003
12	CdTe/TlCl	0.36	7.84	0.05	2000	42.8	0.005	0.029

The calculation results are presented in table 5.1 for various contacts.

In the asymptotic limit of $|z|/R_{nV} \gg 1$, the potential energy takes the form (5.7.6). The ground state of a Hamiltonian with averaged, z-independent effective masses and potential energy is exactly

$$E_0(\beta_0) = -\alpha_{nV}\hbar\omega_{n0} - \frac{m^* e^4(\varepsilon_{n0} - \varepsilon_{\bar{n}0})^2}{512\pi^2\hbar^2\varepsilon_0^2\varepsilon_{n0}^2(\varepsilon_{n0} + \varepsilon_{\bar{n}0})^2}, \tag{5.7.15}$$

where the average distance of the polaron from the surface is

$$\beta_0^{-1} = \frac{16\pi\varepsilon_0\hbar^2\varepsilon_{n0}(\varepsilon_{n0} + \varepsilon_{\bar{n}0})}{m^* e^2(\varepsilon_{n0} - \varepsilon_{\bar{n}0})}. \tag{5.7.16}$$

A comparison of $E_0(\beta_{\min})$ and $E_0(\beta_0)$ indicates that the variational calculation in almost all the considered examples given in table 5.1 gives a deeper level of energy. This is due to the fact that when $\varepsilon_{n0} < \varepsilon_{\bar{n}0}$, regardless of the ratio between ε_n and $\varepsilon_{\bar{n}0}$ in the domain of localization of the wave function (in the vicinity of $\beta_{\min}^{-1}$), the graph of quantum mechanical potential energy passes much lower than the classical one. Even in the case when $\varepsilon_{n0} > \varepsilon_{\bar{n}0}$ (but $\varepsilon_n < \varepsilon_{\bar{n}}$) and in the classical limit there is no attraction to the surface at all, a quantum mechanical potential well exists and gives a local level at distances $\beta_{\min}^{-1}$ from the interface satisfying the accepted approximation $\beta_{\min}^{-1} > R_V$.

The wide difference in the results is explained by the peculiarities of the quantum mechanical formula for potential energy—significant deviations (even in the $|z|/R_V \gg 1$) from the classical limit: the nonmonotonic nature of the $U_T(z)$, change due to different rates of change of the contributions $U_{nS}(z)$ and $U_{nV}(z)$ in the intermediate domain. Thus, the localization conditions on the surface are much more complex than it might seem, and the simple condition $\varepsilon_{n0} < \varepsilon_{\bar{n}0}$ is only sufficient, but not necessary.

5.8 Surface polaronic states in external fields

5.8.1 Surface polaronic states in a homogeneous magnetic field

After the idea of cyclotron resonance of current carriers in a solid was proposed and implemented in practice, magnetic resonance measurements have become widely used in experimental practice and are currently one of the main sources of reliable information about the electronic energy spectrum of solids, in particular, multilayer structures. A large number of studies have been devoted to polaronic states in a magnetic field [40, 51, 57–72, 125], including the contact of two media and quasi-two-dimensional systems. It is of interest to calculate the parameters measured in experiments on cyclotron resonance: cyclotron frequencies $\omega_c = eB/m$ (B—magnetic induction), polaron contributions to the effective mass, etc. Since the connection of the magnetic field to the problems of polaronic states in multilayer structures leads to an unknown number of variants, we will limit ourselves to considering simple systems: the contact of a polar crystal in two variants—internal and external surface polarons and two limiting cases: weak and strong magnetic fields [57, 67].

5.8.2 Weak magnetic field

We will consider a weak magnetic field such that the cyclotron frequency is the lowest of all the frequencies of the system; for example,

$$\omega_c < \frac{E_{\text{loc}}}{\hbar} < \omega_{\text{polariz}}. \tag{5.8.1}$$

When this criterion is fulfilled, according to the previous consideration, the charge carrier is a polaron, the Hamiltonian of which is determined by the formula (5.5.1). Averaging it on the wave functions of the localized state, we obtain a Hamiltonian describing motion in the XOY plane:

$$\hat{H}(\mathbf{P}_\perp) = E_0(\beta_{\min}) + \frac{\hat{p}^2}{2m_\perp^*} + b\left(\frac{\hat{p}^2}{2m_\perp^*}\right)^2, \tag{5.8.2}$$

where $E_0(\beta_{\min})$ is the minimized ground state energy (5.6.13):

$m_\perp^* \approx m(1 + a_V + a_S)$; $b \equiv b_V + b_S$;

$$b_V = -\frac{4\hbar^4}{P_\perp^4 m^2}\sum_{\mathbf{Q}}\frac{|V_{nV}(Q)|^2 |g_{nV}(\mathbf{q}_\perp, q_z, z)|^2 (\mathbf{Q}\mathbf{P}_\perp)^4}{\left(\hbar\omega_{n0} + \frac{\hbar^2 Q^2}{2m^*}\right)^5}; \tag{5.8.3a}$$

$$b_S = -\frac{4\hbar^4}{P_\perp^4 m^2}\sum_{s,\mathbf{Q}}\frac{|V_s(\eta)|^2 |g_{nS}(\eta, z)|^2 (\eta\mathbf{P}_\perp)^4}{\left(\hbar\Omega_s + \frac{\hbar^2\eta^2}{2m^*}\right)^5}. \tag{5.8.3b}$$

The Hamiltonian of a polaron in a magnetic field is obtained by replacing [57, 58, 67]:

$$\mathbf{P}_{\perp} \to \mathbf{P}_{\perp} + e\mathbf{A};\ \mathbf{A} = \frac{1}{2}[\mathbf{B}\boldsymbol{\rho}], \tag{5.8.4}$$

A is the vector potential of the magnetic field, ***B*** is the induction of the magnetic field.

Eigenfunctions of the Hamiltonian

$$\hat{H} = \frac{1}{2m^*}(\hat{\mathbf{P}} + e\mathbf{A})^2 \tag{5.8.5}$$

are well known [73]. They will also be proper for the Hamiltonian (5.8.2). Therefore, the energy spectrum of a polaron in a magnetic field has the form

$$E_n = E(\beta_m) + \hbar\omega_c^*\left(n + \frac{1}{2}\right) - \frac{|b|}{(1 + |a_S| + |a_V|)^2}\left\{\left(n + \frac{1}{2}\right)\hbar\omega_c^*\right\}^2, \tag{5.8.6}$$

where

$$\omega_c^* = \frac{eB}{m_{\perp}^*},\ n = 0, 1, 2, \ldots \tag{5.8.7}$$

The resonant frequencies are found from (5.8.6):

$$\Omega_n = \frac{1}{\hbar(E_n - E_{n-1})} \equiv \omega_c^*(1 - n \cdot \Delta); \tag{5.8.8}$$

$$\Delta = \frac{2\,|b|\hbar\omega_c^*}{(1 + |a_V| + |a_S|)^2}. \tag{5.8.9}$$

The displacement of the resonant frequency proportional to Δ causes the nonequidistance of energy levels in the magnetic field, which should manifest itself in the temperature dependence of the width of the cyclotron resonance line.

5.8.3 Strong magnetic field

Consider polaronic states in a strong magnetic field. We will consider inequality as the criterion of a strong field

$$\frac{E_{\text{loc}}}{\hbar} < \omega_{\text{polariz}} < \omega_c. \tag{5.8.10}$$

In accordance with the accepted criterion, the wave functions of fast motion are found as eigenfunctions of the Hamiltonian (5.8.5). For the ground state

$$\psi_0(\rho) = \pi^{\frac{1}{2}} a_B^{-1} \exp\left(-\frac{p^2}{2a_B^2}\right), \tag{5.8.11}$$

where

$$a_B = \left(\frac{2\hbar}{eB}\right)^{\frac{1}{2}} \tag{5.8.12}$$

is magnetic length.

Averaging the electron–polarization Hamiltonian $\psi_0(\rho)$, we obtain an effective Hamiltonian describing motion in the field created by surface and bulk optical phonons:

$$\begin{aligned}\hat{H}_{eff} = \langle\psi_0|\hat{H}|\psi_0\rangle = \frac{1}{2}\hbar\omega_c + \frac{\hat{P}_{\|}^2}{2m} + U_{SA}(z) + \hat{H}_{ph}^V + \hat{H}_{ph}^S \\ + \sum_{\mathbf{Q}(\mathbf{q}_\perp, q_z)} V_{nV}(Q) g_{nV}(\mathbf{q}_\perp, q_z, z)\langle\psi_0|e^{i\mathbf{q}_\perp\boldsymbol{\rho}}|\psi_0\rangle\left[\hat{b}_{-\mathbf{Q}}^+ + \hat{b}_{\mathbf{Q}}\right] \\ + \sum_{s,\boldsymbol{\eta}} V_s(\eta) g_{ns}(\boldsymbol{\eta}, z)\langle\psi_0|e^{i\boldsymbol{\eta}\boldsymbol{\rho}}|\psi_0\rangle\left[\hat{b}_{s,-\boldsymbol{\eta}}^+ + \hat{b}_{s,-\boldsymbol{\eta}}\right].\end{aligned} \tag{5.8.13}$$

Using the adiabatic approximation and perturbation theory in the second order, we calculate the potential created by surface and bulk optical phonons in the presence of a magnetic field:

$$W_{\text{eff}}(z) = U_{SA}(z) + W_{\text{ph}}(z), \tag{5.8.14}$$

where

$$U_{\text{ph}}(z) = -\sum_{\mathbf{Q}} \frac{|V_{nV}(Q)|^2 |g_{nV}(\mathbf{q}_\perp, q_z, z)|^2}{\hbar\omega_{n0} + \frac{\hbar^2 q_z^2}{2m^*} - \frac{\hbar q_z P_r}{m^*}} - \frac{1}{\hbar\Omega_S}\sum_{s,\boldsymbol{\eta}} |V_S(\eta)|^2 |g_{nS}(\boldsymbol{\eta}, z)|^2 e^{-\eta^2 a_B^2}. \tag{5.8.15}$$

Moving in (5.8.15) from summation to integration, we obtain

$$W_{\text{ph}}(z) = -\alpha_S \hbar\Omega_S \frac{R_S}{a_S}\sqrt{\frac{\pi}{2}} \exp\left(\frac{4z^2}{2a_B^2}\right)\left[1 - \Phi\left(\frac{\sqrt{2}z}{a_B}\right)\right] - \alpha_V \hbar\omega_{n0} J(z, a_B, R_S); \tag{5.8.16}$$

here $\Phi(x)$ is a probability function,

$$J(z, a_B, R_S) = \int_0^\infty \mathrm{d}x\, e^{-\frac{a_B^2 x^2}{2R_S^2}}\left\{\frac{1 + e^{-\frac{2zx}{R_S}}}{1 + x} - \frac{2e^{-\frac{zx}{R_S}}}{1 - x^2}\left(e^{-\frac{zx}{R_S}} - xe^{-\frac{z}{R_S}}\right)\right\}. \tag{5.8.17}$$

For z=0, we get from the expression (5.8.14)

$$\begin{aligned}W_{\text{eff}} = -\alpha_V \hbar\omega_0 \Bigg\{\sqrt{\frac{\pi}{2}}\frac{R_S}{a_B} - \sqrt{\frac{\pi}{2}}\frac{R_S}{a_B}\exp\left(\frac{2z^2}{a_B^2}\right)\left[1 - \Phi\left(\frac{\sqrt{2}z}{a_B}\right)\right] \\ - 2e^{-\frac{z}{a_B}}\frac{R_S^2}{a_B^2}\left[1 - \frac{\sqrt{2\pi}z}{4a_B}\exp\left(\frac{z^2}{2a_B^2}\right)\left[1 - \Phi\left(\frac{z}{\sqrt{2}a_B}\right)\right]\right]\Bigg\}.\end{aligned} \tag{5.8.18}$$

The variational energy of the ground state of the surface polaron in the presence of a strong magnetic field is obtained by substituting (5.8.16) into (5.8.13) and (5.8.14) and using the trial function (5.8.11):

$$\begin{aligned}E(\beta) &= \frac{1}{2}\hbar\omega_c + \frac{\hbar^2\beta^2}{2m^*} + \langle\psi(z)|U_{SA}(z) + W_{ph}(z)|\psi(z)\rangle \\ &= \frac{1}{2}\hbar\omega_c + \frac{\hbar^2\beta^2}{2m^*} + \frac{e^2(\varepsilon_1 - \varepsilon_2)\beta}{16\pi\varepsilon_0\varepsilon_1(\varepsilon_1 + \varepsilon_2)} \\ &- 4\sqrt{\frac{\pi}{2}}\beta^3\alpha_S\hbar\Omega_{10}\frac{R_S}{a_B}\int_0^\infty z^2 e^{-2\beta z + \frac{2z^2}{a_B^2}}\left[1 - \Phi\left(\frac{\sqrt{2}z}{a_B}\right)\right]dz \\ &- 4\beta^3\alpha_V\hbar\omega_{10}\int_0^\infty z^2 e^{-\beta z}dz\int_0^\infty dx\, e^{-\frac{a_B^2x^2}{2R_S^2}}\frac{\left(1 - e^{-\frac{2z}{R_S}x}\right)}{(1 + x)}.\end{aligned} \tag{5.8.19}$$

The result (5.8.19) corresponds to the weak interaction of an electron with a surface. In the case where the bond with the surface is strong

$$\omega_{\text{polariz}} < \frac{E_{\text{loc}}}{\hbar} < \omega_c, \tag{5.8.20}$$

the Hamiltonian (5.8.13) is averaged on the variational wave function

$$\exp\left\{\sum_{\mathbf{Q}}\left(\hat{b}^+_{\mathbf{Q}}f_Q - \hat{b}_{\mathbf{Q}}f_Q^*\right)\right\}\exp\left\{\sum_{s,\eta}\left(\hat{b}^+_{s,\eta}f_{\eta,s} - \hat{b}_{s,\eta}f^*_{\eta,s}\right)\right\}\psi(z)|\,0\rangle; \tag{5.8.21}$$

here, $\psi(z)$ has the form (5.8.11). From the condition of the minimum of the variational energy, we obtain

$$\begin{aligned}E(\beta) &= \frac{1}{2}\hbar\omega_c + \frac{\hbar^2\beta^2}{2m^*} + \sum_{\mathbf{Q}}\hbar\omega_{n0}|f_Q|^2 + \sum_{s,\eta}\hbar\Omega_{10}|f_\eta|^2 \\ &+ \sum_{\mathbf{Q}}V_{nV}(Q)\langle\psi(z)|g_{nV}(\mathbf{q}_\perp, q_z, z)|\psi(z)\rangle\exp\left\{-\frac{\hbar q_\perp^2}{4m^*\omega_c}\right\}(f_Q^* + f_Q) \\ &+ \sum_{s,\eta}V_s(\eta)\langle\psi(z)|g_{ns}(\eta, z)|\psi(z)\rangle\exp\left\{-\frac{\hbar\eta^2}{4m^*\omega_c}\right\}(f_\eta^* + f_\eta),\end{aligned} \tag{5.8.22}$$

where

$$\langle\psi(z)|g_{nV}(\mathbf{q}_\perp, q_z, z)|\psi(z)\rangle = \left(1 + \frac{q_z}{2\beta}\right)^{-3} - \left(1 + \frac{q_\perp}{2\beta}\right)^{-3}; \tag{5.8.23a}$$

$$\langle\psi(z)|g_{ns}(\eta, z)|\psi(z)\rangle = \left(1 + \frac{\eta}{2\beta}\right)^{-3} \tag{5.8.23b}$$

They are determined by the values of the variational parameters f_η, f_Q:

$$f_Q = -\frac{1}{\hbar\omega_{n0}} V_{nV}(Q) \exp\left(-\frac{\hbar\eta^2}{4m^*\omega_c}\right) \langle\psi(z)|g_{nV}(\mathbf{q}_\perp, q_z, z)|\psi(z)\rangle; \tag{5.8.24a}$$

$$f_\eta = -\frac{1}{\hbar\Omega_{10}} V_s(\eta) \exp\left(-\frac{\hbar q_\perp^2}{4m^*\omega_c}\right) \langle\psi(z)|g_{ns}(\boldsymbol{\eta}, z)|\psi(z)\rangle, \tag{5.8.24b}$$

and the variational binding energy takes the form

$$\begin{aligned} E(\beta) &= \frac{1}{2}\hbar\omega_c + \frac{\hbar^2\beta^2}{2m^*} \\ &- \sum_{\mathbf{Q}} \frac{1}{\hbar\omega_{n0}} |V_{nV}(Q)|^2 \left|\langle\psi(z)|g_{nV}(\mathbf{q}_\perp, q_z, z)|\psi(z)\rangle\right|^2 \exp\left\{-\frac{\hbar q_\perp^2}{4m^*\omega_c}\right\} \\ &- \frac{1}{\hbar\Omega_{10}} \sum_{s,\boldsymbol{\eta}} |V_s(\eta)|^2 |\langle\psi(z)|g_{ns}(\boldsymbol{\eta}, z)|\psi(z)\rangle|^2 \exp\left\{-\frac{\hbar\eta^2}{4m^*\omega_c}\right\}, \end{aligned} \tag{5.8.25}$$

Here is the result obtained for the external polaron:

$$E(\beta) = \frac{1}{2}\hbar\omega_c + \frac{\hbar^2\beta^2}{2m^*} - 2\alpha_S\hbar\Omega_{10}R_S J(\beta)\beta + \frac{e^2(\varepsilon_1 - \varepsilon_2)\beta}{16\pi\varepsilon_0\varepsilon_1(\varepsilon_1 + \varepsilon_2)}, \tag{5.8.26}$$

where

$$J(\beta) = \int_0^\infty \frac{dx}{(1+x)^6} \exp\left[-\frac{2\hbar\beta^2 x^2}{m\omega_c}\right] dx. \tag{5.8.27}$$

For a two-dimensional polaron ($\beta \to \infty$) the second and fourth terms in (5.8.26) must be omitted, and the third term gives an energy shift depending on the magnetic field:

$$\lim_{\beta\to\infty} 2\alpha_S\hbar\Omega_{10}R_S J(\beta) = \frac{\sqrt{\pi}}{2}\alpha_S\hbar\Omega_{10}\sqrt{\frac{\omega_c}{\Omega_{10}}}. \tag{5.8.28}$$

The expression (5.8.28) coincides with the result obtained in [66]:

$$E_0^{2D} = \frac{1}{2}\hbar\omega_c - \frac{\sqrt{\pi}}{2}\alpha_S\hbar\Omega_{10}\sqrt{\frac{\omega_c}{\Omega_{10}}}. \tag{5.8.29}$$

In the limit $B \to \infty$, the electron–phonon contribution to the energy of the surface polaron tends to saturation (independent of the magnetic field):

$$E_0 = -\alpha_S^2\hbar\Omega_{10}f_3^2(\varepsilon_1, \varepsilon_{10}, \varepsilon_2), \tag{5.8.30}$$

where

$$f_3(\varepsilon_1, \varepsilon_{10}, \varepsilon_2) = \frac{1}{5} + \frac{(\varepsilon_{10} + \varepsilon_2)(\varepsilon_1 - \varepsilon_2)}{4\varepsilon_2(\varepsilon_{10} - \varepsilon_1)}. \tag{5.8.31}$$

According to the expression (5.8.30), the main level of the polaron localized in a strong magnetic field (minus $\frac{1}{2}\hbar\omega_c$) lies deeper than in the absence of a magnetic field. For example, for the SiO_2–vacuum contact $\varepsilon_{10} = 4$, $\varepsilon_1 = 2$, $f_3(\varepsilon_1, \varepsilon_{10}, \varepsilon_2) = 0.8$, and from the formula (5.6.23b) for $f_2(\varepsilon_1, \varepsilon_{10}, \varepsilon_2)$ we get 0.38 (i.e., the value of the phonon contribution increases by about 4.5 times).

The increase in the phonon contribution with an increase in B is due to a decrease in the radius of the polarization cloud of the electron and the achievement of a certain constant value in extremely strong fields, at which the phonon contribution to energy ceases to depend on the magnitude of magnetic induction.

When considering the problem of a surface polaron at the contact of two polar crystals [63, 72] within the limits of weak and strong magnetic fields, as in the case of the absence of a field [44], the Hamiltonian was used, in which the interaction with surface and bulk optical phonons is described approximately. (So, in the Hamiltonian $\hat{H}^{V}_{\text{e-ph}}$, the dependence on the z-coordinate has the form $\sin q_z z$ instead of $e^{iq_z z} - e^{-q_\perp z}$ in the exact Hamiltonian, and in $\hat{H}^{S_1, S_2}_{\text{e-ph}}$ there is a contribution from only one surface optical mode, so the results obtained in these works need this correction.)

5.8.4 Surface polaronic states in a homogeneous electric field

Stable states of an external polaron located in a nonpolar medium ($n = 2$) are possible if there is a potential barrier at the interface high enough to prevent the penetration of an electron from medium $n = 2$ into medium $n = 1$. This situation is realized, for example, in the case when the band gap of the first substance is larger than the second, and therefore the bottom of the conduction band in the first is located higher, and the ceiling of the valence band is lower than in the second. An example is a MDS structure consisting of silicon wafers with a silicon dioxide film and a metal electrode on a dielectric. In [20, 73, 74], the effect of the electron–phonon interaction on the state of an electron in a semiconductor was taken into account. In particular, in [73] a calculation was performed for a model of a thick polar dielectric layer ($d \sim 10^2 - 10^3$ nm). Recently, systems with extremely thin insulating layers (tunnel-transparent and MDS structures with a dielectric layer thickness of 3.0–10 nm) have been increasingly used. In such systems, dimensional effects begin to play an important role.

Consider the contributions to the potential energy of an electron located in silicon at the interface with a polar dielectric. In the depleted layer of a p-type semiconductor, the potential energy of an electron due to its interaction with the charges of acceptors is determined by the formula [20]

$$U_e(z) = \frac{e^2 N_0}{\varepsilon_0 \varepsilon_2} z \left(1 - \frac{z}{2z_0}\right), \tag{5.8.32}$$

where

$$N_0 = N_A z_0, \quad z_0 = \left(\frac{2\varepsilon_0\varepsilon_2\varphi_0}{eN_A}\right)^{\frac{1}{2}}; \tag{5.8.33}$$

where $e\varphi_0$—bending of the band on the surface of the semiconductor and N_A is the concentration of acceptors.

The potential energy of the interaction of electrons in the inversion layer with each other [20] is

$$U_{ie}(\tilde{z}) = \frac{e^2 N_S \tilde{z}}{\varepsilon_0\varepsilon_2}, \tag{5.8.34}$$

where N_S—two-dimensional electron concentration;

$$\tilde{z} = \int_0^\infty z\ |\psi(z)|^2 dz$$

—the average electron removal of the inversion layer from the interface. To these interactions, which are usually taken into account, we add the potential energy of self-action, recorded taking into account the finite thickness of the polar layer ($n = 1$):

$$U_{SA}(z) = \frac{e^2}{8\pi\varepsilon_0}\int_0^\infty \exp(-2\eta\ |z|)\frac{(\varepsilon_2 - \varepsilon_1 \coth \eta d)}{(\varepsilon_2 + \varepsilon_1 \coth \eta d)}d\eta, \tag{5.8.35}$$

and the potential energy $U_{pS}(z)$ of the interaction of an electron with the surface mode of a polar dielectric, which in the adiabatic approximation described in section 5.5 has the form

$$U_{pS}(z) = -\sum_{s,\eta}\frac{|V_s(\eta)|^2 e^{-2\eta|z|}}{\hbar\Omega_s + \frac{\hbar^2\eta^2}{2m*}}, \tag{5.8.36}$$

where the Hamiltonian (5.2.72)–(5.2.75) is used, obtained for a MDS structure with a polar layer as a dielectric.

Let's consider the polaronic effects in the MDS structure. To calculate the energy spectrum of a polaron, it is necessary to solve the Schrödinger equation with the Hamiltonian:

$$\hat{H} = -\frac{\hbar^2}{2m^*}\frac{d^2}{dz^2} + U_t(z), \tag{5.8.37}$$

where the total potential energy $U_t(z)$ contains the four above contributions:

$$U_t(z) = U_e(\tilde{z}) + U_{ie}(\tilde{z}) + U_{SA}(z) + U_{pS}(z). \tag{5.8.38}$$

By choosing a test ground state wave function in the form of (5.4.57) with a node at the contact of the dielectric with the semiconductor, we obtain the variational energy:

$$E_0(\beta) = \frac{\hbar^2\beta^2}{2m_{||}^*} + \frac{3e^2N_0}{\varepsilon_0\varepsilon_2\beta} - \frac{6e^2N_a}{\varepsilon_0\varepsilon_2\beta^2} - \frac{33e^2N_e}{32\varepsilon_0\varepsilon_2\beta}$$
$$+ \frac{e^2\beta^3d^2}{8\pi\varepsilon_0\varepsilon_2}\int_0^\infty dx\left\{\frac{\varepsilon_2 - \varepsilon_{10}\coth x}{(\varepsilon_2 + \varepsilon_1\coth x)(\beta d + 2x)^3}\right.$$
$$\left. - \frac{2\varepsilon_2(1 + \mathrm{th}^2 x)(\varepsilon_{10} - \varepsilon_1)\coth x}{\left(1 + \frac{R_S^2 x^2}{d^2}\right)(\beta d + 4x)^3(\varepsilon_2 + \varepsilon_{10}\coth x)(\varepsilon_2 + \varepsilon_1\coth x)}\right\}. \quad (5.8.39)$$

Minimization of $E_0(\beta)$ by the β parameter was carried out numerically using the following parameter values: Si: $\varepsilon_2 = 12$, $m^*{=}0.19m_0$, SiO_2: $\varepsilon_{10} = 4$, $\varepsilon_1 = 2$, $\hbar\omega_0 = 60$ meV; at the same time, the concentration of acceptors was considered constant: $N_a = 7 \cdot 10^{14}\ \mathrm{cm}^{-3}$, $N_0 = 1.01 \cdot 10^{11}\ \mathrm{cm}^{-2}$. The values of N_s, for which the displacement of the bottom of the conduction band is calculated, are shown in table 5.2. In the second, third, and fourth columns of table 5.2, the energy values of the ground state of the Hamiltonian (5.8.38) are given, in the potential energy of which the contributions are respectively taken into account: U_e, U_{ie}, U_{ps}, U_{SA} $(d \to \infty)$ and all four terms with a finite value d. The energy $U_t(z \to 0)$ is taken as the zero-reference level. From comparing the energy values in the second and third columns, it follows that the polaron effect additionally lowers the energy level of the ground state of the electron by 7%–8%. The results of calculating the energy at finite d indicate a weakening in comparison with the case $d \to \infty$ of the electron's repulsion from the boundary ($\varepsilon_1 < \varepsilon_2$) due to its attraction to a metal electrode (self-action is due to the electrode).

The 'triangular potential' approximation is of interest. The energy of self-action (5.8.35) can be represented as

Table 5.2. The results of numerical calculation of the variational energy of the ground state. Minimization of $E_0(\beta)$ by the β parameter was carried out numerically using the following parameter values: Si: $\varepsilon_2 = 12$, $m^*{=}0.19m_0$, SiO_2: $\varepsilon_{10} = 4$, $\varepsilon_1 = 2$, $\hbar\omega_0 = 60$ meV; at the same time, the concentration of acceptors was considered constant $N_a = 7 \cdot 10^{14}\ \mathrm{cm}^{-3}$, $N_0 = 1.01 \cdot 10^{11}\ \mathrm{cm}^{-2}$.

$N_0 + N_e$ ($10^{11}\ \mathrm{cm}^{-2}$)	E_0 [50] (meV)	E_0 (meV) ($d \to \infty$)	E_0 (meV) ($d = 1$ nm)	z_0 (Å) [114]	z_0 (Å) ($d \to \infty$)
2	27.3	25.6	21.2	7.73	7.65
3	31.4	29.6	24.9	7.42	7.35
4	35.3	33.4	28.5	7.20	7.05
10	54.8	52.5	46.8	6.28	6.23
20	80.6	77.9	71.6	5.52	5.45
50	140.1	136.8	129.2	4.41	4.38
100	216.6	212.9	204.3	3.56	3.53

$$U_{SA}(z) = -\frac{e^2\delta}{16\pi\varepsilon_0\varepsilon_2 z} - \frac{\varepsilon_1 e^2}{2\pi\varepsilon_0(\varepsilon_1+\varepsilon_2)^2}\int_0^\infty \frac{e^{-2\eta z}d\eta}{e^{2\eta d}+\delta}, \tag{5.8.40}$$

where $\delta = (\varepsilon_1 - \varepsilon_2)/(\varepsilon_1 + \varepsilon_2)$.

Expanding the second term in a series in z, we obtain in a linear approximation

$$\begin{aligned} U_{SA}(z) = &-\frac{e^2\delta}{16\pi\varepsilon_0\varepsilon_2 z} - \frac{e^2\varepsilon_1}{4\pi\varepsilon_0(\varepsilon_1+\varepsilon_2)^2 d}\int_0^\infty \frac{dx}{e^x+\delta} \\ &- \frac{e^2\varepsilon_1 z}{4\pi\varepsilon_0(\varepsilon_1+\varepsilon_2)^2 d^2}\int_0^\infty \frac{x\,dx}{e^x+\delta}. \end{aligned} \tag{5.8.41}$$

Similarly, the phonon contribution $U_{pS}(z)$ can be represented as a 'triangular potential':

$$U_{pS}(z) = U_{pS}^0 + eE_1 z, \tag{5.8.42}$$

where

$$U_{pS}^0 = -\frac{e^2}{8\pi\varepsilon_0 d}\int_0^\infty dx \frac{(1+\mathrm{th}^2 x)}{\left(1+\frac{4R_S^2}{d^2}x^2\right)}\left(\frac{1}{1+\frac{\varepsilon_1}{\varepsilon_2}\coth x} - \frac{1}{1+\frac{\varepsilon_{10}}{\varepsilon_2}\coth x}\right); \tag{5.8.43}$$

$$E_1 = -\frac{e}{2\pi\varepsilon_0 d^2}\int_0^\infty dx \frac{(1+\mathrm{th}^2 x)}{\left(1+\frac{4R_S^2}{d^2}x^2\right)}\left(\frac{1}{1+\frac{\varepsilon_1}{\varepsilon_2}\coth x} - \frac{1}{1+\frac{\varepsilon_{10}}{\varepsilon_2}\coth x}\right). \tag{5.8.44}$$

Substituting (5.8.41) and (5.8.42) into (5.8.38) and excluding terms independent of z, we obtain the resulting potential energy, which increases linearly as we move away from the interface between the semiconductor and the dielectric, and then bends and reaches a constant value (the contribution of the inversion layer $U_B(z)$ becomes constant already at a distance, the corresponding z_{SL}, and the quadratic term in (5.8.32) is small, because $z_0 \gg z_d$, z_d is the thickness of the near-surface domain). Then

$$U_t(z) = eE_t z, \tag{5.8.45}$$

where

$$E_t = \frac{(N_0 + fN_s)}{\varepsilon_0\varepsilon_2} + E_1 + E_2 + E_{\mathrm{ext}}; \tag{5.8.46}$$

E_{ext} is an external homogeneous electric field;

$$\frac{N_s e f}{\varepsilon_2\varepsilon_0}$$

is the average field in the inversion layer.

The solution of the Schrödinger equation with boundary conditions according to which the envelope of the wave function is zero at $z \to 0$ and $z \to \infty$ has the form of the Airy function:

$$\psi_i(z) = C\, Ai\left[\left(\frac{2m_{\|}^* eE_t}{\hbar^2}\right)^{\frac{1}{3}}\left(z - \frac{E_i}{eE_t}\right)\right];\ C = \frac{\left(2m_{\|}^*\right)^{\frac{1}{3}}}{\sqrt{\pi}\,(E_t e)^{\frac{1}{6}}\hbar^{\frac{1}{3}}}; \tag{5.8.47}$$

$$E_i = \left(\frac{\hbar^2}{2m_r^*}\right)^{\frac{1}{3}}\left[\frac{3\pi eE_t}{2}\left(i + \frac{3}{4}\right)\right]^{\frac{2}{3}};\ i = 0, 1, 2, \ldots \tag{5.8.48}$$

where i is the number of the ith subband.

The average value of z in the ith subband is

$$\bar{z} = \frac{2E_i}{3eE_t}. \tag{5.8.49}$$

The total energy of the electron, taking into account the terms independent of z, has the form

$$E_{t,\,i} = U_{pS}(0) - \frac{e^2\varepsilon_1 \ln(1+\delta)}{4\pi\varepsilon_0(\varepsilon_1+\varepsilon_2)^2 d\,\delta} - \frac{e^2}{2d}J(d) + \left(\frac{\hbar^2}{2m_{\|}}\right)^{\frac{1}{3}}\left[\frac{3\pi eE_t}{2}\left(i+\frac{3}{4}\right)\right]^{\frac{2}{3}}, \tag{5.8.50}$$

where

$$J(d) = \int_0^\infty dx \frac{(1+\tanh^2 x)}{\left(1+\frac{4R_S^2}{d^2}x^2\right)}\left(\frac{1}{\varepsilon_2 + \varepsilon_1 \coth x} - \frac{1}{\varepsilon_{20} + \varepsilon_1 \coth x}\right). \tag{5.8.51}$$

The expression (5.8.50) is a satisfactory approximation for the Si – SiO_2–metal structure at $d \geqslant 5\ nm$ (the divergence with the results of the variational calculation is approximately 10ı30%).

5.9 Levitating surface polaronic states

5.9.1 An electron above the surface of a liquid helium film deposited on a polar substrate

The actual contact of two crystals is a complex system with a large number of points and extended, charged, and neutral lattice defects. Free and bound charges accumulate in the contact area. Therefore, the effects of electron–phonon interaction and self-action are difficult to observe in their pure form: they are masked by numerous other electron interactions, for example, described by the potentials U_e and U_{ie} (see section 5.7).

In the late sixties and early seventies of the last century, it was proposed [35, 75–84] an ideal system in which the potential energy of self-action plays a decisive role. We are talking about an electron placed on the surface of a dielectric with negative electron affinity (except for liquid helium, these can be Ne, H_2, D_2—liquid and solid [79]). For helium, the potential barrier at the boundary with air is of the order of 1 eV. The dielectric constant of He_4 $\varepsilon \approx 1.0572$. The forces of self-action create a shallow potential well for an electron near the surface of helium:

$$U_{SA}(z) = -\frac{e^2(\varepsilon - 1)}{16\pi\varepsilon_0(\varepsilon + 1)z}, \tag{5.9.1}$$

and the potential barrier on the surface of helium limits its movement to the domain $z > 0$. In the described one-dimensional potential well, stationary states arise, the energy and radius of which are easily found from the Schrödinger equation with a one-dimensional Coulomb potential and an infinite barrier [35, 76, 79, 85–87]:

$$E_n = -\frac{\kappa m_0 e^4}{2\hbar^2 n^2}, \quad n = 1, 2, 3, \ldots; \tag{5.9.2a}$$

$$\kappa = -\frac{(\varepsilon - 1)^2}{16\pi\varepsilon_0(\varepsilon + 1)^2}; \ \bar{z}_0 = \frac{\hbar^2}{\kappa m_0 e^2}. \tag{5.9.2b}$$

Substituting the value of ε gives an energy of approximately $(6.6/n^2) \cdot 10^{-4}$ eV and radii of states of the order of 10 nm. The difference between energy levels can be significantly increased (i.e., the resonant frequency can be increased), and the radius of the state can be reduced by pressing electrons to the surface of helium with an electric field. An electron 'hanging' above the surface of liquid helium is called levitating. The states of the levitating electron have been studied in great detail theoretically and investigated experimentally. A system with a finite thickness of a helium film is considered, the contribution of the substrate to the potential energy of self-action is taken into account, the influence of electric and magnetic fields is studied, the theory of ripplon interaction is constructed [76], and bubbles filled with electrons in helium are predicted and observed. The results of these studies are presented in the reviews [35, 76, 79].

If a film of liquid helium is deposited on a solid substrate, then the effect of the latter on levitating electrons is equivalent to a pressing electric field. For a homeopolar substrate on which a He_4 c $d \sim 10$ nm is applied, the electric field strength reaches 10^2 CGSE units. Such states have been considered in many theoretical works (see [76, 88, 89]). Since the polarization induced by the levitating electron in the substrate is fast (valence electron plasmons), it follows the motion of the electron inertially and the interaction with it can be described on the basis of classical electrodynamics.

A fundamentally new situation arises when a polar crystal serves as a substrate. In this case, the electron interacts with surface polar vibrations at the boundary of liquid helium and the substrate and turns into a levitating polaron. From the point of view of surface polaronic states, this system is of exceptional interest because it allows direct experimental investigation of the electron–polarization interaction with high accuracy (it is not masked by a large number of other interactions that are always present in the case of contact between two crystals: static charges at the boundary, defect fields, etc); i.e., a direct experimental verification of the theory of surface polaronic states becomes possible. Let's turn to the consideration of levitating surface polaronic states.

The Hamiltonian of the system under consideration includes the kinetic energy of an electron (in the *XOY* plane and in the z direction), the energy of surface optical vibrations of a polar substrate, the energy of interaction of an electron with self-induced inertial polarization in a liquid helium film, and a substrate $U_{SA}(z)$, as well as the energy of interaction of an electron with surface optical phonons of the substrate $\hat{H}^{S}_{\text{e-ph}}$. In addition, the Hamiltonian includes the potential energy of the barrier at the vacuum–film He_4 contact:

$$\hat{H} = \frac{\hat{P}_{\|}^{2}}{2m_{\|}^{*}} + \frac{\hat{P}_{\perp}^{2}}{2m^{*}} + \hat{H}^{V}_{\text{ph}} + \hat{H}^{S}_{\text{e-ph}} + U_{SA}(z), \tag{5.9.3}$$

where in the coordinate system, the XOY-plane of which coincides with the interface: the substrate-liquid helium

$$U_{SA}(z) = \frac{e^2}{8\pi\varepsilon_0}\int_0^{\infty}\frac{e^{-2\eta z}\left[\varepsilon_1 - (\varepsilon_1 - 1)\varepsilon_2 \coth \eta d - \varepsilon_2^2\right]d\eta}{\varepsilon_1 + (\varepsilon_1 + 1)\varepsilon_2 \coth \eta d + \varepsilon_2^2}; \tag{5.9.4}$$

$$\hat{H}^{S}_{ph} = \sum_{\eta} \hbar\Omega_S \hat{b}^{+}_{s,\,\eta}\hat{b}_{s,\,\eta}; \tag{5.9.5}$$

$$\hat{H}^{S}_{e-ph} = \sum_{\eta} V_S(\eta) e^{i\eta\rho} e^{-\eta(z-d)}(\hat{b}^{+}_{-\eta} + \hat{b}_{\eta}), \tag{5.9.6}$$

$$U_B = \begin{cases} U_0 \to \infty, & 0 \leqslant z \leqslant d; \\ 0, & z > d. \end{cases} \tag{5.9.7}$$

The square of the modulus of the electron–phonon interaction constant is given in sections 5.1 (formulas (5.2.63)–(5.2.68)).

Let's move on to calculating the binding energy of a levitating polaron.

Consider a weak coupling polaron. The potential created by the surface optical phonons of the $U_S(z)$ substrate is determined using the second order of perturbation theory and the adiabatic approximation (motion along the axis is considered slow, and in the *XOY* plane–fast). Then the Hamiltonian of the system can be represented as

$$\hat{H} = \frac{\hat{P}_{\|}^{2}}{2m_{\|}^{*}} + \frac{\hat{P}_{\perp}^{2}}{2m^{*}} + U_{SA}(z) + U_S(z) + U_B(z), \tag{5.9.8}$$

where

$$U_S(z, \mathbf{P}_{\perp}) = -\sum_{\eta}\frac{|V_S(\eta)|^2 e^{-2\eta z}}{\hbar\Omega_S + \frac{\hbar^2\eta^2}{2m_{\perp}} - \frac{\hbar\eta\mathbf{P}_{\perp}}{m_{\perp}}}. \tag{5.9.9}$$

The variational energy of the ground state of the levitating polaron is calculated using a trial function:

$$\psi(z) = 2\beta^{\frac{3}{2}}(z - d)e^{-\beta(z-d)}; \tag{5.9.10}$$

where

β is a variation parameter.

As a result, we get

$$\begin{aligned} E_0^{WC}(\beta) = &\frac{\hbar^2\beta^2}{2m_\perp} + \frac{e^2}{8\pi\varepsilon_0 d}\int_0^\infty \frac{dx}{\left(1 + \frac{x}{\beta d}\right)^3}\left[\frac{\varepsilon_1 - (\varepsilon_1 - 1)\varepsilon_2 \coth x - \varepsilon_2^2}{\varepsilon_1 + (\varepsilon_1 + 1)\varepsilon_2 \coth x + \varepsilon_2^2}\right] \\ &- \frac{2\alpha_S\hbar\Omega_S R_S}{d}\int_0^\infty \frac{e^{-2x}dx}{\left(1 + \frac{R_S^2 x^2}{d^2}x^2\right)\left(1 + \frac{x}{\beta d}\right)^3}. \end{aligned} \tag{5.9.11}$$

Consider the intermediate electron–phonon coupling. To solve the problem in the approximation of intermediate electron–phonon coupling, we use the method described in 5.2. We subject the Hamiltonian (5.9.3) to a unitary transformation:

$$\hat{S}_1 = \exp\left\{-i\rho\sum_\eta \eta\hat{b}_\eta^+\hat{b}_\eta\right\}. \tag{5.9.12}$$

In this case, we take the adiabatic approximation, in which the movement of the electron along the z-axis is considered fast, and in the XOY plane it is considered slow. We exclude the z-coordinate by averaging the Hamiltonian $\hat{H}_1 = \hat{S}_1^{-1}\hat{H}\hat{S}_1$ on the wave function (5.9.10).

We perform the second unitary transformation using the operator

$$\hat{S}_2 = \exp\left\{\sum_\eta\left[\hat{b}_\eta^+ f_\eta - \hat{b}_\eta f_\eta^*\right]\right\}e^{i\frac{\mathbf{P}_\perp\rho}{\hbar}} \mid 0\rangle. \tag{5.9.13}$$

For the energy of the ground state, we obtain the expression

$$\begin{aligned} E_0^{(IC)}(\beta) = &\frac{\hbar^2\beta^2}{2m_\perp} + \frac{e^2}{8\pi\varepsilon_0 d}\int_0^\infty \frac{dx}{\left(1 + \frac{x}{\beta d}\right)^3}\left[\frac{\varepsilon_1 - (\varepsilon_1 - 1)\varepsilon_2 \coth x - \varepsilon_2^2}{\varepsilon_1 + (\varepsilon_1 + 1)\varepsilon_2 \coth x + \varepsilon_2^2}\right] \\ &- \frac{2\alpha_S\hbar\Omega_S R_S}{d}\int_0^\infty \frac{e^{-2x}dx}{\left(1 + \frac{R_S^2 x^2}{d^2}x^2\right)\left(1 + \frac{x}{\beta d}\right)^3}. \end{aligned} \tag{5.9.14}$$

In the strong electron–phonon coupling approximation, which takes into account not only the usual localization at the surface, but also localization in the XOY plane, the variational wave function is chosen as

$$\psi_0(\rho, z) = \left(\frac{8\gamma^{2\beta^3}}{\pi}\right)^{\frac{1}{2}} \exp\{-\gamma^2\rho^2 - \beta z\}. \tag{5.9.15}$$

The transformations for calculating the variational energy in this case are performed in the following order: first, the Hamiltonian (5.9.3) is averaged on the wave function (5.9.15), then the unitary transformation is performed:

$$\hat{S}_1 = \exp\left\{\sum_{\eta}\left[\hat{b}_{\eta}^{+} f_{\eta} - \hat{b}_{\eta} f_{\eta}^{*}\right]\right\} \mid 0\rangle. \tag{5.9.16}$$

By performing a standard variational procedure, we obtain the variational energy of the ground state:

$$E_0^{(SC)}(\beta, \gamma) = \frac{\hbar^2(\beta^2 + \gamma^2)}{2m_\perp} + \frac{e^2}{8\pi\varepsilon_0 d}\int_0^\infty \frac{dx}{\left(1 + \frac{x}{\beta d}\right)^3}$$

$$\left[\frac{\varepsilon_1 - (\varepsilon_1 - 1)\varepsilon_2 \coth x - \varepsilon_2^2}{\varepsilon_1 + (\varepsilon_1 + 1)\varepsilon_2 \coth x + \varepsilon_2^2}\right] \tag{5.9.17}$$

$$- \frac{2\alpha_S \hbar\Omega_S R_S}{d}\int_0^\infty dx\left(1 + \frac{x}{2\beta d}\right)^{-6} \exp\left(-2x - \frac{x^2}{4\gamma^2 d^2}\right).$$

5.9.2 Approximation of the triangular potential

Due to two circumstances: (a) the weak interaction of the electron with the helium film ($\varepsilon_{\mathrm{He}_4} - 1 \approx 0.057$) and (b) the electron's distance from the substrate by a finite minimum distance equal to the thickness of the helium film d, with not too small thicknesses of the latter, one can simply approximate the potential energy of the electron in vacuum. Consider successively the terms of potential energy $U_{SA}(z)$ $U_{pS}(z)$. It is convenient to represent the result of the integration of $U_{SA}(z)$ in the form

$$U_{SA}(z) = -\frac{\Lambda_0}{z'} - \Lambda_1 \sum_{n=1}^{\infty} \frac{(-a)^{n-1}}{z' + nd}; \quad z' = z - d, \tag{5.9.18}$$

where

$$\Lambda_0 = \frac{e^2(\varepsilon_2 - 1)}{16\pi\varepsilon_0(\varepsilon_2 + 1)}; \quad \Lambda_1 = \frac{e^2\varepsilon_2(\varepsilon_1 - \varepsilon_2)}{4\pi\varepsilon_0(\varepsilon_2 + 1)^2(\varepsilon_1 + \varepsilon_2)}; \tag{5.9.19a}$$

$$a = \frac{(\varepsilon_1 - 1)(\varepsilon_1 - \varepsilon_2)}{(\varepsilon_1 + \varepsilon_2)(\varepsilon_2 + 1)} > 0. \tag{5.9.19b}$$

From (5.9.19b) it can be seen that $a \ll 1$, so in (5.9.18) it is enough to keep the contribution with n= 1:

$$U_{SA}(z) = -\frac{\Lambda_0}{z'} - \frac{\Lambda_1}{z' + d} \equiv -\frac{\Lambda_0}{z'} - \frac{\Lambda_1}{d} + \frac{\Lambda_1}{d^2}z. \tag{5.9.20}$$

The physical meaning of the terms is obvious. The first of them describes the interaction of an electron with helium, and the second with a substrate. In the integral for potential energy, the interaction of an electron with the surface optical vibrations of the substrate (5.9.9) at $d/R_S > 1$, the main role is played by the domain of small x, therefore:

$$\begin{aligned} U_{pS}(z) &\approx -\alpha_S\hbar\Omega_S \int_0^\infty \exp\left(-\frac{2d}{R_S}x\right)\frac{\left(1 + \frac{2xz}{R_S}\right)}{(1 + x^2)}dx \\ &= -\alpha_S\hbar\Omega_S S_1\left(\frac{2d}{R_S}\right) + \frac{2\alpha_S\hbar\Omega_S}{R_S} z\, S_2\left(\frac{2d}{R_S}\right), \end{aligned} \tag{5.9.21}$$

where

$$S_1(x) = \int_0^\infty \frac{e^{-xt}dt}{1 + t^2} = ci(x)\sin(x) - si(x)\cos(x); \tag{5.9.22a}$$

$$S_2(x) = \int_0^\infty \frac{e^{-xt}tdt}{1 + t^2} = -ci(x)\cos(x) - si(x)\sin(x). \tag{5.9.22b}$$

By adding (5.9.20) and (5.9.21) and connecting the external field F_{ext}, we obtain the potential energy

$$U(z) \approx U_{SA}(z) + U_S(z) = -\frac{\Lambda_0}{z'} - \frac{\widetilde{\Lambda}_1}{d} + F_{\text{eff}}z, \tag{5.9.23}$$

where

$$\widetilde{\Lambda}_1 = \Lambda_1 + \alpha_S\hbar\Omega_S S_1\left(\frac{2d}{R_S}\right); \tag{5.9.24a}$$

$$F_{\text{eff}} = F_{ext} + \frac{\Lambda_1}{d^2} + \frac{2\alpha_S\hbar\Omega_S}{d}S_2\left(\frac{2d}{R_S}\right), \tag{5.9.24b}$$

Since $\varepsilon_{\text{He}_4} - 1 \ll 1$, the first term of the right-hand side (5.9.23) can be neglected. Then the solution of the Schrödinger equation with potential (5.9.23) can be represented as

$$\widetilde{E}_n = -\frac{\widetilde{\Lambda}_1}{d} + \left(\frac{\hbar}{2m}\right)^{\frac{1}{3}} F_{eff}^{\frac{2}{3}}\left[\frac{3\pi}{2}\left(n + \frac{3}{4}\right)\right]^{\frac{2}{3}}; \; n = 0, 1, 2, \ldots; \tag{5.9.25}$$

$$\psi_n = \frac{(2m)^{\frac{1}{3}}}{\sqrt{\pi}F_{\text{эфф}}^{\frac{1}{6}}\hbar^{\frac{2}{3}}}Ai\left[\left(\frac{2mF_{eff}}{\hbar^2}\right)^{\frac{1}{3}}\left(z - \frac{\widetilde{E}_n}{F_{eff}}\right)\right], \tag{5.9.26}$$

$Ai(x)$—the Airy function.

5.9.3 Effective mass of the levitating polaron above the surface of the liquid helium film deposited on the polar substrate

To obtain the effective mass of the surface polaron, we decompose the energy in a series in $\mathbf{P}_\perp$ and limit ourselves to the quadratic terms in

$$E(\mathbf{P}_\perp) = E_0(\beta) + \frac{P_\perp^2}{2m_p}. \tag{5.9.27}$$

In the approximation of weak and intermediate coupling forces for the effective mass of the levitating polaron, we obtain the following expression:

$$\frac{m_p^{(WC)}}{m_0} = 1 + \frac{2\alpha_S \hbar R_S}{m_0 \Omega_S d^3} \int_0^\infty \frac{x^2 e^{-2x} dx}{\left(1 + \frac{R_S^2}{d^2}x^2\right)\left(1 + \frac{x}{\beta d}\right)^3}, \tag{5.9.28a}$$

corresponding to a weak electron—phonon bond and

$$\frac{m_p^{(IC)}}{m_0} = 1 + \frac{2\alpha_S \hbar R_S}{m_0 \Omega_S d^3} \int_0^\infty \frac{x^2 e^{-2x} dx}{\left(1 + \frac{R_S^2}{d^2}x^2\right)\left(1 + \frac{x}{\beta d}\right)^6} \tag{5.9.28b}$$

—in the approximation of an intermediate connection.

Figure 5.11 shows the dependences of the polaron effective mass on the thickness of the helium film d for substrates: 1—LiF, 2—KI, and 3—RbI. The calculation was carried out according to the formula (5.9.28a).

It follows from expressions (5.9.11), (5.9.28a) and (5.9.14), (5.9.28b) that a change in the thickness of the liquid helium film deposited on a polar substrate leads to a change in the parameters of the levitating polaron: the binding energy of the polaron with the substrate and its effective mass, which can be experimentally measured with high accuracy, which allows us to obtain an answer from the experiment to the question of the nature of the change in the binding energy of the polaron during the transition from a weak to an intermediate electron–phonon bond.

This problem has been discussed in the context of levitating polaron phase transitions in the works [40, 48–54, 90, 91].

5.10 Cyclotron resonance of a levitating polaron

In experiments on cyclotron resonance with a levitating electron [80, 81, 92] moving over a helium bath, it was found that the measured cyclotron mass coincides with the mass of a free electron with an accuracy better than 10^{-4}. In clause 5.7.2, the energy spectrum of a surface polaron at the contact of two media in a weak magnetic field was found (see formula (5.8.6)). A similar expression can be obtained for a levitating polaron:

$$E_n = E_0(\beta_m) + \left(n + \frac{1}{2}\right)\hbar\omega_c^* - \frac{4F_2}{(1 + F_1)^2}\left[\left(n + \frac{1}{2}\right)\hbar\omega_c^*\right]^2. \tag{5.10.1}$$

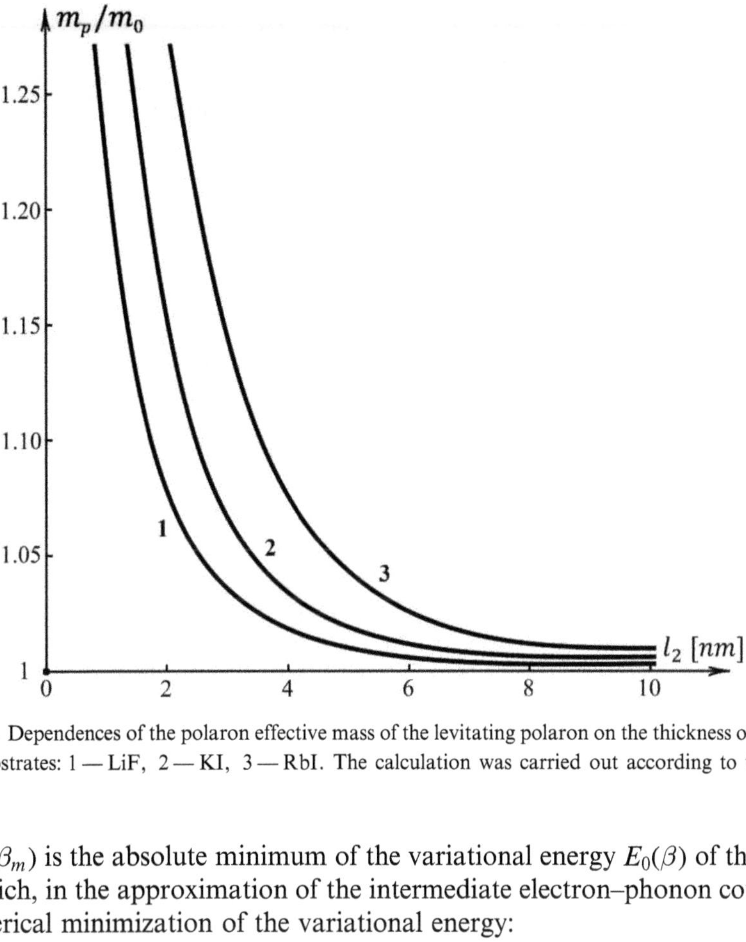

Figure 5.11. Dependences of the polaron effective mass of the levitating polaron on the thickness of the helium film for substrates: 1 — LiF, 2 — KI, 3 — RbI. The calculation was carried out according to the formula (5.9.27a).

Here $E_0(\beta_m)$ is the absolute minimum of the variational energy $E_0(\beta)$ of the ground state, which, in the approximation of the intermediate electron–phonon coupling, is the numerical minimization of the variational energy:

$$E_0(\beta) = \frac{\hbar^2\beta^2}{2m} - 2\alpha_S\hbar\Omega_S R_S \int_0^\infty \frac{\exp(-4Qd)dQ}{\left(1 - R_s^2 Q^2\right)\left(1 + \frac{Q}{2\beta}\right)^6} + \frac{e^2}{8\pi\varepsilon_0}\int_0^\infty \frac{dQ}{\left(1 + \frac{Q}{2\beta}\right)^3}\cdot\frac{\varepsilon_1 - (\varepsilon_1 - 1)\varepsilon_2 \coth \mathrm{Qd} - \varepsilon_2^2}{\varepsilon_1 + (\varepsilon_1 + 1)\varepsilon_2 \coth \mathrm{Qd} + \varepsilon_2^2}; \tag{5.10.2}$$

$$F_1 = 8\alpha_S \int_0^\infty \frac{x^2 \exp\left(-\frac{4d}{R_S}x\right)dx}{(1 + x^2)^3\left(1 + \frac{x}{2\beta_m R_S}\right)^6};$$
$$F_2 = \frac{8\alpha_S}{\hbar\Omega_s}\int_0^\infty \frac{x^4 \exp\left(-\frac{4d}{R_S}x\right)dx}{(1 + x^2)^3\left(1 + \frac{x}{2\beta_m R_S}\right)^6}; \tag{5.10.3a}$$

$$\omega_c^* = \frac{eB}{m_p};\quad m_p = m_0(1 + F_1). \tag{5.10.3b}$$

For resonant frequencies, we obtain an expression similar to (5.7.8):

$$\Omega_n = \omega_c^*(1 - \Delta_n), \tag{5.10.4}$$

with the nonequidistance parameter Δ_n of Landau levels:

$$\Delta_n = \frac{8F_2\hbar\omega_c^* n}{(1 + F_1)^2} \equiv n \cdot \Delta_1. \tag{5.10.5}$$

In the approximation of a triangular potential well, the parameters of the energy spectrum having the form (5.10.1) are obtained by averaging on the Airy function:

$$E_{n,\,i} = \widetilde{\Delta}_i + \hbar\omega_c^*\left(n + \frac{1}{2}\right) - \frac{|b|}{(1 + |a_S|)^2}\left[\hbar\omega_c^*\left(n + \frac{1}{2}\right)\right]^2, \tag{5.10.6}$$

where

$$\widetilde{\Delta}_i = -\frac{e^2\varepsilon_2(\varepsilon_1 - \varepsilon_2)}{4\pi\varepsilon_0(\varepsilon_2 + 1)^2(\varepsilon_1 + \varepsilon_2)d} - \alpha_S\hbar\Omega_S\int_0^\infty \left(1 + x^2\right)^{-1} e^{-\frac{2d}{R_S}x}dx +$$
$$+\left(\frac{\hbar^2}{2m}\right)^{\frac{1}{3}}\left[\frac{3\pi}{2}\left(i + \frac{3}{4}\right)\right]^{\frac{2}{3}} \tag{5.10.7}$$
$$\left\{\frac{e^2\varepsilon_2(\varepsilon_1 - \varepsilon_2)}{4\pi\varepsilon_0(\varepsilon_2 + 1)^2(\varepsilon_1 + \varepsilon_2)d} + \frac{2\alpha_S\hbar\Omega_S}{R_S}\int_0^\infty \frac{xe^{-\frac{2dx}{R_S}}dx}{\left(1 + x^2\right)^2}\right\}^{\frac{2}{3}}.$$

In the case under consideration

$$\frac{m_\perp^*}{m_0} - 1 = |a_S| = 4\alpha_S\zeta^3 J(\zeta); \tag{5.10.8}$$

$$\zeta = 2\beta R_S;\quad J(\zeta) = \int_0^\infty t^2(1 + \zeta^2 t^2)(1 + t)^{-6}dt. \tag{5.10.9}$$

We present an estimate of the value $(m_\perp^*/m_0 - 1)$ and the nonequidistance parameter Δ of Landau levels for various substrates: for a GaAs substrate with a liquid helium film thickness $d_1 = 2$ nm; $|a_S| \approx 0.002$; with $d_2 = 8$ nm; $|a_S| \approx 0.001$; for the substrate CdTe with $d_1 = 2$ nm; $|a_S| \approx 0.016$; with $l_2 = 8$ nm; $|a_S| \approx 0.006$; for the substrate PbI_2 with $l_1 = 2$ nm; $|a_S| \approx 0.105$; with $d_2 = 8$ nm; $|a_S| \approx 0.104$.

For the nonequidistance parameter Δ with $l = 2$ nm for the substrate PbI_2 with $B = 4 \cdot 10^{-3}\,T$ we obtain: $\Delta \approx 0.2$; with $B = 16 \cdot 10^{-3}\,T$; $\Delta \approx 0.9$; with $l = 4$ nm; $B = 16 \cdot 10^{-3}\,T$; $\Delta \approx 0.1$. As follows from the above estimates, $m_\perp^*/m_0 - 1 \equiv |a_S|$ is in the order of magnitude $10^{-2} \div 10^{-3}$ and can be experimentally measured.

5.11 Potential energy of self-action of a charge in a planar structure

The potential energy of a charge located near the interface of various substances contains two main contributions due to: (a) the interaction of this charge with all charges of an ideal undeformed lattice, forming a short-range part of the total potential energy varying within one or two atomic layers of the lattice; (b) correlation effects induced by the electric field of the charge in the lattice (polarization valence electron systems, inertial polarization of ions) and forming an extended part of the total potential energy. The latter is well known as the potential energy of the image forces, which is considered on the basis of a quantum mechanical approach. In the continuum theory developed in this section, we will focus on the extended part, assuming that the short-acting part is given. Consideration of these contact fields is necessary when analyzing phenomena occurring in near-surface areas, the dimensions of which are comparable to the effective penetration length of contact fields into the crystal. These fields shift the energy levels of surface impurity centers and change their spectra, affect the activation energy of 'fast' local centers at the dielectric–semiconductor interface, change the binding energy of impurity complexes located near the interface with metal [36], and also shift the energy levels of excitons in multilayer systems [29, 30, 49]. They have a significant effect on the current-voltage characteristics of barrier structures [27].

Let us proceed to the conclusion of the general expression of the extended part of the contact field in multilayer structures, necessary for the study of polaronic states in these systems.

Let us first consider a special case that is relevant for a large number of experimental situations—a three-layer planar structure [12, 93–95].

5.11.1 Potential energy of self-action in a three-layer anisotropic structure

A three-layer structure is considered, the middle layer of which is enclosed in a domain of space $-l/2 \leqslant z \leqslant l/2$, and the outer layers are in semi-infinite domains $z < l/2$, and $z > l/2$. The dielectric permittivity of all layers is considered to have axial symmetry:

$$\varepsilon_k^{xx} = \varepsilon_k^{yy} \equiv \varepsilon_k^{\perp};\ \varepsilon_k^{zz} = \varepsilon_k^{||};$$

where $k = 1,\ 2,\ 3$ are the layer numbers. The potential due to the bulk charge $\rho_k(\mathbf{r})$ in the $V_k(\mathbf{r})$ in the kth layer of the multilayer structure is derived from the system of Maxwell's electrostatic equations [93]:

$$\mathrm{div}(\overleftrightarrow{\varepsilon}_k \mathrm{grad}\, V_k(\mathbf{r})) = -\varepsilon_0^{-1}\rho_k(\mathbf{r}). \tag{5.11.1}$$

The solution to this problem, taking into account all possible sources of the field (bulk and surface charges, polarization oscillations of the lattice), is given in [18]. In the special case of a point charge located in the middle layer ($k = 2$), $\rho(\mathbf{r}, \mathbf{r}_e) = -e\delta\,(\mathbf{r} - \mathbf{r}_e)\delta_{2,\,k}$ and boundary conditions requiring continuity of the potential and the normal component of the electric displacement vector at the boundaries of the layers, it has the form:

$$V(\mathbf{r}, \mathbf{r}_e) = \int \frac{d^2\eta}{(2\pi)^2} e^{i\eta\rho} V(\eta, z, z_e), \tag{5.11.2}$$

where

$$V(\eta, z, z_e) = \frac{e}{2\varepsilon_0 \overline{\varepsilon}_2 \eta} \Big\{ e^{-\epsilon_2 \eta |z - z_e|} + \frac{2}{e^{2\zeta_2} - \delta_1\delta_3} \Big[\delta_1\delta_3 ch \,\epsilon_2\, \eta(z - z_e) + e^{\zeta_2}(f_1\, ch\, \epsilon_2\, \eta(z + z_e) + f_2\, sh\, \epsilon_2\, \eta(z + z_e))\Big]\Big\}; \tag{5.11.3}$$

$$\delta_j = \frac{\varepsilon_2 - \varepsilon_j}{\varepsilon_2 + \varepsilon_j},\ j = 1, 3;\ \ \epsilon_k = \left(\frac{\varepsilon_k^{\perp}}{\varepsilon_k^{\parallel}}\right)^{\frac{1}{2}};\ \ \varepsilon_k = \left(\varepsilon_k^{\perp} \cdot \varepsilon_k^{\parallel}\right)^{\frac{1}{2}}; \tag{5.11.4a}$$

$$\zeta_k = \epsilon_k \eta l_2;\ f_1 = \frac{\varepsilon_2^2 - \varepsilon_1\varepsilon_3}{(\varepsilon_1 + \varepsilon_2)(\varepsilon_2 + \varepsilon_3)};\ f_2 = \frac{(\varepsilon_1 - \varepsilon_3)\varepsilon_2}{(\varepsilon_1 + \varepsilon_2)(\varepsilon_2 + \varepsilon_3)}. \tag{5.11.4b}$$

Part of the potential

$$V_e(\mathbf{r}, \mathbf{r}_e) = \frac{e}{\varepsilon_0 \varepsilon_2} \int \frac{d^2\eta}{(2\pi)^2 \eta} e^{i\eta\rho} e^{-\epsilon_2 \eta |z - z_e|} \tag{5.11.5}$$

due to the direct action of the electron charge, and the difference

$$V_p(\mathbf{r}, \mathbf{r}_e) = V(\mathbf{r}, \mathbf{r}_e) - V_e(\mathbf{r}, \mathbf{r}_e) \tag{5.11.6}$$

describes the field induced by electronic polarization. The potential energy of the charge of electrons with the same induced polarization field is calculated using the standard electrostatic formula:

$$U_{\mathrm{SA}}(z_e) = \frac{1}{2} \int V_p(\mathbf{r}, \mathbf{r}_e) \rho_2(\mathbf{r}, \mathbf{r}_e) d^3r, \tag{5.11.7}$$

in which the 1/2 multiplier excludes double counting of intercharge interactions. Taking into account the assumption of the pointedness of the bulk charge, as well as using the formulas (5.11.2), (5.11.5) and the definition (5.11.6), we find

$$U_{\mathrm{SA}}(z) = U_{\mathrm{SA}}^0 + U_{\mathrm{SA}}^{\mathrm{even}}(z) + U_{\mathrm{SA}}^{\mathrm{odd}}(z), \tag{5.11.8}$$

where

$$U_{\mathrm{SA}}^0 = \frac{e^2}{4\pi\varepsilon_0 \overline{\varepsilon}_2} \int_0^{\infty} \frac{\delta_1\delta_2 + e^{\zeta_2} f_1}{e^{2\zeta_2} - \delta_1\delta_3} d\eta \tag{5.11.9a}$$

is self-induced displacement of the band edge in the center of the layer,

$$U_{\mathrm{SA}}^{\mathrm{even}}(z) = \frac{e^2 f_1}{2\pi\varepsilon_0 \overline{\varepsilon}_2} \int_0^{\infty} \frac{e^{\zeta_2} \sinh^2(\epsilon_2 \eta z)}{e^{2\zeta_2} - \delta_1\delta_3} d\eta \tag{5.11.9b}$$

is even and

$$U_{\mathrm{SA}}^{\mathrm{odd}}(z) = \frac{e^2 f_2}{2\pi\varepsilon_0 \overline{\varepsilon}_2} \int_0^{\infty} \frac{e^{\zeta_2} \sinh(\in_2 \eta z)\cosh(\in_2 \eta z)}{e^{2\zeta_2} - \delta_1 \delta_3} d\eta \tag{5.11.9c}$$

is an odd part of the potential energy.

From the above explicit form of the functions (5.11.9b) and (5.11.9c), it follows that in the range $-l/2 \leqslant z \leqslant l/2$ and $0 \leqslant z \leqslant l/2$ these functions, together with their derivatives, monotonously depend on z. Because for all values of z, the integral (5.11.9c) is greater than (5.10.c), then in the case of $|f_2| \geqslant |f_1|$ the course of potential energy is determined by an odd function, at $f_2 > 0$ the potential energy of self-action increases monotonously, and at $f_2 < 0$, on the contrary, decreases monotonously. If $|f_2| < |f_1|$, then the course of potential energy is mainly determined by an even function. The odd contribution shifts the extremum of potential energy from the center of the layer (at $f_1 > 0$ the minimum displacement will be a) to the left if $f_2 > 0$, b) to the right if $f_2 < 0$; at $f_1 < 0$ the maximum displacement will be a) at $f_2 > 0$—to the right; b) at $f_2 < 0$—to the left).

Here are examples of numerical calculation of $U_{SA}(z)$ according to the formulas (5.11.8) and (5.11.9a)–(5.11.9, c). In figure 5.12, curves *a, b*, and *c* correspond to the potential energy of self-action in CdTe ($k = 2$), bordering MgF_2 ($k = 3$) and vacuum ($k = 1$). The corresponding parameter values are $\in_2 = 1$, $\varepsilon_1 = 1$, $\varepsilon_2 = 7.1$, $\varepsilon_3 = 5.45$; in this case, $f_1 = 45 > 0$, $f_2 = -31 < 0$. Since $f_1 > f_2$, the minimum potential energy of self-action is shifted to the right. In figure 5.12, one can also see the strong influence of anisotropy on the energy of self-action (curves a', b', c' correspond to the same parameter values, but $\in_2 = 0.1$). In figure 5.13 curves *a, b, c* correspond to the potential energy of self-action in CdTe ($k = 2$), bordering vacuum ($k = 3$) and metal ($k = 1$, $\varepsilon_1 \to \infty$). The curves a', b', c' correspond to $\in_2 = 0, 1$. $U_{SA}(z)$ is a monotone function of z.

Of special interest is the MDS structure ($k = 1$—metal, $\varepsilon_1 \to \infty$, $k = 2$—dielectric, $k = 3$—semiconductor).

From (5.11.9a)–(5.11.9c) with $\varepsilon_1 \to \infty$ we find it for the domain $-l/2 \leqslant z \leqslant l/2$:

$$U_{SA}^{0} = -\frac{e^2}{4\pi\varepsilon_0 \overline{\varepsilon}_2}\left(\delta_3 + \frac{\overline{\varepsilon}_3}{\overline{\varepsilon}_2 + \overline{\varepsilon}_3}\right)\int_0^{\infty} \frac{d\eta}{e^{2\zeta_2} + \delta_3}; \tag{5.11.10a}$$

$$U_{SA}^{even}(z) = -\frac{e^2 \overline{\varepsilon}_3}{4\pi\varepsilon_0 \overline{\varepsilon}_2(\overline{\varepsilon}_2 + \overline{\varepsilon}_3)} \int_0^{\infty} \frac{e^{\zeta_2} \sinh^2(\in_2 \eta z)}{e^{2\zeta_2} + \delta_3} d\eta; \tag{5.11.10b}$$

$$U_{SA}^{odd}(z) = \frac{e^2 \overline{\varepsilon}_2}{2\pi\varepsilon_0 \overline{\varepsilon}_2(\overline{\varepsilon}_2 + \overline{\varepsilon}_3)} \int_0^{\infty} \frac{e^{\zeta_2} \sinh(\in_2 \eta z)\cosh(\in_2 \eta z)}{e^{2\zeta_2} + \delta_3} d\eta. \tag{5.11.10c}$$

Figure 5.14 shows the graphs of $U_{SA}(z)$ for the system: metal ($k = 1$)—dielectric ($k = 2$)—semiconductor ($k = 3$) for three values of the parameter δ_3: –0.9; –0.3; 0. The domain $-l/2 \leqslant z \leqslant l/2$ corresponds to the dielectric; the domain $z < l/2$—metal and $z > l/2$—semiconductor.

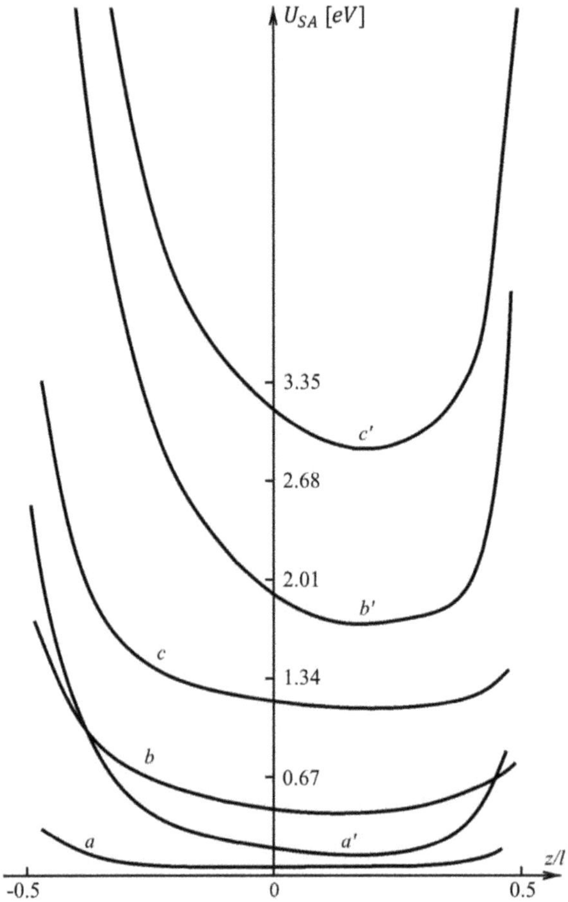

Figure 5.12. Potential energy of interaction in the CdTe layer bordering vacuum and MgF_2. Curves *a, b, c* correspond to the anisotropy parameter $\epsilon_2 = 1$, curves a', b', c'—$\epsilon_2 = 0.1$. The thickness of the CdTe: a, a'—10 nm; b, b'—3 nm; c, c'—2 nm.

$U_{SA}(z)$ in the semiconductor domain can be found from a general solution for a three-layer structure (5.11.3), (5.11.4a), and (5.11.4b):

$$U_{\text{SA}}(z) = -\frac{e^2}{8\pi\varepsilon_0\overline{\varepsilon}_3}\int_0^\infty \frac{e^{-2\eta\left(z+\frac{l}{2}\right)}(\overline{\varepsilon}_3 - \overline{\varepsilon}_2\coth\zeta_2)}{\overline{\varepsilon}_3 + \overline{\varepsilon}_2\coth\zeta_2}d\eta. \tag{5.11.11}$$

Here an expression for $U_{\text{SA}}(z)$ is another relevant case of a symmetric metal–dielectric–metal (MDM) system. Moving to (5.11.8) and (5.11.9a)–(5.11.9c) to the limit $\varepsilon_1 = \varepsilon_3 \rightarrow \infty$, we obtain for the domain $k = 2$

$$U_{\text{SA}}(z)|_{l\rightarrow\infty,\ z\rightarrow z+l/2} \equiv U_{ie}(z) = \frac{e^2}{4\pi\varepsilon_0}\cdot\frac{(\overline{\varepsilon}_2 - \overline{\varepsilon}_1)}{4z\overline{\varepsilon}_2(\overline{\varepsilon}_2 + \overline{\varepsilon}_1)}. \tag{5.11.12}$$

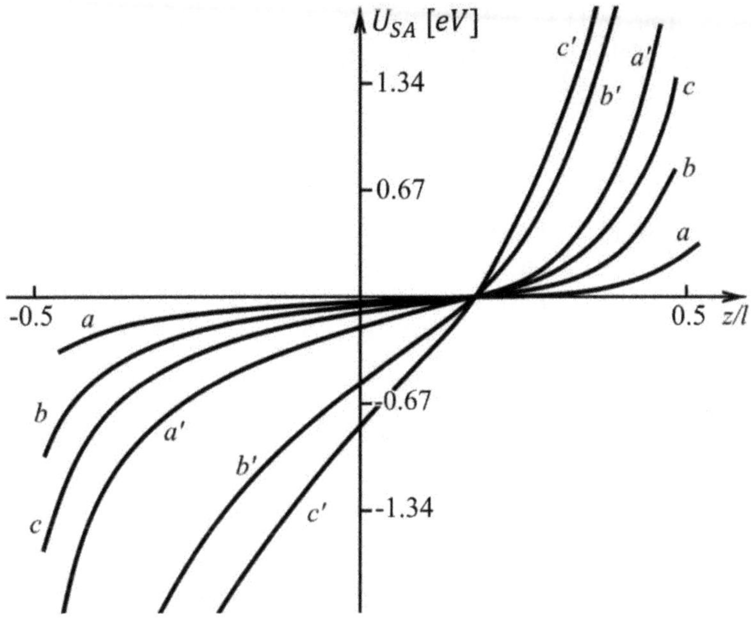

Figure 5.13. The potential energy of self-action in CdTe bordering on vacuum and metal. Curves *a, b, c* correspond to the anisotropy parameter $\epsilon_2 = 1$, curves a', b', c'—$\epsilon_2 = 0.1$. The thickness of the CdTe: a, a'—10 nm; b, b'—3 nm; c, c'—2 nm.

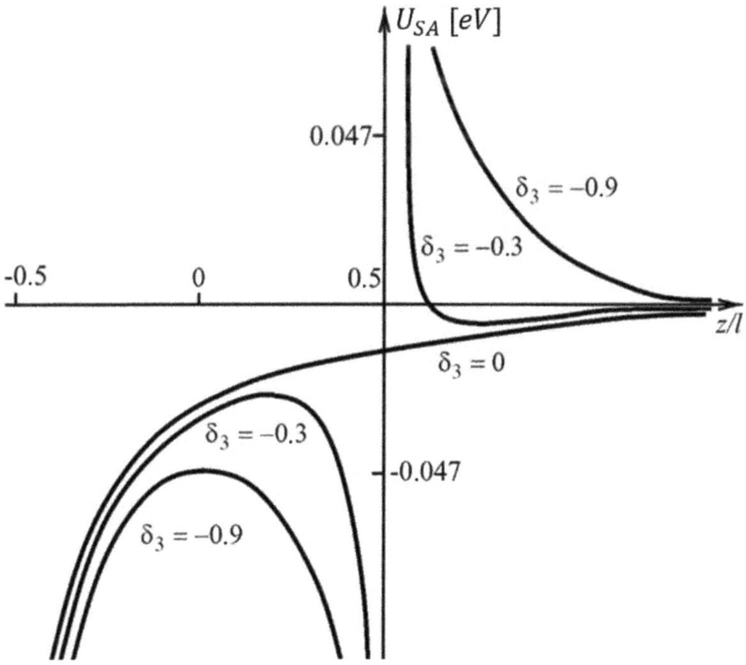

Figure 5.14. The potential energy of self-action in the metal–dielectric–semiconductor structure for three different values of the parameter δ_3.

For $l \to \infty$ from (5.11.8) and (5.11.9a)–(5.11.9c), a known result follows for the potential energy of the interaction of a charge with its image at the contact of two polar media:

$$U_{\rm SA}(z) = \frac{e^2}{4\pi\varepsilon_0\overline{\varepsilon}_2}\int_0^\infty \frac{d\eta}{e^{2\zeta_2}-1}\left[1 - e^{\zeta_2}\cosh(2 \in_2 \eta z)\right]. \tag{5.11.13}$$

It is of interest to find the self-action potential of $U_{SA}(z)$ at small z (small oscillation amplitudes of the particle in the field of self-action potential). Expand (5.11.8) to series of $\in_2\eta z$ up to the quadratic terms:

$$U_{\rm SA}(z) \approx U_0 + K_1 z + K_2 z^2, \tag{5.11.14}$$

where

$$U_0 \equiv U_0(l) = \frac{e^2}{8\pi\varepsilon_0\overline{\varepsilon}_2 \in_2 l_2}\ln\left\{\frac{(1+\sqrt{\delta_1\delta_3})^{\frac{f_1}{\sqrt{\delta_1\delta_3}}-1}}{(1-\sqrt{\delta_1\delta_3})^{\frac{f_1}{\sqrt{\delta_1\delta_3}}+1}}\right\}; \tag{5.11.15a}$$

$$K_1 = K_1(l) = \frac{e^2 f_2 J_1}{2\pi\varepsilon_0\overline{\varepsilon}_2 l^2};\ K_2 = K_2(l) = \frac{e^2 f_1 J_2}{2\pi\varepsilon_0\overline{\varepsilon}_2 l^3}; \tag{5.11.15b}$$

$$J_1 = \int_0^\infty x e^x(e^{2x} - \delta_1\delta_3)^{-1}dx;\ J_2 = \int_0^\infty x^2 e^x(e^{2x} - \delta_1\delta_3)^{-1}dx. \tag{5.11.15c}$$

The expression (5.11.14) can be represented as:

$$U_{\rm SA}(z) \approx U_0 + K_2 z_0^2 + K_2(z + z_0)^2, \tag{5.11.16}$$

where

$$z_0 = \frac{K_1}{2K_2}. \tag{5.11.17}$$

It follows from (5.11.15c) that for $f_1 > 0$, $K_2 > 0$, the potential energy (5.11.14) has a parabolic shape with an offset center. (In a symmetric system $\varepsilon_1 = \varepsilon_3$ and $f_2 = 0$, so the bottom of the parabola lies in the center of the layer $k = 2$.) At $f_1 < 0$, $K_2 < 0$, the potential energy (5.11.14) has the form of a downward-facing parabola with a displaced vertex. In the case of an asymmetric structure with $\int \varepsilon_1 \gg \varepsilon_2 > \varepsilon_3$ (e.g., MDS structures), the quadratic term in (5.11.14) at $\eta(z - l/2) \ll 1$ becomes small compared to the linear one and the expression for $U_{SA}(z)$ can be represented as

$$U_{\rm SA}(z) \approx U_0' + eE_{\it eff}\, z, \tag{5.11.18a}$$

where

$$E_{\rm eff} = \frac{eJ_1'}{2\pi(\overline{\varepsilon}_2 + \overline{\varepsilon}_3)\varepsilon_0 l};\ J_1' = \int_0^\infty x e^x(e^{2x} + \delta_3)^{-1}dx;$$

$$U_0' = -\frac{e^2}{8\pi\varepsilon_0\overline{\varepsilon}_2 \in_2 l}\left\{\ln(1+\delta_3) + \frac{\overline{\varepsilon}_2(\pi - 2\text{artan}\delta_3^{-\frac{1}{2}}}{\delta_3^{\frac{1}{2}}(\overline{\varepsilon}_2 + \overline{\varepsilon}_3)}\right\} \quad (5.11.18b)$$

It can be considered as an effective field of charge-induced polarization (triangular potential approximation).

In conclusion of this section, we note two circumstances: (1) $U_{SA}(z)$ does not depend on the sign of the charge and is the same for an electron and a hole (i.e., the sign of the contribution from self-action to the effective band gap $\widetilde{E}_g$ of a semiconductor film on a substrate depends on the ratio of the permittivity). If $\varepsilon_{\text{substr}} > \varepsilon_{\text{film}}$, then $U_{SA}(z)$ decreases; when $\varepsilon_{\text{substr}} < \varepsilon_{\text{film}}$ increases. In all cases, this contribution significantly depends on the thickness of the film and the anisotropy of its dielectric constant.

Expression for $U_{SA}(z)$ has singularities at the boundaries of the layers. As can be shown, these divergences disappear in the quantum mechanical approach, in which the charge carrier is considered as a quasi-particle (plasmonic polaron) with a finite radius. Thus, in the immediate vicinity of the boundary (at a distance of z of the order R_{pl}), the expression for $U_{SA}(z)$ becomes unfair. For typical values of semiconductor parameters $m^*=0.1m_0$ and $\omega_{pl} \sim 10^{15} - 10^{16}\ \text{s}^{-1}$ the value of R_{pl} is of the order of the lattice constant.

5.11.2 Potential energy of self-action in a superlattice

In a periodic multilayer structure formed by alternating layers of type a and b with charge in layer b, $U_{SA}^{SL}(z)$ has the form [10, 96, 97]

$$\begin{aligned} \mathrm{U}_{\mathrm{SA}}^{\mathrm{SL}}(z) &= \frac{e^2}{4\pi\varepsilon_0\overline{\varepsilon}_b}\int_0^\infty \frac{d\eta}{\sin\mathrm{h}\zeta_b\sqrt{\psi^2-1}} \\ &\left\{e^{-\zeta_b}\sqrt{\psi^2-1} + \frac{\cos\mathrm{h}\zeta_b}{\psi+\sqrt{\psi^2-1}} - \cos\mathrm{h}\zeta_a\right. \\ &\left.+ (\overline{\varepsilon}_b^2 - \overline{\varepsilon}_a^2)(2\overline{\varepsilon}_a\overline{\varepsilon}_b)^{-1}\sin\mathrm{h}\zeta_a \sin\mathrm{h}\zeta_b \cos\mathrm{h}2 \in_b \eta z\right\}, \end{aligned} \quad (5.11.19)$$

where $\overline{\varepsilon}_a$, $\overline{\varepsilon}_b$—dielectric permittivity of layers and l_a, l_b—layer thicknesses,

$$\psi = \cos\mathrm{h}\zeta_a \cos\mathrm{h}\zeta_b + (\overline{\varepsilon}_a^2 - \overline{\varepsilon}_b^2)(2\overline{\varepsilon}_a\overline{\varepsilon}_b)^{-1}\sin\mathrm{h}\zeta_a \sin\mathrm{h}\zeta_b,\ \zeta_{a,b} = \in_{a,b}\eta l_{a,b}.$$

Note the properties of $U_{SA}^{SL}(z)$, which follows from this formula directly: $U_{SA}^{SL}(z)$, firstly, is periodic with a period $L = l_a + l_b$; secondly, it is symmetrical with respect to the center of layers a and b, thirdly, in layer b it has the sign of the difference $\overline{\varepsilon}_b - \overline{\varepsilon}_a$ (for $\overline{\varepsilon}_b = \overline{\varepsilon}_a$ the integrand function vanishes). Denoting $U_a^0(z)$ and $U_b^0(z)$, due to the short-range potential of the electron energy jumps during the transition from the bottom of the conduction band to vacuum, and E_{ga}^0 and E_{gb}^0—the widths of the forbidden bands in an ideal lattice (with undeformed electron shells)—we obtain a change in the level of the bottom of the conduction band $U_a^0(z) - U_b^0(z)$ and the ceiling of the valence band, $E_{ga}^{0} + U_{a}^{0}(z) - E_{gb}^{0} - U_{b}^{0}(z)$, during the transition from layer b to layer a. Note that the chemical affinity also includes the extended potential $\chi_a = U_a^0 + U_{SA}^0$, where U_{SA}^0 is the value of the self-action energy at the interface. Within the framework of classical theory, it is not defined. These jumps form a periodic structure in the superlattice. $U_{SA}^{SL}(z)$ shifts the bottom level of the conduction band in the center of layer b at $\overline{\varepsilon}_b > \overline{\varepsilon}_a$ upwards:

$$\begin{aligned} U_{\mathrm{SA}}^{\mathrm{SL}}(0) &= \frac{e^2}{4\pi\varepsilon_0\overline{\varepsilon}_b}\int_0^{\infty}\frac{d\eta}{sh\zeta_b\sqrt{\psi^2-1}} \\ &\left\{e^{-\zeta_b}\sqrt{\psi^2-1} + \frac{\cos \mathrm{h}\zeta_b}{\psi+\sqrt{\psi^2-1}} - \cos \mathrm{h}\zeta_a\right. \\ &\left. + (\overline{\varepsilon}_b^2 - \overline{\varepsilon}_a^2)(2\overline{\varepsilon}_a\overline{\varepsilon}_b)^{-1}\sin \mathrm{h}\zeta_a \sin \mathrm{h}\zeta_b\right\}. \end{aligned} \tag{5.11.20}$$

The ceiling of the valence band in the same layer is shifted by the same amount, but downwards. For $\overline{\varepsilon}_b - \overline{\varepsilon}_a > 0$, the formula (5.11.20) determines the shift of the bottom of the conduction band and the ceiling of the valence band down and up, respectively. The resulting course of the potential is such that in the case of inequality $\overline{\varepsilon}_b - \overline{\varepsilon}_a > 0$, the band gap of layer b increases and layer a decreases. The potential barrier at the boundary for electrons and holes increases. In the case when $\overline{\varepsilon}_b - \overline{\varepsilon}_a < 0$, the heights of potential barriers at the border decrease.

As shown in figures 5.15 and (b), the width of the band gap in layer a increases, and in layer b it decreases. The potential barrier at the boundary for electrons and holes decreases. In figure 5.15(b), the widths of the forbidden bands change in directions opposite to figure 5.15(a), and the heights of potential barriers at the border increase. Here are some numerical estimates. In the GaAs/Al_xGa_{1-x} ($x = 0.30$) at $l_a = l_b = 5.0$ nm, $\varepsilon_a = 11$, $\varepsilon_b = 13$; $U_{SA}^{SL}(0) \sim 1$ meV. The small value of the self-action energy in this case is explained by the close values of ε_a, ε_b. Taking, for example, $\varepsilon_a = 2$, $\varepsilon_b = 20$, we get for $U_{SA}^{SL}(0)$ the value of 100 meV (i.e., two orders of magnitude more).

In superlattices made of thin layers with very different values of dielectric permittivity for interaction, the corresponding charge self-action in the periodic potential may be significant.

5.12 Potential of bulk charges and self-action potential in a cylindrical wire in a nonpolar medium

5.12.1 Potential of bulk charges in a wire in an infinite dielectric matrix

Consider a wire of radius R and permittivity ε_1, placed in an infinite medium of ε_2. This system can be considered as a special case of a cylindrical multilayer system with two layers ($I = 1$; $K = 2$), with the radius of the second layer directed to infinity. To obtain formulas describing the potential in such a system, it is necessary to put in all formulas of section 2.1:

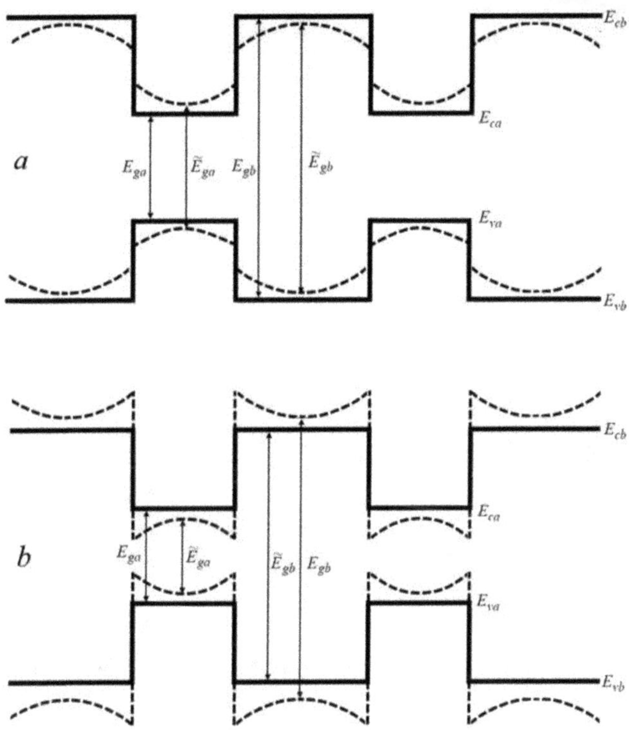

Figure 5.15. The band structure of the superlattice.

$$r_0 \to 0;\ r_1 = R;\ r_2 \to \infty. \tag{5.12.1}$$

Moreover, let the system under consideration be completely isotropic. Then, taking into account the formulas (2.2.19)–(2.2.21), we obtain

$$\eta_1 = \eta_2 = \eta;\ \nu_1 = \nu_2 = |m|. \tag{5.12.2}$$

For a two-layer system, the potential in each of the layers is found by formulas (2.2.29)–(2.2.33) at $n = 1$ and $n = 2$, respectively:

$$\begin{aligned} V_1(r \mid m, \eta) = {} & V(r_1 \mid m, \eta)F_1(r, r_0 \mid m, \eta) + V(r_0 \mid m, \eta)F_1(r_1, r \mid m, \eta) \\ & + \frac{1}{\varepsilon_0}\int_{r_0}^{r_1} G_1(r, r'|m, \eta)\widetilde{\rho}_1(r'|m, \eta)r'dr'; \end{aligned} \tag{5.12.3a}$$

$$\begin{aligned} V_2(r \mid m, \eta) = {} & V(r_2 \mid m, \eta)F_2(r, r_1 \mid m, \eta) + V(r_1 \mid m, \eta)F_2(r_2, r \mid m, \eta) \\ & + \frac{1}{\varepsilon_0}\int_{r_1}^{r_2} G_2(r, r'|m, \eta)\widetilde{\rho}_2(r'|m, \eta)r'dr', \end{aligned} \tag{5.12.3b}$$

where

$$F_1(r, r'|m, \eta) = \frac{S_1(r, r'|m, \eta)}{s_1(m, \eta)}; \tag{5.12.4a}$$

$$F_2(r, r'|m, \eta) = \frac{S_2(r, r'|m, \eta)}{s_2(m, \eta)}; \tag{5.12.4b}$$

$$S_1(r, r'|m, \eta) = I_{\nu_1}(\eta_1 r)K_{\nu_1}(\eta_1 r') - I_{\nu_1}(\eta_1 r')K_{\nu_1}(\eta_1 r); \tag{5.12.5a}$$

$$S_2(r, r'|m, \eta) = I_{\nu_2}(\eta_2 r)K_{\nu_2}(\eta_2 r') - I_{\nu_2}(\eta_2 r')K_{\nu_2}(\eta_2 r); \tag{5.12.5b}$$

$$s_1(m, \eta) = S_1(r_1, r_0|m, \eta); \tag{5.12.6a}$$

$$s_2(m, \eta) = S_2(r_2, r_1|m, \eta); \tag{5.12.6b}$$

$$\begin{aligned} G_1(r, r'|\, m, \eta) = {} & \frac{1}{\varepsilon_1^r s_1(m, \eta)}\{\theta(r > r')S_1(r_1, r|m, \eta)S_1(r', r_0|m, \eta) \\ & + \theta(r < r')S_1(r_1, r'|m, \eta)S_1(r, r_0|m, \eta)\}; \end{aligned} \tag{5.12.7a}$$

$$\begin{aligned} G_2(r, r'|\, m, \eta) = {} & \frac{1}{\varepsilon_2^r s_2(m, \eta)}\{\theta(r > r')S_2(r_2, r \mid m, \eta)S_2(r', r_1 \mid m, \eta) \\ & + \theta(r < r')S_2(r_2, r'|\, m, \eta)S_2(r, r_1 \mid m, \eta)\}. \end{aligned} \tag{5.12.7b}$$

Substituting (5.12.1) into (5.12.4a), (5.12.4b)–(5.12.6a), (5.12.6b), we get simpler expressions:

$$S_1(r, r'|m, \eta) = S_2(r, r'|m, \eta) = I_{|m|}(\eta r)K_{|m|}(\eta r') - I_{|m|}(\eta r')K_{|m|}(\eta r); \tag{5.12.7c}$$

$$s_1(m, \eta) = I_{|m|}(\eta r_1)K_{|m|}(\eta r_0) - I_{|m|}(\eta r_0)K_{|m|}(\eta r_1); \tag{5.12.8a}$$

$$s_2(m, \eta) = I_{|m|}(\eta r_2)K_{|m|}(\eta r_1) - I_{|m|}(\eta r_1)K_{|m|}(\eta r_2); \tag{5.12.8b}$$

Taking into account the isotropy of the media, we redefine

$$\varepsilon_1^r = \varepsilon_1; \varepsilon_2^r = \varepsilon_2. \tag{5.12.9}$$

We will use these designations in the future for the dielectric permittivity of media (in contrast to the definition of the anisotropy parameter (2.2.19) in section 2.1).

The modulation transmission matrix for a two-layer system consists of one element:

$$\begin{aligned} D_{11} = (b_{11})^{-1} = \tilde{\nu}_1 = (\eta R)^{-1} \\ \times \left(\varepsilon_1 \frac{I'_{|m|}(\eta r_1)K_{|m|}(\eta r_0) - I_{|m|}(\eta r_0)K'_{|m|}(\eta r_1)}{I_{|m|}(\eta r_1)K_{|m|}(\eta r_0) - I_{|m|}(\eta r_0)K_{|m|}(\eta r_1)} \right. \\ \left. + \varepsilon_2 \frac{I'_{|m|}(\eta r_1)K_{|m|}(\eta r_2) - I_{|m|}(\eta r_2)K'_{|m|}(\eta r_1)}{I_{|m|}(\eta r_2)K_{|m|}(\eta R) - I_{|m|}(\eta R)K_{|m|}(\eta r_2)} \right). \end{aligned} \tag{5.12.10}$$

The value of the Fourier image of the potential at the thread boundary is determined by the formula

$$V\left(r_1\middle|m,n\right)=\left(\frac{\varepsilon_1}{I_{|m|}(\eta r_1)K_{|m|}(\eta r_0)-I_{|m|}(\eta r_0)K_{|m|}(\eta r_1)}V\left(r_0\middle|m,\eta\right)\right.$$
$$\left.+\frac{\varepsilon_2}{I_{|m|}(\eta r_2)K_{|m|}(\eta r_1)-I_{|m|}(\eta r_1)K_{|m|}(\eta r_2)}V\left(r_2\middle|m,\eta\right)\right)$$
$$+r_1\varepsilon_0^{-1}\Big\{\sigma_1(m,\eta)$$
$$+\frac{1}{r_1}\int_{r_0}^{r_1}\left\{F_1(r',r_0\mid m,\eta)\left[\rho_1(r'\mid m,\eta)-i\eta\,P_1^z(r'\mid m,\eta)\right.\right.$$
$$\left.\left.-\frac{im}{r'}P_1^{\phi}(r'\mid m,\eta)\right]+\frac{\partial F_1\left(r',r_0\middle|m,\eta\right)}{\partial r'}P_1^r\left(r'\middle|m,\eta\right)\right\}r'dr' \tag{5.12.11}$$
$$+\frac{1}{r_1}\int_{r_1}^{r_2}\left\{F_2(r_2,r'\mid m,\eta)\left[\rho_2(r'\mid m,\eta)-i\eta\,P_2^z(r'\mid m,\eta)\right.\right.$$
$$\left.-\frac{im}{r'}P_2^{\phi}(r'\mid m,\eta)\right]$$
$$\left.\left.+\frac{\partial F_2\left(r_2,r'\middle|m,\eta\right)}{\partial r'}P_2^r\left(r'\middle|m,\eta\right)\right\}r'dr'\Big\}\right]\times\tilde{\nu}_1^{-1}.$$

Taking into account (5.12.1) from (5.12.4a)–(5.12.6b), we obtain

$$F_1(r,r_0|m,\eta)\rightarrow\frac{I_{|m|}(\eta r)}{I_{|m|}(\eta R)};\ F_1(r_1,r|m,\eta)\rightarrow 0; \tag{5.12.12a}$$

$$F_2(r_2,r|m,\eta)\rightarrow\frac{K_{|m|}(\eta r)}{K_{|m|}(\eta R)};\ F_2(r,r_1|m,\eta)\rightarrow 0; \tag{5.12.12b}$$

$$\frac{\partial F_1(r,r_0|m,\eta)}{\partial r}\rightarrow\frac{I'_{|m|}(\eta r)}{I_{|m|}(\eta R)};\ \frac{\partial F_2(r_2,r|m,\eta)}{\partial r}\rightarrow\frac{K'_{|m|}(\eta r)}{K_{|m|}(\eta R)}. \tag{5.12.13}$$

In the same limit, the formula (5.12.10) will take the form

$$\tilde{\nu}_1=\frac{\varepsilon_1\frac{I'_{|m|}(\eta R)}{I_{|m|}(\eta R)}-\varepsilon_2\frac{K'_{|m|}(\eta R)}{K_{|m|}(\eta R)}}{\eta R}. \tag{5.12.14}$$

Substituting (5.12.12)–(5.12.14) into (5.12.11), taking into account that the first two terms in the right part vanish, we get

$$\begin{aligned}
&V(R \mid m, \eta) \\
&= \Big[R\varepsilon_0^{-1}\big\{\sigma_1(m, \eta) \\
&+ \varepsilon_0^{-1}\int_0^R \left\{ \frac{I_{|m|}(\eta r')}{I_{|m|}(\eta R)}\Big[\rho_1(r' \mid m, \eta) - \frac{im}{r'}P_1^{\phi}(r' \mid m, \eta) \right. \\
&- i\eta\, P_1^{z}(r' \mid m, \eta)\Big] \\
&\left. + \frac{I'_{|m|}(\eta r')}{I_{|m|}(\eta R)} P_1^{r}\big(r' | m, \eta\big) \right\} r' dr' \\
&+ \varepsilon_0^{-1}\int_R^{\infty} \left\{ \frac{K_{|m|}(\eta r')}{K_{|m|}(\eta R)}\Big[\rho_2(r' \mid m, \eta) - \frac{im}{r'}P_2^{\phi}(r' \mid m, \eta) \right. \\
&- i\eta\, P_2^{z}(r' \mid m, \eta)\Big] \\
&\left.\left. + \frac{K'_{|m|}(\eta r')}{K_{|m|}(\eta R)} P_2^{r}\big(r' | m, \eta\big) \right\} r' dr' \right\}\Big] \cdot \eta R\left(\varepsilon_1\frac{I'_{|m|}(\eta R)}{I_{|m|}(\eta R)} - \varepsilon_2\frac{K'_{|m|}(\eta R)}{K_{|m|}(\eta R)} \right)^{-1}
\end{aligned} \tag{5.12.15}$$

Finally, substituting (5.12.1) into (5.12.7a) and (5.12.7b), we obtain the following expressions for Green's functions in each of the environments:

$$\begin{aligned}
&G_1(r, r' \mid m, \eta) = \big(\varepsilon_1 I_{|m|}(\eta R)\big)^{-1} \\
&\Big[\theta(r > r')\big\{I_{|m|}(\eta R)K_{|m|}(\eta r) - I_{|m|}(\eta r)K_{|m|}(\eta R)\big\}I_{|m|}(\eta r') \\
&+ \theta(r > r')\big\{I_{|m|}(\eta R)K_{|m|}(\eta r') - I_{|m|}(\eta r')K_{|m|}(\eta R)\big\}I_{|m|}(\eta r)\Big];
\end{aligned} \tag{5.12.16a}$$

$$\begin{aligned}
&G_2\big(r, r' | m, \eta\big) = \big(\varepsilon_2 K_{|m|}(\eta R)\big)^{-1} \\
&\Big[\theta(r > r')\big\{I_{|m|}(\eta r')K_{|m|}(\eta R) - I_{|m|}(\eta R)K_{|m|}(\eta r')\big\}K_{|m|}(\eta r) \\
&+ \theta(r < r')\big\{I_{|m|}(\eta r)K_{|m|}(\eta R) - I_{|m|}(\eta R)K_{|m|}(\eta r)\big\}K_{|m|}(\eta r')\Big].
\end{aligned} \tag{5.12.16b}$$

The potential in the thread and the environment is obtained by substituting (5.12.12), (5.12.15), (5.12.16a), (5.12.16b) in formulas (5.12.3a), 5.12.3b), respectively:

$$V_1(r \mid m, \eta)) = -[R\varepsilon_0^{-1}\{\sigma_1(\eta, m)$$

$$+\varepsilon_0^{-1}\int_0^R \left\{\frac{I_{|m|}(\eta r')}{I_{|m|}(\eta R)}\left(\rho_1(r'|m, \eta) - \frac{im}{r'}P_1^{\phi}(r'|m, \eta) - i\eta\, P_1^z(r'|m, \eta)\right)+\right.$$

$$\left.\frac{I'_{|m|}(\eta r')}{I_{|m|}(\eta R)}P_1^r(r'|m, \eta)\right\}r'dr'$$

$$+\,\varepsilon_0^{-1}\int_R^{\infty}\left\{\frac{K_{|m|}(\eta r')}{K_{|m|}(\eta R)}\left(\rho_2(r'|m, \eta) - \frac{im}{r'}P_2^{\phi}(r'|m, \eta) - i\eta\, P_2^z(r'|m, \eta)\right)+\right. \tag{5.12.17a}$$

$$\left.\left.\frac{K'_{|m|}(\eta r')}{K_{|m|}(\eta R)}P_2^r(r'|m, \eta)\right\}r'dr'\right\}\Bigg]\cdot \eta R\left(\varepsilon_1\frac{I'_{|m|}(\eta R)}{I_{|m|}(\eta R)} - \varepsilon_2\frac{K'_{|m|}(\eta R)}{K_{|m|}(\eta R)}\right)^{-1}\cdot\frac{I_{|m|}(\eta r)}{I_{|m|}(\eta R)}$$

$$+\,\varepsilon_0^{-1}\int_0^R(\varepsilon_1 I_{|m|}(\eta R))^{-1}\Big[\theta(r > r')\Big\{I_{|m|}(\eta R)K_{|m|}(\eta r) - I_{|m|}(\eta r)K_{|m|}(\eta R)\Big\}I_{|m|}(\eta r')+$$

$$\theta(r > r')\Big\{I_{|m|}(\eta R)K_{|m|}(\eta r') - I_{|m|}(\eta r')K_{|m|}(\eta R)\Big\}I_{|m|}(\eta r)\Big]\tilde{\rho}_1(r'|\, m, \eta)r'dr';$$

$$V_2(r|m, \eta) = [R\varepsilon_0^{-1}\{\sigma_1(m, \eta)$$

$$+\varepsilon_0^{-1}\int_0^R\left\{\frac{I_{|m|}(\eta r')}{I_{|m|}(\eta R)}\left[\rho_1(r'|\, m, \eta) - \frac{im}{r'}P_1^{\phi}(r'|\, m, \eta) - i\eta\, P_1^z(r'|\, m, \eta)\right]\right.$$

$$\left.+\,\frac{I'_{|m|}(\eta r')}{I_{|m|}(\eta R)}P_1^r(r'|m, \eta)\right\}r'dr'$$

$$+\,\varepsilon_0^{-1}\int_R^{\infty}\left\{\frac{K_{|m|}(\eta r')}{K_{|m|}(\eta R)}\left[\rho_2(r'|\, m, \eta) - \frac{im}{r'}P_2^{\phi}(r'|\, m, \eta) - i\eta\, P_2^z(r'|\, m, \eta)\right]\right.$$

$$\left.\left.+\,\frac{K'_{|m|}(\eta r')}{K_{|m|}(\eta R)}P_2^r(r'|m, \eta)\right\}r'dr'\right\}\Bigg]\cdot\eta R\left(\varepsilon_1\frac{I'_{|m|}(\eta R)}{I_{|m|}(\eta R)} - \varepsilon_2\frac{K'_{|m|}(\eta R)}{K_{|m|}(\eta R)}\right)^{-1} \tag{5.12.17b}$$

$$\cdot\,\frac{K_{|m|}(\eta r)}{K_{|m|}(\eta R)}$$

$$+\,\varepsilon_0^{-1}\int_R^{\infty}(\varepsilon_2 K_{|m|}(\eta R))^{-1}$$

$$[\theta(r > r')\Big\{I_{|m|}(\eta r')K_{|m|}(\eta R) - I_{|m|}(\eta R)K_{|m|}(\eta r')\Big\}K_{|m|}(\eta r)$$

$$+\,\theta(r < r')\Big\{I_{|m|}(\eta r)K_{|m|}(\eta R) - I_{|m|}(\eta R)K_{|m|}(\eta r)\Big\}K_{|m|}(\eta r')\Big]\tilde{\rho}_2(r'|\, m, \eta)r'dr'.$$

5.12.2 Potential of bulk charges in a wire with a metal shield

Let's consider another important special case: a thread in a metal shell. In this case, the electric field is completely shielded inside the metal medium, so the potential is only inside the wire. To move on to this particular case, it is necessary to direct the dielectric constant of the second medium to infinity:

$$\varepsilon_2 \to \infty. \tag{5.12.18}$$

Substituting (5.12.18) into (5.12.17a), we get

$$\begin{aligned}
V_1(r|m,\eta) &= R\varepsilon_0^{-1}\{\sigma_1(m,\eta) \\
&+\varepsilon_0^{-1}\int_0^R \left\{\frac{I_{|m|}(\eta r')}{I_{|m|}(\eta R)}\left(\rho_1(r'\Big|m,\eta) - \frac{im}{r'}P_1^{\phi}(r'\Big|m,\eta) - i\eta P_1^{z}(r'\Big|\eta, m)\right)\right. \\
&\left.+\frac{I_{|m|}(\eta r')}{I_{|m|}(\eta R)}P_1^{r}(r'\Big|m,\eta)\right\}r'dr' \\
&+\varepsilon_0^{-1}\int_0^R \left(\varepsilon_1 I_{|m|}(\eta R)\right)^{-1} \\
&\left[\begin{array}{l}\theta(r>r')\left\{I_{|m|}(\eta R)K_{|m|}(\eta r) - I_{|m|}(\eta r)K_{|m|}(\eta R)\right\}I_{|m|}(\eta r')+ \\ +\theta(r>r')\left\{I_{|m|}(\eta R)K_{|m|}(\eta r') - I_{|m|}(\eta r')K_{|m|}(\eta R)\right\}I_{|m|}(\eta r)\end{array}\right] \\
&\tilde{\rho}_1(r'|\eta, m)r'dr'.
\end{aligned} \tag{5.12.19}$$

5.12.3 Potential energy of self-action of a point charge in a wire in an infinite matrix

Let's find the potential created by a point charge in a two-layer cylindrical system (a quantum wire in a dielectric or metallic medium, figure 5.16). Considering a point charge as a bulk one with a bulk density

$$\rho_k(\mathbf{r}) = e\delta\,(\mathbf{r} - \mathbf{r}_0)\delta_{k,\,k_0}, \tag{5.12.20}$$

where k, $\mathbf{r}$ is the layer number and radius is the vector of the observation point and k_0, $\mathbf{r}_0$ is the layer number and radius is the vector of the point of placement of the point charge (figure 5.16). In cylindrical coordinates for the charge density we get

$$\rho_k(r,\theta,\varphi) = \frac{1}{r}\delta(r - r_0)\delta(\varphi - \varphi_0)\delta(z - z_0)\delta_{k,\,k_0}. \tag{5.12.21}$$

Due to the axial symmetry of the problem, we can put

$$\varphi_0 = 0;\ z_0 = 0, \tag{5.12.22}$$

Then the equation (5.12.21) will take the form

$$\rho_k(r,\theta,\varphi) = \frac{1}{r}\delta(r - r_0)\delta(\varphi)\delta(z)\delta_{k,\,k_0}. \tag{5.12.23}$$

The Fourier image of this function

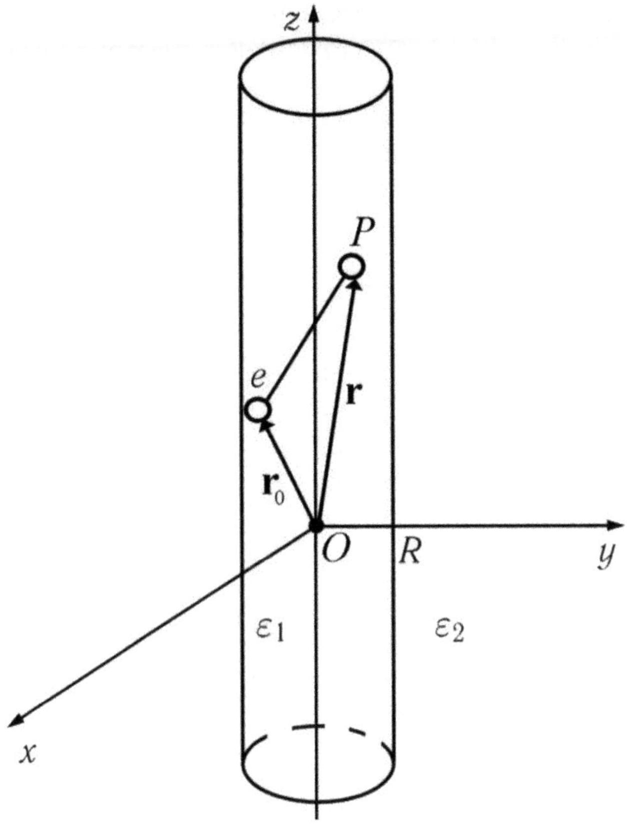

Figure 5.16. A point charge in a quantum wire in a dielectric medium.

$$\rho_k(r|m,\eta) = \frac{1}{(2\pi)^2 r}\delta(r - r_0). \tag{5.12.24}$$

Substituting (5.12.24) into (5.12.17a), (5.12.17b) and assuming that the point charge and the observation point are in the same layer k, we obtain a Fourier image of the potential in each of the layers:

$$\begin{aligned} V_1(r|m,\eta) &= \frac{e}{4\pi^2\varepsilon_0\varepsilon_1}[I_m(\eta r)K_m(\eta r_0)\theta(r<r_0) + I_m(\eta r_0)K_m(\eta r)\theta(r>r_0)] \\ &= \frac{e(\varepsilon_1 - \varepsilon_2)K_m(\eta R)K_m'(\eta R)I_m(\eta r)I_m(\eta r_0)}{4\pi^2\varepsilon_0\varepsilon_1\left(\varepsilon_2 I_m(\eta R)K_m'(\eta R) - \varepsilon_1 I_m'(\eta R)K_m(\eta R)\right)}; \end{aligned} \tag{5.12.25a}$$

$$\begin{aligned} V_2(r|m,\eta) &= \frac{e}{4\pi^2\varepsilon_0\varepsilon_2}[I_m(\eta r)K_m(\eta r_0)\theta(r<r_0) + I_m(\eta r_0)K_m(\eta r)\theta(r>r_0)] \\ &+ \frac{e(\varepsilon_1 - \varepsilon_2)I_m(\eta R)I_m'(\eta R)K_m(\eta r)K_m(\eta r_0)}{4\pi^2\varepsilon_0\varepsilon_2\left(\varepsilon_2 I_m(\eta R)K_m'(\eta R) - \varepsilon_1 I_m'(\eta R)K_m(\eta R)\right)}, \end{aligned} \tag{5.12.25b}$$

where $I_m(x)$, $K_m(x)$ are the modified Bessel function of order m and the MacDonald function, respectively, and $I'_m(x)$, $K'_m(x)$ are the derivatives of these functions by their argument.

The first term in the right-hand sides of both equalities (5.12.25a), (5.12.25b) is the Fourier image of the potential of a point charge placed in an unlimited medium with a dielectric constant ε_k $(k = 1,\ 2)$, and when the self-action potential is found, it must be discarded. Assuming $r = r_0$, we get

$$V_1^{SA}(r\,|\,m,\,\eta) = \frac{e(\varepsilon_1 - \varepsilon_2)K_m(\eta R)K'_m(\eta R)[I_m(\eta r)]^2}{4\pi^2\varepsilon_0\varepsilon_1\left(\varepsilon_2 I_m(\eta R)K'_m(\eta R) - \varepsilon_1 I'_m(\eta R)K_m(\eta R)\right)}; \tag{5.12.26a}$$

$$V_2^{SA}(r\,|\,m,\,\eta) = \frac{e(\varepsilon_1 - \varepsilon_2)I_m(\eta R)I'_m(\eta R)[K_m(\eta r)]^2}{4\pi^2\varepsilon_0\varepsilon_2\left(\varepsilon_2 I_m(\eta R)K'_m(\eta R) - \varepsilon_1 I'_m(\eta R)K_m(\eta R)\right)}. \tag{5.12.26b}$$

An explicit expression for calculating the self-action potential in the system under consideration is obtained by substituting (5.12.26a), (5.12.26b) into the Fourier integral expanding (2.2.14):

$$V_1^{SA}(r,\,\theta,\,\varphi) = \frac{e(\varepsilon_1 - \varepsilon_2)}{4\pi^2\varepsilon_0\varepsilon_1} \sum_{m=-\infty}^{\infty}\int_{-\infty}^{\infty} d\eta\; e^{im\varphi + i\eta z}\frac{K_m(\eta R)K'_m(\eta R)[I_m(\eta r)]^2}{\varepsilon_2 I_m(\eta R)K'_m(\eta R) - \varepsilon_1 I'_m(\eta R)K_m(\eta R)}; \tag{5.12.27a}$$

$$V_2^{SA}(r,\,\theta,\,\varphi) = \frac{e(\varepsilon_1 - \varepsilon_2)}{4\pi^2\varepsilon_0\varepsilon_2} \sum_{m=-\infty}^{\infty}\int_{-\infty}^{\infty} d\eta\; e^{im\varphi + i\eta z}\frac{I_m(\eta R)I'_m(\eta R)[K_m(\eta r)]^2}{\varepsilon_2 I_m(\eta R)K'_m(\eta R) - \varepsilon_1 I'_m(\eta R)K_m(\eta R)}. \tag{5.12.27b}$$

The potential energy of the self-action of a point charge is found by the general formula

$$W_k^{SA}(\mathbf{r}) = \frac{e}{2}V_k^{SA}(\mathbf{r}). \tag{5.12.28}$$

We will get it finally:

$$W_1^{SA}(r,\,\theta,\,\varphi) = \frac{e^2(\varepsilon_1 - \varepsilon_2)}{8\pi^2\varepsilon_0\varepsilon_1} \sum_{m=-\infty}^{\infty}\int_{-\infty}^{\infty} d\eta\; e^{im\varphi + i\eta z}\frac{K_m(\eta R)K'_m(\eta R)[I_m(\eta r)]^2}{\varepsilon_2 I_m(\eta R)K'_m(\eta R) - \varepsilon_1 I'_m(\eta R)K_m(\eta R)}; \tag{5.12.29a}$$

$$V_2^{SA}(r, \theta, \varphi) = \frac{e^2(\varepsilon_1 - \varepsilon_2)}{8\pi^2\varepsilon_0\varepsilon_2}$$
$$\sum_{m=-\infty}^{\infty} \int_{-\infty}^{\infty} d\eta \; e^{im\varphi + i\eta z} \frac{I_m(\eta R) I'_m(\eta R) [K_m(\eta r)]^2}{\varepsilon_2 I_m(\eta R) K'_m(\eta R) - \varepsilon_1 I'_m(\eta R) K_m(\eta R)}. \tag{5.12.29b}$$

Note that the potentials obtained are monotone functions of r, singular at the boundary of the wire and regular at its center. At the same time, in the actual case when the charge is inside the wire, depending on the ratio of dielectric permittivity, the following situations can be realized:

(1) if $\varepsilon_1 < \varepsilon_2$, then the potential at the boundary has the character of hyperbolic attraction, so that the localization of the carrier near the boundary is possible;

(2) if $\varepsilon_1 > \varepsilon_2$, then near the boundary the potential has the character of hyperbolic repulsion, and in the center it has the character of an increasing parabola. In this case, the states are realized near the center of the wire, and to find their energies, you can use the well-known solution of the Schrödinger equation for a two-dimensional harmonic oscillator.

5.13 Point charge potential and self-action potential in spherical structures

Consider the derivation of an expression for the self-action potential of a point charge e in a spherical system from a ball of radius R placed in an infinite matrix. Spherical harmonics for the potential in the nth layer are found by the formula (2.5.33) in the case of $K = 2$:

$$V_n(r; l, m) = \varepsilon_0^{-1} \sum_{k=1}^{2} \int_{R_{k-1}}^{R_k} G_{nk}(r, r') \rho_k(r'; l, m) r'^2 dr', \; (n = 1, \; 2) \tag{5.13.1}$$

where in the case

$$\rho_k(r) = e\delta_{k, k_0} \delta(r - r_0) = \frac{e\delta_{k, k_0}}{r_0^2 \sin\theta_0} \delta(r - r_0)\delta(\theta - \theta_0)\delta(\varphi - \varphi_0); \tag{5.13.2}$$

where $k_0, r_0, \theta_0, \varphi_0$ are the layer numbers and spherical coordinates of the point where the point charge is placed and k, r, θ, φ are the layer numbers and coordinates of the observation point. We decompose the bulk charge density in a row according to spherical functions:

$$\rho_k(r) = \sum_{l=0}^{\infty} \sum_{m=-l}^{l} \rho_k(r; l, m) Y_{lm}(\theta, \varphi); \tag{5.13.3}$$

$$\rho_k(r; l, m) = \int_0^{\pi} \sin\theta d\theta \int_0^{2\pi} d\varphi \; \rho_k(r) Y_{lm}^*(\theta, \varphi) \tag{5.13.4}$$

Substituting (5.13.2) into (5.13.4):

$$\rho_k(r; l, m) = \frac{e\delta_{k,\,k_0}}{r_0^2}\delta(r - r_0)Y^*_{lm}(\theta_0, \varphi_0). \tag{5.13.5}$$

Substituting (5.13.5) into (5.13.1):

$$V_n(r; l, m) = \frac{e}{\varepsilon_0}G_{nk_0}(r, r_0)Y^*_{lm}(\theta_0, \varphi_0). \tag{5.13.6}$$

Substituting (5.13.6) into the decomposition of the potential in spherical harmonics (see (2.4.6a)), we obtain:

$$V_n(\mathbf{r}) = \sum\nolimits_{l=0}^{\infty}\sum\nolimits_{m=-l}^{l}\frac{e}{\varepsilon_0}G_{nk_0}(r, r_0)Y^*_{lm}(\theta_0, \varphi_0)Y_{lm}(\theta, \varphi). \tag{5.13.7}$$

By the addition theorem of spherical functions [98]

$$\sum\nolimits_{m=-l}^{l} Y^*_{lm}(\theta_0, \varphi_0)Y_{lm}(\theta, \varphi) = \frac{2l+1}{4\pi}P_l(\cos\alpha), \tag{5.13.8}$$

where α—the angle between the vectors $\mathbf{n} = \mathbf{r}/r$ and $\mathbf{n}_0 = \mathbf{r}_0/r_0$—we get:

$$V_n(\mathbf{r}) = \frac{e}{4\pi\varepsilon_0}\sum\nolimits_{l=0}^{\infty} G_{nk_0}(r, r_0)(2l+1)P_l(\cos\alpha). \tag{5.13.9}$$

If $n = k_0$, then

$$V_n(\mathbf{r}) = \frac{e}{4\pi\varepsilon_0}\sum\nolimits_{l=0}^{\infty} G_{nn}(r, r_0)(2l+1)P_l(\cos\alpha), \tag{5.13.10}$$

where according to (2.5.38)

$$G_{nn}(r, r_0) = \gamma_n(r, r_0) + \sum\nolimits_{\alpha,\,\beta=1}^{2} D_{nn}^{\alpha\beta}F_n^{\alpha}(r)F_n^{\beta}(r_0). \tag{5.13.11}$$

Here

$$\begin{aligned}\gamma_n(r, r') = \frac{1}{(2l+1)\varepsilon s_n}\Bigg\{&\theta(r > r')\left[\frac{(r_n r')^l}{(r_{n-1}r)^{l+1}} + \frac{(r_{n-1}r)^l}{(r_n r')^{l+1}}\right]\\ &+ \theta(r' > r)\left[\frac{(r_n r)^l}{(r_{n-1}r')^{l+1}} + \frac{(r_{n-1}r')^l}{(r_n r)^{l+1}}\right] - \frac{(r_n r_{n-1})^l}{(rr')^{l+1}} - \frac{(rr')^l}{(r_n r_{n-1})^{l+1}}\Bigg\};\end{aligned} \tag{5.13.12}$$

$$D_{nn}^{\alpha\beta} = \begin{pmatrix} D_{nn} & D_{n,\,n-1} \\ D_{n-1,\,n} & D_{n-1,\,n-1}\end{pmatrix}. \tag{5.13.13}$$

Next, a simple zero-dimensional structure is considered: $r_0 = 0$, $r_1 = R$, $r_2 \to \infty$.

$$\gamma_1(r, r') = \frac{1}{(2l+1)\varepsilon_1}\left\{\theta(r > r')\frac{r'^l}{r^{l+1}} + \theta(r' > r)\frac{r^l}{r'^{l+1}} - \frac{(rr')^l}{R^{2l+1}}\right\}; \tag{5.13.14}$$

$$\gamma_2(r, r') = \frac{1}{(2l+1)\varepsilon_2}\left\{\theta(r > r')\frac{r'^l}{r^{l+1}} + \theta(r' > r)\frac{r^l}{r'^{l+1}} - \frac{R^{2l+1}}{(rr')^l}\right\}; \quad (5.13.15)$$

$$D_{11}^{\alpha\beta} = \begin{pmatrix} \frac{1}{\nu_1} & 0 \\ 0 & 0 \end{pmatrix}; \; D_{12}^{\alpha\beta} = \begin{pmatrix} 0 & \frac{1}{\nu_1} \\ 0 & 0 \end{pmatrix}; \quad (5.13.16a)$$

$$D_{21}^{\alpha\beta} = \begin{pmatrix} 0 & 0 \\ \frac{1}{\nu_1} & 0 \end{pmatrix}; \; D_{22}^{\alpha\beta} = \begin{pmatrix} 0 & 0 \\ 0 & \frac{1}{\nu_1} \end{pmatrix}; \quad (5.13.16b)$$

$$\nu_1 = R[\varepsilon_1 l + \varepsilon_2(l+1)]; \quad (5.13.17)$$

$$\sum\nolimits_{\alpha,\,\beta=1}^{2} D_{11}^{\alpha\beta} F_1^{\alpha}(r) F_1^{\beta}(r) = \frac{r^{2l}}{R^{2l+1}[\varepsilon_1 l + \varepsilon_2(l+1)]}; \quad (5.13.18)$$

$$\sum\nolimits_{\alpha,\,\beta=1}^{2} D_{22}^{\alpha\beta} F_2^{\alpha(r)} F_2^{\beta(r)} = \frac{R^{2l+1}}{r^{2l+2}[\varepsilon_1 l + \varepsilon_2(l+1)]}; \quad (5.13.19)$$

Let's present $G_{nn}(r, r_0)$ as the sum of two terms:

$$G_{nn}(r, r_0) = \gamma_{1,\,0}(r, r_0) + \Delta G_{nn}(r, r_0); \quad (5.13.20)$$

where the first term is

$$\gamma_{n,\,0}(r, r') = \frac{1}{(2l+1)\varepsilon_n}\left\{\theta(r > r')\frac{r'^l}{r^{l+1}} + \theta(r' > r)\frac{r^l}{r'^{l+1}}\right\};$$

after substitution into the expanding (5.13.10), it gives the usual Coulomb potential in an infinite medium with a dielectric constant ε_n, and the second one gives the desired self-action potential (after subtracting the first term and the limit transition $\mathbf{r}_0 \to \mathbf{r}$).

We use the properties of Legendre polynomials:

$$P_l(1) = 1. \quad (5.13.21)$$

Let's denote

$$U_n^{SA}(r) = eV_n^{SA}(r) \quad (5.13.22)$$

as the potential energy of self-action, then

$$U_n^{SA}(r) = \frac{e^2}{4\pi\varepsilon_0}\sum\nolimits_{l=0}^{\infty} \Delta G_{nn}(r, r)(2l+1). \; (n = 1, 2) \quad (5.13.23)$$

We have:

$$\Delta G_1(r, r') = \frac{r^{2l}}{R^{2l+1}} \cdot \frac{(\varepsilon_1 - \varepsilon_2)(l+1)}{\varepsilon_1(2l+1)[\varepsilon_1 l + \varepsilon_2(l+1)]}; \tag{5.13.24}$$

$$\Delta G_2(r, r') = \frac{R^{2l+1}}{r^{2l+2}} \cdot \frac{(\varepsilon_2 - \varepsilon_1)l}{\varepsilon_2(2l+1)[\varepsilon_1 l + \varepsilon_2(l+1)]}; \tag{5.13.25}$$

$$U_1^{SA}(r) = \frac{e^2(\varepsilon_1 - \varepsilon_2)}{4\pi\varepsilon_0\varepsilon_1 R}\sum_{l=0}^{\infty}\left(\frac{r}{R}\right)^{2l} \cdot \frac{l+1}{\varepsilon_1 l + \varepsilon_2(l+1)}; \tag{5.13.26}$$

$$U_2^{SA}(r) = -\frac{e^2(\varepsilon_1 - \varepsilon_2)}{4\pi\varepsilon_0\varepsilon_2 r}\sum_{l=0}^{\infty}\left(\frac{R}{r}\right)^{2l+1} \cdot \frac{l}{\varepsilon_1 l + \varepsilon_2(l+1)}. \tag{5.13.27}$$

It is of interest to consider the limiting cases:

(1) $\varepsilon_1 \gg \varepsilon_2$ (e.g., the first medium is a semiconductor, a polar dielectric or a metal, and the second is a vacuum).

Inside the ball

$$U_1^{SA}(r) \approx \frac{e^2(\varepsilon_1 - \varepsilon_2)}{4\pi\varepsilon_0\varepsilon_1 R}\left\{\frac{1}{\varepsilon_2} - \frac{1}{\varepsilon_1} + \frac{1}{\varepsilon_1}\left[\frac{R^2}{R^2 - r^2} + \ln\frac{R^2}{R^2 - r^2}\right]\right\}; \tag{5.13.28}$$

discarding the nonessential constant:

$$U_1^{SA}(r) \approx \frac{e^2(\varepsilon_1 - \varepsilon_2)}{4\pi\varepsilon_0\varepsilon_1^2 R}\left\{\frac{R^2}{R^2 - r^2} + \ln\frac{R^2}{R^2 - r^2}\right\}. \tag{5.13.29}$$

In the external environment

$$U_2^{SA}(r) \approx -\frac{e^2(\varepsilon_1 - \varepsilon_2)R}{4\pi\varepsilon_0\varepsilon_1\varepsilon_2}\left\{\frac{1}{r^2 - R^2} - \frac{1}{r^2}\right\}. \tag{5.13.30}$$

(2) $\varepsilon_2 \gg \varepsilon_1$ (e.g., the first medium is a vacuum, and the second is a metal, semiconductor, or polar dielectric).

$$U_1^{SA}(r) \approx \frac{e^2(\varepsilon_1 - \varepsilon_2)}{4\pi\varepsilon_0\varepsilon_1\varepsilon_2 R} \cdot \frac{R^2}{R^2 - r^2}; \tag{5.13.31}$$

$$U_2^{SA}(r) \approx \frac{e^2(\varepsilon_2 - \varepsilon_1)}{4\pi\varepsilon_0\varepsilon_2^2 R}\left\{\frac{1}{1 - \left(\frac{R}{r}\right)^2} - 1 - \ln\left(1 - \left(\frac{R}{r}\right)^2\right)\right\}; \tag{5.13.32}$$

discarding the constant:

$$U_2^{SA}(r) \approx \frac{e^2(\varepsilon_2 - \varepsilon_1)}{4\pi\varepsilon_0\varepsilon_2^2 R}\left\{\frac{r^2}{r^2 - R^2} - \ln\frac{r^2}{r^2 - R^2}\right\}. \tag{5.13.33}$$

We transform the obtained general formulas for $U_1^{SA}(r)$ and $U_2^{SA}(r)$ into an integral representation. We have

$$\frac{l+1}{\varepsilon_1 l+\varepsilon_2(l+1)}=\frac{1}{\varepsilon_1+\varepsilon_2}+\frac{\varepsilon_1}{(\varepsilon_1+\varepsilon_2)^2}\cdot\frac{1}{l+\frac{\varepsilon_2}{\varepsilon_1+\varepsilon_2}};$$

Using the formula (5.2.3.4) from [99]

$$\sum\nolimits_{k=0}^{\infty}\frac{x^k}{k+a}=x^{-a}\int_0^x\frac{t^{a-1}}{1-t}dt. \tag{5.13.34}$$

In our case

$$a=\frac{\varepsilon_2}{\varepsilon_1+\varepsilon_2}. \tag{5.13.35}$$

Then from the formulas (5.13.26), (5.13.35) and (5.13.34) we get

$$U_1^{SA}(r)=\frac{e^2(\varepsilon_1-\varepsilon_2)}{4\pi\varepsilon_0\varepsilon_1 R}\left\{\frac{1}{\varepsilon_1+\varepsilon_2}\cdot\frac{1}{1-\left(\frac{r}{R}\right)^2}+\frac{\varepsilon_1}{(\varepsilon_1+\varepsilon_2)^2}\cdot\left(\frac{r}{R}\right)^{-2a}\int_0^{\left(\frac{r}{R}\right)^2}\frac{t^{a-1}}{1-t}dt\right\}. \tag{5.13.36}$$

Let's write (5.13.36) as

$$U_1^{SA}(r)=\frac{e^2}{4\pi\varepsilon_0\varepsilon_1 R}\cdot\frac{\varepsilon_1-\varepsilon_2}{\varepsilon_1+\varepsilon_2}\left[\frac{1}{1-\frac{r^2}{R^2}}+\frac{\varepsilon_1}{\varepsilon_1+\varepsilon_2}\cdot\int_0^1\frac{x^{a-1}}{1-\frac{r^2}{R^2}x}dx\right] \tag{5.13.39}$$

(a formula valid for arbitrary values ε_1, ε_2).

Then we get

$$U_2^{SA}(r)=-\frac{e^2R}{4\pi\varepsilon_0\varepsilon_2 r^2}\cdot\frac{\varepsilon_1-\varepsilon_2}{\varepsilon_1}+\varepsilon_2\left[\frac{1}{1-\frac{R^2}{r^2}}-\frac{\varepsilon_2}{\varepsilon_1+\varepsilon_2}\int_0^1\frac{x^{a-1}}{1-\frac{R^2}{r^2}x}dx\right] \tag{5.13.40}$$

The resulting formulas for $U_{SA}(r)$ can be used to calculate the states of charge carriers in a quantum ball.

5.14 Free charge carriers in a multilayer homeopolar system

We use the formulas derived in the previous paragraph to solve the problem of the energy spectrum of charge carriers in a multilayer homeopolar system. The effect of self-action is fundamental in the sense of its mandatory presence in systems with partition boundaries. Therefore, the state of charge carriers, even in the simple case of the absence of external fields, is described by a z-coordinate-dependent Hamiltonian:

$$\hat{H}_{\|}=\frac{\hat{P}_{\|}^2}{2m_{\|}^*}+U_{SA}(z), \tag{5.14.1}$$

including potential energy.

Assuming that the short-acting part of the potential at the interface of the layers forms sufficiently high rectangular barriers, we subordinate the solutions (5.14.1) to zero boundary conditions.

The motion of the charge in the *XOY* plane is free, and the solution of the Schrödinger equation for it has the form

$$E_{\perp}(k) = \frac{\hbar^2 k_{\perp}^2}{2m_{\perp}^*}; \tag{5.14.2a}$$

$$\psi_{k_{\perp}}(\rho) = C\exp\{i\mathbf{k}_{\perp}\boldsymbol{\rho}\}. \tag{5.14.2b}$$

When calculating the spectrum of the Hamiltonian (5.14.1), we consider various approximations. If the potential energy is small compared to the kinetic energy, in the zero approximation, the second term in the Hamiltonian can be omitted and, taking into account the boundary conditions, write the eigenfunctions and the energy spectrum as

$$\psi_n(z) = C_n \cos\left(\frac{\pi n z}{l}\right), \; -\frac{l}{2} \leqslant z \leqslant \frac{l}{2}; \tag{5.14.3a}$$

$$E_n^0 = \frac{\pi^2\hbar^2 n^2}{2m_{||}^* l^2}, \; n = 1, 2, 3, \ldots, \tag{5.14.3b}$$

where l is the thickness of the layer in question.

It is not difficult to calculate the energy in the zero approximation:

$$E_n = \frac{\pi^2\hbar^2 n^2}{2m_{||}^* l^2} + \langle\psi_n(z)|U_{SA}(z)|\psi(z)\rangle. \tag{5.14.3c}$$

The first significantly positive term describes the effect of the band edge displacement (up for an electron, down for a hole) due to the dimensional quantum effect. The second term, the sign of which depends on the magnitude of the dielectric permittivity of the layers, gives the average potential energy of self-action. The longitudinal part of the energy according to (5.14.3c) and (5.11.2), (5.11.3) has the form

$$E_n = \frac{\pi^2\hbar^2 n^2}{2m_{||}^* l^2} + \frac{e^2}{4\pi\varepsilon_0\varepsilon_2\epsilon_2 l_2}\int_0^{\infty}\frac{dx}{e^{2x} - \delta_1\delta_3}\left\{\delta_1\delta_3 + \frac{\pi^2 n^2 e^x \sinh x}{x(x^2 + \pi^2 n^2)}\right\}. \tag{5.14.3d}$$

Layers $k = 1$ and $k = 3$ are assumed to be infinite ($l_1, \; l_3 \to \infty$). For small z (small oscillation amplitudes of the particle in the field of self-action), the expression for $U_{SA}(z)$ has the form (5.11.16) with U_{SA}^0, K_1 and K_2, defined by the formulas (5.11.15a) and (5.11.15b). In a symmetric system ($\varepsilon_0 = \varepsilon_3$), $\delta_0 = \delta_3$, $K_1 = 0$:

$$U_{SA}(z) = U_0' + K_2 z^2, \tag{5.14.4a}$$

where

$$U_0' = \frac{e^2}{4\pi\varepsilon_0\varepsilon_2\epsilon_2 l_2}\ln\frac{1}{1-\delta^2}; \tag{5.14.4b}$$

$$K_2 = \frac{e^2\delta J_2'}{2\pi\varepsilon_0\varepsilon_2\epsilon_2 l_2}; \; J_2' = \int_0^\infty x^2 e^x(e^{2x} - \delta^2)^{-1}dx. \tag{5.14.4c}$$

If the inequality $E_{s.q.} < E_{SA}$ is satisfied, the intrinsic energy of the operator (5.14.1) with the potential energy (5.14.4a) has the form

$$E_{SA,\,n'} = U_0' + \hbar\omega_{SA}\left(n + \frac{1}{2}\right), \quad n = 0,\, 1,\, 2,\, \ldots. \tag{5.14.5}$$

The last term (5.14.5) (at $n = 0$) describes the cleavage from the edge of the renormalized band of the ground state level by $\frac{1}{2}\hbar\omega_{SA}$, greater $\pi^2\hbar^2/(2m_{||}^* l)^2$ because the criterion for the validity of the spectrum (5.14.5) is the inequality $(\hbar/(m_{||}^*\omega_{SA}))^{1/2} < l$. Included in (5.14.5) is the parameter

$$\omega_{SA} = \left(\frac{e^2 J_2'\delta}{\pi\varepsilon_0\varepsilon_2\epsilon_2 m_{||}^* l^3}\right)^{\frac{1}{2}}. \tag{5.14.6}$$

The wave functions of the Hamiltonian (5.14.1) with potential energy (5.14.4a) are Hermite functions:

$$\psi_n(t) = A_n \exp\left(-\frac{1}{2}t^2\right)H_n(t), \tag{5.14.7}$$

where

$$t = z\left(\frac{m_{||}^*\omega_{SA}}{\hbar}\right)^{\frac{1}{2}}; \; A_n = (2^n\cdot n!)^{-\frac{1}{2}}; \;\; H_n(t) = (-1)^n e^{t^2}\frac{d^n e^{-t^2}}{dt^n}. \tag{5.14.8}$$

In an asymmetric three-layer system, when one of the boundary media has a large dielectric constant (e.g., metal: $\varepsilon_1 \to \infty$, $\varepsilon_1 \gg \varepsilon_2 > \varepsilon_3$), the quadratic term in (5.11.14) becomes small compared to the linear one and the expression for $U_{SA}(z)$) in the intermediate range of z values has the form of 'triangular' potential (formulas (5.11.18a) and (5.11.18b)).

The solution of the Schrödinger equation with the Hamiltonian (5.14.1) and potential (5.11.18a)–c) has the form

$$E_{SA,\,n} = U_0' + \left(\frac{\hbar^2}{2m_{||}^*}\right)^{\frac{1}{3}}\left[\frac{9\pi\varepsilon_{эфф}}{8}\left(n + \frac{3}{4}\right)\right]^{\frac{2}{3}}; \tag{5.14.9}$$

$$\psi_n(z) = \frac{\left(2m_{\|}^{*}\right)^{\frac{1}{3}}}{\sqrt{\pi}\,(eE_{эфф})^{\frac{1}{6}}\hbar^{\frac{2}{3}}} Ai\left[\frac{\left(2m_{\|}^{*}eE_{эфф}\right)^{\frac{1}{3}}}{\hbar^{2}}\left(z - \frac{E_{SA,\,n}}{eE_{eff}}\right)\right], \qquad (5.14.10)$$

where $A_i(x)$ is the the Airy function [100].

Figure 5.17(a) and (b) show the results of numerical calculation using the formulas (5.14.3d), (5.14.5), (5.14.9) of the binding energy of the charge carrier in the field of self-action potential for a CdTe film on two different substrates: a —MgF_2, b—metal. The dotted lines correspond to formulas (5.14.5) and (5.14.9), the

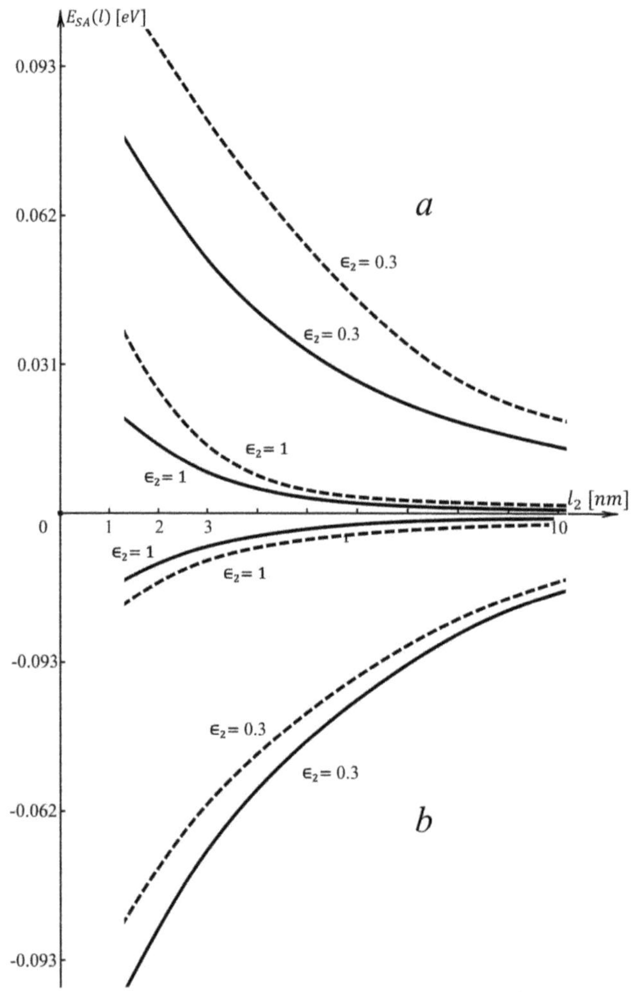

Figure 5.17. The results of numerical calculation using the formulas (5.14.3d), (5.14.5), and (5.14.9) of the binding energy of the charge carrier in the field of self-action potential for CdTe film on two different substrates: (a) MgF_2, (b) metal. Dotted lines correspond to formulas (5.14.5) and (5.14.9); solid lines correspond to formulas (5.14.3d).

solid lines correspond to (5.14.3d). From comparing these results, it follows that the limit formulas (5.14.5) and (5.14.9) at $\epsilon_2 = 1$ give results approximately 5%–10% less than the numerical calculation according to the formula (5.14.3d).

Similar comparisons are made for films and substrates with very different parameters. In all cases, it was found that formulas (5.14.5) and (5.14.9) with $\epsilon_2 = 1$ give a fairly good description (with an accuracy of up to 10%), in the thickness range of the order of 3–10 nm.

The energy spectrum of charge carriers in a superlattice is of great practical interest. A large number of papers and reviews have been devoted to this issue (see, e.g., [101]). Its main feature is its strong anisotropy. The movement of charge carriers in the plane of the layers remains free, whereas in the direction of the superlattice axis it is fundamentally transformed by a system of potential barriers and acquires a mini-band character. The width of the minibands depends on the transparency of the barriers. In the strong coupling approximation, the mini-band energy level is determined by the formula [101]

$$E_S(k_{\|}) = E_S - \Delta_S \cos\left[k_{\|}(l_a + l_b)\right]. \tag{5.14.11}$$

With low transparency of the barriers (i.e., their high height and width), the mini-band spectrum degenerates into a discrete one (e.g., (5.14.3d)). It is of interest to evaluate the role of self-action in the formation of minibands. We will limit ourselves to some qualitative considerations.

Let's take a closer look at the situation when the charge carriers are localized inside the layers (low transparency of the barrier). Then the motion within the layer is described by a wave function satisfying zero boundary conditions. Choosing the origin on the $n-1$ surface and using the formula (5.11.19), we write down the charge energy in the approximation of weak interaction with the self-action potential according to (5.14.3c):

$$\begin{aligned} E_{\|} = {} & \frac{\pi^2\hbar^2}{2m_{\|}l_b^2} + \frac{e^2}{16\pi\varepsilon_0\varepsilon_b l_b}\int_0^\infty \frac{d\eta}{\sinh\zeta_b\sqrt{\psi^2-1}} \\ & \left\{ e^{-\zeta_b}\sqrt{\psi^2-1} + \frac{\cosh\zeta_b}{\psi+\sqrt{\psi^2-1}} \right. \\ & \left. + \frac{\pi^2\left(\varepsilon_b^2-\varepsilon_a^2\right)}{2\varepsilon_a\varepsilon_b}\cdot\frac{\sinh\zeta_a\cdot\sinh\zeta_b}{\zeta_b(\zeta_a^2+\pi^2)} - \cosh\zeta_a \right\}. \end{aligned} \tag{5.14.12}$$

In the approximation of a thin layer b and $\varepsilon_b > \varepsilon_a$ (placing the origin in the middle of layer b), decomposing (5.14.12) by z, we find the potential energy for the Hamiltonian (5.14.1) in the form

$$\widetilde{U}_{SA}(z) \approx \widetilde{U}(0) + \widetilde{A}z^2, \tag{5.14.13}$$

$$\widetilde{U}(0) = \frac{e^2}{2\pi\varepsilon_0}\int_0^\infty d\eta \cdot \eta\left\{\frac{1}{2\eta\varepsilon_b \sinh\zeta_b\sqrt{\psi^2-1}}\right.$$
$$\left[e^{-\zeta_b}\sqrt{\psi^2-1} - \cosh\zeta_a\right. \tag{5.14.14a}$$
$$\left. + \frac{\cosh\zeta_b}{\psi+\sqrt{\psi^2-1}} + \frac{\left(\varepsilon_b^2-\varepsilon_a^2\right)\sinh\zeta_a\sinh\zeta_b}{2\varepsilon_a\varepsilon_b}\right\};$$

$$\widetilde{A} = \frac{e^2}{2\pi\varepsilon_0}\int_0^\infty d\eta \cdot \eta^3\left\{\frac{\sinh\zeta_a\sinh\zeta_b\left(\varepsilon_b^2-\varepsilon_a^2\right)}{4\eta\varepsilon_a\varepsilon_b^2}\right\}. \tag{5.14.14b}$$

The calculation in the approximation of weak and strong electron bonds with self-action potential for a periodic structure is similar to the one already performed for a three-layer system. The results for the periodic structure are obtained by replacing in formulas (5.14.4a), (5.14.4b), (5.14.4c), and (5.14.5) the values $U(0)$, A by $\widetilde{U}(0)$, $\widetilde{A}$, respectively. If $\varepsilon_b < \varepsilon_a$, then the coefficients $\widetilde{U}(0)$ and $\widetilde{A}$ change sign. The bottom of the conduction band shifts downwards, and the parabola wraps its branches downwards, forming a maximum in the center of the layer instead of a minimum.

In the approximation of infinite potential barriers at the boundaries of layers and narrow potential wells, the energy E_{SA}^{SL} in a superlattice can be obtained by averaging (5.11.19) on the wave function (5.14.3a):

$$E_{SA}^{SL} = \frac{e^2}{4\pi\varepsilon_0\varepsilon_b\epsilon_2}\int_0^\infty \frac{dx}{\sinh x\sqrt{\psi^2-1}}$$
$$\left\{e^{-x}\sqrt{\psi^2-1} + \frac{\cosh x}{\psi+\sqrt{\psi^2-1}} - \cosh\left(x\bar{l}\right)\right. \tag{5.14.15}$$
$$\left. + \left[\frac{1}{x} - \frac{x}{x^2+\pi^2}\right]\frac{\left(\varepsilon_b^2-\varepsilon_a^2\right)\sinh x\sinh\left(x\bar{l}\right)}{2\varepsilon_0\varepsilon_b}\right\};$$

where $\bar{l} = 1 + l_a/l_b$. Figure 5.18 shows the ground state energy in the self-action field for two types of superlattices: GaAs/AlGaAs and GaSb/InAs. The contribution from self-action becomes comparable to the value of the barrier at the boundaries of the electron and hole bands at $l \leqslant 5.0$ nm (and is on the order of 5–50 meV), in GaAs the band gap increases, and in GaSb it decreases.

5.15 Polaron in a plate of a polar crystal of finite thickness

Consider the problem of a polaron in a polar crystal layer bordering nonpolar crystals (1|20|3). In [102], the energy of a weakly coupled polaron interacting only

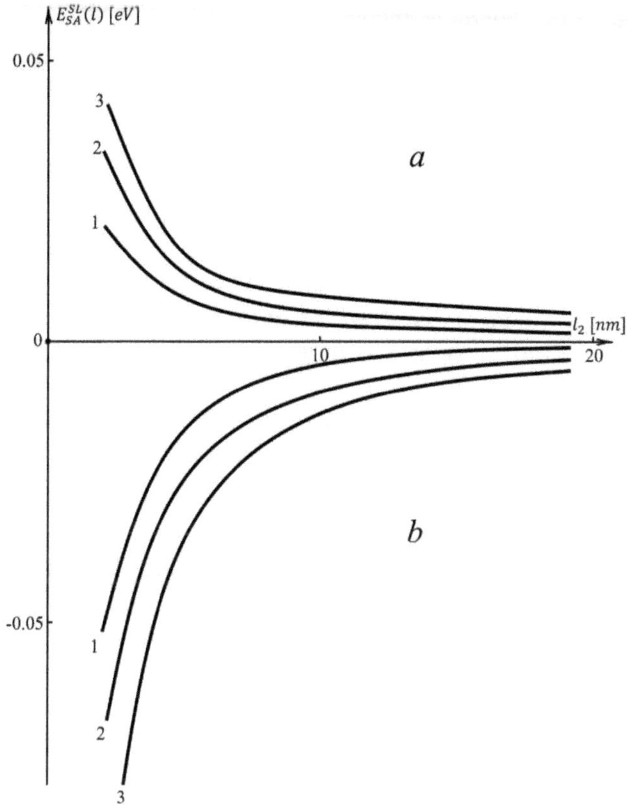

Figure 5.18. Dependence of the ground state energy in the self-action potential field on the period for two types of superlattices (a) GaAs/AlGaAs: 1—$l_b = 3$ nm; 2—$l_b = 5$ nm; 3—$l_b = 10$ nm; (b) GaSb/InAs: 1—$l_b = 3$ nm; 2—$l_b = 10$ nm; 3—$l_b = 20$ nm.

with bulk vibrations in the layer was calculated. Taking into account the attenuation of the interaction at the surface of the layer led to the fact that the polaron energy in absolute value at finite l_2 turned out to be less than in the case of $l_2 \to \infty$, when it is equal to $-\alpha_V \hbar\omega_{20}$.

In [103], the problem of the strong bond polaron was solved; however, as in the bulk case, weak and intermediate bonds are relevant. The interaction with both bulk and surface vibrations in the layer was taken into account by a simple perturbation theory method in [104] using the Hamiltonian of the electron–phonon interaction [7], and in [105] by the combined Lee, Low, and Pines method and perturbation theory. The results of [105] basically coincide with those given in [10], with the difference, however, that in the latter the more general case of dielectric media with $\varepsilon_{1,3} \neq 1$.

5.15.1 Weak electron–phonon coupling

The Hamiltonian of an electron from the conduction band of a polar crystal plate

$$\hat{H} = \frac{\hat{P}_z^2}{2m_{||}^*} + \frac{\hat{P}_\rho^2}{2m_\perp^*} + \hat{H}_{\text{ph}}^{S,\,V} + \hat{H}_{\text{e-ph}}^{S,\,V} + U_{SA}(z) + U_B \tag{5.15.1}$$

includes interaction with all phonon branches $\hat{H}_{\text{e-ph}}^{S,\,V}$ and the potential energy of self-action $U_{SA}(z)$. It was not taken into account in any of the works [102–105]. In section 5.13 the problem of the state of an electron in the field of self-action potential was discussed in detail, so here we will focus only on calculating the polaron part.

Assuming that surface and bulk polarization optical vibrations form a shallow potential well for an electron, it is possible to take into account $\hat{H}_{\text{e-ph}}^{S,\,V}$ as a perturbation (i.e. movement both along the z axis and in the XOY plane is considered slow) and use perturbation theory to solve the problem. The Hamiltonian $\hat{H}_{\text{e-ph}}^{S,\,V}$ was obtained in section 5.1. Here we consider the case of a symmetric three-layer system, $\varepsilon_1 = \varepsilon_3 \equiv \varepsilon$, in order to simplify mathematical calculations.

Let's take the wave function of an electron in the zero approximation in a multiplicative form:

$$\psi_{k_\perp}(\rho, z) = \psi_{k_\perp}(\rho)\psi_n(z), \tag{5.15.2}$$

where

$$\psi_{k_\perp}(\rho) = \frac{1}{\sqrt{L_x L_y}} \exp(i\mathbf{k}_\perp \boldsymbol{\rho}); \tag{5.15.3a}$$

$$\psi_n(z) = \begin{cases} \sqrt{\dfrac{2}{l_2}} \cos\left(\dfrac{\pi n z}{l_2}\right), & n = 1, 3, 5, \ldots; \\ \sqrt{\dfrac{2}{l_2}} \sin\left(\dfrac{\pi n z}{l_2}\right), & n = 2, 4, 6, \ldots; \\ 0, \; z < -\dfrac{l_2}{2}; \; z > \dfrac{l_2}{2}. \end{cases} \tag{5.15.3b}$$

Calculating the contribution from the electron–phonon interaction in the second order of perturbation theory and using the expression (1.40a) for $\hat{H}_{\text{e-ph}}^{V}$ we obtain

$$\Delta E = -\alpha_V \hbar\omega_{20}\frac{d}{R_V}\left\{\frac{8}{\pi^4}\sum_{\substack{m=1,2,3,\ldots\\ n=1,2,3,\ldots}}\left[\ln\left[\frac{(2m-1)^2}{4n(n-1)+\gamma_1(d)}\right]\right.\right.$$

$$\left[\frac{1}{4(n+m-1)^2-1}-\right.$$

$$\left.-\frac{1}{4(n-m)^2-1}\right]^2\left[4(m-n)(m+n-1)+1-\frac{\gamma_1}{d}\right]^{-1}$$

$$+\ln\left[\frac{4m^2}{4n^2-1+\gamma_1(d)}\right]\left[\frac{1}{4(m+n)^2-1}-\frac{1}{4(n+m)^2-1}\right]^2$$

$$\left.\times\left[4(m^2-n^2)+1-\gamma_1(d)\right]^{-1}\right]-\frac{\pi^4}{2}\frac{\sqrt{\varepsilon_0\varepsilon}}{(\varepsilon_{20}-\varepsilon_2)} \tag{5.15.4}$$

$$\times\sum_{n=1,3,\ldots}\left\{(2n-1)^2\right.$$

$$\int_0^\infty \frac{g_1(x)x^2\,\mathrm{th}\,x\,dx}{\left(x^2+\frac{\pi^2n^2}{4}\right)^2\left(x^2+\frac{\pi^2(n-1)^2}{4}\right)^2\left(x^2+\frac{\pi^2n(n-1)^2}{4}+\gamma_2(d)\right)}$$

$$\left.\left.+(2n)^2\int_0^\infty \frac{g_2(x)x^2\coth x\,dx}{\left(x^2+\frac{\pi^2n^2}{4}\right)^2\left(x^2+\frac{\pi^2(2n-1)^2}{16}\right)^2\left(x^2+\frac{\pi^2(2n-1)^2}{16}+\gamma_3(d)\right)}\right\}\right\},$$

where

$$g_{1,2}(x) = \left(\varepsilon_2^{-1}(x)-\varepsilon_{20}^{-1}(x)\right)\left(\frac{\varepsilon_{20}(x)\varepsilon_2}{\varepsilon_2(x)\varepsilon_{20}}\right)^{\frac{1}{2}}; \tag{5.15.5a}$$

$$\gamma_1(d)=\frac{d^2}{\pi^2R_V^2};\ \gamma_{2,3}(d)=\frac{d^2}{4R_{S_{1,2}}^2(x)}. \tag{5.15.5b}$$

Figure 5.19 shows the results of calculating the binding energy ΔE_0 of the polaron at $\mathbf{k}_\perp = 0$ and the GaAs layer bordering nonpolar layers with $\varepsilon = 1$ and $\varepsilon = 10$, respectively. Curve 1 describes the dependence of the kinetic energy of longitudinal motion $\pi^2\hbar^2/(2m_{||}^* l_2^2)$ on l_2, and curve 2 is the binding energy of a bulk polaron. As already noted, at small thicknesses, the decrease in this energy is due to the effect of attenuation of the interaction. Curves 3 and 4 describe, respectively, the contributions of surface optical modes to the polaron energy for $\varepsilon = 10$ and $\varepsilon = 1.0$, growing in absolute magnitude with decreasing l_2, and the total energy (curves 5 and 6) has a minimum in magnitude of rapid growth of the dimensional quantization energy at

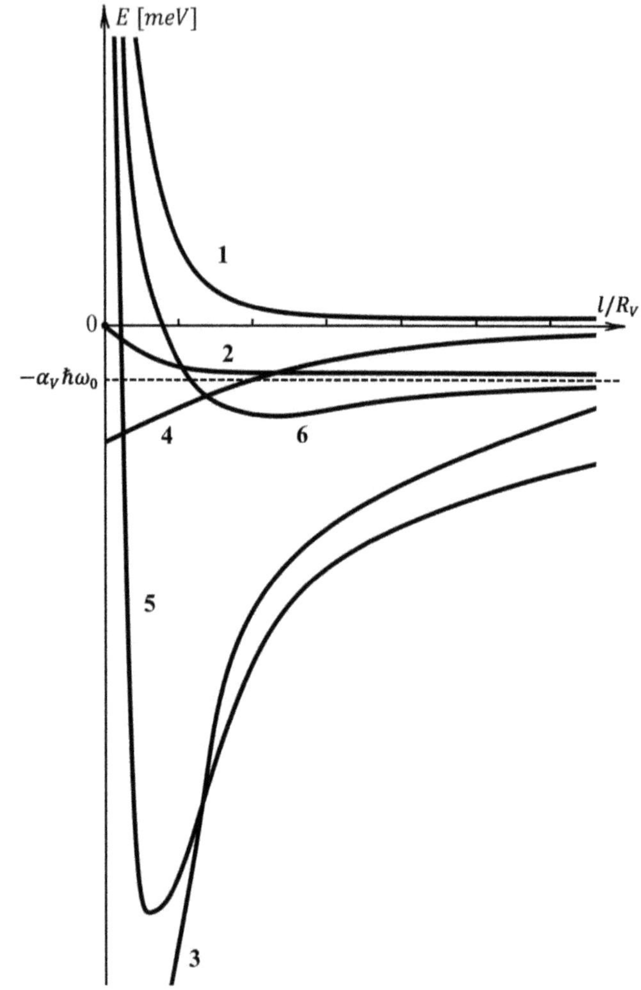

Figure 5.19. The results of calculating the binding energy of the polaron at $\mathbf{k}_\perp = 0$ and the GaAs, layer bordering nonpolar layers with $\varepsilon = 1$ and $\varepsilon = 10$, respectively. Curve *1* describes the dependence of the kinetic energy of longitudinal motion $\pi^2\hbar^2/(2m_{||}^* l_2^2)$ on l_2, and curve 2 is the binding energy of a bulk polaron. Curves 3 and 4 describe, respectively, the contributions of surface optical modes to the polaron energy for $\varepsilon = 10$ and $\varepsilon = 1.0$, and curves 5 and 6 describe the total energy for $\varepsilon = 10$ and $\varepsilon = 1.0$ respectively.

small l_2 ($\sim l_2^{-2}$). The depth of the minimum significantly depends on ε and, as follows from figure 5.19, increases with large ε. The minimum is achieved with still sufficiently large layer thicknesses $l_2 > R_S$ ($R_S \cong 4.0$ nm) for GaAs. Consequently, interaction with surface modes begins to noticeably deepen the polaron states at l_2, significantly exceeding R_S, especially when $\varepsilon \gg 1$.

5.15.2 Intermediate electron–phonon coupling with weak coupling elements

Now let's consider the case when the movement along the z-axis is slow, and in the plane of the layer it is fast. Then, as in the polaron problem at the contact of two crystals, adiabatic convergence can be applied and the potential energy of the electron can be calculated as a function of the z-coordinate, which together with $U_{SA}(z)$ forms the total potential energy.

Let's introduce the Hamiltonian (5.15.1) as

$$\hat{H} = \hat{H}_{||} + \hat{H}_{\perp}, \tag{5.15.6}$$

$$\hat{H}_{||} = \frac{\hat{P}_z^2}{2m_z^*} + U_{SA}(z) + U_B; \tag{5.15.7a}$$

$$\hat{H}_{\perp} = \frac{\hat{P}_{\perp}^2}{2m_{\perp}^*} + \hat{H}_{\text{ph}}^{S,\,V} + \hat{H}_{\text{e-ph}}^{S,\,V}, \tag{5.15.7b}$$

where $\hat{H}_{||,\,\perp}$—Hamiltonians of the slow (||) and fast ($\perp$) subsystems. In clause 5.4.1, we already considered the solution of the polaronic problem in the case of weak localization along the z-axis. Since the mathematical procedure remains the same as given in clause 4.2.1, we will write down the results immediately after performing the first unitary transformation:

$$\hat{S}_1 = \exp\left\{i\rho\left[\mathbf{k} - \sum_{\mathbf{Q}}\mathbf{q}_{\perp}\hat{b}_{\mathbf{Q}}^{+}\hat{b}_{\mathbf{Q}} - \sum_{s,\boldsymbol{\eta}}\boldsymbol{\eta}\hat{b}_{s,\,\boldsymbol{\eta}}^{+}\hat{b}_{s,\,\boldsymbol{\eta}}\right]\right\}, \tag{5.15.8}$$

where $\mathbf{k}_{\perp}$ is the projection of the total polaron momentum in the XOY plane.

It is convenient to represent the Hamiltonian of the system after the first transformation as

$$\widetilde{H} = \hat{S}_1^{-1}\hat{H}\hat{S}_1 = \hat{H}_0 + \hat{H}_1 + \hat{H}_2, \tag{5.15.9}$$

where

$$\begin{aligned}\hat{H}_0 = {} & \frac{\hat{P}_z^2}{2m_{||}^*} + \frac{\hbar^2K^2}{2m_{\perp}^*} + \frac{\hbar^2}{2m_{\perp}^*}\left\{\sum_{\mathbf{q}_{\perp},m,i}\hat{b}_{m,\,i}^{+}\hat{b}_{m,\,i}\left(q_{\perp}^2 + \beta_V^2\right) + \sum_{s,\boldsymbol{\eta}}\hat{b}_{s,\,\boldsymbol{\eta}}^{+}\hat{b}_{s,\,\boldsymbol{\eta}}\left(\eta^2 + \beta_S^2\right)\right\} \\ & + \sum_{\mathbf{q}_{\perp}}\left\{C_V\left[\sum_{m=1,2,3}^{\frac{N}{2}}\frac{\cos\left(\frac{m\pi z}{2d}\right)}{\left[q_{\perp}^2 + \left(\frac{m\pi}{2d}\right)^2\right]^{\frac{1}{2}}}\hat{b}_{m,\,+}^{+} + \sum_{m=2,4,6}^{\frac{N}{2}}\frac{\sin\left(\frac{m\pi z}{2d}\right)}{\left[q_{\perp}^2 + \left(\frac{m\pi}{2d}\right)^2\right]^{\frac{1}{2}}}\hat{b}_{m,\,-}^{-}\right]\right. \\ & \left. + \sum_{s,\boldsymbol{\eta}}C_s(\eta)\begin{Bmatrix}\cosh \eta z, & s = 1 \\ \sinh \eta z, & s = 2\end{Bmatrix} + h.\,c.\right\} + U_{SA}(z) + U_B)z(;\end{aligned} \tag{5.15.10a}$$

$$\begin{aligned}\hat{H}_1 = {} & \frac{\hbar^2}{2m_{\perp}^*}\left\{\left[\sum_{\mathbf{q}_{\perp},m,i}\hat{b}_{m,\,i}^{+}\hat{b}_{m,\,i}\mathbf{q}_{\perp}\right]^2 - \left[\sum_{\mathbf{q}_{\perp},m,i}\hat{b}_{m,\,i}^{+}\hat{b}_{m,\,i}q_{\perp}^2\right]\right\} + \\ & + \frac{\hbar^2}{2m_{\perp}^*}\left[\left(\sum_{\boldsymbol{\eta},s}\hat{b}_{s,\,\boldsymbol{\eta}}^{+}\hat{b}_{s,\,\boldsymbol{\eta}}\boldsymbol{\eta}\right)^2 - \sum_{\boldsymbol{\eta},s}\hat{b}_{s,\,\boldsymbol{\eta}}^{+}\hat{b}_{s,\,\boldsymbol{\eta}}\eta^2\right] + \\ & + \frac{\hbar^2}{2m_{\perp}^*}\sum_{\mathbf{q}_{\perp},m,i}\hat{b}_{m,\,i}^{+}\hat{b}_{m,\,i}\mathbf{q}_{\perp}\cdot\sum_{\boldsymbol{\eta},s}\hat{b}_{s,\,\boldsymbol{\eta}}^{+}\hat{b}_{s,\,\boldsymbol{\eta}}\boldsymbol{\eta};\end{aligned} \tag{5.15.10b}$$

$$\hat{H}_2 = -\frac{\hbar^2}{m_\perp^*}\left[\sum_{\mathbf{q}_\perp, m, i} \hat{b}^+_{m,i}\hat{b}_{m,i}\mathbf{K}_\perp\mathbf{q}_\perp + \sum_{\boldsymbol{\eta}, s}\hat{b}^+_{s,\boldsymbol{\eta}}\hat{b}_{s,\boldsymbol{\eta}}\mathbf{K}_\perp\boldsymbol{\eta}\right]; \tag{5.15.10c}$$

here

$$\beta_V = \left(\frac{2m_\perp^*\omega_0}{\hbar}\right)^{\frac{1}{2}}; \ \beta_S = \left(\frac{2m_\perp^*\Omega_S}{\hbar}\right)^{\frac{1}{2}}. \tag{5.15.10d}$$

The Hamiltonian $\hat{H}_0$ is further considered as unperturbed, and $\hat{H}_1$ and $\hat{H}_2$ as perturbations.

Performing the second unitary transformation over the Hamiltonian (5.15.9)–(5.15.10d)

$$\hat{S}_2 = \exp\left\{\sum_{\mathbf{Q}}\left[\hat{b}^+_{m,i}(\mathbf{q}_\perp)f_{m,i}(\mathbf{q}_\perp) - \hat{b}_{m,i}(\mathbf{q}_\perp)f^*_{m,i}(\mathbf{q}_\perp)\right] + \sum_{\boldsymbol{\eta},s}\left[\hat{b}^+_{s,\boldsymbol{\eta}}f_s(\boldsymbol{\eta}) - \hat{b}_{s,\boldsymbol{\eta}}f^*_s(\boldsymbol{\eta})\right]\right\}, \tag{5.15.11}$$

we get

$$\widetilde{H} = \hat{S}_2^{-1}\hat{H}\hat{S}_2 = \hat{H}_0 + \hat{H}_1 + \hat{H}_2, \tag{5.15.12}$$

where

$$\begin{aligned}
\hat{H}_0 \ = \ & \frac{\hat{P}_z^2}{2m_{\|}^*} + \frac{\hbar^2 K_\perp^2}{2m_\perp^*} \\
& + \frac{\hbar^2}{2m_\perp^*}\left\{\sum_{\mathbf{q}_\perp, m, i}\left[\hat{b}^+_{m,i}\hat{b}_{m,i} + \left|f_{m,i}(\mathbf{q}_\perp)\right|^2\right]\left(q_\perp^{\ 2} + \beta_V^2\right) + \right. \\
& \left. \sum_{\boldsymbol{\eta},s}\left[\hat{b}^+_{s,\boldsymbol{\eta}}\hat{b}_{s,\boldsymbol{\eta}} + \left|f_s(\boldsymbol{\eta})\right|^2\right]\left(\eta^2 + \beta_s^2\right)\right\} \\
& + \sum_{\mathbf{q}_\perp}\left\{C_V \sum_{m=1,3,\ldots}\frac{\cos\left(\frac{m\pi z}{2d}\right)}{\left[q_\perp^2 + \left(\frac{m\pi}{2d}\right)^2\right]}\cdot\left(f^*_{m,+} + \hat{b}^*_{m,+}\right)\right. \\
& \left. + \sum_{m=2,4,\ldots}\frac{\sin\left(\frac{m\pi z}{2d}\right)}{\left[q_\perp^2 + \left(\frac{m\pi}{2d}\right)^2\right]^{\frac{1}{2}}}\cdot\left(f^*_{m,-} + \hat{b}^*_{m,-}\right) + h.\ c.\right\} \\
& + \sum_{\boldsymbol{\eta},s}\left\{C_s(\eta)\begin{Bmatrix}\cosh \eta z, & s = 1 \\ \sinh \eta z, & s = 2\end{Bmatrix}\left(\hat{b}^+_{s,\boldsymbol{\eta}} + f_s(\boldsymbol{\eta})\right) + \text{h. c.}\right\} \\
& + U_{SA}(z) + U_B.
\end{aligned} \tag{5.15.13}$$

Averaging (5.15.13) over the phonon vacuum and applying the variational principle, we find the amplitudes $f_{m,\pm}(\mathbf{q}_\perp)$ and $f_{s,\boldsymbol{\eta}}$:

$$f_{m,\pm}(\mathbf{q}_\perp, z) = -C_V \begin{Bmatrix} \cos\left(\dfrac{m\pi z}{2d}\right), & m = 1, 3, 5, \ldots \\ \sin\left(\dfrac{m\pi z}{2d}\right), & m = 2, 4, 6, \ldots \end{Bmatrix} \tag{5.15.14a}$$

$$\left[q_\perp^2 + \left(\frac{m\pi}{2d}\right)^2\right]^{-\frac{1}{2}} \frac{2m_\perp^*}{\hbar^2}\left(q_\perp{}^2 + \beta_V^2\right)^{-1};$$

$$f^*_{\eta,\, s}(\eta, z) = -C_s(\eta)\begin{Bmatrix} \cosh \eta z, & s = 1 \\ \sinh \eta z, & s = 2 \end{Bmatrix} \frac{2m_\perp^*}{\hbar^2}\left(\eta^2 + \beta_s^2\right)^{-1}. \tag{5.15.14b}$$

Substituting (5.15.14a), (5.15.14b) and (5.15.13), after averaging over the phonon vacuum, we obtain the effective Hamiltonian of the system:

$$\hat{H}_{\text{eff}} = \frac{\hat{P}_z^2}{2m_{||}^*} + \frac{\hbar^2 K_\perp^2}{2m_{\text{eff}}^*} + W_{eff}(z), \tag{5.15.15}$$

where

$$\begin{aligned} W_{\text{eff}}(z) \;=\; & -\frac{2m_\perp^*}{\hbar^2}\sum_{\mathbf{q}_\perp}\frac{|C_V|^2}{\left(q_\perp^2 + \beta_V^2\right)} \\ & \left\{\sum_{m=1,3,\ldots}\frac{\cos^2\left(\frac{m\pi d}{2d}\right)}{\left[q_\perp^2 + \left(\frac{m\pi}{2d}\right)^2\right]} + \sum_{m=2,4,\ldots}\frac{\sin^2\left(\frac{m\pi d}{2d}\right)}{\left[q_\perp^2 + \left(\frac{m\pi}{2d}\right)^2\right]}\right\} \\ & -\frac{2m_\perp^*}{\hbar^2}\sum_{\eta,s}\frac{|C_s(\eta)|^2\begin{Bmatrix}\cosh^2 \eta z, & s = 1 \\ \sinh^2 \eta z, & s = 2\end{Bmatrix}}{\hbar\Omega_s + \frac{\hbar^2\eta^2}{2m_\perp^*}} + U_{SA}(z) + U_B, \end{aligned} \tag{5.15.16}$$

and the effective mass of the polaron is

$$m_{\text{eff}}^* = m^*\left\{1 - \frac{8}{\beta_V d}\alpha_V L(z) + \frac{2}{d^2}[M_1(z)] + M_2(z)\right\}^{-1}; \tag{5.15.17a}$$

here

$$L(z) = \sum_{n=1,3,5,\ldots}\ln\left(\frac{\pi n}{d\beta_V}\right)\cos^2\left(\frac{\pi nz}{d}\right) + \sum_{n=2,4,6,\ldots}\ln\left(\frac{\pi n}{d\beta_V}\right)\sin^2\left(\frac{\pi nz}{d}\right). \tag{5.15.17b}$$

$$M_s = \int_0^\infty d\eta\;\alpha_s\frac{\beta_s\eta^2}{\left(\eta^2 + \beta_s^2\right)^3}\begin{Bmatrix}\cosh^2 \eta z, & s = 1 \\ \sinh^2 \eta z, & s = 1\end{Bmatrix}. \tag{5.15.17c}$$

Averaging (5.15.15)–(5.15.16) on the wave function of the quantum well (5.15.3b), we obtain

$$E_0(d) = E_{\text{р.кв.}} + E^V_{\text{ph}} + E^S_{\text{ph}} + E_{SA}, \tag{5.15.18}$$

where

$$E_{\text{р.кв.}} = \frac{\pi^2\hbar^2 n^2}{8m^*_{||}d^2} = \frac{\pi^2\hbar^2 n^2}{8m^*_{||}(Na)^2};\ n = 1, 2, 3, \ldots; \tag{5.15.19a}$$

where a is the lattice constant:

$$E^V_{\text{ph}} = -\frac{2\alpha_V\hbar\omega_0\beta_V}{Na} \left\{\sum_{n,m=1}^{N/2}\left[\left(\frac{m\pi}{Na}\right)^2 - \beta_V{}^2\right]^{-1}\ln\left(\frac{m\pi}{Na\beta_V}\right)+\frac{1}{2}\left[\left(\frac{n\pi}{Na}\right)^2 - \beta_V^2\right]^{-1} \ln\left(\frac{n\pi}{Na\beta_V}\right)\right\}; \tag{5.15.19b}$$

$$\begin{aligned} E^S_{\text{ph}} = & -\frac{e^2\beta_V}{16\pi\varepsilon_0\xi}\int_0^\infty dx\left(\left[\varepsilon_{20}\tanh x+\varepsilon\right]^{-1} - \left[\varepsilon_2\tanh x+\varepsilon\right]^{-1}\right)\frac{\tanh x}{\cosh^2 x} \\ & \times\frac{2x\left(\pi^2 + 4x^2\right) + \pi^2\sinh 2x}{2x\left(\pi^2 + 4x^2\right)\left(1 + \frac{\omega_0}{\Omega_{S_1}}\frac{x^2}{\xi^2}\right)}\frac{e^2\beta_V}{16\pi\varepsilon_0\xi} \\ & \int_0^\infty dx\left(\left[\varepsilon_2 + \varepsilon\tanh x\right]^{-1} - \left[\varepsilon_{20} + \varepsilon\tanh x\right]^{-1}\right)\frac{1}{\cosh^2 x} \\ & \times\left[-2x\left(\pi^2 + 4x^2\right) + \pi^2\sinh 2x\right] \\ & \left[2x\left(x^2 + 4\pi^2\right)\left(1 + \frac{\omega_0 x^2}{\Omega_{S_2}\varepsilon^2}\right)\right]^{-1}; \end{aligned} \tag{5.15.19c}$$

$$E_{SA} = \frac{e^2\beta_V}{4\pi\varepsilon_0\xi}\int_0^\infty dx(\mathrm{e}^{2x} - \delta^2)^{-1}[\delta^2 + \mathrm{e}^x\delta\pi^2\sinh x\,(x^{-1}[\pi^2 + x^2]^{-1})]. \tag{5.15.19d}$$

Figures 5.20 and 5.21 show the dependences $E_{s.q.}$, E^V_{ph}, E^S_{ph}, E_{SA} on d for a GaAs, plate bordering a vacuum. The following parameter values were used in the calculation $\varepsilon_{20} = 12.8$; $\varepsilon_2 = 10.9$; $\hbar\omega_0 = 36$ meV; $m^*_e = 0.066m_0$; $a = 5.65$ Å, $\alpha_V = 0.068$; $-\alpha\hbar\omega_0 = -2.45$ meV. From figure 5.21 it can be seen that with $d \approx 500$ Å, the polaron energy in the layer becomes equal to the binding energy of the bulk polaron.

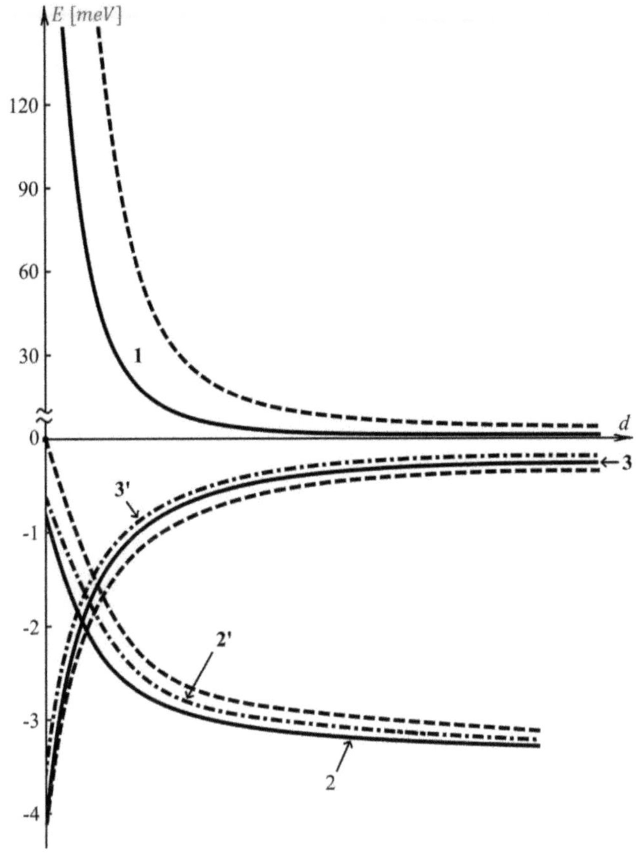

Figure 5.20. The dependences of the contributions to the energy of the polaron on the thickness of the GaAs layer bordering the vacuum: curve 1—$E_{s.q.} + E_{SA}$; curve 2—E^V_{ph}; curve 3—E^S_{ph}; and curve 4—E_0. The following parameter values were used in the calculation: $\varepsilon_{20} = 12.8$; $\varepsilon_2 = 10.9$; $\hbar\omega_0 = 36$ meV; $m_e^* = 0.066m_0$; $a = 5.65$ Å, $\alpha_V = 0.068$ ($\alpha_V\hbar\omega_0 = 2.45$ meV). Solid lines correspond to n=1; dashed lines correspond to $n = 2$.

In the case when the media bordering the GaAs layer have a greater dielectric constant than ε_{20} ($\varepsilon > \varepsilon_{20}$), then $E_{SA} < 0$ and $|E_0| > |\alpha_V \hbar\omega_0|$ in the domain $d \leqslant R_{\text{ph}} \sim 40$ Å.

5.15.3 Intermediate communication

Finally, consider the case when the movements of the charge carrier along the z-axis and in the plane of the layer are fast. As shown in clause 5.4.3, the displacements of the oscillators in this case are determined not by the z-coordinate, but by the quantum state of the electron. The x, y coordinates are eliminated from the Hamiltonian (5.15.1) using the former unitary transformation (5.15.8). But now, before performing the second unitary transformation, it is necessary to perform averaging on the wave function describing movement along the z-axis. As a result, for variational amplitudes we obtain the expressions:

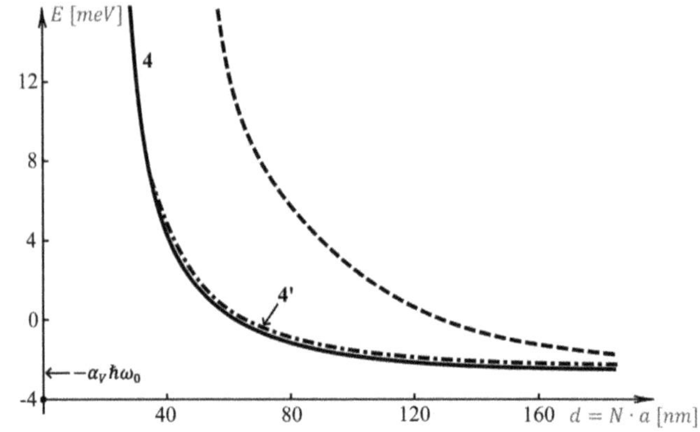

Figure 5.21. Dependences of the polaron energy on the thickness of the GaAs, layer bordering the vacuum: curve 4. The following parameter values were used in the calculation: $\varepsilon_{20} = 12.8$; $\varepsilon_2 = 10.9$; $\hbar\omega_0 = 36$ meV; $m_e^* = 0.066 m_0$; $a = 5.65$ Å, $\alpha_V = 0.068$ ($\alpha_V \hbar\omega_0 = 2.45$ meV). Solid lines correspond to $n = 1$, dashed lines correspond to $n = 2$.

$$f^*_{m,\,i}(\mathbf{q}_\perp) = -\,C_V \left\langle \begin{Bmatrix} \cos\left(\frac{m\pi z}{2d}\right), & m = 1, 3, 5, \ldots \\ \sin\left(\frac{m\pi z}{2d}\right), & m = 2, 4, 6, \ \ldots \end{Bmatrix} \right\rangle \left[q_\perp^2 + \left(\frac{m\pi}{2d}\right)\right]^{-\frac{1}{2}} \times$$

$$\times \left[\hbar\omega_0 - (1 - \tilde{\eta})\frac{\hbar^2 \mathbf{K}_\perp \mathbf{q}_\perp}{m_\perp^*} + \frac{\hbar^2 q_\perp^2}{2m_\perp^*}\right]^{-1}; \tag{5.15.20a}$$

$$f_s^*(\eta) = -C_s(\eta)\left\langle \begin{matrix} \cosh \eta z, & s = 1 \\ \sinh \eta z, & s = 2 \end{matrix} \right\rangle \left[\hbar\Omega_s - \frac{(1 - \tilde{\eta})\hbar^2 \eta \mathbf{K}}{m_\perp^*} + \frac{\hbar^2 \eta^2}{2m_\perp^*}\right]^{-1}, \tag{5.15.20b}$$

where $\langle \ldots \rangle$ means averaging on the wave function $\psi(z)$.

For the phonon contribution to the energy of the ground state and the effective mass, we obtain the following expressions:

$$\Delta E_{\text{ph}}^{S,\,V} = -\sum_{\mathbf{q}_\perp} \frac{\left|\widetilde{C}_{m,\,i}(\mathbf{q}_\perp)\right|^2}{\hbar\omega_0 + \frac{\hbar^2 q_\perp^2}{2m_\perp^*}} - \sum_{\eta, s} \frac{|\widetilde{C}_s(\eta)|^2}{\hbar\Omega_s + \frac{\hbar^2 \eta^2}{2m_\perp^*}}; \tag{5.15.21}$$

$$\frac{m^*_{\text{eff}}}{m^*} = 1 + \frac{2\hbar^2}{m_\perp^*} \left\{ \sum_{\mathbf{q}_\perp} \frac{q_\perp^2 \left|\widetilde{C}_{m,\,i}(\mathbf{q}_\perp)\right|^2}{\left(\hbar\omega_0 + \frac{\hbar^2 q_\perp^2}{2m_\perp^*}\right)^3} + \sum_{\eta, s} \frac{\eta^2 |\widetilde{C}_s(\eta)|^2}{\left(\hbar\Omega_s + \frac{\hbar^2 \eta^2}{2m_\perp^*}\right)^3} \right\}; \tag{5.15.22}$$

with symbols:

$$|\widetilde{C}_{m,i}(\mathbf{q}_\perp)|^2 = |C_V|^2 \left| \left\langle \begin{Bmatrix} \cos\left(\dfrac{m\pi z}{2d}\right), & m = 1, 3, 5, \ldots \\ \sin\left(\dfrac{m\pi z}{2d}\right), & m = 2, 4, 6, \ldots \end{Bmatrix} \right\rangle \right|^2 \qquad (5.15.23a)$$

$$\left[q_\perp^2 + \left(\frac{m\pi}{2d}\right)^2\right]^{-1};$$

$$|\widetilde{C}_s(\eta)|^2 = |C_s(\eta)|^2 \left| \left\langle \begin{Bmatrix} \cosh \eta z, & s = 1 \\ \sinh \eta z, & s = 2 \end{Bmatrix} \right\rangle \right|^2. \qquad (5.15.23b)$$

Switching from summation to integration in (5.15.21), (5.15.22), we obtain an expression for the binding energy and effective mass in a plate of a polar crystal of finite thickness in plates of nonpolar crystals. In figure 5.20, curves 2 and 3 correspond to the calculation according to the formula (5.15.22). As can be seen from the comparison of the results, a weak bond for a GaAs crystal gives a deeper level of energy, which is explained by the fulfillment of the criteria of weak and nonfulfillment—for an intermediate electron–phonon bond. Indeed, for a ZnO crystal plate ($\varepsilon_{20} = 8.5$; $\varepsilon_2 = 4.0$; $\hbar\omega_0 = 73.0$ meV; $\alpha_V = 0.85$), the parameters of which lie in the intermediate bond domain, the intermediate bond theory gives an energy level 10% deeper than the weak electron–phonon bond in the thickness range $R_{ph}^V - 10R_{ph}^V$.

A comparison of the results of numerical calculations obtained by various methods, weak, intermediate with elements of weak, and intermediate electron–phonon bonds, allows us to formulate the following conclusions:

1. The binding energy of a polaron in a plate of a polar crystal can significantly exceed the corresponding energy in a massive crystal. The decrease in the energy level of the polaron is influenced by the media bordering the crystal, the parameters of which determine the energy of self-action and the polaron constants of interactions with surface and bulk optical phonons.
2. A specific feature of the dimensional dependence of the polaron energy is its nonmonotonicity when certain criteria are met. The nonmonotonicity of the behavior is explained by the fact that in the thickness range $d < R_{ph}$ the main contribution is the contribution from spatial quantization, which decreases as d^{-2} with increasing d. While the polaron energy and self-action energy, which have a negative sign, decrease in modulus as d^{-1} and become predominant at certain thicknesses.
3. With decreasing d, the contribution from bulk optical phonons decreases and in the domain of $d \approx 10 - 15$ Å for GaAs becomes negligible in comparison with the contribution from surface phonons, which increases with decreasing d.
4. An essential distinguishing property of a polaron in a limited crystal in comparison with a polaron in a massive crystal is the dependence of its binding energy not only on the constant of the electron–phonon interaction,

but also on the dielectric permittivity (as well as on the frequencies of optical vibrations, if the bordering media are polar) of neighboring media.

5.16 Surface spatially extended optical phonons. Polaronic states in composite superlattices

We have already discussed the problem of potentials and vibrational spectra in superlattices (chapter 3) formed by alternating homeopolar layers of two types. If the superlattice is formed by polar layers, then it is necessary, as in the case of a single polar layer (section 5.14), to take into account the interaction of the charge carrier with polarization surface and bulk optical vibrations. The problem of the Hamiltonian of the electron–phonon interaction in two-layer periodic systems was first considered in [97, 106], and its generalization to the case of an arbitrary multilayer system, from which the Hamiltonian of the electron–phonon interaction for superlattices follows as a special case, was performed in [22]. A complete derivation of the Hamiltonian of the electron–phonon interaction of the Pekar–Fröhlich type for arbitrary multilayer structures and composite superlattices is given in chapter 3.

Let's turn to the consideration of the specific features of the electron–phonon interaction in superlattices. One of the most important features is the emergence of new elementary excitations, which were first predicted theoretically in the works of Pokatilov and Beril [97, 106] and discovered experimentally by the authors of the works [107–114].

We present an elementary theory of surface spatially extended optical vibrations for a composite superlattice consisting of alternating polar layers of two types: '1' and '2' with thicknesses d_1, d_2 and permittivity $\varepsilon_1(\omega)$, $\varepsilon_2(\omega)$. The length of the period of the superlattice is denoted by $c = (d_1 + d_2)/2$.

5.16.1 Law of dispersion of interacting surface modes

We solve the problem in a continuous approximation, assuming that d_1 and d_2 are significantly larger than the lattice constant and the dielectric permittivity $\varepsilon_{1,2}(\omega)$ have the same values as in an unlimited crystal. We will place the origin of the coordinate system in the middle of one of the layers, for example, layer 1. The d axis is perpendicular to the plane of the layers. We assume that N layers are arranged symmetrically relative to the center ($N/2$ layers each to the left and right of the central layer). The numbering of the layers n is indicated by Arabic numerals: to the right of the central layer—positive, to the left—negative. Even numbers correspond to the layers of crystal 1, and odd numbers correspond to crystal 2. The z_n coordinate of the center of the nth layer is defined as

$$z_n = n \cdot c; n = 0, +1, +2, \ldots; \tag{5.16.1}$$

$$z = z_n + z', \tag{5.16.2}$$

where $-d_1/2 \leqslant z' \leqslant d_1$—for the crystal 1 and $-d_2/2 \leqslant z' \leqslant d_2$– for the crystal 2.

Electric fields caused by fluctuations in the polarization of crystals (ignoring magnetic effects) are described by Maxwell's equations:

$$\text{rot}\,\mathbf{E} = 0,\ \ \text{div}\,\mathbf{D} = 0; \tag{5.16.3}$$

$$\mathbf{D} = \varepsilon_0\varepsilon(\omega)\mathbf{E}; \tag{5.16.4}$$

where $\boldsymbol{E}$ and $\boldsymbol{D}$ are the vectors of electric field strength and induction ($\mathbf{E} = -\nabla\Phi$).

Considering the dielectric function $\varepsilon(\omega)$ independent of the spatial coordinates within each layer, we obtain the equation for the potential of the electric field of polarization oscillations:

$$\varepsilon(\omega)\Delta\Phi(\mathbf{r}, t) = 0. \tag{5.16.5}$$

Due to the spatial homogeneity of the system under consideration in the XOY plane, the potential $\Phi(\mathbf{r}, t)$ can be represented as

$$\Phi(\mathbf{r}, t) = \sum_{\mathbf{q}_\perp}\Phi_{q_\perp}(z, t)e^{i\mathbf{q}_\perp\boldsymbol{\rho}}, \tag{5.16.6}$$

where

$$\Phi_{q_\perp}(z, t) = \Phi_{q_\perp}(z)e^{i\omega t}; \tag{5.16.7}$$

$\boldsymbol{\rho}$—radius is a vector in the plane of the layer.

Substituting the expression (5.16.6) in (5.16.5), we obtain the equation for determining $\Phi_{q_\perp}(z)$:

$$\varepsilon(\omega)\left[\frac{d^2\Phi_{q_\perp}(z)}{dz^2} - q_\perp^2\Phi_{q_\perp}(z)\right] = 0. \tag{5.16.8a}$$

Using the transformation (5.16.2) to the 'local' coordinate system for the nth layer, we obtain

$$\varepsilon^{(1)}(\omega)\left[\frac{d^2\Phi^{(1)}_{q_\perp n}(z')}{dz'^2} - q_\perp^2\Phi^{(1)}_{q_\perp n}(z')\right] = 0,\ -\frac{d_1}{2} \leqslant z' \leqslant \frac{d_1}{2}; \tag{5.16.8b}$$

$$\varepsilon^{(2)}(\omega)\left[\frac{d^2\Phi^{(2)}_{q_\perp n}(z')}{dz'^2} - q_\perp^2\Phi^{(2)}_{q_\perp n}(z')\right] = 0,\ -\frac{d_2}{2} \leqslant z' \leqslant \frac{d_2}{2}. \tag{5.16.8c}$$

The boundary conditions for potentials and fields, for example, for the central layer ($n = 0$), have the form

$$\begin{cases} \Phi^{(1)}_{q_\perp,\,n=0}\left(z' = \dfrac{d_1}{2}, t\right) = \Phi^{(2)}_{q_\perp,\,n=1}\left(z' = -\dfrac{d_2}{2}, t\right); \\ \Phi^{(1)}_{q_\perp,\,n=0}\left(z' = -\dfrac{d_1}{2}, t\right) = \Phi^{(2)}_{q_\perp,\,n=1}\left(z' = \dfrac{d}{2}, t\right); \end{cases} \tag{5.16.9}$$

$$\begin{cases} \varepsilon^{(1)}(\omega)\dfrac{d}{dz'}\Phi^{(1)}_{q_\perp,\,n=0}\left(z'=\dfrac{d_1}{2},\,t\right)=\varepsilon^{(2)}(\omega)\dfrac{d}{dz'}\Phi^{(2)}_{q_\perp,\,n=1}\left(z'=-\dfrac{d_2}{2},\,t\right); \\ \varepsilon^{(2)}(\omega)\dfrac{d}{dz'}\Phi^{(1)}_{q_\perp,\,n=0}\left(z'=-\dfrac{d_1}{2},\,t\right)=\varepsilon^{(2)}(\omega)\dfrac{d}{dz'}\Phi^{(2)}_{q_\perp,\,n=-1}\left(z'=\dfrac{d_2}{2},\,t\right). \end{cases} \tag{5.16.10}$$

Equations (5.16.8a) and (5.16.8b) are satisfied in two cases:

$$\varepsilon^{(l)}(\omega)=0,\ l=1,\,2; \tag{5.16.11a}$$

or

$$\frac{d^2\Phi^{(l)}_{q_\perp,\,n}(z')}{dz'^2}-q_\perp^2\Phi^{(l)}_{q_\perp,\,n}(z')=0. \tag{5.16.11b}$$

Consider the case defined by the condition (5.6.11a). To find the roots of this equation (ω_V), we specify the type of the dielectric function $\varepsilon^{(l)}(\omega)$, taking it in the oscillatory approximation:

$$\varepsilon^{(l)}(\omega)=\varepsilon^{(l)}_\infty+\frac{\varepsilon^{(l)}_0-\varepsilon^{(l)}_\infty}{1-\dfrac{\omega^2}{\left(\omega^{(l)}_{TO}\right)^2}}, \tag{5.16.12}$$

where $\varepsilon^{(l)}_0$, $\varepsilon^{(l)}_\infty$ are static and high-frequency dielectric permittivity of crystals. Then from equation (5.16.11a) we find

$$\omega^{(l)}_{LO}=\omega^{(l)}_{TO}\left(\frac{\varepsilon^{(l)}_0}{\varepsilon^{(l)}_\infty}\right)^{\frac{1}{2}}. \tag{5.16.13}$$

These are the frequencies of bulk longitudinal optical vibrations. Since equation (5.16.8a) is satisfied for any values $\Phi^{(l)}_{q_\perp,\,n}(z)$, that is, the most general kind of Fourier components $\Phi^{(l)}_{q_\perp,\,n}$ is

$$\Phi^{(l)}_{q_\perp,\,V}(z')=a^{(l)}_{q_\perp}e^{iq_z z'}+b^{(l)}_{q_\perp}e^{-iq_\perp z'}; \tag{5.16.14}$$

The 'V' index indicates that the potential corresponds to bulk fluctuations.

Due to various time dependencies ($\omega^{(1)}_V\neq\omega^{(2)}_V$) of the potentials of bulk polarization oscillations, the boundary conditions (5.16.9)–(5.16.10) for $\Phi^{(l)}_{q_\perp V}(z',\,t)$ are satisfied at any time if the potential $\Phi^{(l)}_{q_\perp V}(z',\,t)$ vanishes at $z'=\pm d_{1,\,2}/2$. Therefore, the potentials of the bulk modes have nodes at the boundaries of the sections of the layers, i.e.,

$$\Phi^{(l)}_{q_\perp,\,V}(z')=a^{(l)}_{q_\perp}\begin{cases}\sin\left(\dfrac{\pi kz'}{d^{(l)}}\right);\ k=2,\,4,\,6,\,\ldots; \\ \cos\left(\dfrac{\pi kz'}{d^{(l)}}\right);\ k=1,\,3,\,5,\,\ldots.\end{cases} \tag{5.16.15}$$

In the case defined by equation (5.16.11b), we find the potentials of local surface polarization modes:

$$\Phi^{(l)}_{q_\perp, n}(z') = A^{(l)}_{q_\perp, n} e^{q_\perp z'} + B^{(l)}_{q_\perp, n} e^{-q_\perp z'}. \tag{5.16.16}$$

Taking into account the periodicity of the structure along the z direction, it is possible from a system of $(2N + 1)$ equations to determine the coefficients $A^{(l)}_{q_\perp, n}$, $B^{(l)}_{q_\perp, n}$ using the Fourier transform

$$X^{(l)}_{q_\perp, n} = \sum_\kappa X^{(l)}_\kappa(q_\perp) e^{i\kappa z_n}, \tag{5.16.17}$$

where

$$X^{(l)}_{q_\perp, n} = (A^{(l)}_{q_\perp, n}; B^{(l)}_{q_\perp, n}),$$

go to the system of four equations for four unknowns $A^{(1, 2)}_\kappa$ and $B^{(1, 2)}_\kappa$:

$$\begin{cases} A^{(1)}_\kappa e^{\frac{q_\perp d_1}{2}} + B^{(1)}_\kappa e^{-\frac{q_\perp d_1}{2}} = A^{(2)}_\kappa e^{-\frac{q_\perp d_2}{2} + i\kappa c} + B^{(2)}_\kappa e^{\frac{q_\perp d_2}{2} + i\kappa c}; \\ A^{(1)}_\kappa e^{-\frac{q_\perp d_1}{2}} + B^{(1)}_\kappa e^{\frac{q_\perp d_1}{2}} = A^{(2)}_\kappa e^{\frac{q_\perp d_2}{2} - i\kappa c} + B^{(2)}_\kappa e^{\frac{q_\perp d_2}{2} - i\kappa c}; \\ \varepsilon^{(1)}(\omega)\left(A^{(1)}_\kappa e^{\frac{q_\perp d_1}{2}} - B^{(1)}_\kappa e^{-\frac{q_\perp d_1}{2}}\right) = \varepsilon^{(2)}(\omega)\left(A^{(2)}_\kappa e^{-\frac{q_\perp d_2}{2} + i\kappa c} - B^{(2)}_\kappa e^{\frac{q_\perp d_2}{2} + i\kappa c}\right); \\ \varepsilon^{(1)}(\omega)\left(A^{(1)}_\kappa e^{-\frac{q_\perp d_1}{2}} - B^{(1)} e^{\frac{q_\perp d_1}{2}}\right) = \varepsilon^{(2)}(\omega)\left(A^{(2)}_\kappa e^{\frac{q_\perp d_2}{2} - i\kappa c} - B^{(2)}_\kappa e^{-\frac{q_\perp d_2}{2} - i\kappa c}\right). \end{cases} \tag{5.16.18}$$

From the condition of existence of nontrivial solutions of the system of homogeneous equations (5.16.18), we find the equation for frequencies:

$$[\varepsilon^{(1)}(\omega)]^2 + [\varepsilon^{(2)}(\omega)]^2 + 2\varepsilon^{(1)}(\omega)\varepsilon^{(2)}(\omega)[D_1 + D_2 \sin^2(\kappa c)] = 0; \tag{5.16.19}$$

The designations are introduced here:

$$D_1 = \frac{\cosh x_1 \cosh x_2 - 1}{\sinh x_1 \sinh x_2}; \quad D_2 = \frac{2}{\sinh x_1 \sinh x_2}; \quad x_{1, 2} = q_\perp d_{1, 2}. \tag{5.16.20}$$

Equation (5.16.19) is of the fourth order with respect to ω^2 and has all real roots. Consequently, optical surface vibrations in the structure under consideration, interacting, form four types of new vibrations, which can be conditionally called spatially extended.

The solutions of equation (5.16.19) have the form

$$\begin{aligned} \Omega^2_{s, j=1, 3} = {} & [2(\varepsilon_{\infty 2} + F_+\varepsilon_{\infty 1})]^{-1}\Big\{\omega^2_{TO1}(\varepsilon_{\infty 2} + F_+\varepsilon_{01}) + \omega^2_{TO2}(\varepsilon_{02} + F_+\varepsilon_{\infty 1}) \pm \\ & \pm \Big[\omega^2_{TO1}(\varepsilon_{\infty 2} + F_+\varepsilon_{01}) + \omega_{TO2}{}^2(\varepsilon_{02} + F_+\varepsilon_{\infty 1})\Big]^2 - \\ & - 4(\varepsilon_{\infty 2} + \varepsilon_{\infty 1}F_+)(\varepsilon_{02} + \varepsilon_{01}F_+)\omega^2_{TO1}\omega^2_{TO2}]\Big\}^{1/2}; \end{aligned} \tag{5.16.21}$$

$$\Omega^2_{s,j=2,4} = [2(\varepsilon_{\infty 2} + F_{-}\varepsilon_{\infty 1})]^{-1}\{\omega^2_{TO1}(\varepsilon_{\infty 2} + F_{-}\varepsilon_{01}) + \omega^2_{TO2}(\varepsilon_{02} + F_{-}\varepsilon_{\infty 1}) \pm$$
$$\left[\left[\omega^2_{TO1}(\varepsilon_{\infty 2} + F_{-}\varepsilon_{01}) + \omega^2_{TO2}(\varepsilon_{02} + F_{-}\varepsilon_{\infty 1})\right]^2 \quad (5.16.22)$$
$$- 4\omega^2_{TO1}\omega^2_{TO2}(\varepsilon_{\infty 2} + F_{-}\varepsilon_{\infty 1})(\varepsilon_{02} + F_{-}\varepsilon_{01})]^{\frac{1}{2}},$$

where

$$F_{\pm} = D_1 + D_2 \sin^2(\kappa c) \pm [(D_1 + D_2 \sin^2(\kappa c) \pm D_1 + D_2 \sin^2(\kappa c))^2 - 1]^{\frac{1}{2}}. \quad (5.16.23)$$

Figure 3.1 shows graphs of the functions $\Omega_{s,j}(\kappa)$ for the structure of the AlSb and GaSb layers. In general, there is a significant variance of the functions $\omega_j(\kappa)$, the more pronounced the smaller $q_\perp d$. At large $q_\perp d$ ($q_\perp d \gg 1$) the surface spatially extended optical vibrations turn into ordinary surface modes. The spectrum of $\Omega_{s,j}(\kappa)$ lies in the interval between the highest and lowest of the frequencies $\omega^{(l)}_{LO}$ and $\omega^{(l)}_{TO}$. For $d_1 = d_2$ and $\kappa = 0$ the frequencies of modes with different j coincide in pairs.

Here we have considered a simple case of substances with one resonant frequency ω (5.16.12). However, in the developed theory there are no restrictions on the use of dielectric functions of another type, in particular, with two or more resonant frequencies. The only advantage of (5.16.12) over other models is that (5.16.12) allowed us to obtain solutions in an analytical form (formulas (5.16.22)–(5.16.23)).

Surface spatially extended optical phonons also have dispersion along the wave vector $\mathbf{q}_\perp$, as follows from the formulas for frequencies (5.16.22)–(5.16.23). According to the latter, four bands are obtained, the lower and upper edges of which correspond to the values of $\kappa c = \pi$. The widths of all bands decrease with increasing $q_\perp$ and at large values of $q_\perp$ the upper and lower edges of each band practically merge (figure 3.2), asymptotically approaching the values of the frequencies of the surface modes:

$$\Omega^2_{S_1} = \omega^2_{TO1}\left(\frac{\varepsilon_{10} + \varepsilon_2}{\varepsilon_1 + \varepsilon_2}\right); \; \Omega^2_{S_2} = \omega^2_{TO2}\left(\frac{\varepsilon_{20} + \varepsilon_1}{\varepsilon_2 + \varepsilon_1}\right). \quad (5.16.24)$$

Areas of phonons are shown in figure 3.1 and were built using the parameters of the superlattice layers on GaAs (1): $d_1 = 5,887$ nm, $\omega_{LO1} = 295$ cm^{-1}, $\omega_{LO_1} = 273$ cm^{-1} and AlAs (2): $d_2 = 1,805$ nm, $\omega_{LO2} = 407$ cm^{-1}, $\omega_{TO2} = 364$ cm^{-1}. As can be seen from figure 3.2, the bands are combined in pairs into: (a) the frequency range $\omega_{LO1} - \omega_{TO1}$—gallium–arsenic-like and (b) the frequency range $\omega_{LO2} - \omega_{TO2}$—aluminum–arsenic-like; moreover, the outer edges (corresponding to $\kappa c = \pi$) of the pairs of bands in the limit $q_\perp = 0$ are adjacent to the frequencies ω_{TO1}, ω_{LO1} and ω_{TO2}, ω_{LO2}, respectively, and, consequently, their position is stable: it does not depend on the thicknesses of the layers. However, the latter have a strong influence on the position of the inner edges. When $d_1 = d_2$, the inner edges of the same type of pairs of bands close.

It is of interest to calculate the density of phonon states $P_s(\omega)$, $s = 1, \ldots, 4$—the band number. The frequency distribution of $P_s(\omega)$ is in accordance with the general formula

$$P_s(\omega) = \frac{c}{\pi}\frac{\partial\kappa(\omega)}{\partial\omega}. \tag{5.16.25}$$

This derivative can be calculated analytically for the initial participants of the dispersion frequencies at $q_\perp \to 0$. Here are the expressions for gallium–arsenic-like modes:

$$P_1(\kappa)|_{\kappa(\Omega_{S1})} = \frac{\sqrt{M_1(q_\perp)}}{\pi(\Omega_{S1} - \omega_{TO1})\sqrt{\Omega_{S1} - \Omega_{S1}^{\min}(q_\perp)}}; \tag{5.16.26a}$$

$$P_2(\kappa)|_{\kappa(\Omega_{S2})} = \frac{\sqrt{M_2(q_\perp)}}{\pi(\omega_{LO1} - \Omega_{S2})\sqrt{\Omega_{S2}^{\max}(q_\perp) - \Omega_{S2}}}. \tag{5.16.26b}$$

In these formulas, Ω_{S1}, Ω_{S2} are variables and $\Omega_{S1}^{\max}$, $\Omega_{S2}^{\max}$ are the lower edge of the first and upper edge of the second bands of the 'gallium–arsenic-like' type, respectively. The densities of states in aluminum–arsenic-like bands are recorded in a similar way. In the numerators of formulas (5.16.26a), (5.16.26b) square functions of the wave vector, which, with the above parameters, have the form $M_1(q_\perp) = 4.27 \cdot 10^{-17} q_\perp^2$, $M(q_\perp) = 6.49 \cdot 10^{-17} q_\perp^2$. It follows from these expressions for P_s that the maximum density of states occurs at those edges that meet in the limit $q_\perp \to 0$ with bulk frequency values, and half of the total number of vibrational modes falls on the area of $|\Omega_s - \omega(\pi/2; q_\perp)| = M_s(q_\perp)$ close up $\Omega_s(\pi/2; q_\perp)$.

The obtained results can be given a simple qualitative interpretation: at the contact of two semi-infinite polar crystals, normal optical modes are represented by two types: (a) bulk-enclosed inside each of the substances and nonpenetrating into another substance; (b) surface-localized at the interface, with amplitudes exponentially decreasing on both sides of the interface.

In a periodic system of alternating polar layers of two types of sufficiently small thickness, the surface vibrational modes localized at different surfaces overlap, forming spatial vibrations propagating through the entire structure in a direction perpendicular to the planes of the layers.

In the language of interactions between structural elements of layers (atoms, molecules, etc), the fundamental transformation of the essence of surface vibrations and their transformation from localized to propagating ones can be explained as follows: these forces have a short-acting character, therefore there is no direct interaction between atoms of the same type of layers. This eliminates the possibility of the formation of bulk modes common to the entire superlattice. They are isolated in each layer and are called confined—'captive'. But for surface vibrations in superlattices, there is a fundamentally different situation: surface vibrations localized at the boundaries of each layer interact inside it. Through this interaction, surface vibrations are 'collectivized' and delocalized, acquiring the ability to freely propagate in any direction of the superlattice. At the same time, as it was shown, their frequency spectrum is split into bands. The effectiveness of 'collectivization' strongly depends on the thickness of the layers; short-wave oscillations rapidly decreasing from the boundary deep into the layers remain nonlocalized.

Finally, experimental evidence of the existence of surface spatially extended optical phonons is of interest.

The first report on their observation in experiments on Raman scattering of light was published by M. Cardona and his colleagues [113]. The experiment was carried out on a GaAs (a)/AlAs (b) superlattice in the backscattering geometry. The parameters of the superlattice were as follows: sample A—$l_a = 2$ nm, $l_b = 2$ nm; sample B: $l_a = 2$ nm, $l_b = 6$ nm; sample C: $l_a = 6$ nm, $l_b = 2$ nm.

In superlattices of a different type, GaSb/AlSb, surface spatially extended phonons were observed in experiments by another group of researchers from the Bell Laboratory (New Jersey, USA) [111, 112]. Experimental data were also interpreted on the basis of theory [97, 106].

The observation of surface spatially extended optical phonons in GaAs/AlAs superlattices was reported in the works of M. Klein [108] (University of Illinois, USA).

To date, there are more than a hundred references to the works [97, 106] in the world scientific literature, in which the existence of surface spatially extended optical phonons is experimentally and theoretically confirmed.

5.16.2 Hamiltonian of the electron–phonon interaction in a superlattice

It follows from the previous consideration that the electron–phonon interaction should be represented as an interaction with bulk and surface spatially extended optical phonons. Since the bulk optical vibrations of the individual layers of the superlattice do not interact, the corresponding electron–phonon interaction Hamiltonian has the same form as in the layer (see section 5.1).

The Hamiltonian of interaction with surface spatially extended surface optical modes was first obtained in [22, 97] (its conclusion is also contained in the monograph [18]) and has the form

$$\begin{aligned}\hat{H}^S_{e-ph} = {} & \sum_{\eta} e^{i\eta\rho} \sum_{\kappa} e^{\frac{i}{2}n\kappa L} \frac{e\sqrt{\hbar}}{4\sqrt{\varepsilon_0 \eta L_x L_y \sinh\left(\frac{\varsigma_b}{2}\right)}} \cdot \\ & \sum_{s=1}^{4} \frac{1}{\sqrt{\Omega_s}} \sum_{r'=1}^{4} \frac{\omega_c \sqrt{\varepsilon_{c0} - \varepsilon_c}}{\sqrt{\tanh\left(\frac{\varsigma_c}{2}\right)}} \\ & \times \left[K_{1b,r\prime}(\kappa)\cosh \eta z + K_{2b,r\prime}(\kappa)\sinh \eta z\right] \\ & \cdot\left[\delta_{r\prime,s} - \left(1 - \delta_{r\prime,s}\right)F_{r\prime,s}\right] \\ & \times \left(1 + \sum_{r''\neq s} \left| F_{r\prime\prime,s} \right|^2\right)^{-\frac{1}{2}} \left[\hat{b}^{+}_{s}(-\kappa, -\eta, 0) + \hat{b}_s(\kappa, \eta, 0)\right].\end{aligned} \tag{5.16.27}$$

The expression (5.16.27) uses the same notation as in [18, 22]: Ω_s are the frequencies of surface spatially extended oscillations ($s = 1, \ldots, 4$) of the superlattices found above. The explicit form $K_{1b,\,r'}$, $K_{1b,\,r'}$, $F_{r',\,s}$ is described by the following expressions:

$$F_{rs} = \sum_{r' \neq s} N_{rr'}^{-1}(\ldots, \Omega_s) N_{r's}(\ldots, \Omega_s); \tag{5.16.28}$$

N_{rs}^{-1}—the matrix is the inverse of

$$\left\| N_{jb,\,lc}(\kappa, \Omega_c) \right\| = \left\| \left(\Omega_s^2 - \omega_b^2\right)\delta_{j,\,l}\delta_{b,\,c} + M_{jb,\,lc} \right\|; \tag{5.16.29}$$

$$\left\| N_{jb,\,lc}(\kappa) \right\| = \frac{1}{2}\omega_b \sqrt{\frac{\varepsilon_{b0} - \varepsilon_b}{\tanh\left(\frac{\zeta_b}{2}\right)}}\,\omega_c \sqrt{\frac{\varepsilon_{c0} - \varepsilon_c}{\tanh\left(\frac{\zeta_c}{2}\right)}}\,K_{jb,\,lc}(\kappa); \tag{5.16.30}$$

r, r', r'', s—paired indexes satisfying the rule: $1 \to (1, a)$, $2 \to (2, a)$, $3 \to (3, b)$, $4 \to (2, b)$, $(j, b) = r$, $(lc) = r'$;

$$K_{1b,\,1c}(\kappa) = -\eta \tanh\left(\frac{\zeta_b}{2}\right) \tanh\left(\frac{\zeta_c}{2}\right) A_{bc}(\kappa, ++++); \tag{5.16.31a}$$

$$K_{1b,\,2c}(\kappa) = \eta \tanh\left(\frac{\zeta_b}{2}\right) A_{bc}(\kappa, ++--); \tag{5.16.31b}$$

$$K_{2b,\,1c}(\kappa) = -\eta \tanh\left(\frac{\zeta_c}{2}\right) A_{bc}(\kappa, +-+-); \tag{5.16.31c}$$

$$K_{2b,\,2c}(\kappa) = \eta A_{bc}(\kappa, +--+), \tag{5.16.31d}$$

where

$$\left\| A_{bc}(\kappa, +\pm\pm+) \right\| = \frac{2}{\Delta(\kappa)} \left\| \begin{matrix} \pm\nu + \mu_a + \mu_b \cos(\kappa L) & (\nu \pm \mu_a \pm \mu_b)\cos\left(\frac{\kappa L}{2}\right) \\ (\nu \pm \mu_a \pm \mu_b)\cos(\kappa L) & \pm\nu + \mu_b + \mu_b \cos(\kappa L) \end{matrix} \right\|; \tag{5.16.32a}$$

$$\left\| A_{bc}(\kappa, \pm\mp-) \right\| = \frac{2i}{\Delta(\kappa)} \left\| \begin{matrix} \mu_b \sin(\kappa L) & (\nu \pm \mu_a \mp \mu_b)\sin\left(\frac{\kappa L}{2}\right) \\ (\nu \mp \mu_a \pm \mu_b)\sin\left(\frac{\kappa L}{2}\right) & \mu_a \sin(\kappa L) \end{matrix} \right\|; \tag{5.16.32b}$$

$$\nu = \eta(\zeta_a \coth \zeta_a + \zeta_b \coth \zeta_b);\ \mu_k = \frac{\varepsilon_k \eta}{\sinh \zeta_k};\ \Delta(\kappa) = \nu^2 - \mu^2 - 2\mu_a \mu_b \cos(\kappa L). \tag{5.16.32c}$$

The Hamiltonian of the electron–phonon interaction (5.16.27)–(5.16.32) describes the interaction with four surface spatially extended optical modes of a composite superlattice formed by alternating two polar layers of type ‘a’ and ‘b’. It satisfies all the necessary limiting transitions: (a) a polar crystal layer in a lining of nonpolar semi-infinite crystals or vacuum, (b) contact of two polar crystals, and (c) contact of a polar crystal with a nonpolar (or vacuum). This Hamiltonian is necessary to solve the problem of the Fröhlich polaron in a superlattice.

5.17 Polaronic states in a composite superlattice. Weak connection

Let us consider the problem of a polaron in a superlattice, when the movement of a free charge carrier is limited by the limits of one layer, at the boundaries of which potential barriers are assumed to be infinite. Then the solution to this problem contains two stages: (1) isolation of a seed quasi-particle—an electron polaron—formed due to the interaction of an electron with rapid polarization (plasmons of valence electrons). Due to the smallness of the radius of the electron polaron compared with the thicknesses of the layers at typical parameter values, this problem can be solved within the framework of the classical electrodynamic approach, which is developed in section 5.13; (2) taking into account the interaction of the electron polaron with surface and bulk optical phonons of the superlattice, the Hamiltonian of which, as in the case of a polar crystal plate, can be represented in an additive. The bulk part of this Hamiltonian is the same as in the isolated layer, and the surface part is described by the formulas (5.16.27)–(5.16.32).

Here we will consider only the intermediate version with weak electron–phonon coupling elements, analyzed in detail in clause 5.14.2. Since the problems are mathematically identical, we will immediately write down the final results.

The effective polaron Hamiltonian in a superlattice can be represented as

$$\hat{H}_{\text{eff}} = \frac{\hat{P}_z^2}{2m_{\|}^*} + \frac{\hat{P}_\perp^2}{2m_{\text{eff}}^*} + W_{\text{eff}}, \tag{5.17.1}$$

where

$$W_{\text{eff}}(z) = U_{SA}^{SL}(z) + U_B - \frac{2m_\perp^*}{\hbar^2} \sum_{q_\perp} \frac{|C_V^{(b)}|^2}{\left(q_\perp^2 + \beta_{Vb}^2\right)} \left\{ \sum_{m=1,3,\ldots} \frac{\cos^2\left(\frac{m\pi z}{2d}\right)}{\left[q_\perp^2 + \left(\frac{m\pi}{2d}\right)^2\right]} + \sum_{m=2,4,6,\ldots} \frac{\sin^2\left(\frac{m\pi z}{2d}\right)}{\left[q_\perp^2 + \left(\frac{m\pi}{2d}\right)^2\right]} - \sum_{\eta,x} \frac{|\gamma_{s,\eta}(z)|^2}{\left(\hbar\Omega_s + \frac{\hbar^2\eta^2}{2m_{\perp b}^*}\right)} \right\}; \tag{5.17.2}$$

here

$$\gamma_{\eta,\,s}(z) = \sum_{\kappa}\frac{e\sqrt{\hbar}\exp\left(\frac{i}{2}n\kappa L\right)}{4\sqrt{\varepsilon_0\eta L_x L_y \sinh\left(\frac{\varsigma_b}{2}\right)}}\sum_{s=1}^{4}\frac{1}{\sqrt{\Omega_s}}$$
$$\times\sum_{r'=1}^{4}\frac{\omega_c\sqrt{\varepsilon_{c0}-\varepsilon_c}}{\sqrt{\mathrm{th}\left(\frac{\varsigma_c}{2}\right)}}\left[K_{1b,\,r'}(\kappa)\cosh\eta z + K_{2b,\,r'}(\kappa)\sinh\eta z\right]; \tag{5.17.3}$$

where $U_{SA}^{SL}(z)$ is the potential energy of self-action in a layer of type 'b', having the form (5.11.19), and U_B are potential barriers at the boundaries of layer 'b'.

The potential energy corresponding to bulk optical phonons is calculated on the Hamiltonian [7] (formula (5.2.32)).

Assuming that layer 'b' is narrow enough for the inequality to hold:

$$E_{s.q.} >> E_{\mathrm{ph}} + E_{SA},$$

where E_{ph}, E_{SA} is the energy of interactions with phonons and in the field of self-action potential, we average the Hamiltonian (5.17.1) on the wave functions (5.15.2)–(5.15.3b). As a result, for the energies of the ground state of the system, we obtain

$$E_t^{SL}(l_b) = \frac{\pi^2\hbar^2}{2m_{\|}^{*}l_b^2} +$$
$$+\frac{e^2}{4\pi\varepsilon_0\varepsilon_b\epsilon_b}\int_0^{\infty}\frac{dx}{\sinh x\sqrt{\psi^2(x)-1}}$$
$$\left\{e^{-x}\sqrt{\psi^2(x)-1} + \frac{\cosh x}{\psi(x)+\sqrt{\psi^2(x)-1}} - \cosh\left(x\bar{l}\right)\right.$$
$$\left.+\frac{\left(\varepsilon_b^2+\varepsilon_a^2\right)}{2\varepsilon_a\varepsilon_b} + \sinh\left(x\bar{l}\right)\left[\frac{1}{x} - \frac{x}{x^2+\pi^2}\right]\right\} + \frac{2\alpha_{V_b}\hbar\omega_{0b}\beta_{Vb}}{(2l_b)}$$
$$\times\left\{\sum_{\substack{m=1\\n=1}}^{\frac{N}{2}}\frac{\ln\left(\frac{m\pi}{2l_b\beta_{Vb}}\right)}{\left[\left(\frac{m\pi}{2l_b}\right)^2-\beta_V^2\right]} + \frac{1}{2}\frac{\ln\left(\frac{n\pi}{(2l_b)\beta_{Vb}}\right)}{\left[\left(\frac{n\pi}{2l_b}\right)^2-\beta_V^2\right]}\right\} \tag{5.17.4}$$
$$-\frac{e^2}{\varepsilon_0}\sum_{s=1}^{4}\int_0^{\infty}dx\int_{-\pi}^{\pi}\frac{dy}{\varsigma_b^2\left(\varsigma_b^2+4\pi^2\right)^2}$$
$$\times\frac{1}{\Omega_s}\left\{\sum_{r'=1}^{4}\frac{\omega_c\sqrt{\varepsilon_{c0}-\varepsilon_c}}{\sqrt{\mathrm{th}\left(\frac{\varsigma_c}{2}\right)}}K_{1b,r\prime}(y)\left[\delta_{r\prime,s}-\left(1-\delta_{r\prime\prime,s}\right)F_{r\prime,s}(y)\right]\left(1+\sum_{r''\neq s}\left|F_{r\prime\prime,s}\right|^2\right)^{-\frac{1}{2}}\right\}^2$$
$$\times\left[1+R_e^2(\Omega_s)x^2\right]^{-1}.$$

From the expression (5.17.4), it is possible to determine the dimensional dependence of the energy of the ground state of the charge carrier in periodic structures with quantum wells, taking into account the effect of self-action and the polaron effect. All contributions to $E_t^{SL}(l_b)$were already evaluated earlier for the GaAs/AlGaAs. structure. The contribution from the surface spatially extended modes (the odd mode fell out after averaging on the wave function of the quantum well) has the same specificity as in the case of a separate layer: it contains the parameters of the entire structure, therefore it is possible to change the absolute value of this contribution by varying the parameters of neighboring media. In order of magnitude, this contribution is the same as that of subsurface optical phonons in a three-layer structure.

5.18 Polaronic states in nanoscale cylindrical and spherical structures

Let's calculate the ground state energy and the effective mass of the polaron, as well as the electron scattering velocity on surface (interface) optical phonons in a wireous cylindrical MDS structure, the cross-section of which is shown in figure 5.22.

If an excess of electrons is created in a *p*-type semiconductor, which violates the electroneutrality of the structure, then an electric field appears, pressing the electrons to the semiconductor–dielectric interface. As a result, an inversion layer is formed near the boundary. In the inversion layers of the MDS structure, the interaction of charge carriers with surface optical phonons is often the dominant scattering mechanism [73]—firstly, due to an increase in the ratio of the surface area to the volume in which the carriers are localized, and secondly, due to a decrease in the relative concentration of impurities and defects per electron compared with the main domain of the semiconductor in wireous semiconductor and dielectric structures, as shown in [115], the relative contribution of the interaction of charge carriers with surface and interface phonons to the polaron parameters turns out to be even greater than in planar systems.

5.18.1 Electron–phonon interaction in a quantum cylindrical structure

The natural frequencies of optical phonons and the amplitudes of the electron–phonon interaction are calculated within the framework of the dielectric continuum model. The eigenmode frequencies and the Hamiltonian of the electron–phonon interaction for a quantum wire bordering an infinite nonpolar dielectric and for a multilayer cylindrical structure are found in chapter 2 and in [115]. In this section, we will consider a special case of formulas from [116] related to the specific structure under consideration. It has a set of interface modes with certain values of the wavenumber q and the quantum number of the moment projection on the direction of motion m. These modes are characterized by natural frequencies associated with the frequencies of longitudinal $\omega_{2,\,LO}$ and transverse $\omega_{2,\,TO}$ oscillations in the dielectric by the ratio

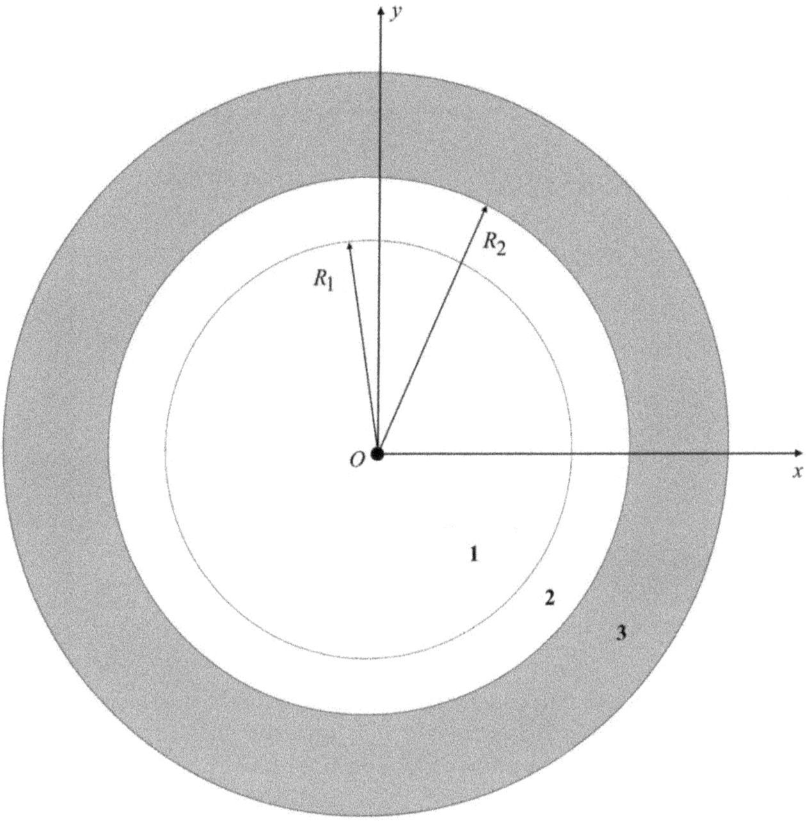

Figure 5.22. Diagram of a cylindrical MDS structure; 1—semiconductor, 2—insulator, 3—electrode.

$$\omega_1(q,\, m) = \sqrt{\frac{\omega^2_{2,\, TO}\varepsilon_1\chi_1(q,\, m) + \omega^2_{2,\, LO}\varepsilon_2\chi_2(q,\, m)}{\varepsilon_1\chi_1(q,\, m) + \varepsilon_2\chi_2(q,\, m)}}, \tag{5.18.1}$$

where ε_1, ε_2 are the high-frequency dielectric permittivity of a semiconductor and a dielectric, respectively, and the coefficients $\chi_1(q,\, m)$ and $\chi_2(q,\, m)$ are expressed in terms of modified cylindrical functions and their derivatives:

$$\psi_1(q,\, m) = \frac{I'_{|m|}(|q|R_1)}{I_{|m|}(|q|R_1)}; \tag{5.18.2}$$

$$\psi_2(q,\, m) = \frac{I'_{|m|}(|q|R_1)K_{|m|}(|q|R_2) - K'_{|m|}(|q|R_1)I_{|m|}(|q|R_2)}{I_{|m|}(|q|R_2)K_{|m|}(|q|R_1) - I_{|m|}(|q|R_1)K_{|m|}(|q|R_2)}. \tag{5.18.3}$$

The Hamiltonian of the electron–phonon interaction in the particular case under consideration is obtained from the general expression [115] and has the form

$$\hat{H}_{e-ph} = \sum_{q,m}\left[\gamma_{q,\,r}(r)\hat{a}_{q,\,m} + \gamma^{*}_{q,\,r}(r)\hat{a}^{+}_{q,\,m}\right]. \tag{5.18.4}$$

It is convenient to represent the amplitudes of the interaction in the form

$$\gamma_{q,\,r}(r) = \Gamma_{q,\,m}(\rho)\frac{1}{\sqrt{2\pi L}}\exp\left[i(qz + m\varphi)\right], \tag{5.18.5}$$

where L is the thread length and $\Gamma_{q,\,m}(\rho)$ is the radial part of the amplitude:

$$\Gamma_{q,\,m}(\rho) = \hbar\omega_{2,\,LO}\sqrt{\frac{2\sqrt{2}\pi\alpha_1 R_p \hbar\omega_{2,\,LO}}{|q|R_1\omega_1(q,\,m)}}\cdot\frac{\varepsilon_2\sqrt{\chi_2}}{[\varepsilon_1\chi_1 + \varepsilon_2\chi_2]}\cdot\frac{I_{|m|}(q\,|\rho|)}{I_{|m|}(qR_1)}, \tag{5.18.6}$$

where

$$R_p = \left(\frac{\hbar}{2m_1^*\omega_{2,\,LO}}\right)^{\frac{1}{2}}; \tag{5.18.7}$$

$$\alpha_1 = \frac{e^2}{\sqrt{2}\hbar\omega_{2,\,LO}R_p}\left(\frac{1}{\varepsilon_0} - \frac{1}{\varepsilon_{20}}\right). \tag{5.18.8}$$

The dependence of the amplitudes on the thickness of the dielectric $l_2 = R_2 - R_1$ under the condition $\varepsilon_1 > \varepsilon_2$ is nonmonotonic: with the growth of l_2, the increase is replaced by a decrease. In the opposite case, $\varepsilon_1 < \varepsilon_2$ the amplitudes of the interaction monotonously increase with increasing dielectric thickness.

A semiconductor (cylinder 1) is considered nonpolar, and a dielectric (cylindrical layer 2) is considered polar.

5.18.2 Intrinsic energy of the polaron and the electron scattering velocity

Further, it is assumed that the inversion layer is thin enough to neglect the electron transitions between the levels of dimensional quantization. Let's choose the wave function of the electron without taking into account the interaction with phonons [116]:

$$\psi_{k,\,m}(r) = \sqrt{\frac{b^3}{2\pi L}}\frac{(R_1 - \rho)}{\sqrt{\rho}}\exp\left[-\frac{b}{2}(R_1 - \rho) + i(kz - m\varphi)\right], \tag{5.18.9}$$

which is applicable if the thickness of the inversion layer $d_1 \equiv \langle R_1 - \rho\rangle_\psi = 3/b$ is small compared to the radius of the semiconductor domain. In this case, the centrifugal potential arising in the Schrödinger equation for the radial wave function is taken into account as a perturbation. The electron energy levels in the considered approximation have the form

$$E_0(k,\,m) = \frac{\hbar^2}{2m_1^*}\left\{k^2 - \frac{b^2}{4} + \left(m^2 - \frac{1}{4}\right)\langle\rho^{-2}\rangle_\psi\right\}. \tag{5.18.10}$$

The intrinsic energy of the polaron is calculated in the second order of perturbation theory by the amplitudes of the electron–phonon interaction:

$$\Delta E_p(k, m) = \sum_{k',m'}\sum_{q,\mu} \frac{\left|\left\langle \psi_{k,m} \middle| \gamma_{q,\mu} \middle| \psi_{k',m'} \right\rangle\right|^2}{[E_0(k, m) - E_0(k', m') - \hbar\omega_1(q, \mu)]}. \tag{5.18.11}$$

The first two terms of the decomposition of the intrinsic energy by degrees of the wavenumber give, respectively, polaronic corrections to the ground state energy (ΔE_{p0}) and to the kinetic energy, which leads to a change in the effective mass by Δm_p compared with the mass of the band electron. The results of numerical calculations of the polaron characteristics are shown in figures 5.23 and 5.24.

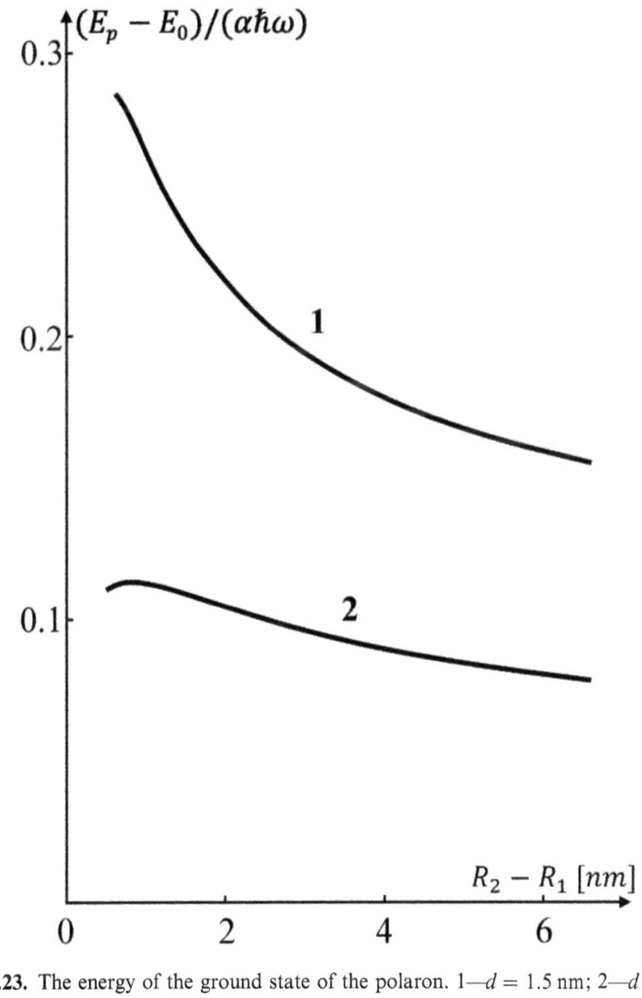

Figure 5.23. The energy of the ground state of the polaron. 1—d = 1.5 nm; 2—d = 4.5 nm.

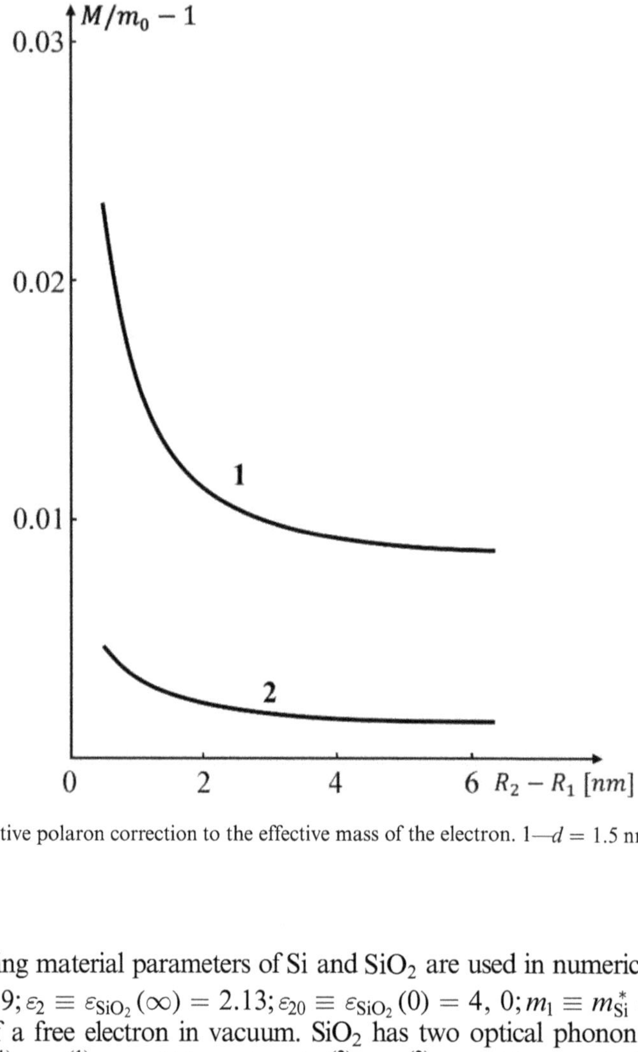

Figure 5.24. Relative polaron correction to the effective mass of the electron. 1—$d = 1.5$ nm; 2—$d = 4.5$ nm.

The following material parameters of Si and SiO_2 are used in numerical calculations: $\varepsilon_1 \equiv \varepsilon_{Si} = 11.9; \varepsilon_2 \equiv \varepsilon_{SiO_2}(\infty) = 2.13; \varepsilon_{20} \equiv \varepsilon_{SiO_2}(0) = 4,\ 0; m_1 \equiv m_{Si}^* = 0,\ 33m_0; m_0$ is the mass of a free electron in vacuum. SiO_2 has two optical phonon branches with frequencies $\omega_{20}^{(1)} \equiv \omega_{2,\,LO}^{(1)} = 509\ \text{cm}^{-1}$ and $\omega_{20}^{(2)} \equiv \omega_{2,\,LO}^{(2)} = 1235\ \text{cm}^{-1}$. (Data are taken from [117].) The calculation was carried out for the radius of the semiconductor $R_1 = 30$ nm. Based on the above data, the following values of polaron radii and coupling constants are obtained: $R_p^{(1)} = 1.91$ nm; $R_p^{(2)} = 1.23$ nm; $\alpha_1^{(1)} = 0.86; \alpha_2^{(2)} = 2.18$.

It can be seen from the figures that with a decrease in the thickness of the inversion layer, both the absolute value of the ground state energy and the effective mass of the polaron increase. In this case, the range of values d_i, in which the polaronic characteristics noticeably change their magnitude, turns out to be of the order of the polaron radius R_p.

The electron scattering velocity is calculated using the same approximations as the intrinsic energy of the polaron, using the formula:

$$W(k, m) = \frac{2\pi}{\hbar} \sum_{k',m'} \sum_{q,\mu} \left| \left\langle \psi_{k,m} \middle| \gamma_{q,\mu} \middle| \psi_{k',m'} \right\rangle \right|^2 \delta\left[E_0(k, m) - E_0(k', m') - \hbar\omega_1(q, \mu) \right], \tag{5.18.12}$$

where only the processes of phonon emission are taken into account (a low-temperature approximation, which, of course, is performed for SiO_2 at room temperature). Figures 5.25 and 5.26 show the results of calculating the electron scattering velocity (a) for different thicknesses of the inversion layer and (b) for different thicknesses of the dielectric layer.

Just like the polaron characteristics, the scattering velocity is a sharply decreasing function of the thickness of the inversion layer and nonmonotonically depends on the thickness of the dielectric. This behavior of these physical quantities turns out to

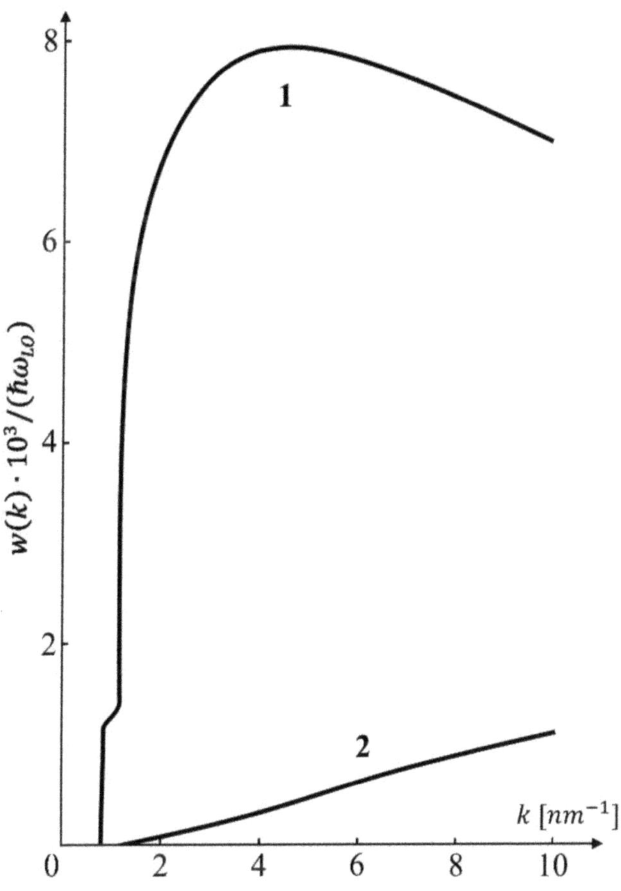

Figure 5.25. The electron scattering velocity as a function of the electron wavenumber: 1—$d = 1.5$ nm; 2—$d = 15$ nm.

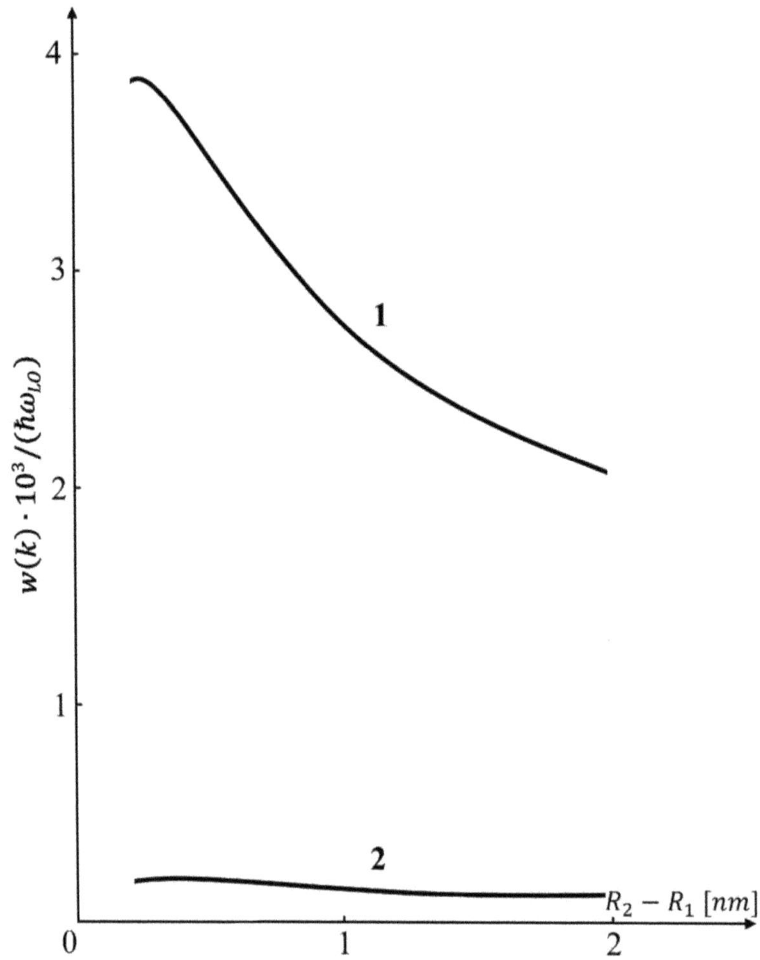

Figure 5.26. The electron scattering velocity as a function of the thickness of the dielectric layer: 1—$d = 3$ nm; 2—$d = 9$ nm.

be common to all polaronic effects and is due to the specific dependence of the amplitudes of the electron–phonon interaction on the geometric parameters of the structure. The nontrivial effect of the metal electrode on the magnitude of the polaron effect should be noted. In a certain range of dielectric thickness values, the electrode does not shield, but strengthens the electron–phonon bond. Since the Coulomb repulsion between electrons near the metal surface is always weakened, the discovered effect should contribute to the appearance of bipolarons and Cooper pairs and, thus, improve the conditions of the superconducting phase transition.

Quantum dots with sizes smaller than 3 nm were obtained in [118, 119]. It is of interest to study the polaron effect in thin nanostructures. It was shown in [120] that in quasi-one-dimensional cylindrical and planar structures, the polaron effect increases as the cross-section of the quantum well decreases. In the case of a quantum wire, its noticeable increase begins when the radius of the wire becomes equal to or less than the polaron radius, and the greater the constant of the electron–phonon interaction, the lower the curve $\Delta E_p(R, \alpha)$, where ΔE_p passes, is the polaron displacement of the level of dimensional quantization; R is the radius of the wire.

5.18.3 Polaronic states in quantum dots of various geometries

Consider a 'flat' quantum dot in which the motion of an electron is limited as follows: a parabolic potential acts in the plane (x, y), and along the z-axis there is a one-dimensional potential well of almost arbitrary symmetrical shape, but narrow enough to provide an adiabatic approximation. We use the Feynman variational method for the calculation. As it was shown in [121] for wide ranges of α and R values, the possibilities of the Feynman variational method are strictly limited by potentials of the parabolic form, therefore, we will assume the confinement at the quantum dot to be parabolic. A possible alternative is to exclude the fast motion of an electron in one or two directions by averaging the Hamiltonian on the wave functions describing this motion (adiabatic approximation). In this section, we will use both calculations and compare their results.

Let's write the Hamiltonian in the form

$$\hat{H} = \frac{P_\perp^2}{2m} + \frac{P_z^2}{2m} + V(z) + \frac{m\Omega^2(x^2 + y^2)}{2} + \hat{H}_{\mathrm{ph}} + \hat{H}_{\mathrm{e-ph}}; \qquad (5.18.13)$$

where $V(z)$ is the potential well along the z-axis; Ω is the parameter of the potential well in the plane (x, y); and $\hat{H}_{\mathrm{ph}}$, $\hat{H}_{\mathrm{e-ph}}$ are phonon and electron–phonon Hamiltonians, respectively.

The problem is solved in a continuous approximation, where m is the band effective mass of the electron. We average the Hamiltonian (5.17.13) on the wave function-limiting motion along the z-axis. The easiest way to perform calculations is using a Gaussian type wave function:

$$\psi(z) = \left(\frac{\beta}{\pi^{1/2}}\right)^{1/2} \exp\left(-\frac{\beta^2 z^2}{2}\right), \qquad (5.18.14)$$

where β is the variational parameter.

For z-dependent averages of operators and functions, we obtain

$$V = \int \omega^2 V(z) dz; \qquad (5.18.15a)$$

$$\left\langle \psi \left| \frac{p_{||}^2}{2m} \right| \psi \right\rangle = \int \psi \frac{p_{||}^2}{2m} \psi dz = \frac{\hbar^2 \beta^2}{4m}; \qquad (5.18.15b)$$

$$\int \psi^2(z) \exp(\pm i k_{||} z) dz = \exp\left(-\frac{k_{||}^2}{4\beta^2}\right) = \gamma_0(k_{||}). \qquad (5.18.16)$$

Substituting (5.18.14)–(5.18.16) into (5.18.13), we find the averaged Hamiltonian:

$$\widehat{\widetilde{H}} = \frac{\hbar^2\beta^2}{2m} + \widetilde{V} + \frac{p_\perp^2}{2m_\perp} + \frac{m\Omega^2(x^2 + y^2)}{2} \\ + \frac{\hbar}{\sqrt{V}}\sum_k \gamma_0(k_\parallel)\left(C_k e^{ik_\perp\rho}\hat{b}_k + C_k^* e^{ik_\perp\rho}\hat{b}_k^+\right) + \sum \hbar\omega_k \hat{b}_k^+ b_k; \tag{5.18.17}$$

$$C_k = \frac{\sqrt{4\pi a}}{k}\omega_0\sqrt{\frac{\hbar}{2m\omega_0}}. \tag{5.18.18}$$

Here ω_0—frequency of polar optical oscillations;

$$k = \sqrt{k_\parallel^2 + k_\perp^2}. \tag{5.18.19}$$

Corresponding to the Hamiltonian (5.18.17) Lagrangian (without constant terms $\hbar^2\beta^2/(2m) + V$) has the form

$$L = \frac{m(\dot{x}^2 + \dot{y}^2)}{2} + \frac{m\Omega^2(x^2 + y^2)}{2} + \frac{1}{2}\sum_k\left(\dot{q}_k^2 - \omega_0 q_k^2\right) + \sum_k \gamma_k(x, y)q_k, \tag{5.18.20}$$

where

$$\gamma_k(x, y) = \frac{\sqrt{2\hbar}}{V}\gamma_0 C_k \begin{cases} \sin(\mathbf{k}_\perp \boldsymbol{\rho}), & k_x > 0; \\ \cos(\mathbf{k}_\perp \boldsymbol{\rho}), & k_x \leqslant 0; \end{cases} \tag{5.18.21}$$

$\boldsymbol{\rho}$ is a radius-vector in the plane (x, y).

We choose the trial Lagrangian in the standard form:

$$L_0 = \frac{m\dot{\rho}^2}{2} + \frac{m\dot{R}^2}{2} - \frac{m\Omega^2\rho^2}{2} - \frac{M\Omega_0^2 R^2}{2} - \frac{k}{2}(\mathbf{R} - \boldsymbol{\rho})^2, \tag{5.18.22}$$

where M, k are Feynman variational parameters. Let's move to L_0 to the normal variables $\xi_1(\omega_1)$, $\xi_2(\omega_2)$, where ω_1, ω_2 are the normal frequencies found from the system of equations of motion for $\boldsymbol{\rho}$, $\mathbf{R}$:

$$\omega_{1,2}^2 = \frac{1}{2}\left(\frac{k}{\mu} + \Omega^2\right) \pm \frac{1}{2}\left[\left(\frac{k}{\mu} + \Omega^2\right)^2 - 4\frac{k\Omega^2}{M}\right]^{1/2}, \tag{5.18.23a}$$

where ω_1, ω_2 are the normal frequencies related by the ratio

$$\omega_1^2\omega_2^2 = \omega_f^2\Omega^2; \; \omega_1^2 + \omega_2^2 = \frac{k}{\mu} + \Omega^2 \tag{5.18.23b}$$

with symbols:

$$\omega_f^2 = \frac{k}{\mu}; \; \mu = \frac{mM}{m + M}. \tag{5.18.23c}$$

Natural variables $\boldsymbol{\rho}$ and $\mathbf{R}$ expressed in terms of normal formulas

$$\boldsymbol{\rho} = C_1\xi_1 + C_2F_2(\omega_2)\xi_2;\ \mathbf{R} = C_1F_1(\omega_1)\xi_1 + C_1\xi_2; \tag{5.18.24a}$$

where

$$F_1(\omega_1) = \frac{k}{k - M\omega_1^2};\ F_2(\omega_2) = \frac{k}{k + M\Omega^2 - m\omega_2^2};\ F_2 = -\frac{M}{m}F_2; \tag{5.18.24b}$$

$$C_1^2 = \frac{1}{m + MF_1^2};\ C_2^2 = \frac{1}{M + mF_2^2};\ C_2^2 = \frac{m}{M}C_1^2. \tag{5.18.24c}$$

In normal variables, the Lagrangian has the form

$$L_0 = \frac{1}{2}\left(\dot{\xi}_1^2 + \dot{\xi}_2^2\right) - \frac{1}{2}\omega_1^2\xi_1^2 - \frac{1}{2}\omega_2^2\xi_2^2. \tag{5.18.25}$$

In accordance with the Feynman variational principle, the sum of the polaron states is expressed by the formula

$$\ln Z = \ln Z_0 + \langle S - S_0\rangle_{S_0}, \tag{5.18.26}$$

where Z_0 is the sum of states for a system with a Lagrangian $L_0$and S and S_0 are actions for exact and trial systems after excluding phonon variables and effective particle variables, respectively.

To find individual terms in the expression (5.18.26), we use the generating function

$$\psi_{\mathbf{k}_\perp}(\xi, \eta) \equiv \langle e^{i\mathbf{k}_\perp(\xi\boldsymbol{\rho}_\tau - \eta\boldsymbol{\rho}_\sigma)}\rangle_{S_0}. \tag{5.18.27}$$

The energy of the polaron is determined by the formula

$$E_p = -\frac{1}{\lambda}\ln Z = -\frac{1}{\lambda}\ln Z_0 - \frac{1}{\lambda}\langle S - S_0\rangle_{S_0}, \tag{5.18.28}$$

where $\lambda = 1/T$; T is the absolute temperature.

After some transformations in (5.18.28) using (5.18.27) in the limit $\lambda \to \infty$ $(T \to 0)$ we find

$$\begin{aligned}\Delta E_p &= E_p - \Omega_\perp - \frac{\Omega_\parallel}{2} = \frac{\beta^2}{4} + \frac{\Omega_\parallel^2}{4\beta^2} + \frac{\left(\Omega - \omega_1\right)^2\left(\Omega - \omega^2\right)^2}{2\Omega^2\left(\omega_1 + \omega_2\right)} - \\ &- \frac{\alpha\beta_0}{\sqrt{\pi}}\int_0^\infty \frac{d\tau\, e^{-\tau}}{\left|1 - 2\beta_0^2A_0\right|^{1/2}} \\ &\begin{cases}\operatorname{arcsinh}\left(\dfrac{1}{2\beta_0A_0} - 1\right)^{\frac{1}{2}}, & \text{if } \dfrac{1}{2\beta_0A_0} > 1; \\ \arcsin\left(1 - \dfrac{1}{2\beta_0A_0}\right)^{\frac{1}{2}}, & \text{if } \dfrac{1}{2\beta_0A_0} < 1,\end{cases}\end{aligned} \tag{5.18.29}$$

where all values are written in dimensionless form using the Feynman system of units: $\hbar\omega_0$ for energy and $(\hbar/m\omega_0)^{1/2}$—for length;

$$A_0 = \frac{1}{2}\frac{d_1^2}{\omega_1}(1 - e^{-\omega_1 T}) + \frac{1}{2}\frac{d_2^2}{\omega_2}(1 - e^{-\omega_2 T}); \; d_1^2 = \frac{\omega_1^2 - \omega_f^2}{\omega_1^2 - \omega_2^2}; \; d_2^2 = \frac{\omega_f^2 - \omega_2^2}{\omega_1^2 - \omega_2^2}.$$

Consider the polaronic states in an ellipsoidal quantum dot.
We replace the potential well $V(z)$ in (5.18.13) with a parabolic one:

$$\hat{H} = \frac{p_\perp^2}{2m} + \frac{p_z^2}{2m} + \frac{m\Omega_\parallel^2 z^2}{2} + \frac{m\Omega_\perp^2 \rho^2}{2} + \hat{H}_{\text{ph}} + \hat{H}_{\text{e-ph}}. \tag{5.18.30}$$

In this section, the Hamiltonian is not averaged and the corresponding Lagrangian is immediately written:

$$L = \frac{m\dot{\rho}^2}{2} + \frac{m\dot{z}^2}{2} - \frac{m\Omega_\perp^2 \rho^2}{2} - \frac{m\Omega_\parallel^2 z^2}{2} + \frac{1}{2}\sum_k \left(\dot{q}_k^2 - \omega_0^2 q_k^2\right) + \sum \gamma_k(x, y) q_k, \tag{5.18.31}$$

where

$$\gamma_k(\mathbf{r}) = \frac{\sqrt{2\hbar}}{V}\gamma_0 C_k \begin{cases} \sin(\mathbf{k}_\perp \mathbf{r}), & k_x > 0; \\ \cos(\mathbf{k}_\perp \mathbf{r}), & k_x \leqslant 0. \end{cases} \tag{5.18.32}$$

The trial Lagrangian is taken as

$$L_0 = \frac{m\dot{\rho}^2}{2} + \frac{M_\perp \dot{R}^2}{2} - \frac{m\Omega_\perp^2 \rho^2}{2} - \frac{k_1}{2}(\mathbf{R} - \boldsymbol{\rho})^2 + \frac{m\dot{z}^2}{2} + \frac{M_\parallel \dot{Z}^2}{2} - \frac{m\Omega_\parallel^2 z^2}{2} - \frac{k_2}{2}(Z - z)^2. \tag{5.18.33}$$

The variational variables are $M_\perp$, $M_\parallel$, k_1, k_2. The four equations of motion break up into two independent systems of two equations each. The normal frequencies of transverse motion are described by the formula (5.18.23a), in which all variables should now be marked with the index «$\perp$». The normal frequencies of longitudinal motion are expressed by a similar formula with the index «$\perp$» replaced by the index«$\parallel$».

The formula for energy is derived according to the same scheme as in the previous case. As a result, the polaron shift of the basic level of dimensional quantization is expressed by the formula

$$E_p = E_p - \Omega_\perp - \frac{\Omega_{||}}{2}$$
$$= \frac{(\Omega_\perp - \omega_{\perp,1})^2(\Omega_\perp - \omega_{\perp,2})^2}{2\Omega_\perp^2(\omega_{\perp,1} + \omega_{\perp,2})} + \frac{(\Omega_{||} - \omega_{||,1})^2(\Omega_{||} - \omega_{||,2})^2}{2\Omega^2(\omega_{||,1} + \omega_{||,2})}$$
$$- \frac{a}{\sqrt{\pi}} \int_0^\infty \frac{d\tau\, e^{-\tau}}{|A_\perp - A_{||}|^{1/2}} \begin{cases} \operatorname{arcsinh}\left(\frac{A_{||}}{A_\perp} - 1\right)^{\frac{1}{2}}, & \text{if } \frac{A_{||}}{A_\perp} > 1; \\ \arcsin\left(1 - \frac{A_{||}}{A_\perp}\right)^{\frac{1}{2}}, & \text{if } \frac{A_{||}}{A_\perp} < 1, \end{cases} \tag{5.18.34}$$

where

$$A_{\perp,||} = \frac{1}{2}\frac{d_{1,\perp,||}^2}{\omega_{1,\perp,||}}(1 - e^{-\omega_{1,\perp,||}\tau}) + \frac{1}{2}\frac{d_{2,\perp,||}^2}{\omega_{2,\perp,||}}(1 - e^{-\omega_{2,\perp,||}\tau});$$
$$d_1^2 = \frac{\omega_1^2 - \omega_f^2}{\omega_1^2 - \omega_2^2};\ d_2^2 = \frac{\omega_f^2 - \omega_2^2}{\omega_1^2 - \omega_2^2}.$$

All variables in (5.18.34) are taken in dimensionless form using the Feynman system of units. It is convenient to vary variables using the variables $\omega_{1\perp}$, $\omega_{2\perp}$, $\omega_{1||}$, $\omega_{2||}$, instead of the original $M_\perp$, $M_{||}$, k_1, k_2.

Let's note the main patterns.

For all α, a decrease in the radius of the parabolic well leads to an increase in the polaron effect. When R changes by about 30 times, the energy changes by about 3 times, and this proportion is maintained for all α.

As the width of the pit decreases along the z-axis (greater Ω_0) for the same R, the polaron level also deepens, but more slowly than with decreasing R. This is due to the fact that in the first case, the reduction in size occurs in two directions at once, whereas in the second—in one direction. Based on these results, it can be concluded that any reduction in the size of the quantum well leads to an increase in the polaron binding energy.

The same patterns follow from the results for the ellipsoidal pit. Calculations show an increase in the polaron energy both with a decrease in length and with a reduction in the transverse dimensions of the ellipsoid. However, it is important to note that the energy of a polaron with a smaller α may lie deeper than with a larger one, given the sufficiently small size of the pit in the first case. We come to an important conclusion that it is possible to control the mode of polaronic communication by changing the size of the quantum well. The quasi-one-dimensional case is considered in [120].

5.19 Magnetopolaron in a cylindrical quantum wire in a dielectric medium

Let's consider magnetopolaronic states in a cylindrical quantum wire with an infinite potential on its surface and a two-dimensional parabolic potential inside it. To calculate the polaron parameters, we use the electron–phonon Hamiltonian, which includes interaction with bulk-like and interface phonons, derived in [115] and used to calculate large-radius polaron states.

5.19.1 Wave function and electron energy

The wave function and the energy of an electron in a cylindrical well with a parabolic bottom with a magnetic field $\mathcal{H}$ directed along the z-axis of the cylinder are found from the solution of the Schrödinger equation with a Hamiltonian in a cylindrical coordinate system:

$$\hat{H}_e = \frac{\hat{P}^2}{2M} - \frac{c\hbar}{2Mc}\mathcal{H}\hat{l}_z + \frac{e^2\mathcal{H}^2}{8Mc^2}\rho^2 + U(\rho), \tag{5.19.1}$$

where $\hat{l}_z = -i\partial/\partial\varphi$ is the moment projection operator.

Full potential U includes a barrier on the cylinder surface ($\rho = R$):

$$U_B = \begin{cases} 0, & \rho < R; \\ \infty, & \rho > R \end{cases} \tag{5.19.2}$$

and the parabolic potential inside the cylinder ($\rho < R$):

$$U_0 = \frac{M\Omega^2}{2}\rho^2. \tag{5.19.3}$$

The eigenfunctions of the Hamiltonian (5.19.1) have the form

$$\psi_{m,k}(\rho, \varphi, z) = \frac{A}{2\pi}e^{i(kz+m\varphi)}\,e^{-\rho^2/(2a^2)}\left(\frac{\rho^2}{a^2}\right)^{|m|/2} F_1\left(-\nu_r^{|m|}, |m|+1, \frac{\rho^2}{a^2}\right), \tag{5.19.4}$$

where $F_1(\nu, \gamma, \rho)$ is the degenerate hypergeometric function,

$$\frac{1}{a^4} = \frac{1}{4a_{\mathcal{H}}^4} + \frac{1}{a_0^4}, \quad a_{\mathcal{H}} = \sqrt{\frac{\hbar}{M\omega_{\mathcal{H}}}}, \quad a_0 = \sqrt{\frac{\hbar}{M\Omega}}, \quad \omega_{\mathcal{H}} = \frac{|e|\mathcal{H}}{Mc}. \tag{5.19.5}$$

The eigenvalues of E_γ of the Hamiltonian (5.19.1) are determined by the expression

$$E_\gamma - \frac{\hbar^2k^2}{2M} = E_{\nu,m} = \hbar\varpi\left(\nu_r^{|m|} + \frac{|m|+1}{2}\right) + \hbar\omega_{\mathcal{H}}\frac{m}{2}; \; \gamma = \left(m, \nu_r^{|m|}, k\right); \tag{5.19.6}$$

$$\varpi = \left[\omega_{\mathcal{H}}^2 + (2\Omega)^2\right]^{1/2}$$

It is found from the condition of vanishing at the boundary $\rho = R$ of a degenerate hypergeometric function

$$F_1\left(-\nu_r^{|m|}, |m|+1, \frac{R^2}{a^2}\right) = 0. \tag{5.19.7}$$

In the limit $z = 0$, $R \to \infty$, $\Omega \neq 0$, the results of [122] follow from (5.19.5) and (5.19.7).

In the limit $R/a \to 0$, the asymptotics of the roots of equation (5.19.7) has the form $v_r^{|m|} \approx j^2_{|m|,\,r}(a^2/4R^2)$, where $j_{|m|,\,r}$ is the r th zero of the Bessel function $J_m(x)$, and energy levels are determined by dimensional quantization.

At $\mathcal{H} = 0$ in the part of the spectrum lying significantly below the level

$$U_0(R) = \frac{MR^2\Omega^2}{2} \gg E\left(\nu_r^{|m|}, |m|, \Omega\right), \tag{5.19.8}$$

energy takes on an oscillatory appearance:

$$E_{n_0}(\Omega) = \hbar\Omega(n_0 + 1), \tag{5.19.9}$$

with a degeneracy multiplicity of $n_0 + 1$, where $n_0 = 2(r - 1) + |m|$ is the quantum number of a two-dimensional oscillator. In the area (above and below) $U_0(R)$ there is a splitting of oscillatory levels by the barrier potential U_B into $n_0/2 + 1$ in the case of even n_0 and $(n_0 + 1)/2$—in the case of an odd n_0.

Finally, at very high energies $E_{m,\,r} \gg U_0(R)$, we have a dimensional quantization spectrum with doubly sign-degenerate m levels. The general case is described by the formula (5.19.6). In the domain of weak magnetic fields ($a_{\mathcal{H}} \gg a_0$, $\Omega_{\mathcal{H}} \ll \Omega$) splitting of all levels from $n_0 \neq 0$ and complete removal of degeneracy takes place. In this limit, the contribution of $\omega_{\mathcal{H}}$ в $\widetilde{\omega}$ is small and the splitting value is determined by the term $\hbar\omega_{\mathcal{H}}m/2$ in the general formula (5.19.6). In the opposite limit of strong magnetic fields, when $\omega_H \gg \Omega$ and the magnetic radius $\omega_{\mathcal{H}} \gg \Omega$, the roots of the $a_{\mathcal{H}} \ll a_0$, $\nu_r^{|m|}$ equation (5.19.7), as indicated, degenerate, and the equation for energy has the form

$$E = \hbar\omega_{\mathcal{H}}\left(n + \frac{1}{2}\right), \tag{5.19.10}$$

where $n \equiv r - 1 + (m + |m|)/2$—the quantum Landau number.

5.19.2 Energy and effective mass of the polaron

The Hamiltonian of the electron–phonon interaction and the cylindrical wire was obtained in [115]. To simplify numerical calculations, we will limit ourselves to the case of dimensional magnetic quantization ($\Omega = 0$).

The Hamiltonian of the polaron problem has the form

$$\hat{H} = \hat{H}_1 + \hat{H}_{\text{ph}} + \hat{H}_{\text{e-ph}}; \tag{5.19.11}$$

here $\hat{H}_1 = \hat{H}_e + \hat{H}_{SA}$ is the electronic part of the Hamiltonian, including the Hamiltonian (5.19.1) with potential $U = U_B$ and the potential energy of self-action (potential energy of image forces), which for an electron inside the cylinder has the form [115]

$$U_{SA}(\rho) = \frac{\alpha\hbar\omega_L(\varepsilon_2 - \varepsilon_1)\varepsilon_{10}}{R(\varepsilon_{10} - \varepsilon_1)\varepsilon_1}\frac{1}{\pi}\sum_{m=-\infty}^{+\infty}\int_{-\infty}^{+\infty} d\eta \, \frac{\eta K_m'(\eta)K_m(\eta)I_m^2(\eta\rho)}{1 - \left(\frac{\varepsilon_1}{\varepsilon_2} - 1\right)K_m'(\eta)I_m(\eta)\eta}, \tag{5.19.12}$$

where

$$\alpha = \frac{1}{2}\frac{e^2}{4\pi\varepsilon_0 R_p \hbar\omega_L}\left(\frac{1}{\varepsilon_1} - \frac{1}{\varepsilon_{10}}\right)$$

is the constant of the electron–phonon interaction of Pekar–Fröhlich, $I_m(x)$, $K_m(x)$, are cylindrical functions of the imaginary argument, and $I_m'(x)$, $K_m'(x)$ are their derivatives [52]. The function $U_{SA}(\rho)$ is monotonically decreasing or monotonously increasing from the axis of the cylinder, depending on the sign $\varepsilon_2 - \varepsilon_1$. The contribution to the electron energy from self-action decreases rapidly as $1/R$. Figure 5.27 shows the dependence on the magnetic field of the energy $\Delta E_{\gamma,\,SA} = \langle\psi_\gamma\,|U_{SA}|\psi_\gamma\rangle$ for $\gamma = (0,\,1)$, $(\pm 1,\,1)$. The decrease $|\Delta E_{\gamma,\,SA}|$ is explained by the increased localization of the electron near the center of the cylinder as the magnetic field increases.

The Hamiltonian of free phonons is

$$\hat{H}_{\text{ph}} = \sum_{l,\eta,q}\hbar\omega_b \hat{a}^{+}_{lq\eta}\hat{a}_{lq\eta} + \sum_{l,\eta}\hbar\omega_s(l,\,\eta)\hat{b}^{+}_{l\eta} + \hat{b}_{l\eta}. \tag{5.19.13}$$

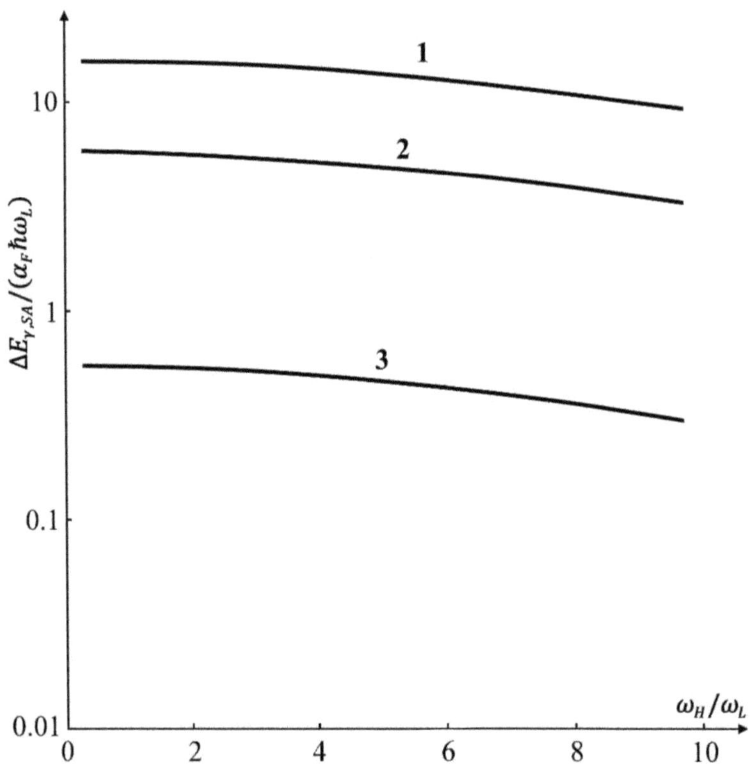

Figure 5.27. Dependence of the self-action energy on the magnetic field. The dielectric constant of the substance inside the cylinder is $\varepsilon_1 = 10.6$ (GaAs), and the external medium is ε_2: 1 – 1; 2 – 3, 3 – 9(AlAs).

It includes the bulk and surface branches of vibrations. The bulk and interface parts of the electron–phonon interaction Hamiltonian have the form

$$H^{V}_{\text{e-ph}} = \sum_{\chi_V}[\Gamma_V(\rho,\ \eta,\ l,\ q_l)\mathrm{e}^{i(\eta z+l\varphi)}\ \hat{a}_{l,\ q,\ \eta} + \text{к. с.}]; \tag{5.19.14}$$

$$H^{S}_{\text{e-ph}} = \sum_{\chi_S}[\Gamma_S(\rho,\ \eta,\ l)\mathrm{e}^{i(\eta z+l\varphi)}\ \hat{b}_{l,\ \eta} + \text{к. с.}], \tag{5.19.15}$$

where $\chi_V = (l,\ q,\ \eta)$ and $\chi_S = (l,\ \eta)$ are complete sets of quantum numbers for bulk (V) and surface (S) modes. The amplitudes of the electron–phonon interaction $\Gamma_V(\rho,\ l,\ q_l,\ \eta)$ and $\Gamma_S(\rho,\ l,\ \eta)$ are greatly simplified for the considered case of a polar wire in a nonpolar medium.

In the approximation of the Wigner–Brillouin perturbation theory for the polaron energy, we have the equation

$$E = E^{0}_{\gamma_0,\ n^0_\chi} - \sum_{\gamma,n_\chi}\frac{\left|\left\langle \psi_{\gamma,\ n_\chi}\left|\hat{H}_{\text{e-ph}}\right|\psi_{\gamma_0,\ n^0_\chi}\right\rangle\right|^2}{E^{0}_{\gamma,\ n_\chi} - E}, \tag{5.19.16}$$

where $\psi_{\gamma,\ n_\chi} = \psi(\rho;\gamma)\mid n_\chi\rangle$ with $\gamma = (\nu_r,\ m,\ k)$ is the wave function of the zero-order approximation, in which $\psi(\rho;\gamma)$ and $\mid n_\chi\rangle$ are eigenfunctions of the Hamiltonians $\hat{H}_e$ and $\hat{H}_{\text{ph}}$and n_χ is phonon filling state χ. The phonon wave function describing the initial state will be considered vacuum is $n^0_\chi = 0$. Since the operator $\hat{H}_{\text{e-ph}}$ is single phonon, there must be one phonon in the perturbed state $n_\chi = 1$: $E^0_{\gamma,\ 1_\chi} = E^0_\gamma + \hbar\omega_\chi$. This electron–phonon level will be called virtual. The desired energy E is represented as the sum of $E = E^0_{\gamma_0,\ 0_\chi} + \Delta E_p(\gamma)$, where ΔE_p is the polar correction. The energy difference in the denominator of the formula (5.19.16) is written as

$$\Delta E^{\gamma_0 0_\chi}_{\gamma 1_\chi} = E^0_\gamma - E^0_{\gamma_0} + \hbar\omega_\chi + \frac{\hbar^2\eta^2}{2M} - \frac{\hbar^2\eta k_0}{M} + \Delta E_p(\gamma). \tag{5.19.17}$$

Expanding the polaron part in the formula (5.19.16) according to k_0 and limiting ourselves to terms proportional to k_0^2, we obtain at $k_0 = 0$ the equation for polaron shifts, and the part proportional to k_0^2 gives corrections to the translational effective mass.

When calculating the polaron corrections to the ground state energy, they can be neglected in the denominators of the formula (5.19.16), which corresponds to the use of Rayleigh–Schrödinger perturbation theory. Some results of such calculation for R/R_p using the parameters GaAs (m^*=0.06624m; $\omega_L = 5.5\cdot 10^{13}\ \text{s}^{-1}$; $\varepsilon_0 = 12.87$; $\varepsilon_\infty = 10.9$; $\alpha_F = 0,\ 07$) are shown in figure 5.28, where the dependences of the surface contributions $|\Delta E^{(S)}_p|/(\alpha\hbar\omega_L)$ (dashed lines s, $2s$, $3s$ for $\varepsilon_2 = 1,\ 3,\ 9$ respectively), the bulk contribution $|\Delta E^{(V)}_p|/(\alpha\hbar\omega_L)$ (thin line V), as well as the total polaron shift $|\Delta E^{(V)}_p + \Delta E^{(S)}_p|/(\alpha\hbar\omega_L)$ (thick lines $1(V+S),\ 2(V+S),\ 3(V+S)$) from the magnetic field are taken in units of $\omega_{\mathcal{H}}/\omega_L$. The surface contribution, mainly due to interaction with the zero-interface mode ($l = 0$), practically does not depend on

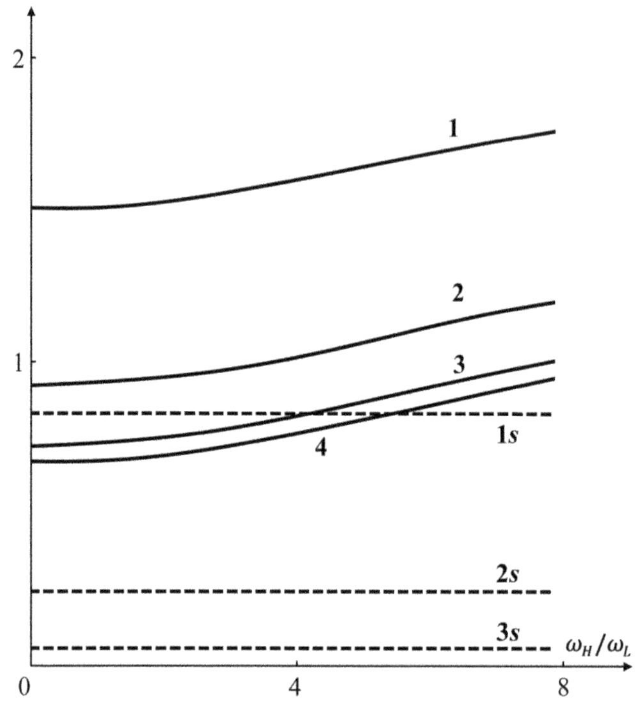

Figure 5.28. Dependence of the polaron energy on the magnetic field. 1*s*, 2*s*, 3*s* are the surface polaron contributions at $\varepsilon_2 = 1,\ 3,\ 9$ (AlAs), respectively; 4 is the bulk contribution; and 1, 2, 3 are the corresponding total polaron energies.

the magnetic field and decreases sharply with an increase in the dielectric constant of the external medium ε_2. At large R, the bulk contribution increases, tending to the limit $\Delta E_p^{(\mathrm{M})}/(\alpha\hbar\omega_L) = 1$, and the surface decreases rapidly. Note that depending on the values of ε_2, R, $\omega_{\mathcal{H}}$ $|\Delta E_p^{(V)} + \Delta E_p^{(S)}|/(\alpha\hbar\omega_L)$ can be either greater or less than one. It is known that as the dimension of the electron gas decreases, the energy increases when interacting with 3D phonons. When taking into account the dimensional quantization of phonons, a similar comparison can be made for bulk polaron contributions $\Delta E_p^{(V)}$. regardless of the choice of the external environment. If for a layer (2D) of thickness L $\Delta E_p^{(V)}|_{L\to 0} \to 0$, then for a cylinder (1D) of radius R $\Delta E_p^{(V)}|_{R\to 0} \to$ const and for a ball (0d) of radius R $\Delta E_p^{(V)}|_{R\to 0} \to \infty$ [115].

Passing in the formula (5.19.16) from the sum with respect to the integral η and performing integration, we obtain two different equations for calculating the polaron displacements ΔE_{p1} and ΔE_{p2} depending on the sign of the resonant term $E_\gamma^0 = \hbar\omega_\chi - E$. The splitting value, defined as the difference of the polaron shifts $\Delta E_p = \Delta E_{p1} - \Delta E_{p2}$ at the resonance point $\beta_{\gamma_1}^{\gamma_0} = E_{\gamma 1} + \hbar\omega_\chi - E_{\gamma_0} = 0$, calculated based on the formula (5.19.16), gives $\Delta E_p \sim \alpha^{2/3}$ for the bulk contribution. This result coincides with the one obtained for the problem with 3D electrons and phonons in a magnetic field, since in both cases the electron has one translational degree of freedom. In contrast, for a 2D electron (in a layer) in an inclined magnetic field, due to the full dimensional quantization of the electronic spectrum, $\Delta E_p \sim \alpha^{1/2}$ [64].

To construct resonant curves, it is necessary to solve two equations obtained after integration with respect to η, in which the desired value ΔE_p is stored only in resonant terms:

$$\sum_{q_j^l} \frac{\left|\left\langle \psi_{m_1,\,\nu_1} \left| H_{\text{e-ph}} \right| \psi_{m_0,\,\nu_0} \right\rangle\right|^2}{\left(\beta_\gamma^{\gamma_0} - \Delta E_{p1}\right)\left(\beta_\gamma^{\gamma_0} - \Delta E_{p1} + q_j^l\right) q_j^l} + S_+ = -\Delta E_{p1}, \;\; \beta_\gamma^{\gamma 0} - \Delta E_{p1} > 0, \quad (5.19.18)$$

$$-\sum_{q_j^l} \frac{\left|\left\langle \psi_{m_1,\,\nu_1} \left| H_{\text{e-ph}} \right| \psi_{m_0,\,\nu_0} \right\rangle\right|^2}{\left[\left(\beta_\gamma^{\gamma_0} - \Delta E_{p2}\right)^2 + q_j^{l2}\right] q_j^l} + S_- = -\Delta E_{p2}, \;\; \beta_\gamma^{\gamma 0} - \Delta E_{p2} \leqslant 0, \quad (5.19.19)$$

where $S_\pm$ is the sum of all non-resonant terms, including the surface part.

The results of numerical calculations of splits using the parameters of GaAs crystals, AlAs $\omega_L = (7.61 \cdot 10^{13}\,\text{s}^{-1},\ \varepsilon_\infty = 9)$ at the resonance of the level (–1.1) with (0.1) (intersection of the electronic level (–1.1) with the virtual level (0.1) (relatively strong dimensional quantization) and at the resonances of the electronic level (1.1) with (–1.1) and (0.1) (relatively weak dimensional quantization) are shown in figures 5.29, 5.30. Thin lines show the electronic levels $E_\gamma^0/\hbar\omega_L$, dashed virtual $(E_\gamma^0 + \hbar\omega_L)/\hbar\omega_L$, and the thick ones are polar $(E_\gamma^0 + \Delta E_p)/\hbar\omega_L$ depending on $\omega_\mathcal{H}/\omega_L$ for $R/R_p = 2$ (a) and $R/R_p = 4$ (b). A feature of the right resonance in figure 5.31 is the fixity of its position $\omega_\mathcal{H}^0 = \omega_L$ regardless of the values of the dimensional quantization parameters $\Omega,\ R$.

The calculation of the translational polaron effective mass M_{eff} is performed in the described way in accordance with its definition:

$$E_{\gamma_0} = E_{m_0\nu_0} + \Delta E_p + \frac{\hbar^2 k_0^2}{2M_{\text{eff}}}; \quad (5.19.20)$$

$$\frac{1}{M_{\text{eff}}} = \frac{1}{M}\left(1 - \frac{\alpha S}{6}\right), \quad (5.19.21)$$

where $\alpha S/6$ is the coefficient at k_0^2 in the decomposition of the formula (5.19.16) by k_0.

The calculation results [123] for the ground state at $R/R_p = 2$ are shown in figure 5.30 (right), where the dashed lines $1s,\ 2s,\ 3s$ correspond to the functions $S^{(S)}(\omega_\mathcal{H},\ R)$ for surface phonons at $\varepsilon_2 = 1,\ 3,\ 9$, the thin line V gives the function $S^{(V)}(\omega_H,\ R)$ for bulk phonons, and thick lines $1(V+S),\ 2(V+S),\ 3(V+S)$ are their sums $S^{(S)} + S^{(V)}$.

The bulk part increases, and the surface part decreases with increasing $\omega_\mathcal{H}$ at a fixed R.

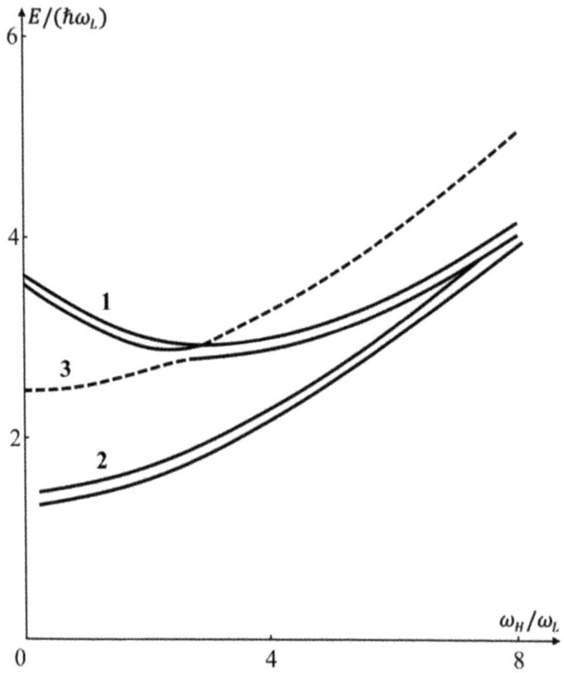

Figure 5.29. Splitting of polaron levels: resonance of levels (0, 1)–(−1, 1). Curve 1—level (−1, 1); curve 2—(0,1); and curve 3—(0, 1) + ph.

Thus, it can be argued that the dimensional effect weakens the influence of the magnetic field on the polaron contributions and the energy and effective mass of the electron, especially in the domain of strong quantization (small R, large Ω), but in a certain range of values of the radius R facilitates the achievement of magnetophonon resonance $0 \ll \omega_{\mathcal{H}}^0 \ll \omega_L$.

5.20 Conclusion

The theory of surface polarons in planar systems, based on the derived exact Hamiltonians of the electron–polarization interaction, was applied to describe surface, near-surface, film, external, levitating, electronic, and others. Criteria for the existence of surface polaronic states were established. The quantum theory of image potential and forces was presented. Based on the polaronic theory, the dielectric function of a quantum dielectric was derived for planar systems, and levitating polaronic states were investigated. The potential energy of self-action was found and the polaronic states in multilayer cylindrical and spherical structures considered. Polaron states in homogeneous electric and magnetic fields in planetary, cylindrical, and spherical nanostructures and in composite superlattices were studied.

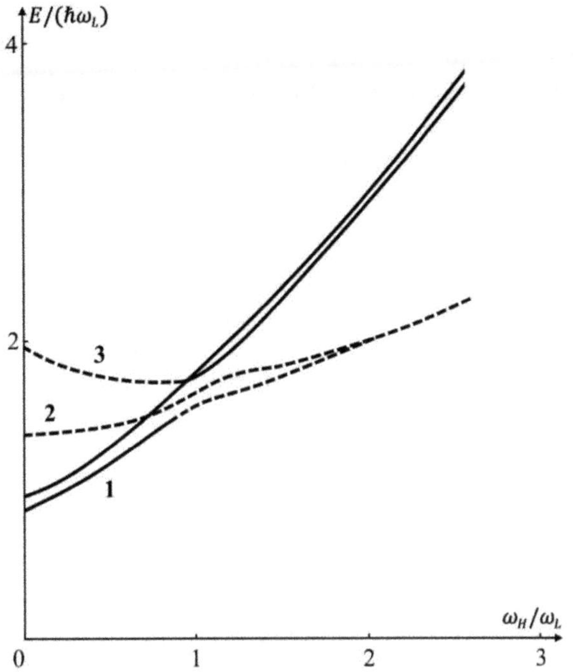

Figure 5.30. Splitting of polaron levels: resonance of levels (0, 1)–(1, 1) and (−1, 1). Curve 1 is level (1,1); curve 2—(0, 1) + ph; and curve 3—(−1, 1) + ph.

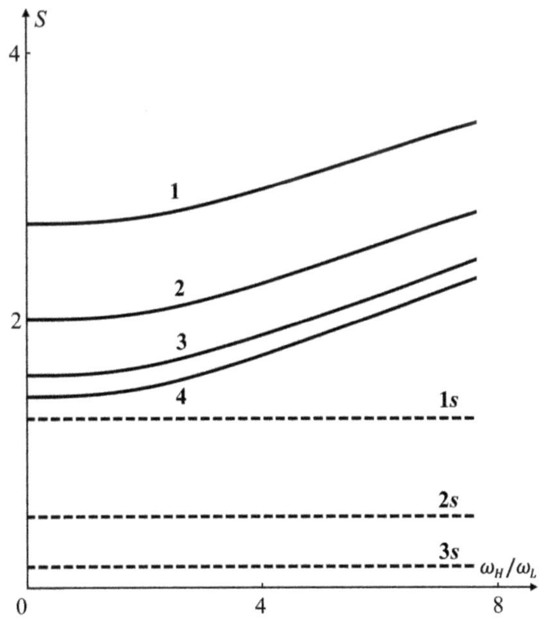

Figure 5.31. Dependence of the effective mass parameter S (on the right) on the magnetic field. $1s$, $2s$, $3s$ are the surface polaron contributions at $\varepsilon_2 = 1, 3, 9$ (AlAs) respectively; 4 is the bulk contribution; and 1, 2, 3 are the corresponding total polaron energies.

References

[1] Landau L D 1933 Über die bewegung der elektronen in kristallgitter *Phys. Z. Sowjetunion* **3** 664–5

[2] Pekar S I 1959 *Research on the Electronic Theory of Crystals* (Moscow: GITTL) p 256

[3] Bryksin V V, Mirlin D N and Firsov Y A 1973 Surface optical phonons in ionic crystals *UFN* **113** 30–63

[4] Evans E and Mills D Z 1972 Interaction of slow electrons by long-wavelength surface optical phonons *Phys. Rev.* B **5** 1238–4126

[5] Evans E and Mills D Z 1973 Theory of inelastic scattering of slow electrons with a surface of a dielectric: theory of surface polarons *Phys. Rev.* B **8** 4004–18

[6] Evans E and Mills D Z 1972 Theory of surface polarons *Solid State Commun.* **11** 1093–98

[7] Licary J J and Evrard R 1977 Electron–phonon interaction in a dielectric slab: effect of the electronic polarizability *Phys. Rev.* B **15** 2254–64

[8] Lucas A A, Kartheuser L and Badro R 1970 Electron–phonon interaction in dielectric films. Application to electron energy loss and again spectra *Phys. Rev.* **2** 2488–99

[9] Sac I 1972 Theory of surface polarons *Phys. Rev.* B **6** 3981–86

[10] Beril S I, Pokatilov E P and Movile V 1984 Polaronic states in dimensionally limited crystals *Surface* 5–8

[11] Beril S I and Pokatilov E P 1978 Surface states in a quantum dielectric *FTT* **12** 2030–33

[12] Beril S I and Pokatilov E P 1977 Surface polaron in ionic crystals *FTT* **19** 1627–31

[13] Beril S I and Pokatilov E P 1981 Surface polaron at the contact of two crystal *FTT* **23** 1181–84

[14] Beril S I and Pokatilov E P 1978 Surface weak bond polaron in ionic crystals *FTT* **20** 2386–90

[15] Beril S, Pokatilov E P and Cheban I S 1985 Polaron at the contact of two media and its phase diagrams *Physics of Semiconductors and Semiconductor Microelectronics* (Chisinau: Stiinza) pp 85–94

[16] Pokatilov E P, Beril S I, Fomin V M and Ryabukhin G Y 1988 Surface polaron at the contact of two polar kristallov *Chisinau* 18c Dept. in MoldNIINTI. 12/21/88. No. 1063

[17] Pokatilov E P, Beril S I, Fomin V M and Riabukhin G Y 1989 Surface polarons at the contact of two polar crystals *Phys. Status Solidi* B **156** 225–34

[18] Pokatilov E P, Fomin V M and Beril S I 1990 *Vibrational Excitations, Polarons and Excitons in Multilayer Systems and Superlattices* (Chisinau: Stiinza) 280 p

[19] Bryksin V V and Firsov Y A 1971 Interaction of an electron with surface phonons in an ion crystal plate *FTT* **13** 496–503

[20] Ando M, Fowler A and Stern F 1985 *Electronic Properties of Two–Dimensional Systems* (Moscow: Mir) p 415

[21] Pokatilov E P and Fomin V M 1984 *Theory of Potential in Multilayer Systems* No. 500M-05.85 (Chisinau: MoldNIINTI) p 19
Pokatilov E P and Fomin V M 1984 *Interaction of an Electron with Polarization Optical Vibrations in Periodic Structures* No. 508M-05.85 (Chisinau: MoldNIINTI) p 14

[22] Fomin V M and Pokatilov E P 1985 Phonons and the electron–phonon interaction in multilayer systems *Phys. Status Solidi* B **132** 69–82

[23] Camley R E and Mills D Z 1984 Collective excitations of semi-infinite superlattice structures: surface plasmons, bulk plasmons and the electron-energy-loss spectrum *Phys. Rev.* B **29** 1695–706

[24] Toyozawa Y 1954 Theory of electronic polaron and ionization of trapped electron by excited *Prog. Theor. Phys.* **12** 421–42

[25] Hermanson J 1972 Simple model of electronic correlation in insulators *Phys. Rev.* B **6** 2427–32

[26] Hermanson J 1974 The self-energy problem on quantum dielectrics *Elementary Excitations in Solids, Molecules and Atoms: Part* B (New York: Springer) pp 199–211

[27] Pokatilov E P, Beril S and Fomin V M 1988 Potentials and image forces of the electron polaron model *Surface* **5** 5–12

[28] Pokatilov E P, Beril S I and Fomin V M 1988 Image potentials and image forces in the polaron theory *Phys. Status Solidi* B **147** 163–72

[29] Gabovich A M, Ilchenko L G and Pashitsky E A 1980 The effect of spatial dispersion effects on image forces and the energy spectrum of electrons above the surface liquid helium *JETF* **29** 665–71

[30] Gabovich A M, Ilchenko L G and Pashitsky E A 1979 Electrostatic interaction of charges with the surface of metals and semiconductors *FTT* **21** 1683–9

[31] Ilchenko L G, Pashitsky E A and Romanov Y A 1980 The electrostatic potential of charges in layered systems with spatial dispersion *FTT* **22** 2700–10

[32] Inkson I C 1971 The electrostatic image potential in metal–semiconductor junctions *J. Phys.* **4** 591–7

[33] Schulze K R and Unger K 1974 The linear dielectric response of a new analytic form for the dielectric function *Phys. Status Solidi* B **66** 491–8

[34] Kiselev V A 1979 Exciton reflection of light in the presence of a Schottky barrier *FTT* **21** 1069–74

[35] Gabovich A M, Ilchenko L G and Pashitsky E A 1981 Image forces and electron spectrum above the surface of liquid helium (Preprint No. 10) Institute of Physics of the Academy of Sciences of the Ukrainian SSR

[36] Born M and Huang K 1958 *The Dynamic Theory of Crystal Lattices* (Moscow: ILL)

[37] Davydov A S 1972 *Theory of a Solid Body* (Moscow: Nauka) 640 p

[38] Lee T D, Low F E and Pines D 1953 Interaction of a nonrelativity particle with a scalar field with application to slow electron in polar crystal *Phys. Rev.* B **92** 883–91

[39] Lee T D, Low F E and Pines D 1953 The motion of slow electrons in a polar crystals *Phys. Rev.* B **90** 297–302

[40] Tokuda N and Kato H 1987 Strong-coupled polarons in a magnetic field *J. Phys. C: Solid State Phys.* **20** 3021–7

[41] Huybrecht W I 1977 Internal excited state of the optical polaron *J. Phys. C: Solid State Phys.* **10** 3761–7

[42] Jin-Sher P 1985 The surface or interface polaron in polar crystals *Phys. Status Solidi* B **127** 307–18

[43] Jin-Sher P 1985 The surface or interface polaron in polar crystals *Phys. Status Solidi* B **128** 292–7

[44] Matsuura M 1976 Polaron effects in surface electron in polar semiconductor *J. Phys. Soc. Jpn.* **41** 394–9

[45] Fuchs R, Kliewer K Z and Pardey W J 1966 Optical properties of an ionic crystal slab *Phys. Rev.* **150** 589–96

[46] Kliewer K Z and Fuchs R 1965 Optical modes of vibration in an ionic crystal slab *Phys. Rev.* **140A** 2076–88

[47] Liang X X and Gu S W 1984 The polarons and their dead layers in semi-infinity polar crystals *Solid State Commun* **50** 505–8
Liang X X 1985 The interface polaron in polar-polar crystals *Solid State Commun.* **55** 215–8
[48] Peeters F M and Devreese J T 1982 On the existence of a transition for Fröhlich Polaron *Phys. Status Solidi* B **112** 219–29
[49] Gorikov S N, Rodriguez K and Fedyanin V K 1985 On the question of phase transitions in the theory of the polaron *The 3rd Int. Symp. on the Problems of Statistical Mechanics* vol 1 *(Dubna, 1984)* pp 227–33
[50] Gerlach B and Lowen H 1987 Proof of the nonresistance of formal phase transitions in polaron systems *Phys. Rev.* B **35** 4291–6
[51] Gerlach B and Lowen H 1988 Proof of the nonresistance of formal phase transitions in polaron systems, exposed to a homogenous magnetic field *Phys. Scr.* **37** 925–9
[52] Lepine Y and Matz D 1979 Mean field theory of a single Fröhlich Polaron *Phys. Status Solidi* B **96** 797–806
[53] Manka R 1978 The first-order phase transition on the large polaron ground state *Phys. Rev. Lett.* **67A** 311–2
[54] Matz D and Burkey B C 1971 Dynamical theory of the large polaron: fock approximation *Phys. Rev.* B **3** 3487–97
[55] Tokuda N 1982 The surface polaron and its phase diagram *J. Phys. C: Solid State Phys.* **15** 1953–60
[56] Gerlach B, Kalina F and Smondyrev M 2003 On the LO-phonon dispersion in dimensions *Phys. Status Solidi* B **237** 204–14
[57] Beril S I and Pokatilov E P 1978 Surface polaron in a magnetic field *FTP.* **12** 1184–6
[58] Bajaj K K 1968 Polaron in a magnetic field *Phys. Rev.* **170** 694–8
[59] Chen C Y, Ding T Z and Lin D Z 1987 Interface polaron in a strong magnetic field *Phys. Rev.* B **35** 4398–403
[60] Wei C-W, Kong X-J and Gu S-W 1989 Cyclotron resonance of a interface polaron in a polar-polar crystals *Phys. Rev.* B **39** 3230–8
[61] Devreese J T 1989 Polaron physics in 2D and 3D *Phys. Scr.* **25** 309–15
[62] Ercelebie A and Sualp G 1987 Variational treatment of a two-dimensional polaron *J. Phys. Chem.* **48** 739–42
[63] Gu S W, Kong X J and Wei C W 1987 Properties of a magnetopolaron of the interface of polar-polar crystals *Phys. Rev.* B **36** 7977–83
[64] Haupt R and Wendler L Effects of the electron–phonon interaction on the cyclotron resonance of parabolic quantum wells in a tilted magnetic field *Ann. Phys.* **233** 214–47
[65] Larsen D 1987 Fourth-order perturbation calculation of the cyclotron resonance frequency of the two-dimensional polaron *Phys. Rev.* B **35** 4427–34
[66] Larsen D M 1986 Perturbation theory of the two-dimensional polaron of the magnetic field *Phys. Rev.* **33** 799–806
[67] Peeters F M, Xiaoguang W and Devreese J T 1986 Landau levels above the optical phonon continuum in two and three dimensions *Phys. Rev.* **33** 4338–40
[68] Suh E-K, Bartolomev D U, Ramdas A K and Rodriguez S 1987 Raman scattering from superlattices of diluted magnetic semiconductors *Phys. Rev.* **36** 4316–31
[69] Wei C W, Kong X J and Gu S W 1989 Cyclotron resonance of a magnetic polaron in a semiconductor quantum well *Phys. Rev.* B **39** 3230–8

[70] Xiaoquang W, Peeters F M and Devreese J T 1986 Theory of the cyclotron resonance spectrum of a polaron in two dimensions *Phys. Rev.* B **34** 8800–9

[71] Xiaoquang W, Peeters F M and Devreese J T 1985 Two-dimensional polaron in a magnetic field *Phys. Rev.* B **32** 7964–9

[72] Trallero-Giner C, Santigo-Perez D G and Fomin V M 2023 New magneto-polaron resonances in a monolayer of a transition metal dichalcogenide *Sci. Rep.* https://www.nature.com/articles/s41598-023-27404-x

[73] Hess K and Vogl P 1979 Remote polar phonon scattering in silicon inversion layers *Solid State Commun* **30** 797–9

[74] Rahman T S, Mills D L and Riseborough P S 1981 Electron–phonon interaction and electron inversion layers on polar semiconductors *Phys. Rev.* B **23** 4081–88

[75] Zhakin A I 2013 Electrodynamics of charged surfaces *UFN* **183** 153–77

[76] Shikin V B and Monarkha Y P 1975 Surface charges in helium *FNT* **1** 957–83

[77] Shikin V B 1970 On the movement of helium ions near the vapor-liquid boundary *JETF* **58** 1748–56

[78] Shikin V B 1999 On the supersaturation state of a 2D electronic system on the surface of liquid helium *Lett. JETF* **70** 274–8

[79] Edelman V S 1980 Levitating electrons *UFN* **130** 675–704

[80] Edelman V S 1977 Nonlinear resonance on surface electrons in liquid helium *Lett. JETF* **24** 510–13

[81] Edelman V S 1977 On cyclotron resonance on surface electrons in liquid helium *Lett. ZhETF* **24** 510–13

[82] Brown T R and Grimes C C 1972 Observations of cyclotron resonance in surface-bound electrons on liquid helium *Phys. Rev. Lett.* **29** 1233–6

[83] Cole M W and Cohen M H 1969 Image-potential induced surface bands in insulators *Phys. Rev. Lett.* **23** 1238–41

[84] Cole M W 1970 Properties of image-potential induced surface states of insulators *Phys. Rev.* **2** 4239–52

[85] Jackson S A and Platzman P M 1981 Polaronic aspects of two-dimensional-electrons on film liquid He *Phys. Rev.* B **24** 499–502

[86] Jackson S A and Platzman P M 1982 Temperature dependent effective mass of a self-trapped electron on the surface of a liquid-helium film *Phys. Rev.* B **25** 4886–9

[87] Kajita K 1983 Stability of electrons on thin helium film solid neon system-surface electrons and bubble electrons *J. Phys. Soc. Jpn.* **52** 372–5

[88] Peeters F M and Jackson S A 1985 Frequency-dependent response of an electron on a liquid helium film *Phys. Rev.* B **31** 7098–108

[89] Sokolov S S 2004 Variational approach to the problem of energy spectrum of surface electrons over liquid-helium film *Low Temp. Phys.* **30** 271–5

[90] Adamenko I N, Zhukov A V and Nemchenko K E 2000 Influence of electron–electron interaction on electron mobility over liquid helium *Phys. Low Temp.* **26** 631–7

[91] Beril S, Fomin V M and Starchuk A S 2020 *Theory of Polarons, Excitons, Bipolarons and Kinetic Effects in Multilayer Structures of Various Geometries and Superlattices* (Tiraspol: Pridnestrovian University Publishing House) p 696

[92] Edelman V S 1976 Effective mass of electrons localized above the surface of liquid helium *Lett. JETF* **24** 510–13

[93] Landau L D and Lifshits E M Theoretical physics: in 10 volumes *Electrodynamics of Continuous Media: A Textbook* **vol 8** (Moscow: Fizmatlit,) 2005 p 656

[94] Litovchenko V G, Zuev V A, Korbutyak D V and Veitz V V 1974 Investigation of the GaAs–Si_3 N_4 interface by the surface PL method *DAN of the Ukrainian SSR. Series* A No. 1 69–72

[95] Litovchenko V G 1980 *Fundamentals of Physics of Semiconductor Layered Systems* (Kiev: Naukova Dumka) p 282

[96] Pokatilov E P, Beril S and Fomin V M 1985 Manifestation of the self-action effect in exciton spectra of multilayer structures *FTT* **27** 1892–5

[97] Pokatilov E P and Beril S I 1983 Electron–phonon interaction in periodic two layer structures *Phys. Status Solidi* B **118** 567–73

[98] Nikiforov A F and Uvarov V B 1984 *Special Functions of Mathematical Physics* (Moscow: Nauka) p 344

[99] Prudnikov A P, Brychkov Y A and Marichev O I 1981 *Integrals and Series* (Moscow: Nauka) p 800

[100] Gradstein I S and Ryzhik I M 1963 *Tables of Integrals, Sums, Series and Products* (Moscow: GIFML) p 1100

[101] Silin A P and Semiconductor S R 1985 *UFN* **147** 485–521

[102] Licari J J 1979 Polaron self-energy in a dielectric slab *Solid State Commun.* **29** 625–8

[103] Sherman A V 1981 Dependence of the polaron binding energy and effective mass in a crystal layer on its thickness *Solid State Commun.* **39** 273–7

[104] Liang X X, Gu S and Lin D L 1986 Polaronic states in slab of a polar crystal *Phys. Rev.* B **34** 2807–14

[105] Gu S W, Li Y C and Zheng L F 1989 Intermediate-coupling polaron in polar–polar crystal slab *Phys. Rev.* B **39** 1346–56

[106] Pokatilov E P and Beril S I 1982 Spatially extended optical modes in two layer periodical structures *Phys. Status Solidi* B **110** K75–78

[107] Esaki L A 1986 Berd's-Eye view on the evolution of semiconductor superlattices and quantum wells *IEEE J. Quantum Electron.* **QE-22** 1611–24

[108] Klein M V 1986 Phonons in semiconductor superlattices *IEEE J. Quantum Electron.* **QE-22** 1760–70

[109] Lambin P, Vigneron J P, Lucas A A, Thiry P A *et al* 1986 Observation of long-wavelength interface phonons in a GaAs–AlGaAs superlattices *Phys. Rev. Lett.* **56** 1842–5

[110] Maciel A C, Campelo L C and Ryan J F 1987 Resonant Raman scattering from confine phonons and interface phonon in GaAs/GaAlAs superlattices *J. Phys. C: Solid State Phys* **20** 3041–6

[111] Schwartz G P, Gualtieri G J, Sunder W A and Farrow L A 1987 Light scattering from quantum confine and interface optical vibrational modes in strained-layer GaSb/AlSb superlattices *Phys. Rev.* B **36** 4868–77

[112] Schwartz G P, Gualtieri G J, Sunder W A and Farrow L A 1987 Raman scattering from periodic and nonperiodic GaSb/AlSb strained-layer superlattices *Superlattices Microstruct.* **3** 523–33

[113] Sood A K, Menendez J, Cardona M and Ploog K 1985 Interface vibrational modes in GaAs–AlAs superlattices *Phys. Rev. Lett.* **54** 2115–8

[114] Stern F and Howard W E 1967 Properties of semiconductor surface inversion layers in the electric quantum limit *Phys. Rev.* **163** 816–35

[115] Klimin S N, Pokatilov E P and Fomin V M 1994 Bulk and interface polarons in quantum wires and dots *Phys. Status Solidi* B **184** 373–83

[116] Klein M C, Hache F, Ricard D and Flytzanis C 1990 Size dependence of electron–phonon coupling in semiconductor nanospheres: the case of CdSe *Phys. Rev.* B **42** 11123–32

[117] Broser I *et al* 1982 *Physics of II-VI and I-VII Compounds, Semimagnetic Semiconductors* Landolt-Börnstein: Numerical Data and Functional Relationships in Science and Technology (Berlin: Springer) 543 p

[118] Duan J, Bishop G G, Gillman E S, Safron S A and Skofronick J G 1992 Homoepitaxial growth investigated by high-resolution He atom scattering: NaCl onto NaCl (001) *J. Vac. Sci. Technol.* A **10** 1999–2005

[119] Safron S A, Duan J, Bishop G G, Gillman E S and Skofronick J G 1993 Investigation of epitaxial growth via high-resolution helium atom scattering: potassium bromide onto rubidium chloride (001) *J. Phys. Chem.* **97** 1749–57

[120] Pokatilov E P, Fomin V M, Balaban S N, Klimin S N, Fai L C and Devreese J T 1998 *Superlattices Microstruct.* **23** 331–6

[121] Pokatilov E, Croitoru M, Beril S and Balaban S 1998 Polaronic states in nanoscale structures *Anale Ştiintice* 63–70

[122] Wendler L, Chaplik A V, Haupt R and Hipolito O 1993 Magnetopolarons in quantum dots: comparison of polaronic effects from three to quasi-zero dimensions *J. Phys.: Condens. Matter* **5** 8031–46

[123] Pokatilov E P, Klimin S N, Balaban S N and Beril S I 1996 Magnetopolaron in a cylindrical quantum wire *FTP* **30** 641–50

[124] Cotrufo M, Sun L, Choi J, Alù A and Li X 2019 Enhancing functionalities of atomically thin semiconductors with plasmonicnanostructures *Nanophotonics* **8** 577–98

[125] Trallero-Giner C, Santiago-Pérez D G, Tkachenko D V, Marques G E and Fomin V M– 2024 Raman scattering owing to magneto-polaron states in monolayer transition metal dichalcogenides *Sci. Rep.* **14** 12857

IOP Publishing

Vibrational Excitations in Multilayer Nanostructures

Properties and manifestations

Stepan I Beril, Vladimir M Fomin and Alexander S Starchuk

Chapter 6

Wannier–Mott excitons in homeopolar multilayer structures. Polaronic excitons at the contact of two media, in dimensionally limited crystals and in quantum wires

6.1 Introduction

The sixth chapter is devoted to the study of electronic and exciton states in multilayer structures of homeopolar crystals. The general theory of Wannier–Mott exciton states in composite superlattices is developed.

The effective Hamiltonian of the electron–hole interaction at the contact of two polar crystals is derived and the polaronic exciton states within the limits of Haken ($R_{ex} > R_p$) and Mayer ($R_{ex} < R_p$) are investigated. The biexciton states and exciton complexes on the surface of a polar crystal are considered. The effect of a strong magnetic field on a polaron exciton in a quantum well is investigated. A theoretical study of the Coulomb interaction and Wannier–Mott excitons in polar quantum wires and a comparison of theory and experiment is carried out.

6.2 General description of the exciton problem in a quantum well made of a nonpolar semiconductor

Among the most advanced heterostructures, heterostructures based on GaAs and $Al_xGa_{1-x}As$ compounds can be distinguished due to the proximity of their permanent lattices. The perfection of the resulting contacts makes it possible to create structures of various types: three-layer ones containing a thin layer of GaAs in plates made of thick layers of $Al_xGa_{1-x}As$and multilayer ones from a large number of thin layers of both types. In experimental work [1–6] the exciton luminescence spectra of three-layer and multilayer structures were studied. To analyze the experimental data, the authors of [2, 4] proposed an exciton model in which the

doi:10.1088/978-0-7503-6164-4ch6

self-action effect is completely neglected, and the interaction of an electron and a hole is described by a bulk Coulomb potential with an effective dielectric constant $\varepsilon_{эфф} = \sqrt{\varepsilon_1 \cdot \varepsilon_2}$, where ε_1 is the dielectric constant of the $Al_xGa_{1-x}As$; and ε_2—compound; and 2 is the dielectric constant of GaAs. The proximity of values ε_1 and ε_2 of these compounds (GaAs: $\varepsilon_2 = 13.1$; $Al_xGa_{1-x}As$: $\varepsilon_1 = 13.4$) somewhat mitigates the rough approximation of the proposed method. It is of interest to analyze exciton states using exact expressions of potential energy. The difficulty of such studies lies in the need to use forms of potential energy with implicit dependence on coordinates, even for thin layers, since obtaining explicit forms is due to the implementation of strict criteria inequalities that greatly narrow the permissible parameter intervals. Therefore, these calculations are feasible only by approximate methods, for example, variational ones [7]. In experiments, the GaAs layer containing exciton had small thicknesses of l_2 ($\leqslant$3 nm). Since the dimensional effects are particularly pronounced precisely in the domain of small thicknesses of the exciton layer, we will assume in the calculation that $l_2 \leqslant 10$ nm. Since there is a difference of $(\varepsilon_2 - \varepsilon_1) \ll \varepsilon_1,\ \varepsilon_2$, in the structure under consideration, we will not take into account the effect of self-action ($\Delta E_{SA} < 1$ meV).

The exciton Hamiltonian in a quantum well in a center-of-mass system for motion in a plane ($\boldsymbol{\rho} = \boldsymbol{\rho}_e - \boldsymbol{\rho}_h$; $\mathbf{R} = (m_e^* \boldsymbol{\rho}_e + m_h^* \boldsymbol{\rho}_h)/M_{ex}$, $M_{ex} = m_e^* + m_h^*$) has the view

$$\hat{H} = \frac{\hat{P}_{e\|}^2}{2m_{e\|}} + \frac{\hat{P}_{h\|}^2}{2m_{h\|}} + \frac{\hat{P}_R^2}{2M_{ex}} + \frac{\hat{P}_\rho^2}{2\mu} + U(\rho, z_e, z_h) + U_B(z_e, z_h); \tag{6.2.1}$$

where $U(\rho, z_e, z_h)$ is the exact potential of the electron–hole interaction in a layer of finite thickness, which is obtained from solving the Poisson equation and which has the form (5.11.2)–(5.11.4). In the case of a symmetric structure considered here ($\varepsilon_1 = \varepsilon_3$) $U(\rho, z_e, z_h)$ has the view

$$U(\rho, z_e, z_h) = -\frac{e^2}{4\pi\varepsilon_0\varepsilon_2}\int_0^\infty J_0(\eta\rho)\Phi(\eta, z_e, z_h)\mathrm{d}\eta; \tag{6.2.2a}$$

$$\Phi(\eta, z_e, z_h) = e^{-|z_e - z_h|} + \frac{2}{e^{2\zeta} - \delta^2}[\delta \cosh \eta(z_e - z_h) + e^{\zeta_2} \cosh \eta(z_e + z_h)], \tag{6.2.2b}$$

where $\delta = (\varepsilon_2 - \varepsilon_1)/(\varepsilon_2 + \varepsilon_1)$.

Taking into account the small thickness of the layer, we will select the test wave function of the ground state in the form

$$\Psi_{1s}(\eta, z_e, z_h) = C\cos\left(\frac{\pi z_e}{l_2}\right)\cos\left(\frac{\pi z_h}{l_2}\right)\exp\left(-\frac{\rho}{\lambda}\right); \tag{6.2.3}$$

here C is the normalization constant, λ is a variational parameter. Note that according to calculations from [8], the wave function (6.2.3) with a bulk Coulomb potential, but taking into account boundary conditions, gives the same results for exciton energy as a more complex wave function, including dependence on the

$(z_e - z_h)$ difference. The variational energy calculated on the Hamiltonian (6.2.1) using the wave function (6.2.3) [7] is equal to

$$E_{ex}^{1s}(\lambda) = \frac{\pi^2\hbar^2}{2\mu l_2^2} + \frac{\hbar^2}{2\mu\lambda^2} - \frac{8e^2}{\pi\varepsilon_0 l_2}F(\lambda, l_2), \tag{6.2.4a}$$

where

$$F(\lambda, l_2) = \int_0^\infty \frac{dx}{\left(4 + \frac{\lambda^2}{l_2^2}\right)^{\frac{3}{2}}} \left\{ \frac{3x^2 + 8\pi^2}{4x(x^2 + 4\pi^2)} - \frac{8\pi^4(1 - e^{-x})}{x^2(x^2 + 4\pi^2)} + \frac{32\pi^4\delta \sinh^2(x/2)}{x^2(e^x - \delta)(x^2 + 4\pi^2)^2} \right\}. \tag{6.2.4b}$$

The values of the parameters are used in the numerical calculation: $\varepsilon_1 = 11.4$; $\varepsilon_2 = 13.1$; $m_e = 0.067m_0$; $m_{lh} = 0.08m_0$; $m_{hh} = 0.38m_0$. The given effective exciton masses were determined by the formula (41) from [8]:

$$\frac{1}{\mu_\pm} = \frac{1}{m_e} + \frac{1}{2m_{hh}}\left(1 \mp \frac{1}{2}\right) + \frac{1}{2m_{lh}}\left(1 \pm \frac{1}{2}\right) \tag{6.2.5}$$

(the upper sign is for heavy, the lower one is for light excitons).

Figure 6.1 shows the results of the numerical calculation of $E_{1s}^{ex}(\lambda)$ for various l_2 values in the three-layer $GaAs/Al_xGa_{1-x}As$ ($x = 0.37$), from which it follows that the calculation clarifies the values of the energies of the 1s-state of light and heavy excitons by 10…15% compared with approximate calculations [8] on the bulk Coulomb potential. To illustrate the effect of weakening the electron–hole coupling by specific conditions on the outer surfaces of the structure (metal electrodes), the 1s state of a heavy exciton was calculated using the formula (6.2.4a) at $l_1 = l_3 = 650$ nm: $E_{ex}^{1s} = -13.84$ meV; at $l_1 = l_3 = 10$ nm, $E_{ex}^{1s} = -6.41$ meV; with $l_1 = l_3 = 3$ nm, $E_{ex}^{1s} = -0.14$ meV. In all cases, $l_2 = 5$ nm. This implies the possibility of changing the exciton-binding energy by selecting neighboring media with certain material and geometric parameters.

To calculate the energy of the excited exciton state (2s-state), the test function was taken in the same form as in [8]:

$$\Psi_{2s}(\rho, z_e, z_h) = C_2 \cos\left(\frac{\pi z_e}{l_2}\right)\cos\left(\frac{\pi z_h}{l_2}\right)\left(1 - \frac{2\rho}{\beta}\right)e^{-\frac{\rho}{\beta}}, \tag{6.2.6}$$

where C_2 is the normalization constant and β is a variational parameter. The calculation results are shown in figure 6.2 in the form of plots of the dependence of $|E_{ex}^{1s} - E_{ex}^{2s}|$ on the thickness l_2 of the layer. Here, a comparison is given with experimental data and calculations from [8]. It should be noted that the energy values of the 2 s-state obtained in the described calculation turned out to be significantly different from those in [8]. For example, at $l_2 = 5$ nm, $l_1 = l_3 = 650$ nm, $E_{ex}^{2s} = -2.6$ meV, while in [8] $E_{ex}^{2s} = -1,9$ meV was obtained. The significant difference between these values reflects the fact that the model used in [8] becomes rougher the larger the radius of the exciton

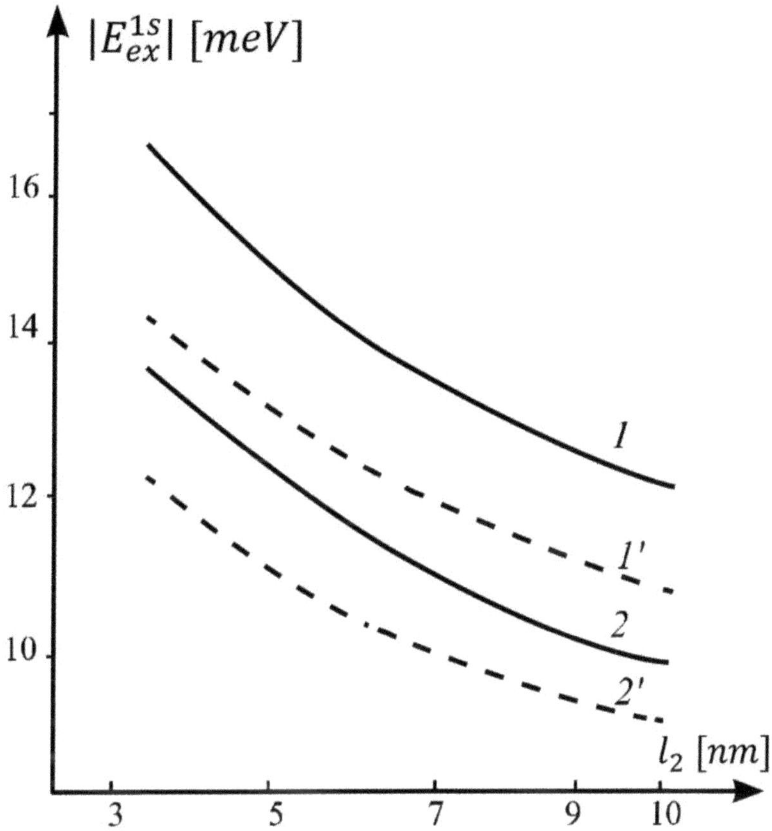

Figure 6.1. Dependence of the exciton energy in the 1s-state on the thickness of the inner layer in the three-layer GaAs/Al_xGa_{1-x}As (x = 0, 30). Curves 1 and 2 are the results of the calculation according to the formulas (6.2.4a), (6.2.4b), respectively; curves 1′ and 2′ are the results of the work [8].

state. The agreement of the $|E_{ex}^{1s} - E_{ex}^{2s}|$ differences calculated in [8] with the experimental values is obviously explained by the mutual compensation of errors of individual terms of the differences. It has already been noted that in this task the effects of self-action are weak. For comparison with the experiment, their calculation is not necessary at all, since they are mutually reduced in the difference $|E_{ex}^{1s} - E_{ex}^{2s}|$.

Exciton luminescence from individual quantum wells is also observed in multilayer GaAs/Al_xGa_{1-x}As structures with GaAs quantum wells at $l_2 \geqslant 30$ nm well thicknesses. It is of interest to compare the results for the 1s state of an isolated quantum well, given above, with the results for the periodic structure, for which the described variational procedure was also used [9].

Averaging the Hamiltonian (6.2.1) with the potential energy of the electron–hole interaction

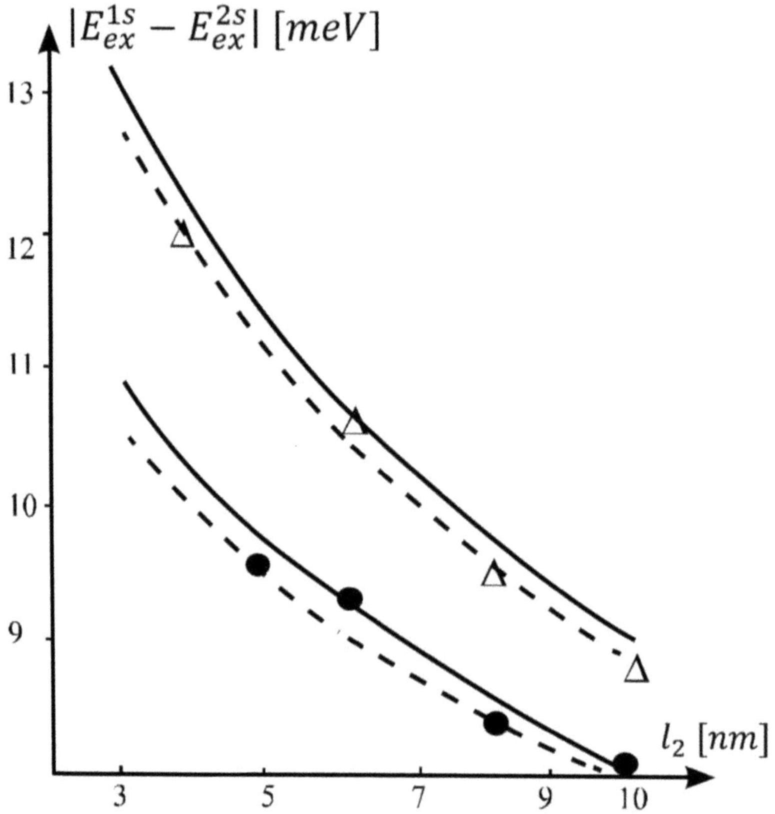

Figure 6.2. Dependence $|E_{ex}^{1s} - E_{ex}^{2s}|$ on the thickness of the GaAs layer: Curves 1 and 2—calculation results according to the formulas (6.2.4a), (6.2.4b), respectively; curves 1′ and 2′ are the results of [8]; Δand are experimental results for a light exciton; •—for a heavy exciton.

$$U(\rho, z_e, z_h) = -\frac{e^2}{4\pi\varepsilon_0\varepsilon_b}\int_0^\infty d\eta\, J_0(\eta\rho)\left\{e^{-\eta|z_e - z_h|} + \frac{1}{\sinh\zeta_b\sqrt{\psi^2 - 1}}\right.$$

$$\left[(e^{-\zeta_b}\sqrt{\psi^2 - 1} + \frac{\cosh\zeta_b}{\psi + \sqrt{\psi^2 - 1}} - \cosh\zeta_b)\cosh\eta(z_e - z_h)\right. \quad (6.2.7)$$

$$\left.\left.+\frac{(\varepsilon_b{}^2 - \varepsilon_a{}^2)}{2\varepsilon_a\varepsilon_b}\sinh\zeta_a\sinh\zeta_b\cosh\eta(z_e + z_h)\right]\right\},$$

where

$$\psi = \frac{(\varepsilon_b{}^2 + \varepsilon_a{}^2)}{2\varepsilon_a\varepsilon_b}\sinh\zeta_a\sinh\zeta_b + \cosh\zeta_a\cosh\zeta_b,$$

obtained in [9] for the considered structure of alternating layers of type *a* and *b* (electron and hole are in a layer of type b) on the wave function (6.2.3), we find the variational energy in dimensionless variables:

$$\widetilde{E}_{ex}^{1s} = \widetilde{\lambda}^{-2} - \frac{16}{\pi\varepsilon_0\widetilde{l}_b}\int_0^\infty \mathrm{d}x\left(4 + \frac{\widetilde{\lambda}^2 x^2}{\widetilde{l}_b^2}\right)^{-\frac{3}{2}}$$

$$\left\{\left[\frac{1}{2x} + \frac{x}{4\left(x^2 + 4\pi^2\right)} - \frac{8\pi^4(1 - e^{-x})}{x^2\left(x^2 + 4\pi^2\right)^2}\right] + \frac{16\pi^4 \sinh\left(\frac{x}{2}\right)}{x^2\left(x^2 + 4\pi^2\right)^2}\right.$$

$$\left[\frac{1}{\sinh x\sqrt{\psi^2 - 1}}\left(e^{-x}\sqrt{\psi^2 - 1} + \frac{\cosh x}{\psi + \sqrt{\psi^2 - 1}} - \cosh\left(x\frac{l_a}{l_b}\right)\right)\right. \tag{6.2.8}$$

$$\left. + \frac{\left(\varepsilon_b{}^2 - \varepsilon_a{}^2\right)}{2\varepsilon_a\varepsilon_b}\frac{\sinh\left(x\frac{\widetilde{l}_a}{\widetilde{l}_b}\right)}{\sqrt{\psi^2 - 1}}\right\},$$

where $\widetilde{E}_{ex}^{1s} = E_{ex}^{1s}/E_0$; $\widetilde{l}_{a,\,b} = l_{a,\,b}/a_0$; $\widetilde{\lambda} = \lambda/a_0$; E_0, a_0 are the ground state energy and the radius of the bulk exciton in material b. The results of the variational calculation according to the formula (6.2.8) are shown in figure 6.3. The exciton energy in layer b decreases monotonously with increasing thickness of this layer. Taking into account the final thickness of layers 1 and 3 shows that large values of the thicknesses of the outer layers with $\varepsilon_a < \varepsilon_b$ correspond to large exciton energies. For example, for fixed values $l_b = 3$ nm, $E_{ex}^{1s}(l_a = 3\text{ nm}) = -2.73E_0$; $E_{ex}^{1s}(l_a = 650\text{ nm}) = -3.74E_0$. In figure 6.3 it can be seen that even at very small l_b ($l_b = 3$ nm), the calculation according to the formula (6.2.7) with an exact potential gives energy values significantly different from the limit values obtained on a two-dimensional Coulomb potential:

$$U^{2D}(\rho) = -\frac{e^2}{4\pi_0 S\rho}\left(\frac{l_a}{\varepsilon_a^{\|}} + \frac{l_b}{\varepsilon_b^{\|}}\right);\ S = l_a + l_b, \tag{6.2.9a}$$

which has a spectrum

$$E_{ex}^{2D} = -4\left(1 + \lambda\frac{\varepsilon_b^{\|}}{\varepsilon_a^{\|}}\right)\left(1 + \lambda\frac{\varepsilon_a^{\perp}}{\varepsilon_b^{\perp}}\right)^{-1}(2n + 1)^{-2}E_V;\ E_V = \frac{\mu_\perp e^4}{32\pi^2\varepsilon_0^2\varepsilon_2^2\hbar^2}. \tag{6.2.9b}$$

The use of the wave function (6.2.3) is justified when the motion of both the electron and the hole along z is limited by the thickness of the film. This condition can be written as an inequality

$$a_V^{(e,\,h)} > l_2, \tag{6.2.9c}$$

where $a_V^{(e,\,h)}$ are the radii of the electron (e) and hole (h) clouds in the ground state of the exciton. For an electron with an effective mass $m_e^* = 0.068m_0$, the inequality

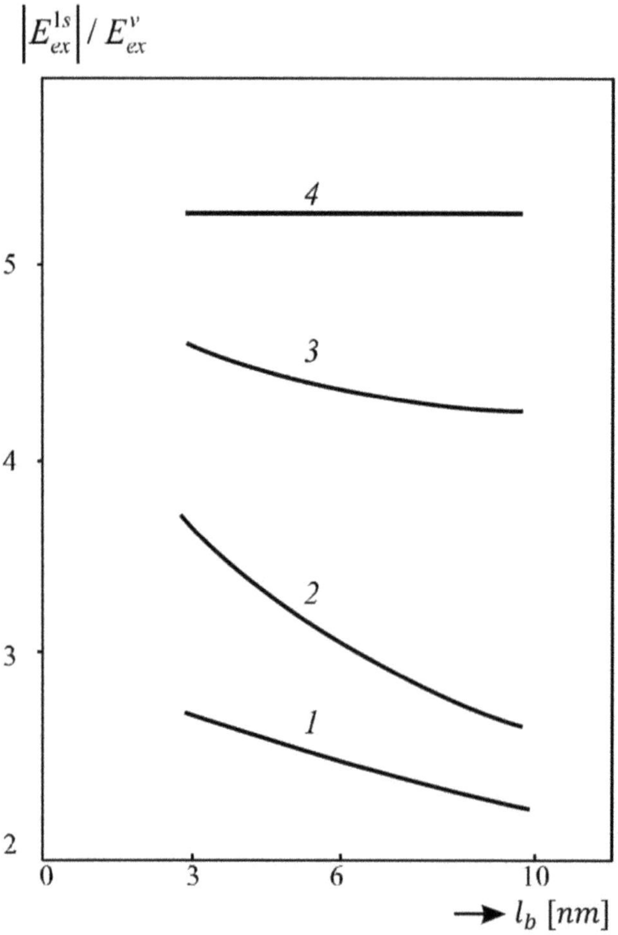

Figure 6.3. Dependence of the energy of the 1s state of the exciton in the GaAs layer of the GaAs/$Al_xGa_{1-x}As$ ($x = 0, 37$) with $1 - l_b = 3$ nm; $2 - l_b = 650$ nm; curves 3 and 4 describe the energy of $1s$ state calculated on the 2D Coulomb potential.

(6.2.9c) is performed well enough already at $l \leqslant 10$ nm. However, for a hole with an effective mass $m_h^* = 0, 38m_0$, the criterion (6.2.9c) is fulfilled at $l < 5$ nm. In fact, the criterion in the form of (6.2.9c) is hard, since the interaction energy with the electron charge smeared inside the layer is significantly less than with the Coulomb center. This, apparently, explains the proximity to the experimental values of the energy values obtained by the formula (6.2.4a), derived using the wave function (6.2.3).

6.2.1 Exciton formed by a heavy hole

Consider an intermediate situation in which the motion of a light particle is mainly determined by the boundaries of the layer (dimensionally quantized), and the motion of a heavy particle is determined by its interaction with the charge cloud of a light particle and is concentrated near the center of the layer. Intermediate layer thicknesses adjacent to the domain of small values correspond to this case. Under these conditions, an adiabatic approximation can be used to describe the interaction of light and heavy particles.

As an example, consider a three-layer system with the same ($\varepsilon_1 = \varepsilon_3$) infinite layers l_1 and l_3 (three-layer symmetric structure). The potential energy $U_0(\rho, z_e, z_h)$ has the form (6.2.2). Taking the assumption of the dimensionally quantized motion of an electron, the wave function of the fast longitudinal motion of an electron and the variational function of the transverse motion of an electron–hole system are written as

$$\psi_1(\rho, z_e) = N_1 \cos\left(\frac{\pi z_e}{l_2}\right) \exp\left(-\frac{\rho}{\lambda}\right), \tag{6.2.10}$$

where $N_1^2 = 4/(\pi\lambda^2 l_2)$, λ is the variation parameter.

Let's average the exciton Hamiltonian (6.2.1), taken without the kinetic energy of the longitudinal motion of the hole, on the wave function (6.2.10) and write down the equation of motion of the slow subsystem, including the energy of the fast one as a potential:

$$\left\{\frac{\hat{P}_{\|h}^2}{2m_{h\|}} + U(z_h)\right\}\psi(z_h) = E_{ex}(\lambda, l_2)\psi(z_h), \tag{6.2.11a}$$

where

$$E_{ex}(\lambda,\ l_2) = E_{ex} - \frac{\pi^2\hbar^2}{2m_{\|e} l_2^2} - \frac{\hbar^2}{2\mu_\perp \lambda^2}; \tag{6.2.11b}$$

$$\begin{aligned} U(z_h) \;=\; & -\frac{e^2 N_1^2}{2\varepsilon_0\varepsilon_2}\int_0^\infty \mathrm{d}\eta\left(4 + \lambda^2\eta^2\right)^{-\frac{3}{2}} \\ & \left\{\frac{1}{\eta} + \frac{\eta\cos\left(\frac{2\pi z_h}{l_2}\right)}{\eta^2 + \frac{4\pi^2}{l_2^2}} - \frac{4\pi^2 e^{-\frac{\eta l_2}{2}}}{\eta l_2^2\left(\eta^2 + \frac{4\pi^2}{l_2^2}\right)}\right. \\ & + \frac{2\left(4\pi^2/l_2^2\right)\sinh\left(\frac{\eta l_2}{2}\right)}{\left(e^{2\eta l_2} - \delta_1\delta_3\right)\eta\left(\eta^2 + 4\pi^2/l_2^2\right)} \\ & \left.\left[\left(\delta_1\delta_3 + e^{\eta l_2} f_1\right)\cosh\eta z_h + e^{\eta l_2} f_2 \sinh\eta z_h\right]\right\}. \end{aligned} \tag{6.2.11c}$$

Assuming that the motion of the heavy hole is concentrated in the central domain of the layer (i.e., ηz_h is small), we expand (6.2.11c) series, limiting ourselves only to quadratic terms:

$$U(z_h) \approx U(0) + A_1 z_h + A_2 z_h^2, \tag{6.2.12a}$$

where

$$U(0) = -\frac{e^2 N_1^2}{2\varepsilon_0\varepsilon_2}\int_0^\infty \frac{d\eta}{(4+\lambda^2\eta^2)^{\frac{3}{2}}}$$
$$\left\{\frac{1}{\eta} + \frac{\eta}{\eta^2 + \frac{4\pi^2}{l_2^2}} - \frac{4\pi^2 e^{-\frac{\eta l_2}{2}}}{\eta l_2^2\left(\eta^2 + \frac{4\pi^2}{l_2^2}\right)} \right. \tag{6.2.12b}$$
$$\left. + \frac{2\left(\frac{4\pi^2}{l_2^2}\right)\sinh\left(\frac{\eta l_2}{2}\right)}{(e^{\eta l_2} - \delta_1\delta_3)\eta\left(\eta^2 + \frac{4\pi^2}{l_2^2}\right)}\left[\delta_1\delta_3 + e^{\eta l_2} f_1\right]\right\};$$

$$A_1 = -\frac{2\pi^2 e^2 N_1^2}{\varepsilon_0\varepsilon_2 l_2^2}\int_0^\infty \frac{d\eta\, e^{-\frac{\eta l_2}{2}}\sinh\left(\frac{\eta l_2}{2}\right)}{(4+\lambda^2\eta^2)^{\frac{3}{2}}\left(\eta^2 + \frac{4\pi^2}{l_2^2}\right)}; \tag{6.2.12c}$$

$$A_2 = -\frac{e^2 N_1^2}{2\varepsilon_0\varepsilon_2}\int_0^\infty \frac{d\eta}{(4+\lambda^2\eta^2)^{\frac{3}{2}}}\left\{-\frac{2\pi^2\eta}{l_2^2\left(\eta^2 + \frac{4\pi^2}{l_2^2}\right)} - \frac{2\pi^2\eta e^{-\frac{\eta l_2}{2}}}{l_2^2\left(\eta^2 + \frac{4\pi^2}{l_2^2}\right)} \right. \tag{6.2.12d}$$
$$\left. + \frac{4\pi^2\eta\sinh\left(\frac{\eta l_2}{2}\right)}{(e^{2\eta l_2} - \delta_1\delta_3)\left(\eta^2 + \frac{4\pi^2}{l_2^2}\right)}(\delta_1\delta_3 + e^{\eta l_2} f_1)\right\}.$$

For small values of f_2 the linear term (describing the deviation from the center of the hole oscillations) can be neglected. In general, this deviation is equal to $z_0 = A_1/2A_2$, and the potential energy (6.2.12a) can be represented as

$$U(z_h) \approx U_0 - A_2 z_0^2 + A_2(z_h + z_0)^2. \tag{6.2.13}$$

Substituting (6.2.13) into (6.2.11) and averaging on the oscillator function, we obtain the variational energy:

$$E_{ex}(\lambda, l_2) = \frac{\hbar^2}{2\mu l_2^2} - A_2 z_0^2 + \hbar\omega(\lambda, l_2)\left(n + \frac{1}{2}\right); \tag{6.2.14a}$$

here the parameter $\omega_n(\lambda, l_2)$ is defined by the expression

$$\omega_n(\lambda, l_2) = \left\{\frac{2A_2(\lambda, l_2)}{m_{h\parallel}}\right\}^{\frac{1}{2}}. \tag{6.2.14b}$$

Minimizing $E_{ex}(\lambda)$ by λ gives the exciton energy in the case under consideration when the hole motion is limited by a parabolic potential with a minimum in the center of the layer. Note that in the case of $\varepsilon_2 > \varepsilon_1 = \varepsilon_3$, the contribution to the potential from the self-action of the hole also has the form $\sim z_h^2$. Therefore, taking into account self-action is reduced to converting the parameters: $U_0 \to U_0 + U_{SA}(0)$ and $A_2 \to A_2 + A$.

The criterion for the validity of the oscillatory approximation is the inequality

$$\left(\frac{\hbar}{m_{h\|}^* l_2}\right)^{\frac{1}{2}} < \frac{l_2}{2}. \tag{6.2.15}$$

In [10], a method for accounting for finite film thicknesses is proposed by averaging the difference of approximate two-dimensional and exact potentials on the wave functions of a fast subsystem. The approximate bulk Coulomb potential was used as an exact one in [10].

In another limit: for large film thicknesses $l_2 > a_0$, the translational motion of the exciton as a whole is quantized [10]. The energy of the absorbed layer contains the terms

$$\hbar\omega_{ex} = \frac{\pi^2\hbar^2}{2M_{ex}l_2^2} + \widetilde{E}_g + E_{ex}, \tag{6.2.16}$$

where the first term is the kinetic energy of an exciton in a quantum well (the energy of the lowest level of size quantization),

$$\widetilde{E}_g = E_{g0} + U_{SA}^{(e)}(l_2) + U_{SA}^{(h)}(l_2) \tag{6.2.17}$$

is the band gap renormalized by self-action and E_{ex} is the internal exciton energy calculated in the same way as in a massive crystal.

Based on the study of exciton states in a quantum well in cases of three-layer and multilayer periodic structures, it can be concluded that: (a) two-dimensional exciton states are not realized even in cases of very narrow quantum wells (i.e., accurate electron–hole potentials should be used to calculate the exciton energy spectrum; (b) the self-action potential and the additional potential associated with a heavy hole—the adiabatic potential—can lead to a change in the character of quantization of the electron spectrum in a quantum well (in a rectangular well $E_{s.q.} \sim n^2$, and in a parabolic one—the energy spectrum is equidistant).

6.2.2 Coulomb interaction and coupled electron–hole states in ultrathin homeopolar films

The case of an ultrathin homeopolar film ($\eta l_2 \ll 1$) is of independent interest. The Coulomb interaction in this limiting case has been discussed in the works [11–20]. Let's consider a symmetric system. The potential of the electron–hole interaction [21, 22] can be found from the general expressions (6.2.2a) and (6.2.2b). From (6.2.2c) in the limit $\eta l_2 \ll 1$ we obtain

$$\Phi(\eta) = \frac{e^{\eta l_2} + \delta}{e^{\eta l_2} - \delta}. \tag{6.2.18}$$

Substituting (6.2.18) into (6.2.2a), we find

$$U(\rho) = -\frac{\pi e^2(1+\delta)}{4\pi\varepsilon_0\varepsilon_2 l_2}\left[H_0\left(\frac{2\rho(1-\delta)}{l_2}\right) - N_0\left(\frac{2\rho(1-\delta)}{l_2}\right)\right], \tag{6.2.19}$$

where $H_0(x)$, $N_0(x)$ are the Struve and Neumann functions, respectively.

Let's consider two ranges of values of the function argument (6.2.19):

$$2\rho(1-\delta)/l_2 \gg 1;\ \rho \gg l_2, \tag{6.2.20a}$$

$$H_0(x) - N_0(x) \approx \frac{2}{\pi x}, \tag{6.2.20b}$$

then from (6.2.19) we get:

$$U(\rho) = -\frac{e^2(1+\delta)}{\varepsilon_2(1-\delta)} = -\frac{e^2}{4\pi\varepsilon_0\varepsilon_1\rho} \tag{6.2.20c}$$

is two-dimensional Coulomb potential with dielectric permittivity of media adjacent to the film;

$$2\rho(1-\delta)/l_2 \ll 1;\ \rho \gg l_2, \tag{6.2.21a}$$

$$H_0(x) - N_0(x) = -\frac{2}{\pi}[\ln x - C], \tag{6.2.21b}$$

where C is the Euler constant; then from (6.2.19) we get:

$$U(\rho) = -\frac{e^2}{2\pi\varepsilon_0\varepsilon_2 l_2}\left[\ln\left(\frac{\varepsilon_2 l_2}{\varepsilon_1\rho}\right) - C\right]. \tag{6.2.21c}$$

The results presented here are given in [12, 13, 15, 16] and generalized in [7, 23] for the case of an anisotropic structure.

In cases where conditions (6.2.20a) and (6.2.21a) are met, the exciton energy can be found in analytical form. It is convenient to write down the solutions of the Schrodinger equation with potential energy (6.2.20c) and (6.2.21c) as

$$E_n^{2D} = -\frac{\mu e^4}{16\pi^2\varepsilon_0^2\varepsilon_1^2\hbar^2\left(n+\frac{1}{2}\right)^2};n = 0,\ 1,\ 2,\ \ldots; \tag{6.2.22a}$$

$$E_n = -\frac{e^2}{4\pi\varepsilon_0\varepsilon_2 l_2}\left\{\ln\left[\frac{4\varepsilon_2 l_2}{\varepsilon_1 a_{ex}}\right] - 2C - 2\gamma_n\right\}; \tag{6.2.22b}$$

here γ_n is found from the solution of the differential equation

$$\Delta_{\xi}\,\psi_n(\xi) - \ln|\xi|\psi_n(\xi) = \gamma_n\psi_n(\xi) \tag{6.2.22c}$$

and in the order of magnitude of the order of 1.

Comparing these results with the results for a 3D exciton, whose energy is $E_n^{3D} = -\mu e^4/(32\pi^2\varepsilon_0^2\varepsilon_2^2\hbar^2 n^2)$; $n = 1, 2, \ldots$, we can draw a general conclusion that in the structure with $\varepsilon_2 > \varepsilon_1$, with a decrease in the thickness of the l_2 homeopolar film, the exciton binding energy monotonously increases from the energy of the 3D exciton to the energy of the 2D exciton when fulfilling the criterion inequality (6.2.21a), passing through the area in which the logarithmic law (6.2.22b) is fulfilled. The reason for the increase in the exciton binding energy is the displacement of electric field lines in an area with lower permittivity values and a geometric effect (the Keldysh effect [13]). As will be shown below, another channel for increasing the exciton binding energy appears in polar semiconductor films—the dimensional effect of losing part of the inertial shielding [24].

6.3 General theory of Wannier–Mott exciton states in composite superlattices

In the previous sections, the problem of Coulomb interaction and exciton states in thin films, quantum wells, and three-layer systems was discussed. Generalization of this theory to the case of a periodic system is not trivial, as evidenced by attempts to develop it for composite superlattices [1, 10, 17, 25]. A common disadvantage of these studies is the use of phenomenological potentials of the electron–hole interaction (or a bulk Coulomb potential with effective dielectric constant), which ultimately do not allow for a qualitative comparison with the experiment. In some works [10, 17], the starting point is the spectrum of a two-dimensional exciton, to which corrections are calculated taking into account the final thickness of the layer in which the charge carriers are located. This approach is also not entirely correct, since it is very important already at the initial stage to take into account the final thickness of the layers.

Based on the approach outlined in section 6.1, we developed a general theory of Wannier–Mott exciton states in two-layer periodic structures. At the first stage, based on the theory of potential in multilayer systems [26], the seed potential of the electron–hole interaction is found, having the form (6.2.7). At the second stage, the problem of dynamic shielding of the electron–hole interaction by polarization optical vibrations of the structure is solved—the confinement phonons of this layer and the surface spatially extended phonons of the entire structure.

The Hamiltonian of interaction with spatially extended surface optical phonons has the form (5.16.27)–(5.16.32).

In the limiting case of Haken and the approximation of a rectangular quantum well $E_{s.q.} > E_{ex}$ the effective potential of the electron–hole interaction has the form

$$U_{\text{eff}}(\rho) = \widetilde{U}_0(\rho) + \delta W^{(V)}_{ph,\,1}(\rho) + \delta W^{(S)}_{ph,\,1}(\rho), \tag{6.3.1}$$

where

$$\widetilde{U}_0(\rho) = \left\langle \psi(z_e, z_h) \,\middle|\, U_0(\rho, z_e, z_h) \,\middle|\, \psi(z_e, z_h) \right\rangle = -\frac{8e^2}{\pi\varepsilon_0\varepsilon_b l_b}$$

$$\int_0^\infty \mathrm{d}x\, J_0(x\rho)\left\{\frac{1}{2x} + \frac{x}{4\left(x^2 + 4\pi^2\right)} - \frac{8\pi^4\left(1 - e^{-x}\right)}{x^2\left(x^2 + 4\pi^2\right)^2}\right.$$

$$+\frac{16\pi^4 \sinh^2\left(\frac{x}{2}\right)}{x^2\left(x^2 + 4\pi^2\right)^2}\left[\frac{1}{\sinh x\,\sqrt{\psi^2 - 1}}\right. \tag{6.3.2}$$

$$\left(e^{-x}\sqrt{\psi^2 - 1} + \frac{\cosh x}{\psi + \sqrt{\psi^2 - 1}} - \cosh\left(x\frac{l_a}{l_b}\right)\right)$$

$$\left.-\frac{\left(\varepsilon_b^2 - \varepsilon_a^2\right)}{2\varepsilon_a\varepsilon_b}\frac{\sinh\left(x\frac{l_a}{l_b}\right)}{\sqrt{\psi^2 - 1}}\right\};$$

the contributions from shielding by bulk and surface spatially extended optical phonons are equal, respectively:

$$\delta W_{ph,1}^{(V)}(\rho) = \frac{128 l_b}{\pi^4\rho^2\varepsilon_0}\left(\frac{1}{\varepsilon_{b0}} - \frac{1}{\varepsilon_b}\right)\sum_{m=1,3,5,\ldots}\left[m^4\left(m^2 - 4\right)^2\right]^{-1}$$

$$\times \int_0^\infty dy \cdot y\left(J_0(y) - 1\right)\left[1 + \frac{1}{(m\pi)^2}\left(\frac{l_b}{\rho}\right)^2\right]^{-1} \tag{6.3.3a}$$

$$\left\{\left[1 + \frac{R_e^2}{\rho^2}y^2\right]^{-1} + \left[1 + \frac{R_h^2}{\rho^2}y^2\right]^{-1}\right\};$$

$$\delta W_{ph,2}^{(V)}(\rho) = \frac{128 e^2 l_b \hbar^2}{2\pi^4\varepsilon_0\mu\hbar\omega_{b0}\rho^4}\left(\frac{1}{\varepsilon_{b0}} - \frac{1}{\varepsilon_b}\right)$$

$$\sum_{m=1,3,5,\ldots}\left[m^4\left(m^2 - 4\right)^2\right]^{-1}$$

$$\times \int_0^\infty dy \cdot y^3 J_0(y)\left[1 + \frac{y^2}{(m\pi)^2}\left(\frac{l_b}{\rho}\right)^2\right]^{-1} \tag{6.3.3b}$$

$$\left[1 + \frac{R_e^2}{\rho^2}y^2\right]^{-1} \cdot \left[1 + \frac{R_h^2}{\rho^2}y^2\right]^{-1};$$

$$\delta W^{(S)}_{ph,1}(\rho) = \frac{e^2}{\varepsilon_0 \rho}\sum_{s=1}^{4}\int_0^{\infty} dy \int_{-\pi}^{\pi} \frac{dx\left[J_0(y) - 1\right]}{\zeta_b^2\left(\zeta_b^2 + 4\pi^2\right)^2 \Omega_s^2} \times$$
$$\left\{\sum_{r'=1}^{4} \frac{\omega_b \sqrt{\varepsilon_{b0} - \varepsilon_b}}{\sqrt{\tanh\left(\frac{\zeta_b}{2}\right)}} K_{1b,r'}(x) \times \right.$$
$$\left[\delta_{r',s} - \left(1 - \delta_{r'',s}\right)F_{r',s}\right]$$
$$\left[1 + \sum_{r'' \neq s} \left| F_{r'',s} \right|^2\right]^{-\frac{1}{2}2} \tag{6.3.4a}$$
$$\left. \times \left\{\left[1 + \frac{R_e^2}{\rho^2} y^2\right]^{-1} + \left[1 + \frac{R_h^2}{\rho^2} y^2\right]^{-1}\right\}\right\};$$

$$\delta W^{(S)}_{ph,2}(\rho) = \frac{2\hbar^2 e^2}{\mu_\perp \varepsilon_0 \rho^3}\sum_{s=1}^{4}\int_0^{\infty} dy \cdot y^2 \int_{-\pi}^{\pi} \frac{J_0(y) dx}{\hbar \Omega_s^3 \zeta_b^2\left(\zeta_b^2 + 4\pi^2\right)} \times$$
$$\times \left\{\sum_{r'=1}^{4} \frac{\omega_b \sqrt{\varepsilon_{b0} - \varepsilon_b}}{\sqrt{\tanh\left(\frac{\zeta_b}{2}\right)}} K_{1b,r}[\delta_{r',s} - (1 - \delta_{r',s})F_{r',s}]\left(1 + \sum_{r' \neq s} |F_{r'',s}|^2\right)^{-\frac{1}{2}} \times \right. \tag{6.3.4b}$$
$$\left. \times \left[1 + \frac{R_e^2}{\rho^2} y^2\right]^{-1}\left[1 + \frac{R_h^2}{\rho^2} y^2\right]^{-1}\right\}$$

(the electron and the hole are localized in a layer of type b and with parameters ε_{b0}, ε_b, l_b, ω_b).

The following known limiting results can be obtained from expression (6.3.1):

(1) two-dimensional potential ($l_a \to \infty$, $l_b \to 0$) in the Haken approximation;

(2) the potential of the electron–hole interaction in a single quantum well ($l_a \to \infty$);

(3) the potential of the electron–hole interaction at the contact of two crystals: (a) polar–polar crystals ($l_a \to \infty$, $l_b \to \infty$); (b) polar–nonpolar crystals ($\varepsilon_{a0} = \varepsilon_0$).

The exciton binding energy is determined by minimizing the variational energy:

$$E(\beta, l_a, l_b) = \langle \psi(\rho)| - \frac{\hbar^2}{2\mu_\perp}\Delta_\rho + \sum_{c=e,h} \frac{\pi^2 \hbar^2}{2m_{cz} l_b^2} + E_{SA,c}(l) + U_{\text{eff}}(\rho)|\psi(\rho)\rangle \tag{6.3.5}$$

and it includes zone and intraexiton energies. Numerical calculations of the exciton binding energy for specific superlattices were carried out for the Mayer limit case, which was discussed in detail in the exciton problem at the contact of two polar media in the previous chapter. We will give here only the final expressions for the variational binding energy and the effective mass of the exciton:

$$E(\beta,\ l_a,\ l_b) = \frac{\hbar^2}{2\mu_{\perp}\beta^2} + \sum_c \left(\frac{\pi^2\hbar^2 N_c^2}{2m_{cz}l_b^2} + E_{SA,c}(l_{a,b}) \right) - \frac{e^2}{\pi\varepsilon_0\varepsilon_b l_b}$$

$$\int_0^\infty \mathrm{d}x \left[1 + \frac{\beta^2 x^2}{4l_b^2}\right]^{-\frac{3}{2}} \left\{ \frac{1}{x} + \frac{x}{4(x^2+4\pi^2)} - \frac{8\pi^4(1-e^{-x})}{x^2(x^2+4\pi^2)^2} \right.$$

$$+ \frac{16\pi^4 \sinh^2\left(\frac{x}{2}\right)}{x^2(x^2+4\pi^2)^2} \times \frac{1}{\sinh x\sqrt{\psi^2-1}} \qquad (6.3.6)$$

$$\left(e^{-x}\sqrt{\psi^2-1} + \frac{\cosh x}{\psi + \sqrt{\psi^2-1}} - \cosh\left(x\frac{l_a}{l_b}\right) \right.$$

$$\left.\left. + \frac{(\varepsilon_b{}^2 - \varepsilon_a{}^2)\sinh\left(x\frac{l_a}{l_b}\right)}{2\varepsilon_a\varepsilon_b\sqrt{\psi^2-1}} \right)\right\} + \delta W_{ph}^{(V)}(\beta) + \delta W_{ph}^{(S)}(\beta),$$

where

$$\delta W_{\mathrm{ph}}^{(V)}(\beta) = -\sum_{m=1,3,5,\ldots} \alpha_V \hbar\omega_0 (l_b R_V^{-1}) \tilde{g}_m^2 \int_0^\infty dx \left[1 + \left(\frac{R_V^{-1} l_b}{\pi m}\right)^2 x^2\right]^{-1} \times$$

$$\left[\left[1 + \left(\frac{s_h R_V^{-1} x\beta}{2}\right)^2\right]^{-\frac{3}{2}} - \left[1 + \left(\frac{s_e R_V^{-1} x\beta}{2}\right)^2\right]^{-\frac{3}{2}} \frac{x}{(1+x^2)} \times \right. \qquad (6.3.7a)$$

$$\left.\left\{1 - \frac{1}{2}(1-\tilde{\eta})^2 \left(\frac{2x}{1+x^2}\right)^2 \frac{K^2}{R_V^{-2}}\right\}\right];$$

$$\delta W_{\mathrm{ph}}^{(s)}(\beta) = \frac{e^2}{64\pi^2\varepsilon_0} \int_0^\infty dx \int_{-\pi}^{\pi} dy \sum_{s=1}^{4} \frac{|\rho_\eta|^2 R_s^{-1}}{\Omega_s^2} \left\{ \sum_{r'=1}^{4} \frac{\omega_b\sqrt{\varepsilon_{b0} - \varepsilon_b}}{\sqrt{\tanh\left(\frac{\zeta_b}{2}\right)}} K_{1f,\,r'}(x) \times \right.$$

$$[\delta_{r',\,s} - (1-\delta_{r',\,s})F_{r',\,s}] \left(1 + \sum_{r'' \neq s} |F_{r'',\,s}|^2\right)^{-\frac{1}{2}} (1+x^2)^{-1} \qquad (6.3.7b)$$

$$\left. \times \left\{1 - \frac{1}{2}(1-\tilde{\eta})^2 \left(\frac{2x}{1+x^2}\right)^2 \frac{K^2}{R_s^{-2}}\right\}\right\}.$$

The effective mass of the exciton can be represented as:

$$M_{\text{eff}} = M_{ex\perp}\left(1 + A_{\text{ph}}^{V} + A_{\text{ph}}^{S}\right); \tag{6.3.8}$$

here, the contributions from bulk and surface spatially extended optical phonons have the form, respectively:

$$A_{\text{ph}}^{V} = 2\sum_{m=1,3,5,\dots} \alpha_V\left(l_b R_V^{-1}\right)\tilde{g}_m^2 \int_0^{\infty} dx \cdot x^3(1+x^2)^{-3}\left[1 + \left(\frac{l_b}{m\pi\, R_V}\right)^2 x^2\right]^{-1}$$
$$\times\left\{\left[1 + \left(\frac{s_h x\beta}{2R_V}\right)^2\right]^{-\frac{3}{2}} - \left[1 + \left(\frac{s_e x\beta}{2R_V}\right)^2\right]^{-\frac{3}{2}}\right\}; \tag{6.3.9a}$$

$$A_{\text{ph}}^{S} = \frac{e^2}{16\pi\varepsilon_0}\int_0^{\infty} \mathrm{d}x \int_{-\pi}^{\pi} \frac{|\rho|^2 M_{\perp} x^2 R_S}{\Omega_S^2 \hbar^2 (1+x^2)^3}\left\{\sum_{r'=1}^{4} \frac{\omega_b\sqrt{\varepsilon_{b0} - \varepsilon_b}}{\sqrt{\tanh\left(\frac{\varsigma}{2}\right)}} K_{1b,\, r'}(x) \times\right.$$
$$\left.\left[\delta_{r',\,s} - (1 - \delta_{r',\,s})F_{r',\,s}\right]\left[1 + \sum_{r''\neq s} |F_{r'',\,s}|^2\right]^{-\frac{1}{2}}\right\}. \tag{6.3.9b}$$

In the expressions (6.3.9a and (6.3.9b), the notation is introduced:

$$\tilde{g}_m = \frac{2}{m\pi}\left\{\frac{4}{\pi}\frac{1}{m^2 - 4}\sin\left[(m-2)\frac{\pi}{2}\right] + \frac{4}{\pi}\frac{1}{m(m+2)}\sin\left[\left(\frac{m\pi}{2}\right)\right]\right\};$$

$$R_V^{-1} = \left(\frac{2M_{\perp}\omega_{b0}}{\hbar}\right)^{\frac{1}{2}};\ R_S^{-1} = \left(\frac{2M_{\perp}\Omega_S(\kappa, \eta)}{\hbar}\right)^{\frac{1}{2}}.$$

6.3.1 Wannier–Mott excitons in $GaAs/Al_xGa_{1-x}As$ superlattices

Calculations of the Wannier–Mott exciton binding energy were performed numerically for $GaAs/Al_xGa_{1-x}As$ superlattices. Figures 6.4(a) and (b) show the dependences of the binding energies of heavy and light excitons in GaAs, layers obtained for the following values of the structure parameters: $\varepsilon_b = 10.9$; $\omega_{b0} = 5.34 \cdot 10^{13}\,\text{s}^{-1}$; $E_{ex}^{V} = -4.58$ meV; $m_e^* = 0.067m_0$; $m_h^{lh} = 0.197m_0$; $m_h^{hh} = 0.099m_0$; $\mu_{\perp}^{lh} = 0.05m_0$; $\mu_{\perp}^{hh} = 0.04m_0$. Curve 1 corresponds to the calculation of E_{ex} without taking into account the shielding by optical phonons (obtained using the potential (6.3.2)); curve 2—taking into account the contribution from confinement and surface spatially extended optical phonons (expressions (6.3.7a) and (6.3.7b)); curves 3 and 4 correspond to the results of work [27] without taking into account shielding by bulk optical phonons (curves 3 and 4, respectively). The phonon contributions in [27] were calculated using the bulk Fröhlich Hamiltonian of the electron–phonon

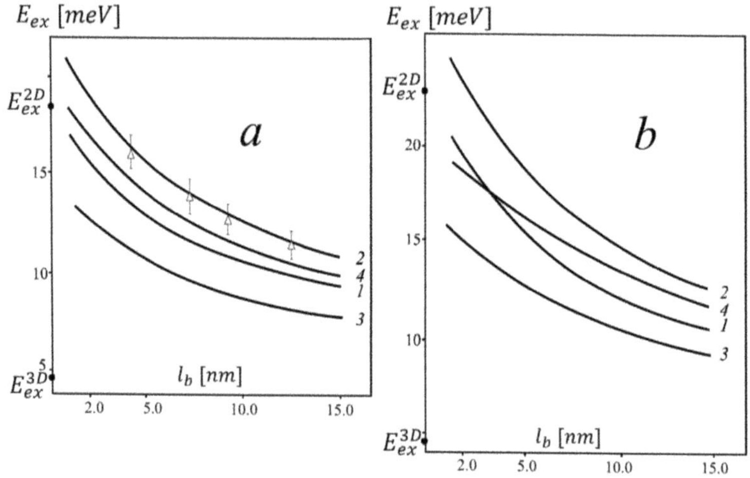

Figure 6.4. Dependences of the binding energies of heavy (a) and light (b) excitons in GaAs, layers obtained for the following values of structure parameters: $\varepsilon_b = 10.9$; $\omega_{b0} = 5.34 \cdot 10^{13} s^{-1}$; $E_{ex}^{V} = -4.58$ meV; $m_e^* = 0.067 m_0$; $m_h^{lh} = 0.197 m_0$; $m_h^{hh} = 0.099 m_0$; $\mu_\perp^{lh} = 0.05 m_0$; $\mu_\perp^{hh} = 0.04 m_0$. Curve 1 corresponds to the calculation of E_{ex} without taking into account the shielding by optical phonons (obtained using the potential (6.3.2)); curve 2—taking into account the contribution from confinement and surface spatially extended optical phonons (expressions (6.2.7a, b)); Curves 3 and 4 correspond to the results of work [27] without taking into account and taking into account shielding by bulk optical phonons (curves 3 and 4, respectively). The phonon contributions in [27] were calculated using the bulk Fröhlich Hamiltonian of the electron–phonon interaction without taking into account the restructuring of the optical oscillation spectrum.

interaction without taking into account the restructuring of the optical oscillation spectrum.

From the comparison of the results presented in figures 6.4(a) and (b), the following conclusions can be drawn:

(1) the use of the V_{eh}^{3D} potential in calculations of light and heavy excitons underestimates the absolute value of E_{ex} by about 15%–20% in comparison with the calculation results based on the exact potential of the electron–hole interaction;
(2) the correct two-dimensional limit for $E_{ex}(l_b)$ is obtained only when using the exact potential of the electron–hole interaction (6.3.2). The latter takes into account not only the geometric effect leading to the growth of $E_{ex}(l)$, but also the change in the nature of inertial and inertialess shielding;
(3) with a change in l_b, the role of contributions from optical phonon shielding changes: with a decrease in l_b, the contribution from bulk phonons decreases and at $l_b \to 0$ $\delta W_{\text{ph}}^{V} \to 0$, and from surface phonons increases. The total contribution of $\delta W_{\text{ph}}^{S} + \delta W_{\text{ph}}^{V}$ has a weak dimensional dependence.

In conclusion, we will make a remark about comparing theory with experiment. In superlattices, the situation is similar to that considered in [28] for a single quantum well, in which it was established that theory and experiment should be compared not in terms of the E_{ex} value, but in terms of the photon energy $\hbar\omega_{ex}$, corresponding to the recombination of an electron–hole pair. The latter contains two groups of contributions: (a) renormalizing the band gap widths of the superlattice layers; (b) renormalizing the electron–hole interaction. In the $GaAs/Al_xGa_{1-x}As$ superlattice, the main contribution to the size dependence of $\hbar\omega_{ex}(l)$ is made by the dimensional quantization energies of the electron and hole, the energy of the electron–hole interaction, and the polaron energies, which are included in the optical band gap. The contribution from self-action in this structure does not exceed 0.5 meV. The experimental E_{ex} data shown in figures 6.4(a) and (b) are in good agreement with those calculated on the basis of this theory. This allows us to conclude that the proposed model of the polaron exciton in a composite superlattice with polar composites is correct.

6.3.2 Coulomb interaction and exciton states in a superlattice in the limit of narrow quantum wells

The bulkiness of the results obtained in the previous section for the electron–hole interaction in a superlattice makes it advisable to derive formulas that would make it possible to discuss the problem and predict the analysis of experiments, similar to how it was in section 1.2 for an ultrathin nonpolar film. The initial expressions for such an analysis are (6.2.7).

In the domain of relatively small values of the radius of exciton states ($\rho \gg l_a,\ l_b$), we obtain from formula (1.7)

$$U(\rho) = -\frac{e^2}{4\pi\varepsilon_0 s\rho}\left(\frac{l_a}{\varepsilon_a^{\parallel}} - \frac{l_b}{\varepsilon_b^{\parallel}}\right), \tag{6.3.10a}$$

where

$$s = \left[\epsilon_a^2 l_a^2 + \epsilon_b^2 l_b^2 + \frac{\epsilon_a l_a \epsilon_b l_b\left(\bar{\varepsilon}_a^2 + \bar{\varepsilon}_b^2\right)}{\bar{\varepsilon}_a\bar{\varepsilon}_b}\right]^{\frac{1}{2}}. \tag{6.3.10b}$$

In this domain, the energy and radius of the exciton are equal, respectively:

$$E_{ex,\,n} = -4\left(1 + \lambda\frac{\varepsilon_b^{\parallel}}{\varepsilon_a^{\parallel}}\right)\left(1 + \lambda\frac{\varepsilon_a^{\perp}}{\varepsilon_b^{\perp}}\right)(2n+1)^{-2}E_V;\ n = 0,\ 1,\ \ldots; \tag{6.3.11a}$$

$$a_n = \frac{1}{2}\left(1 + \lambda\frac{\varepsilon_b^{\parallel}}{\varepsilon_a^{\parallel}}\right)^{-\frac{1}{2}}\left(1 + \lambda\frac{\varepsilon_a^{\perp}}{\varepsilon_b^{\perp}}\right)^{\frac{1}{2}}(2n+1)a_V; \tag{6.3.11b}$$

where μ is the reduced mass of the electron and the hole in the XOY plane and $\lambda = l_a/l_b$ is the ratio of the thicknesses of the layers.

The criterion of applicability of the Coulomb law follows from the inequality adopted in its derivation and is reduced to the requirement that the effective period length of the periodic system is small compared to the radius of the exciton state, which can be written as

$$x \gg x_n = 2\sqrt{\epsilon_b}\left(1 + \lambda\frac{\varepsilon_b^{\|}}{\varepsilon_a^{\|}}\right)(2n + 1)^{-1}, \tag{6.3.11c}$$

where $x = a_V / l_b$ is the dimensionless inverse layer thickness b.

In the intermediate range of values ρ

$$\max\{\epsilon_a l_a,\ \epsilon_b l_b\} < \rho < s \tag{6.3.12a}$$

formulas for exciton energy follow from formula (5.13.7):

$$E_{ex,\,n}^{\perp} = -\left[\epsilon_b\left(1 + \lambda\frac{\varepsilon_a^{\perp}}{\varepsilon_b^{\perp}}\right)\right]^{-1} x\left\{\ln\left[\frac{8\sqrt{\epsilon_b}}{x}\left(1 + \lambda\frac{\varepsilon_b^{\|}}{\varepsilon_a^{\|}}\right)\right] - 2C - 2\gamma_n\right\}E_V; \tag{6.3.12b}$$

and for the radius of the ground state

$$a_0 = \left\{\kappa\frac{\sqrt{\epsilon_b}}{2x}\left(1 + \lambda\frac{\varepsilon_a^{\perp}}{\varepsilon_b^{\perp}}\right)\right\}^{\frac{1}{2}} a_V, \tag{6.3.12c}$$

fair, as follows from the inequality (6.3.12a), in the range of parameter values determined by inequalities:

$$\max\{x_n,\ x_b\} < x < x_e, \tag{6.3.12d}$$

where

$$x_e = \frac{\sqrt{\epsilon_b}}{\kappa}\left(1 + \lambda\frac{\varepsilon_b^{\|}}{\varepsilon_a^{\|}}\right);\ \{x_a,\ x_b\} = \frac{2}{\kappa\sqrt{\epsilon_b}}\left(1 + \lambda\frac{\varepsilon_a^{\perp}}{\varepsilon_b^{\perp}}\right)\{\varepsilon_a\lambda^2,\ \varepsilon_b\}. \tag{6.3.12e}$$

In conclusion of this section, we describe some features of excitons in periodic structures and compare them with the corresponding features of excitons in finite-dimensional structures. By absolute value, the energy of the main exciton value in a periodic structure is less than in a symmetric three-layer structure with infinite outer layers of type a ($\lambda \to \infty$). This is explained by the fact that in a three-layer structure, the lines of force coming out of layer b intersect medium a with a lower permittivity value, whereas in a periodic structure they also pass through layers of type b. With a decrease in $\lambda = l_a / l_b$, the role of the b layers increases, and the absolute value of the exciton energy decreases. Note also that as the left margin λ of applicability of the logarithmic law increases, it moves towards the larger ones faster than the right one, and the area of applicability narrows.

A comparison of the results of calculating the exciton binding energies in the GaAs/AlGaAs superlattice for the exact Green function (1.7) and according to formulas (6.3.11a), (6.3.12b) indicates that even in extremely thin layers of $l_b \leqslant 3,0$ nm, the discrepancy is approximately 1.5–2 times (i.e., the quasi-two-dimensional description reflects only a qualitative trend E_{ex} changes with l_b change).

The exciton binding energy in a superlattice containing polar layers should be given taking into account the shielding of the electron–hole interaction by spatially extended phonons [29–31].

6.4 Hydrogen-like impurity states in multilayer systems

Studies of the optical spectra of structures with quantum wells indicate that along with exciton [3, 8], impurity [32] states are observed in them. The theoretical description of impurity states in these structures was carried out in [32] on the basis of a simple hydrogen-like model in which the potential energy of the interaction of an electron with a charged center in a three-layer symmetric system has the form

$$U(\rho,\ z_e, z_i) = -\frac{e^2}{\varepsilon_{\text{eff}}\sqrt{\rho^2 + \left(z_e^2 - z_i^2\right)^2}}, \tag{6.4.1}$$

$\varepsilon_{\text{eff}} = \sqrt{\varepsilon_1\varepsilon_2}$; ε_1, ε_2 is dielectric permittivity of the layers of the structure.
The potential barriers at the boundaries of the quantum well are assumed to be infinite, and the zones are assumed to be parabolic.

The improvement of the model [32] was carried out in several directions. In [33, 34], the finite depth of the quantum well was taken into account and it was shown that the binding energy of a hydrogen-like impurity center depends on the height of the barrier. Thus, for an impurity in the center of the GaAs pit in structure GaAs/AlGaAs, an increase in the height of the barrier from $V_B = 25R_0$ to $V_B = 50R_0$ leads to an increase in the binding energy from $E_i = 2.3R_0$ to $E_i = 2.5R_0$ at the pit width $l = 0.35a_0$, where $R_0 = m_e^* e^4/(2\hbar^2\varepsilon_{\text{eff}}^2)$, $a_0 = \hbar^2\varepsilon_{\text{eff}}/(2m_e^* e^2)$ (i.e., using the approximation of infinite barriers introduces an error in the value E_i around 8% at $l < a_0$ and 1%–3%—at $l > a_0$).

Another source of error is the use of a bulk Coulomb potential instead of the exact potential of the electron–impurity interaction. It was shown in [7] that a similar approximation for the Wannier–Mott exciton in the GaAs/AlGaAs structure gives an error in calculating E_{ex} of about 10%–15%. (In structures with very different dielectric permittivity layers, the error in the calculation will be much greater [14, 28].)

It is of interest to perform calculations based on the exact potential of the electron–impurity interaction. In addition, neither [32] nor other theoretical studies of the impurity center took into account the effect of self-action of charge carriers, which, as will be shown below, affects the position of the impurity peak in the optical spectrum.

In this section, in the model of a quantum well with infinite barriers at the boundaries, the energy of the ground state of a fine impurity center is calculated as a function of its position and the width of the well using the exact potential of the electron–impurity interaction. Consider the impurity center in the middle layer of a three-layer structure with identical outer layers. The Hamiltonian of an impurity electron has the form

$$\hat{H}_i = \hat{H}_1(z_e, z_i) + \hat{H}_2(\rho, z_e, z_i), \tag{6.4.2}$$

where

$$\hat{H}_1(z_e, z_i) = \hat{K}_{||} + U_{SA}(z_e) + U_{SA}(z_i) + U_B(z_e); \tag{6.4.3}$$

$$\hat{H}_2(\rho, z_e, z_i) = \hat{K}_{\perp} + U(\rho, z_e, z_i). \tag{6.4.4}$$

In formulas (6.4.1)–(6.4.4), the notation is adopted: $\hat{K}_{||}$, $\hat{K}_{\perp}$ are the components of the electron kinetic energy operator, which, in disregard of the nonparabolicity effect, have the form

$$\hat{K}_{||} = \frac{\hat{P}_{||}^2}{2m_{e||}^*}; \; \hat{K}_{\perp} = \frac{\hat{P}_{\perp}^2}{2m_{e\perp}^*}; \tag{6.4.5}$$

$U(\rho, z_e, z_i)$ is the energy of interaction of an electron with an impurity center:

$$U(\rho, z_e, z_i) = -\frac{e^2}{4\pi\varepsilon_0\varepsilon_2}\int_0^{\infty} d\eta \, J_0(\eta, \rho)\times$$
$$\times\left\{e^{-\eta|z_e - z_i|} + \frac{2\delta}{e^{2\varsigma_2} - \delta^2}[\delta \cosh \eta(z_e - z_i) + e^{\varsigma_2}\cosh \eta(z_e + z_i)]\right\}; \tag{6.4.6}$$

where z_e, z_i are the coordinates of the electron and the impurity center, respectively, measured from the center of the layer.

In the future, we will talk about the donor, although all the results can be applied to the acceptor if the valence band is parabolic and nondegenerate; $U_{SA}(z_i)$ is the potential energy of self-action describing the interaction of an electron and an impurity center with inertia-free polarization induced by them.

The solution of the Schrodinger equation with potential energy (6.4.5) will be sought by the variational method, choosing a trial ground state function in the form [32]

$$\psi_{1s}(\rho, z_e, z_i) = \begin{cases} C\cos\left(\frac{\pi z_e}{l_2}\right)\exp\{-\lambda^{-1}\sqrt{\rho^2 + (z_e + z_i)^2}\}; \; |z_e| \leqslant \frac{l_2}{2}; \\ 0, \; z_e < -\frac{l_2}{2}; \; z_e > \frac{l_2}{2}; \end{cases} \tag{6.4.7}$$

where λ is variational parameter and C is the normalization constant:

$$C^{-2} = \frac{\pi\lambda^3}{2}\left\{1 + \frac{\cos 2bz_i}{(1 + b^2\lambda^2)^2}\left[\frac{z_i}{\lambda}\frac{b^2\lambda^2}{(1 + b^2\lambda^2)}\sinh\left(\frac{2z_i}{\lambda}\right) + \right.\right.$$
$$\left.\left.([1 + b^2\lambda^2]^{-2} - b^2 l_2\lambda[2(1 + b^2\lambda^2)]^{-1} - 1)\cosh\left(\frac{2z_i}{\lambda}\right)\right]e^{-\frac{l_2}{\lambda}}\right\}. \tag{6.4.8}$$

The wave function (6.4.7) has limits that are exact solutions of the Schrodinger equation with the Hamiltonian (6.4.1) at $l_2 \to 0$ and $l_2 \to \infty$. Averaging the Hamiltonian (6.4.1) on the wave function (6.4.7), we obtain the variational energy of the system:

$$E_{1s}(\lambda, l, z_i) = E_1(\lambda, l_2, z_i) + E_2(\lambda, l_2, z_i), \tag{6.4.9}$$

where

$$E_1(\lambda, l_2, z_i) = \langle \psi_{1s}(\rho, z_e, z_i) | \hat{H}_1(z_e, z_i) | \psi_{1s}(\rho, z_e, z_i) \rangle; \tag{6.4.10}$$

$$E_B \equiv E_2(\lambda, l_2, z_i) = \left\langle \psi_{1s}(\rho, z_e, z_i) \middle| \hat{H}_2(\rho, z_e, z_i) | \psi_{1s}(\rho, z_e, z_i) \right\rangle. \tag{6.4.11}$$

In the case of a thin layer (narrow quantum well), that is, $\eta l_2 \ll 1$ or $l_2/\lambda \ll 1$, $E_1(\lambda, l_2, z_i)$ weakly depends on the parameter λ and determines the band size effects, including the energy of dimensional quantization, as well as the shift of the bottom of the conduction band relative to the energy level of the impurity due to self-action:

$$E_1(l_2, z_i) = \frac{\pi^2 \hbar^2}{2m_{\|}^* l_2^2} + \frac{e^2 \delta}{8\pi\varepsilon_0 \varepsilon_2 l_2} \int_0^\infty \frac{dx}{e^{2x} - \delta^2} \left\{ 2\delta + \frac{\pi^2 e^{-x} \sinh x}{x(x^2 + \pi^2)} + e^x \cosh\left(x \frac{2z_i}{l_2} \right) \right\}. \tag{6.4.12}$$

Value $E_2(\lambda, l_2, z_i)$ determines the internal energy of the electron in the impurity center:

$$\begin{aligned} E_2(\lambda, l_2, z_i) = {} & \frac{\hbar^2}{2m_\perp^* \lambda^2} - \frac{e^2 C^2 \lambda^2}{8\varepsilon_0 \varepsilon_2} \\ & \left\{ 1 + \frac{\cos(2bz_i)}{1 + b^2\lambda^2} - \frac{b^2\lambda^2}{1 + b^2\lambda^2} \exp\left(-\frac{l_2}{\lambda} \right) \cosh\left(\frac{2z_i}{\lambda} \right) \right\} \\ & + \Delta E_2(\lambda, l_2, z_i). \end{aligned} \tag{6.4.13}$$

$$\lim_{l_2 \to \infty, z_i = 0} E_2(l_2, z_i) = -\frac{m_e^* e^4}{32\pi^2 \varepsilon_0^2 \hbar^2 \varepsilon_2^2}; \tag{6.4.14}$$

$$\lim_{l_2 \to \infty, z_i = \pm l_2/2} E_2(l_2, z_i) = -\frac{m_e^* e^4}{32\pi^2 \varepsilon_0^2 \hbar^2 (\varepsilon_1 + \varepsilon_2)^2}; \tag{6.4.15}$$

$$\lim_{l_2 \to 0} E_2(l_2, z_i) = -\frac{m_e^* e^4}{8\pi^2 \varepsilon_0^2 \hbar^2 \varepsilon_1^2}. \tag{6.4.16}$$

In the limit of the thin layer ($\eta l_2 \ll 1$) from (6.4.13) we get

$$E_2(\lambda, l_2, z_i) = \frac{\hbar^2}{2m_\perp^* \lambda^2} - e^2 C^2 \lambda^2 (\varepsilon_0 \varepsilon_2)^{-1} \times \int_0^\infty \mathrm{d}x \left[1 + \frac{x^2 \lambda^2}{l_2^2}\right]^{-\frac{3}{2}}$$
$$\left\{\frac{1}{x} + \frac{x \cos(2bz_i)}{x^2 + 4b^2 l^2} - \frac{4b^2 l_2^2 \cosh\left(\frac{z_i x}{\lambda}\right)}{x\left(x^2 + 4b^2 l^2\right)} \left(e^{-\frac{x}{2}} - \frac{2\delta \sinh\left(\frac{x}{2}\right)}{e^x - \delta}\right)\right\}. \tag{6.4.17}$$

The results of numerical calculation of the energy $E_2(\lambda, l_2, z_i)$ according to the formula (6.4.13) for the GaAs/$Al_xGa_{1-x}As$ ($x = 0.30$) are shown in figure 6.5. Curves 1 and 2 correspond to impurities in the center and at the boundaries of the layer: 1—$E_2(l_2, z_i = 0)$; 2—$E_2(l_2, z_i = \pm l_2/2)$; and curves 3 and 4 are the results of [32] (energy and length units are used here): $E_0 = m_e^* e^4/(32\pi^2\varepsilon_0^2\hbar^2\varepsilon_2^2)$; $a_0 = 4\pi\varepsilon_0\hbar^2/(m_e^* e^2)$.

A comparison of the results indicates that for an impurity located in the center of the layer, the ground state energy at $l_2/a_0 \leqslant 1$ is about $(\varepsilon_2/\varepsilon_1)$ times deeper than obtained in [32]. This is approximately 15% of the ground state energy. With an increase in (l_2/a_0), the results converge and coincide with $l_2/a_0 \geqslant 6$. For an impurity

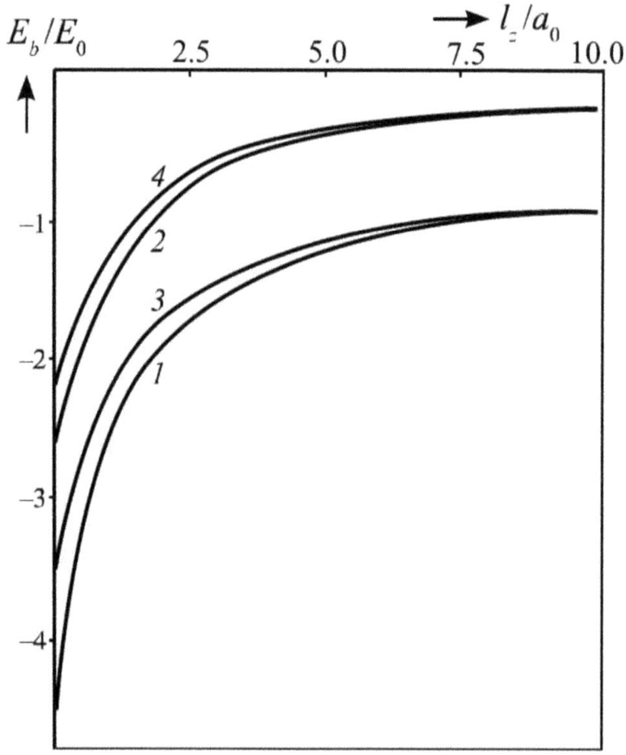

Figure 6.5. The results of numerical calculation of the energy $E_2(\lambda, l_2, z_i)$ according to the formula (6.4.13) for the GaAs/$Al_xGa_{1-x}As$ ($x = 0.30$) are shown in figure 6.5. Curves 1 and 2 correspond to impurities in the center and at the boundaries of the layer: 1—$E_2(l_2, z_i = 0)$; 2—$E_2(l_2, z_i = \pm l_2/2)$; curves 3 and 4 are the results of [32] (energy and length units are used here): $E_0 = m_e^* e^4/(32\pi^2\varepsilon_0^2\hbar^2\varepsilon_2^2)$; $a_0 = 4\pi\varepsilon_0\hbar^2/(m_e^* e^2)$.

located at the boundary at $l_2/a_0 < 1$, the ground state energy is $(\varepsilon_2/\varepsilon_1)$ times deeper, and at $l_2/a_0 \gg 1$, the difference between the results decreases to 1% (at $l_2/a_0 \approx 10$, a rough estimate gives $(\varepsilon_1 + \varepsilon_2)^2/(2\varepsilon_{\text{eff}})^2[1 + (a/l_2)^2]$).

The impurity zone additionally shifts relative to the bottom of the conduction band due to self-action by an amount equal to

$$\Delta E_g(l_2) = \frac{e^2\delta}{4\pi\varepsilon_0\varepsilon_2 l_2}\int_0^\infty \frac{dx}{e^{2x} - \delta^2}\{\delta + e^x\},\ z_i = 0; \tag{6.4.18a}$$

$$\Delta E_g(l_2) = \frac{e^2\delta}{4\pi\varepsilon_0\varepsilon_2 l_2}\int_0^\infty \frac{dx}{e^{2x} - \delta^2}\{\delta + e^x \cosh x\},\ z_i = \pm\frac{l_2}{2}. \tag{6.4.18b}$$

In the structure of $GaAs/Al_xGa_{1-x}As$ $\Delta E_g \sim 3$ meV. With significant differences in the dielectric permittivity of neighboring layers, the contribution from the self-action effect plays a more significant role.

6.5 Effective Hamiltonian in the problem of a polaron exciton at the contact of two crystals

This chapter discusses the states of a large-radius exciton (Wannier–Mott) in a polar crystal. The polarization of the crystal induced by the charges of the electron and the hole affects the state of the exciton and, in turn, depends on this state [35–37]. Thus, there is a problem of matching the movement of charges in the exciton and the polarization of the lattice. A large number of papers have been devoted to this problem, and multiparametric variational approaches have been developed [38, 39], which require a large amount of numerical calculations for their implementation. In the extreme case

$$a_{ex} \leqslant R_{ex} \tag{6.5.1}$$

(Mayer, [40]), when the exciton radius is of the order of and less than the polaron one, the polarizations caused by the dissimilar charges of the electron and the hole, overlapping (due to this inequality), are largely compensated and have a weak effect on the exciton state. In this limit, the exciton–polarization interaction can be taken into account according to the perturbation theory [40]. In the opposite extreme case

$$a_{ex} > R_{ex} \tag{6.5.2}$$

(Haken, [38, 39, 41]), the polarization state of the lattice is close to that due to the independent electron and hole. Therefore, it can be assumed that the Haken exciton is formed from electron and hole polarons, the polarization states of which are deformed by their interaction with each other. Since the electron–phonon interaction constant is $\alpha < 6$, for most ionic crystals, the Lee, Lowe, and Pines method can be used to derive the effective exciton Hamiltonian (i.e., to exclude phonon variables) (section 5.5, clause 5.4.2).

To obtain the initial Hamiltonian, it is necessary to add up the electron and hole polaron Hamiltonians of a general form, connect the potential energy of the electron–hole interaction, and switch to the variables of the center of mass and relative motion. For the same reason as in the polar chapter, we will follow the path of increasing the complexity of the system. Let's start with the simplest system—the contact of a polar crystal with a nonpolar one (10|2)—and having one surface optical mode:

$$\begin{aligned}\hat{H} = &\frac{\hat{P}^2_{\|e}}{2m^*_{e\|}} + \frac{\hat{P}^2_{\|h}}{2m^*_{h\|}} + \frac{\hat{P}^2_{R_\perp}}{2M_\perp} + \frac{\hat{P}^2_{\rho_\perp}}{2\mu_\perp} \\ &+U_0(\rho, z_e, z_h) + U_{SA}(z_e, z_h) \\ &+\sum_Q \hbar\omega_{10}\left[\hat{b}^+_Q\hat{b}_Q + \frac{1}{2}\right] + \sum_\eta \hbar\Omega_s\left[\hat{b}^+_\eta\hat{b}_\eta + \frac{1}{2}\right] \\ &+\sum_Q\Big\{V_{1B}(Q)e^{iq_\perp R_\perp}\Big[g_V(q_\perp, q_z, z_e)e^{i\sigma_1 q_\perp\rho + iq_z z} \\ &-g_V(q_\perp, q_z, z_h)e^{-i\sigma_2\sigma_1 q_\perp\rho + iq_z z}\Big]\hat{b}^+_{-Q}\Big\} \\ &+h.\,c. + \sum_\eta\Big\{V_s(\eta)e^{i\eta R_\perp}\Big[g_s(\eta, z_e)e^{i\sigma_1\eta\rho} - g_s(\eta, z_h)e^{-i\sigma_2\eta\rho}\Big]\hat{b}^+_{-\eta}\Big\} \\ &+h.\,c.\end{aligned} \tag{6.5.3}$$

$$\sigma_1 = \frac{m^*_{\perp h}}{M_\perp};\ \sigma_2 = \frac{m^*_{\perp e}}{M_\perp};\ M_\perp = m^*_{e\perp} + m^*_{h\perp}. \tag{6.5.4}$$

6.5.1 Haken limit

By a transformation similar to (5.4.15), it is possible to switch to the coordinate system associated with the center of mass of the exciton. However, in the future we will take an approximation in which we will consider the exciton to be stationary and located at a point with a radius vector $\mathbf{R}_\perp = 0$. After this simplification, following [39], we choose the exciton wave function in the form

$$\Psi(\rho, z_e, z_h, \{\hat{b}_Q\}, \{\hat{b}_\eta\}) = C\,\Psi(\rho, z_e, z_h)e^{\hat{S}_S}e^{\hat{S}_V}\,|0>, \tag{6.5.5a}$$

with definitions

$$\hat{S}_S = \sum_\eta\Big\{\hat{b}_\eta f^*_{s,\eta}(\rho, z_e, z_h) - \hat{b}^+_{-\eta}f_{s,\eta}(\rho, z_e, z_h)\Big\}; \tag{6.5.5b}$$

$$\hat{S}_V = \sum_Q\Big\{\hat{b}_Q f^*_Q(\rho, z_e, z_h) - \hat{b}^+_{-Q}f_Q(\rho, z_e, z_h)\Big\}. \tag{6.5.5c}$$

The structure of the wave function (6.5.5a) is similar to that used in polaronic problems, with the essential feature that the amplitudes of the displacements of the polarization oscillators of the lattice $f_{s,\eta}$, f_Q are generally assumed to depend not only on the position of the electron and the hole relative to the interface of the layers but also on the relative position of the electron and the hole. In accordance with [38, 39], the Hamiltonian (6.5.3) is averaged over the wave function (6.5.5a). The function obtained after averaging is differentiated by $f_{s,\eta}$, f_Q and $\Psi(\rho, z_e, z_h)$ and derivatives are assumed to be equal to zero:

$$-\frac{\hbar^2}{2M_\perp}\Delta_\rho f_{s,\eta}-\frac{\hbar^2}{2m_{e\|}^*}\Delta_{z_e} f_{s,\eta}-\frac{\hbar^2}{2m_{h\|}^*}\Delta_{z_h} f_{s,\eta}$$

$$+f_{s,\eta}\left\{\hbar\Omega_s+\frac{\hbar^2\eta^2}{2M_\perp}+\frac{1}{2}\frac{\hbar^2}{2M_\perp}\frac{\Delta_\rho(\psi\psi^*)}{|\psi|^2}+\frac{1}{2}\frac{\hbar^2}{2m_{e\|}^*}\frac{\Delta_z(\psi\psi^*)}{|\psi|^2}+\right. \tag{6.5.6a}$$

$$\left.\frac{1}{2}\frac{\hbar^2}{2m_{h\|}^*}\frac{\Delta_{z_h}(\psi\psi^*)}{|\psi|^2}+V_s^*(\eta)(e^{\eta z_e+i\sigma_1\eta\rho}-e^{\eta z_h-i\sigma_2\eta\rho})\right\}=0;$$

$$-\frac{\hbar^2}{2\mu_\perp}\Delta_\rho f_Q-\frac{\hbar^2}{2m_{e\|}}\Delta_{z_e} f_Q-\frac{\hbar^2}{2m_{h\|}}\Delta_{z_h} f_Q$$

$$+f_Q\left\{\hbar\omega_0+\frac{\hbar^2Q^2}{2M_\perp}\right.$$

$$+\frac{1}{2}\frac{\hbar^2}{2\mu_\perp}\frac{\Delta_\rho(\psi\psi^*)}{|\psi|^2}+\frac{1}{2}\frac{\hbar^2}{2m_{e\|}^*}\frac{\Delta_{z_e}(\psi\psi^*)}{|\psi|^2}+\frac{1}{2}\frac{\hbar^2}{2m_{h\|}^*}\frac{\Delta_{z_h}(\psi\psi^*)}{|\psi|^2}+ \tag{6.5.6b}$$

$$V_{1B}^*\left\{\left(1-e^{q_\perp z_e+iq_\perp z_e}\right)e^{i\sigma_1\eta\rho+iq_z z_e}\right.$$

$$\left.\left.-\left(1-e^{q_\perp z_h+iq_z z_h}\right),e^{-i\sigma_2 q_\perp\rho+iq_z z_h}\right\}\right\}=0;$$

$$-\left\{\frac{\hbar^2}{2\mu_\perp}\Delta_\rho\psi+\frac{\hbar^2}{2m_{e\|}}\Delta_{z_e}\psi+\frac{\hbar^2}{2m_{h\|}}\Delta_{z_h}\psi\right\}$$

$$+\left[U(\rho,z_e,z_h)+U_{SA}(z_e)+U_{SA}(z_h)\right]\psi$$

$$-\frac{1}{2}\frac{\hbar^2}{2\mu_\perp}\sum_\eta\left(f_{s,\eta}\Delta_\rho f_{s,\eta}^*+f_{s,\eta}^*\Delta_\rho f_{s,\eta}-\Delta_\rho\left|f_{s,\eta}\right|^2\right)$$

$$-\frac{1}{2}\sum_c\frac{\hbar^2}{2m_{c\|}}\left(f_{s,\eta}\Delta_{z_c} f_{s,\eta}^*+f_{s,\eta}^*\Delta_{z_c} f_{s,\eta}-\Delta_\rho\left|f_{s,\eta}\right|^2\right) \tag{6.5.6c}$$

$$+\sum_\eta\left|f_{s,\eta}\right|^2\left(\hbar\Omega_s+\frac{\hbar^2\eta^2}{2M_\perp}\right)$$

$$+\sum_\eta\left\{V_s(\eta)\left[e^{\eta z_e+i\sigma_1\eta\rho}-e^{\eta z_h-i\sigma_2\eta\rho}\right]f_{s,\eta}+h.\,c.\right\}$$

$$+Idem\{S\to V;\,\eta\to Q\}=E\Psi.$$

The system of equations (6.5.6a) and (6.5.6b) describes the self-consistent motion of the electron and the hole and the polarization of the lattice. To determine the polarization states and calculate the potential energy of charges in the polarization field, it is necessary to solve the equations (6.5.6a and (6.5.6b) for $f_{s,\,\eta}$ and f_Q, which take into account the exciton states. In general, this is a task of high complexity that cannot be solved analytically. However, the problem can be significantly simplified if we take advantage of the inequality (6.5.2). Expand $|\psi|^2$ into a Fourier series:

$$|\psi|^2 = \sum_{\mathbf{k}_\perp, k_{e\|}, k_{h\|}} C(\mathbf{k}_\perp, k_{e\|}, k_{h\|}) \exp\{i\mathbf{k}_\perp \boldsymbol{\rho} + ik_{e\|}z_e + ik_{h\|}z_h\}, \tag{6.5.7}$$

where $\mathbf{k}_\perp$ is the wave vector of the relative motion of the electron and the hole in the exciton, and $k_{e\|,h\|}$ are the components of the wave vectors of the longitudinal motion of the electron and the hole. Let's estimate the contribution from ψ in (6.5.6a and (6.5.6b):

Maximum value of $C(k_\perp, k_{e\|}, k_{h\|})$ falls on the value $k_\perp \sim a_{ex\perp}^{-1}$, that's why

$$\Delta_\rho(\psi\psi^*)/|\psi|^2 \sim k_\perp^2 \sim a_{ex\perp}^{-2}; \tag{6.5.8}$$

similarly

$$\Delta_{z_c}(\psi\psi^*)/|\psi|^2 \sim a_{\|e,\,h\,ex}^{-2}. \tag{6.5.9}$$

In this approximation, the expression in curly brackets for $f_{s,\,\eta}$, f_Q equations (6.5.6a) and (6.5.6b) of the order

$$1 + R_\perp^2 Q^2 + \frac{1}{2}R_\perp^2/a_{ex\perp}^2 + \frac{1}{2}R_{e\|}^2/a_{e\|}^2 + \frac{1}{2}R_{h\|}^2/a_{h\|}^2.$$

Criterion (6.5.2) allows us to limit ourselves to only the first two terms. In [39], an explicit form of the wave function $\psi \sim \exp(-\beta r)$ was used to justify this simplification. The criterion of the Haken approximation was discussed in [42]. Since, according to the virial theorem, there is a proportional relationship between the kinetic energy of a system and its total energy with a proportionality coefficient determined by an explicit type of potential, criterion (6.5.2) means that of the three energies:, phonon ($\hbar\omega_{1V}$, $\hbar\,\Omega_s$), polaron $\hbar^2Q^2/(2M_\perp) \sim |E_{polar.}|$, and exciton $P_\perp^2/(2\mu_\perp) \sim |E_{ex}|$, the exciton energy is considered to be the lowest. The inequality $a_{\|ex} > R_{e,\,h}$ for longitudinal motion means that the attenuation depth of the exciton wave function deep into the crystal is significantly greater than the polaron radii. After these simplifications, the equations (6.5.6a, b) are easily solved:

$$f_{s,\eta} = -\frac{V_s^*(\eta)e^{\eta z_e + i\sigma_1\eta\rho}}{\hbar\Omega_s + (1-\gamma_e)\hbar^2\eta^2/(2m_{e\perp})} + \frac{V_s^*(\eta)e^{\eta z_h + i\sigma_2\eta\rho}}{\hbar\Omega_s + (1-\gamma_h)\hbar^2\eta^2/(2m_\perp)}; \tag{6.5.10a}$$

$$f_Q = \frac{V_{1B}^*\sin(q_z z_e)e^{i\sigma_1\mathbf{q}_\perp\boldsymbol{\rho}}}{\hbar\omega_{10} + \frac{\hbar^2 q_\perp^2}{2m_{e\perp}^*} + \frac{\hbar^2 q_z^2}{2m_{e\|}^*}} - \frac{V_{1B}^*\sin(q_z z_h)e^{-i\sigma_2\mathbf{q}_\perp\boldsymbol{\rho}}}{\hbar\omega_{10} + \frac{\hbar^2 q_\perp^2}{2m_{h\perp}^*} + \frac{\hbar^2 q_z^2}{2m_{h\|}^*}}. \tag{6.5.10b}$$

In equations (6.5.6a, b), the interaction attenuation factor $g_V = 1 - \exp(q_\perp z - iq_z z)$ to simplify integration, we replace it with the expression $\sin(q_z z_e)\exp(-iq_z z_{e,h})$ (in this form it was obtained in [43]).

Substituting (6.5.10a, b) into (6.5.6a, b), we obtain the effective potential energy of the interaction of the exciton with the polarization of the lattice:

$$W_{\mathrm{eff}}(\rho, z_e, z_h) = W_S^{(1)} + W_S^{(2)} + W_V^{(1)} + W_V^{(2)}, \tag{6.5.11a}$$

where

$$W_S^{(1)} = \frac{1}{2}\sum_{\boldsymbol{\eta}}\left(f_{S,\eta}\widetilde{V}_S(\eta) + f_{S,\eta}^*\widetilde{V}_S^*(\eta)\right)$$
$$= \sum_{\boldsymbol{\eta},c=e,h}\frac{|V_S(\eta)|^2 e^{2\eta z_c}}{\hbar\Omega_S + \hbar^2\eta^2/(2m_{c\perp})} + \sum_{\boldsymbol{\eta},c=e,h}|V_S(\eta)|^2\cos(\boldsymbol{\eta\rho})\left[\hbar\Omega_S + \frac{\hbar^2\eta^2}{2m_{e\perp}}\right]^{-1}; \tag{6.5.11b}$$

$$W_S^{(2)} = \frac{1}{2}\frac{\hbar^2}{2\mu_\perp}\sum_{\boldsymbol{\eta}}\Delta_\rho\left|f_{S,\eta}\right|^2 + \frac{1}{2}\frac{\hbar^2}{2}\sum_{\boldsymbol{\eta},c=e,h}\frac{1}{m_{c\|}}\Delta_{z_c}\left|f_S(\eta)\right|^2$$
$$= \sum_{\boldsymbol{\eta},c=e,h}\frac{(\hbar^2\eta^2/m_{c\|})|V_S(\eta)|^2 e^{-2\eta z_c}}{\hbar\Omega_S + (1-\gamma_c)\hbar^2\eta^2/(2m_{c\perp})} \tag{6.5.11c}$$
$$+\sum_{\boldsymbol{\eta}}\frac{|V_S(\eta)|^2\cos(\boldsymbol{\eta\rho})\left(\mu_\perp^{-1} - m_{e\|}^{-1} - m_{h\|}^{-1}\right)\hbar^2\eta^2/2}{(\hbar\Omega_s + (1-\gamma_e)\hbar^2\eta^2/(2m_{e\perp}))(\hbar\Omega_S + (1-\gamma_h)\hbar^2\eta^2/(2m_{h\perp}))};$$

$$W_V^{(1)} = \frac{1}{2}\sum_{\mathbf{Q}}\left(f_Q\widetilde{V}_{1B}(Q) + f_Q^*\widetilde{V}_{1B}^*\right)$$
$$= -\sum_{\mathbf{Q},c=e,h}\frac{|V_{1B}|^2\sin^2(q_z z_c)}{\hbar\omega_{10} + \frac{\hbar^2 q_\perp^2}{2m_{c\perp}} + \frac{\hbar^2 q_z^2}{2m_{c\|}}} + \sum_{\mathbf{Q},c=e,h}\frac{|V_{1B}|^2\sin(q_z z_e)\sin(q_z z_h)\cos(\mathbf{q}_\perp\boldsymbol{\rho})}{\hbar\omega_{10} + \frac{\hbar^2 q_\perp^2}{2m_{c\perp}} + \frac{\hbar^2 q_z^2}{2m_{c\|}}}; \tag{6.5.11d}$$

$$W_V^{(2)} = \frac{1}{2}\frac{\hbar^2}{2\mu_\perp}\sum_{\mathbf{Q}}\Delta_\rho\left|f_Q\right|^2 + \frac{1}{2}\frac{\hbar^2}{2}\sum_{\mathbf{Q},c}\frac{1}{m_{c\|}}\Delta_{z_c}\left|f_Q\right|^2$$
$$= \sum_{\mathbf{Q}}\frac{|V_{1B}|^2\sin(q_z z_e)\cos(\mathbf{q}_\perp\boldsymbol{\rho})\left[\frac{\hbar^2 q_\perp^2}{2\mu_\perp} + \frac{\hbar^2 q_z^2}{2\mu_\perp}\left(m_{e\|}^{-1} + m_{h\|}^{-1}\right)\right]}{\left[\hbar\omega_{10} + \frac{\hbar^2}{2}\left(q_\perp^2 m_{e\perp}^{-1} + q_z^2 m_{e\|}^{-1}\right)\right]\left[\hbar\omega_{10} + \frac{\hbar^2}{2}\left(q_\perp^2 m_{h\perp}^{-1} + q_z^2 m_{e\perp}^{-1}\right)\right]} \tag{6.5.11e}$$
$$+\sum_{\mathbf{Q},c=e,h}|V_{1B}|^2\cos(2q_z z_e)\frac{\hbar^2 q_z^2}{2m_{c\|}}\left[\hbar\omega_{10} + \frac{\hbar^2}{2}\left(q_\perp^2 m_{c\perp}^{-1} + q_z^2 m_{c\|}^{-1}\right)\right]^{-2};$$

$$\widetilde{V}_S(\boldsymbol{\eta}) = V_S(\eta)[\exp(\eta z_e - i\sigma_1\boldsymbol{\eta\rho}) - \exp(\eta z_h - i\sigma_2\boldsymbol{\eta\rho})]; \tag{6.5.12a}$$

$$\widetilde{V}_{1B}(Q) = V_{1B}(Q)[\sin(q_z z_e)e^{-i\sigma_1\mathbf{q}_\perp\boldsymbol{\rho}} - \sin(q_z z_h)e^{-i\sigma_2\mathbf{q}_\perp\boldsymbol{\rho}}]. \tag{6.5.12b}$$

The contributions $W_S^{(2)}$ and $W_V^{(2)}$ at values of the m_e/m_h ratio within the order are corrective in nature. For example, in the area of large z_c, the calculation using the formulas (6.5.11c) and (6.5.11d) gives $W^{(2)} \sim z_c^{-3}$.

In another limit, when $m_h \to \infty$ (impurity center), the sum

$$W_i(\rho, z_e, z_h) = U(\rho, z_e, z_h) + W_S^{(1)} + W_S^{(2)} + W_V^{(1)} + W_V^{(2)} \tag{6.5.13a}$$

gives a well-known formula for the potential energy of an impurity center located at a point with the z_h coordinate and shielded by a static dielectric constant ε_{10}:

$$W_i = -\frac{e^2}{4\pi\varepsilon_0\varepsilon_{10}\sqrt{\rho^2 + (z_e - z_i)^2}} - \frac{e^2(\varepsilon_{10} - \varepsilon_2)}{4\pi\varepsilon_0\varepsilon_{10}(\varepsilon_{10} + \varepsilon_2)\sqrt{\rho^2 + (z_e + z_i)^2}} + \frac{e^2(\varepsilon_{10} - \varepsilon_2)\left(z_e^{-1} + z_i^{-1}\right)}{4\pi\varepsilon_0\varepsilon_{10} \cdot 4(\varepsilon_{10} + \varepsilon_2)}. \tag{6.5.13b}$$

The given formulas (6.5.6a), (6.5.6b), (6.5.10a), (6.5.10b), (6.5.11a), and (6.5.13a) solve the problem of deriving the effective Hamiltonian for the exciton at the interface. In their preparation, the ideology of a 'deep' potential well at the surface was used, forming the exciton state to a greater extent than the electron–hole interaction: $\underline{z}_e, \underline{z}_h < a_{ex}$. The limiting situation is the 'flat exciton': $\bar{z}_e, \bar{z}_h \to 0$, considered in detail in [27]. In particular, for the crystal–vacuum contact for $W_{\text{eff}}(\rho)$ the expression was obtained in [27]

$$W_{\text{eff}}(\rho) = W_0(\rho) + W_S^{(1)}(\rho) + W^{(2)}(\rho), \tag{6.5.14}$$

where $W_0(\rho)$ is the seed potential of the electron–hole interaction, taking into account the shielding by rapid polarization and having the form

$$W_0(\rho) = -\frac{2e^2}{(\varepsilon_1 + 1)\rho}; \tag{6.5.15a}$$

$W_S^{(1)}(\rho)$, $W^{(2)}(\rho)$ are phonon contributions to the shielding of the electron–hole interaction, having the form

$$\begin{aligned} W_S^{(1)}(\rho) &= \pi\alpha_{Se}\hbar\Omega_S \frac{m_h^*}{\left(m_e^* - m_h^*\right)} \\ &\left\{\left(\frac{m_e^*}{m_h^*}\right)^{\frac{1}{2}}\left[I_0\left(\frac{\rho}{R_h}\right) - L_0\left(\frac{\rho}{R_h}\right)\right] - \left[I_0\left(\frac{\rho}{R_e}\right) - L_0\left(\frac{\rho}{R_e}\right)\right]\right\} \\ &- \frac{\pi}{2}(\alpha_{Se} + \alpha_{Sh})\hbar\Omega_S, \end{aligned} \tag{6.5.15b}$$

where

$$\alpha_{Se} = \frac{e^2(\varepsilon_{10} - \varepsilon_1)}{4\pi\varepsilon_0(\varepsilon_{10} + 1)(\varepsilon_1 + 1)\hbar\Omega_S R_S}; \tag{6.5.15c}$$

$$\begin{aligned} W_S^{(2)}(\rho) = \frac{\pi}{4}\alpha_{Se}\hbar\Omega_S\left(\frac{m_e^* + m_h^*}{m_h^* - m_e^*}\right)&\left\{\left(\frac{m_h^*}{m_e^*}\right)^{\frac{1}{2}}\left[I_0\left(\frac{\rho}{R_h}\right) - L_0\left(\frac{\rho}{R_h}\right)\right]\right. \\ &\left.-\left[I_0\left(\frac{\rho}{R_e}\right) - L_0\left(\frac{\rho}{R_e}\right)\right]\right\}. \end{aligned} \tag{6.5.15d}$$

Here $I_0(x)$, $L_0(x)$ are modified Bessel and Struve functions, respectively.

In the limit $\rho \gg R_{e,\,h}$ from (6.5.14) and (6.5.15a), (6.5.15b), (6.5.15c), (6.5.15d) follows:

$$W_{\rm eff}(\rho) = -\frac{\pi}{2}(\alpha_{se} + \alpha_{sh})\hbar\Omega_s - \frac{2e^2}{(\varepsilon_{10} + 1)\rho}. \tag{6.5.16a}$$

In the limit $\rho \ll R_{e,\,h}$ from (6.5.14), (6.5.15a), (6.5.15b), (6.5.15c) follows:

$$W_{\rm eff}(\rho) = -\frac{\pi}{2}(\alpha_{Se} + \alpha_{Sh})\hbar\Omega_S - \frac{2e^2}{(\varepsilon_1 + 1)\rho}. \tag{6.5.16b}$$

In the opposite case, an exciton quasi-particle moving along a potential well should be considered. This problem was solved in [44] and using a simplified form of electron–hole interaction in [45].

For the considered case of contact of a polar crystal with a vacuum in [10], it was obtained:

$$\begin{aligned}
W_{\rm eff}(\rho, z_e, z_h) = {}& \frac{e^2(\varepsilon_1 - 1)}{16\pi\varepsilon_0\varepsilon_1(\varepsilon_1 + 1)z_e} + \frac{e^2(\varepsilon_1 - 1)}{16\pi\varepsilon_0\varepsilon_1(\varepsilon_1 + 1)z_h} - \\
& \frac{e^2}{4\pi\varepsilon_0\varepsilon_1\sqrt{\rho^2 + (z_e - z_h)^2}} - \frac{e^2(\varepsilon_1 - 1)}{4\pi\varepsilon_0\varepsilon_1(\varepsilon_1 + 1)\sqrt{\rho^2 + (z_e + z_h)^2}} - \\
& \alpha_s\hbar\Omega_s\left\{F_1\left(\frac{2z_e}{R_e}\right) + \left(\frac{R_e}{R_h}\right)F_1\left(\frac{2z_h}{R_h}\right) - 2R_eF_2(\rho, z_e, z_h)\right\} - \\
& (\alpha_{Ve} + \alpha_{Vh})\hbar\omega_{10} + \frac{e^2(\varepsilon_{10} - \varepsilon_1)}{16\pi\varepsilon_0\varepsilon_{10}\varepsilon_1}\left\{\frac{1}{z_e}\left[1 - \exp\left(-\frac{2z_e}{R_e}\right)\right]\right. \\
& \left. + \frac{1}{z_h}\left[1 - \exp\left(-\frac{2z_h}{R_h}\right)\right]\right\} - \alpha_V\hbar\omega_{10}R_{Ve} \\
& \left\{\frac{2 - \exp\left(-R_e^{-1}\sqrt{\rho^2 + (z_e - z_h)^2}\right)}{\sqrt{\rho^2 + (z_e - z_h)^2}}\right. \\
& \left. -\exp\left(-R_h^{-1}\sqrt{\rho^2 + (z_e - z_h)^2}\right)\right\} + \frac{1}{\sqrt{\rho^2 + (z_e + z_h)^2}} \\
& \left\{2 - \exp\left(-R_e^{-1}\sqrt{\rho^2 + (z_e + z_h)^2}\right)\right. \\
& \left. - \exp\left(-R_h^{-1}\sqrt{\rho^2 + (z_e + z_h)^2}\right)\right\},
\end{aligned} \tag{6.5.17}$$

where

$$F_1(x) = ci(x)\sin(x) - si(x)\cos(x); \tag{6.5.18a}$$

$$F_2(\rho, z_e, z_h) = \int_0^\infty dy\, e^{-y(z_e + z_h)} J_0(y\rho)\left(\frac{1}{1 + R_e^2 y^2} - \frac{1}{1 + R_h^2 y^2}\right). \quad (6.5.18b)$$

As in the planar case, the expression (6.5.17) follows the well-known classical limits in the cases of $\rho, z_e, z_h \gg R_{e,h}$ и $\rho, z_e, z_h < R_{e,h}$, which have the structure of the seed Coulomb potential.

6.5.2 The Mayer limit

In the Mayer approach for a massive crystal, the problem was investigated in [40, 46, 47]. As mentioned earlier, as a first step in these works, an exciton is constructed, and then the interaction with the polarization surrounding it is taken into account, that is, the hydrogen-like motion in the exciton is separated, which is considered as a quasi-particle interacting with the polarization vibrations of the lattice during its translational motion. Since the exciton as a whole is neutral, it interacts with lattice vibrations more weakly than in the limiting case of Haken. Let's move on to the consideration of this limit. From the Hamiltonian (6.5.3) by a unitary transformation

$$\hat{S}_1 = \exp\left\{-i\mathbf{R}_\perp\left[\sum_{\boldsymbol{\eta}} \boldsymbol{\eta}\hat{b}_{\boldsymbol{\eta}}^+\hat{b}_{\boldsymbol{\eta}} + \sum_{\mathbf{Q}} \mathbf{q}_\perp \hat{b}_{\mathbf{Q}}^+\hat{b}_{\mathbf{Q}}\right]\right\} \quad (6.5.19)$$

we remove the coordinate of the exciton's center of mass $\mathbf{R}_\perp$ in the contact plane, after which we perform the second unitary transformation over the Hamiltonian:

$$\hat{S}_2 = \exp\left\{-\sum_{\boldsymbol{\eta}}\left[f_\eta^*\hat{b}_{\boldsymbol{\eta}} - f_\eta \hat{b}_{-\boldsymbol{\eta}}^+\right] - \sum_{\mathbf{Q}}\left[f_Q^*\hat{b}_Q - f_Q\hat{b}_{-\mathbf{Q}}^+\right]\right\}, \quad (6.5.20)$$

averaging the Hamiltonian beforehand $\hat{H}_1 = S_1^{-1}\hat{H}\hat{S}_1$ on the wave function 1s-state $\varphi(\rho, z_e, z_h)$. Then, applying the variational principle, we find the amplitudes of the displacements f_η and f_Q:

$$f_\eta = \frac{V^*(\eta)\rho(\eta)}{\hbar\Omega_S + \frac{\hbar^2\eta^2}{2M_\perp}}; \quad (6.5.21a)$$

$$f_Q = \frac{C_Q^*\rho(Q)}{\hbar\omega_{10} + \hbar^2Q^2/(2M_\perp)}, \quad (6.5.21b)$$

where

$$\rho(\eta) = \langle\varphi(\rho, z_e, z_h)|e^{-\eta z_e + is_1\boldsymbol{\eta}\boldsymbol{\rho}} - e^{-\eta z_h + is_2\boldsymbol{\eta}\boldsymbol{\rho}}|\varphi(\rho, z_e, z_h)\rangle; \quad (6.5.22a)$$

$$\rho(Q) = \left\langle\varphi(\rho, z_e, z_h)\right|\sin(q_z z_e)e^{is_1\mathbf{q}_\perp\boldsymbol{\rho}} - \sin(q_z z_h)e^{-is_2\mathbf{q}_\perp\boldsymbol{\rho}}|\varphi(\rho, z_e, z_h)\rangle. \quad (6.5.22b)$$

As in the Haken limit, an effective Hamiltonian can also be obtained here, the averaging of which is the energy of the ground state of the exciton:

$$E_{ex}^0 = \langle \varphi(\rho, z_e, z_h) | \hat{H}_{\text{eff}} | \varphi(\rho, z_e, z_h) \rangle, \tag{6.5.23}$$

where $\hat{H}_{\text{eff}}$ is the effective Hamiltonian of the electron–hole–phonon system:

$$\begin{aligned} \hat{H}_{\text{eff}} = {} & \frac{\hat{P}_\rho^2}{2M_\perp} + \frac{\hat{P}_{z_e}^2}{2m_{e\parallel}^*} + \frac{\hat{P}_{z_h}^2}{2m_{h\parallel}^*} + U(\rho, z_e, z_h) + U_{SA}(z_e) + U_{SA}(z_h) \\ & - \sum_{\boldsymbol{\eta}} |V_\eta|^2 \rho^*(\eta)[e^{-\eta z_e + is_1\boldsymbol{\eta\rho}} - e^{-\eta z_h - is_2\boldsymbol{\eta\rho}}]\left(\hbar\Omega_S + \frac{\hbar^2\eta^2}{2M_\perp}\right)^{-1} \\ & - \sum_{\mathbf{Q}} |V_Q|^2 \rho^*(\eta)[\sin(q_z z_e)e^{is_1\mathbf{q}_\perp\boldsymbol{\rho}} - \sin(q_z z_h)e^{-is_2\mathbf{q}_\perp\boldsymbol{\rho}}]\left(\hbar\omega_0 + \frac{\hbar^2Q^2}{2M_\perp}\right)^{-1}. \end{aligned} \tag{6.5.24}$$

In the special case of a 'flat' surface exciton, the effective Hamiltonian of the electron–hole–phonon interaction has the form

$$\begin{aligned} \hat{H}_{\text{eff}} = {} & \frac{\hat{P}_\rho^2}{2\mu_\perp} + U_0(\rho) - \sum_{\eta} \frac{|V_\eta|^2 \rho^*(\eta)\left(e^{is_1\eta\rho} - e^{-is_2\eta\rho}\right)}{\hbar\Omega_S + \frac{\hbar^2\eta^2}{2M_\perp}} \\ & - \frac{\pi}{2}(\alpha_{Se} + \alpha_{Sh})\hbar\Omega_S, \end{aligned} \tag{6.5.25}$$

where

$$\rho^*(\eta) = \int |\varphi(\rho)|^2 (e^{-is_1\boldsymbol{\eta\rho}} - e^{is_2\boldsymbol{\eta\rho}}) d\boldsymbol{\rho}. \tag{6.5.26a}$$

If as a trial function $1s$-states take a hydrogen-like function

$$\varphi(\rho) = \left(\frac{2}{\pi}\right)^{\frac{1}{2}} \lambda e^{-\lambda\rho},$$

λ is the variational parameter, then for $\rho^*(\eta)$ we obtain

$$\rho^*(\eta) = \left[1 + \left(\frac{s_1\eta}{2\lambda}\right)^2\right]^{-\frac{3}{2}} - \left[1 + \left(\frac{s_2\eta}{2\lambda}\right)^2\right]^{-\frac{3}{2}}. \quad (6.4.26b) \tag{6.5.26b}$$

Substituting (6.5.26b) into (6.5.25) and performing integration, we obtain:

$$\hat{H}_{\text{eff}} = \frac{\hat{P}_\rho^2}{2\mu_\perp} + U_0(\rho) - \alpha_S \hbar\Omega_S R_S F(\rho, R_S); \tag{6.5.27a}$$

here

$$F(\rho, R_S) = \int_0^{\infty} d\eta \, [J_0(s_1\eta\rho) - J_0(s_2\eta\rho)]\left(1 + R_S^2\eta^2\right)^{-1} \times$$
$$\times \left\{\left[1 + \left(\frac{s_1\eta}{2\lambda}\right)^2\right]^{-\frac{3}{2}} - \left[1 + \left(\frac{s_2\eta}{2\lambda}\right)^2\right]^{-\frac{3}{2}}\right\}. \quad (6.5.27b)$$

6.5.3 Effective Hamiltonian of the electron–hole interaction at an arbitrary ratio $a_{ex}/R_{e,h}$

For the first time, an attempt was made to develop a method suitable for calculating the energy of a polaron exciton at an arbitrary ratio $a_{ex}/R_{e,h}$ in [48], in which a variational method was constructed leading to an exact limit of the energy of the ground state of the exciton. The authors [48] obtained satisfactory agreement with the experiment for a number of polar crystals. Various aspects of the polaron exciton problem have been discussed in a number of other papers [49–52], in which the results of the Haken and Mayer theories are improved. Noteworthy is the work [53], the authors of which deduced the effective interaction potential of a system of electrons and holes, taking into account polarization oscillations, suitable for calculating not only the ground state, but also excited states with an arbitrary ratio $a_{ex}/R_{e,h}$.

Using the concept of work [53], it is possible to derive the effective Hamiltonian of the surface polaron exciton at an arbitrary ratio $a_{ex}/R_{e,h}$. Let's average the Hamiltonian (7.2.3) on the variational function (7.2.5), in which the amplitudes $f_s(\eta)$ and $f_V(Q)$ (formula (7.2.10)) contain additional variational parameters $\alpha_1, \alpha_2, \beta_1, \beta_2$:

$$f_S(\eta) = -\frac{\alpha_1\beta_1 V_S^*(\eta)e^{-\eta z_e - is_2\eta\rho}}{\hbar\Omega_S + \beta_1^2\hbar^2\eta^2/\left(2m_{h\perp}^*\right)} + \frac{\alpha_2\beta_2 V_S^*(\eta)e^{-\eta z_h + is_1\eta\rho}}{\hbar\Omega_S + \beta_2^2\hbar^2\eta^2/\left(2m_{e\perp}^*\right)}; \quad (7.1.28a)$$

$$f_V(Q) = -\frac{\alpha_1\beta_1 V^*(Q)\sin(q_z z_e)e^{-is_2 q_\perp \rho}}{\hbar\omega_0 + \beta_1^2\hbar^2Q^2/\left(2m_{h\perp}^*\right)} + \frac{\alpha_2\beta_2 V^*(Q)\sin(q_z z_h)e^{is_1\eta\rho}}{\hbar\omega_0 + \beta_2^2\hbar^2Q^2/\left(2m_{e\perp}^*\right)}. \quad (7.1.28b)$$

These parameters are determined by minimizing the variational energy of the system, the structure of which is the same as in the Haken approach.

6.6 Binding energy and effective mass of a polaron exciton at the contact of two crystals

The effective Hamiltonians of the electron–hole–phonon system, derived in sections 6.3.2 and 6.3.3, are used in this section for variational calculation of the binding energy of the surface exciton.

The binding energy ΔE_{ex}^{S} of a surface exciton, by definition, is equal to the difference between the ground state energy of the surface exciton ΔE_{ex}^{0S} and the intrinsic energies of the electron and hole polarons at the contact:

$$\Delta E_{ex}^{S} = \Delta E_{ex}^{0S} - (E_{pe}^{S} + E_{\text{ph}}^{S}), \tag{6.6.1}$$

where E_{ex}^{0S}. It is found from minimizing the variational energy of the surface exciton:

$$E_{ex}^{0S}(\lambda, \beta_e, \beta_h) = \langle \varphi(\rho, z_e, z_h) | \hat{H}_{\text{eff}} | \varphi(\rho, z_e, z_h) \rangle; \tag{6.6.2}$$

E_{ex}^{S} is the binding energy of the electron (e) and hole (h) surface polarons, including their self-action energies (λ, β_e, β_h are variational parameters).

$$\varphi(\rho, z_e, z_h) = \varphi(\rho)\varphi(z_e)\varphi(z_h), \tag{6.6.3a}$$

where

$$\varphi(\rho) = (2\pi^{-1})^{\frac{1}{2}} \lambda e^{-\lambda\rho}; \tag{6.6.3b}$$

$$\varphi(z_i) = 2\beta_i^{\frac{3}{2}} z_i e^{-\beta_i z_i}; \; i = e, \; h, \tag{6.6.4}$$

which satisfies the boundary condition

$$\varphi(\rho, z_e, z_h)\Big|_{z_e, z_h = 0} = 0, \tag{6.6.5}$$

the corresponding assumption of an infinite potential barrier on the contact.

A more complex trial function is a function of the form

$$\varphi(\rho, z_e, z_h) = \varphi(r)\varphi(z_e)\varphi(z_h), \tag{6.6.6a}$$

where

$$\varphi(r) = C \cdot \exp\left(-\lambda\sqrt{\rho^2 + (z_e - z_h)^2}\right). \tag{6.6.6b}$$

In some special cases, however, the binding energy of the surface exciton can be found analytically, as, for example, in the case of a 'flat' surface exciton in the limiting cases determined by the inequalities $\rho \ll R_{e,h}$ and $\rho \gg R_{e,h}$. The expression for the potential energy of the electron–hole interaction has the form (3.15a) and (3.16a), respectively. The Schrodinger equation with a two-dimensional Coulomb potential of the form

$$W(\rho) = -\frac{e^2}{\varepsilon^* \rho} \tag{6.6.7}$$

It has an analytical solution [54, 55]:

$$E_{n,m}^{ex} = -\frac{\mu_{\perp} e^4}{2\hbar^2 \varepsilon^{*2}} \cdot \frac{1}{\left(n + |m| + \frac{1}{2}\right)^2}; \tag{6.6.8}$$

$$\Psi_{n,\,m}(\rho,\,\varphi) = C_{nm}\left(\frac{4\rho}{a_0(2n+2\,|m|+1)}\right)^{|m|} \times \exp\left(im\varphi - \frac{2\rho}{(2n+2\,|m|+1)}\right)L_{|2m|+n}^{|2m|}\left(\frac{4\rho}{a_0(2n+2\,|m|+1)}\right), \tag{6.6.9}$$

where $n = 0,\ 1,\ 2,\ \ldots; m = 0,\ \pm1,\ \pm2,\ \ldots(|m| < n)$—the main and magnetic quantum numbers and $a_0 = \hbar^2\varepsilon^*/(\mu_\perp e^2)$, $L_p^v(x)$ are attached Laguerre polynomials.

Thus, the binding energy of the polaron surface exciton in the limit of $\rho \gg R_{e,\,h}$ is equal to

$$\Delta E_{ex}^S = -\frac{2\mu_\perp e^4}{\hbar^2\left(\varepsilon_{10}^*\right)^2},\ \varepsilon_{10}^* = \frac{\varepsilon_{10}+1}{2}. \tag{6.6.10}$$

Comparing it with the binding energy of an exciton in a massive crystal in this limit ($\Delta E_{ex}^V = -\mu_\perp e^4/(2\hbar^2\varepsilon_{10}^2)$), it can be seen that ΔE_{ex}^S exceeds $16(\varepsilon_{10}/(\varepsilon_{10}+1))^2$ times, that is, by more than an order of magnitude, which is due to two reasons: (a) the geometric effect; (b) a change like the shielding of the electron–hole interaction at the contact of two media.

In the opposite extreme case, when $\rho < R_{e,\,h}$, there is a further deepening of the energy levels of the surface exciton due to the weakening of the electron–hole shielding by inertial polarization (optical phonons):

$$\Delta E_{ex}^S = -\frac{2\mu e^4}{\hbar^2\varepsilon_1^{*2}} + \frac{\pi\hbar\Omega_s}{m_e^* - m_h^*}(\alpha_{Sh} - \alpha_{Se}). \tag{6.6.11}$$

In the general case of a 'flat' surface exciton, when the electron–hole interaction has the form (3.14)–(3.15c), the variational energy does not allow analytical minimization and is found by numerical integration.

We have calculated the binding energy of the surface exciton at the GaAs–vacuum contact when the attenuation depth of the electron and hole wave functions along the z-axis is much less than the exciton radius. Then the flatness criterion is well fulfilled, and the expressions (3.14)–(3.15b)—in the Haken version and (3.27a and (3.27b)—in the case of the Mayer approach can be used for the binding energy of the surface exciton. The following parameter values were used in the calculation: $\varepsilon_{10} = 12.8$; $\varepsilon_1 = 10.9$; $m_e^* = m_h^* \cong m_0$; $\hbar\omega_0 = 36$ meV. As a result for ΔE_{ex}^S we obtain

$\Delta E_{ex}^S(H) = -360$ meV, $R_{ex}(H) = 1.2$ nm—the Haken method,

$\Delta E_{ex}^S(M) = -400$ meV, $R_{ex}(M) = 0.8$ nm—the Mayer method.

(at the same time $R_{e,\,h} = 1.0$ nm).

These results are in good agreement with the experimental ones presented in [56], the authors of which observed the Wannier–Mott surface exciton states at the GaAs–vacuum contact. The exciton binding energy reached approximately 0.4 eV and was about two orders of magnitude higher than the exciton binding energy in the GaAs volume. (In this case, the criteria for the applicability of the continuum approximation are fulfilled, although at the limit.) Since the self-action forces of the electron and the hole have a repulsive character, in principle, a 'dead zone' for

Wannier–Mott excitons should arise in GaAs upon contact with vacuum [57]. However, an electron and a hole can be localized on surface Tamm states having a small (on the order of 1 – 3 Å) attenuation depth; or, due to the excitation of an electron–hole pair on the GaAs surface, they can be trapped in a potential well having a polarization nature (see clause 5.3.3), which also has a small width ($R_{pl} \sim a$) and in which the states of the charge carriers are separated from the band states in the crystal volume by a potential barrier.

In [58, 59], surface exciton states at the GaAs–dielectric contact were experimentally observed. A Si_3N_4 (GeN_4) film was used as a dielectric. The GaAs surface, covered with a thin Si_3N_4, film, was irradiated with light, which, penetrating into the crystal, generated excitons near the GaAs surface. At the same time, the emission band of free bulk excitons of 1.5160 eV and a new emission band of 1.5137 eV, shifted by 2.3 meV relative to the bulk one, were observed. This band was interpreted [59, 60] as being caused by surface excitons. The formulas obtained in this work allow us to analyze the experimental data in detail. Since the dielectric constant of the film depends on the conditions of its application, various values are used in the calculations given. The results of numerical calculation of the binding energy of surface exciton states are shown in table 6.1 for the following GaAs and parameters: $m_e^*/m_0 = 0.07$; $m_h^*/m_0 = 0.40$; $\varepsilon_{10} = 12.8$, $\varepsilon_1 = 10.9$, $\hbar\omega_0 = 36$ meV. According to these data, at relatively low values of the dielectric constant of the film, the calculated binding energy of the surface exciton $(\Delta E_{ex}^S)_{\text{theor}}$ was less than the experimental $(E_{ex}^S)_{\text{exp}}$. The experimental energy can be represented as the sum of the shift of the energy of the surface electronic state relative to the bulk (2.3 meV) and the binding energy of the bulk exciton (4.7 meV):

$$\left(E_{ex}^S\right)_{\text{exp}} = \left(\Delta E_{ex}^S\right) + \Delta E_0^V = 2.3\text{ meV} + 4.7\text{ meV} = 7.0\text{ meV}.$$

Theoretical significance

$$\left(E_{ex}^S\right)_{\text{theor}} = 3.1\text{ meV} + 4.5\text{ meV} = 7.6\text{ meV}.$$

This value was obtained under the assumption that the dielectric constant of the Si_3N_4 (Ge_3N_4) film is infinitely high ($\varepsilon \to 0$). The possibility of such conditions of the GaAs surface when applying Si_3N_4 (Ge_3N_4) film was noted by the authors of experimental

Table 6.1. The results of numerical calculations of the binding energy of the surface exciton at the GaAs–vacuum contact. The following parameter values are used in the calculation: $\varepsilon_{10} = 12.8$; $\varepsilon_1 = 10.9$; $m_e^*/m_0 = 0.07$; $m_h^*/m_0 = 0.40$; $\hbar\omega_0 = 36$ meV.

ε_2	$(\Delta E_{ex}^S)_T'$	λ_1^{-1}, Å	β_e^{-1}, Å	β_h^{-1}, Å	$(\Delta E_{ex}^S)_T''$	λ_2^{-1}, Å	β_{e2}^{-1}, Å	β_{h2}^{-1}, Å	$(\Delta E_{ex}^S)_{\text{exp}}$
5	–2.1	154	167	134	–0.5	168	178	138	—
10	–2.6	155	164	127	–1.1	170	177	135	—
15	–2.9	156	161	121	–1.4	172	177	133	–2.3
20	–3.1	156	159	117	–1.6	172	176	131	—
	–5.1	155	137	34	–3.1	169	177	49	—

works [58, 61]. According to the table, the contribution of phonons $(E_{ex}^{S})_{\text{theor}}$ to its spectrum is very significant. Indeed, without taking into account phonons

$$\left(E_{ex}^{S}\right)_{\text{theor}} = 5.1\ \text{meV} + 4.5\ \text{meV} = 9.6\ \text{meV},$$

where $(E_{ex}^{S})'_{\text{theor}} = 5.1$ meV is a shift in the energy level of the surface electronic state relative to the bulk one, without taking into account the contribution of optical phonons.

Note that the wave function we have chosen (6.6.3a) assumes that the interaction with the surface is greater or of the same order as the Coulomb one (formulas (3.17) and (3.18)). The calculated values of β_e, β_h, λ confirm the conclusion.

In conclusion of this section, we will discuss the question of the effective mass of the polaron exciton.

Since in the limiting case of Haken, the surface exciton is formed by electron and hole surface polarons, the upper limit of the effective mass of the exciton will be the sum of the effective masses of the electron (m_{pe}^{*}) and hole (m_{ph}^{*}) polarons, which for the contact of a polar crystal with a nonpolar one were obtained in section 5.4 (formula (6.5.4.12))

$$M_{ex}^{*}(\text{X}) = m_{pe}^{*} + m_{\text{ph}}^{*}. \tag{6.6.12}$$

In the Mayer limit case, to determine the effective mass of the surface exciton, we expand the functional of the energy of the surface exciton

$$\begin{aligned} E^{S}(\boldsymbol{P}_{\perp}) = & \left\langle \varphi(\rho) \left| \frac{P_{\rho}^{2}}{2\mu_{\perp}} - \frac{e^{2}}{\varepsilon_{1}^{*}\rho} \right| \varphi(\rho) \right\rangle \\ & + \frac{P_{\perp}^{2}}{2M_{ex}} + \frac{1}{2M_{ex}} \left(\sum_{\eta} \hbar\eta \left| f_{s}(\eta) \right|^{2} \right)^{2} \\ & + \sum_{\eta} \left[\hbar\Omega_{s} + \frac{\hbar^{2}\eta^{2}}{2M_{ex}} - \frac{\hbar\eta \boldsymbol{P}_{\perp}}{M_{ex}} \right] \left| f_{s}(\eta) \right|^{2} \\ & + \sum_{\eta} \left[V_{s}^{*}(\eta) f_{s}(\eta) \rho^{*}(\eta) + \text{c. c.} \right] \end{aligned} \tag{6.6.13}$$

in a series of powers of $\mathbf{P}_{\perp}$ up to and including quadratic terms, and substitute the exciton energy in the form

$$E_{ex}^{S}(\mathbf{P}_{\perp}) = E_{ex,\,0}^{S} + \frac{P_{\perp}^{2}}{2M_{ex}^{*}}, \tag{6.6.14}$$

where M_{ex}^{*} is the effective mass of the surface exciton, determined by the expression

$$M_{ex}^{*}(M) = M_{ex} \left\{ 1 + \frac{2\hbar^{2}}{M_{ex}} \sum_{\eta} \frac{\eta^{2}\, |\rho(\eta)|^{2}\, |V_{s}(\eta)|^{2}}{\left(\hbar\Omega_{s} + \frac{\hbar^{2}\eta^{2}}{2M_{ex}} \right)^{3}} \right\}. \tag{6.6.15}$$

Estimates of the effective mass of the surface exciton were carried out for a flat exciton on a GaAs, surface bordering a vacuum using formulas (6.6.12) and (6.6.15):

$$M^*_{ex}(H) = 3,4\,M_{ex};\;\; M^*_{ex}(M) = 2,5\,M_{ex};\;\; M_{ex} = m^*_{e\perp} + m^*_{h\perp}.$$

6.7 Biexciton states on the crystal surface

At high excitation levels, when the exciton density is high ($\geqslant 10^{17}\,\text{cm}^{-3}$), and the temperature remains low, collective exciton states can occur in semiconductors: biexcitons [60, 62], trions [62], electron–hole metallic liquid [63], and Bose–Einstein condensate of excitons or biexcitons [64–66]. The realization of certain states depends on the band structure of the crystal, the nature of the interaction between excitons, the concentration of excitons, and temperature.

Theoretical calculations have shown that the biexciton is stable at all values $0 \leqslant \sigma \leqslant 1$ ($\sigma = m_e^*/m_h^*$). In particular, it was shown in [67] that a two-dimensional system of four particles, two electrons, and two holes is stable in the most unfavorable case $\sigma = 1$, and the binding energy in such a system is greater than that in the three-dimensional case.

In a number of physical situations, which, for example, were discussed when discussing plane exciton states, electronic states can be highly localized near the surface of a polar crystal. In this case, there is a fast movement of the electron and the hole along the z-axis and a slow movement in the XOY plane. Therefore, the wave functions of electrons and phonons can be represented as a wave function depending on z and a wave function depending on x, y. Suppose that the radii of the electron bound states (exciton and biexciton) in the surface plane are greater than their attenuation depths into the crystal, that is, the flatness criterion is fulfilled. Then, when averaging the electronic part of the Hamiltonian over the z-dependent wave function, $z = 0$ can be put in the potential of the Coulomb interaction of an electron and a hole, which only leads to a shift of all energy levels (in the future, energy is counted from this shift). At $z = 0$, the interaction with bulk optical phonons disappears, and with surface phonons it does not depend on the z-coordinate.

When deriving the effective Hamiltonian of a biexciton, from which the interaction of an electron and a hole (or excitons) with phonons is excluded, we will assume the fulfillment of the condition $a_{ex} > R_{e,h}$ (Haken limit).

The Hamiltonian of interacting electrons, holes, and phonons has the form

$$\hat{H} = \hat{K} + W_0(\rho) + \hat{H}_{\text{ph}} + \hat{H}_{ex}, \tag{6.7.1}$$

where $\hat{K}$ is the kinetic energy operator of two electrons and two holes:

$$\hat{K} = -\frac{\hbar^2}{2m_e^*}(\Delta_{\rho_{e_1}} + \Delta_{\rho_{e_2}}) - \frac{\hbar^2}{2m_h^*}(\Delta_{\rho_{h_1}} + \Delta_{\rho_{h_2}}); \tag{6.7.2}$$

$W_0(\rho)$ is the potential energy of the interparticle interaction in the biexciton, which has the form

$$W_0 = \frac{e^2}{\varepsilon_1^*}\left(\frac{1}{\rho_{e_1e_2}} + \frac{1}{\rho_{h_1h_2}} - \frac{1}{\rho_{e_1h_1}} - \frac{1}{\rho_{e_1h_2}} - \frac{1}{\rho_{e_2h_1}} - \frac{1}{\rho_{e_2h_2}}\right). \tag{6.7.3}$$

Here $\varepsilon_1^* = (\varepsilon_1 + 1)/2$ is effective inertialess dielectric constant at the polar crystal–vacuum contact.

The expression (6.7.3) is obtained in the Haken limit when taking into account the dynamic shielding of the electron–hole interaction by surface plasma vibrations of valence electrons when $\rho_{ij} \gg R_{i,j}$.

The Hamiltonians of the surface optical phonons and the electron–(hole–)phonon interaction have the form

$$\hat{H}_{\text{ph}} = \sum_{\boldsymbol{\eta}} \hbar\Omega_s \hat{b}_{\boldsymbol{\eta}}^{+} \hat{b}_{\boldsymbol{\eta}}; \tag{6.7.4}$$

$$\hat{H}_{\text{e-ph}}^{(i)} = \sum_{\boldsymbol{\eta}} V_s(\eta) e^{i\boldsymbol{\eta}\boldsymbol{\rho}_i}\left(\hat{b}_{-\boldsymbol{\eta}}^{+} + \hat{b}_{\boldsymbol{\eta}}\right). \tag{6.7.5}$$

In the coordinates of the center of mass and relative motion of the electron and the hole in the exciton.

$$\mathbf{R}_j = \alpha_e \boldsymbol{\rho}_{e_j} + \beta \boldsymbol{\rho}_{h_j};\ R^2 = x^2 + y^2; \tag{6.7.6a}$$

$$\boldsymbol{\rho}_j = \boldsymbol{\rho}_{ej} - \boldsymbol{\rho}_{hj}; \tag{6.7.6b}$$

here $\alpha_e = m_e^*/M;\ \beta = 1 - \alpha;\ M = m_e^* + m_h^*$.

Then the Hamiltonian of the system in question can be written as

$$\begin{aligned}\hat{H} = &-\frac{\hbar^2}{2M_{ex}}(\Delta_{\mathbf{R}_1} + \Delta_{\mathbf{R}_2}) - \frac{\hbar^2}{2\mu}(\Delta_{\boldsymbol{\rho}_1} + \Delta_{\boldsymbol{\rho}_2}) + \sum_{\boldsymbol{\eta}} \hbar\Omega_s \hat{b}_{\boldsymbol{\eta}}^{+}\hat{b}_{\boldsymbol{\eta}} + W_0(\rho)\\ &+ \sum_{\boldsymbol{\eta}}\{V_\eta[F_\eta(\rho_1)e^{i\boldsymbol{\eta}\mathbf{R}_1} + F_\eta(\rho_2)e^{i\boldsymbol{\eta}\mathbf{R}_2}]\hat{b}_{\boldsymbol{\eta}} + h.\ c.\};\end{aligned} \tag{6.7.7}$$

$$F_\eta(\rho) = e^{i\beta\,\boldsymbol{\eta}\boldsymbol{\rho}} - e^{-i\alpha\,\boldsymbol{\eta}\boldsymbol{\rho}}. \tag{6.7.8}$$

To derive W_{eff}, which takes into account the contribution from inertial polarization, we calculate the average value of the Hamiltonian (6.7.7) on the wave function of two excitons and phonons [68, 69]:

$$\begin{aligned}\Psi &= \exp\{i(\mathbf{K}_1\mathbf{R}_1 + \mathbf{K}_2\mathbf{R}_2)\}\varphi(\rho_1, \rho_2, \rho)\cdot\Phi_0\\ &\times \exp\left\{\sum_{\boldsymbol{\eta}}[f_\eta(\rho_1)e^{i\boldsymbol{\eta}\mathbf{R}_I} + f_\eta(\rho_2)e^{i\boldsymbol{\eta}\mathbf{R}_2}]\hat{b}_{\boldsymbol{\eta}} + h.\ c.\right\};\end{aligned} \tag{6.7.9a}$$

$$\varphi(\rho_1, \rho_2, \rho) = \varphi(\rho_1)\varphi(\rho_2) + \varphi(|\boldsymbol{\rho} + \alpha\boldsymbol{\rho}_2 + \beta\boldsymbol{\rho}_I|)\varphi(|\boldsymbol{\rho} - \alpha\boldsymbol{\rho}_I - \beta\boldsymbol{\rho}_2|). \tag{6.7.9b}$$

In formulas (6.7.8a,b), the notation is introduced: $\boldsymbol{\rho} = \mathbf{R}_1 - \mathbf{R}_2$; $\mathbf{K}_j$ ($j = 1, 2$) is the wave vector of the translational motion of the jth exciton; $\varphi(\rho)$ is the wave function of the relative motion of the electron and the hole in the exciton; f_η is the variational amplitudes of the displacement of the phonon mode operators; and Φ_0 the basic state of phonons.

Consider the case $\mathbf{K}_1 = \mathbf{K}_2 = 0$ and assume that both excitons are in the ground state:

$$\varphi(\rho) = \sqrt{\frac{2}{\pi}}\, a_{ex}^{-1} \exp\left(-\frac{\rho}{a_{ex}}\right), \tag{6.7.10}$$

then

$$\sum_{\eta} \left| f_\eta^* (\boldsymbol{\rho}_1, \boldsymbol{\rho}_2, \boldsymbol{\rho}) \right|^2 \eta = 0; \tag{6.7.11a}$$

$$\sum_{\eta} \left[f_\eta (\boldsymbol{\rho}_1, \boldsymbol{\rho}_2, \boldsymbol{\rho})\, \nabla_{\boldsymbol{\rho}} f_\eta^* (\boldsymbol{\rho}_1, \boldsymbol{\rho}_2, \boldsymbol{\rho}) - f_\eta^* (\boldsymbol{\rho}_1, \boldsymbol{\rho}_2, \boldsymbol{\rho})\, \nabla_{\boldsymbol{\rho}_1} f_\eta (\boldsymbol{\rho}_1, \boldsymbol{\rho}_2, \boldsymbol{\rho}) \right] = 0, \tag{6.7.11b}$$

and the energy functional of the ground state of the surface biexciton has the form

$$\begin{aligned} E \int d^2\rho_1 d^2\rho_2\, |\varphi(\boldsymbol{\rho}_1, \boldsymbol{\rho}_2, \boldsymbol{\rho})|^2 &= E_0 + \iint d^2\rho_1 d^2\rho_2\, |\varphi(\boldsymbol{\rho}_1, \boldsymbol{\rho}_2, \boldsymbol{\rho})|^2 \\ &\times \sum_{\eta} |f_\eta (\boldsymbol{\rho}_1, \boldsymbol{\rho}_2, \boldsymbol{\rho})|^2 \left(\hbar\Omega_s + \frac{\hbar^2\eta^2}{2M_{ex}} \right) - \frac{\hbar^2}{2\mu} \iint d^2\rho_1 d^2\rho_2\, |\varphi(\boldsymbol{\rho}_1, \boldsymbol{\rho}_2, \boldsymbol{\rho})|^2 \\ &\times \sum_{\eta} \left[f_\eta (\boldsymbol{\rho}_1, \boldsymbol{\rho}_2, \boldsymbol{\rho})(\Delta_{\boldsymbol{\rho}_1} + \Delta_{\boldsymbol{\rho}_2}) f_\eta (\boldsymbol{\rho}_1, \boldsymbol{\rho}_2, \boldsymbol{\rho}) - \frac{1}{2}(\Delta_{\boldsymbol{\rho}_1} + \Delta_{\boldsymbol{\rho}_2}) |f_\eta (\boldsymbol{\rho}_1, \boldsymbol{\rho}_2, \boldsymbol{\rho})|^2 \right] \\ &- \iint d^2\rho_1 d^2\rho_2\, |\varphi(\boldsymbol{\rho}_1, \boldsymbol{\rho}_2, \boldsymbol{\rho})|^2 \sum_{\eta} \left\{ F_\eta(\rho_1) f_\eta (\boldsymbol{\rho}_1, \boldsymbol{\rho}_2, \boldsymbol{\rho})(\rho_1) \right. \\ &\left. + F_\eta(\rho_2) f_\eta (\boldsymbol{\rho}_1, \boldsymbol{\rho}_2, \boldsymbol{\rho}) e^{-i\eta\boldsymbol{\rho}} \right] + c.\ c. \Big\}; \end{aligned} \tag{6.7.12}$$

$$\begin{aligned} E_0 &= \iint d^2\rho_1 d^2\rho_2 \varphi^*(\rho_1, \rho_2, \rho) \\ &\left[-\frac{\hbar^2}{2\mu}\left(\Delta_{\rho_1} + \Delta_{\rho_2}\right) + W_0(\rho_1, \rho_2, \rho) \right] \varphi(\rho_1, \rho_2, \rho); \end{aligned} \tag{6.7.13a}$$

$$f_\eta (\boldsymbol{\rho}_1, \boldsymbol{\rho}_2, \boldsymbol{\rho}) = f_\eta (\rho_1) + f_\eta (\rho_2) e^{i\eta\boldsymbol{\rho}}. \tag{6.7.13b}$$

The variation of the energy functional (6.7.12) by the function (6.7.13b) leads to

$$\begin{aligned} &\left\{ -\frac{\hbar^2}{2\mu}\left(\Delta_{\boldsymbol{\rho}_1} + \Delta_{\boldsymbol{\rho}_2}\right) + \frac{\hbar^2\eta^2}{2M_{ex}} + \hbar\Omega_S \right. \\ &\left. + \frac{\hbar^2}{2\mu}\left[\frac{\left(\Delta_{\boldsymbol{\rho}_1} + \Delta_{\boldsymbol{\rho}_2}\right) |\varphi(\boldsymbol{\rho}_1, \boldsymbol{\rho}_2, \boldsymbol{\rho})|^2}{2\, |\varphi(\boldsymbol{\rho}_1, \boldsymbol{\rho}_2, \boldsymbol{\rho})|^2} \right] \right\} f_\eta (\boldsymbol{\rho}_1, \boldsymbol{\rho}_2, \boldsymbol{\rho}) \\ &- F_\eta^*(\rho_1) \exp(i\eta\boldsymbol{\rho}) = 0. \end{aligned} \tag{6.7.14}$$

In the Haken approximation, the terms in the square bracket of equation (6.7.14) can be neglected. In essence, this approximation is equivalent to the requirement:

$$\hbar\Omega_S > E_{ex}^S. \tag{6.7.15}$$

In this case, the solution of equation (6.7.14) has a simple form:

$$f_\kappa(\rho_1, \rho_2, \rho) = V_S^*(\eta)\left\{\frac{e^{i\beta\,\eta\rho_1} + e^{-i\beta\,\eta\rho_2 + i\eta\rho}}{\hbar\Omega_S + \dfrac{\hbar^2\eta^2}{2m_e^*}} - \frac{e^{i\alpha\,\eta\rho_1} + e^{-i\alpha\,\eta\rho_2 + i\eta\rho}}{\hbar\Omega_S + \dfrac{\hbar^2\eta^2}{2m_h^*}}\right\}. \tag{6.7.16}$$

Using equation (6.7.14), it is possible to simplify the expression for the energy of the system (6.7.12):

$$\begin{aligned}
&E\iint d^2\rho_1 d^2\rho_2 \left|\varphi(\rho_1, \rho_2, \rho)\right|^2 = E_0\\
&-\frac{1}{2}\iint d^2\rho_1 d^2\rho_2 \left|\varphi(\rho_1, \rho_2, \rho)\right|^2 \sum_\eta\\
&\left\{\left[f_\eta(\rho_1, \rho_2, \rho)F_\eta(\rho_1) + f_\eta(\rho_1, \rho_2, \rho)F_\eta(\rho_2)e^{-i\eta\rho}\right] + \text{c. c.}\right\}\\
&+\frac{1}{2}\frac{\hbar^2}{2\mu}\iint d^2\rho_1 d^2\rho_2 \left|\varphi(\rho_1, \rho_2, \rho)\right|^2\\
&\sum_\eta\left(\Delta_{\rho_1} + \Delta_{\rho_2}\right)\left|\varphi(\rho_1, \rho_2, \rho)\right|^2.
\end{aligned} \tag{6.7.17}$$

Here E_0 describes the undisturbed energy of a biexciton with potential W_0.

The additional members of the right-hand side (6.7.17) turn this potential into an effective one, that is,

$$E = \frac{\iint d^2\rho_1 d^2\rho_2 \varphi^*(\rho_1, \rho_2, \rho)\left[-\frac{\hbar^2}{2\mu}(\Delta_{\rho_1} + \Delta_{\rho_2}) + W_{\text{eff}}^{(i)}\right]\varphi(\rho_1, \rho_2, \rho)}{\iint d^2\rho_1 d^2\rho_2 \left|\varphi(\rho_1, \rho_2, \rho)\right|^2}, \tag{6.7.18}$$

where

$$W_{\text{eff}}^{(1)} = W_0 + \delta W_1 + \delta W_2. \tag{6.7.19}$$

Expressions for phonon contributions to the effective potential of interparticle interaction have the form

$$\delta W_1 = \frac{1}{2}\sum_\eta\{[f_\eta(\rho_1, \rho_2, \rho)\left[F_\eta(\rho_1) + F_\eta(\rho_2)e^{-i\eta\rho}\right] + c.\ c.]\}; \tag{6.7.20a}$$

$$\delta W_2 = \frac{1}{2}\frac{\hbar^2}{2\mu}\sum_{\boldsymbol{\eta}}(\Delta_{\rho_1} + \Delta_{\rho_2})|f_\eta(\boldsymbol{\rho}_1, \boldsymbol{\rho}_2, \boldsymbol{\rho})|^2. \tag{6.7.20b}$$

Varying the energy functional (6.7.17) by $\varphi(\boldsymbol{\rho}_1, \boldsymbol{\rho}_2, \boldsymbol{\rho})$, we obtain an expression for the effective Coulomb potential of the interaction of electrons and holes, taking into account the shielding by inertial polarization:

$$\begin{aligned} W_{\text{eff}}^{(1)} &= \frac{e^2}{\varepsilon_1^*}\left(\frac{1}{\rho_{e_1e_2}} + \frac{1}{\rho_{h_1h_2}} - \frac{1}{\rho_{e_1h_1}} - \frac{1}{\rho_{e_1h_2}} - \frac{1}{\rho_{e_2h_1}} - \frac{1}{\rho_{e_2h_1}}\right) \\ &- \pi\alpha_S\hbar\Omega_S\frac{m_h^*}{(m_e^* - m_h^*)}\left\{\left(\frac{m_e^*}{m_h^*}\right)^{\frac{1}{2}}\left[I_0\left(\frac{\rho_{e_1e_2}}{R_e}\right) - L_0\left(\frac{\rho_{e_1e_2}}{R_e}\right)\right]\right. \\ &\left.+\left[I_0\left(\frac{\rho_{h_1h_2}}{R_h}\right) - L_0\left(\frac{\rho_{h_1h_2}}{R_h}\right)\right]\right\} \\ &- 2\cdot\frac{\pi}{2}\hbar\Omega_S(\alpha_{Se} + \alpha_{Sh}) + \frac{\pi}{2}\alpha_{Se}\hbar\Omega_{S_1}\frac{m_h^*}{(m_e^* - m_h^*)} \\ &\times \sum_{j=e,h;\ l=e_1,e_2;\ m=h_1h_2}\left(\frac{m_j}{m_h}\right)^{\frac{1}{2}}\left[I_0\left(\frac{\rho_{lm}}{R_j}\right) - L_0\left(\frac{\rho_{lm}}{R_j}\right)\right]. \end{aligned} \tag{6.7.21}$$

Let's move on to calculating the dissociation energy of the biexciton. The most detailed studies of the stability of the bulk biexciton in the entire range of values $0 \leqslant \sigma \leqslant 1$ are contained in [70–72]. In contrast to the works [71, 72], where the adiabatic mass approximation is used, in [70] it is assumed that the dissociation energy of the biexciton is much less than the ionization potential of the exciton. We will use this model of a biexciton consisting of two weakly deformable excitons. In units of length $a_0 = \hbar^2\varepsilon_0^*/(2\mu e^2)$ and energy $E_0 = 2\mu e^4/(\hbar^2\varepsilon_0^{*2})$, the Hamiltonian of the biexciton has the form

$$\begin{aligned} \hat{H} &= -\Delta_{\rho_1} - \Delta_{\rho_2} - \frac{\sigma\Delta_{\mathbf{R}}}{2(1+\sigma)^2} - \frac{2\sigma\Delta_{\boldsymbol{\rho}}}{(1+\sigma)^2} \\ &+ \left[\frac{1}{|\boldsymbol{\rho} + \beta(\boldsymbol{\rho}_1 - \boldsymbol{\rho}_2)|} + \frac{1}{|\boldsymbol{\rho} - \alpha(\boldsymbol{\rho}_1 - \boldsymbol{\rho}_2)|} - \frac{1}{\rho_1} - \frac{1}{\rho_2}\right. \\ &\left.- \frac{1}{|\boldsymbol{\rho} + \beta\boldsymbol{\rho}_1 + \alpha\boldsymbol{\rho}_2|} - \frac{1}{|\boldsymbol{\rho} + \alpha\boldsymbol{\rho}_1 + \beta\boldsymbol{\rho}_2|}\right]. \end{aligned} \tag{6.7.22}$$

The energy of the biexciton E_m is defined by the expression

$$E_m = \langle\Psi_m|\hat{H}|\Psi_m\rangle, \tag{6.7.23}$$

where the variational function is

$$\Psi_m = \frac{\exp(i\boldsymbol{KR})}{\sqrt{2(1+S)}}\{\Phi(\rho)\varphi(\rho_1)\varphi(\rho_2) + \Phi(|(\beta-\alpha)\boldsymbol{\rho} - 2\alpha\beta(\boldsymbol{\rho}_1 - \boldsymbol{\rho}_2)|) \varphi(|\boldsymbol{\rho} + \beta\boldsymbol{\rho}_1 + \alpha\boldsymbol{\rho}_2|)\varphi(|\boldsymbol{\rho} - \alpha\boldsymbol{\rho}_1 - \beta\boldsymbol{\rho}_2|)\}; \tag{6.7.24}$$

$$\Phi(\rho) = \sqrt{\frac{2}{\pi}}a_m^{-1}e^{-\frac{\rho}{a_m}};\ \varphi(\rho) = \sqrt{\frac{2}{\pi}}a_{ex}^{-1}e^{-\frac{\rho}{a_{ex}}}; \tag{6.7.25}$$

a_m and a_{ex} are variational parameters that make sense of the radii of the biexciton and exciton, respectively;

$$S = \int\Phi(\rho)\Phi(|(\beta-\alpha)\boldsymbol{\rho} - 2\alpha\beta(\boldsymbol{\rho}_1 - \boldsymbol{\rho}_2)|)\varphi(\rho_1)\varphi(\rho_2) \Phi(|\boldsymbol{\rho} + \beta\boldsymbol{\rho}_1 + \alpha\boldsymbol{\rho}_2|) \times \varphi(|\boldsymbol{\rho} - \alpha\boldsymbol{\rho}_1 - \beta\boldsymbol{\rho}_2)d\rho d\rho_1 d\rho_2. \tag{6.7.26}$$

All integrals included in (6.7.23) can be calculated analytically if the Slater orbitals are represented as a linear combination of Gaussian orbitals:

$$\exp(-\xi\rho) = \sum_{i=1}^{4} C_i\lambda_i^{\frac{1}{2}}\exp(-\lambda_i\xi^2\rho^2) \tag{6.7.27}$$

(C_i, λ_i are given in the work [73]).

The binding energy of a flat surface biexciton is determined by the expression

$$\Delta E_m(\sigma) = E_m(\sigma) - 2. \tag{6.7.28}$$

The numerical calculation of $\Delta E_m(\sigma)$ was performed in [74, 75], and its results are shown in figure 6.1 together with the binding energy of the biexciton calculated in [71].

The surface biexciton is stable at all values σ. For =0 $\Delta E_m = -0.468E_0$; при $\sigma = 1$ $\Delta E_m = -0.042E_0$. For comparison, we present the binding energy of a bulk biexciton at the same values σ:

$$\sigma = 0,\ \Delta E_m^V = -0.0298\frac{\mu e^4}{2\hbar^2\varepsilon_0^2};\ \sigma = 1,\ \Delta E_m^V = -0.0273\frac{\mu e^4}{2\hbar^2\varepsilon_0^2}.$$

Thus, a change in the geometry of the biexciton, as well as a change in the parameters of its interaction with lattice vibrations ($\varepsilon_0 \to \varepsilon_0^*$), as in the case of an exciton on the surface of a polar crystal, leads to a significant increase in its binding energy.

6.8 Surface exciton complexes

For the first time, the idea of the existence of charged complexes was expressed in [62]. Their formation becomes favorable at a high density of excitons, when free electrons and holes appear in the system due to the decay of some of them. As a result of the interaction of an electron or hole with an exciton, bound states of band carriers with an exciton, called three-particle exciton complexes, can arise. In [21, 62], it was pointed out the possibility of the existence of various types of such complexes and an analogy was established between the structure of the energy levels of complexes and their analogues—well–studied theoretically and experimentally—atomic and molecular systems (ionized hydrogen molecule H_2^+ and hydrogen atom ion H^-).

In [76–79], the stability of such complexes with respect to decay into an exciton and a quasi-free charge carrier was investigated. The authors of [76, 79] established the stability of the H_2^+ and H^-complexes at any values of the parameter $\sigma = m_e^*/m_h^*$. Exciton complexes can make a significant contribution to the luminescence and conductivity of crystals. Experimentally, such complexes were discovered by the authors of works [80, 81] on the measurement of luminescence at high excitation levels in the range $T \sim 5 - 15$ K in Ge crystals and in experiments on cyclotron absorption at low temperatures.

Taking into account the dynamic shielding of the Coulomb interaction by polar optical phonons, as shown in [82, 83], does not change the conclusions about the stability of exciton complexes.

Along with H_2^+ and H^--type systems, complexes formed by an exciton and charged by a donor and acceptor are of great theoretical and experimental interest. For the first time, the coupling of such a complex was calculated in [84] on a simple Coulomb potential of interparticle interaction, and in [85, 86] various effective potentials describing electron–hole and electron–impurity interactions in polar crystals were used in calculations.

In this section, the issue of the binding energy of an exciton complex at the contact of a polar crystal with a nonpolar one is discussed when the particles forming the complex are localized in a narrow near-surface domain of the polar crystal. This assumption, as shown in [87, 88], significantly simplifies the mathematical calculations of the physical model of the complex. Since the theory of the surface exciton complex is similar to the theory of the biexciton on the surface of a polar crystal developed in the previous paragraph, we will present here only the final results and their discussion.

6.8.1 Effective Hamiltonian and binding energy of the complex 'exciton–neutral donor'

The impurity ion is considered heavy ($M \gg m_e^*,\ m_h^*$), so the center of mass of the system is located in it. The problem of deriving the interaction potential of the particles that make up the complex is solved in two stages: first, the screening of the Coulomb interaction of the particles of the complex by inertial polarization is taken into account. At the second stage, the potential found in this way is considered as a seed potential and the contribution from the screening of the Coulomb interaction by inertial polarization is found.

The energy functional of the ground state of a flat exciton complex can be represented by the formula:

$$E^S_{\text{compl}} = \langle \Psi(\rho) | \hat{H}_{\text{eff}} | \Psi(\rho) \rangle, \tag{6.8.1}$$

where

$$\hat{H}_{\text{eff}} = -\frac{\hbar^2}{2m^*_{e_1}}\Delta_{\rho_1} - \frac{\hbar^2}{2m^*_{e_2}}\Delta_{\rho_2} - \frac{\hbar^2}{2m^*_h}\Delta_{\rho_3} + W_{\text{eff}}; \tag{6.8.2}$$

$$\begin{aligned} W_{\text{eff}} &= \frac{e^2}{\varepsilon^*_1}\left(\frac{1}{\rho_3} - \frac{1}{\rho_1} - \frac{1}{\rho_2}\right) + \frac{e^2}{\varepsilon^*_1}\left(\frac{1}{\rho_{12}} - \frac{1}{\rho_{13}} - \frac{1}{\rho_{23}}\right) \\ &+ e^2\left[\frac{1}{\varepsilon^*_S(\rho_{12})\rho_{12}} - \frac{1}{\varepsilon^*_S(\rho_{13})\rho_{13}} - \frac{1}{\varepsilon^*_S(\rho_{23})\rho_{23}}\right]. \end{aligned} \tag{6.8.3}$$

In (6.8.3), the designation is introduced:

$$\begin{aligned} \left[\varepsilon^*_S(\rho)\right]^{-1} \equiv \frac{1}{2}\Bigg\{ & \frac{1}{\varepsilon^*_1} - \frac{\pi\rho}{2(\sigma-1)\varepsilon^* R_s} \\ & \left[\sqrt{\sigma}\left(I_0\left(\frac{\rho}{R_h}\right) - L_0\left(\frac{\rho}{R_h}\right)\right) - \left(I_0\left(\frac{\rho}{R_e}\right) - L_0\left(\frac{\rho}{R_e}\right)\right)\right]\Bigg\}; \end{aligned} \tag{6.8.4a}$$

$$(\varepsilon^*)^{-1} = \left(\varepsilon^*_1\right)^{-1} - \left(\varepsilon^*_{10}\right)^{-1}, \tag{6.8.4b}$$

where $\boldsymbol{\rho}_1$, $\boldsymbol{\rho}_2$, $\boldsymbol{\rho}_3$ are the radius-vectors of the particles and ρ_{12}, ρ_{13}, ρ_{23} are their relative distances.

From expressions (6.8.2)–(6.8.4), one can obtain the Hamiltonian of the exciton-ionized donor complex, consisting of one heavy particle of infinite mass (donor) and two light particles forming an exciton. Figure 6.6(a) shows the results of numerical

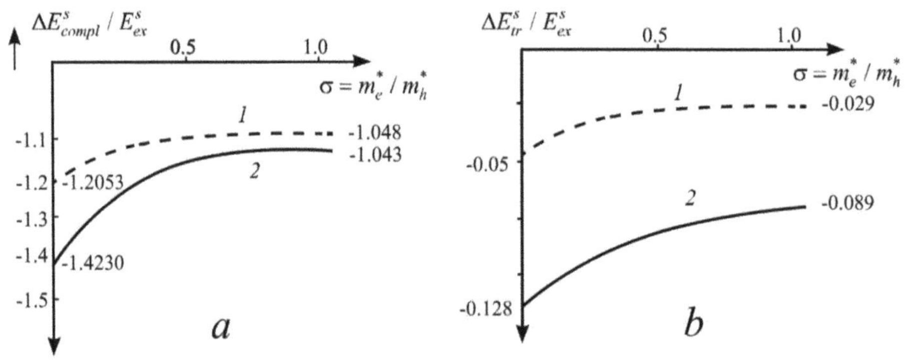

Figure 6.6. The binding energy of the complex as a function of a parameter at the contact of a CdS crystal with a nonpolar crystal ($\varepsilon_2 = 12.83$): (a) an exciton-ionized donor; (b) a planar trion. Curves 1 correspond to 3D complexes, and curves 2 correspond to 2D complexes.

calculation of the binding energy of the exciton-ionized donor complex (E^{S}_{compl}) as a function of the parameter at the contact of a CdS crystal with a nonpolar crystal ($\varepsilon_2 = 12, 83$):

$$E^{S}_{compl} = E^{0S}_{compl} + \frac{\pi}{2}(\alpha_{Se} + \alpha_{Sh})\hbar\Omega_S, \tag{6.8.5}$$

where E^{0S}_{compl} is the energy of the ground state of the exciton complex:

$$E^{0S}_{compl}(\lambda, \ \beta, \ \gamma) = \int \Psi^*(\rho_e, \ \rho_h, \ \rho_{eh})\hat{H}_{\text{eff}}\Psi(\rho_e, \ \rho_h, \ \rho_{eh})d\tau. \tag{6.8.6}$$

The trial wave function is taken as

$$\Psi(\rho_e, \ \rho_h, \ \rho_{eh}) = A \exp\{-\lambda\rho_{eh} - \beta\rho_e - \gamma\rho_h\}; \tag{6.8.7}$$

where $\lambda, \ \beta, \ \gamma$ are variational parameters.

The following parameter values of CdS are used in the calculation: $\varepsilon_{10} = 9.7$; $\varepsilon_1 = 5.35$; $m_e^* = 0.18m_0$; $\hbar\omega_0 = 33$ meV.

6.8.2 Effective Hamiltonian and the binding energy of a planar trion (*ehh*)

In contrast to the complex discussed above the 'exciton–neutral donor' trion is formed by three moving band charge carriers, and, consequently, the center of mass of this system performs translational movement through the crystal. Taking into account the dynamic shielding of the Coulomb interaction of particles, the effective Hamiltonian of the trion has the form

$$\begin{aligned}\hat{H}_{\text{eff}} &= -\frac{\hbar^2}{2m_{e1}^*}\Delta_{\rho_{e1}} - \frac{\hbar^2}{2m_h^*}(\Delta_{\rho_{h_1}} + \Delta_{\rho_{h_2}}) \\ &\quad -\frac{e^2}{\varepsilon_S^*\left(\rho_{eh_1}\right)\rho_{eh}} - \frac{e^2}{\varepsilon_S^*\left(\rho_{eh_2}\right)\rho_{eh_2}} + \frac{e^2}{\varepsilon_S\rho_{h_1h_2}}.\end{aligned} \tag{6.8.8}$$

In the center of mass system (6.8.8) it has the form

$$\begin{aligned}H_{\text{eff}} &= -\frac{\hbar^2}{2\mu}\left(\nabla_\xi^2 + \nabla_\eta^2\right) - \frac{\hbar^2}{m_e^*}\nabla_\xi\nabla_\eta - \frac{\hbar^2}{2M_{compl}}\nabla_{\boldsymbol{R}}^2 \\ &\quad -\frac{e^2}{\varepsilon_S^*(\zeta)\zeta} - \frac{e^2}{\varepsilon_S^*(\eta)\eta} + \frac{e^2}{\varepsilon_S^*(|\zeta - \eta|)|\zeta - \eta|},\end{aligned} \tag{6.8.9}$$

where the third term of the right part (6.8.9) describes the translational motion of the center of mass:

$$\boldsymbol{\zeta} = \boldsymbol{\rho}_e - \boldsymbol{\rho}_{h_1}; \ \boldsymbol{\eta} = \boldsymbol{\rho}_e - \boldsymbol{\rho}_{h_2}; \tag{6.8.10a}$$

$$\mathbf{R} = M_{compl}^{-1}\left\{m_h^*(\boldsymbol{\rho}_{h1} + \boldsymbol{\rho}_{h2}) + m_e^*\boldsymbol{\rho}_e\right\}, \tag{6.8.10b}$$

and ε_S^* has the form (6.8.4).

The method of 'coordinate stretching' is used to derive the variational binding energy of the considered complex. As a result, for the binding energy of the complex, we obtain

$$E_{0S}^{tr} = -\left[\int |\Psi|^2 W_{\text{eff}} d\tau\right]^2 \times \left\{4\left[\int |\Psi|^2 d\tau\right]\left[\int\left\{\frac{\sigma}{1+\sigma}\left[\left(\nabla_{\rho_{h_1}}\Psi\right)^2 + \left(\nabla_{\rho_{h_2}}\Psi\right)^2\right] + \frac{\left(\nabla_{\rho_e}\Psi\right)}{1+\sigma}\right\}d\tau\right]\right\}^{-1} \tag{6.8.11}$$

.

The variational function $\Psi(\rho_{e_{h_1}}, \rho_{e_{h_2}}, \rho_{h_1h_2})$ is chosen in the form proposed by the authors of the work [89]:

$$\Psi(s, t, u) = \exp\left(-\frac{1}{2}s\right)(1 + C_1u + C_2t^2), \tag{6.8.12}$$

where

$$s = \rho_{h_1} + \rho_{h_2};\ t = \rho_{h_1} - \rho_{h_2};\ u = \rho_{h_1h_2}. \tag{6.8.13}$$

Calculations are performed in the Hylleraas coordinate system on the plane

$$\int F\left(\rho_{eh_1}, \rho_{eh_2}, \rho_{h_1h_2}\right)d\tau = 2\pi\int_0^{\infty} ds \int_0^{s} \frac{udu}{\sqrt{s^2 - u^2}} \int_0^{\infty} \frac{F(s, t, u)\left(s^2 - t^2\right)dt}{\sqrt{u^2 - t^2}}. \tag{6.8.14}$$

As a result, we obtain the variational energy in the form of

$$E_{0S}^{tr} = -\frac{1}{8}\left\{4 - \frac{3\pi}{8} + C_1^2\left(32 - \frac{15\pi}{8}\right) + C_1C_2(45\pi - 16) + C^2\left(288 - \frac{495\pi}{64}\right) + C_1(6\pi - 4) + C_2\left(32 - \frac{21\pi}{16}\right)\right\}^2 \times \left\{1 + 12C_1^2 + 72C_2^2 + \frac{15\pi}{8}C_1 + 8C_2 + \frac{405\pi}{32}C_1C_2\right\}^{-1} \times \left\{1 + 8C_1^2 + 136C_2^2 + \frac{\pi C_1}{(1+\sigma)}\left(1,5\sigma + \frac{9}{8}\right) + 8C_2. + \frac{\pi C_1C_2}{1+\sigma}\left(\frac{45\sigma}{4} + \frac{411}{32}\right)\right\}^{-1} \tag{6.8.15}$$

The results of the numerical calculation of E_{0S}^{tr} are shown in figure 6.6(b). Based on the above study, the following conclusions can be drawn:

(a) planar complexes: exciton-ionized donor and trion are stable with respect to the decay process into a free exciton and a donor center and into three band carriers, respectively, for any parameter values in the range $0 \leqslant \sigma \leqslant 1$;

(b) the described exciton complexes are more stable than their three-dimensional counterparts (curves 1 in figures 6.6(a) and (b) correspond to 3D complexes, and curves 2 correspond to 2D complexes).

6.9 Surface polaron exciton in a strong magnetic field

In general, the problem of a magnetic polaron exciton is much more complicated than the problems of a polaron exciton and exciton complexes in the absence of a field discussed in the previous paragraphs. However, due to at least the theoretical possibility of changing the magnitude of the magnetic field induction B from zero to arbitrarily large values, it is possible to implement limiting cases in which the problem with a magnetic field is solved quite simply. For example, in the limit of very strong magnetic fields directed along the z-axis, at which $\omega_c > \omega_0$, the exciton Hamiltonian can first be averaged on the wave function describing motion in the XOY plane, then approximations of strong or weak bonds with the surface can be used. Since the change introduced by taking into account the magnetic field in the electron–phonon Hamiltonian consists of replacing the exponent exp $(i\eta\rho)$ by the factor $\langle\Psi \exp (i\eta\rho)\Psi\rangle$, the calculations of the exciton binding energy with the surface are similar to those described in sections 6.8.1–6.8.3. In chapter 5, it was shown that for all values of the electron–phonon interaction constant, the binding energy of surface polaron states increases with an increase in the magnetic field. Obviously, this should lead to a change in the nature of the dynamic shielding of the electron–hole interaction by polar optical phonons in the presence of a magnetic field. In [90], the effect of a strong magnetic field on the electron–hole interaction in the framework of a simple Coulomb model and on the energy spectrum of the Wannier–Mott exciton in a quasi-two-dimensional system was theoretically investigated. In this section, we will consider the problem of a polaron exciton at the contact of two crystals (in a plane approximation) in the limit of a strong magnetic field and investigate the dependence of contributions to the energy and effective mass of the exciton on magnetic induction at arbitrary exciton pulses.

The exciton Hamiltonian at the contact of two crystals in a homogeneous magnetic field has the form (7.2.3), in which substitutions must be performed:

$$\mathbf{P}_{c\perp} \to \mathbf{P}_{c\perp} \pm \mathbf{A}_c; \; c = e, \; h; \tag{6.9.1}$$

$\mathbf{A}_c$ is the vector potential of a homogeneous magnetic field, having the form

$$\mathbf{A}_c = \frac{1}{2}[\mathbf{B}\boldsymbol{\rho}_c]. \tag{6.9.2}$$

The criterion of a strong magnetic field is the inequality:

$$\hbar\omega_c > Ry; \tag{6.9.3}$$

where $\omega_c = eB/\mu$ is cyclotron frequency, $Ry \equiv E_{ex}^S$ is exciton Rydberg, and μ is the reduced weight.

The effective exciton–phonon Hamiltonian in the Mayer limit case (formula (6.8.1)) for a strong magnetic field is determined using the method of operation [91]. In the case under consideration, the role of the operator of a two-dimensional exciton pulse in a magnetic field is played by the value

$$\mathbf{P}_\perp = (-i\hbar\ \nabla_{\rho_e} + e\mathbf{A}_e) + (-i\hbar\ \nabla_{\rho_h} + e\mathbf{A}_h) - e[\mathbf{B},\ \rho_e - \ \rho_h]. \qquad (6.9.4)$$

The wave function of the exciton–phonon system has the form

$$\Psi(\rho_e,\ \rho_h) \equiv \Psi(\rho) = \exp\{i(\mathbf{P}_\perp + \frac{e}{2}[\mathbf{B}\rho])\frac{\mathbf{R}_\perp}{\hbar}\}\Psi_1(\rho,\ \eta); \qquad (6.9.5a)$$

$$\Psi_1(\rho,\ \eta) = \Phi(\rho - \rho_0,\ \eta)\exp(\frac{i}{2}\gamma\rho\mathbf{P}_\perp); \qquad (6.9.5b)$$

$$\Phi(\rho - \rho_0,\ \eta) = \Phi_1(\rho - \rho_0)\exp\{\sum_\eta f_\eta^*\hat{b}_\eta - f_\eta\hat{b}_{-\eta}^+\}|0\rangle; \qquad (6.9.5c)$$

here $\mathbf{R}_\perp = (m_e^*\rho_e + m_h^*\rho_h)/(m_e^* + m_h^*)$, $\rho = \rho_e - \rho_h$ are the coordinate of the center of mass and the relative coordinate, respectively; $M = m_e^* + m_h^*$ is exciton mass; $\gamma = (m_h^* - m_e^*)/(m_h^* + m_e^*)$; $\rho_0 = (eB^2)^{-1}[\mathbf{B}\rho]$ is the vector defining the local center of the orbit in the classical analogue of this problem; f_η, f_η^* is the variational amplitudes of displacement of phonon mode operators $\hat{b}_\eta$, $\hat{b}_\eta$; and $|0\rangle$ is the basic state of the phonon subsystem.

The wave function $\Phi_1(\rho - \rho_0)$ relative motion satisfies the Schrodinger equation

$$\hat{H}_{\text{eff}}\Phi_1(\rho - \rho_0) = E\Phi_1(\rho - \rho_0) \qquad (6.9.6)$$

with an effective Hamiltonian

$$\begin{aligned}\hat{H}_{\text{eff}} = &-\frac{\hbar^2}{2\mu}\Delta_\rho - \frac{ie\hbar}{2\mu}\gamma B[\rho\nabla_\rho] + \frac{e^2}{8\mu}B^2\rho^2 - \frac{e^2}{4\pi\varepsilon_0\varepsilon_1^*\,|\rho + \rho_0|}\\ &+\frac{\hbar^2}{2M}\sum_\eta \eta^2\,|f_\eta|^2 + \frac{1}{2M}\left\{\sum_\eta \hbar\eta\,|f_\eta|^2\right\}^2 - \frac{1}{M}(\mathbf{P}_\perp + 2e\mathbf{A})\sum_\eta \hbar\eta\,|f_\eta|^2\\ &+\sum_\eta \hbar\Omega_S\,|f_\eta|^2 - \sum_\eta[V_\eta(e^{is_2\eta(\rho+\rho_0)} - e^{-is_1\eta(\rho+\rho_0)})f_\eta + c.\,c.].\end{aligned} \qquad (6.9.7)$$

Applying the variational principle for the total energy, we find the amplitudes f_η, f_η^*:

$$f_\eta^* = \frac{V_\eta^* A_\eta}{\hbar\Omega_S + \frac{\hbar^2\eta^2}{2M} - (1 - \zeta)\frac{\hbar\eta\mathbf{P}_\perp}{M}}; \qquad (6.9.8)$$

here

$$A_\eta = \exp\left(is_2\boldsymbol{\eta}\boldsymbol{\rho}_0 - \frac{1}{2}s_2^2\eta^2R_B^2\right) - \exp\left(-is_1\boldsymbol{\eta}\boldsymbol{\rho}_0 - \frac{1}{2}s_1^2\eta^2R_B^2\right); \tag{6.9.9}$$

where $R_B = (\hbar/(eB))^{1/2}$ is magnetic length, $s_1 = m_e^*/M$, $s_2 = m_h^*/M$. Parametr ζ is introduced according to the definition

$$\zeta\mathbf{P}_\perp = \sum_{\boldsymbol{\eta}} \hbar\boldsymbol{\eta}\, |f_\eta|^2. \tag{6.9.10}$$

(For small values of $\mathbf{P}_\perp$, the parameter ζ does not depend on $\mathbf{P}_\perp$.)
Substituting (6.9.8), (6.9.9) into (6.9.7), we obtain

$$\hat{H}_{\text{eff}} = -\frac{\hbar^2}{2\mu}\Delta_\rho + \frac{ie\gamma}{2\mu}[\mathbf{B}\boldsymbol{\rho}]\,\nabla_\rho + \frac{e^2}{8\mu c}B^2\rho^2 + W_{\text{eff}}. \tag{6.9.11}$$

The effective potential of the electron–hole interaction introduced in (6.9.11) takes into account the contributions to the shielding from two types of polarization —inertial and inertialess—and has the form

$$W_{\text{eff}} = -\frac{e^2}{4\pi\varepsilon_0\varepsilon_1^*\,|\boldsymbol{\rho} + \boldsymbol{\rho}_0|} - \frac{\zeta^2P_\perp^2}{2M} - \sum_{\boldsymbol{\eta}}\frac{|V_\eta|^2|A_\eta|^2}{\hbar\Omega_S + \frac{\hbar^2\eta^2}{2M} - (1-\zeta)\frac{\hbar\boldsymbol{\eta}\mathbf{P}_\perp}{M}}. \tag{6.9.12}$$

In the W_{eff} zero approximation, equation (6.9.6) will transform into the Schrodinger equation for a free particle of mass μ in the field [90] $\mathbf{B}$, if $\gamma = 1$, which corresponds to the motion of one particle with mass $\mu = m_e^*$ (or m_h^*) around a stationary impurity.

The wave function of relative motion, however, does not depend on γ and is given by the expression

$$\Phi_{nm}(\rho) = \left(\frac{n!}{2^{|m|+1}(n + |m|!\pi)}\right)^{\frac{1}{2}}\frac{e^{-im\varphi}}{R_B}\left(\frac{\rho}{R_B}\right)^{|m|}L_n^{|m|}\left(\frac{\rho^2}{2R_B^2}\right)\exp\left(-\frac{\rho^2}{4R_B^2}\right); \tag{6.9.13}$$

here $L_n^{|m|}(x)$ are Laguerre polynomials.

The energy of the zero approximation is a function of γ:

$$E_{nm}^{(0)} = \hbar\omega_c\left[n + \frac{1}{2}(|m| - \gamma m + 1)\right]. \tag{6.9.14}$$

It is obvious that this energy is degenerate in terms of the moment m only at $\gamma = 1$. For exciton $|\gamma| < 1$ (i.e., there is no degeneracy in the moment of relative motion).

The spectrum (6.9.14) is completely discrete (unlike the three-dimensional case, where the spectrum is continuous in the z-component), and the correction to the exciton energy can be considered according to the usual perturbation theory.

The functional of the surface exciton energy in the zero order of perturbation theory has the form

$$\left\langle n, m \left| \hat{H}_{\text{eff}} \right| n, m \right\rangle = E_{nm}^{(0)} - \left\langle n, m \left| \frac{e^2}{4\pi\varepsilon_0\varepsilon_1^* \left| \rho + \rho_0 \right|} \right| n, m \right\rangle \\ - \sum_{\eta} \frac{\left| V_\eta \right|^2 \left| A_{nm}(\eta) \right|^2}{\hbar\Omega_S + \frac{\hbar^2\eta^2}{2M}}. \tag{6.9.15}$$

The energy corrections corresponding to the second term of the right-hand side (6.9.15) were determined in [90]:

$$- \langle n, m | \frac{e^2}{4\pi\varepsilon_0\varepsilon_1^* \left| \mathbf{\rho} + \mathbf{\rho}_0 \right|} | n, m \rangle \\ = - \frac{e^2}{4\pi\varepsilon_0\varepsilon_1^*} \int_0^{2\pi} d\varphi \int_0^{\infty} \rho d\rho \frac{n!}{2^{|m|+1}\left(n + \left| m \right|!\right) R_B^2} \times \\ \times \left(\frac{\rho}{R_B}\right)^{2|m|} \left| \mathbf{\rho} + \mathbf{\rho}_0 \right|^{-1} \exp\left(-\frac{\rho^2}{2R_B^2}\right) \left[L_n^{|m|}\left(\frac{\rho^2}{2R_B^2}\right)\right]^2. \tag{6.9.16a}$$

In general, it is impossible to calculate integrals in (6.9.16). Here are the results for some special cases:

a) $n = m = 0$:

$$\langle 0, 0 | - \frac{e^2}{4\pi\varepsilon_0\varepsilon_1^* |\mathbf{\rho} + \mathbf{\rho}_0|} |0, 0\rangle = - \frac{e^2}{8\pi^2\varepsilon_0\varepsilon_1^* R_B^2} \int_0^{2\pi} \int_0^{\infty} \frac{e^{-\frac{\rho^2}{2R_B^2}} \rho d\rho d\varphi}{|\mathbf{\rho} + \mathbf{\rho}_0|} \\ = - \frac{e^2}{4\pi\varepsilon_0\varepsilon_1^* R_B^2} \exp\left(-\frac{\rho_0^2}{4R_B^2}\right) I_0\left(\frac{\rho_0^2}{4R_B^2}\right); \tag{6.9.16b}$$

here $I_\nu(x)$ is a modified Bessel function $(\rho_0^2/(4r_0^2) = P_\perp^2/(4e\hbar))$. For arbitrary n and m, the integral included in the expression for $A_{nm}(\eta)$ is not calculated. In the special case, when $mm = 0$, $n \neq 0$, we have

$$A_{n0}(\eta) = R_B^{-2} \int_0^{\infty} \left[J_0(s_2\eta\rho) - J_0(s_1\eta\rho) \right] \\ \exp\left(-\frac{\rho^2}{2R_B^2}\right) \left[L_n^0\left(\frac{\rho^2}{2R_B^2}\right)\right]^2 \rho d\rho \\ = \frac{1}{\pi} \sum_{l=0}^{n} \frac{(-1)^l \Gamma\left(n - l + \frac{1}{2}\right) \Gamma\left(l + \frac{1}{2}\right)}{l!(n - l)!} \\ \times \left\{ \exp\left(-\frac{s_2^2\eta^2 R_B^2}{2}\right) L_{2l}^0\left(s_2^2\eta^2 R_B^2\right) - \exp\left(-\frac{s_1^2\eta^2 R_B^2}{2}\right) \right. \\ \left. L_{2l}^0\left(s_1^2\eta^2 R_B^2\right) \right\}. \tag{6.9.17a}$$

If $n = 0$, $m \neq 0$, then

$$A_{0m}(\eta) = \left[2^{|m|}m!R_B^2\right]^{-1}\int_0^\infty \left[J_0(s_2\eta\rho) - J_0(s_1\eta\rho)\right]\left(\frac{\rho}{R_B}\right)^{2|m|} \exp\left(-\frac{\rho^2}{2R_B^2}\right)\rho d\rho$$

$$= \exp\left\{-\frac{s_2^2R_B^2\eta^2}{2}\right\}L_{|m|}^0\left(\frac{s_2^2R_B^2\eta^2}{2}\right) - \exp\left\{-\frac{s_1^2R_B^2\eta^2}{2}\right\} L_{|m|}^0\left(\frac{s_1^2R_B^2\eta^2}{2}\right). \quad (6.9.17b)$$

Substituting (6.9.17a and (6.9.17b) into (6.9.12), we calculate the corrections to the energy and effective mass from the interaction of an electron and a hole with surface optical phonons. Let's write down a general expression for the energy in the Landau zero zone:

$$E_{0m} = \frac{1}{2}\hbar\omega_c(|m| - \gamma m + 1) - \frac{e^2\Gamma\left(|m| + \frac{1}{2}\right)}{4\pi\varepsilon_0\varepsilon_1^*\sqrt{2}R_B|m|!} - C\int_0^\infty \left\{\hbar\Omega_S + \frac{\hbar^2\eta^2}{2M}\right\}^{-1} \left\{e^{-\frac{s_2^2\eta^2R_B^2}{2}}L_{|m|}^0\left(\frac{s_2^2\eta^2R_B^2}{2}\right) - e^{-\frac{s_1^2R_B^2\eta^2}{2}}L_{|m|}^0\left(\frac{s_1^2R_B^2\eta^2}{2}\right)\right\}d\eta. \quad (6.9.18)$$

Expressions for the ground state energy and effective mass included in the formula:

$$E_{nm}(\mathbf{P}_\perp) = E_{nm}(0) + \frac{P_\perp^2}{2M_{nm}^*}, \quad (6.9.19)$$

for a special case $n = m = 0$:

$$E_{0,0} = \frac{2\hbar^2\lambda^2}{\mu} - \frac{2\sqrt{2\pi}\lambda e^2}{4\pi\varepsilon_0\varepsilon_1^*} - \frac{MC}{2\sqrt{2M\Omega_S\hbar^3}} \times\left\{\exp\left(\frac{M\Omega_S s_2^2}{2\lambda^2\hbar}\right)\left[1 - \Phi\left(\frac{s_2}{\lambda}\sqrt{\frac{M\Omega_s}{2\hbar}}\right)\right] + \exp\left(\frac{M\Omega_S s_1^2}{2\lambda^2\hbar}\right)\left[1 - \Phi\left(\frac{s_1}{\lambda}\sqrt{\frac{M\Omega_S}{2\hbar}}\right)\right]\right\} - 2\exp\left(\frac{M\Omega_S(s_1^2 + s_2^2)}{4\lambda^2\hbar}\right)\left[1 - \Phi\left(\frac{1}{2\lambda}\sqrt{\frac{M\Omega_S(s_1^2 + s_2^2)}{\hbar}}\right)\right]. \quad (6.9.20)$$

Here $\Phi(x)$ is a probability function.

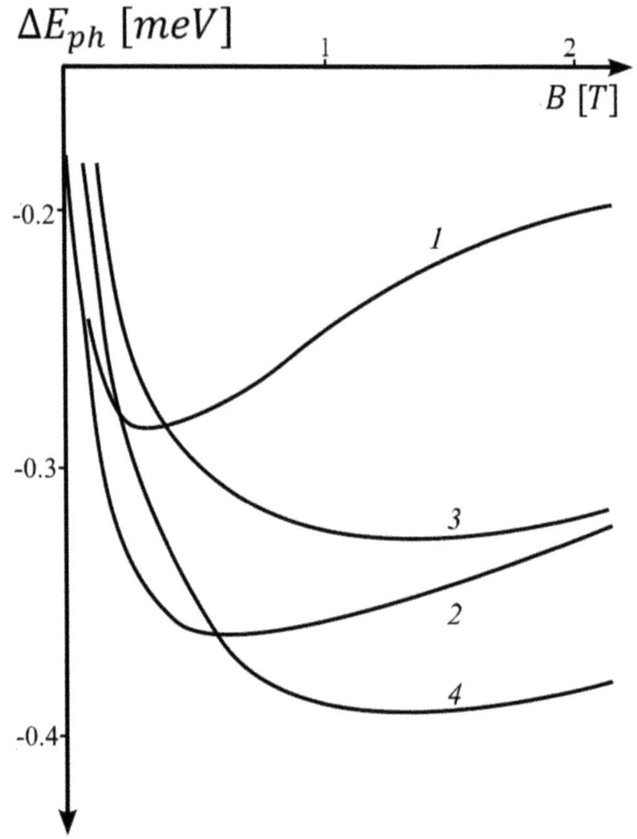

Figure 6.7. The dependences of $(\Delta E_{ph})_{nm}$ on the magnitude of magnetic induction, numerically calculated using the formulas (6.9.15) (the last term), (6.9.17a) and (6.9.17b) [92].

It follows from formula (6.9.16a) that the effective radius of the state is proportional to R_B and tends to zero with increasing B, which leads to an increase in the Coulomb interaction energy in proportion to $\sqrt{B}$. The dependence of the phonon contribution to the ground state energy (6.9.20) turns out to be nonmonotonic. Figure 6.7 shows the dependences of $(\Delta E_{ph})_{nm}$ on the magnitude of magnetic induction, numerically calculated using the formulas (6.9.15)—the last term, (6.9.17a) and (6.9.17b) [92]. The phonon correction to the effective mass, which was first obtained in [92] and is determined by the expression, which also turns out to be a nonmonotonic function of the field:

$$(\Delta M_{ph})_{00} = \frac{B_1}{B_2(B_2 - B_1)}; \tag{6.9.21}$$

here

$$B_1^{-1} = \frac{8\pi\varepsilon_0\varepsilon_1^*}{e^2}\left(\frac{2eB}{\pi}\right)^{\frac{1}{2}}. \tag{6.9.22}$$

Contribution to the effective exciton mass from inertialess polarization

$$B_2^{-1} = \frac{M(1 + F_1)}{F_1(1 + F_2) + F_2}, \tag{6.9.23}$$

where

$$F_1 = 2R_S R_{ex}^{-1} f_1(\varepsilon_{10}, \varepsilon_1, \varepsilon) \int_0^\infty x^2(1 + x^2)^{-3}\left(e^{-\zeta_1^2 x^2} - e^{-\zeta_2^2 x^2}\right)dx; \tag{6.9.24a}$$

$$F_2 = \frac{1}{2} R_B^4 R_{ex}^{-1} R_S^{-1} f(\varepsilon_{10}, \varepsilon_1, \varepsilon) \int_0^\infty x^2(1 + x^2)^{-1} e^{-(\zeta_1^2 + \zeta_2^2)x^2} dx; \tag{6.9.24b}$$

$$f(\varepsilon_{10}, \varepsilon_1, \varepsilon) = \frac{\varepsilon_{10} - \varepsilon_1}{(\varepsilon_{10} + \varepsilon)(\varepsilon_1 + \varepsilon)}; \; R_S = \left(\frac{\hbar}{2M\Omega_S}\right)^{\frac{1}{2}}; \; R_{ex} = \frac{\hbar^2 \varepsilon_1}{Me^2}; \; \zeta_{1,2} = \frac{m_{e,h}^* R_B}{\sqrt{2} M R_S}.$$

In a nonpolar crystal, the contribution from phonons to the effective mass disappears from the general expression for the effective mass, which can be obtained by decomposing into a series of $E_{n,m}(\mathbf{P}_\perp)$ (formula (6.9.16b)) up to quadratic terms:

$$M_{ex} = \varepsilon_1^* \sqrt{\frac{B}{\pi}} \left[\frac{4\sqrt{\pi\varepsilon_0}\hbar}{e}\right]^{\frac{3}{2}}. \tag{6.9.25}$$

Figure 6.8 shows the dependence of the phonon contribution to the effective mass on the induction of a magnetic field. Figures 6.7 and 6.8 show that the dependences

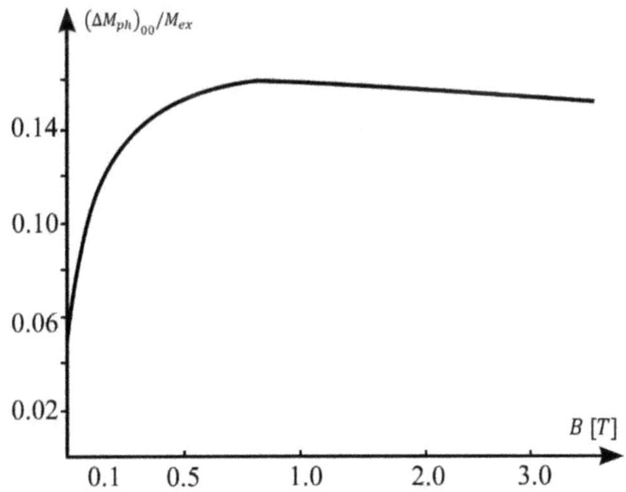

Figure 6.8. Dependence of the phonon contribution to the effective mass on the induction of the magnetic field.

of the phonon contributions to the exciton energy and its effective mass are nonmonotonic functions of the field. The reason for nonmonotonicity is as follows: when the magnetic length R_B is greater than the polaron radii $R_{e,\,h}$ of the electron and the hole

$$R_B > R_{e,\,h}\,(R_{e,\,h} = \left(\hbar/2m^*_{e,\,h}\Omega_s\right)^{1/2}), \tag{6.9.26}$$

the magnitude of the phonon contribution increases (in absolute value) proportionally to $\sqrt{B}$ due to the increased polarizing effect of the electron and hole charges. In the case when

$$R_B \sim R_{e,\,h}(\sim R_{ex}). \tag{6.9.27}$$

The polarization clouds of the electron and hole polarons overlap strongly, which significantly reduces the polarizing force of these charges, and the phonon contribution decreases proportionally to $1/\sqrt{B}$.

The domain of magnetic fields at which the phonon contribution to the energy of a given exciton state is maximal is determined from the condition

$$R_B \sim R_{e,\,h}(\sim R_{ex}). \tag{6.9.28}$$

For example, for the ground state ($n = m = 0$), the field value corresponding to the maximum phonon contribution is

$$B \sim \frac{\mu\Omega_S}{2e}\,(\sim 0,\, 2\;T\;(\mathrm{InSb})).$$

Figure 6.7 also shows that higher excited exciton states correspond to large phonon contributions to energy, which has a simple physical interpretation—an increase in the polarizing effect of particles with an increase in the exciton radius. Note that in order of magnitude $(\Delta E_{ph})_{n,\,m} = (0.01 - 0.1)E^{(0)}_{n,\,n}$ depending on the contact materials.

6.10 General approach to the Wannier–Mott exciton problem in a polar film

To obtain an exciton–phonon Hamiltonian, we write the initial Hamiltonian in the center-of-mass system:

$$
\begin{aligned}
\hat{H} = & -\frac{\hat{P}_\rho^2}{2\mu_\perp} + \frac{\hat{P}_{\|e}^2}{2m_{e\|}} + \frac{\hat{P}_{\|h}^2}{2m_{h\|}} + U(\rho,\, z_e,\, z_h) + U_{SA}(z_e) + U_{SA}(z_h) \\
& + \sum_{\eta, s_j} \hbar\Omega_{S_j} \hat{b}^+_{\eta,j} \hat{b}_{\eta,j} + \sum_{Q} \hbar\omega_{20} \hat{b}^+_Q \hat{b}_Q \\
& + \sum_{\eta, s_j} \Big\{ V^j_{\left\{\begin{smallmatrix}1\\3\end{smallmatrix}\right\}} \left[\cosh \eta z_e e^{i\sigma_1\eta\rho} - \cosh \eta z_h e^{-i\sigma_2\eta\rho}\right] \\
& + V^j_{\left\{\begin{smallmatrix}2\\4\end{smallmatrix}\right\}} \left[\sinh \eta z_e e^{i\sigma_1\eta\rho} - \sinh \eta z_h\, e^{-i\sigma_2\eta\rho}\right] \hat{b}^+_j(-\eta) + h.\, c.\Big\} \\
& + \sum_{Q, m=1,3,5,\ldots} \Big\{ V_{q_{\perp},m} \Big[\cos\left(\frac{m\pi\ z_e}{l_2}\right) e^{i\sigma_1 q_\perp \rho} - \cos \\
& \left(\frac{m\pi\ z_h}{l_2}\right) e^{-i\sigma_2 q_\perp \rho} \Big] \hat{b}^+_m(-Q) + h.\, c.\Big\} \\
& + \sum_{Q, m=2,4,6,\ldots} \Big\{ V_{q_{\perp},m} \Big[\sin\left(\frac{m\pi\ z_e}{l_2}\right) e^{i\sigma_1 q_\perp \rho} - \sin \\
& \left(\frac{m\pi\ z_h}{l_2}\right) e^{-i\sigma_2 q_\perp \rho} \Big] \hat{b}^+_m(-Q) + h.\, c.\Big\}.
\end{aligned} \tag{6.10.1}
$$

The derivation of the effective Hamiltonian from (6.10.1) by the Haken method is preceded by the following remark. In a plate, the movement of charges along is limited by its surfaces. If $l_2 \sim R_{e,\, h}$, then criterion (7.1.2) is violated. However, at small plate thicknesses, at which the level of dimensional quantization $\pi\ {}^2\hbar/(2m_{\square\square} l_2^2)$ reaches phonon energy values $\hbar\omega_{20}$, $\hbar\Omega_{S_{1,\,2}}$, it can be assumed that polarization is determined not by the electron z-coordinate, but by the density distribution of the electron cloud. For this reason, $f_{s,\,\eta}$, f_Q can be considered independent of z_e, z_h. In the future, we will limit ourselves to this simple situation. In accordance with the noted simplification, we first average the exciton Hamiltonian (6.10.1) on the wave functions describing the motion of an electron and a hole along the z-axis (in the special case, these are the wave functions of a particle in a box), and then on the wave function:

$$
\Psi(\rho,\ \{\hat{b}_{\nu_j}\}) = \varphi(\rho) \exp\left\{ \sum_{\nu_j} \hat{b}_{\nu_j} f^*_{\nu_j}(\rho) - h.\, c. \right\} |0\rangle, \tag{6.10.2}
$$

where $\nu_j = \mathbf{Q}$ or η. The sequence of further operations is described in section 7.1. After completing them, we find the displacement amplitudes:

$$= \frac{\widetilde{V}^{j*}_{\begin{Bmatrix} i_1 \\ i_2 \end{Bmatrix}}(\eta) e^{-i\sigma_1 \eta \rho}}{\hbar \Omega_{S_j} + \frac{\hbar^2 \eta^2}{2m_{e\perp}}} - \frac{\widetilde{V}^{j*}_{\begin{Bmatrix} i_1 \\ i_2 \end{Bmatrix}}(\eta) e^{i\sigma_2 \eta \rho}}{\hbar \Omega_{S_j} + \frac{\hbar^2 \eta^2}{2m_{h\perp}}}, \ j = 1, 2; \qquad (6.10.3a)$$

$$f_V = \frac{\widetilde{V}^*(\mathbf{q}_\perp) e^{-i\sigma_1 \mathbf{q}_\perp \rho}}{\hbar \omega_{20} + \frac{\hbar^2 q_\perp^2}{2m_{e\perp}}} - \frac{\widetilde{V}^*(\mathbf{q}_\perp) e^{i\sigma_2 \mathbf{q}_\perp \rho}}{\hbar \omega_{20} + \frac{\hbar^2 q_\perp^2}{2m_{h\perp}}}, \qquad (6.10.3b)$$

where

$$\widetilde{V}^{j}_{\begin{Bmatrix} i_1 \\ i_2 \end{Bmatrix}}(\eta) = \widetilde{V}^{j}_{\begin{Bmatrix} i_1 \\ i_2 \end{Bmatrix}}(\eta) \begin{Bmatrix} \langle \cosh \eta z_k \rangle_{e,h} \\ \langle \sinh \eta z_k \rangle_{e,h} \end{Bmatrix}; \ \widetilde{V}(q_\perp)$$

$$= \sum_m V_m \left[q_\perp^2 + \left(\frac{\pi m}{2 l_2} \right)^2 \right]^{-\frac{1}{2}} \begin{Bmatrix} \left\langle \cos\left(\frac{m\pi z_k}{l_2} \right) \right\rangle_{e,h} \\ \left\langle \sin\left(\frac{m\pi z_k}{l_2} \right) \right\rangle_{e,h} \end{Bmatrix}.$$

The brackets $\langle \ldots \rangle_{e,h}$ have the meaning of averaging on the wave functions of an electron and a hole, which in general can be different. Similarly:

$$\widetilde{V}(\mathbf{q}_\perp) = \sum_m \frac{V_m}{\left[q_\perp^2 + (\pi m/(2 l_2))^2 \right]^{1/2}} \begin{Bmatrix} \left\langle \cos\left(\frac{m\pi z_k}{l_2} \right) \right\rangle_{e,h}, m = 1, 3, 5, \ldots; \\ \left\langle \sin\left(\frac{m\pi z_k}{l_2} \right) \right\rangle_{e,h}, m = 2, 4, 6, \ldots. \end{Bmatrix} \qquad (6.10.4)$$

With their help, the surface and volume polarization contributions to the potential energy of the exciton are calculated:

$$W_S(\rho) = -\sum_{\eta, c=e,h} \frac{\left| \widetilde{V}^{(c)}_{\begin{Bmatrix} 1 \\ 3 \end{Bmatrix}}(\eta) \right|^2}{\hbar \Omega_{S_j} + \frac{\hbar^2 \eta^2}{2m_{c\perp}}} + \sum_{\eta, c} \frac{\left| \widetilde{V}^{(c)}_{\begin{Bmatrix} 1 \\ 3 \end{Bmatrix}}(\eta) \right|^2 \cos(\eta \rho)}{\hbar \Omega_{S_j} + \frac{\hbar^2 \eta^2}{2m_{c\perp}}} + \qquad (6.10.5a)$$

$$+ \sum_{\eta, j} \left| \widetilde{V}^{j}_{\begin{Bmatrix} 1 \\ 3 \end{Bmatrix}}(\eta) \right|^2 \cos(\eta \rho) \frac{\hbar^2 \eta^2}{2\mu_\perp} \prod_{c=e,h} \left[\left(\hbar \Omega_{S_j} + \frac{\hbar^2 \eta^2}{2m_{c\perp}} \right)^{-1} \right];$$

$$W_V(\rho) = -\sum_{\mathbf{q}_\perp,\, c=e,\, h}\left(\hbar\omega_{20} + \frac{\hbar^2 q_\perp^2}{2m_{c\perp}}\right)^{-1}\left[\sum_{m,m'=1,3,5,\dots}\widetilde{V}_m^{(c)}(\mathbf{q}_\perp)\widetilde{V}_{m'}^{(c)}(\mathbf{q}_\perp)\right]$$
$$+\sum_{\mathbf{\eta},c=e,h}\left(\hbar\omega_{20} + \frac{\hbar^2 q_\perp^2}{2m_{c\perp}}\right)^{-1}\left[\sum_{m,m'=1,3,5,\dots}\widetilde{V}_m(\mathbf{q}_\perp)\widetilde{V}_{m'}(\mathbf{q}_\perp)\right]\cos(\mathbf{q}_\perp\boldsymbol{\rho}) \quad (6.10.6)$$
$$+\sum_{\mathbf{\eta}}\left(\sum_{m,m'=1,3,5,\dots}\widetilde{V}_m(\mathbf{q}_\perp)\widetilde{V}_{m'}(\mathbf{q}_\perp)\right)\cos(\mathbf{q}_\perp,\boldsymbol{\rho})\frac{\hbar^2 q_\perp^2}{2m_\perp}\prod_{c=e,h}\left[\hbar\omega_{20} + \frac{\hbar^2 q_\perp^2}{2m_{c\perp}}\right]^{-1}.$$

The effective Hamiltonian averaged over z_e and z_h has the form

$$\overline{H}_{ex} = \frac{\pi^2\hbar^2}{2m_{e\|}} + \frac{\pi^2\hbar^2}{2m_{h\|}} + \frac{\hat{P}_\rho^2}{2\mu_\perp} + \overline{U}(\rho) + 2\overline{U}_{SA} + W_S(\rho) + W_V(\rho). \quad (6.10.7)$$

To find the exciton wave function $\varphi(\rho)$ let's use the variational method. Let's average (6.10.7) on the function $\varphi(\rho) = C\exp(-\lambda\rho)$:

$$\langle\varphi\,|\overline{H}_{ex}|\varphi\rangle = \frac{\pi^2\hbar^2}{2\mu_{\|}l_2^2} + \mathcal{K}(\lambda) + U(\lambda,\ l_2) + 2U_{SA}(l_2)$$
$$+W_V(\lambda,\ l_2) + W_V(l_2) + W_S(\lambda,\ l_2) + W_s(l_2) \equiv +E_1(\lambda,\ l_2) + E_2(l_2) \quad (6.10.8)$$

and we obtain the variational energy of the system, which includes terms of two types: with λ and without λ, that is, depending on the internal state of the exciton and independent of it. We will write them out explicitly when the wave functions of an electron and a hole describing motion along z are taken as cosines, that is, for particles in a rectangular potential well. The part $E_1(\lambda,\ l_2)$ should be interpreted as the exciton energy dimensionally dependent on the film thickness, and $E_2(\lambda,\ l_2)$ as describing the dimensional band effects. The components are included in $E_1(\lambda,\ l_2)$:

$$\mathcal{K}(\lambda) = \frac{\hbar^2\lambda^2}{2\mu_\perp} \quad (6.10.9)$$

is kinetic energy of internal transverse motion in an exciton;

$$U(\rho,\ \lambda_2) = -\frac{e^2}{2\pi\varepsilon_0\varepsilon_2}\int_0^\infty J_0\left(\frac{2\rho}{l_2}x\right)\left\{\frac{1}{x} + \frac{x}{2(x^2+\pi^2)} - \frac{\pi^4 e^{-x}\sinh x}{x^2(x^2+\pi^2)^2} + \right.$$
$$\left. + \frac{2\pi^4\sinh^2 x}{x^2(x^2+\pi^2)(e^{4x}-\delta_1\delta_3)}\left[\delta_1\delta_3 + \frac{e^{4x}\left(\varepsilon_2^2 - \varepsilon_1\varepsilon_3\right)}{(\varepsilon_2+\varepsilon_1)(\varepsilon_2+\varepsilon_3)}\right]\right\}dx \quad (6.10.10)$$

is the energy of the interaction of charges, describing two-dimensional effects having different origins. The first of them is due to the convergence of the electron and the hole with a decrease in l_2: $r \sim \sqrt{\rho^2 + (z_e - z_h)^2} \sim \sqrt{\rho^2 + l^2}$; which leads to an increase in the exciton binding energy. In the $l_2 \to 0$ limit, this effect gives a fourfold increase in the exciton Rydberg (geometric effect). The second is due to the fact that with a decrease in l_2, the lines of force connecting the hole to the electron are displaced into media 1 and 3, which at $\varepsilon_1, \varepsilon_3 > \varepsilon_2$ leads to a decrease in the exciton binding energy, and at $\varepsilon_1, \varepsilon_3 < \varepsilon_2$ to its increase (Keldysh effect) [93]. The functions $W_{S,V}(\lambda, l_2)$ describe the shielding of the electron–hole interaction by lattice polarization. This is the term of the exciton energy:

$$E_{ex} = E_{ex}^0(\lambda, l_2) + W_{S,V}(\lambda, l_2), \tag{6.10.11a}$$

where the energy of the unshielded exciton is

$$E_{ex}^0(\lambda, l_2) = \mathcal{K}(\lambda) + U(\lambda, l_2). \tag{6.10.11b}$$

The spatial quantization of the kinetic energy of the longitudinal motion of the exciton, leading to an effective increase in the band gap $\sim l_2^{-2}$, is given by the sum of the kinetic energies of the electron and the hole:

$$\mathcal{K}(l_2) = \frac{\pi^2\hbar^2}{2l_2^2}\left(\frac{1}{m_{e\|}} + \frac{1}{m_{h\|}}\right) = \frac{\pi^2\hbar^2}{2\mu_{\|}l_2^2}. \tag{6.10.12}$$

With small film thicknesses, this dimensional effect is the main one. The zone size effect due to self-action [94] is described by the term

$$\begin{aligned} 2U_{SA}(l_2) = &\frac{e^2}{2\pi\varepsilon_0\varepsilon_2 l_2}\int_0^\infty \frac{dx}{e^{2x} - \delta_1\delta_3} \\ &\left\{\delta_1\delta_3 + \frac{(\varepsilon_2^2 - \varepsilon_1\varepsilon_3)4\pi^2 e^x \sinh x}{(\varepsilon_2 + \varepsilon_1)(\varepsilon_2 + \varepsilon_3)x(x^2 + 4\pi^2)}\right\}. \end{aligned} \tag{6.10.13}$$

It plays an important role at low l_2, shifting the edges of the band gap in opposite directions (narrowing or expanding it). For small l_2 $U_{SA} \sim l_2^{-1}$. The dimensionally dependent polaronic displacements of the edges are described by the terms $W_S(l_2)$ and $W_V(l_2)$ from (6.10.8). The explicit form of phonon contributions is not given due to their bulkiness.

In another situation, when the longitudinal motion of a heavy particle is limited by the electron–hole interaction, its potential energy is found by averaging the electron–hole potential on the wave function of a light particle. Averaging on the oscillator function gives the variational energy:

$$\begin{aligned} E_{ex}(\lambda, l_2) = &\frac{\hbar^2}{2\mu_{\|}l_2^2} - A_2 z_0^2 + \hbar\omega_n(\lambda, l_2)\left(n + \frac{1}{2}\right) \\ &+ W_S(\lambda, l_2) + W_V(\lambda, l_2), \end{aligned} \tag{6.10.14}$$

where $W_S(\lambda, l_2)$ and $W_V(\lambda, l_2)$ are phonon contributions calculated with averaging on the wave functions (6.2.3) and (6.2.10) for an electron and a hole, respectively.

In the general case of arbitrary electron–hole and electron–phonon interactions, it is impossible to determine the explicit form of wave functions describing motion along z, and therefore it is impossible to perform averaging λ, in (6.10.3) and (6.10.4). However, in the extreme case of oscillatory motion of a heavy particle, calculations can be completed in a variational procedure with one variational parameter, using the wave functions (6.2.3) and (6.2.10) to calculate the averages in (6.10.3), (6.10.4). If necessary, for specific cases, calculations will be carried out in these two alternative approximations of both light masses or one light and one heavy mass.

6.10.1 Coulomb interaction in the limit of a thin polar film

As already mentioned in clause 6.1.2, in a thin film of a polar crystal, there may be a specific effect of strengthening the electron–hole bond, which consists of reducing the contributions to the shielding of the electron–hole interaction of inertial polarization with a decrease in l_2. Inertial vibrations actively shield the interaction in the case when the exciton frequency $\omega_{ex} = E_{ex}/\hbar$ is significantly lower than the frequency of longitudinal optical vibrations ω_0, Ω_{S_j}:

$$\omega_{ex} \ll \omega_0 \tag{6.10.15}$$

(adiabaticity of optical vibrations).

However, as l_2 decreases, the exciton frequency increases (clause 6.1.2) and the inequality (6.10.15) weakens. It is also possible that the sign of inequality changes to the opposite (this is the case, for example, in a CdTe film on a substrate: at $l_2 \gg a_{ex} \sim 10$ nm $E_{ex} = 10$ meV, $\hbar\omega_0 = 22$ meV; при $l_2 \approx 3$ nm $E_{ex} = 100$ meV).

The indicated effect of the loss of inertial shielding in turn leads to an additional strengthening of the bond and is manifested in the fact that with a decrease in l_2 terms with ε_{20}appear in the formulas for the potential describing the electron–hole interaction, deepening the potential, and with a further decrease in l_2, the terms with ε_{20} disappear and are replaced by similar ones, but with ε_2 instead ε_{20}.

From the general expression for the potential energy of the electron–hole interaction, which is the sum of $U(\rho, \lambda_2)$ (formula (6.10.10)), $W_S(\rho)$, $W_V(\rho)$ (formulas (6.10.5a), (6.10.6)); in the $l_2 \ll a_{ex}$ limit, we obtain

$$U(\rho) = W_{pl}(\rho) + W_{\text{ph}}(\rho), \tag{6.10.16}$$

where

$$W_{pl}(\rho) = -\frac{e^2}{4\varepsilon_0\varepsilon_2 l_2}\left[H_0\left(\frac{2\varepsilon_1\rho}{\varepsilon_2 l_2}\right) - N_0\left(\frac{2\varepsilon_1\rho}{\varepsilon_2 l_2}\right)\right]; \tag{6.10.17a}$$

$$W_{\text{ph}}(\rho) = 2[F(\varepsilon_2, R_S) - F(\varepsilon_{20}, R_S)]; \tag{6.10.17b}$$

$$F(\varepsilon_i, R_S) = \frac{e^2}{4\varepsilon_0\varepsilon_i l_2}\left[1 + \left(\frac{2\varepsilon_1 R_s}{\varepsilon_i}\right)^2\right]^{-1}\left\{\frac{\pi}{2}\left[H_0\left(\frac{2\varepsilon_1\rho}{\varepsilon_i l_2}\right) - N_0\left(\frac{2\varepsilon_1\rho}{\varepsilon_i l_2}\right)\right] - K_0\left(\frac{\rho}{R_S}\right) + \frac{\pi\varepsilon_1 R_S}{\varepsilon_i l_2}\left[I_0\left(\frac{\rho}{R_S}\right) - L_0\left(\frac{\rho}{R_S}\right)\right]\right\};\ i = 2, 20; \tag{6.10.17c}$$

where $K_0(x)$ is the MacDonald function (here, for simplicity, we consider the case when $m_e^* = m_h^*$).

Let's enter the parameter $\kappa_i = \varepsilon_1/\varepsilon_i$ and consider a situation where the condition is met:

$$\kappa_i \ll 1. \tag{6.10.18}$$

If the inequality is satisfied:

$$l_2 < a_{ex} < \rho < l_{2\mathrm{eff,i}}\ (l_{2\mathrm{eff,i}} = l_2/\kappa_i), \tag{6.10.19a}$$

then from formulas (6.10.16), (6.10.17a–b) and using decompositions of Struve, Neumann, Bessel, and MacDonald functions [95], we obtain:

$$U(\rho) \approx -\frac{e^2}{4\varepsilon_0\varepsilon_{20} l_2}\left[\ln\left(\frac{l_{2\mathrm{eff},\,20}}{\rho}\right) - C\right] + O\left(\frac{\rho}{l_{2\mathrm{eff},\,20}}\right). \tag{6.10.19b}$$

When performing an inequality

$$l_2 < R_S < l_{2\mathrm{eff},\,i} < \rho \tag{6.10.19c}$$

from the formulas (6.10.16)–(6.10.17a–c) we get

$$U(\rho) = -\frac{e^2}{4\pi\varepsilon_0\varepsilon_1\rho} + \frac{e^2}{4\pi\varepsilon_0\varepsilon_1\rho}\left(\frac{\varepsilon_{20} l_2}{2\varepsilon_1\rho}\right)^2 + O\left(\mathrm{e}^{-\frac{\rho}{R_S}}\right). \tag{6.10.19d}$$

Finally, when moving to inequality

$$l_2 < l_{2eff,\,i} < R_S < \rho \tag{6.10.19e}$$

from expressions (6.10.16)–(6.10.17) we get

$$U(\rho) = -\frac{e^2}{4\pi\varepsilon_0\varepsilon_1\rho} + \frac{e^2}{4\pi\varepsilon_0\varepsilon_1\rho}\left(\frac{\varepsilon_2 l_2}{2\varepsilon_1\rho}\right)^2 + O\left(\mathrm{e}^{-\frac{\rho}{R_S}}\right). \tag{6.10.19f}$$

Formulas (6.10.19b, 6.10.19d, 6.10.19f) describe the dynamics of deepening the potential of the electron–hole interaction with a decrease in l_2 both due to the effect of displacement of force lines into a medium with a smaller one ε, and the effect of loss of inertial shielding (the appearance of ε_2 in formula (6.10.19f) instead of ε_{20} in (6.10.19e)). The second effect is significant at intermediate values of l_2, at which the potential has not yet reached its asymptotic value $U^{2D}(\rho) = -e^2/(4\pi\varepsilon_0\varepsilon_1\rho)$. In section 7.7 we discuss the manifestation of this effect in real experimental situations.

6.11 Excitons in thin films of PbI_2 and CdTe

6.11.1 Exciton in a thin film of PbI_2 on various substrates

For semiconductors in which excitons have a high oscillator strength, the effect on the optical absorption spectra of the substrate material and dimensional exciton effects can be observed in ultrathin films ($l < 10.0$ nm), obtained by vacuum spraying. An example is the PbI_2, films deposited by evaporation in vacuum [82, 96–98] on various substrates. To obtain a linear distribution of thicknesses along the wedge, modulating the evaporated flow of matter using a disk was used [99]. The quality of the obtained films was determined by the electronogram. As noted, the light absorption coefficient in the exciton domain of these films is quite high ($7 \cdot 10^5$ cm) [100]; the minimum transmission can be confidently fixed even at small thicknesses of $l_2 \sim 5.0$ nm. Let's describe the main results of the experiment [97, 98, 101]. With a decrease in the film thickness, a short-wave shift in the energy position of the transmission minimum was observed. To study the effect of media bordering the film on the energy position of the exciton level, transmission spectra of thin PbI_2 films deposited on different substrates under the same technological conditions were measured. Figure 6.9 shows the dependences of the energy position of the exciton level on the thickness of the PbI_2 films deposited on a glass ($\varepsilon_3 = 4$) and mica ($\varepsilon_3 = 8$) substrate. The displacement of the exciton peak into the short-wavelength domain of the spectrum was observed with a decrease in thickness, and for a film deposited on a substrate with a lower dielectric constant (glass), a stronger displacement was observed than for a substrate with a higher dielectric constant (mica). The experiment also noted a discrepancy in the energy values of the bulk exciton in layers on different substrates (figure 6.9, curves 1 and 2), possibly due to some uncontrolled influence of technology. Figure 6.10 shows the energy of a photon generating an exciton in a PbI_2 film on a silver substrate ($\varepsilon_3 \to \infty$), calculated from the dimensional dependence of the reflection minimum (the minimum offset was determined by its position in the spectrum of the thickest film on the same substrate). It follows from this figure that the dependence of the energy position of the exciton line on l_2 is nonmonotonic in nature and has a pronounced minimum at $l_2 = 10$ nm.

These measurements relate to a three-layer structure (1|20|3) in which layer 1 is liquid helium, 2 is PbI_2, and 3 is glass, mica or silver. The thicknesses of the layers l_1, l_3 can be considered infinite.

To describe the experimental results reflected in the plots in figures 6.9 and 6.10, we will use the theory set out in section 7.6 in both versions. The minimization of the variational energy was carried out by numerical methods at the following values of the structure parameters: $\varepsilon_{20} = 26.4$; $\varepsilon_2 = 6.1$ [26], $m_e|| = 2.1m_0$; $m_{e\perp} = 0.48m_0$; $m_h|| = 0.195m_0$; $\hbar\omega_0 = 13$ meV; $\varepsilon_1 = 1$ (liquid helium), $\varepsilon_3 = 4$ (quartz glass), $\varepsilon \to \infty$ (metal). The energy position of the exciton level in the massive sample ($l_2 \to \infty$) $\hbar\omega_{ex} = 2.498$ meV, and the band gap of the massive crystal PbI_2 $E_{g0} = 2.525$ eV.

The energy of the photon generating the exciton is directly determined from the experiment:

$$\hbar\omega_{ex} = E_{g0} + \mathcal{K}(l_2) + 2U_{SA}(l_2) + W_S(l_2) + W_V(l_2) + E_{ex}[\lambda(l), l_2]; \qquad (6.11.1)$$

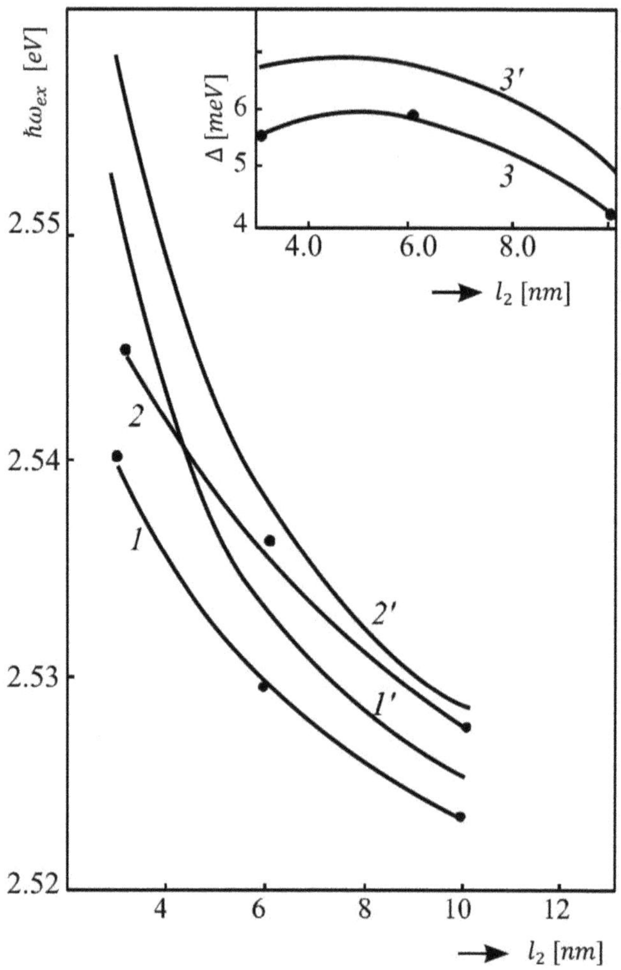

Figure 6.9. (a) The dependence of the energy position of the exciton level on the thickness of the PbI_2films deposited on a glass (curves 1, 1′) and mica (curves 2, 2′) substrate. Curves 1, 2 are constructed according to experimental data, and curves 1′, 2′ are calculated theoretically. In the inset: the difference in the values of exciton levels on mica and glass substrates. Curve 3 is based on experimentally measured values; curve 3′ is based on theoretically calculated values.

where

$E_{ex}[\lambda(l), l_2]$ is the minimized on λ exciton energy.

The calculation results are summarized in table 6.2, which allows us to draw the following conclusions: from comparing the values of the exciton energy $E^0_{ex} = K(\lambda) + U(\lambda, l)$, calculated without taking into account the electron–phonon interaction (column 3 of the table) with $E_{ex}(\lambda, l)$, including the electron–phonon interaction (column 6 of the table), it follows that the latter plays an important role in the formation of exciton states in PbI_2 films. Shielding of the electron–hole

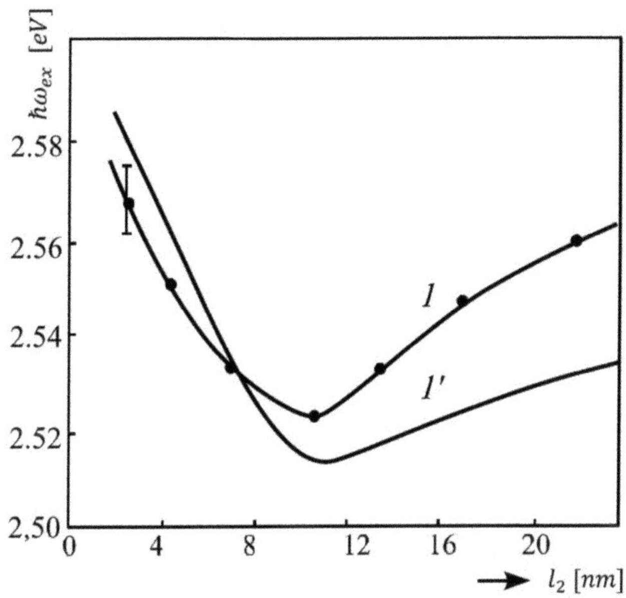

Figure 6.10. The energy of a photon generating an exciton in a PbI_2 film on a silver substrate ($\varepsilon_3 \to \infty$), calculated from the dimensional dependence of the minimum reflection. Curve 1 is based on experimental data; curve 1′ is based on theoretical data.

interaction by bulk and surface vibrations reduces the binding energy of the polaron exciton by about three times for all l_2 and substrates $\varepsilon_3 = 4$ and $\varepsilon_3 = 8$. With a decrease in l_2, the shielding of the electron–hole interaction by bulk vibrations decreases rapidly, and by surface vibrations it also increases rapidly, as a result of which the total effect remains almost constant (columns 4 and 5). Among the band size effects, the most important for small l_2 is the dimensional quantization of the kinetic energy $K(l_2)$ of longitudinal motion, which increases the band gap width and with it the energy of the exciton-generating photon. With intermediate l_2, the energy of self-action plays an important role (column 10). In the case of a substrate made of glass ($\varepsilon_3 = 4$) and helium ($\varepsilon_1 = 1$) $\varepsilon_2 > \varepsilon_3,\ \varepsilon_1$ the self-action energy of both surfaces (1|2) and (2|3) has a positive value, therefore it is the highest; in the case of mica plates, it becomes negative and is generally less than that of glass. Finally, in the case of a metal lining at the surface (2|3), the self-action is large and has a negative sign, so $U_{SA}(l_2)$ as a whole also becomes negative.

Summarizing, it can be concluded that the dimensional effects fall into two groups: (a) related to the renormalization of zones (band effects)—dimensional quantization of the kinetic energy of charge carriers; the effect of self-action of an electron and a hole; dependences of the polaronic energies of an electron and a hole on l_2 (b) intraexiton dimensional effects, consisting of dependence of the electron–hole interaction on polarization contributions to the shielding of electron–hole interaction by bulk and surface phonons (the dimensional effect of the change in

Table 6.2. The result of calculating the exciton energy.

Substrate material	d nm	E_{ex}^{0} eV	$U^{S_1} + U^{S_2}$ eV	U^{V} eV	E_{ex} eV	$U_{ph}^{S_1} + U_{ph}^{S_2}$ eV	U_{ph}^{V} eV	K eV	U_{SA} eV
glass	3	−0.152	0.084 + 0.014	0.006	−0.048	−0.054 − 0.009	−0.002	0.234	0.086
	6	−0.145	0.061 + 0.014	0.031	−0.039	−0.037 − 0.008	−0.017	0.059	0.043
	10	−0.144	0.043 + 0.013	0.053	−0.034	−0.025 − 0.006	−0.045	0.021	0.026
mica	3	−0.115	0.046 + 0.024	0.004	−0.041	−0.029 − 0.014	−0.002	0.234	0.042
	6	−0.108	0.034 + 0.012	0.021	−0.033	−0.021 − 0.012	−0.017	0.059	0.021
	10	−0.106	0.021 + 0.012	0.043	−0.029	−0.014 − 0.009	−0.045	0.021	0.018
metal	3	−0.091	0 + 0.052	0.003	−0.036	−0.020 − 0	−0.002	0.234	−0051
	6	−0.088	0 + 0.036	0.024	−0.028	−0.015 − 0	−0.017	0.059	−0.021
	10	−0.075	0 + 0.020	0.030	−0.025	−0.011 − 0	−0.045	0.021	−0.015

shielding [1] on the thickness of the film). At the same time, for a quantitative description of the experiment, it is necessary to correctly take into account all contributions to the energy of the photon generating the exciton, which is clearly demonstrated in table 6.2.

6.11.2 Dimensionally quantized states of the Wannier–Mott exciton in ultrathin CdTe films

Of great interest for the application of the theory are recent experiments on the observation of large-radius excitons in thin films of cadmium telluride deposited on a polar substrate of magnesium fluoride [102–104]. Exciton absorption peaks were observed for transitions with the formation of light and heavy excitons for different levels of dimensional quantization.

The general theory developed in section 6.1 cannot be directly used to describe these experiments and needs some clarifications and additions [19]:

(1) the valence band of CdTe is fourfold degenerated. In the experimentally studied samples with thicknesses from 25.0 to 3.0 nm, the valence band is split into two doubly degenerate zones of light and heavy holes;
(2) the transverse masses of the electron and the light hole turn out to depend on the thickness of the film due to the effect of the parabolicity of the zones;
(3) exciton states with electrons and holes appear in the optical absorption spectrum of a dimensionally quantized *CdTe* film at higher levels of dimensional quantization ($N_e = N_h = 2$). To calculate them, it is necessary to take into account the dependence of the longitudinal effective mass on the number of the dimensional quantization level [35, 105];
(4) in thick films ($l_2 > a_{ex}$), the energy of dimensional quantization of an electron and a hole becomes less than the Coulomb energy. In this case, the motion of the exciton as a whole is quantized, with a slight change in its binding energy [106].

This section provides a theoretical description of these effects and, based on this theory, an interpretation of the experimental absorption spectra of ultrathin CdTe films [102, 103].

The exciton–phonon Hamiltonian of the system under consideration differs from the Hamiltonian (6.10.1) by the presence of additional terms describing the influence of the polar substrate, taking into account the nonparabolicity of the zones and changing the type of operators of interactions with surface optical phonons.

The operators of the kinetic energies of an electron and a hole can be represented as

$$\hat{H}_i = \frac{\hat{P}_\rho^2}{2\mu_\perp} + \frac{\hat{P}_z^2}{2\mu_{||}}, \tag{6.11.2}$$

where $\mu_{\square\square} = \mu_{N\square\square}(l)$ is the reduced longitudinal mass of the electron and the hole, taking into account renormalization due to nonparabolicity; and $\mu_{\perp} = \mu_{\perp}(N)$ is the reduced transverse mass, which depends on the number of the subband of the dimensional quantization N.

For the longitudinal effective masses of an electron and a light hole renormalized due to nonparabolicity, the formulas obtained in [107] can be used:

$$\frac{m}{m_e^*} = 1 + \frac{\mathcal{P}^2}{2D}\left[\frac{D + D_h}{3}\left(\frac{2}{E_g} + \frac{1}{E_g + \Delta_{SO}}\right) - \frac{D - D_h}{D - E_g}\right]; \tag{6.11.3a}$$

$$\frac{m}{m_{lh}^*} = \frac{2\mathcal{P}^2}{3E_g} - 1. \tag{6.11.3b}$$

The following numerical parameter values were used for numerical calculations: $\mathcal{P}^2 = 21$ eV; $\Delta_{SO} = 0.92$ eV; $D = E(\Gamma_{15C}) - E(\Gamma_{15V}) = 5.2$ eV; $D_h = E(\Gamma_{15}) - E(\Gamma_{25}) = 3.2$ eV; $E_g = E_g(D)$ is the width of the forbidden zone, which takes into account all contributions renormalizing its value:

$$E_g(d) = E_{g0} + E_{s.q.} + E_{SA}^{(l)}(d) + E_{SA}^{(h)}(d) + E_{ph}^{(e)}(d) + E_{ph}^{(h)}(d); \tag{6.11.4}$$

where E_{g0} is the band gap width of a massive crystal and $E_{s.q.}$ is the contribution from the effect of size quantization.

The components of the effective masses in the film plane can be calculated using the formulas obtained in [19]:

$$\frac{m_0}{m_{N_\perp}^{hh,lh}} = A \mp B \mp \frac{C^2}{2B} + \frac{\operatorname{sign}(A \mp B)\sqrt{A^2 - B^2}\left[(-1)^{n+1} + \cos\left(\frac{A \mp B}{A \pm B}\right)^{\frac{1}{2}}\pi n\right]}{\pi n B^2 \sin\left[\frac{A - B}{A + B}\right]^{\frac{1}{2}}\pi n}, \tag{6.11.5}$$

where $A = -4.11$; $B = -3.44$; $C^2 = -6.33$; $\operatorname{sign}(x) = 1,\ x > 0$; $\operatorname{sign}(x) = -1,\ x < 0$.

The upper signs correspond to *hh*, the lower ones to *lh*.

For the first zone of dimensional quantization from the expression (6.11.5) we obtain the following values: $m_{1\perp}^{lh} = 0.197m_0$; $m_{1\perp}^{hh} = 0.446m_0$; $m_{2\perp}^{lh} = 0.106m_0$; $m_{2\perp}^{hh} = 0,\ 208m_0$.

The interaction of an electron and a hole with bulk and surface optical phonons of a film and substrate is described by the Hamiltonian (6.5.1.28) and (6.5.1.31). Since ω_{TO_1} (CdTe) $< \omega_{TO_2}$ (MgF_2), the substrate can be considered as nonpolar with a static dielectric constant of $\varepsilon_0(MgF_2) = 5.45$, which apparently does not introduce a noticeable error. For a light exciton and both quantization levels ($N = 1$; $N = 2$)

the wave function can be selected in the form (6.2.3). Averaging all the contributions to the exciton energy on it, we find:

1. For $N = 1$—the same formulas as for PbI_2 (clause 6.10.1), taking into account paragraphs 1 and 2 of this section;
2. For $N = 2$, the energy of the 'naked' exciton is found from minimizing the variational energy

$$[E_{ex}(\lambda, l)]_{N=2} = \frac{\hbar^2}{2\mu_\perp l^2} - \frac{e^2}{2\pi\varepsilon_0\varepsilon_2}\int_0^\infty dx\left(1 + \frac{\lambda^2}{l^2}x^2\right)^{-\frac{3}{2}} \times$$
$$\times\left\{\frac{1}{x} + \frac{x}{2(x^2+4\pi^2)} - \frac{8\pi^4(1-e^{-2x})}{x^2(x^2+4\pi^2)^2} + \frac{32\pi^4\sinh^2 x}{e^{4x}-\delta_1\delta_3}[\delta_1\delta_3 + e^{2x}f_1]\right\}. \quad (6.11.6)$$

The energy of the exciton–phonon interaction is described by the formulas:

$$W_{S_{1,2}}(\lambda, l) = B\int_0^\infty dx\,\varphi_{1,2}(x, l)\left(1 + \frac{\lambda^2}{l^2}x^2\right)^{-\frac{3}{2}}, \quad B = \frac{4\pi^3 e^2}{\varepsilon_0 l}; \quad (6.11.7a)$$

$$W_V(\lambda, l) = \frac{128B}{\pi^8}\left(\varepsilon_2^{-1} - \varepsilon_{20}^{-1}\right)\int_0^\infty dx$$
$$\sum_{m=1,3,5,\ldots}[x + m^2\pi^2]^{-1}\left\{\left[1 + \frac{R_e^2}{l^2}(x + m^2\pi^2)\right]^{-1}\right.$$
$$+\left[1 + \frac{R_h^2}{l^2}(x + m^2\pi^2)\right]^{-1} \quad (6.11.7b)$$
$$\left.\left(1 + \frac{x^2}{4\lambda^2}\right)^{-\frac{3}{2}}\left(8 - m^2\right)^2 m^{-2}\left(16 - m^2\right)^{-1}\right\},$$

где

$$\varphi_{1,2}(x, l) = \left(\tilde{\varepsilon}_2^{-1} - \tilde{\varepsilon}_{20}^{1,2}\right)^{-1}\frac{\tanh^2 x}{x^2}\frac{F_{1,3}^2}{\left(x^2 + 4\pi^2\right)^2}\left[\left(1 + \frac{4R_e^2}{l^2}x^2\right)^{-1}\right.$$
$$\left. + \left(1 + \frac{4R_h^2}{l^2}x^2\right)\right]. \quad (6.11.7c)$$

The sum of expressions (6.11.6)–(6.11.7c) gives the exciton energy taking into account phonon shielding:

$$[E_{ex}(\lambda, l)]_{N=2} = \left[E_{ex}^0(\lambda, l) + W_{S_1}(\lambda, l) + W_{S_2}(\lambda, l) + W_V(\lambda, l)\right]_{N=2}. \quad (6.11.8)$$

The energy of self-action has the form

$$U_{SA}(l) = \frac{B}{4\pi^4\varepsilon_2}\int_0^\infty dx(e^{2x} - \delta_1\delta_3)^{-1}\left\{\delta_1\delta_3 + f_1\frac{4\pi^2 e^x \sinh x}{x(x^2 + 4\pi^2)}\right\}. \qquad (6.11.9)$$

The polaron contributions are equal:

$$W_{S_{1,2}}(l) = B\int_0^\infty dx\,\varphi_{1,2}(x, l); \qquad (6.11.10a)$$

$$W_V(l) = -\sum_{c=e,h}\alpha_c\hbar\omega_{02}\frac{4l}{\pi^5 R_c}$$
$$\sum_{m=1,3,5,\ldots}\left\{\ln\left[\frac{(2m-1)^2\left[(4(2m+1)^2-1)^{-1}-(4(2m)^2-1)\right]}{4(m-2)(m+1)+1-\gamma_c(l)}\right]^{-1}\right. \qquad (6.11.10b)$$
$$+\ln\left[\frac{4m^2}{15+\gamma_c(l)}\right]\left[\frac{1}{4(2+m)^2-1}-\frac{1}{4(2-m)^2-1}\right]$$
$$\frac{1}{\left[4(m^2-4)+1-\gamma_c(l)\right]}.$$

The energy of the exciton-generating quantum, with $N = 2$, is equal to

$$[\hbar\omega_{ex}]_{N=2} = [E_{ex}(l)]_{N=2} + 2U_{SA}(l) + W_{S_1}(l) + W_{S_2}(l) + W_V(l). \qquad (6.11.11)$$

For a heavy exciton at $N = 2$ the function (6.2.3) can also be used. For $N = 1$, in parallel with the calculation of the wave function (6.2.3), it is necessary to perform a calculation with the wave function of a light particle and the oscillatory wave function of a heavy particle. The calculation results are shown in the plots of figures 6.11 and 6.12 and in table 6.3.

1. The calculation of the binding energy of light and heavy excitons was performed numerically. The following values of the system parameters were used in the calculation: liquid helium–cadmium telluride film–magnesium fluoride substrate—$\varepsilon_{He} \approx 1$; $\varepsilon_{20} \equiv \varepsilon_0 = 10.6$; $\varepsilon_2 \equiv \varepsilon_\infty = 7.13$; $\varepsilon_{MgF_2} = 5.45$; $\hbar\omega_0 = 21.7$ meV; $m_e^* = 0.096m_0$; $m_{lh}^* = 0.133m_0$; $m_{hh} = 1.493m_0$. Figure 6.12 shows the dependences of the reduced transverse effective masses of the electron and the hole on the thickness of the film, calculated using the formulas (6.11.3a and (6.11.3b). Taking into account the nonparabolicity of the zones significantly reduces the energy of dimensional quantization for both light and heavy excitons. Figure 6.13 shows the dependence of the binding energies of light (curve 1—$N = 1$; 2—$N = 2$) and heavy (curve 3—$N = 1$; 4—$N = 2$) excitons on $\ln l/l$. For comparison, the results obtained using the Keldysh formula [108] (curve E_{ex}^{1e-1lh}, 6—E_{ex}^{1e-1hh}):

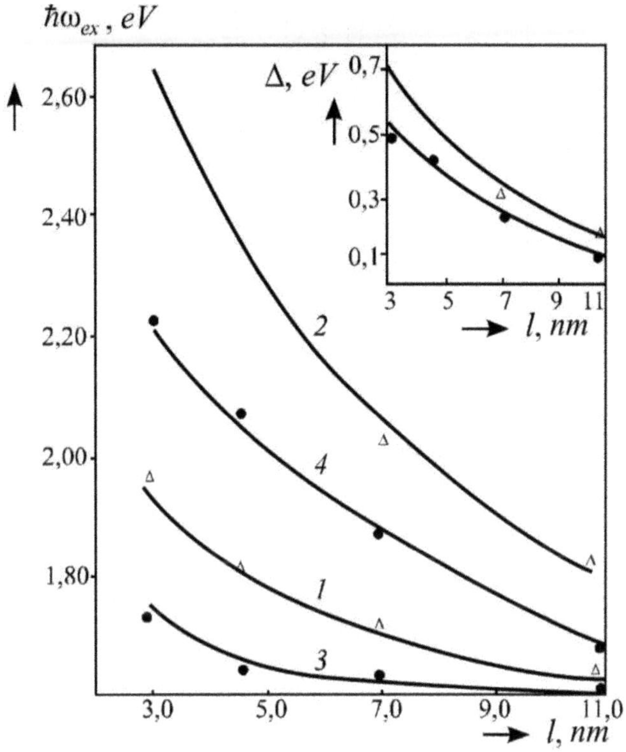

Figure 6.11. The dependences of $\hbar\omega_{ex}$ on l for light (curve 1—$N = 1$; 2—$N = 2$) and heavy (curve 3—$N = 1$; 4—$N = 2$) excitons calculated according to the formula (33) of [82]. In the box: the difference in the energy values of the quantum generating heavy and light excitons.

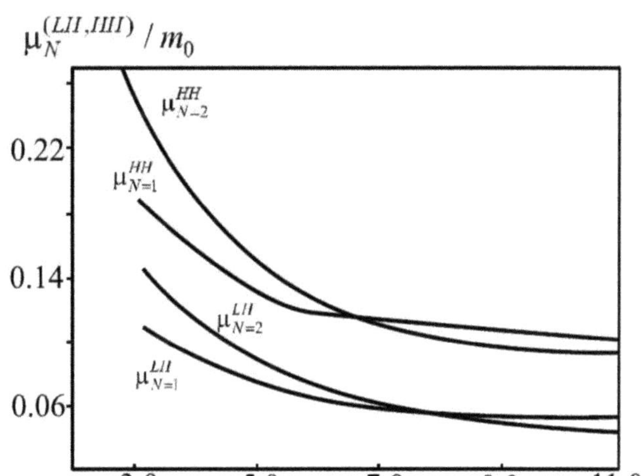

Figure 6.12. The dependences of the reduced transverse effective masses of the electron and the hole on the thickness of the film, calculated by the formulas (6.10.3a, b).

Table 6.3. Results of calculation of the energy of the exciton-generating quantum (formula (6.11.11)).

	l (nm)	E_{ex}^0 (meV)	λ (nm)	$U^{S_1} + U^{S_2}$ meV	U^V (meV)	E_{ex} (meV)	$U_{ph}^{S_1} + U_{ph}^{S_2}$ (meV)	U_{ph}^V (meV)	K (meV)	U_{SA} (meV)	$\hbar_{ex}^{theor}$ (meV)	$\hbar_{ex}^{\exp}$ (meV)
$1e - 1hh$	3.1	−105	3.5	15+5	4	−81	−10−4	−2	204	63	1770	1739
	4.5	−88	3.9	13+4	5	−66	−9−4	−5	127	44	1687	1671
	7.0	−68	4.7	11+4	9	−44	−7−3	−13	67	23	1629	1646
	11.0	−52	5.6	9+3	16	−24	−6−2	−19	39	18	1608	1620
$2e - 2hh$	3.1	−62	3.9	8+2	2	−50	−8−2	−1	560	91	2190	2193
	4.5	−54	4.7	6+2	6	−40	−7−2	−5	470	63	2081	2087
	7.0	−44	5.3	5+1	10	−29	−5−1	−8	300	41	1898	1894
	11.0	−35	6.4	4+1	13	−17	−4−1	−13	120	26	1711	1709

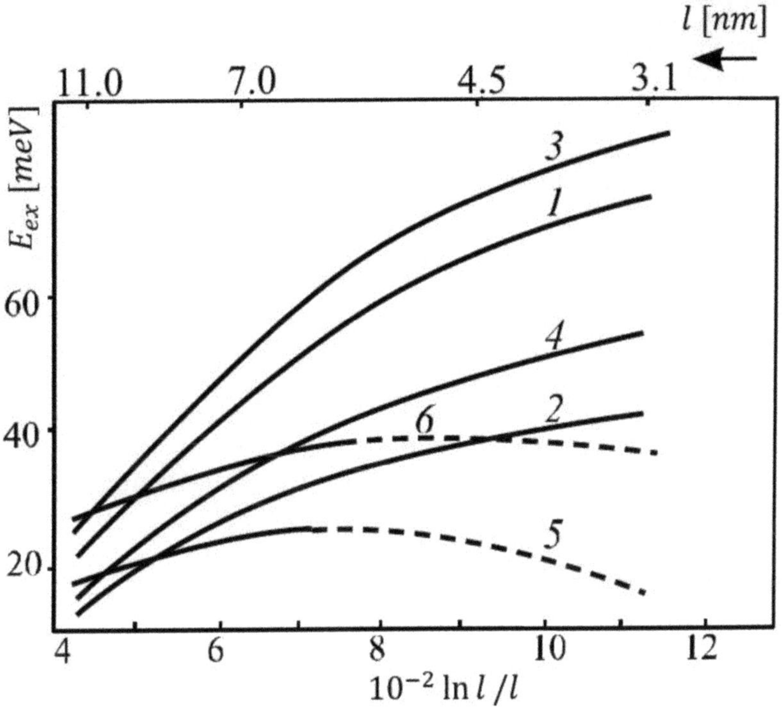

Figure 6.13. Dependence of the binding energies of light (curve 1—$N = 1$; 2—$N = 2$) and heavy (curve 3 —$N = 1$; 4—$N = 2$) excitons on $\ln l/l$. For comparison, the results obtained using the Keldysh formula [108] (curve 5—E_{ex}^{1e-1lh}, 6—E_{ex}^{1e-1hh}) are presented.

$$E_{ex}(l) = -\frac{e^2}{\varepsilon_{20} l}\left[\ln\left(\frac{4l}{\delta^2 a_{ex}}\right) - 2.2\right], \qquad (6.11.12)$$

where $\delta = (\varepsilon_{10} + 1)/(2\varepsilon_{20})$.

From the comparison of the results shown in figure 6.13, it follows that the variational calculation gives significantly higher values of the binding energy for both heavy and light excitons than the corresponding results obtained by the formula (6.11.12). This is due to the manifestation of the dimensional effect of the loss of inertial shielding [100], which in the thickness range $l \sim 3.0$ nm gives values $E_{ex}(l)$ B $\varepsilon_{20}/\varepsilon_2$ times greater for all l. Indeed, the volume radius of the exciton in cadmium telluride is $\sim$10.0 nm, and its binding energy is $E_{ex} \sim 10$ meV. Since the condition $\omega_0 > \omega_{ex} = E_{ex}/\hbar$ for a bulk exciton is well fulfilled, the shielding of the electron–hole interaction is carried out by inertial polarization. For small film thicknesses ($l \leqslant 5.0$ nm), the binding energy becomes greater than the energy of the longitudinal optical phonon, as a result of which part of the inertial shielding is 'turned off'. Note also that the dependence of E_{ex} on $l^{-1} \ln l$ is not linear, that is, the logarithmic law of electron–hole interaction does not hold in the specified thickness range [108, 109].

2. A fundamental circumstance when comparing theory with experiment is the impossibility of directly measuring the exciton binding energy from the optical spectra of ultrathin films, as is the case in the case of a massive crystal. This is due to the complex renormalization of the band gap $E_g(l)$. The experimentally measured value is the energy of the quantum $\hbar\omega_{ex}(l)$, which generates an exciton, and contains $E_{ex}(l)$ as one of the contributions. In figure 6.11 the dependences of $\hbar\omega_{ex}(l)$ for light and heavy excitons are given, taking into account additional factors affecting the values of individual contributions to $\hbar\omega_{ex}(l)$. A comparison of the results allows us to draw an important conclusion: the best quantitative agreement of the theory with the experiment, which is demonstrated in table 6.3, is achieved only by taking into account all the numerous dimensional effects indicated in [82] and in this section.

3. The observation of exciton peaks corresponding to higher levels of dimensional quantization in the absorption spectrum of ultrathin cadmium telluride films [103] makes it possible to theoretically analyze the question of the fulfillment of the quantization law for a single quantum well. In particular, calculating the energy gap

$$\Delta = \hbar\omega_{ex}^{2e-2lh,\, hh} - \hbar\omega_{ex}^{1e-1lh,\, 1hh}$$

and comparing it with the experimental one, we can answer this question. A good agreement of the theory with the experiment (box to figure 6.11) allows us to conclude that the law of $E_{s.q.} \sim N^2/l^2$ is fulfilled up to thicknesses of the order of 3.0 nm.

In conclusion, we note one more important circumstance. In [110, 111], a decrease in the binding energy of the Wannier–Mott exciton in a single quantum well in the thickness range $l \leqslant 3$ nm was observed. In a CdTe quantum well on an MgF_2 substrate, an increase in binding energy is observed up to thicknesses of the order of 2, 0 nm. This is due to the existence of large potential barriers at both film boundaries (approximately 4 – 5 eV), which is much larger than all the size-dependent contributions included in $E_{ex}(l)$ and $\hbar\omega_{ex}(l)$. In the $GaAs/Al_xGa_{1-x}As$, quantum well, as is known, the potential barriers for $l = 0.30$ are 0.30 eV and 0.20 eV, and for $l = 3.0$ nm and at $l = 3.0$ nm are of the same order as the kinetic energies of the electron and the hole, which leads to delocalization of the exciton from the quantum well. For the maximum $E_{ex}(l)$, magnification that can be achieved in $GaAs/Al_xGa_{1-x}As$, quantum wells is approximately 3.5 times the volume value. At the same time, l is on the order of 3.0 nm. A further decrease in l leads to a decrease in $E_{ex}(l)$ due to delocalization of the exciton from the quantum well. In a CdTe quantum well on an MgF_2 substrate, $E_{ex}(l)$ can be increased by about 6–8 times compared to the volume value, and an increase in $E_{ex}(l)$ is observed throughout the domain of experimentally achievable thicknesses.

6.12 Magnetic polaron exciton in a quantum well structure

The problem of a plane magnetic polaron exciton in the Mayer limit has already been discussed in section 6.8. In this section, we will consider the case of an exciton in a quantum well of finite thickness. The Coulomb interaction is added to the dimensional quantization of the electron and hole states, which leads to the formation of bound states and at the same time causes the free movement of the exciton as a whole [91].

The main task of this section is to study the exciton–phonon interaction in a magnetic field in a relatively simple system (1|20|1) with an exciton in the middle layer. The plates will be considered symmetrical, $\varepsilon_1 = \varepsilon_3 = \varepsilon$; $l_1 = l_3 \to \infty$, and all three layers are isotropic. Let's limit ourselves to two well-known limiting cases—Haken and Mayer. We will write the exciton Hamiltonian using polaron Hamiltonians for holes and electrons:

$$\begin{aligned}\hat{H}_{ex} &= \hat{H}_e + \hat{H}_h + U(\rho,\ z_e,\ z_h) + \hat{H}_{B,S}\\ &\quad + \hat{H}_{ex-B} + \hat{H}_{ex-S} + \hat{U}_{SA}(z_e,\ z_h) + U,\end{aligned} \tag{6.12.1}$$

where

$$\hat{H}_c = \frac{1}{2m_c}(\mathbf{P}_{c\perp} \pm e\mathbf{A}_c)^2 + \frac{\hat{P}_{c\|}^2}{2m_{c\|}^*};\ c = e,\ h; \tag{6.12.2}$$

where $\mathbf{A}$ is the vector potential of a homogeneous magnetic field, determined by the formula (6.11.5.2) (the sign '+' is taken for $c = e$; '−'—for $c = e$); $\hat{H}_{B,\ S} = \hat{H}_B + \hat{H}_S$ is the sum of the energy operators of polar bulk and surface optical vibrations in the second layer; $\hat{H}_S$ takes into account two surface modes $s = 1,\ 2$; $\hat{H}_{ex-B} = \hat{H}_{e-B} + \hat{H}_{h-B}$, where $\hat{H}_{c-B}$ is the energy operator of the interaction of the charge with bulk optical vibrations; $U(\rho,\ z_e,\ z_h)$ is the potential energy of the electron–hole interaction; and U_0 is the potential barrier on the surfaces of the layers.

Consider the case of strong dimensional quantization and the Haken limit for transverse motion:

$$\mathcal{K}(l) > \hbar\omega_{20} > \hbar\omega_c > E_{ex},\ (R_c < a_{ex},\ a_B) \tag{6.12.3}$$

where a_B is the magnetic length ($a_B = (\hbar/(eB))^{1/2}$).

Taking into account (6.11.3), we average the Hamiltonian (6.12.1) on the wave functions of the quantum well. By introducing a coordinate system associated with the center of mass of the exciton, we obtain

$$\begin{aligned}\hat{H}_1 &= \left\langle \Psi(z_e,\ z_h) \,\middle|\, \hat{H} \,\middle|\, \Psi(z_e,\ z_h) \right\rangle\\ &= \frac{1}{2m_{e_\perp}}\left(\mathbf{P}_{e\perp} + e\mathbf{A}_e\right)^2 + \frac{1}{2m_{h\perp}}\left(\mathbf{P}_{h\perp} - e\mathbf{A}_h\right)^2\\ &+ \frac{\pi^2\hbar^2}{2\mu l_2^2} - \frac{e^2}{2\pi\varepsilon_0\varepsilon_2 l}\int_0^\infty J_0\left(\frac{2\rho x}{l_2}\right)B_1(x)dx + \frac{e^2\delta}{2\pi\varepsilon_0\varepsilon_2 l}\int_0^\infty B_2(x)dx\\ &\pm \sum_{\eta,s,c}\tilde{V}^c(\eta)e^{i\eta\rho_c}\left(\hat{b}_{-\eta}^+ + \hat{b}_\eta\right) + \sum_\eta \hbar\Omega_S\hat{b}_S^+\hat{b}_S \pm\\ &\pm \sum_{\mathbf{Q},c}\tilde{V}^{(c)}(Q)e^{i\mathbf{q}_\perp\rho_c}\left(\hat{b}_{-\mathbf{Q}}^+ + \hat{b}_\mathbf{Q}\right) + \sum_\mathbf{Q}\hbar\omega_{20}\hat{b}_\mathbf{Q}^+\hat{b}_\mathbf{Q},\end{aligned} \tag{6.12.4}$$

where the designations introduced are

$$\widetilde{V}^c(\eta) = V^c(\eta)\langle\Psi(z_e, z_h)|\cosh \eta z_c|\Psi(z_e, z_h)\rangle;$$

$$\widetilde{V}^c(Q) = V^c(Q)\left\langle\Psi(z_e, z_h)\middle|g_V(\mathbf{q}_\perp, q_z, z_c)|\Psi(z_e, z_h)\right\rangle.$$

The total momentum of the system in the XOY plane is the value

$$\mathcal{P} = \mathbf{P}_{R_\perp} - \frac{e}{2}[\mathbf{B}\boldsymbol{\rho}] + \sum_{\eta}\hbar\eta\hat{b}_\eta^+\hat{b}_\eta + \sum_{\eta}\hbar\mathbf{q}_\perp\hat{b}_\mathbf{Q}^+\hat{b}_\mathbf{Q}. \tag{6.12.5}$$

The eigenfunctions of the total momentum operator are

$$\Psi(\boldsymbol{\rho}, \mathbf{R}_\perp) = \exp\left\{\frac{1}{\hbar}\left(\mathcal{P} - \frac{e}{2}[\mathbf{B}\boldsymbol{\rho}]\right) - \sum_{\eta}\hbar\eta\hat{b}_\eta^+\hat{b}_\eta - \sum_{\mathbf{Q}}\hbar\mathbf{q}_\perp\hat{b}_\mathbf{Q}^+\hat{b}_\mathbf{Q}\right\}. \tag{6.12.6}$$

Following the works [90, 91], we perform two unitary transformations on the Hamiltonian (6.11.4): (1) transition to a full pulse; (2) displacement of the amplitudes of phonon oscillators:

$$\hat{U}_1 = \exp\left\{\frac{1}{\hbar}\left(\mathcal{P}_\perp + \frac{e}{2}[\mathbf{B}\boldsymbol{\rho}]\right)\mathbf{R}_\perp\right\}\exp\left\{-i\sum_{\eta}\eta\mathbf{R}_\perp\hat{b}_\eta^+\hat{b}_\eta - \sum_{\mathbf{Q}}\mathbf{q}_\perp\mathbf{R}_\perp\hat{b}_\mathbf{Q}^+\hat{b}_\mathbf{Q}\right\}; \tag{6.12.7}$$

$$\hat{U}_2 = \exp\left\{\sum_{\eta}f_\eta^*(\rho)\hat{b}_\eta - f_\eta(\rho)\hat{b}_\eta^+\right\}\exp\left\{\sum_{\mathbf{Q}}f_\mathbf{Q}^*(\rho)\hat{b}_\mathbf{Q} - f_\mathbf{Q}(\rho)\hat{b}_\mathbf{Q}^+\right\}; \tag{6.12.8}$$

and after averaging on the wave function of relative motion, we get the variational energy:

$$\begin{aligned}
\overline{E} = &-\frac{\hbar^2}{2\mu_\perp}\int\Psi^*\Delta_\rho\Psi d\tau - \frac{\hbar^2}{2\mu_\perp}\int\Psi^*\left[i\hbar\gamma\cdot 2eA_\rho\nabla_\rho - \gamma^2e^2A_\rho^2\right]\Psi d\tau\\
&+\left\{\frac{1}{2}\frac{\hbar^2}{2\mu_\perp}\int|\Psi|^2\sum_{\eta}\left[\Delta_\rho|f_\eta|^2 - f_\eta^*\Delta_\rho f_\eta - f_\eta\Delta_\rho f_\eta^*\right]d\tau\right.\\
&+\int|\Psi|^2\sum_{\eta}\left(\hbar\Omega_s + \frac{\hbar^2\eta^2}{2M_\perp}\right)|f_\eta|^2 d\tau\\
&-\int|\Psi|^2\sum_{\eta}\left[\widetilde{V}(\eta)f_\eta a(\boldsymbol{\rho}) + c.\,c.\right]d\tau + \left\{Idem\begin{bmatrix}\eta\to\mathbf{Q}\\ \Omega_S\to\omega_0\end{bmatrix}\right\}\\
&+\frac{\pi^2\hbar^2}{2\mu_{\|}l^2}\\
&\left.-\frac{e^2}{2\pi\varepsilon_0\varepsilon_2 l_2}\iint|\Psi|^2 J_0\left(\frac{2\rho x}{l_2}\right)B(x)dx\right\};
\end{aligned} \tag{6.12.9}$$

where

$$a(\boldsymbol{\rho}) = \exp(i\sigma_1 \boldsymbol{\eta\rho}) - \exp(-i\sigma_2 \boldsymbol{\eta\rho});\ \gamma = (m_{h_\perp} - m_{e_\perp})/(m_{h_\perp} + m_{e_\perp}).$$

Minimizing it by variational functions $f_\eta(\rho)$, $f_Q(\rho)$, $\Psi(\rho)$ we obtain a system of equations for these functions similar to those considered in the polaron exciton problem. Denoting $\mathbf{p} = \{\boldsymbol{\eta},\ \mathbf{Q}\}$, we write down the equation $f_p(\rho)$:

$$-\frac{\hbar^2}{2\mu_\perp}\Delta_\rho f_p + \left\{\hbar\Omega_p + \frac{\hbar^2 p^2}{2M_\perp} + \frac{1}{2}\frac{\hbar^2}{2\mu_\perp}\frac{\Delta_\rho |\Psi|^2}{|\Psi|^2}\right\} f_p - \widetilde{V}^*_p a^*_p(\rho) = 0; \tag{6.12.10}$$

where

$$\Omega_p = \begin{cases} \Omega_S, & \mathbf{p} = \boldsymbol{\eta}; \\ \omega_{20}, & \mathbf{p} = \mathbf{Q}; \end{cases} \quad \widetilde{V}^*_p = \begin{cases} \widetilde{V}^*(\boldsymbol{\eta}), & \mathbf{p} = \boldsymbol{\eta}; \\ \widetilde{V}^*(\mathbf{Q}), & \mathbf{p} = \mathbf{Q}. \end{cases}$$

We will look for an approximate partial solution of the equation (6.12.10) in the form

$$f_p = \frac{\widetilde{V}^*_p \exp(-i\sigma_1 \mathbf{p\rho})}{\Pi(m_{e\perp},\ \mathbf{p})} - \frac{\widetilde{V}^*_p \exp(-i\sigma_2 \mathbf{p\rho})}{\Pi(m_{h\perp}, \mathbf{p})}. \tag{6.12.11}$$

Substituting (6.12.11) into (6.12.10) and averaging on Gaussian-type wave functions

$$\Psi(\rho) = C\exp(-\beta\rho^2),$$

we get

$$\left\langle \Psi \left| \frac{1}{2}\frac{\hbar^2}{2\mu_\perp}\frac{\Delta_\rho |\Psi|^2}{|\Psi|^2} \right| \Psi \right\rangle = -\frac{\hbar^2\beta}{\mu_\perp}\left(1 + \frac{\eta^2}{8\beta}\right) \left[1 + \frac{1}{4}\frac{\left(\kappa_1^2 + \kappa_2^2\right)}{\kappa_1\kappa_2}\exp\left(\frac{\eta^2}{8\beta}\right)\right]^{-1}, \tag{6.12.12}$$

where

$$\kappa_{1,2} \equiv \left|\widetilde{V}_p\right|^2 \Pi^{-2}(m_{c\perp},\ \mathbf{p}).$$

At $\eta^2 < 8\beta$ and $\kappa_1 \approx \kappa_2$

$$\left\langle \Psi \left| \frac{1}{2}\frac{\hbar^2}{2\mu_\perp}\frac{\Delta_\rho |\Psi|^2}{|\Psi|^2} \right| \Psi \right\rangle \approx -\frac{2\hbar^2\beta}{3\mu_\perp}. \tag{6.12.13}$$

The described approximate transformations can only conditionally be considered the justification for the estimate (6.12.13); however, the formula (6.12.11) gives a qualitatively correct description of the influence of the magnetic field on $f_p(\rho)$. Substituting (6.12.11) into the equation for the wave function, we can write

$$\hat{H}_{\text{eff}}\Psi(\boldsymbol{\rho} - \boldsymbol{\rho}_0) = E\Psi(\boldsymbol{\rho} - \boldsymbol{\rho}_0), \tag{6.12.14}$$

where $\boldsymbol{\rho}_0 = (eB^2)^{-1}[\mathbf{BP}_\perp]$ is the vector of the local center of the orbit.

The effective Hamiltonian of the exciton–phonon system

$$\hat{H}_{\text{eff}} = -\frac{\hbar^2}{2\mu_\perp}\Delta_\rho - i\gamma\,\hbar\frac{e}{2\mu_\perp}\mathbf{B}[\boldsymbol{\rho}\nabla_\rho] + \frac{e^2B^2\rho^2}{8\mu_\perp} + W_{\text{eff}}(|\boldsymbol{\rho} + \boldsymbol{\rho}_0|, l) \tag{6.12.15}$$

includes an effective potential with phonon contributions:

$$W_{\text{eff}}(|\boldsymbol{\rho} + \boldsymbol{\rho}_0|, l_2) = U(|\boldsymbol{\rho} + \boldsymbol{\rho}_0|, l_2) + W_{\text{ph}}(|\boldsymbol{\rho} + \boldsymbol{\rho}_0|, l_2) \tag{6.12.16}$$

and with the polaron energies of an electron and a hole:

$$W(l) = -\sum_{\mathbf{p}, c=e,h} \left| \tilde{V}_p^c \right|^2 (\Pi(\mathbf{p}, m_{C\perp}))^{-1}. \tag{6.12.17}$$

The contributions from the shielding of the electron–hole interaction by optical phonons have the form

$$W(|\boldsymbol{\rho} + \boldsymbol{\rho}_0|, l_2) = \sum_{\mathbf{p},\, C=e,h} \left| \tilde{V}_p^C \right|^2 \cos[\mathbf{p}(\boldsymbol{\rho} + \boldsymbol{\rho}_0)](\Pi(m_{C\perp}, \mathbf{p}))^{-1}. \tag{6.12.18}$$

From the general view of the screening solution

$$W_S(|\boldsymbol{\rho} + \boldsymbol{\rho}_0|, l_2) = \sum_{c=e,h} \left| 1 - \frac{\omega_c}{\Omega_{S_1}} \right|^{-1} \{F(\varepsilon_2, R_{Sc}) - F(\varepsilon_{20}, R_{Sc})\}; \tag{6.12.19}$$

here

$$F(\varepsilon_i, \Omega_S) = \left[\frac{e^2}{4\pi\varepsilon_0\varepsilon_i l_2}\right]\left[1 + \left(\frac{2\varepsilon\tilde{R}_s}{l_2}\right)^2\right]^{-1}\left\{\frac{\pi}{2}\left[H_0\left(\frac{2\varepsilon|\boldsymbol{\rho}+\boldsymbol{\rho}_0|}{\varepsilon_i l_2}\right) - N_0\left(\frac{2\varepsilon|\boldsymbol{\rho}+\boldsymbol{\rho}_0|}{\varepsilon_i l_2}\right)\right] - K_0\left(\frac{|\boldsymbol{\rho}+\boldsymbol{\rho}_0|}{\tilde{R}_S}\right) + \frac{\pi\varepsilon\tilde{R}_S}{\varepsilon_i l_2}\left[I_0\left(\frac{|\boldsymbol{\rho}+\boldsymbol{\rho}_0|}{\tilde{R}_S}\right) - L_0\left(\frac{|\boldsymbol{\rho}+\boldsymbol{\rho}_0|}{\tilde{R}_S}\right)\right]\right\}.$$

it follows that it vanishes with a decrease in the thickness of the film and with an increase in the magnetic field. For example, with a fixed magnetic field, but a decreasing effective value of the thickness of the quantum well $l_{\text{eff}} = l/\delta_0\,(\varepsilon/\varepsilon_{20})$ it is possible to trace the dynamics of changes in the potential energy of the electron–hole interaction.

When performing an inequality

$$l < \widetilde{R}_{Sc} < a_{ex} < l_{\text{eff}},$$

using the asymptotic expansions in (6.12.19), we obtain

$$W_s(|\rho + \rho_0|, l_2) = \frac{e^2}{4\pi\varepsilon_0}\left(\frac{2\widetilde{R}_{Sc}\varepsilon}{l^2}\right)\left\{\frac{1}{\varepsilon_2^2}\left[\ln\left(\frac{|\rho + \rho_0|\varepsilon}{\varepsilon_2 l}\right) + \gamma_0\right] - \frac{1}{\varepsilon_{20}^2}\left[\ln\left(\frac{|\rho + \rho_0|\varepsilon}{\varepsilon_{20} l}\right) + \gamma_0\right]\right\}. \text{ (6.11.20)} \tag{6.12.20}$$

When the condition is met

$$l < \widetilde{R}_{Sc} < l_{\text{eff}} < a_{ex}$$

from (6.12.19) we get

$$W_S(|\rho + \rho_0|, l_2) \approx \frac{e^2}{4\pi\varepsilon_0\varepsilon\,|\rho + \rho_0|} + O\left[\exp\left(-\frac{|\rho + \rho_0|}{\widetilde{R}_{Sc}}\right)\right]. \tag{6.12.21}$$

Finally, at the limit

$$l < l_{\text{eff}} < \widetilde{R}_{Sc} < a_{ex}$$

the effective electron–hole–phonon potential takes the form

$$W_S(|\rho + \rho_0|, l) \approx \frac{e^2}{4\pi\varepsilon_0\varepsilon\,|\rho + \rho_0|}\left(\frac{\varepsilon_2 l}{2\varepsilon\,|\rho + \rho_0|}\right)^2 + O\left[\exp\left(-\frac{|\rho + \rho_0|}{\widetilde{R}_{Sc}}\right)\right]. \tag{6.12.22}$$

Now, by fixing the thickness, it is possible to follow the dynamics of changes in the electron–hole interaction when the magnetic field changes. Since the relative coordinate is included in the potential of the electron–hole interaction in the combination $(1 - \omega_c/\Omega_{Sc})^{1/2}|\rho + \rho_0|$, an increase of the ω_c is equivalent to a decrease in the radius of the exciton state, which can be represented as:

(a) In the case when the condition is fulfilled

$$l < \widetilde{R}_{Sc} < l_{\text{eff}} < \tilde{a}_{ex},\ \tilde{a}_{ex} = a_{ex}(1 - \omega_c/\Omega_S)^{1/2},$$

from the expression (6.12.19) we get

$$W_S(|\rho + \rho_0|, l) \approx \frac{e^2}{4\pi\varepsilon_0\varepsilon\,|\rho + \rho_0|} - \frac{2e^2}{4\pi\varepsilon_0\varepsilon_2 l}\left(\frac{\pi\widetilde{R}_{Sc}}{2\,|\rho + \rho_0|}\right)\exp\left(-\frac{|\rho + \rho_0|}{\widetilde{R}_{Sc}}\right)\left(1 - \frac{\varepsilon}{\varepsilon_{20}}\right). \tag{6.12.23}$$

(b) When the condition is fulfilled $l < R_{Sc} < \tilde{a}_{ex} < l_{\text{eff}}$ the effective potential of the electron–hole–phonon interaction is expressed by the formula (6.12.20).

(c) Finally, in the limit $l < \tilde{a}_{ex} < R_{Sc} < l_{\text{eff}}$ for $W_S(|\boldsymbol{\rho} + \boldsymbol{\rho}_0|, l)$ we get the following expression:

$$W_S(|\rho + \rho_0|, l) \approx \frac{e^2}{2\pi\varepsilon_0\varepsilon_2 l}\left[\ln\left(\frac{|\rho + \rho_0|}{l/\delta} + \tilde{\gamma}_0\right)\right] - \frac{e^2}{4\pi\varepsilon_0\varepsilon\,|\rho + \rho_0|}\left(\frac{|\rho + \rho_0|}{l/\delta}\right)^{-2}; \tag{6.12.24}$$

$$\tilde{\gamma}_0 = C + \left(1 - \frac{\varepsilon_2}{\varepsilon_{20}}\right)\ln\left(\frac{l}{\delta R_{Sc}}\right) - \left(\frac{\varepsilon_2}{\varepsilon_{20}}\right)\ln\left(\frac{\varepsilon_{20}}{\varepsilon_2}\right).$$

Thus, the growth of the magnetic field reduces the phonon contribution to the shielding of the electron–hole interaction and, in the limit of a strong magnetic field, generally turns off inertial polarization from the shielding.

6.12.1 Strong magnetic field and Haken limit

Let's consider first a light exciton. When fulfilling criterion (6.12.3), to calculate the binding energy and the effective mass of the exciton in the presence of a magnetic field, we use the method of perturbation theory [90], taking as a perturbation the effective potential of the electron–hole interaction $W_{\text{eff}}(\boldsymbol{\rho}, l)$, which can be written as

$$W_{\text{eff}}(|\boldsymbol{\rho} + \boldsymbol{\rho}_0|, l) \approx -\frac{e^2}{2\pi\varepsilon_0\varepsilon_2 l}\int_0^\infty dx\, J_0\left(\frac{2\,|\boldsymbol{\rho} + \boldsymbol{\rho}_0|}{l}x\right)B_1(x)dx + W_S(|\boldsymbol{\rho} + \boldsymbol{\rho}_0|, l). \tag{6.12.25}$$

In the W_{eff} zero approximation, the Schrodinger equation (6.12.14) with the Hamiltonian (6.12.15) transforms into the equation for a free particle with mass $\mu_\perp$ in a magnetic field with induction B. The wave function of relative motion does not depend on γ and is given by the expression (6.12.13). The energy of the zero approximation has the form (6.12.14). Coulomb corrections to the levels (6.12.14) are found according to the usual perturbation theory for nondegenerate systems. We present here the results for the lowest Landau state in the thin layer limit ($l < a_{ex}$), when the effective potential of the electron–hole interaction can be represented as

$$W_{\text{eff}}\left(\left|\rho+\rho_0\right|,l\right)=-\frac{e^2}{4\pi\varepsilon_0\varepsilon_2}\left(\frac{\varepsilon_2}{\varepsilon}-e^{-\frac{\left|\rho+\rho_0\right|}{l}}\right)\frac{1}{\left|\rho+\rho_0\right|}+\frac{e^2\left(\varepsilon_{20}^2-\varepsilon_2^2\right)}{64\pi\varepsilon_0\varepsilon_2\varepsilon^2l^2}\sum_{c=e,h}\frac{l^3}{R_{Sc}^3}\left(1-\frac{\omega_c}{\Omega_S}\right)^{\frac{1}{2}}\left(\frac{\pi R_{Sc}}{2\left|\rho+\rho_0\right|}\right)^{\frac{1}{2}}\exp\left\{-\frac{\rho+\rho_0}{R_S\left(1-\frac{\omega_c}{\Omega_S}\right)^{-\frac{1}{2}}}\right\} \tag{6.12.26}$$

when performing an inequality: $l_{\text{eff}} \ll |\boldsymbol{\rho}+\boldsymbol{\rho}_0|$ and

$$W_{\text{eff}}(|\boldsymbol{\rho}+\boldsymbol{\rho}_0|,l)\approx-\frac{e^2}{\pi\varepsilon_0(\varepsilon_2+\varepsilon)l}\left[\ln\left(\frac{(\varepsilon_2+\varepsilon)l}{4\varepsilon\,|\boldsymbol{\rho}+\boldsymbol{\rho}_0|}\right)-\gamma_0\right]-\frac{e^2\left(\varepsilon_{20}^2-\varepsilon_2\right)}{64\pi\varepsilon_0\varepsilon_2\varepsilon^3l}\sum_{c=e,h}\frac{l^3}{R_{Sc}^3}\left(1-\frac{\omega_c}{\Omega_S}\right)^{\frac{1}{2}}\left[\ln\left(\frac{2R_{Sc}\left(1-\frac{\omega_c}{\Omega_S}\right)^{-\frac{1}{2}}}{|\boldsymbol{\rho}+\boldsymbol{\rho}_0|}\right)-\gamma_0\right], \tag{6.12.27}$$

when the condition is met: $l<|\boldsymbol{\rho}+\boldsymbol{\rho}_0|\ll l_{\text{eff}}$.

Using the procedure described in section 6.8, we obtain the following expressions for the binding energy and effective mass of a magnetic polaron exciton:

$$\Delta E_{00}(B)=-\frac{e^2}{2\sqrt{2\pi}\,\varepsilon_0\varepsilon_2a_B}\left\{\frac{\varepsilon_2}{\varepsilon}-\exp\left(-\frac{a_B^2}{2l^2}\right)erfc\left(\frac{a_B}{\sqrt{2}l}\right)\right\}+\frac{e^2\left(\varepsilon_{20}^2-\varepsilon_2\right)}{64\pi\varepsilon_0\varepsilon_2\varepsilon^3l}\sum_{c=e,h}\frac{l^3}{R_{Sc}^3}\left(1-\frac{\omega_c}{\Omega_S}\right)^{\frac{1}{2}}\left(\frac{\pi R_{Sc}}{2a_B}\right)^{\frac{1}{2}}\exp\left(\frac{\beta_c^2a_B^2}{4}\right)D_{-\frac{3}{2}}(\beta_ca_B); \tag{6.12.28}$$

$$M_{eH}^{-1}(B)=\frac{1}{8\sqrt{2\pi}\,\varepsilon_0\varepsilon_2a_B^3B^2}\left\{\frac{\varepsilon_2}{\varepsilon}+\exp\left(\frac{a_B^2}{2l^2}\right)\left[erfc\left(\frac{a_B}{\sqrt{2}\,l}\right)-\frac{a_B^2}{l^2}erfc\left(\frac{a_B}{\sqrt{2}\,l}\right)-\frac{a_B^2}{l^2}\right]\right\}+\frac{e^2\left(\varepsilon_{20}^2-\varepsilon_2^2\right)\hbar^{-1}}{128\pi\varepsilon_0\varepsilon^2\varepsilon_2B}\sum_{c=e,h}\frac{\pi l^3}{R_S^3}\left(1-\frac{\omega_c}{\Omega_S}\right)^{\frac{1}{4}}\left(\frac{R_{Sc}}{2a_B}\right)^{\frac{1}{2}}\exp\left(\frac{\beta_c^2a_B^2}{4}\right)\times\left[\frac{1}{4}D_{-\frac{7}{2}}(\beta_ca_B)-D_{-\frac{3}{2}}(\beta_ca_B)\right], \tag{6.12.29}$$

where $\beta_c=(1-\omega_c/\Omega_S)^{1/2}R_{Sc}^{-1}$, $D_n(x)$ are functions of the parabolic cylinder. The first terms of the right parts (6.12.28) and (6.12.29) coincide with the expressions obtained in [91] (formulas (11) and (12)) for a two-dimensional exciton in a strong magnetic field. The second and third terms take into account the extent of the wave functions of the electron and the hole in the quantum well and the contributions from inertial polarization, respectively. Taking into account the finite thickness of the quantum well leads to a decrease in the binding energy of the exciton and an increase in its effective mass. The exciton binding energy is maximal in the quasi-two-dimensional limit and is equal to

$$\Delta E_0^{2D}(B) = -\frac{e^2}{4\sqrt{2\pi}\,\varepsilon_0\varepsilon a_B}. \tag{6.12.30}$$

At values of l, for which the condition begins to be fulfilled $\hbar\omega_c \geqslant E_{s.q.}$ (i.e., $a_B < l$), the exciton binding energy has the form

$$\Delta E_0^{2D}(B) = -\frac{e^2}{4\pi\varepsilon_0\varepsilon_2 l} + \frac{e^2 a_B}{8\sqrt{2\pi}\,\varepsilon_0\varepsilon l}. \tag{6.12.31}$$

In the case of the logarithmic limit of the Coulomb interaction, the corresponding expressions for the binding energy and effective mass have the form

$$\begin{aligned}\Delta E_0(B) = &-\frac{e^2}{\pi\varepsilon_0(\varepsilon_2+\varepsilon)l}\left[\ln\left(\frac{(\varepsilon_2+\varepsilon)l}{4\varepsilon a_B}\right)-\gamma\right]\\ &+\frac{e^2\left(\varepsilon_{20}^2-\varepsilon_2^2\right)}{64\pi\varepsilon_0\varepsilon_2\varepsilon^2 l}\sum_c\frac{l^3}{R_{Sc}^3}\left(1-\frac{\omega_c}{\Omega_S}\right)^{\frac{1}{2}}\left[\ln\left(\frac{2R_S}{a_B\left(1-\frac{\omega_c}{\Omega_S}\right)^{\frac{1}{2}}}\right)-\gamma_0\right];\end{aligned} \tag{6.12.32}$$

$$M_{\text{eff}}^{-1} = \frac{e^2 a_B^2}{4\pi\varepsilon_0\hbar^2 l}\left\{\frac{1}{\varepsilon_2+\varepsilon}-\frac{\left(\varepsilon_{20}^2-\varepsilon_2^2\right)}{64\varepsilon_2\varepsilon^2}\sum_{c=e,h}\frac{l^3}{R_{Sc}^3}\left(1-\frac{\omega_c}{R_{Sc}}\right)^{\frac{1}{2}}\right\}. \tag{6.12.33}$$

It follows from expressions (6.12.32) and (6.12.33) that with an increase in the magnetic field, the binding energy increases as $\ln B$, in contrast to $B^{1/2}$, obtained in [91], that is, in quantum wells of finite sizes, the dependence of the binding energy on the magnitude of the magnetic field is weaker than predicted by the two-dimensional exciton model [91]. This is related to a stronger dependence of the effective mass on the magnitude of the field: instead of $B^{1/2}$ for the two-dimensional exciton model, we get B in a quantum well of finite dimensions.

The above analysis of the behavior of the binding energy and the effective mass is qualitative, since it requires the fulfillment of criterion inequalities, which cannot always be fulfilled for real systems.

Now let's perform a similar analysis for a heavy exciton. As has already been established, in the case of a heavy exciton, the motion of the hole is weakly quantized ($E_{s.q.}^{(h)} \leqslant E_{ex}$). This leads to the appearance of an additional potential acting on the hole and leading to its localization, in the case of a symmetric three-layer system, in the center of the quantum well. Let's imagine the potential energy of the electron–hole interaction as

$$U\left(\left|\rho+\rho_0\right|, z_h\right) = -\frac{e^2}{4\pi\varepsilon_0\varepsilon\left|\rho+\rho_0\right|} + \left\{\frac{e^2}{4\pi\varepsilon_0\varepsilon\left|\rho+\rho_0\right|} - U\left(\left|\rho+\rho_0\right|, z_h\right)\right\}. \tag{6.12.34}$$

In the limit of an ultrathin film (a narrow quantum well), the expression in square brackets can be considered a perturbation. Averaging it on a wave function describing the relative motion of an electron and a hole in the XOY plane and their movement along z, we obtain the desired additional potential in which the hole moves:

$$U(z_h) = U_{SA}(z_h) + U_{n,\,m}^{N_e}(z_h), \tag{6.12.35}$$

where $U_{SA}(z_h)$ is the potential energy of the self-action of the hole, and $U_{n,\,m}^{N_e}(z_h)$ has the form

$$\begin{aligned} U_{n,m}^{N_e}(z_h) &= \frac{e^2}{4\pi\varepsilon_0\varepsilon_2}\int d\rho\int dz_e\left|\Psi_{n,m}(\rho)\right|^2\left(\frac{2}{l}\right)\cos^2\left(\frac{\pi N_e z_e}{l}\right)\left\{\frac{\varepsilon_2}{\varepsilon}\left|\rho+\rho_0\right|^{-1}\right. \\ &- \int_0^\infty d\eta\, J_0\left(\left|\rho+\rho_0\right|, \eta\right)\left[e^{-\eta\left|z_e-z_h\right|} + \frac{2\delta}{e^{2\eta l}-\delta^2}\times\right. \\ &\times\left.\left.\left[\delta\cosh\eta(z_e-z_h) + e^{\eta l}\cosh(z_e-z_h)\eta\right]\right]\right\}. \end{aligned} \tag{6.12.36}$$

For the lowest state of Landau ($n = m = 0$):

$$U_{0,\,0,\,1}(z_h) \approx \frac{e^2 l}{16\pi\varepsilon_0\varepsilon_2 a_B^2}\left\{\frac{z_h^2}{l^2} + \left[\frac{1}{4} + \frac{1}{2}\left(\frac{\varepsilon_2^2}{\varepsilon^2} - 1\right)\right] - \frac{1}{\pi^2}\cos\left(\frac{\pi z_h}{l}\right)\right\}. \tag{6.12.37}$$

Then for the $U(z_h)$ we get the following expression:

$$\begin{aligned} U(z_h) &\approx \frac{e^2 l}{4\pi\varepsilon_0\varepsilon_2 a_B^2}\left[\frac{1}{4} + \frac{1}{2}\left(\frac{\varepsilon_2^2}{\varepsilon^2} - 1\right)\right] + \frac{e^2\ln(1-\delta)^{-1}}{4\pi\varepsilon_0\varepsilon_2 l} \\ &+ \frac{z_h^2}{l^2}\left[\frac{e^2 l}{4\pi\varepsilon_0\varepsilon_2 a_B^2} + \frac{2e^2\delta J(\delta)}{4\pi\varepsilon_0\varepsilon_2 l}\right]. \end{aligned} \tag{6.12.38}$$

The solution of the Schrödinger equation with potential energy (6.12.38) has the form

$$\Delta E_0 \approx \hbar\omega_c^*\left(n+\frac{1}{2}\right)+\frac{e^2 l}{4\pi\varepsilon_0\varepsilon_2 a_B^2}\left[\frac{1}{4}+\frac{1}{2}\left(\frac{\varepsilon_2^2}{\varepsilon^2}-1\right)\right]+\frac{e^2\ln\left(1-\delta\right)^{-1}}{4\pi\varepsilon_0\varepsilon_2 l}, \tag{6.12.39}$$

where $n = 0,\ 1,\ 2,\ \ldots$;

$$\omega_c^* = \left\{\frac{2}{m_{h\|}l^2}\left[\frac{e^2 l}{4\pi\varepsilon_0\varepsilon_2 a_B^2}+\frac{2e^2\delta J(\delta)}{4\pi\varepsilon_0\varepsilon_2 l}\right]\right\}^{\frac{1}{2}}. \tag{6.12.40}$$

6.12.2 Strong magnetic field and the Mayer limit

If the parameters of the system are such that the polarization clouds of the electron and the holes overlap (a situation is possible when the Haken limit is realized at $B = 0$ and the Mayer limit at large), then the interaction of the exciton with polarization is considered as a weak perturbation. The exciton is formed mainly by the electron–hole interaction, relatively weakly shielded by high-frequency dielectric constant ε. Therefore, the frequency of internal motion in the exciton may be higher than ω_{n0}. Due to this inequality, the polarization state is determined by the electron and hole clouds. Therefore, in the Mayer limit, the displacements of f_p should be considered independent of the coordinates of the electron and the hole, which greatly simplifies the task of calculating the amplitudes of f_p. Since the procedure for calculating the phonon contributions to the ground state energy and the effective mass of the exciton is completely similar to that given in section 6.8 of this chapter, we present the final results here:

$$f_p = -\frac{\widetilde{V}_p^* A_p}{\hbar\Omega_p+\frac{\hbar^2 p^2}{2M_\perp}-(1-\kappa_p)\frac{\hbar \mathbf{p}\mathbf{P}_\perp}{M_\perp}}, \tag{6.12.41}$$

$$A_p = \exp\left(i\sigma_2\mathbf{p}\boldsymbol{\rho}_0-\frac{1}{2}\sigma_2^2 p^2 a_B^2\right)-\exp\left(i\sigma_1\mathbf{p}\boldsymbol{\rho}_0-\frac{1}{2}\sigma_1^2 p^2 a_B^2\right), \tag{6.12.42}$$

where

$$\kappa\mathbf{P}_\perp = \sum_{\mathbf{p}}\hbar\mathbf{p}\left|f_p\right|^2; \tag{6.12.43}$$

$$\Delta E_{00} = -\frac{e^2}{4\pi\varepsilon_0 l}\int_0^\infty \frac{dx\left[\exp\left(-\frac{2\sigma_1 a_B^2 x^2}{2}\right) - \exp\left(-\frac{2\sigma_2 a_B^2 x^2}{2}\right)\right]}{1 + \left(\frac{2R_S}{l}\right)^2 x^2} \times \tag{6.12.44a}$$

$$\times\left[(\varepsilon + \varepsilon_2 \tanh x)^{-1} - (\varepsilon + \varepsilon_{20} \tanh x)^{-1}\right],$$

$$(M_{ph})_{00} = M_{00}\left[2F_5 + M_\perp^{-1}F_6(1 + F_6)^{-1}\right]^{-1}. \tag{6.12.44b}$$

The factor $[\exp(-2\sigma_1 a_B^2 x^2/2) - \exp(-2\sigma_2 a_B^2 x^2/2)]$ in terms of energy and effective mass describes the disappearance of polarization: (1) with equal masses of the electron and the hole ($\sigma_1 = \sigma_2$); (2) in the limit of infinite magnetic fields, for which $a_B \to 0$.

For small $l \ll a_B$ the expression (6.12.41) can be represented as

$$\Delta E_{00} = -\frac{e^2(\varepsilon_{20}^2 - \varepsilon_2^2)}{8\pi\varepsilon_0 l\varepsilon^2}\left(\frac{l}{2R_S}\right)^2\left\{\mathrm{e}^{-\frac{\sigma_2^2 a_B^2}{2R_S^2}} Ei\left(-\frac{\sigma_2^2 a_B^2}{2R_S^2}\right) - e^{-\frac{\sigma_1^2 a_B^2}{2R_S^2}} Ei\left(-\frac{\sigma_1^2 a_B^2}{2R_S^2}\right)\right\}, \tag{6.12.45}$$

from which it follows that the growth of the magnetic field leads to nonmonotonicity of the phonon contribution to the binding energy and the effective mass of the exciton in the field values determined from the conditions: (a) $a_B \geqslant R_S$—phonon contributions increase with field growth; (b) $a_B < R_S$—phonon contributions decrease with field growth. The nature of nonmonotonicity was discussed in detail in section 6.8 in the problem of a surface polaron exciton in a magnetic field. Note that in the case of an exciton in a quantum well, these effects become more pronounced and can be more easily investigated experimentally. Shown in figure 7.8.1 the dependences of phonon contributions to the binding energy and the effective mass of the exciton for quantum wells of indium antimonide and cadmium telluride as a function of the magnitude of the magnetic field and the thickness of the quantum wells confirm this conclusion.

6.13 Coulomb interaction and Wannier–Mott excitons in polar semiconductor quantum wires

Quantum semiconductor and semimetallic wires (quasi-one-dimensional structures) are of great interest among structures of reduced dimension due to the possibility of their use to create new electronic and optoelectronic devices. In the works [61, 92, 112–117] some properties of such structures have been experimentally investigated.

It was shown in [93, 118, 119] that the Coulomb interaction in a thin wire can significantly increase as a result of a decrease in the dimension of the medium (geometric effect), as well as as a result of dielectric amplification due to the 'displacement' of the Coulomb field lines into neighboring domains with low dielectric permeability.

In [83, 120, 121], the Wannier–Mott theory of exciton states for a semiconductor quantum wire was developed, taking into account both effects, used to analyze and interpret experimental spectra of quantum wires made of cadmium selenide, lead iodide, and gallium arsenide in a dielectric matrix [111, 122].

In [1], it was found that for a planar structure made of a polar semiconductor, along with the effects of geometric and dielectric amplifications of the electron–hole interaction established in [11, 58], another effect may occur due to the loss of part of the inertial shielding of the Coulomb interaction of an electron and a hole due to the overlap of their polarization clouds.

Since in the experiments [121, 122] quantum wires were made of polar semiconductors, when calculating the binding energy of the Wannier–Mott exciton, it is also necessary to take into account the effect of shielding the electron–hole interaction by polar optical phonons.

In this section, the contribution to the exciton binding energy from the dimensional effect of the loss of inertial shielding is calculated, taking into account the approximations adopted in [93, 118, 121]:

(1) the continuum approximation

$$R \gg a_0, \tag{6.13.1}$$

where R *is* the radius of the wire and a_0 is the interatomic distance;

(2) approximation of the effective mass;

(3) an adiabatic approximation that allows to separate the transverse and longitudinal movements of an electron and a hole

$$E_{s.q.} \gg V_{e-h}, \tag{6.13.2}$$

where $E_{s.q.}$ is the energy of dimensional quantization and V_{e-h} is the average energy of the electron–hole interaction;

(4) the approximation of an infinite potential barrier on the surface of the wire and the environment.

6.13.1 Exciton Hamiltonian

The Hamiltonian of a system consisting of two charges (an electron and a hole) moving in the conduction band and the valence band, respectively, in a polar semiconductor wire of radius R, bordering a semi-infinite polar medium (figure 6.14) has the form

$$\begin{aligned}\hat{H} &= \hat{K}_e + \hat{K}_h + \hat{H}_{\mathrm{ph}}^{S_{1,2}} + \hat{H}_{\mathrm{ph}}^{V} + \hat{H}_{\mathrm{e,\,h-ph}}^{S_{1,2}} + \hat{H}_{\mathrm{e-ph}}^{S} \\ &+ \hat{H}_{h-ph}^{S} + V_{e-h} + V_{SA}(\rho_e) + V_{SA}(\rho_h) + V_0,\end{aligned} \tag{6.13.3}$$

where $\hat{K}_i$ are operators of kinetic energies of an electron ($i = e$) and a hole ($i = h$):

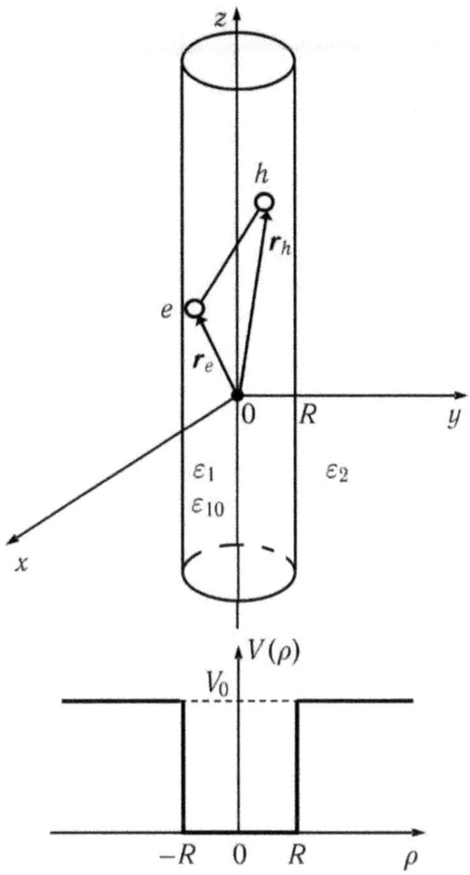

Figure 6.14. An electron (e) and a hole (h) in the wire. Below: the shape of a potential barrier at the wire boundary

$$\hat{K}_i = \frac{\hat{p}_{z_i}^2}{2m_{II_i}^*} + \frac{\hat{p}_{\rho_i}^2}{2m_{\perp_i}^*}, \tag{6.13.4}$$

$m_{II_i}^*$, $m_{\perp_i}^*$ are the longitudinal and transverse components of the effective mass of the electron and the hole, and $\hat{H}_{ph}^j$ are the Hamiltonians of the surface ($j = S$) and bulk ($j = V$) optical phonons:

$$\hat{H}_{ph}^j = \sum\nolimits_{\chi_j} \hbar\Omega_{\chi_j} \hat{a}_{\chi_j}^+ \hat{a}_{\chi_j}, \tag{6.13.5}$$

where $\Omega_{j=V} \equiv \omega_{10}$ is the frequency of longitudinal optical phonons; $\Omega_{j=S} = \Omega_S$ is the frequency of surface optical phonons; and $\hat{H}_{i-ph}^j$ are the Hamiltonians of electron and hole interactions with surface ($j = S$) and bulk ($j = V$) optical phonons obtained in [20, 123].

$$\hat{H}^{j}_{i-ph} = \sum\nolimits_{\chi_j,\, n}[\Gamma_{\chi_j} e^{i(\eta_j z_i + l_i\phi_i)}\hat{a}_{\chi_j,\, n} + \text{к. с.}]. \tag{6.13.6}$$

Here:

$$\chi_j = \begin{cases} (m,\, q_1,\, \eta) \text{ при } j = V; \\ (m,\, \eta) \text{ при } j = S; \end{cases} \tag{6.13.7}$$

are complete sets of quantum numbers for volume and surface modes: q_1—radial quantum number; m—quantum number for the projection of the moment on the z axis; η—wave number of motion along z; and $\hat{a}^{+}_{\chi_j}$, $\hat{a}_{\chi_j}$ are operators of the birth and destruction of surface ($j = S$) and bulk ($j = V$) phonons in the polar wire:

$$\Gamma_{\chi_j} = \begin{cases} V_{\chi_v}\psi_1(\rho_i) \text{ при } j = V; \\ V_{\chi_s} F_1^1(\rho_i) \text{ при } j = S; \end{cases} \tag{6.13.8}$$

V_{χ_V} is the amplitude of the electron–phonon interaction for bulk polar optical phonons in a wire [124]:

$$|V_{\chi_V}|^2 = \frac{e^2}{4\pi\varepsilon_0}\left(\frac{1}{\varepsilon_1} - \frac{1}{\varepsilon_{10}}\right)\frac{\hbar\omega_{10}}{\left(q_1^2 + q^2\right)R^2}; \tag{6.13.9}$$

$$\psi_1(\rho_i) = \frac{\sqrt{2}\, J_m(q_1\rho_i)}{|J_{m+1}(q_1 R)|}, \tag{6.13.10}$$

ε_1, ε_{10} are high-frequency and static dielectric permittivity of the wire and $J_m(x)$ is the Bessel function; the value q_1 is found from the equation:

$$J_m(q_1 R) = 0; \tag{6.13.11}$$

V_{χ_S} is the amplitude of the electron–phonon interaction for surface phonons in a wire:

$$|V_{\chi_S}|^2 = \frac{e^2\hbar\omega_{10}(\varepsilon_{10} - \varepsilon_1)^{\frac{1}{2}}}{2\pi\varepsilon_0\eta R(\chi_1 + \chi_2)^{\frac{3}{2}}(\varepsilon_1\chi_2 - \varepsilon_{10}\chi_1)^{\frac{1}{2}}}; \tag{6.13.12}$$

$$F_1^1(\rho_i) = \frac{I_m(\eta\rho_i)}{I_m(\eta R)}; \tag{6.13.13}$$

$$\chi_1 = \varepsilon_1\frac{I'_{|m|}(\eta R)}{I_{|m|}(\eta R)};\ \chi_2 = -\varepsilon_2\frac{K'_{|m|}(\eta R)}{K_{|m|}(\eta R)}. \tag{6.13.14}$$

ε_2 is the dielectric constant of the environment. $I_m(x)$, $K_m(x)$ are modified Bessel functions and MacDonald functions, respectively; and V_{e-h} is the potential energy of the interaction of an electron and a hole, obtained in [44, 100, 125] and having the form:

$$V_{e-h} = -e\int_{-\infty}^{+\infty} dk \cdot e^{ik(z_e - z_h)}\sum_{n=-\infty}^{\infty} e^{in\gamma}\, \Phi_n(k, \rho_e, \rho_h); \tag{6.13.15}$$

(Since V_{e-h} does not depend on the angle γ, only one term with $n = 0$ remains in the expression (6.13.15).)

$$\begin{gathered}\Phi_0(k, \rho_e, \rho_h) = \frac{e^2}{4\pi^2\varepsilon_0\varepsilon_1}\left\{\frac{(\varepsilon_1 - \varepsilon_2)K_0(kR)K_0'(kR)I_0(k\rho_e)I_0(k\rho_h)}{\varepsilon_2 I_0(kR)K_0'(kR) - \varepsilon_1 I_0'(kR)K_0(kR)} + \right.\\ \left. + \theta(\rho_h < \rho_e)I_0(k\rho_h)K_0(\rho_e) + \theta(\rho_h > \rho_e)I_0(k\rho_e)K_0(\rho_h)\right\},\ \rho,\ \rho_e \leqslant R,\end{gathered} \tag{6.13.16}$$

$$\theta(\rho) = 1,\ \ \rho > 0;\ \theta(\rho) = 0,\ \ \rho < 0.$$

$V_{SA}(\rho_i)$ is the potential energy of the self-action of an electron (hole) due to the interaction of the carrier with the inertia-free polarization of the medium induced by it [121, 122]:

$$V_{SA}(\rho_i) = -\frac{1}{2}e\int_{-\infty}^{+\infty} \mathrm{d}k\sum_{n=-\infty}^{\infty} \Phi_m(k, \rho_i), \tag{6.13.17}$$

where

$$\Phi_0(k, \rho_i) = \frac{e}{4\pi^2\varepsilon_0\varepsilon_1}\frac{(\varepsilon_1 - \varepsilon_2)K_0(kR)K_0'(kR)I_0^2(k\rho_i)}{\varepsilon_2 I_0(kR)K_0'(kR) - \varepsilon_1 I_0'(kR)K_0(kR)},\quad \rho_i \leqslant R; \tag{6.13.18}$$

V_0 is a potential barrier at the boundary of the wire and its environment:

$$V_0 \to \infty,\ \ \rho \geqslant R;\ V_0 = 0,\ \ \rho < R. \tag{6.13.19}$$

6.13.2 Effective potential of the electron–hole interaction

Considering V_{e-h} as a seed potential that takes into account the shielding of the electron–hole interaction by fast polarization (plasmons of valence electrons), we apply the method [20, 126] to obtain the effective potential of the electron–hole interaction, including shielding by slow polarization (surface and bulk optical phonons of the polar wire).

The wave function of the system, taking into account the accepted approximations, can be represented in a multiplicative form:

$$\psi(\rho_e, \rho_h, \varphi_e, \varphi_h, z) = \psi(z)\,\psi(\rho_e, \varphi_e)\,\psi(\rho_h, \varphi_h), \tag{6.13.20}$$

where $\psi(z)$ is a wave function describing the relative motion of an electron and a hole along the z-axis; $\psi(\rho_i, \varphi_i)$ is a wave function describing the motion of an electron ($i = e$) and a hole ($i = h$) inside a wire perpendicular to its axis (in a cylindrical quantum well with infinite barriers) and satisfying the Schrödinger equation:

$$\frac{1}{\rho}\frac{\partial}{\partial\rho}\left(\rho\frac{\partial}{\partial\rho}\Phi(\rho_i)\right)+\left(\xi^2-\frac{m^2}{\rho^2}\right)\Phi(\rho_i)=0; \tag{6.13.21}$$

in the field of $\rho < R$,

$$\xi^2=\beta^2-k^2;\beta^2=\frac{2m^*_{\perp,\,i}E_i}{\hbar^2}. \tag{6.13.22}$$

The solution of equation (6.13.21) is given in [121].
The wave function $\psi(\rho_i,\,\varphi_i)$ can be represented as

$$\psi(\rho_i,\,\varphi_i)=\frac{1}{\sqrt{2\pi}}\Phi\,(\rho_i)e^{im_i\varphi_i}. \tag{6.13.23}$$

Considering the motion of the electron and the hole perpendicular to the z-axis to be fast, we average the Hamiltonian (6.13.3)–(6.13.18) on the wave function (6.13.20), (6.13.23) and, switching to the center of mass system for movement along z, we obtain:

$$\begin{aligned}\widetilde{H}_1&=\left\langle\Phi(\rho_e)\Phi(\rho_h)\,|\hat{H}|\,\Phi(\rho_e)\Phi(\rho_h)\right\rangle\\&=\frac{\hat{p}_Z^2}{2M_{II}}+\frac{\hat{p}_z^2}{2\mu_{II}}+\frac{\hbar^2\xi_e^2(m,\,k)}{2m^*_{e\perp}}+\frac{\hbar^2\xi_h^2(m,\,k)}{2m^*_{h\perp}}\\&+\widetilde{V}_{e-h}(R,\,z)+\widetilde{V}_{SA}(R)+\sum\nolimits_{\mathbf{k}(\eta,\,q_1)}\hbar\omega_{10}a^+_{\mathbf{k}}a_{\mathbf{k}}+\sum\nolimits_{\eta}\hbar\Omega_s a^+_{\eta}a_{\eta}\\&+\sum\nolimits_{\mathbf{k}(\eta,\,q_1)}\widetilde{V}_{\chi_V}e^{iqZ}\left\{\left[e^{i\sigma_1qz+im\varphi_e}-e^{-i\sigma_2qz+im\varphi_h}\right]\hat{a}^+_{\mathbf{k}}+\text{к.с.}\right\}\\&+\sum\nolimits_{\eta}\widetilde{V}_{\chi_S}e^{i\eta Z}\left\{\left[e^{i\sigma_1\eta z+im\varphi_e}-e^{-i\sigma_2\eta z+im\varphi_h}\right]\hat{a}^+_{\eta}+\text{к. с.}\right\},\end{aligned} \tag{6.13.24}$$

where

$$\begin{aligned}\widetilde{V}_{e-h}(R,\,z)&=-e\int_{-\infty}^{+\infty}dk\cdot e^{ikz}\int_0^R\int_0^R|\,\varphi(\rho_e)\,|^2|\,\varphi(\rho_h)\,|^2\\&\quad\Phi_0(k,\,\rho_e,\,\rho_h)\rho_e d\rho_e\rho_h d\rho_h,\end{aligned} \tag{6.13.25}$$

$$\begin{aligned}\widetilde{V}_{SA}(R)&=-\frac{e}{2}\sum\nolimits_{i=e,h}\int_{-\infty}^{+\infty}dk\int_0^R\int_0^R|\,\varphi(\rho_e)\,|^2|\,\varphi(\rho_h)\,|^2\\&\quad\Phi_0(k,\,\rho_i)\rho_e d\rho_e\rho_h d\rho_h.\end{aligned} \tag{6.13.26}$$

The designations are introduced here:

$$\widetilde{V}_{\chi_V}=\frac{\sqrt{2}\,V_{\chi_V}\int_0^R|\varphi(\rho_i)|^2J_m(q_1\rho_i)\rho_i d\rho_i}{J_{m+1}(q_1R)}; \tag{6.13.27}$$

$$\widetilde{V}_{\chi_S} = \frac{\sqrt{2}\, V_{\chi_S} \int_0^R |\varphi(\rho_i)|^2 I_m(\eta\rho_i)\rho_i d\rho_i}{I_{m+1}(\eta R)}; \tag{6.13.28}$$

$$z = z_e - z_h;\ Z = \frac{m_{IIe}^* z_e + m_{IIh}^* z_h}{m_{IIe}^* + m_{IIh}^*}; \tag{6.13.29}$$

$$\mu^{-1} = m_{\|e}^{*-1} + m_{\|h}^{*-1};\ M_{\|} = m_{\|e}^* + m_{\|h}^*; \tag{6.13.30}$$

$$\sigma_1 = \frac{m_{IIh}^*}{M_{II}};\ \sigma_2 = \frac{m_{IIe}^*}{M_{II}}. \tag{6.13.31}$$

In the limit of a thin wire ($kR \ll 1$), the wave functions $\Phi(\rho_i)$ in (6.13.23) can be taken as [121]

$$\Phi(\rho_i) = \begin{cases} \dfrac{1}{\sqrt{\pi}R},\ \rho_i \leqslant R; \\ 0,\ \rho_i > R. \end{cases} \tag{6.13.32}$$

In this case, the integrals (6.13.25)–(6.13.26) are calculated analytically:

$$\begin{aligned} \widetilde{V}_{e-h}(R, z) = &-\frac{e^2}{\pi^2\varepsilon_0\varepsilon_1}\int_{-\infty}^{+\infty} \mathrm{d}k \cdot \frac{e^{ikz}}{k^2} \\ &\times \left\{ \frac{(\varepsilon_1 - \varepsilon_2)K_0'(kR)K_0(kR)I_1^2(kR)}{\varepsilon_2 I_0(kR)K_0'(kR) - \varepsilon_1 I_0'(kR)K_0(kR)} - I_1(kR)K_1(kR) + \frac{1}{2}\right\}; \end{aligned} \tag{6.13.33}$$

$$\begin{aligned} \widetilde{V}_{SA}(R, z) &= -\frac{e^2}{4\pi^2\varepsilon_0\varepsilon_1}\frac{(\varepsilon_1 - \varepsilon_2)}{\pi R^2} \\ &\int_{-\infty}^{+\infty} dk \frac{K_0'(kR)K_0(kR)\int_0^R I_0^2(k\rho_i)\rho_i d\rho_i}{\varepsilon_2 I_0(kR)K_0'(kR) - \varepsilon_1 I_0'(kR)K_0(kR)}. \end{aligned} \tag{6.13.34}$$

Following the Haken method [38, 39], we average the Hamiltonian (6.13.24) on the wave function

$$\psi(z, \{\hat{a}_\kappa\}, \{\hat{a}_\eta\}) = \psi(z)\, e^{\hat{U}_1} e^{\hat{U}_2}\, |0\rangle_V\, |0\rangle_S, \tag{6.13.35}$$

where $\hat{U}_1$ и $\hat{U}_2$ are operators of unitary transformations defined by expressions:

$$\hat{U}_1 = \sum_\kappa \left\{\hat{a}_\kappa f_\kappa^*(z) - \hat{a}_{-\kappa}^+ f_\kappa(z)\right\};\ \hat{U}_2 = \sum_\eta \left\{\hat{a}_\eta f_\eta^*(z) - \hat{a}_{-\eta}^+ f_\eta(z)\right\}, \tag{6.13.36}$$

where $|0\rangle_S\ |0\rangle_V$ is the wave function of the ground state of the phonon subsystem.

In the center of mass system ($Z = 0$), for the variational amplitudes of the displacements of the phonon mode operators $f_\kappa(z)$, $f_\eta(z)$ we obtain the following expressions:

$$f_\kappa(z) = \frac{\sqrt{2}\,\widetilde{V}_V^{*} e^{i\sigma_1\eta z}}{\hbar\omega_{10} + \frac{\hbar^2\eta^2}{2m^*_{\parallel e}} + \frac{\hbar^2 q_1^2}{2m^*_{\perp e}}} - \frac{\sqrt{2}\,\widetilde{V}_V^{*} e^{-i\sigma_2\eta z}}{\hbar\omega_{10} + \frac{\hbar^2\eta^2}{2m^*_{\parallel h}} + \frac{\hbar^2 q_1^2}{2m^*_{\perp h}}}, \quad \kappa \equiv \kappa(\eta, q_1); \tag{6.13.37a}$$

$$f_\eta(z) = f_\kappa(z) = \frac{\sqrt{2}\,\widetilde{V}_V^{*} e^{i\sigma_1\eta z}}{\hbar\Omega_S + \frac{\hbar^2\eta^2}{2m^*_{\parallel e}}} - \frac{\sqrt{2}\,\widetilde{V}_V^{*} e^{i\sigma_2\eta z}}{\hbar\Omega_S + \frac{\hbar^2\eta^2}{2m^*_{\parallel h}}}. \tag{6.13.37b}$$

The effective potential of the electron–hole interaction $W_{\text{eff}}(R, z)$ is the sum of the seed Coulomb potential $\widetilde{V}_{e-h}(R, z)$ and the contributions from the shielding of the Coulomb interaction by bulk and surface phonons:

$$W_{\text{eff}}(R, z) = \widetilde{V}_{e-h}(R, z) + W_V(R, z) + W_S(R, z), \tag{6.13.38}$$

where

$$\widetilde{V}_{e-h}(R, z) = -\frac{2e^2(1-\kappa)}{\pi^2 R\varepsilon_0\varepsilon_1} F_1(R, z); \tag{6.13.39}$$

$$W_V(R, z) = \frac{e^2}{\pi^3 R\varepsilon_0}\left(\frac{1}{\varepsilon_0} - \frac{1}{\varepsilon_{10}}\right)\frac{F_2(q_1 R)}{\left|J_{m+1}(q_1 R)\right|^2} \left\{F_3\left(z, R_{\parallel e}, R_{\perp e}\right) - F_3\left(z, R_{\parallel h}, R_{\perp h}\right)\right\}; \tag{6.13.40}$$

$$W_S(R, z) = \frac{e^2(\varsigma - 1)}{\pi^2 R\varepsilon_0\varepsilon_1} F_1(R, z). \tag{6.13.41}$$

Notation is introduced in formulas (6.13.40)–(6.13.42):

$$F_1(R, z) = \int_0^\infty \frac{\cos\left(\frac{z}{R}x\right)}{x^2}\left\{\frac{K_0(x)\,K_1(x)\,I_1^2(x)}{I_1(x)K_0(x) + \kappa I_0(x)K_1(x)} - I_1(x)K_1(x) + \frac{1}{2}\right\}dx; \tag{6.13.42}$$

$$F_2(q_1 R) = \int_0^1 J_m(q_1 Rx)dx; \tag{6.13.43}$$

$$F_3(z, R_{\parallel i}, R_{\perp i}) = \int_0^\infty \frac{dx\cos\left(\frac{z}{R_{S\parallel i}}x\right)}{\left(x^2 + q_1^2 R_{\parallel i}^2\right)\left(1 + x^2 + q_1^2 R_{\perp i}^2\right)}; \tag{6.13.44}$$

$$F_4(R, z) = \int_0^\infty \frac{K_0^2(x)\, I_1^3(x) \cos\left(\frac{z}{R}x\right)\left\{\left(1 + \frac{R_{\parallel e}^2}{R^2}x^2\right)^{-1} - \left(1 + \frac{R_{\parallel h}^2}{R^2}x^2\right)^{-1}\right\}}{x^3 I_0(x)\left\{\varsigma I_1(x)K_0(x) + \kappa I_0(x)K_1(x)\right\}} \tag{6.13.45}$$

$$\left[\frac{\omega_{10}}{\Omega_S(x)}\right] dx;$$

$$\varsigma = \frac{\varepsilon_{10}}{\varepsilon_1};\ \kappa = \frac{\varepsilon_2}{\varepsilon_1};\ R_{S\parallel i} = \left(\frac{\hbar^2}{2m_{\parallel i}^*\Omega_S}\right)^{\frac{1}{2}};\ R_{\parallel i} = \left(\frac{\hbar^2}{2m_{\parallel i}^*\omega_{10}}\right)^{\frac{1}{2}}. \tag{6.13.46}$$

For the polaron contributions to the energy of self-action from bulk $E^V_{p,\, i=e,\, h}(R)$ and surface phonons, we obtain the following expressions:

$$E^V_{p,\, i=e,\, h}(R) = -\frac{e^2 R_{\parallel i}}{4\pi^3 \varepsilon_0 R^2}\left(\frac{1}{\varepsilon_0} - \frac{1}{\varepsilon_{10}}\right)\frac{F_2^2(q_1 R) F_5(q_{\perp i}, R_{\parallel i}, R_{\perp i})}{|J_{m+1}(q_1 R)|^2}, \tag{6.13.47}$$

$$E^S_{p,\, i=e,\, h}(R) = -\frac{e^2}{\pi^3 \varepsilon_0 R}(\varepsilon_{10} - \varepsilon_1)^{\frac{1}{2}}\frac{F_6(R, R_{S\parallel i})}{|J_{m+1}(q_1 R)|^2}, \tag{6.13.48}$$

where the designations accepted are:

$$F_5(z, R_{\parallel i}, R_{\perp i}) = \int_0^\infty \frac{dx}{\left(x^2 + q_1^2 R_{\parallel i}^2\right)\left(1 + x^2 + q_1^2 R_{\perp i}^2\right)}; \tag{6.13.49}$$

$$F_6\left(\frac{R}{R_{S\parallel i}}\right) = \int_{-\infty}^\infty \frac{dx}{x}\frac{\left[\tilde{I}\left(\frac{R}{R_{S\parallel i}}\right)\right]}{\left(\frac{\Omega_S}{\omega_{10}}\right)(\chi_1 + \chi_2)^{\frac{3}{2}}(\varepsilon_1\chi_2 - \varepsilon_{10}\chi_1)^{\frac{1}{2}}\left|I_m\left(\frac{R}{R_{S\parallel i}}\right)\right|^2\left(1 + x^2\right)}; \tag{6.13.50}$$

$$\frac{\Omega_S}{\omega_{10}} = \left[\left(\frac{\omega_1}{\omega_{10}}\right)^2 + \frac{\chi_1}{\chi_1 + \chi_2}\right]^{\frac{1}{2}}; \tag{6.13.51}$$

$$\chi_1 = \varepsilon_1 \frac{I'_{|m|}\left(x\frac{R}{R_{S\parallel}}\right)}{I|m|\left(x\frac{R}{R_{S\parallel}}\right)};\ \chi_2 = -\varepsilon_2\frac{K'_{|m|}\left(x\frac{R}{R_{S\parallel}}\right)}{K_{|m|}\left(x\frac{R}{R_{S\parallel}}\right)};\ \left[\tilde{I}\left(\frac{R}{R_{S\parallel}}\right)\right] \tag{6.13.52}$$

$$= \int_0^1 J_m\left(\frac{R}{R_{S\parallel}}x\right)x dx.$$

6.13.3 Exciton binding energy in a quantum wire

To calculate the energy of the exciton ground state in a polar semiconductor quantum wire, we use the effective Hamiltonian:

$$\hat{H}_{\text{eff}} = -\frac{\hbar^2}{2\mu_{\parallel}}\frac{d^2}{dz^2} + W_{\text{eff}}(R, z), \tag{6.13.53}$$

where $W_{\text{eff}}(R, z)$ is defined by the expressions (6.13.38)–(6.13.46).

Let's choose a trial wave function in the form

$$\psi(z) = \sqrt{\frac{\beta}{\sqrt{\pi}}}\exp\left\{-\frac{1}{2}\beta^2(z_e - z_h)^2\right\}, \tag{6.13.54}$$

where β is the variation parameter.

By matrixing the Hamiltonian (6.13.53) on the trial wave function (6.13.54), we obtain a variational functional for calculating the energy of the exciton ground state:

$$\begin{aligned} E_0(R) &= \frac{\hbar^2\beta^2}{4\mu_{\parallel}} - \frac{2e^2(1-k)}{\pi^2\varepsilon_0\varepsilon_1 R}\widetilde{F}_1(\beta, R) \\ &+ \frac{e^2}{\pi^3\varepsilon_0 R}\left(\frac{1}{\varepsilon_1} - \frac{1}{\varepsilon_{10}}\right)\frac{F_2(q_1 R)}{\left|J_{m+1}(q_1 R)\right|^2}\left\{\widetilde{F}_3\left(\beta, R_{\parallel e}, R_{\perp e}\right) - \widetilde{F}_3\left(\beta, R_{\parallel h}, R_{\perp h}\right)\right\} \\ &+ \frac{e^2(\varsigma - 1)}{2\pi^2\varepsilon_0\varepsilon_1 R}\widetilde{F}_4(\beta, R), \end{aligned} \tag{6.13.55}$$

where

$$\widetilde{F}_1(\beta, R) = \int_0^\infty \frac{\exp\left(-\frac{x^2}{4\beta^2}\right)}{x^2}\left\{\frac{K_0(x)\, K_1(x)\, I_1^2(x)}{I_1(x)K_0(x) + \kappa I_0(x)K_1(x)} - I_1(x)K_1(x) + \frac{1}{2}\right\}dx; \tag{6.13.56}$$

$$\widetilde{F}_3(\beta, R_{\parallel i}, R_{\perp i}) = \int_0^\infty \frac{dx\, \exp\left\{-\frac{x^2}{4\beta^2 z^2}\right\}}{\left(x^2 + q_1^2 R_{\parallel i}^2\right)\left(1 + x^2 + q_1^2 R_{\perp i}^2\right)}; \tag{6.13.57}$$

$$\begin{aligned} \widetilde{F}_4(\beta, R) = \int_0^\infty &\frac{dx\, K_0^2(x)\, I_1^3(x)\exp\left\{-\frac{x^2}{4\beta^2 z^2}\right\}\left[\frac{\omega_{10}}{\Omega_S(x)}\right]}{x^3 I_0(x)\{\varsigma I_1(x)K_0(x) + \kappa I_0(x)K_1(x)\}} \\ &\times\left\{\left(1 + \frac{R_{\parallel e}^2}{R^2}x^2\right)^{-1} - \left(1 + \frac{R_{\parallel h}^2}{R^2}x^2\right)^{-1}\right\}. \end{aligned} \tag{6.13.58}$$

In the limit of a thin quantum wire, when the criterion is fulfilled

$$\left(\frac{\varepsilon_2}{\varepsilon_1}\right)(\beta R)^2 \ln|\beta R| \ll 1, \tag{6.13.59}$$

from the expressions (6.13.55)–(6.13.58) we get

$$E_0(\beta) = \frac{\hbar^2\beta^2}{4\mu_\parallel} + \frac{2e^2}{4\pi\varepsilon_0\varepsilon_2}\frac{\beta}{\sqrt{\pi}}\left[\ln\left(\frac{\beta R_{\text{eff}}}{2}\right) + \frac{C}{2}\right] + F_V(\beta) + F_S(\beta). \tag{6.13.60}$$

Here

$$R_{\text{eff}} = R\exp\left\{\frac{\varepsilon_2}{\varepsilon_1}S(R) + S(\infty) - S(R)\right\}; \tag{6.13.61}$$

$$S(x) = (2\pi)^2\int_0^x \rho d\rho \ln\left(\frac{\rho}{R}\right)\int_0^R \rho' d\rho'\left\{\Phi_e^2(\rho)\Phi_h^2(\rho') + \Phi_e^2(\rho')\Phi_h^2(\rho)\right\}. \tag{6.13.62}$$

Let's choose the values as units of energy and length in the formula (6.13.60):

$$Ry^* = \frac{\mu_\parallel e^4}{2(4\pi\varepsilon_0)^2\hbar^2\varepsilon_1^2}, \quad r_B^* = \frac{\hbar^2\varepsilon_1}{\mu_{II}e^2}. \tag{6.13.63}$$

Then the variational functional (6.13.60) can be represented as:

$$\widetilde{E}(\widetilde{\beta}) = \frac{\widetilde{\beta}^2}{2} + \frac{4\widetilde{\beta}}{\varepsilon_2\sqrt{\pi}}\left[\ln\left(\frac{\widetilde{\beta}\widetilde{R}_{\text{eff}}}{2}\right) + \frac{C}{2}\right] + \widetilde{F}_V(\widetilde{\beta}) + \widetilde{F}_S(\widetilde{\beta}), \tag{6.13.64}$$

where

$$\widetilde{\beta} = \beta r_B^*; \; \widetilde{E} = \frac{E}{Ry^*}; \; \widetilde{R}_{\text{eff}} = \frac{R_{\text{eff}}}{r_B^*}. \tag{6.13.65}$$

In the general case, when the criterion (6.13.59) is not fulfilled, the exciton binding energy should be calculated using the functional (6.13.55)–(6.13.58).

Tables 6.4 and 6.5 show the results of numerical calculation of the exciton binding energy for CdSe and GaAs quantum wires in dielectric matrices made of chrysotile asbestos, excluding and taking into account the contribution from phonons, calculated by the approximate formula (6.13.60) (columns 2 and 4) and by exact formulas (6.13.55)–(6.13.58) (columns 3 and 5) depending on the radius of the quantum wire. As can be seen from table 6.4, for a more polar CdSe semiconductor, the phonon contribution ranges from 4 to 15% in the R range from 0.01 r_B^* to 10 r_B^*, while for a less polar GaAs phonons contribute from 2 to 6% in the same ranges of the radius of the quantum wire (table 6.5).

Table 6.4. The results of numerical calculation of the exciton binding energy for a CdSe quantum wire in a chrysotile asbestos dielectric matrix.

R, r_B^*	E_{cl}^0 (R), Ry^*	E_{cl} (R), Ry^*	$E_{\text{ph}}^0(R)$, Ry^*	$E_{\text{ph}}(R)$, Ry^*	a_0^{-1}, r_B^*	a^{-1}, r_B^*
0.01	116.87	114.5	114.63	109.6	0.0959	0.0964
0.03	61.42	59.62	60.00	55.67	0.151	0.154
0.1	26.15	25.84	25.42	23.45	0.293	0.274
0.3	10.47	10.98	10.13	9.74	0.644	0.496
1	3.439	3.973	3.322	3.457	1.812	0.990
3	1.182	1.493	1.141	1.282	5.123	1.922
10	0.359	0.489	0.346	0.416	16.69	4.218

Table 6.5. Results of numerical calculation of exciton binding energy for GaAs quantum wire in a chrysotile asbestos dielectric matrix.

R, r_B^*	E_{cl}^0 (R), Ry^*	E_{cl} (R), Ry^*	$E_{\text{ph}}^0(R)$, Ry^*	$E_{\text{ph}}(R)$, Ry^*	a_0^{-1}, r_B^*	a^{-1}, r_B^*
0.01	282.0	261.2	280.6	255.2	0.06637	0.07394
0.03	137.1	123.5	136.3	119.4	0.1136	0.1264
0.1	53.33	48.73	52.95	46.64	0.2476	0.2435
0.3	20.01	19.35	19.85	18.39	0.5987	0.4605
1	6.323	6.631	6.271	6.266	1.812	0.9596
3	2.143	2.403	2.125	2.263	5.267	1.912
10	0.6467	0.7665	0.6412	0.7199	17.36	4.237

6.13.4 Comparison of theory and experiment

In the experiment [122, 127], the directly measured value is the energy of a photon generating an exciton:

$$\hbar\omega_{ex} = E_g^0 + \sum\nolimits_{i=e,\,h}(\mathrm{E}_{s.q.}^i + E_{SA}^i + W_{\text{ph},\,i}^{S,\,V}) - E_b. \tag{6.13.66}$$

Here E_g^0 is the band gap of a massive semiconductor crystal (wire substance). The second term in (6.13.66) determines the contribution to the renormalization of the band gap width due to the dimensional quantum effect, the self-action effect, and the polaron effect, respectively, and E_b is the exciton binding energy.

In accordance with the approach developed in [122], the $\hbar\omega_{ex}$ takes into account the effects of nonparabolicity of zones, as well as the splitting of the valence band into zones of light and heavy holes and the formation of 'light' and 'heavy' excitons, which are observed independently in the optical spectrum of a CdSe and GaAs quantum wire.

Table 6.6. The values of the calculated parameters of excitons in *GaAs* and *CdSe A*, quantum wires crystallized in chrysotile-asbestos nanotubes and experimental values of the energy of exciton transitions. $\hbar\omega_{ex}$ is the value of the energy of the exciton-generating quantum calculated by the authors of this book, $\hbar\omega_{ex}^{cl}$, in [122]. The remaining values are also taken from [122].

Type of wire and exciton	Model parameters							Theory		Experiment
	$\frac{m_{\perp h}}{m_0}$	$\frac{m_{\parallel h}}{m_0}$	$\frac{\mu}{m_0}$	E_g^0, eV	ε_1	ε_2	d, nm	$\hbar\omega_{ex}^{cl}$, eV	$\hbar\omega_{ex}$, eV	$\hbar\omega_{ex}$, eV
GaAs $e - hh$	0.067	0.050	0.034	1.426	10.9	2.2	4.8	1.840	1.824	1.82 ± 0,04
							6.0	1.710	1.693	1.69 ± 0,04
GaAs $e - lh$	0.067	0.068	0.059	1.426	10.9	2.2	4.8	2.118	2.111	2.11 ± 0,05
							6.0	1.910	1.899	1.89 ± 0,05
CdSe *A*	0.12	0.45	0.107	1.751	5.8	2.2	4.8	1.992	1.995	1.98 ± 0,08
							6.0	1.905	1.908	

Table 6.6 shows the size dependences of $\hbar\omega_{ex}$: theoretical and experimental values calculated for the quantum wire GaAs ($e - hh$), GaAs ($e - lh$), CdSe (subzone of light holes A).

As can be seen from the comparison of theoretical and experimental data given in table 6.6, taking into account the contributions of optical phonons improves the agreement of theory and experiment. At the same time, it should be taken into account that this has a weaker effect on the value of $\hbar\omega_{ex}$, since contributions come in with opposite signs.

6.14 Conclusion

The general theory of Wannier–Mott exciton states in multilayer structures with quantum wells and in composite superlattices was applied to the study of Wannier–Mott excitons in single quantum wells and in $GaAs/Al_xGa_{1-x}As$ superlattices.

A generalization of the theory of polaron excitons in semibounded polar crystals and at the contact of two polar crystals within the limits of Haken $a_{ex} > R_p$, Mayer $a_{ex} < R_p$, and at an arbitrary value of the ratio a_{ex}/R_p, where a_{ex} and R_p are the radii of the exciton and the polaron, respectively, was given.

The theory and experiment for excitons in thin polar films of lead iodide and cadmium telluride were compared.

References

[1] Bastard B, Mendez E E, Chang L L and Esaki L 1982 Exciton binding energy in quantum wells *Phys. Rev.* B **26** 1974–9

[2] Miller R C, Kleinmann D A and Gossard A C 1984 Energy gap discontinuities and effective masses for GaAs – Al_xAs quantum well *Phys. Rev.* B **29** 7085–91

[3] Miller R C, Kleinmann D A, Munteanu O and Tsang W T 1981 New transitions in the photoluminescence of GaAs quantum wells *Appl. Phys. Lett.* **39** 1–3

[4] Miller R C, Kleinmann D A, Nordland N A and Gossard A C 1984 Luminescence studies of optically pumped GaAs − Al_xAs quantum well *Phys. Rev.* B **29** 7085–91

[5] Vojak V A, Holonjak N, Laiding M D *et al* 1980 The exciton recombination in GaAs − $Al_xGa_{1-x}As$ quantum well-heterostructures *Solid State Commun.* **35** 477–81

[6] Weisbuch C, Dingle R, Gossard A C and Wiegmann W 1980 Optical properties and interface disorder of GaAs − $Al_xGa_{1-x}As$ MQW structures *J. Vac. Sci. Tecnol* **17** 1128–9

[7] Beril S I, Pokatilov E P, Fomin V M and Pogorilko G A 1985 Wannier–Mott exciton in multilayer systems *FTP* **19** 412–7

[8] Miller R C, Kleinmann D A, Tsang W T and Gossard A C 1981 Observation of the exciton level in GaAs quantum wells *Phys. Rev.* B **24** 1134–6

[9] Pokatilov E P, Beril S I, Fomin V M and Pogorilko G A 1985 Wannier–Mott exciton states in two-layers periodic structures *Phys. Status Solidi* B **130** 278–88

[10] Efros A A 1986 Excitons in structures with quantum wells *FTP* **20** 1987–97

[11] Andryushin E A and Silin A P 1980 Excitons in thin semiconductor films *FTT* **22** 2676–80

[12] Ginzburg V L and Kelle V V 1973 On electron–hole type surface excitons and related collective phenomena *Lett. JETF* **17** 428–31

[13] Keldysh L V 1979 Coulomb interaction in thin films of semiconductors and semimetals *Lett. JETF* **29** 716–9

[14] Pokatilov E P, Beril S I, Fomin V M, Kalinovsky V V, Litovchenko V G, Korbutyak D V *et al* The dimensionally quantized states of the Wannier–Mott exciton in film structures *Part III Dep. hands. MoldNIINTI* 12/21/88. No. 1064 18

[15] Rytova N S 1967 The shielded potential of a point charge in a thin film *Bull. Mosc. State Univ* **3** 30–7

[16] Chaplik A V and Entin M V 1971 Charged impurities in very thin layers *JETF* **61** 2496–503

[17] Kartiel J 1984 Spectrum of two-dimensional excitons in heterojunction superlattices *Phys. Rev. Lett.* **101A** 158–60

[18] Shinada M and Sugano S 1966 Interband optical transitions in extremely anisotropic semiconductors *J. Phys. Soc. Jpn* **21** 1936–46

[19] Pokatilov E P, Beril S I, Fomin V M, Kalinovsky V V, Litovchenko V G, Korbutyak D V *et al* The dimensionally quantized states of the Wannier–Mott exciton in film structures *Part III Dep. hands. MoldNIINTI* 12/21/88. No. 1064 18

[20] Pokatilov E P, Fomin V M and Beril S I 1990 *Vibrational Excitations, Polarons and Excitons in Multilayer Systems and in Superlattices* (Chisinau: Stiinza) p 278

[21] Molas M R 2023 Excitons and phonons in two-dimensional materials: from fundamental to applications *Nanomaterials* **13** 3047

[22] Glutsch S 2004 *Excitons in Low-Dimensional Semiconductors. Theory Numerical Methods Applications* (Berlin: Springer) p 298

[23] Beril S I, Pokatilov E P, Fomin V M and Pogorilko G A 1986 The effect of anisotropy and self–action on the energy spectrum of the Wannier–Mott exciton in a three-layer structure *Optical Properties of Semiconductors* (Chisinau: Stiinza)

[24] Beril S I, Pokatilov E P and Cheban I S 1984 The dimensional effect of the loss of inertial shielding by the Wannier–Mott exciton in a polar crystal film *FTT* **26** 3698–700

[25] Gu S W and Shen M J 1987 Exciton in a slab of polar crystal *Phys. Rev.* B **35** 9817–29

[26] Bryksin V V and Firsov Y A 1971 Interaction of an electron with surface phonons in an ion crystal plate *FTT* **13** 496–503

[27] Degani M and Hipolito O 1987 Polaron effects on exciton in GaAs – $Ga_{1-x}Al_xAs$ quantum wells *Phys. Rev.* B **35** 4507–10

[28] Pokatilov E P, Beril S I, Fomin V M *et al* 1988 The size-quantized states of the Wannier–Mott exciton in structures with superthin films *Phys. Status Solidi* B **145** 535–44

[29] Pokatilov E P and Beril S I 1983 Electron–phonon interaction in periodic two-layer structures *Phys. Status Solidi* B **118** 567–73

[30] Pokatilov E P and Beril S I 1982 Spatially extended optical modes in two-layer periodical structures *Phys. Status Solidi* B **110** k75–8

[31] Pokatilov E P *et al* 2008 Excitons in wurzite AlGaN/GaN quantum-well heterostructures *Phys. Rev.* B **77** 125328

[32] Bastard G 1981 Hydrogenic impurity states in quantum wells: a simple model *Phys. Rev.* B **24** 4714–22

[33] Greene R L and Bajaj K K 1983 Energy levels of hydrogenic impurity states in GaAs – $Al_xGa_{1-x}As$ quantum well structures *Solid State Commun.* **45** 825–9

[34] Greene R L and Bajaj K K 1985 Shallow impurity center in semiconductor quantum well structures *Solid State Commun.* **53** 1103–8

[35] Dyakonov M I and Khayetsky A V 1982 Dimensional quantization of holes in a semiconductor with a complex valence band and carriers in a slit-free semiconductor **82** 1584–90

[36] Maslov A Y and Proshina O V 2010 The role of interface phonons in the formation of polaronic states in quantum wells *Phys. Technol. Semicond* **44** 200–4

[37] Peter Y and Cardona M 2010 *Fundamentals of Semiconductors. Physics and Material Properties* (Berlin: Springer) p 775

[38] Haken G 1959 Theory of excitons in crystals *UFN* **68** 566–619

[39] Haken H 1956 Zur quantumtheorie des mehrelectronen systems im swingenden Gitter *Z. Phys.* **146** 527–54

[40] Meyer H I G 1956 Interaction of the excitons with lattice vibrations in polar crystals. 1. General theory *Physica* **22** 109–21

[41] Pekar S I 1951 *Research on the Electronic Theory of Crystals* (Moscow: Gostekhizdat)

[42] Bobrysheva A J and Vybornov V I 1978 The biexciton in polar crystals *Phys. Status Solidi* B **88** 315–9

[43] Licary J J and Evrard R 1977 Electron–phonon interaction in a dielectric slab: effect of the electronic polarizability *Phys. Rev.* B **15** 2254–64

[44] Beril S I, Pokatilov E P and Cheban I S 1982 Polar exciton at the contact of two polar crystals *UFN* **27** 585–90

[45] Lozovik Y E and Nishanov V N 1976 Vante–Mott excitons in layered structures and near the boundary of two media *FTT* **18** 3267–72

[46] Ipatova I P 1956 On the theory of exciton in ionic crystals *ZhTF* **26** 2787–92

[47] Tulub A V 1957 On the theory of interaction of an electron with lattice vibrations *Izv. Leningrad Univ. Ser. Phys.* **4** 53–8

[48] Pollmann J and Buttner H 1975 Upper bounds the ground state energy of the exciton–phonon systems *Solid State Commun.* **17** 1171–5

[49] Aldrich C and Bajaj K 1977 Binding energy of a Mott–Wannier exciton in a polarizable medium *Solid State Commun.* **22** 157–60

[50] Bajaj K and Aldrich C 1977 Effective electron–hole interaction for intermediate and strong electron–phonon coupling *Phys. Status Solidi* B **2** 663–6

[51] Barentzen H 1971 Effective electron–hole interaction for intermediate and strong electron–phonon coupling *Phys. Status Solidi* B **71** 245–50

[52] Pollmann J and Buttner H 1977 Effective Hamiltonian and binding energies of Wannier excitons in polar semiconductors *Phys. Rev.* B **16** 4480–5

[53] Bednarek S, Adamowski J and Suffczynski M 1977 Effective Hamiltonian for few-particle systems in polar semiconductors *Solid State Commun.* **21** 1–3

[54] Kohn W and Luttinger J M 1955 Theory of donor states in silicon *Phys. Rev.* B **98** 915–22

[55] Shinada M and Sugano S 1966 Interband optical transitions in extremely anisotropic semiconductors *J. Phys. Soc. Jpn* **21** 1936–46

[56] Lapeyere G and Anderson J R 1975 Evidence for a surface-states exciton on GaAs (110) *Phys. Rev.* B **35** 117–9

[57] Deigen M F and Glinchuk M D 1963 Excitons near the surface of a homeopolar crystal *FTT* **5** 3250–8

[58] Zuev V A, Korbutyak D V, Litovchenko V G *et al* 1975 Collective effects on the surface of GaAs semiconductors *JETF* **69** 1289–300

[59] Zuev V A, Korbutiak D V and Litovchenko V G 1975 Surface radiative recombination in GaAs with surphon participation *Surf. Sci.* **50** 215–28

[60] Moskalenko S A 1958 On the theory of Mott exciton in alkali-galloid crystals *Opt. Spectrosc* **5** 147–55

[61] Weman H, Potemski M, Lazzouni M E, Miller M S and Merz J L 1996 Magneto-optical determination of exciton binding energies in quantum-wire superlattices *Phys. Rev.* B **53** 6959–62

[62] Lambert M A 1958 Mobile and immobile effective-mass-particle complexes in nonmetallic solids *Phys. Rev. Lett.* **1** 450–3

[63] Keldysh L V 1969 *Proc. of the IX Int. Conf. on Semiconductor Physics* (Moscow: Nauka) Vol. 2 1384–92

[64] Moskalenko S A 1962 Reversible optical-hydrodynamic phenomena in an imperfect exciton gas *FTT* **4** 271–84

[65] Blatt J M, Boer K W and Brandt W 1962 Bose-Einstein condensation of excitons *Phys. Rev.* **162** 1691–2

[66] Casella R C 1963 A criterion for exciton binding in dense electron–hole systems—application to line narrowing observed in GaAs *J. Appl. Phys.* **34** 1703–5

[67] Ivanov L M, Lozovik Y E and Musin D R 1978 On the ground state of the two- and three-dimensional excitonic molecules *J. Phys. C: Solid State Phys.* **11** 2527–34

[68] Pollmann J and Buttner H 1973 Phonon interaction potential in polar crystals *Solid State Commun.* **12** 1105–8

[69] Tran Thoai D B 1977 Exciton interaction potential in polar crystals *Z. Phys.* **26** 115–23

[70] Moskalenko S A *et al* 1974 *The Interaction of Excitons in Semiconductors* (Chisinau: Stiinza)

[71] Akimoto O and Hanamura E 1972 Excitonic molecule. 1. Calculation of the binding energy *J. Phys. Soc. Jpn* **32** 1537–44

[72] Brinkman W F, Rice T M and Bell B 1973 The excitonic molecules *Phys. Rev.* B **8** 1570–80

[73] O-hata K, Takata H and Huzinaga S 1966 Gaussian expansion of atomic orbitals *J. Phys. Soc. Jpn.* **21** 2306–13

[74] Bobrysheva A I, Beryl S I, Moskalenko S A, Pokatilov E P and Cheban I S 1962 Biexcitonic states on the surface of a polar crystal *Excitons and Biexcitons in Semiconductors* (Chisinau: Stiinza) pp 182–95

[75] Bobrysheva A I, Beril S I, Moskalenko S A and Pokatilov E P 1980 The dissociation energy of the surfae biexciton *Phys. Status Solidi* B **100** 281–8

[76] Insepov Z A and Norman G A 1975 Three-particle charged electron–hole complexes in semiconductors *JETF* **69** 1321–4

[77] Atzmüller H, Fröschi F and Schröder U 1979 Theory of excitons bound to neutral impurities in polar semiconductors *Phys. Rev.* B **19** 3118–29

[78] Gerlach B 1974 Bound states in electron–exciton collisions *Phys. Status Solidi* B **63** 459–63

[79] Stebe B and Munchy G 1975 Binding energies of the excitonic molecule and of the excitonic ion *Solid State Commun.* **17** 1051–4

[80] Kawabata T, Muro K and Narita S 1977 Observation of cyclotron resonance absorption due to excitonic ion and excitonic molecule ion in silicon *Solid State Commun.* **23** 267–70

[81] Thomas G A and Rice T M 1977 Trions, molecules and excitons above the Mott density in Ge *Solid State Commun.* **23** 359–63

[82] Pokatilov E P, Beril S I, Fomin V M, Kalinovsky V V, Litovchenko V G, Korbutyak D V *et al* 1987 The dimensionally quantized states of the Wannier–Mott exciton in film structures *Part I Dep. hands. MoldNIINTI* No. 845. 8 p. No. 846. part II. 1987 13

[83] Beril S I and Starchuk A S 2009 Coulomb interaction and Wannie–Mott excitons in polar semi-conducting quantum wires *J. Nanoelectron. Optoelectron* **4** 159–64

[84] Elkomoss S C and Amer A S 1975 Binding energy for some atomic and excitonic complex systems *Phys. Rev.* B **11** 2925–32
Elkomoss S C and Amer A S 1975 New method for the calculation of the binding energy of exciton complex *Phys. Rev.* B **11** 2222–8

[85] Petelenz P and Smith V 1980 Binding energies of the lowest exited states of Wannier exciton and ionized donor complex *J. Phys. C: Solid State Phys.* **13** 47–56

[86] Smith V and Petelenz P 1978 Effective potentials of electrons and holes. Binding energy for exciton complex system *Phys. Rev.* B **15** 3253–62

[87] Beril S I, Pokatilov E P and Kabine S 1982 Exciton complexes at the contact of two polar semiconductors *Excitons and Biexcitons in Semiconductors* (Chisinau: Stiinza) p 286

[88] Bobrysheva A I, Zyukov V T, Bilinkis P G and Gorodetsky M V 1982 Excited states of a biexciton and an exciton ion on a crystal surface *Excitons and Biexcitons in Semiconductors* (Chisinau: Stiinza) pp 195–204

[89] Hylleraase E A 1947 Electron affinity of positronium *Phys. Rev.* **71** 491–3

[90] Lerner I V and Lozovik Y E 1980 Wannier–Mott excitons in quasi-two-dimensional semiconductors in a strong magnetic field *JETF* **78** 1167–75

[91] Gorkov L P and Dzyaloshinsky I E 1967 On the Wannier–Mott theory in a strong magnetic field *JETF* **53** 717–22

[92] Beril S I and Pokatilov E P 1980 Exciton states on the crystal surface *FTP* **14** 37–42

[93] Keldysh L V, Silin A P and Babichenko V S 1980 Coulomb interaction in thin semiconductor and semi-metallic wires *FTT* **22** 1238–40

[94] Pokatilov E P, Beril S I, Semenavskaya N N and Fahood M 1990 Charge energy spectrum in multilayer structures and superlattices in a field of self-action potentials *Phys. Status Solidi* B **158** 165–74

[95] Janke E, Emde F and Lesh F 1964 *Special Functions* (Moscow: Nauka) p 344

[96] Litovchenko V G, Beril S I, Korbutyak D V, Lashkevich E G, Pokatilov E P and Mikhailovskaya E V 1988 Exciton states in dimensionally quantized film structures *DAN of the Ukrainian SSR. – Ser. A. – Phys.-Math. Sci.* 57–61

[97] Litovchenko V G, Korbutyak D V and Zuev V A 1982 Detection and investigation of a quasi-two-dimensional electron–hole condensate *Izv. Akad. Nauk SSSR–Ser. Fiz* **46** 1452–62

[98] Pokatilov E P, Beril S I, Fomin V M *et al* 1988 The size-quantized states of the Wannier–Mott exciton in structures with superthin films *Phys. Status Solidi* B **145** 535–44

[99] Mirkin L I 1961 *Handbook of X-ray Diffraction Analysis* (Moscow: State Publishing House of Physical and Mathematical Literature) p 863

[100] Beril S I, Pokatilov E P and Cheban I S 1984 The dimensional effect of the loss of inertial shielding by the Wannier–Mott exciton in a polar crystal film *FTT* **26** 3698–700

[101] Litovchenko V G, Zuev V A and Korbutyak D V 1976 Investigation of collective effects in layered semiconductors using photoluminescence spectra *Izv. Akad. Nauk SSSR–Ser. Phys.* **40** 1833–6

[102] Babaev N A, Bagaev V S, Gaponov S V *et al* 1983 Dimensional quantization in cadmium telluride thin films *Lett. JETF* **37** 524–7

[103] Babaev N A, Bagaev V S, Garin F V *et al* 1984 Dimensional quantization of excitons in CdTe *Lett. JETF* **40** 190–3

[104] Babaev N A, Bagaev V S, Kopylovsky A G *et al* 1984 Measurement of features in the optical absorption of thin semiconductor films *FTT* **26** 3611–7

[105] Nedorezov S S 1970 Spatial quantization in semiconductor films *FTT* **12** 2269–76

[106] Efros A A 1986 Excitons in structures with quantum wells *FTP* **20** 1987–97

[107] Cardona M 1963 Band parameters of semiconductors with zinc blende, wurzite and Germanium structure *J. Phys. Chem. Solids* **24** 1543–55

[108] Keldysh L V 1979 Coulomb interaction in thin films of semiconductors and semimetals *Lett. JETF* **29** 716–9

[109] Andryushin E A and Silin A P 1980 Excitons in thin semiconductor films *FTT* **22** 2676–80

[110] Koteles E S and Chi J Y 1988 Experimental exciton binding quantum wells a junction of well width *Phys. Rev.* B **37** 6332–5

[111] Miller R C, Kleinmann D A, Tsang W T and Gossard A C 1981 Observation of the exciton level in GaAs quantum wells *Phys. Rev.* B **24** 1134–6

[112] Kapon E, Kash K, Clousen E M, Hwang D M and Colas E 1992 Luminescence characteristics of quantum wires grown by organometallic chemical vapor deposition on nonplanar substrates *Appl. Phys. Lett.* **60** 477–9

[113] Nagamune Y, Arakawa Y, Tsukamoto S, Nishioka M, Sasaki S and Miura N 1992 Photoluminescence spectra and anisotropic energy shift of GaAs quantum wires in high magnetic fields *Phys. Rev. Lett.* **69** 2963–6

[114] Schooss D, Mews A, Eychmüller A and Weller H 1994 Quantum-dot quantum well CdS/HgS/CdS: theory and experiment *Phys. Rev.* B **49** 17072–8

[115] Someya T, Akiyama H and Sakaki H 1996 Enhanced binding energy of one-dimensional excitons in quantum wires *Phys. Rev. Lett.* **76** 2965–8

[116] Tsukamoto S, Nagamune Y, Nishioka M and Arakawa Y 1993 *Appl. Phys. Lett.* **63** 355

[117] Wegscheider W, Pfeiffer L N, Dignam M M, Pinczuk A, West K W, McCall S L and Hull R 1993 Lasing from excitons in quantum wires *Phys. Rev. Lett.* **71** 4071–4

[118] Andryushin E A and Silin A P 1993 Excitons in quantum wells and quantum wires *FTT* **35** 1947–57

[119] Lisachenko M G and Timoshenko V Y u 1999 The effect of the dielectric environment on the exciton spectrum of silicon quantum wires *Bull. Mosc. Univ. – Series 3. – Physics. Astronomy* 30–3

[120] Beril S I, Pokatilov E P and Starchuk A S 2006 Coulomb interaction and Wannier–Mott excitons in polar semiconductor quantum wires *Bulletin of the Moscow University. – Series 3. – No. 5. Physics. Astronomy* 33–9

[121] Mulyarov E A and Tichodeev S G 1997 Dielectric amplification of excitons in semiconductor quantum wires *JETF* **111** 274–82

[122] Dneprovsky V S, Zhukov E A, Mulyarov E A and Tichodeev S G 1998 Linear and nonlinear absorption of excitons in semiconductor quantum wires crystallized in a dielectric matrix *JETF* **114** 700–10

[123] Klimin S N, Pokatilov E P and Fomin V M 1994 Bulk and interface polarons in quantum wires and dots *Phys. Status Solidi* B **184** 373–83

[124] Fomin V M and Pokatilov E P 1985 Phonons and the electron–phonon interaction in multilayer systems *Phys. Status Solidi* B **132** 69–82

[125] Beril S I, Pokatilov E P and Cibotaru L F 1982 Surface exciton in a magnetic field *FTT* **24** 663–9

[126] Pokatilov E P, Klimin S N, Balaban S N and Fomin V M 1995 Polarons in a cylindrical quantum wire with Finite–Barrier well *Phys. Status Solidi* B **191** 311–23

[127] Muljarov E A, Zhukov E A, Dneprovskii V S and Masumoto Y 2000 Dielectrically enhanced excitons in semiconductor-insulator quantum wires: theory and experiment *Phys. Rev.* B **62** 7420–32

Chapter 7

Bipolaronic states of large radius in multilayer planar and cylindrical structures. High-temperature bipolaronic superconductivity in multilayer structures

7.1 Introduction

In chapter 7, large-radius bipolaronic states in multilayer polar planar and cylindrical structures are investigated. The Hamiltonians of the electron–phonon interaction are derived for two cases: (a) in a quantum layer (δ-layer) separating semi-infinite polar crystals and (b) in quantum layers (δ-layers) spatially separated by a polar layer (interlayer bipolaronic states).

The Hamiltonian of the electron–phonon interaction in a quantum wire in a polar medium is obtained, based on which the effective potential energy of the electron–electron interaction (attraction) and the binding energy of a bipolaron in a quantum wire are found.

A study of bipolaronic states in inhomogeneous systems was carried out for cases (a) and (b), and it was found that significantly more favorable conditions for the formation of bipolarons arise in the considered composite multilayer structures due to the possibility of an independent choice of parameters of neighboring media describing their polarization properties responsible for the formation of bipolarons.

The possibility of the occurrence of high-temperature bipolaronic superconductivity in structures of the Ginzburg sandwich type, FeSe (monolayer)/$SrTiO_3$, FeSe (monolayer)/$SrTiO_3$/FeSe (monolayer), is investigated. The critical temperature T_c in the studied multilayer structures is estimated.

In [1, 2], the nature and mechanism of superconductivity were established, based on the phenomenon of the occurrence of elementary excitations from paired electrons (Cooper pairs). However, in homogeneous and isotropic metals, the temperature of the superconducting transition (critical temperature T_c) turned out

doi:10.1088/978-0-7503-6164-4ch7

to be very low, compared with the boiling point of liquid nitrogen. As was shown in several works, the beginning of which was the research carried out in [3–6], the inhomogeneity and anisotropy of the medium could contribute to an increase in the critical temperature T_c [7, 8]. In [3–5, 9–11]; structures consisting of metallic and dielectric layers with exciton, plasmon, and phonon mechanisms for the formation of electron pairs located in both the same and different layers of the structure were considered.

7.2 Bipolaronic states in a monolayer (δ-layer) separating semi-infinite polar crystals

The problem associated with the enhancement of electron–phonon interaction due to the separation of regions in which electrons forming Cooper pairs move with regions in which polarization clouds are induced can be solved by selecting optimal geometric (thickness) and material (dielectric permittivity, optical frequencies) parameters. In turn, this allows us to provide conditions for the formation of bipolaronic states with sufficiently large values of binding energy.

The possibility of the formation of bipolaronic quasi-particles due to the interaction of polarons was considered in the works [12–15]. In [3, 16, 17], bipolaronic states were considered in the study of superfluidity.

It should be noted that although the mechanisms of bipolaron formation and electron pairing into Cooper pairs differ in the method of description, in fact, the basis of both effects is the electron–phonon interaction. In [12, 17–19], the conditions of stability of a bipolaron in a homogeneous medium at different values of the strength of the electron–phonon interaction were studied, which allowed us to establish that the criterion for the existence of a stable bipolaron is quite complex.

In particular, the following criteria were obtained in [16]:

$$\frac{\varepsilon_0}{\varepsilon_\infty} > 20, \quad \alpha \geqslant 10, \tag{7.2.1}$$

where ε_0—static, ε_∞—high-frequency dielectric constants, and α—Frelich constant of electron–phonon coupling. In [17], the inequality (7.2.1) was somewhat mitigated:

$$\frac{\varepsilon_0}{\varepsilon_\infty} > 10, \quad \alpha > 7.3. \tag{7.2.2}$$

But even these conditions turned out to be impossible for most semiconductor compounds. Therefore, the existence of stable bipolaronic states in homogeneous media remained problematic.

In [15, 20, 21], we showed that significantly more favorable conditions for the formation of polaron pairs can occur in layered structures, especially in composite multilayer structures with quantum wells, for the reason mentioned above: to ensure an independent choice of parameters of the medium in which the electrons are located, with parameters describing the polarization properties of neighboring media that cause the appearance of polarons and their pairing into bipolarons.

In this section, the interaction of two electrons moving in a monolayer (δ-layer) [22] with polarization vibrations of polar crystals adjacent to it on both sides will be studied [23].

7.2.1 The Hamiltonian of the electron–phonon interaction in a three-layer structure

A three-layer system (10|2|30) is considered, consisting of two semi-infinite polar (10, 30) crystals and a nonpolar (2) semiconductor layer located between them and having a thickness d and a dielectric constant of ε_2; $\varepsilon_{i\infty} \equiv \varepsilon_i$ $(i = 1, 3)$ is a high-frequency dielectric permittivity of polar crystals; ε_{i0} $(i = 1, 3)$ is a static dielectric permittivity. The solution to the problem of the spectrum of vibrational excitations and electron–phonon interaction in multilayer structures with an arbitrary number of polar and nonpolar layers is given in [24] and in the monoplot [25].

The Hamiltonian of the considered system with two electrons in layer 2 has the form

$$\begin{aligned}\hat{H} &= \hat{K}_1 + \hat{K}_2 + V_{e_1-e_2}(|\rho_2 - \rho_2|, z_1, z_2) \\ &+ V_{SA}(z_1, z_2) + \hat{H}^S_{\text{ph}} + \hat{H}^S_{\text{e-ph}} + V_B(z_1, z_2),\end{aligned} \tag{7.2.3}$$

where $\hat{K}_{1,2}$ are kinetic energy operators:

$$\hat{K}_n = \frac{\hat{P}^2_{||n}}{2m_{||}} + \frac{\hat{P}^2_{\perp n}}{2m_{\perp}}, \quad n = 1, \quad 2. \tag{7.2.4}$$

The direction of the z-axis is chosen so that it is perpendicular to the boundaries of the layer, and the origin is placed in the center of layer 2. The designations '||' and '$\perp$' refer respectively to the directions parallel and perpendicular to the z-axis.

The potential energy of the electrons, due to both the direct electron–electron interaction and the interaction of each of the electrons with rapid polarization of the layer and the crystals bordering it, has the form [26]

$$\begin{aligned}V_{e_1-e_2}(\rho, z_1, z_2) &= \frac{e^2}{4\pi\varepsilon_0\varepsilon_2}\int_0^\infty d\eta\, J_0(\eta\rho)\{e^{-\eta|z_1-z_2|} \\ &+ \frac{2}{e^{2\eta d} - \delta_1\delta_3}\left[\delta_1\delta_3\cosh\eta(z_1 - z_2) + e^{\eta d}(f_1\cosh\eta(z_1 + z_2) + f_2\sinh\eta(z_1 + z_2)\right\},\end{aligned} \tag{7.2.5}$$

where the following designations are introduced:

$$\delta_{1,3} = \frac{\varepsilon_2 - \varepsilon_{1,3}}{\varepsilon_2 + \varepsilon_{1,3}}; \; f_1 = \frac{\varepsilon_2^2 - \varepsilon_1\varepsilon_3}{(\varepsilon_2 + \varepsilon_1)(\varepsilon_2 + \varepsilon_3)}; \; f_2 = \frac{(\varepsilon_1 - \varepsilon_3)\varepsilon_2}{(\varepsilon_2 + \varepsilon_1)(\varepsilon_2 + \varepsilon_3)}; \tag{7.2.6}$$

where $J_0(x)$ is the Bessel function of zero order and ε_0 is the electrical constant.

The potential energy of each electron with its induced rapid polarization of the layer and neighboring crystals (the so-called self-action energy) can be written as

$$V_{SA}(z_n)|_{n=1,2}$$

$$=\frac{e^2}{4\pi\varepsilon_0\varepsilon_2}\int_0^{\infty} d\eta(e^{2\eta d} - \delta_1\delta_3)^{-1}\{\delta_1\delta_3 + e^{2\eta d}(f_1 \cosh 2\eta z_n + f_2 \sinh 2\eta z_n)\}. \quad (7.2.7)$$

In the three-layer system under consideration, there are two surface polarization optical modes, the energy of which is described by the operator [21]

$$\hat{H}_{ph} = \sum_{\mathbf{q},s=1,2} \hbar\Omega_s \hat{b}^{+}_{s,\mathbf{q}}\hat{b}_{s,\mathbf{q}}, \quad (7.2.8)$$

where

$$\Omega^2_{1,2} = \frac{1}{2}p_1 \pm \sqrt{\frac{1}{4}p_1^2 - p_2}; \quad (7.2.9)$$

$$p_1 = \frac{1}{\widetilde{B}}\left\{\omega_1^2\left[\varepsilon_2^2 + \hat{\varepsilon}_2(\varepsilon_3 + \varepsilon_{10}) + \varepsilon_3\varepsilon_{10}\right] + \omega_3^2\left[\varepsilon_2^2 + \hat{\varepsilon}_2(\varepsilon_1 + \varepsilon_{30}) + \varepsilon_{30}\varepsilon_1\right]\right.; \quad (7.2.10a)$$

$$p_2 = \frac{\omega_1^2\omega_3^2}{\widetilde{B}}\left[\varepsilon_2^2 + \hat{\varepsilon}_2(\varepsilon_{10} + \varepsilon_{30}) + \varepsilon_{30}\varepsilon_{10}\right]; \quad (7.2.10b)$$

$$\widetilde{B} = \varepsilon_2^2 + \hat{\varepsilon}_2(\varepsilon_1 + \varepsilon_3) + \varepsilon_1\varepsilon_3; \quad (7.2.11a)$$

$$\hat{\varepsilon}_2 = \varepsilon_2 \coth 2\eta d. \quad (7.2.11b)$$

By the boundary conditions obtained in [21, 24], electrons interact in the layer only with surface optical modes of polar media, the interaction operator with which has the form:

$$\begin{aligned}\hat{H}^{s}_{e-ph} = \sum_{\eta,n=1,2} C(\eta)e^{i\eta\rho}\Big\{&[B_1K_{13,22} + B_2K_{13,24}F_{21}]\cosh \eta z_n\\ &+ [B_1K_{23,22} + B_2K_{23,34}F_{21}]\sinh \eta z_n\left(\hat{b}^{+}_{1,-\eta} + \hat{b}_{1,\eta}\right)\\ &+ [B_3K_{13,12} + B_4K_{13,24}F_{21}]\cosh \eta z_n\\ &+ [B_3K_{23,22}F_{12} + B_4K_{23,34}]\sinh \eta z_n\left(\hat{b}^{+}_{2,-\eta} + \hat{b}_{2,\eta}\right)\Big\},\end{aligned} \quad (7.2.12)$$

with

$$C(\eta) = \frac{e\sqrt{\hbar}}{4\sqrt{\varepsilon_0\eta L_x L_y \sinh\left(\frac{\eta d}{2}\right)}}; \quad (7.2.13)$$

$$B_{1,2} = \frac{\omega_{1,3}\sqrt{2(\varepsilon_{1,30} - \varepsilon_{1,3})}}{\sqrt{\Omega_1\left(1 + F_{21}^2\right)}}; \; B_{3,4} = \frac{\omega_{1,3}\sqrt{2(\varepsilon_{1,30} - \varepsilon_{1,3})}}{\sqrt{\Omega_2\left(1 + F_{12}^2\right)}}; \quad (7.2.14)$$

$$K_{13,22} = -\frac{\text{th}\frac{\eta d}{2}}{\widetilde{B}}\left(\varepsilon_3 \coth\frac{\eta d}{2} + \varepsilon_3\right);\ K_{23,22} = \frac{\varepsilon_2 \tanh\frac{\eta d}{2} + \varepsilon_3}{\widetilde{B}}; \quad (7.2.15)$$

$$K_{13,24} = \frac{\text{th}\frac{\eta d}{2}}{\widetilde{B}}\left(\varepsilon_2 \coth\frac{\eta d}{2} + \varepsilon_1\right);\ K_{23,24} = \frac{\varepsilon_2 \tanh\frac{\eta d}{2} + \varepsilon_1}{\widetilde{B}}; \quad (7.2.16)$$

$$F_{12} = -\frac{\omega_1\omega_3\varepsilon_2\sqrt{\varepsilon_{10} - \varepsilon_1} \cdot \sqrt{\varepsilon_{30} - \varepsilon_3}}{\sinh \eta d \quad [\widetilde{B}\Omega_2^2 - \omega_1^2[\varepsilon_2^2 + \hat{\varepsilon}_2(\varepsilon_3 + \varepsilon_{10}) + \varepsilon_{10}\varepsilon_{30}]}. \quad (7.2.17)$$

$$F_{21} = -\frac{\omega_1\omega_3\varepsilon_2\sqrt{\varepsilon_{10} - \varepsilon_1} \cdot \sqrt{\varepsilon_{30} - \varepsilon_3}}{\sinh \eta d \quad [\widetilde{B}\Omega_1^2 - \omega_3^2[\varepsilon_2^2 + \hat{\varepsilon}_2(\varepsilon_1 + \varepsilon_{30}) + \varepsilon_{10}\varepsilon_{30}]}. \quad (7.2.18)$$

The potential energy of electrons in a rectangular well with infinitely high barriers is equal to

$$V_B(z_n)|_{n=1,2} = \left\{0,\ -\frac{d}{2} < z_n < \frac{d}{2};\ \infty,\ z_n \leqslant -\frac{d}{2}; z_n \geqslant \frac{d}{2}.\right. \quad (7.2.19)$$

7.2.2 The energy of the ground state of the bipolaron

Let's consider a quasi-two-dimensional situation as an alternative to a three-dimensional (three-dimensional) one. Let's assume that the thickness of d layer 2 is so small that the energy of dimensional quantization is much greater than the energy of phonons and electron–phonon interaction. In this approximation, the motion of an electron along the *z-axis* can be considered as fast and the wave functions can be chosen in the form describing the basic dimensional quantized state:

$$\psi(z_1, z_2) = \frac{2}{d}\cos\left(\frac{\pi z_1}{d}\right)\cos\left(\frac{\pi z_2}{d}\right). \quad (7.2.20)$$

After averaging the Hamiltonian (7.2.3) on the wave function (7.2.20), we exclude the variables z_1 and z_2 and obtain a 'quasi-two-dimensional' problem with the Hamiltonian

$$\hat{H}_1(\rho) = \langle\psi(z_1, z_2)|\hat{H}_1(\rho, z_1, z_2)|\psi(z_1, z_2)\rangle. \quad (7.2.21)$$

Let's place the origin of the XOY coordinate system in the center of mass of the two-electron system (the radius vectors of the electrons are respectively denoted $\boldsymbol{\rho}_1$ and $\boldsymbol{\rho}_1$). Thus, the developed theory describes the localized states of the bipolar.

Let's perform a unitary transformation:

$$\hat{H}_2(\rho) = U_1^{-1}\hat{H}_1 U_1, \quad (7.2.22)$$

where the transformation operator has the form

$$U_1 = \exp\left\{\sum_{\eta,s} f(\eta,\rho_1, \rho_2)\hat{b}^+_{s,\mathbf{q}}\hat{b}_{s,\mathbf{q}}\right\}. \quad (7.2.23)$$

Here $f(\eta, \rho_1, \rho_2)$ are the variational amplitudes of the displacement of the phonon mode operators.

By averaging the obtained Hamiltonian $\hat{H}_2$ over the phonon vacuum $\hat{H}_2$, we obtain an effective Hamiltonian that does not contain phonon variables:

$$\hat{H}_s(\rho_1, \rho_2) = \langle \varphi_0 \, | \hat{H}_2 | \varphi_0 \rangle . \tag{7.2.24}$$

Let's choose the variational amplitudes of the displacement in formula (7.2.23) in a form similar to [17]:

$$f(\eta, \rho_1, \rho_2) = \frac{C(\eta)F_s(\eta)}{\hbar\Omega_s}\left\{\frac{e^{i\eta\rho_1} + e^{i\eta\,\rho_2}}{1 + \eta^2 R_s^2} + \frac{\lambda}{\left(1 + \eta^2 R_s^2 \beta^2\right)^2}\right\}, \tag{7.2.25}$$

where λ and β are the variational parameters determining the optimal values of the induced charge and the effective radius of the polaron, respectively.

At $\lambda = 0$, the function $f(\eta, \rho_1, \rho_2)$ takes the form of a Haken amplitude, taking into account the specific nature of the system under consideration. The multiplier F_s is found from this condition and has the form

$$F_1(\eta) = \frac{\pi^2 \sinh\left(\frac{\eta d}{2}\right)[B_1 K_{13,22} + B_2 F_{21} K_{13,24}]}{\frac{\eta d}{4}\left(\frac{\eta^2 d^2}{16} + \pi^2\right)\sqrt{\hbar\Omega_1}}; \tag{7.2.26a}$$

$$F_2(\eta) = \frac{\pi^2 \sinh\left(\frac{\eta d}{2}\right)[B_3 K_{13,22} F_{12} + B_4 K_{13,24}]}{\frac{\eta d}{4}\left(\frac{\eta^2 d^2}{16} + \pi^2\right)\sqrt{\hbar\Omega_2}}. \tag{7.2.26b}$$

The second term in the formula (7.2.25) was introduced in [17, 27, 28] according to the theory of volumetric bipolaronic and exciton states to more accurately describe the distribution of slow polarization induced by the field of a two-charge system: an electron and a hole in an exciton or two electrons in a bipolaron. It was shown in [28] that the variational parameter λ describes the magnitude of the polarization charge arising in a two-electron system in addition to the polaron and localized near the center of mass, and the variational parameter β takes into account the size of the region in which the charge is distributed. Now let's write down the effective Hamiltonian:

$$\begin{aligned}&\hat{H}_{\text{eff}}(\rho, \rho_1, \rho_2)\\ &= E_0(d) - \frac{\hbar^2}{2m_\perp}\Delta_1 - \frac{\hbar^2}{2m_\perp}\Delta_2 + W_{\text{eff}}(\rho, d) + W_p(\rho_1, \rho_2, d) + E_p(d).\end{aligned} \tag{7.2.27}$$

The summand

$$E_0(d) = \frac{\pi^2\hbar^2}{m_{||}d^2} + \langle V_{SA}\rangle$$

includes the energy of the basic level of dimensional quantization:

$$\langle\psi(z_1, z_2)|\left[-\frac{\hbar^2}{2m_{||}}\left(\frac{\partial^2}{\partial z_1^2}+\frac{\partial^2}{\partial z_2^2}\right)\right]|\varphi(z_1, z_2)\rangle = \frac{\pi^2\hbar^2}{m_{||}d^2} \tag{7.2.28}$$

and the energy of self-action:

$$\begin{aligned}\langle V_{SA}\rangle &\equiv \langle\varphi(z_1, z_2)|V_{SA}(z_1, z_2)|\varphi(z_1, z_2)\rangle \\ &= \frac{1}{4\pi\varepsilon_0}\cdot\frac{\mathrm{e}^2}{\varepsilon_2 d}\int_0^\infty \mathrm{d}x\frac{1}{\mathrm{e}^{2x}-\delta_1\delta_3}\left\{\delta_1\delta_3+f_1\frac{\pi^2 e^x\sinh x}{x(x^2+\pi^2)}\right\}.\end{aligned} \tag{7.2.29}$$

The term $W_{\text{eff}}(\rho, d)$ includes the energy $V_{e_1, e_2}(\rho, d)$ (expression (7.2.5)) averaged over the wave function (7.2.20), and the contribution of the electron–phonon interaction resulting from the interaction of each of the electrons with polaron polarization (independent of λ) the other electron and the interaction of polaron polarizations between each other:

$$\begin{aligned}W_{\text{eff}}(\rho,\quad d) = V_{e_1, e_2}(\rho, d) + \frac{e^2}{16\pi\varepsilon_0 d}\int_0^\infty J_0\left(\frac{x\rho}{d}\right)\ \Big\{F_1^2(x)a_{11}(a_{11}-2) \\ + F_2^2(x)a_{12}(a_{12}-2)\Big\}\mathrm{d}x,\end{aligned} \tag{7.2.30}$$

where

$$a_{1s}(x) = \frac{1}{1+\frac{R_s^2x^2}{d^2}}, \qquad s = 1, 2. \tag{7.2.31}$$

Let's consider various limiting cases that allow us to obtain $W_{\text{eff}}(\rho, d)$ in an analytical form:

(1) In the limit $d \to 0$, $\varepsilon_{30} = \varepsilon_3$ (one of the crystals is nonpolar) from the expression (7.2.30) we obtain:

$$\begin{aligned}W_{\text{eff}}^{2D}(\rho, d \to 0) &= \frac{e^2}{2\pi\varepsilon_0(\varepsilon_1+\varepsilon_3)\rho} \\ &+ \frac{e^2}{2\pi\varepsilon_0 R_s}\left(\frac{1}{\varepsilon_1+\varepsilon_3}-\frac{1}{\varepsilon_{10}+\varepsilon_3}\right)\int_0^\infty J_0\left(\frac{\rho x}{R_s}\right)\left(\frac{x^2-1}{x^2+1}\right)\mathrm{d}x.\end{aligned} \tag{7.2.32}$$

Integrating the expressions (7.2.32) we obtain:

$$\begin{aligned}W_{\text{eff}}^{2D} &= \frac{e^2}{2\pi\varepsilon_0(\varepsilon_1+\varepsilon_3)\rho} \\ &+ \frac{e^2}{4\pi\varepsilon_0 R_s}\left(\frac{1}{\varepsilon_1+\varepsilon_3}-\frac{1}{\varepsilon_{10}+\varepsilon_3}\right)\left\{-\frac{3\pi}{4}\left[I_0\left(\frac{\rho}{R_s}\right)-L_0\left(\frac{\rho}{R_s}\right)\right]\right. \\ &\left.+\frac{\rho}{R_s}-\frac{\pi\rho}{8R_s}\left[I_1\left(\frac{\rho}{R_s}\right)-L_1\left(\frac{\rho}{R_s}\right)\right]\right\},\end{aligned} \tag{7.2.33}$$

where $I_n(x)$, $L_n(x)$; $n = 0,\ 1$ are the modified Bessel and Struve functions, respectively, and $R_s = (\hbar/(2m_\perp \Omega_s))^{1/2}$ is the radius of the polaron.

This limit was investigated by us in [15, 23].

(2) Of particular interest is the case $d \to 0$, when a polaron is formed at the contact of two polar crystals in a monolayer (δ-layer). In this case, a strong coupling of localized δ-layer electrons with surface optical vibrations is realized.

Assuming both polar crystals to be the same, $\varepsilon_1 = \varepsilon_3 \equiv \tilde{\varepsilon}$, $\varepsilon_{10} = \varepsilon_{30} \equiv \tilde{\varepsilon}_0$, we obtain the following expression for the potential energy of the electron–electron interaction, taking into account its dynamic shielding by surface polarization optical vibrations of polar crystals:

$$W_{\text{eff}}^{2D} = \frac{e^2}{4\pi\varepsilon_0\tilde{\varepsilon}\,\rho} + \frac{e^2}{4\pi\varepsilon_0 R_s}\left(\frac{1}{\tilde{\varepsilon}} - \frac{1}{\tilde{\varepsilon}_0}\right)\int_0^\infty J_0\left(\frac{\rho x}{R_s}\right)\left(\frac{x^2-1}{x^2+1}\right)\mathrm{d}x. \quad (7.2.34)$$

Integrating, we get:

$$W_{\text{eff}}^{2D} = \frac{e^2}{4\pi\varepsilon_0\tilde{\varepsilon}\,\rho} + \frac{e^2}{8\pi\varepsilon_0 R_S}\left(\frac{1}{\tilde{\varepsilon}} - \frac{1}{\tilde{\varepsilon}_0}\right)\left\{-\frac{3\pi}{4}\left[I_0\left(\frac{\rho}{R_s}\right) - L_0\left(\frac{\rho}{R_S}\right)\right]\right.$$
$$\left. + \frac{\rho}{2R_S} - \frac{\pi\rho}{8R_S}\left[I_1\left(\frac{\rho}{R_s}\right) - L_1\left(\frac{\rho}{R_s}\right)\right]\right\}. \quad (7.2.35)$$

Figure 7.1 shows plots of the function (7.2.33) constructed for contacts of $SrTiO_3$, TiO2, and BaO crystals (table 7.1) with vacuum for $m_e^* = m_0$—the mass of a free electron.

From figure 7.1, in particular, it can be seen that within the limits of $\rho \ll R_s$ and $\rho \gg R_s$, the interaction between electrons has a repulsive character, but in a certain range of values of the ratio ρ/R_S, repulsion is replaced by attraction. An analysis of a similar expression for a massive crystal was obtained by Haken in [29]:

$$W_{\text{eff}}^{3D}(r) = \frac{e^2}{\tilde{\varepsilon}\ r} - \frac{e^2}{r}\left(\frac{1}{\tilde{\varepsilon}} - \frac{1}{\tilde{\varepsilon}_0}\right)\left\{1 - \frac{1}{2}\left(\mathrm{e}^{-\frac{r}{R_{V_1}}} + \mathrm{e}^{-\frac{r}{R_{V_2}}}\right)\right\} \quad (7.2.36)$$

which shows that at no parameter values ε, ε_0 (inertia-free and static permittivity, respectively), $W_{\text{eff}}^{3D}(r)$ as a function of r does not change its sign ($R_{V_{1,2}} = (\hbar/(2m_{e1,2}^*\omega_0))^{1/2}$ are the radii of the volumetric polarons).

Thus, the occurrence of attraction between free charges of the same name through induced slow polarization takes place only in spatially inhomogeneous systems in which it is possible to separate regions with charge carriers and regions with polarizations, through interaction with which carriers are paired and bipolarons are formed.

(3) In the general case, when the polar crystals are different, and the layer between them has a finite thickness d, the calculation of the potential profile should be performed using general formulas (7.2.30) and (7.2.31).

The appearance of a tendency to decrease repulsion at distances between electrons of the order of their polaron radii is an important factor in the formation of bipolarons.

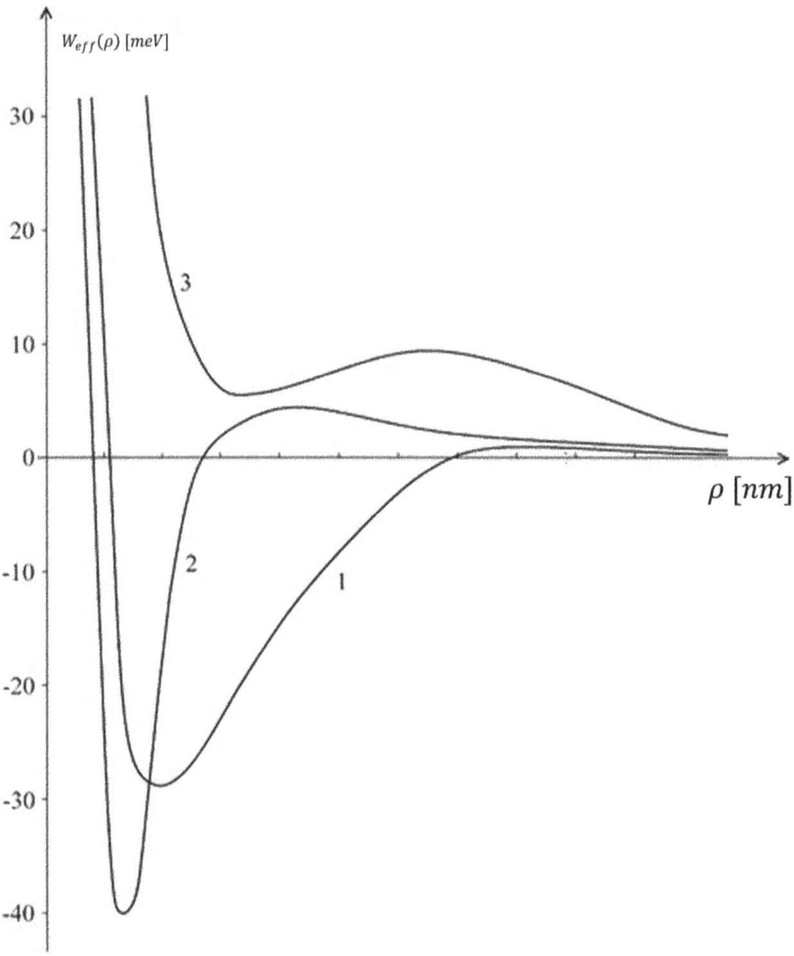

Figure 7.1. Plots of $W_{\text{eff}}^{2D}(\rho)$ dependencies for contacts of $SrTiO_3$ crystals (curve 1), TiO_2 (curve 2), and BaO (curve 3) with vacuum $m_e^* = m_0$. The material parameters of crystalline media are presented in table 7.1.

Table 7.1. The values of the parameters of the media forming the contact used for numerical calculations.

Material Parameters	$SrTiO_3$	TiO_2	BaO
ε_{10}	1000	170	34
ε_1	5	6	4
ω_1, 10^{13} s^{-1}	1.42	2.42	3.3

The summand

$$W_p(\rho_1, \rho_2, d) = \sum_{I=1,2} |C(\eta)F_s(\eta)|^2(2a_{1s} - 1)2a_{2s}\cos(\eta\rho_i), \qquad (7.2.37)$$

which is propotional to a_{2s} is determined by the second term in the formula (7.2.25). It describes the potential energy of the interaction of an electron with an additional polarization charge induced by the joint action of both electrons (and absent in a single-electron system) and located in the region of the center of mass of a two-electron system.

Since the sign of the additional charge is opposite to the sign of the electrons, and the distance between it and the electrons is less than the distance between the electrons, $W_p(\rho_1, \rho_2, d)$ makes a certain contribution to the effect of polaron pairing and the formation of a bipolaron.

For the contact of a polar crystal with a vacuum, the density of the total surface charge can be found by the formula:

$$\sigma(\rho) = 2\varepsilon_0 E_z|_{z=0} = -2\varepsilon_0 \frac{\partial V_p}{\partial z}\bigg|_{z=0}$$

$$= \frac{e(\varepsilon_{10} - \varepsilon_1)}{2\pi(\varepsilon_1 + 1)(\varepsilon_{10} + 1)} \int_0^\infty \eta d\eta\, e^{-\eta z_0} \left(\frac{\lambda_1}{\left(1 + \beta_1^2 R_s^2 \eta^2\right)} \left[J_0(\eta\, |\rho_1 - \rho| \right.\right. \qquad (7.2.38)$$

$$\left.\left. + J_0(\eta\, |\rho_2 - \rho|) \right] + \frac{\lambda_2 J_0(\eta\rho)}{(1 + \eta^2\rho^2)^2} \right),$$

where V_p is the potential of the polarization field, z_0 is the distance from the plane in which the electrons are located to the contact, $\rho = (x, y)$ is the radius vector in the contact plane, measured from the center of mass of the electrons, ρ_1, ρ_2 are two-dimensional radius vectors of electrons, and λ_1, λ_2, β_1, β_2 are variational parameters.

Assuming $\lambda_1 = \lambda_2 = \beta_1 = \beta_2 = 1$, $z_0 = 0$ and performing integration, we obtain an explicit expression for $\sigma(\rho)$:

$$\sigma(\rho) = \frac{e(\varepsilon_{10} - \varepsilon_1)}{4\pi(\varepsilon_1 + 1)(\varepsilon_{10} + 1)R_s^2} \left\{ 2\left[K_0\left(\frac{|\rho_1 - \rho|}{R_s}\right) + K_0\left(\frac{|\rho_2 - \rho|}{R_s}\right) \right] + \frac{\rho}{R_s} K_1\left(\frac{\rho}{R_s}\right) \right\}; \qquad (7.2.39)$$

where $K_0(x)$, $K_1(x)$ are McDonald's functions.

Figure 7.2 shows the dependencies of the induced charge density on the x coordinate taken along the axis on which the electrons are located. As can be seen from figure 7.2, additional induced positive charges arise at the surface of polar crystals (figure 7.3).

Integration of the expression (7.2.39) by giving ρ the value of the effective electric charge e^* induced by electrons:

$$e^* = \frac{e(\varepsilon_{10} - \varepsilon_1)}{(\varepsilon_1 + \varepsilon_3)(\varepsilon_{10} + \varepsilon_3)}. \qquad (7.2.40)$$

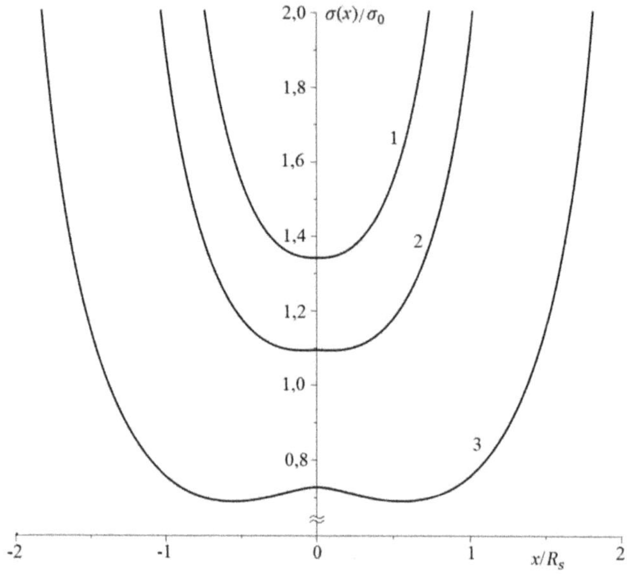

Figure 7.2. Dependence of the surface density of the induced charge on the coordinate along the straight line on which the electrons are located. The induced charge density is calculated by the formula (7.2.39) and expressed in units $\sigma_0 = e(\varepsilon_{10} - \varepsilon_1)/(4\pi(\varepsilon_1 + 1)(\varepsilon_{10} + 1)R_S^2)$, where the x coordinate is in units R_S. Curve 1 corresponds to a distance between electrons twice as large as R_S, curve 2 corresponds 2,5R_s, and curve 3 corresponds 4R_s.

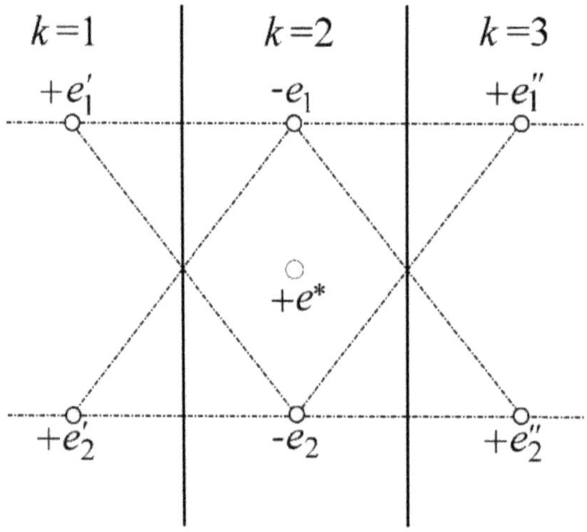

Figure 7.3. Formation of induced charges in a three-layer system. Here $-e_1$, $-e_2$ are electrons located in the middle layer, $+e_1'$, $+e_1''$, $+e_2'$, $+e_2''$ are image charges induced by electrons in neighboring media, and $+e^*$ is the center of the induced positive charge.

A coordinate-independent part of the potential energy

$$E_p(d) = \sum_{\eta,s} |C(\eta)F_s(\eta)|^2\left(2a_{1s}^2 - 4a_{1s} + 2a_{1s}^2R_s^2\eta^2 + a_{2s}^2\right) \tag{7.2.41}$$

takes into account the polaron effect (i.e., the energy of interaction of each of the electrons with the polarization created by it and the energy of an elastically deformed lattice).

The variational wave function of the electron pair is selected as

$$\varphi(\rho) = 2(3\pi)^{-1/2}\gamma^2\rho e^{-\gamma\rho}, \tag{7.2.42}$$

where γ is the variational parameter.

The energy of the ground state of the bipolaron is found by minimizing the variational function of the energy:

$$E(\gamma, \beta, \lambda) = \langle\varphi(\rho)\big|\hat{H}(\rho, \rho_1, \rho_2, d)|\varphi(\rho)\rangle. \tag{7.2.43}$$

The binding energy of a bipolaron is determined by the expression

$$W_b = E_b - 2E_p, \tag{7.2.44}$$

where E_b is the minimum value of $E(\gamma, \beta, \lambda)$ (i.e., the energy of the bipolaronic system) and E_p is the energy of a single polaron. For $d \to 0$ and $\gamma \to \infty$ for E_p, we obtain the expression

$$E_p = -\frac{\pi}{2}\sum_{s=1,2}\alpha_s\hbar\Omega_s. \tag{7.2.45}$$

7.2.3 Discussion

Let's discuss the results of the calculations. It should be noted that the fulfillment of the criteria of the continuous approximation adopted in the developed theory imposes certain restrictions on the values of the material parameters of crystals and variational parameters γ, β, R_s. With a decrease in the parameter β varying the radius of the polaron, the interaction energy of the induced charge e^* with electrons increases due to increased localization of the additional charge (decrease βR_s). The decrease in the binding energy of the bipolaron W_b is limited by an increase in elastic energy proportional to a_S^2. The calculation shows that the passage through the minimum W_b occurs at $\beta \approx 0, 1 - 0, 2$ (i.e., when the radius of the additional polarization charge βR_s becomes much smaller than the radius of the polaron R_S, therefore, at R_S nm, the criterion of applicability of the continuous approximation is violated). In this case, to make qualitative estimates of W_b, $\beta = a_0/R_S$, where a_0 the lattice constant can be fixed, and minimization can be carried out only by two parameters: γ and λ.

We calculate the binding energy W_b of the bipolaron (7.2.44) at the contacts using the formula (7.2.43) on the variational function (7.2.42):

1. $SrTiO_3$/δ-layer ($d \to 0$)/vacuum, at $m_e^* = m_0$; $R_S = 5.6$ nm

$$W_b = -196 \text{ meV};$$

2. BaO/δ-layer ($d \to 0$)/BaO, at $m_e^* = m_0$; $R_S = 8.1$ nm

$$W_b = -29.1 \text{ meV};$$

3. $SrTiO_3$/δ-layer ($d \to 0$)/BaO, at $m_e^* = m_0$; $R_S = 6.0$ nm

$$W_b = -186.9 \text{ meV}.$$

It follows from the above estimates that for the real values of the parameters of two- and three-layer structures, the binding energy of bipolarons W_b ranges from tens to hundreds of millielectronvolts (i.e., bipolarons can exist at temperatures in the range from 10^2 to 10^3 K). The interaction of electrons with an additional polarizing charge located near the center of mass of the system plays an important role in the formation of bipolaronic states. Electron–electron pairing is facilitated by the weakening of repulsion between electrons, which, with a favorable ratio of the parameters of the contact crystals, turns into attraction. The binding energy of the bipolaron increases with decreasing layer thickness ($d/R_S \leqslant 1$). High W_b values are obtained even at $d/R_S \sim 0, 5$. The maximum value of the binding energy is obtained for the polaron arising in the δ-layer ($d \rightarrow 0$).

Since the effective value of the electron–phonon interaction constant is $S \approx 1$, the developed theory relating to the field of intermediate coupling will remain satisfactory and give good results.

It is of interest to draw an analogy between the crystal structure of layered high-temperature superconductors, on the one hand, and the crystal structure of a composite superlattice, on the other. It consists of the fact that the CuO_2 layers in the superconductor correspond to the conductive layers in the superlattice, and the BaO layers (e.g., in the high-temperature superconducting compound $YBa_2Cu_3O_{7-\delta}$) play the role of polar crystalline layers in the superlattice. Within the framework of this analogy, based on the calculations and estimates carried out, it can be assumed that such oscillation modes in high-temperature superconducting materials, which in the macroscopic limit are 'surface modes', should play an important role in the effect of electron–electron pairing and the formation of a bipolaron. To substantiate such a pairing mechanism in compounds with high-temperature superconductivity, special experimental and theoretical studies of the vibration spectra of these compounds are necessary to identify vibrational excitations with specific limiting properties in them. Finally, although experimental studies of models of multilayer structures of a promising type have been carried out are limited, there are experimental confirmations that during the transition from homogeneous samples (PbTe, PbS, PbSe) to multilayer structures and superlattices based on them [30], an increase in the critical transition temperature to the superconducting state is seen.

7.3 Bipolaronic states in spatially separated monolayers (δ-layers) in multilayer structures with quantum wells

Large-radius bipolaronic states in multilayer structures with δ-layers and quantum wells were considered in section 7.2 for the case when the pairing of identical charge carriers (electrons or holes) occurred in the same monolayer (δ-layer). It has been shown that when choosing the geometric (layer thickness) and material parameters (dielectric permittivity, optical phonon frequencies, effective mass of charge carriers) of multilayer structures, the conditions for the occurrence of bipolaronic states can become significantly more favorable than in bulk crystals. The possibility to improve the criteria for the occurrence of bipolaronic states appears in multilayer structures due to the possibility of separating regions with charge carriers located in them and regions in which inertial (slow) polarization is induced. In [5, 7, 31, 32], studies of multilayer structures were carried out in which the pairing of charge carriers of the same name from different layers was observed. Based on such interlayer pairing, various models of high-temperature superconductivity have been developed [8]. Such a system differed from purely two-dimensional ones, which were widely discussed in a number of papers, in that as a result of this interlayer interaction, fluctuations in the order parameter are suppressed.

In section 7.2, the analogy of the multilayer structure under consideration with δ-layers separating quantum wells from polar semiconductor (dielectric) layers with superconducting ceramics (having a layered structure) was described: $Y-Ba-Cu-O$, in which conductive two-dimensional CuO_2 layers are separated by polar dielectric layers BaO. Since the thicknesses of the BaO layers are quite small (up to several nanometers), charge carriers moving in different CuO_2, layers will interact with each other both directly (electrostatic repulsion) and indirectly (attraction, due to interaction with the surface optical phonons of the polar layers). It is of interest to study the electron–electron interaction of charge carriers moving in different δ-layers when their attraction is carried out through slow polarization induced by them in the polar layer separating them (interlayer pairing).

In this section, based on the approach developed in section 7.2, the potential energy of the electron–electron interaction and the energy of interlayer bipolaronic states are calculated as functions of the material and geometric parameters of the system. To describe the interlayer states, a model is used in which a layer of a polar dielectric separates the interacting electrons.

7.3.1 The Hamiltonian of the interlayer bipolaron

The system under consideration is a three-layer structure (1|20|3) (unlike the structure (10|2|30), discussed in section 7.2), where 1 and 3 are identical nonpolar quantum layers (δ_1-, δ_3-layers) with dielectric permittivity $\varepsilon_1 = \varepsilon_2 \equiv \varepsilon$, separated by a layer of polar dielectric (20) with a thickness d, occupying a region of space $-d/2 \leqslant z \leqslant d/2$.

The static and optical permittivity are denoted as ε_{20} and ε_2, respectively, and the frequencies of longitudinal and transverse optical vibrations are ω_{20} and ω_2, respectively. The charge carriers are located in layers 1 and 3 at a distance ⍰z⍰ from the origin (which is in the middle of the second layer); $z_i \geqslant d/2$ (figure 7.4).

The problem of the spectrum of vibrational excitations and electron–phonon interaction in multilayer structures with an arbitrary number of polar and nonpolar layers was solved in [21, 24].

The Hamiltonian of the structure under consideration has the form

$$\begin{aligned} \hat{H} &= \hat{K}_1 + \hat{K}_2 + V_{e_1-e_2}\left(\left| \rho_1 - \rho_2 \right|, z_1, z_2\right) \\ &\quad + V_{SA}(z_1, z_2) + \hat{H}^S_{\text{ph}} + \hat{H}^S_{\text{e-ph}} + V_B(z_1, z_2), \end{aligned} \tag{7.3.1}$$

where $\hat{K}_{1,2}$ are the operators of the kinetic energies of electrons:

$$\hat{K}_i = \frac{\hat{P}^2_{\|i}}{2m_{\|}} + \frac{\hat{P}^2_{\perp i}}{2m_{\perp}}, \quad i = 1, \quad 2. \tag{7.3.2}$$

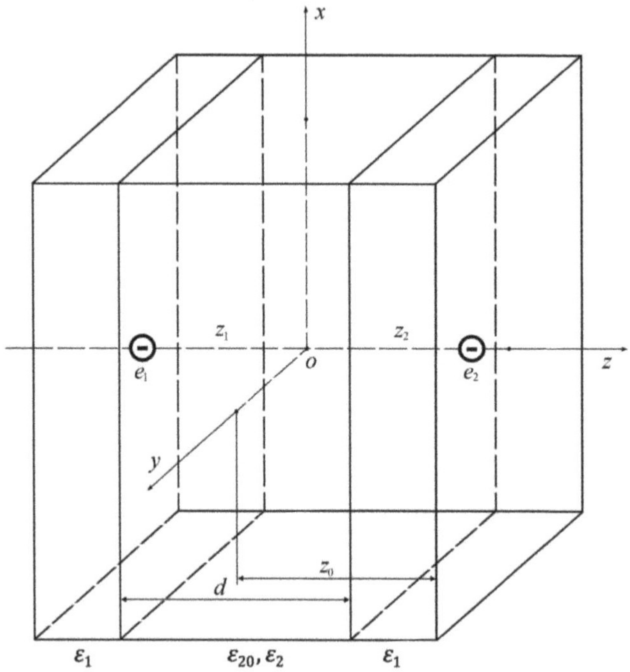

Figure 7.4. A three-layer structure (1|20|3) consisting of a polar layer (20) separating nonpolar quantum layers (δ_1-, δ_2-layers) 1 and 3.

(The designations || and $\perp$ are directions parallel and perpendicular to the z-axis, respectively.)

Initial potential energy

$$W_0(\rho_{12}, z_1, z_2) = V_{e_1-e_2}(\rho_{12}, z_1, z_2) + V_{SA}(z_1, z_2) \tag{7.3.3}$$

includes the energy of direct electron–electron interaction and the energy of interaction of electrons with rapid polarization (plasmons of valence electrons) of the polar layer.

By [15, 21], $V_{e_1-e_2}(|\rho_2 - \rho_1|, z_1, z_2)$ can be presented in the following form:

$$V_{e_1-e_2}\left(\left|\boldsymbol{\rho}_2 - \boldsymbol{\rho}_1\right|, z_1, z_2\right) = \frac{e^2}{2\pi\varepsilon_0}\int_0^\infty \mathrm{d}\eta\, J_0(\eta\rho_{12}) \frac{\varepsilon_2 \mathrm{e}^{\eta\left(z_1+\frac{d}{2}\right)} e^{\eta\left(\frac{d}{2}-z_2\right)}}{\sinh\ \xi_2\left[\varepsilon_2^2 + 2\varepsilon_1\varepsilon_2 \coth \xi_2 + \varepsilon_1\right]}, \tag{7.3.4}$$

where $J_0(x)$ is the zero-order Bessel function and ε_0 is the electric constant.

The potential energy of self-action of an electron

$$V_{SA}(z_1, z_2) = -\frac{e^2}{4\pi\varepsilon_0}\cdot\frac{\left(\varepsilon_2^2-\varepsilon_1^2\right)}{2\varepsilon_1}\int_0^\infty \frac{e^{-2\eta\left|z_1+\frac{d}{2}\right|}d\eta}{\varepsilon_2^2+2\varepsilon_1\varepsilon_2\coth\xi_2+\varepsilon_1} - \frac{e^2}{4\pi\varepsilon_0}\cdot\frac{\left(\varepsilon_2^2-\varepsilon_1^2\right)}{2\varepsilon_1}\int_0^\infty \frac{e^{-2\eta\left|z_2-\frac{d}{2}\right|}d\eta}{\varepsilon_2^2+2\varepsilon_1\varepsilon_2\coth\xi_2+\varepsilon_1}. \quad (7.3.5)$$

In formulas (7.3.4) and (7.3.5) we imply that

$$-\infty < z_1 \leqslant -\frac{d}{2};\ \frac{d}{2} \leqslant z_2 < \infty.$$

In the three-layer structure under consideration, there are two surface optical modes, which are described by the Hamiltonian

$$\hat{H}^S_{\text{ph}} = \sum_{\eta,s=1,2} \hbar\Omega_s\hat{b}^+_{s,\eta}\hat{b}_{s,\eta}, \quad (7.3.6)$$

where

$$\Omega_1^2 = \omega_{TO}^2\left\{\frac{\varepsilon_{20}+\varepsilon_1\coth\left(\frac{\eta d}{2}\right)}{\varepsilon_2+\varepsilon_1\coth\left(\frac{\eta d}{2}\right)}\right\};\ \Omega_2^2 = \omega_{TO}^2\left\{\frac{\varepsilon_{20}+\varepsilon_1\tanh\left(\frac{\eta d}{2}\right)}{\varepsilon_2+\varepsilon_1\tanh\left(\frac{\eta d}{2}\right)}\right\}. \quad (7.3.7)$$

By the boundary conditions adopted in [21], the electrons in layers 1 and 3 (δ_1-, δ_2-layers) interact only with the optical surface modes of the polar medium.

The Hamiltonian of the interaction of electrons with surface vibrations in layers 1 and 3 can be written, following formula (14.16c) from [21] and formula (58) from [24], as

$$\hat{H}^S_{e_1-\text{ph}} = \sum_{\eta_1} C_{\eta_1}e^{-\eta_1\left(z_1-\frac{d}{2}\right)}e^{i\eta_1\rho_1}\left(\hat{b}_{\eta_1}+\hat{b}^+_{-\eta_1}\right) + \sum_{\eta_2} C_{\eta_2}e^{-\eta_2\left(z_1-\frac{d}{2}\right)}e^{i\eta_2\rho_1}(\hat{b}_{\eta_2}+\hat{b}^+_{-\eta_2}); \quad (7.3.8a)$$

$$\hat{H}^S_{e_2-\text{ph}} = \sum_{\eta_1} C_{\eta_1}e^{-\eta_1\left(z_2-\frac{d}{2}\right)}e^{i\eta_1\rho_2}\left(\hat{b}_{\eta_1}+\hat{b}^+_{-\eta_1}\right) + \sum_{\eta_2} C_{\eta_2}e^{-\eta_2\left(z_2-\frac{d}{2}\right)}e^{i\eta_2\rho_2}(\hat{b}_{\eta_2}+\hat{b}^+_{-\eta_2}); \quad (7.3.8b)$$

$$\left|C_{\eta_1}\right|^2 = \frac{1}{L^2}\cdot\frac{2\pi\alpha_{S_1}}{\beta_{S_1}\eta_1}\cdot\frac{(\hbar\omega_{S_1})^2}{th\left(\frac{\eta d}{2}\right)};\ \beta_{S_1} = \left(\frac{2m_e^*\omega_{S_1}}{\hbar}\right)^{\frac{1}{2}}; \quad (7.3.9a)$$

$$\alpha_{S_1} = \frac{e^2}{\hbar}\left\{\frac{1}{\varepsilon_2^{(1)}}-\frac{1}{\varepsilon_{20}^{(1)}}\right\}\left(\frac{m_e^*}{2\hbar\omega_{S_1}}\right)^{\frac{1}{2}};\ \omega_{S_1} = \omega_{TO}\left(\frac{\varepsilon_{20}^{(1)}}{\varepsilon_2^{(1)}}\right)^{\frac{1}{2}}; \quad (7.3.9b)$$

$$\varepsilon_{20,2}^{(1)} = \varepsilon_{20,2} + \varepsilon_1 \coth\left(\frac{\eta d}{2}\right);\ R_{S_j} = \left(\frac{\hbar}{2m_e^* \omega_{S_j}}\right)^{\frac{1}{2}}; \tag{7.3.9c}$$

$$\left|C_{\eta_2}\right|^2 = \frac{1}{L^2}\cdot\frac{2\pi\alpha_{S_2}(\hbar\omega_{S_2})^2}{\beta_{S_2}\eta_2}\tanh\left(\frac{\eta d}{2}\right);\ \beta_{S_2} = \left(\frac{2m_e^*\omega_{S_2}}{\hbar}\right)^{\frac{1}{2}}; \tag{7.3.9d}$$

$$\alpha_{S_2} = \frac{e^2}{\hbar}\left\{\frac{1}{\varepsilon_2^{(2)}} - \frac{1}{\varepsilon_{20}^{(2)}}\right\}\left(\frac{m_e^*}{2\hbar\omega_{S_2}}\right)^{\frac{1}{2}};\ \omega_{S_2} = \omega_{TO}\left(\frac{\varepsilon_{20}^{(2)}}{\varepsilon_2^{(2)}}\right)^{\frac{1}{2}}; \tag{7.3.9e}$$

$$\varepsilon_{20,2}^{(2)} = \varepsilon_{20,2} + \varepsilon_1 \tanh\left(\frac{\eta d}{2}\right). \tag{7.3.9f}$$

It is assumed that there are infinite potential barriers at the interface between polar and nonpolar layers that prevent electrons from entering the polar plate:

$$V_B(z_i) = 0,\ |z_i| \geqslant \frac{d}{2};\ V_B(z_i) \to \infty,\ |z_i| \leqslant \frac{d}{2}. \tag{7.3.9g}$$

7.3.2 Effective potential of electron–electron interaction

The total potential of the electrostatic interaction of the charge carriers of the same name located in different layers of the structure (figure 7.4) can be represented as

$$W_{\text{eff}}(\rho_{12}, z_1, z_2) = W_0(\rho_{12}, z_1, z_2) + \delta W_{\text{ph}}(\rho_{12}, z_1, z_2), \tag{7.3.10}$$

where $W_0(\rho_{12}, z_1, z_2)$ is seed potential and is determined by the formula (7.3.4) and $\delta W_{\text{ph}}(\rho_{12}, z_1, z_2)$ is the contribution to the interaction potential of electrons from their interaction with slow polarization. There is no interaction with volumetric optical phonons in regions 1 and 3, therefore $\delta W_{\text{ph}}(\rho_{12}, z_1, z_2)$ determines the contribution from interaction only with surface optical phonons.

The movement of charge carriers along the z-axis is considered fast, so the potential energy (7.3.10) can be averaged along the z-coordinate. To describe the potential energy profile, we select the wave function for electrons in layers 1 and 3 when their thicknesses satisfy the condition $d \ll \lambda$, where λ is the de Broglie wavelength of the electron (δ-layer) in the form of a δ-function:

$$\psi(z_i) = C\delta\,(z_{0i} - z_i),\quad i = 1, 2. \tag{7.3.11}$$

The choice of a δ-like z-dependence of the wave function is valid if the electrons are localized at the surface (e.g., at the Tamm levels), as well as within the framework of the model considered in this chapter, in which the conduction electrons are in thin δ-layers (similar to the case of CuO_2 layers in materials with YBaCuO). The use of the wave function (7.3.11) leads to the replacement of the z-coordinate in the formulas (7.3.4), (7.3.5), (7.3.8a), and (7.3.8b) for fixed values z_{01} and z_{02}. Due to the symmetry of the structure, these values can be defined as

$$z_{02} = -z_{01} \equiv z_0;\ |z_0| > \frac{d}{2}. \tag{7.3.12}$$

Taking into account (7.3.12) for the Hamiltonian of the electron–phonon interaction it can be written after averaging it on wave functions (7.3.11):

$$\left\langle \psi(z_1)\psi(z_2)\right| \hat{H}^{S}_{\text{e-ph}} \left| \psi(z_1)\psi(z_2)\right\rangle$$
$$= \sum_{\eta_1} C_{\eta_1} e^{-\eta_1\left(z_0 - \frac{d}{2}\right)}\{e^{i\eta_1\rho_1} + e^{i\eta_1\rho_2}\}\left(\hat{b}_{\eta_1} + \hat{b}^{+}_{-\eta_1}\right) \tag{7.3.13}$$
$$+ \sum_{\eta_2} C_{\eta_2} e^{-\eta_2\left(z_1 - \frac{d}{2}\right)}\{e^{i\eta_2\rho_1} + e^{i\eta_2\rho_2}\}(\hat{b}_{\eta_2} + \hat{b}^{+}_{-\eta_2});$$

The procedure for determining the expression $\delta W_{\text{ph}}(\rho_{12}, z_1, z_2)$ is similar to that described in [15].

As a result, we get:

$$\delta W_{\text{ph}}(\rho_{12}, \boldsymbol{\rho}_1, \boldsymbol{\rho}_2) = \sum_{\eta_j, j=1,2} |A_j(\eta_j)|^2 \Big\{2a_{1j}^2(1 + \cos[\boldsymbol{\eta}_j(\boldsymbol{\rho}_1 - \boldsymbol{\rho}_2)])$$
$$+ 2a_{1j}a_{2j}(\cos \boldsymbol{\eta}_j\boldsymbol{\rho}_1 + \cos \boldsymbol{\eta}_j\boldsymbol{\rho}_2) + a_{2j}^2\Big\}$$
$$- 2 \sum_{\eta_j, j=1,2} |A_j(\eta_j)|^2 \{2a_{1j}(1 + \cos[\boldsymbol{\eta}_j(\boldsymbol{\rho}_1 - \boldsymbol{\rho}_2)]) \tag{7.3.14}$$
$$+ a_{2j}(\cos \boldsymbol{\eta}_j\boldsymbol{\rho}_1 + \cos \boldsymbol{\eta}_j\boldsymbol{\rho}_2)\} + 2 \sum_{\eta_j, j=1,2} |A_j(\eta_j)|^2 \eta_j a_{1j} R_{S_j}^2,$$

where

$$\left|A_j(\eta_j)\right|^2 = \left|C_{Q_j}\right|^2 e^{-2\eta_j\left(z_0 - \frac{d}{2}\right)}. \tag{7.3.15}$$

The term describing the shielding of the electron–electron interaction includes the coordinate $\rho_{12} = |\,\boldsymbol{\rho}_1 - \boldsymbol{\rho}_2\,|$. By isolating these terms and replacing the sums with integrals taken by the variable $x = \eta d$, we can, by some transformations, obtain an effective electrostatic interaction in the following form:

$$W_{\text{eff}}(\rho_{12}, z_0, d) = \frac{e^2}{4\pi\varepsilon_0} \cdot \frac{2}{d} \int_0^\infty dx \frac{J_0\left\{\frac{\rho_{12}}{d}x\right\} \varepsilon_2^2 e^{\left(1 - \frac{2z_0}{d}\right)x}}{\sinh x \;\; f(\varepsilon)}$$
$$+ \frac{1}{4\pi\varepsilon_0} \cdot \frac{1}{d} \int_0^\infty dx \;\; \{|F_1|^2 \;\; J_0\left(\frac{\rho_{12}}{d}x\right) a_{11}(a_{11} - 2) \tag{7.3.16}$$
$$+ |F_2|^2 \;\; J_0\left(\frac{\rho_{12}}{d}x\right) a_{12}(a_{12} - 2)\Big\},$$

where

$$f(\varepsilon) = \varepsilon_2^2 + \varepsilon_1\varepsilon_3 \coth x + \varepsilon_1^2; \tag{7.3.17}$$

$$|F_1|^2 = \frac{\alpha_{S_1}}{\beta_{S_1} \tanh\left(\frac{x}{2}\right)} e^{-\left(\frac{2z_0}{d} - 1\right)}; \; |F_2|^2 = \frac{\alpha_{S_2} \tanh\left(\frac{x}{2}\right)}{\beta_{S_2}} e^{-\left(\frac{2z_0}{d} - 1\right)}; \tag{7.3.18a}$$

$$a_{2j} = \frac{\lambda}{\left(1 + R_{S_j}^2\left(\frac{x^2}{d^2}\right)\right)^2}; \qquad (7.3.18b)$$

where λ, β are variational parameters.

Following (7.3.16), the dependence of the effective potential energy of the electrostatic interaction W_{eff} on the parameters $\rho_{12} = |\rho_1 - \rho_2| z_0$ and d are shown in figures 7.5 and 7.6.

In figure 7.5, curve 1 corresponds to the direct Coulomb repulsion of electrons ($V_{e_1 - e_2}(\rho_{12})$ is the first term in (7.3.16)). Curve 2 describes the total effective potential energy (W_{eff} from formula (7.3.16)). As follows from figure 7.5, slow polarization at certain values of the parameters of the polar layer not only greatly reduces the

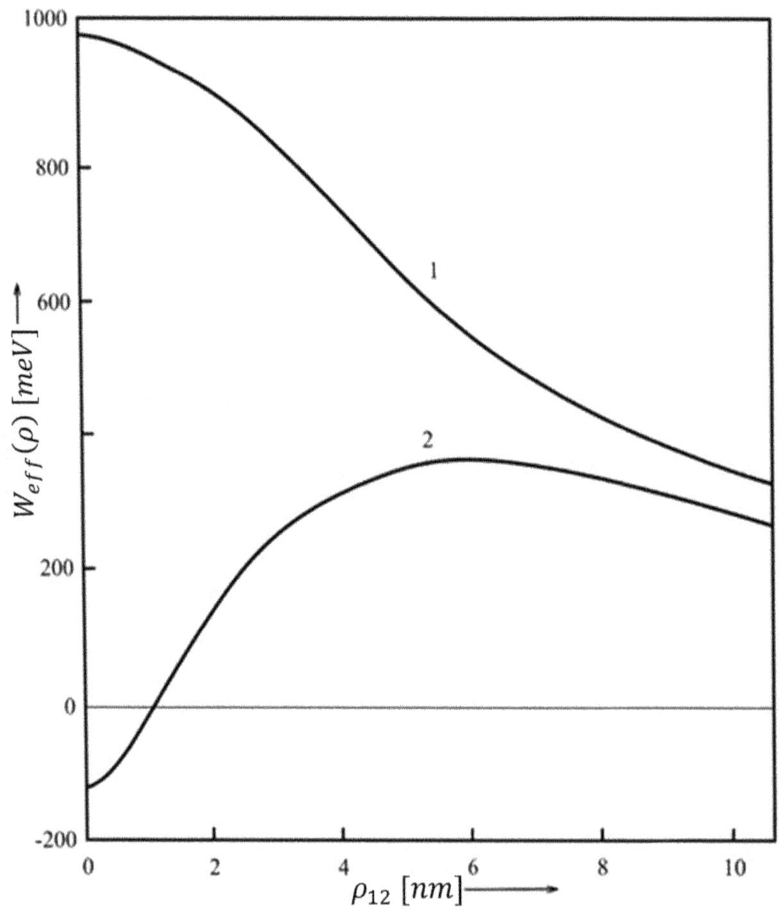

Figure 7.5. Effective potential of electron–electron interaction. Curve 1 describes the direct Coulomb repulsion (the first term in equation (7.3.16)). Curve 2 is the full effective potential of W_{eff} calculated by the formula (7.3.16). The following parameters of the low-dimensional structure were used for calculation: for polar layers BaO $d = 5$ nm; for nonpolar δ-layers $\varepsilon_1 = \varepsilon_3$; $m_e^* = m_0$; and their thickness is 0.1 nm.

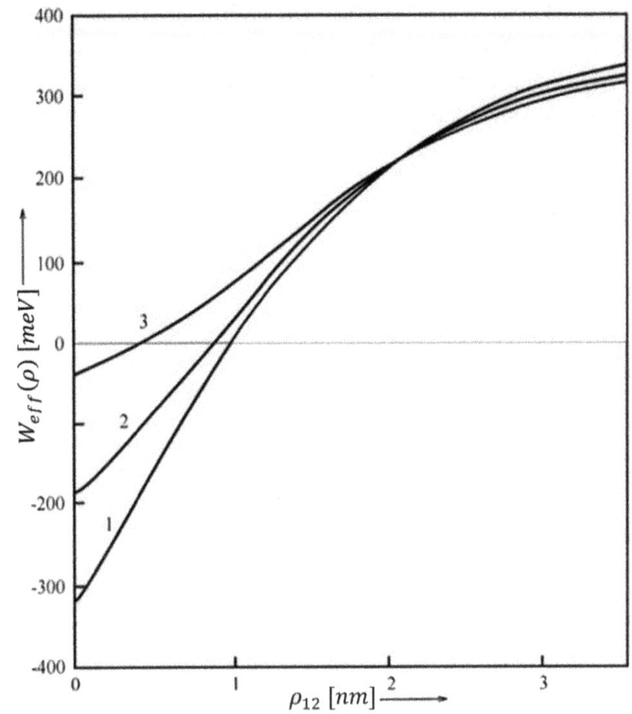

Figure 7.6. The effective potential of the electron–electron interaction. Curves 1–3 are constructed for different thicknesses of nonpolar δ layers: $d/2 = 0.05; 0.13$ and $0, 3$ nm, respectively. The other parameters are the same as in figure 7.5.

electrostatic repulsion as a result of its shielding by slow polarization, but also changes the sign of interaction from repulsion to attraction.

Figure 7.6 shows the dependence of W_{eff} on the distance between the electrons and the layers of the structure containing the polar layers of BaO.

7.3.3 Binding energy of the bipolaron

The binding energy of a bipolaron is calculated similarly to the case of a bipolaron in a quantum well (see section section 7.2). Averaging the Hamiltonian (7.3.1) on the wave function (7.3.11) leads to the problem of the two-dimensional case with the Hamiltonian:

$$\hat{H}_1(\rho_{12}) = \langle \psi(z_1)\psi(z_2) \big| \hat{H}(\rho_{12}) | \psi(z_1)\psi(z_2) \rangle . \tag{7.3.19}$$

Let's place the origin at the center of mass of the two-electron system, then ρ_1 and ρ_2 are the radius vectors of the electrons. We will perform a unitary transformation to shift the amplitudes of the phonon modes:

$$\hat{H}_2 = U_1^{-1} \hat{H}_1 U_1, \tag{7.3.20}$$

where the expression for U_1 is given in [15].

Following the work [15, 23, 25], after averaging on the phonon vacuum wave function, we obtain an effective Hamiltonian that depends only on variables ρ_1, ρ_2 and the variational parameter λ. Section 7.2 shows that the parameter λ describes the charge induced by polarization. The effective Hamiltonian is obtained in the form:

$$\hat{H}_3(\rho_{12}, \rho_1, \rho_2) = E_0(d) - \frac{\hbar^2}{2m_1}\Delta_1 - \frac{\hbar^2}{2m_2}\Delta_2 + W_{\text{eff}}(\rho_{12}, d) + W_p(\rho_1, \rho_2) + E_p(d, z_0). \quad (7.3.21)$$

The energy $E_0(d) = \langle V_{SA} \rangle$ includes the energy of self-action:

$$\langle \psi_1(z_1)\psi_2(z_2)| V_{SA}(z_1, z_2)|\psi_1(z_1)\psi_2(z_2)\rangle = \frac{e^2}{4\pi\varepsilon_0} \cdot \frac{1}{d}\left(\varepsilon_2^2 - \varepsilon_1^2\right) \int_0^\infty \mathrm{d}x \frac{e^{-\left|1-\frac{2z_0}{d}\right|x}}{f(\varepsilon)}; \quad (7.3.22)$$

$W_{\text{eff}}(\rho_{12}, d)$ is described by the expression (7.3.16).

Term

$$W_p(\rho_1, \rho_2) = \sum_{i=1,2} W_p(\rho_i, d) = \frac{1}{d} \sum_{i,j=1,2} \int_0^\infty \mathrm{d}x \left\{ |F_j|^2 J_0\left(\frac{\rho_i}{d}x\right) a_{2j}(a_{1j} - 2) \right\} \quad (7.3.23)$$

is proportional to λ and describes the potential energy of the interaction between electrons and the additional charge induced by the joint action of both electrons (which is absent in a single-electron system).

A coordinate-independent part of the energy

$$E_p(d, z_0) = \frac{1}{d} \sum_{j=1,2} \int_0^\infty dx \, |F_j|^2 \left\{ 2a_{2j}^2 - 4a_{1j} + 2a_{1j}^2 R_{S_j}^2 \left(\frac{x^2}{d^2}\right) + a_{2j}^2 \right\}, \quad (7.3.24)$$

corresponding to the polaron effect, it contains the energy of interaction between each electron and the polarization induced by it, proportional to a_{1j}, and the energy of an elastically deformed lattice, proportional to a_{1j}^2, a_{2j}^2.

Since W_{eff} as a function of ρ_{12} changes its sign from repulsion to attraction, we choose the normalized variational wave function of the ground state of the electron pair as

$$\varphi(\rho_{12}) = \sqrt{\frac{\gamma}{\pi}} e^{-\gamma\rho_{12}}, \quad (7.3.25)$$

where γ is a variational parameter that takes into account the spatial correlation of electrons.

Minimization of variational energy

$$H(\gamma, \lambda) = \left\langle \varphi(\rho_{12}) \middle| \hat{H}_3(\rho_{12}, \rho_1, \rho_2, d, z_0) | \varphi(\rho_{12}) \right\rangle \quad (7.3.26)$$

it is performed numerically.

The binding energy of a bipolaron is the difference:

$$W_b = E_b - 2E_p, \qquad (7.3.27)$$

where E_b is the minimum value of $H(\gamma, \lambda)$ (ground state energy) and E_p is the ground state energy of the polaron.

The estimation of the binding energy W_b of an interlayer bipolaron formed by electrons in CuO_2 layers separated by a layer of polar dielectric BaO of a three-layer structure monolayer CuO_2/BaO/monolayer CuO_2 with the thickness of the layer BaO $d \sim 7.5$ nm gives: $W_b \approx -70$ meV, which is by the order of magnitude comparable to the binding energy of a bipolaron in a monolayer FeSe on a polar substrate $SrTiO_3$.

Based on the conducted research, the formation of interlayer bipolarons, their condensation, and the occurrence of high-temperature superconductivity is possible. As in the case of the formation of bipolarons in monolayers, their binding energy significantly depends on the geometric and material parameters of the layers of the multilayer structure.

7.4 Bipolaronic states in a quantum wire in a polar medium

As noted in sections 7.2 and 7.3, the problem of bipolaronic states in dimensionally limited systems attracts great interest due to the more favorable possibility of fulfilling the criteria for their experimental observation and the possible explanation of the occurrence of high-temperature superconductivity based on the bipolaronic mechanism [11, 13, 16].

Obviously, in structures with a lower dimension than planar ones, favorable conditions for the emergence of stable bipolaronic states should also be expected. This problem aroused particular interest after the appearance of Little's work [6] on the possible creation of an 'exciton superconductor' based on organic compounds by constructing a long conductive (metallic) organic molecule surrounded by side 'polarizers' (i.e., Little indicated a possible way to increase T_c even to room and higher temperatures).

Consider this system: a quantum polar wire in a polar medium (similar to the one shown in figure 6.14, only for a pair of electrons, and the external environment is polar).

The solution to the problem of a bipolaron in a quantum wire in a polar medium decays into two stages: at the first stage, the exact Green function of the Poisson equation is derived by the method of potential theory, which allows us to obtain the potential of the electron–electron interaction $W_{e_1-e_2}$, taking into account the shielding by rapid polarization (plasma of valence electrons). In the second stage, based on the modified Haken method [13, 29], developed in the problem of charge interaction in polarizing media, the effective potential W_{eff} of the electron–electron interaction is found, taking into account the direct Coulomb interaction $W_{e_1-e_2}$ as a seed (the case of charges on the axis of the wire).

Further, in the case of the existence of a region of effective carrier attraction, the energy of the ground state of a bipolaronic pair can be calculated based on a variational approach.

7.4.1 Hamiltonian of the electron–phonon interaction in a quantum wire

The Hamiltonian of the electron–phonon interaction for a polar quantum wire in a polar medium (figure 7.7) is derived in [33] and has the form:

$$\hat{H}_{\text{e-ph}, n} = \sum\nolimits_{\chi} [\Gamma_{\chi, n}(\rho) \cdot e^{i(m\cdot\phi+\eta\cdot z)} \cdot \hat{b}_{\chi, n} + \text{h. c.}], \tag{7.4.1}$$

where $n = 1, 2$ numbers the parameters of the thread and the medium: 1 is the thread, 2 is the medium; R is the radius of the thread; χ is a set of quantum numbers of phonons: $\chi = \chi(q, m, n)$, where q is the radial quantum number, m is the projection of the moment on the z, η is the wave number of motion along the z-axis; and $\hat{b}^{+}_{\chi, n}$, $\hat{b}_{\chi, n}$ are phonon creation and an destruction operators in the medium n.

The amplitude of the electron–phonon interaction for bulk phonons in medium n is represented as

$$\Gamma^{bn}_{\chi} = V^{bn}_{\chi} \psi_n(\rho); \tag{7.4.2}$$

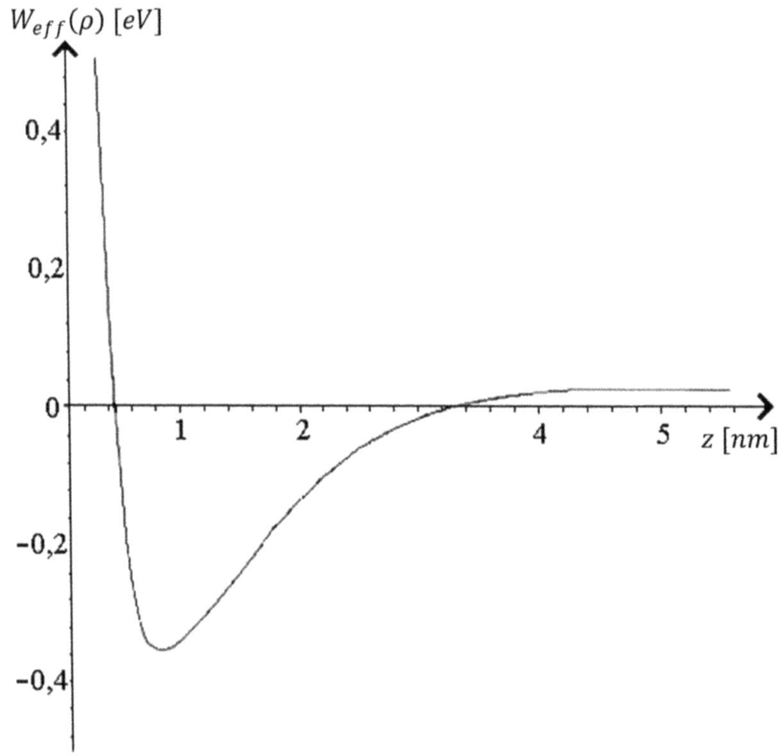

Figure 7.7. Dependence of the interaction energy of two electrons in a nonpolar wire in a polar medium on the distance between electrons at the values of the wire radius $R \approx a_0$ ($a_0 = \hbar^2\varepsilon_1/(m_e^* e^2)$); dielectric permittivity of the first medium $\varepsilon_1 = 1$ and high-frequency and low-frequency dielectric permittivity of the environment $\varepsilon_2 = 2$; $\varepsilon_{20} = 20$, respectively.

$$\left| V_{\chi}^{bn} \right|^2 = \alpha_{Fn}\hbar\omega_{nL}\frac{2\hbar\omega_{nL}R_{pn}}{\left(q_n^2+\eta^2\right)R^2};\ R_{pn}=\left(\frac{\hbar}{2m_n\omega_{nL}}\right)^{\frac{1}{2}}; \tag{7.4.3}$$

$$\psi_1(\rho)=\frac{\sqrt{2}J_m(q_1\rho)}{\left|J_{m+1}(q_1R)\right|}, \tag{7.4.4}$$

where q_1 is found from the equation:

$$J_m(q_1R)=0. \tag{7.4.5}$$

Further, limiting ourselves to the ground state of the dimensional state, we assume $m = 0$ and take the smallest positive root of the equation (7.4.5).

For surface modes, the amplitude of the electron–phonon interaction in the medium n has the following form:

$$\Gamma_{\sigma\chi}^{Sn}=\sqrt{\frac{\hbar}{2\Omega_\sigma}}\cdot\frac{e\Omega_\sigma^0(m,\eta)}{(2\pi\varepsilon_0\,|\eta|R(\chi_1+\chi_2))^{\frac{1}{2}}}\cdot F_n^n(\rho); \tag{7.4.6}$$

where σ =1, 2 are the numbers of the two surface modes:

$$F_1^1=\frac{I_m(\eta\rho)}{I_m(\eta R)};\ F_1^1=\frac{I_m(\eta\rho)}{I_m(\eta R)}; \tag{7.4.7}$$

where $\Gamma_{\sigma=1,\,\chi}^{S,\,n=1}$ is the amplitude of the first surface mode in the first medium, $\Gamma_{\sigma\chi}^{Sn}$ is the amplitude of the first surface mode in the second medium, $\Gamma_{\sigma\chi_{\sigma=2,\,\chi}}^{Sn\,S,\,n=1}$ is the amplitude of the second surface mode in the first medium, and $\Gamma_{\sigma\chi_{\sigma=2,\,\chi}}^{Sn\,S,\,n=2}$ is the amplitude of the second surface mode in the second medium.

The expressions for the frequencies of the surface modes have the form

$$\Omega_{1,2}(\eta)=\frac{1}{\sqrt{2}}\left\{B_{11}+B_{22}+\sqrt{(B_{11}-B_n)^2+4B_{12}^2}\cdot(B_{11}-B_{22})\right\}; \tag{7.4.8}$$

where $B_{nk}(n,\,k=1,\;2)$ have a form

$$B_{nk}=\omega_n^2\delta_{nk}+\frac{\omega_{n0}\omega_{k0}\sqrt{\chi_n\chi_k}}{\chi_1+\chi_2},\ \omega_{n0}^2=\omega_n^2-\omega_n^2=\omega_n^2\cdot\frac{\varepsilon_{n0}-\varepsilon_n}{\varepsilon_n}; \tag{7.4.9}$$

$$\chi_1(\eta R)=\varepsilon_1\frac{I'_{|m|}(\eta\rho)}{I_{|m|}(\eta R)};\ \chi_2(\eta R)=-\varepsilon_2\frac{K'_{|m|}(\eta\rho)}{K_{|m|}(\eta R)}; \tag{7.4.10}$$

$I_{|m|}(x)$ and $K_{|m|}(x)$ are the modified Bessel functions:

$$I'_{|m|}(x)=\frac{dI_m(x)}{dx};\ K'_{|m|}(x)=\frac{dK_m(x)}{dx}; \tag{7.4.11}$$

$$\Omega_{\sigma}^{0}(\eta, m) = \sum\nolimits_{n=1}^{2}\left(\frac{\chi_n}{\chi_1 + \chi_2}\right)^{\frac{1}{2}} \omega_{n0} F_{n\sigma}; \tag{7.4.12}$$

$$F_{11} = F_{22} = \left\{\frac{1}{2} + \frac{B_{11} - B_{22}}{2\left[(B_{11} - B_{22})^2 + 4B_{12}^2\right]^{\frac{1}{2}}}\right\}^{\frac{1}{2}}; \tag{7.4.13}$$

$$F_{21} = -F_{12} = \left\{\frac{1}{2} - \frac{B_{11} - B_{22}}{2\left[(B_{11} - B_{22})^2 + 4B_{12}^2\right]^{\frac{1}{2}}}\right\}^{\frac{1}{2}} \text{sign}(\quad B_{11} - B_{22}). \tag{7.4.14}$$

If we consider only the external environment to be polar, then in the formulas obtained it is necessary to take

$$\varepsilon_{10} = \varepsilon_1;\ B_{11} = \omega_1{}^2;\ B_{12} = B_{21} = 0; \tag{7.4.15}$$

$$\Omega_{1,2}(\eta) = \frac{1}{\sqrt{2}} \cdot \{B_{11} + B_{22} \pm |B_{11} - B_{22}|\}^{\frac{1}{2}}; \tag{7.4.16}$$

$$\Omega_1(\eta) = \frac{1}{\sqrt{2}} \cdot \{B_{11} + B_{11}\}^{\frac{1}{2}} = \omega_1; \tag{7.4.17}$$

$$\Omega_2(\eta) = \frac{1}{\sqrt{2}} \cdot \{B_{22} + B_{22}\}^{\frac{1}{2}} = (B_{22})^{\frac{1}{2}}; \tag{7.4.18}$$

$$\Omega_{\sigma}^{0}(\eta, m) = \left(\frac{\chi_2}{\chi_1 + \chi_2}\right)^{\frac{1}{2}} \cdot \omega_2\left(\frac{\varepsilon_{20}}{\varepsilon_2} - 1\right)^{\frac{1}{2}} \cdot F_{2\sigma}; \tag{7.4.19}$$

Then

$$\begin{aligned}\hat{H}_{\text{e-ph}}^{s} = &\sum_{\eta}\frac{A}{2\pi}\sqrt{\frac{\hbar}{2\Omega_1}}\cdot\frac{e\Omega_1^0(m, \eta)}{\sqrt{2\pi\varepsilon_0\eta}\cdot R\cdot(\chi_1 + \chi_2)} \times F_1^1(\rho)\cdot e^{i(m\varphi + \eta z)}[\hat{b}_{\chi,\,1} + h.\,c.]\\ &+\sum\nolimits_{\eta}\sqrt{\frac{\hbar}{2\Omega_2}}\cdot\frac{e\Omega_2^0(m, \eta)}{\sqrt{2\pi\varepsilon_0\eta}\cdot R\cdot(\chi_1 + \chi_2)} \times F_1^1(\rho)\cdot e^{i(m\varphi + \eta z)}[\hat{b}_{\chi,\,2} + h.\,c.];\end{aligned} \tag{7.4.20}$$

$$\Omega_1^0(\eta, m) = \sum\nolimits_{n=1}^{2}\left(\frac{\chi_n}{\chi_1 + \chi_2}\right)^{\frac{1}{2}} \omega_{n0} F_{n1}; \tag{7.4.21}$$

$$\Omega_2^0(\eta, m) = \sum\nolimits_{n=1}^{2}\left(\frac{\chi_n}{\chi_1 + \chi_2}\right)^{\frac{1}{2}} \omega_{n0} F_{n2}; \tag{7.4.22}$$

$$\Omega_1^0(\eta, m) = \left(\frac{\chi_2}{\chi_1 + \chi_2}\right)^{\frac{1}{2}} \cdot \omega_2 \left(\frac{\varepsilon_{20}}{\varepsilon_2} - 1\right)^{\frac{1}{2}} \cdot F_{21}; \tag{7.4.23}$$

$$\Omega(\eta, m) = \left(\frac{\chi_2}{\chi_1 + \chi_2}\right)^{\frac{1}{2}} \cdot \omega_2 \left(\frac{\varepsilon_{20}}{\varepsilon_2} - 1\right)^{\frac{1}{2}} \cdot F_{22}. \tag{7.4.24}$$

7.4.2 Effective Hamiltonian of the electron–phonon interaction and the effective potential energy of the electron–electron interaction

Next, we consider a system consisting of a nonpolar quantum wire with a dielectric constant of ε_1 and radius R located in a polar medium with high-frequency and low-frequency dielectric permittivity of ε_2 and ε_{20}, respectively (figure 6.14).

The Hamiltonian of such a system with two electrons in the wire following [13]

$$\hat{H} = \hat{K}_{e1} + \hat{K}_{e2} + \hat{H}_{\text{ph}}^{S} + \hat{H}_{\text{e-ph}} + W_{e1-e2}(\rho, z, \varphi), \tag{7.4.25}$$

$$\hat{K}_{ei} = \frac{\hat{p}_{zi}^2}{2m_{\parallel i}^*} + \frac{\hat{p}_{\rho i}^2}{2m_{\perp i}^*} \tag{7.4.26}$$

are the operators of the kinetic energy of electrons,

$$\hat{H}_{\text{ph}}^{S} = \sum\nolimits_{\eta,\, i=1,2} \hbar \Omega_{Si} \hat{b}_{\chi i}^{+}(\eta) \hat{b}_{\chi i}(\eta); \tag{7.4.27}$$

where $\hat{H}_{\text{e-ph}}$ is the Hamiltonian of the electron–phonon interaction (7.4.1).

The energy of the Coulomb interaction of two electrons located inside the wire is [13]

$$W_{e1-e2}(\rho, z, \varphi) = \frac{e^2}{(4\pi\varepsilon_0)\varepsilon_1\sqrt{\rho^2 + z^2}} + \frac{2e^2\left(\frac{\varepsilon_1}{\varepsilon_2} - 1\right)}{(4\pi\varepsilon_0)\varepsilon_1\pi} \cdot J(z); \tag{7.4.28}$$

where

$$J(z) = \int_0^\infty dx \cos(xz) \cdot \frac{K_0(xa)K_1(xa)I_0(x\rho)}{\delta K_0(xa)I_1(xa) + I_0(xa)K_1(xa)}. \tag{7.4.29}$$

The following designations are accepted here:

$$\delta = \frac{\varepsilon_1}{\varepsilon_2};\; a = 2R. \tag{7.4.30}$$

We will be interested in a situation close to one-dimensional. Therefore, we take the radius of the wire to be small and, accordingly, the energy of dimensional quantization to be large (more than all other contributions of the system).

As a result of averaging the Hamiltonian (7.4.25) on the wave function of the ground state of dimensional quantization, we obtain

$$\begin{aligned}\hat{H}_1 &= \left\langle \Phi(\rho,\varphi) \,\middle|\, \hat{H} \,\middle|\, \Phi(\rho,\varphi) \right\rangle = E_{\text{s.q.}} + \frac{\hat{p}_{z1}^2}{2m_\parallel^*} + \frac{\hat{p}_{z2}^2}{2m_\parallel^*} + \widetilde{W}_0(R,z) \\ &+ \sum_{\eta,i=1,2} \hbar\Omega_{Si}\hat{b}_{\chi i}^{+}(\eta)\hat{b}_{\chi i}(\eta) + \sum_{\eta}\sqrt{\frac{\hbar}{2\Omega_1}}\cdot\frac{e\Omega_1^0(m,\eta_1)}{\sqrt{2\pi\varepsilon_0\eta R(\chi_1+\chi_2)}} \\ &\times \left\langle \Phi(\rho\varphi) \,\middle|\, F_1^1(\rho) \,\middle|\, \Phi(\rho\varphi) \right\rangle \cdot e^{i\eta\cdot z_{e1}} \cdot \left[\hat{b}_{\chi,1} + h.\,c.\right] \\ &+ \sum_{\eta_2}\sqrt{\frac{\hbar}{2\Omega_2}}\cdot\frac{e\Omega_1^0(m,\eta_2)}{\sqrt{2\pi\varepsilon_0\eta R(\chi_1+\chi_2)}} \\ &\times \left\langle \Phi(\rho\varphi) \,\middle|\, F_1^1(\rho) \,\middle|\, \Phi(\rho\phi) \right\rangle \cdot e^{i\eta\cdot z_{e2}} \cdot \left\{\hat{b}_{\chi,2} + \text{h. c.}\right\};\end{aligned} \tag{7.4.31}$$

$$\begin{aligned}\widetilde{W}_0(R,z) &= \langle \Phi(\rho,\varphi)|\, W_{e1-e2}(\rho,z,\varphi)|\Phi(\rho,\varphi)\rangle \\ &= \frac{2\pi A_0^2}{4\pi\varepsilon_0}\int_0^R \frac{e^2 J_0^2(\sqrt{\alpha(R)}\cdot\rho)}{\varepsilon_1\sqrt{\rho^2+z^2}}\cdot\rho d\rho + \frac{4\pi e^2\left(\frac{\varepsilon 1}{\varepsilon 2}-1\right)}{4\pi\varepsilon_0\varepsilon_1\pi}\widetilde{J}(z).\end{aligned} \tag{7.4.32}$$

In the last term, it is assumed that $m = 0$, and

$$\begin{aligned}\widetilde{J}(z) &= A_0^2\int_0^\infty dx\cos(x\,z)\cdot\frac{K_0(xa)K_1(xa)I_0(x\rho)}{\alpha K_0(xa)K_1(xa)+I_0(xa)K_1(xa)} \\ &\times \int_0^R J_0^2(\sqrt{\alpha(R)}\cdot\rho)I_0(x\rho)\,\rho d\rho.\end{aligned} \tag{7.4.33}$$

Let's take the following notation:

$$\begin{aligned}\widetilde{F}_1^1(\eta R) &= \left\langle \Phi(\rho\varphi) \,\middle|\, F_1^1(\rho) \,\middle|\, \Phi(\rho\varphi) \right\rangle \\ &= 2\pi A_0^2\int_0^R J_0^2\left(\sqrt{\alpha(R)}\rho\right)\cdot\frac{I_0(\eta\rho)}{I_0(\eta R)}\cdot\rho d\rho.\end{aligned} \tag{7.4.34}$$

Let's introduce a center of mass system:

$$z_1 = Z + \tilde{z}_1;\; z_2 = Z + \tilde{z}_2, \tag{7.4.35}$$

then

$$\exp(i\eta_i z_1) + \exp(i\eta_i z_2) = \exp(i\eta Z)\cdot[\exp(i\eta_i\tilde{z}_1) + \exp(i\eta_i\tilde{z}_2)]; \tag{7.4.36}$$

$$E_{s.q.}(R) = 2\cdot[\widetilde{E}_{\text{р.кв.}}(R) + E_{\text{SA}}(R)]; \tag{7.4.37}$$

$$\begin{aligned}H_{\text{e-ph}}^S &= \sum\nolimits_{\eta_j,s=1,2} A_j(m,\eta)\widetilde{F}_1^1(\eta R)\cdot \\ &\left[e^{i\eta_j Z}\left(\exp\left(i\eta_j\tilde{z}_1\right) + \exp\left(i\eta_j\tilde{z}_2\right)\right)\cdot\left[\hat{b}_{-\eta_j}^{+} + \text{h. c.}\right]\right].\end{aligned} \tag{7.4.38}$$

Let's make a unitary transformation over the Hamiltonian (7.4.31):

$$\hat{H}_2 = \hat{U}^{-1}\hat{H}_1\hat{U}^1;\; U = \exp\left\{\sum\nolimits_{\eta_j}\left(f_{\eta_j}(R, z)\cdot\hat{b}_{\eta_j}^{\pm f_{\eta_j}^{+}(R, z)}\cdot\hat{b}_{\eta_j}\right)\right\}; \tag{7.4.39}$$

$$f_{\eta_j}(R, z) = \frac{A_j(m, \eta)F_1^1(\eta_j R)}{\hbar\Omega_{Sj}}\cdot\left\{\frac{\lambda_{1j}e^{i\eta_j Z}(e^{i\eta_j z_1}+e^{i\eta_j \tilde{z}_2})}{\beta_{Sj}^{2R_{Sj}^{2\eta^2}}+1}+\frac{\lambda_{2j}}{\left(\beta_{Sj}^2R_{Sj}^2\eta^2+1\right)^2}\right\}; \tag{7.4.40}$$

$$\hat{H}_{eff} = E_{\text{s.q.}}^{(0)}(R) + \frac{\hbar^2}{2m_\parallel^*}\Delta_{\tilde{z}_1} - \frac{\hbar^2}{2m_\parallel^*}\Delta_{\tilde{z}_2} + W_{\text{eff}}(|z_2 - z_1|, R) + \sum\nolimits_{j=1,2}W_p(|z_j|, R), \tag{7.4.41}$$

where $E_{s.q.}^{(0)}(R)$ is the energy of the lowest dimensional quantum state [13], and

$$W_{\text{eff}}(|z|, R) = \widetilde{W}_{\text{cl}}(R, z) + \delta W_{\text{ph}}(R, z); \tag{7.4.42}$$

is the effective potential energy of the interaction of electrons, and $\widetilde{W}_{\text{cl}}(R, z)$ is the energy of the Coulomb interaction of two electrons [13]:

$$\widetilde{W}_{cl}(R, z) = 2\pi A_m^2[J_1(R, z) + J_2(R, z)]; \tag{7.4.43}$$

where

$$J_1(R, z) = \frac{e^2}{4\pi\varepsilon_0\varepsilon_1}\int_0^R \frac{J_0^2(\sqrt{\alpha(R)}\cdot\rho)}{\sqrt{\rho^2+z^2}}\rho d\rho; \tag{7.4.44}$$

$$J_2(R, z) = \frac{2e^2\left(\frac{\varepsilon_1}{\varepsilon_2}-1\right)}{4\pi^2\varepsilon_0\varepsilon_1}\int_0^\infty \mathrm{d}x\cos(xz)\cdot\frac{K_0(xa)K_1(xa)I_0(x\rho)}{\alpha K_0(xa)I_1(xa)+I_0(xa)K_1(xa)} \times\int_0^R J_0^2(\sqrt{\alpha(R)}\cdot\rho)I_0(x\rho)\,\rho d\rho; \tag{7.4.45}$$

$$\delta W_{\text{ph}}(R, z) = \sum\nolimits_{\eta_j}\left|\frac{A_j(m, \eta)}{2\pi}\cdot\widetilde{F}_1^1(\eta R)\right|^2\{2a_{2j}\cdot\cos(\eta_j\tilde{z}_1)\cdot(2a_{1j}-1)\}; \tag{7.4.46}$$

where $\sum_{j=1,2}W_p(|z_j|, R)$ is the energy of free (noninteracting) polarons.

$$a_{1j} = \frac{\lambda_{1j}}{\beta_{1j}^2R_{Sj}^2\eta^2+1};\; a_{2j} = \frac{\lambda_{2j}}{\left(\beta_{2j}^2R_{Sj}^2\eta^2+1\right)^2}; \tag{7.4.47}$$

$$\delta W_{\text{ph}}(R, z) = \frac{1}{2\pi}\int_0^\infty \frac{\hbar}{2\Omega_1}\cdot\frac{\left|e\Omega_1^0(m, \eta)\right|^2}{2\pi\varepsilon_0 R(\chi_1+\chi_2)}\left|\widetilde{F}_1^1(R)\right|^2 \times\left\{\frac{2\lambda_{1j}^2}{\left(\beta_{1j}^2R_{Sj}^2\eta^2+1\right)^2}+\frac{\lambda_{2j}}{\left(\beta_{2j}^2R_{Sj}^2\eta^2+1\right)^2}\right\}\cos(\eta z)\mathrm{d}\eta; \tag{7.4.48}$$

Let us place both electrons on the axis of the quantum wire and move in expression (7.4.46) from summation to integration. Assuming

$\lambda_{1j} = \lambda_{2j} = \beta_{1j} = \beta_{2j} = 1$, we obtain the final expression for the effective interaction potential, which consists of the energy of the Coulomb interaction and the contribution from inertial polarization:

$$W_{\text{eff}}(z, R) = \frac{e^2}{(4\pi\varepsilon_0)\varepsilon_1 z} + \frac{e^2\left(\frac{1}{\varepsilon_2} - \frac{1}{\varepsilon_1}\right)}{(4\pi\varepsilon_0)\pi R}$$
$$\times \int_0^\infty dx \cdot \frac{\cos\left(\frac{xz}{2R}\right) \cdot K_0(x)K_1(x)}{\left[\delta K_0(x)I_1(x) + I_0(x)K_1(x)\right]} + \frac{2e^2\left(\frac{1}{\varepsilon_{20}} - \frac{1}{\varepsilon_2}\right)}{(4\pi\varepsilon_0)\pi R} \tag{7.4.49}$$
$$\times \int_0^\infty dx \cos\left(\frac{xz}{R}\right) \cdot \frac{K_0(x)\left\{\frac{2}{1+a^2x^2} + \frac{2}{\left(1+a^2x^2\right)^2}\right\}}{xI_0(x) \cdot \left[\delta K_0(x)I_1(x) + I_0(x)K_1(x)\right]},$$

where $a = R_S/R$.

The plot of $W_{\text{eff}}(z, R)$ is shown in figure 6.14 for next parameter values: $\varepsilon_1 = 1$; $\varepsilon_2 = 2$; $\varepsilon_{20} = 20$; $R \approx a_0$ ($a_0 = \hbar^2\varepsilon_1/(m_e^* e^2)$).

The behavior of $W_{\text{eff}}(z, R)$, as a function of the distance between electrons z, has the form of a molecular potential with a potential well depth corresponding to the region of effective electron attraction and a component of $\sim$0.3 eV.

For wires of radius $R \approx (2-5)$ nm, the interaction of electrons decreases by one or two orders of magnitude and in the limit $z \gg R$ passes into the Coulomb repulsion potential.

In the case when $z \gg R$, the classical limit follows from the expression (7.4.49):

$$W_{\text{eff}}(z)\,\Big|_{z\gg R} = W_{\text{cl}}(z) + W_{\text{ph}}(z) \approx \frac{e^2}{4\pi\varepsilon_0\varepsilon_2 z}$$
$$- \frac{e^2}{4\pi\varepsilon_0\varepsilon_2 z}\left(\frac{1}{\varepsilon_2} - \frac{1}{\varepsilon_{20}}\right) = \frac{e^2}{4\pi\varepsilon_0\varepsilon_{20} z}, \tag{7.4.50}$$

which corresponds to the situation with the disappearance of the region of effective attraction of electrons for large values of the radius of the wire R.

7.4.3 Binding energy of the bipolaron in the quantum wire

Let's calculate the binding energy of a bipolaron in a nonpolar quantum wire in a polar medium.

The variational wave function of the bipolaron is chosen in the form

$$\psi = 2\beta^{3/2} z e^{-\beta z}. \tag{7.4.51}$$

Averaging the effective Hamiltonian of the system on the wave function (7.4.51), we obtain the variational functional for the binding energy (E_b) of the bipolaron:

$$E(\beta) = \langle \psi(z) | \hat{H}_{\text{eff}} | \psi(z) \rangle; \tag{7.4.52}$$

where

$$\hat{H}_{\text{eff}} = -\frac{\hbar^2}{2m}\frac{d^2}{dz^2} + W_{\text{eff}}(z, R), \tag{7.4.53}$$

where

W_{eff} is the effective potential energy of the interaction of electrons, determined by the expression (7.4.49).

We will estimate the binding energy E_b in a quantum wire for several parameter values:

$$\eta = \frac{\varepsilon_{20}}{\varepsilon_2}; \varepsilon_1 = 2; \varepsilon_2 = 5.$$

$$\eta_2 = 15 \quad E_b \approx -2.20 \text{ meV}$$

$$\eta_3 = 20 \quad E_b \approx -2.78 \text{ meV}$$

1. $R_1 = 1$ nm $\eta_1 = 10 \quad E_b \approx -4.21$ meV
   $\eta_2 = 15 \quad E_b \approx -5.15$ meV
   $\eta_3 = 20 \quad E_b \approx -5.70$ meV

2. $R_2 = 1, 5$ nm $\eta_1 = 10 \quad E_b \approx -1.51$ meV
   $\eta_2 = 15 \quad E_b \approx -2.20$ meV
   $\eta_3 = 20 \quad E_b \approx -2.78$ meV

As follows from the above estimates:

(a) an increase in the radius of the wire reduces the binding energy of the bipolaron by more than two times;

(b) an increase in the parameter leads to an increase in the E_b energy.

It should be noted that the absolute value of the binding energy of the bipolaron in the quantum wire is one to two orders of magnitude less than the binding energy of the bipolaron in the FeSe quantum layer in the structures: FeSe (monolayer)/$SrTiO_3$; $SrTiO_3$/FeSe (monolayer)/$SrTiO_3$ ('Ginzburg sandwiches'), which is in agreement with the Ginzburg justification [34] for the preference for replacing a quasi-one-dimensional conductive wire (Little's model [6]) with a quasi-two-dimensional structure due to the significantly smaller role of fluctuations and the possibility of creating layered compounds—a stack of 'Ginzburg sandwiches'.

7.5 High-temperature bipolaronic superconductivity in structures of the 'Ginzburg sandwiches' type: $FeSe/SrTiO_3$; $SrTiO_3/FeSe/SrTiO_3$

7.5.1 Introduction

The discovery of high-temperature superconductivity (HTSC) in layered iron chalcogenides has stimulated great interest among researchers, since the nature of

HTSC in these materials, as well as a number of their physical properties, differs significantly from Y – Ba – Cu – O, compounds in which HTSC $T_c > 92$ K was first discovered [35, 36]. Detailed experimental and theoretical studies of the systems, monolayer FeSe, film deposited on a $SrTiO_3$ substrate, intercalated compounds based on FeSe, and multilayer systems based on FeSe, TiO_2, SrO, $SrTiO_3$, were aimed at developing new models of HTSC, since existing theories proved to be inapplicable to explain the reasons for such a high growth of T_c ($\sim$80 – 100 K) [37, 38] in comparison with T_c for bulk FeSe, crystals, the critical temperature of which did not exceed values in in the range $T_c \sim (8 - 10)$ K. HTSC studies in the system, a monolayer FeSe film on a substrate of a highly polar semiconductor $SrTiO_3$, indicated the important role of the substrate in a significant increase in T_c in the FeSe monolayer to record temperatures of $\gtrsim$100 K.

In the review [8], a systematic review and analysis of the results of theoretical and experimental studies of HTSC in structures of the $FeSe/SrTiO_3$ type and in other compounds and possible mechanisms for increasing T_c in these systems, representing typical 'Ginzburg sandwiches' [4, 9], was carried out.

Let's highlight the most relevant models from the work [8], which provide the possibility of implementing the exciton mechanism of the Ginzburg HTSC:

1. A model of T_c increase in the FeSe monolayer due to the interaction of charge carriers moving in the conduction band of the FeSe layer with elementary excitations in the $SrTiO_3$, substrate, following the picture of the Ginzburg exciton mechanism [4, 5, 9].
2. The exciton mechanism of Allender–Bray–Bardin [10], which, as shown in the review [8], was considered when explaining high T_c in FeSe monolayers on $SrTiO_3$ ($BaTiO_3$) substrates, but turned out to be not entirely effective.
3. A phonon mechanism in which the interaction of electrons in a metal monolayer FeSe with optical phonons of the $SrTiO_3$ substrate is considered a variant of the exciton mechanism in the geometry of the Ginzburg sandwich;
4. Nonadiabatic phonon superconductivity due to a significant excess of the energy of the optical phonon $SrTiO_3$ of the Fermi energy E_F [39]. In this case, the interaction of electrons with high-energy optical phonons is considered as a possible mechanism for increasing T_c in the $FeSe/SrTiO_3$. system.

Based on the analysis of these mechanisms, it was concluded that the high T_c growth in the FeSe monolayer on the $SrTiO_3$ substrate in comparison with the bulk FeSe is apparently due to an additional pairing mechanism resulting from the interaction of electrons with $SrTiO_3$, optical phonons, realizing a kind of 'pseudo-exciton' pairing mechanism.

Thus, in all the selected models, the interaction with the optical phonons of the substrate to one degree or another plays a decisive role in the formation of high T_c values.

Experiments on high-resolution electron spectroscopy of energy losses have confirmed the presence of strong electron–phonon interaction at the $FeSe/SrTiO_3$ [40]. An experimental evaluation of the interaction constant of an electron in a monolayer of FeSe with a surface optical phonon $\omega_{SLO} = 92$ meV of the $SrTiO_3$ substrate confirmed the presence of a strong electron–phonon bond with $\alpha \gtrsim 1$ (it should be noted that the electron in FeSe interacts with the field of surface optical phonons of the $SrTiO_3$ substrate since volumetric optical phonons of $SrTiO_3$ do not create fields outside the substrate due to the confinement effect).

In [41], experimental studies were conducted on the role of electron–phonon interaction in enhancing superconductivity in FeSe films on $SrTiO_3$ substrates, including in the structure: a monolayer of FeSe deposited on an unalloyed and alloyed $SrTiO_3$ substrate.

In experiments on high-resolution electron energy loss spectroscopy (HREELS) with two-dimensional energy and momentum mapping, surface phonon excitations were studied both in the FeSe/$SrTiO_3$ monolayer and separately on the pure surface (001) $SrTiO_3$.

The electronic structure of the FeSe/$SrTiO_3$ monolayer was determined by angular resolution photoemission spectroscopy (ARPES). It is established that the electrons of the FeSe layer interact with the surface optical vibrations of $SrTiO_3$, which leads to the formation of dynamic interphase polarons. In the microscopic model [42], it was shown that the 'polaron-polaron' interaction in the FeSe layer can cause additional attraction between electrons, which, in turn, leads to an increase in superconductivity (an increase in T_c).

Thus, it is of interest to carry out a consistent theoretical description of polaronic and bipolaronic states in a monolayer FeSe film in the FeSe/$SrTiO_3$ system, based on the basic principles and its basis to estimate the critical temperature of Bose condensation of bipolarons. Since the polaronic and bipolaronic states in the monolayer film of the FeSe/$SrTiO_3$ structure are formed as a result of the interaction of electrons from the FeSe conduction band with surface optical phonons $SrTiO_3$, the resulting bipolaronic states should be considered as surface (it was previously mentioned that volumetric optical vibrations do not create electric fields outside the $SrTiO_3$ surface in the FeSe monolayer).

It should be noted that the path proposed by Ginzburg [3, 4, 9] to enhance the electron–phonon interaction and achieve HTSC by separating the regions where electrons are located (forming Cooper pairs or bipolarons) with regions in which excitons are excited (or inertial polarization is induced) made it possible to implement criteria for the formation of bipolaronic states in multilayer structures with high binding energy, due to the possibility of selecting optimal geometric and material parameters (layer thicknesses, dielectric permittivity, optical frequencies, effective masses) [42].

The formation of bipolarons in ionic crystals was first investigated by Pekar [12] in the framework of the continuum theory of polarons. Ginzburg [3] considered the Bose condensate of bipolarons in the study of superfluidity. In [16, 34], the superfluidity of a charged Bose gas and the bipolaronic mechanism of superconductivity were studied. Even though the mechanisms of formation of bipolarons and Cooper pairs differ in the methods of description, in essence, the basis of both effects is the electron–phonon interaction.

We apply the theory of large-radius bipolaronic states developed by us using the exact Hamiltonian of electron–phonon interaction for arbitrary multilayer structures to the study of the processes of interaction of polarons and the formation of bipolaronic states in Ginzburg sandwiches: a monolayer of FeSe on an $SrTiO_3$ substrate and in a monolayer of FeSe in a three-layer structure of $SrTiO_3$/FeSe/$SrTiO_3$. In the systems under consideration, we will evaluate the critical temperature of condensation of bipolarons and the formation of a superfluid bipolaronic condensate [43].

7.5.2 Basic principles of the theory of large-radius bipolarons in a three-layer structure with a quantum well

Considered a three-layer structure $\langle 10 \mid 2 \mid 30 \rangle$ consisting of two semi-infinite polar crystals $\langle 10 \rangle$, $\langle 20 \rangle$ and nonpolar semiconductor quantum layer $\langle 2 \rangle$ with thickness d ($d \sim \lambda$, λ is de Broglie wavelength of the electron in the layer $\langle 2 \rangle$) between them and having a dielectric constant ε_2.

The theory was based on the exact Hamiltonian of the electron–phonon interaction for arbitrary multilayer systems [15, 21, 25]. For the considered system, the Hamiltonian of the electron–phonon interaction of electrons in layer (2) with the surface optical phonons of polar crystals (10) and (30) has the form (7.2.12)–(7.2.18).

The effective Hamiltonian of the electron–electron interaction (7.2.27), based on which the profile of the effective potential $W_{\text{eff}}(\rho, d)$ of the interaction of two electrons in structures of the Ginzburg sandwich type FeSe/$SrTiO_3$; FeSe/$SrTiO_3$/FeSe, is calculated in this section. In clause 7.5.4, the mechanism of formation of large-radius bipolarons is investigated and the binding energy is calculated in these structures, in which high-temperature superconductivity is experimentally realized.

7.5.3 Investigation of the effective potential of electron–electron interaction

We investigate the profile of the effective potential of the electron–electron interaction $W_{\text{eff}}(\rho, d)$, defined by formulas (7.2.30), (7.2.31), for two actual cases:

(a) in a monolayer of FeSe deposited on a massive polar substrate $SrTiO_3$ (one interface FeSe/$SrTiO_3$), in which HTSC was observed experimentally ($d \to 0$, $\varepsilon_{30} = \varepsilon_3$);

(b) in the FeSe monolayer separating the $SrTiO_3$ crystal layers (two FeSe/$SrTiO_3$ interfaces) ($d \to 0$, a bipolaron is formed at the contact of two polar crystals in the monolayer (δ-layer)).

Both cases under consideration are typical structures—'Ginzburg sandwiches', in which two electrons from the conduction band of a monolayer FeSe film interact with the surface optical phonons of the $SrTiO_3$ substrate (case a) and a monolayer FeSe film separating polar $SrTiO_3$ crystals (case b)).

As it was shown in [17, 21, 25], due to boundary conditions, bulk longitudinal optical vibrations of $SrTiO_3$ do not create electric fields in neighboring media (confinement effect); in this case, in the FeSe monolayer, in which free electrons will interact only with surface optical phonons of $SrTiO_3$. Let's assume that the FeSe monolayer is a quasi-two-dimensional quantum well with infinite walls.

For case (a), the effective potential of the electron–electron interaction can be obtained from the formulas (7.2.30) and (7.2.31) in the limit $d \to 0$:

$$W^{2D}(\rho) = \frac{e^2}{2\pi\varepsilon_0(\varepsilon_1+1)\rho} + \frac{e^2}{2\pi\varepsilon_0 R_s}\left(\frac{1}{\varepsilon_1+1} - \frac{1}{\varepsilon_{10}+1}\right) \int_0^\infty J_0\left(\frac{\rho x}{R_s}\right)\left(\frac{x^2-1}{x^2+1}\right)dx, \tag{7.5.1a}$$

where

$$R_S = \left(\frac{\hbar}{2m_e^*\Omega_S}\right)^{\frac{1}{2}};\ \Omega_S = \omega_1\sqrt{\frac{\varepsilon_{10}+1}{\varepsilon_1+1}}; \tag{7.5.1b}$$

where Ω_S is the frequency of the longitudinal surface optical phonon.

By carrying out the integration, we get:

$$W_{\text{eff}}^{2D}(\rho) = \frac{e^2}{2\pi\varepsilon_0\rho}\left(\frac{2}{\varepsilon_1+1} - \frac{1}{\varepsilon_{10}+1}\right) - \frac{e^2}{2\varepsilon_0 R_S}\left(\frac{1}{\varepsilon_1+1} - \frac{1}{\varepsilon_{10}+1}\right)\left\{I_0\left(\frac{\rho}{R_S}\right) - L_0\left(\frac{\rho}{R_S}\right)\right\}. \tag{7.5.2}$$

Here $I_0(x)$, $L_0(x)$ are the modified Bessel and Struve functions are of zero order, respectively.

Then the effective Hamiltonian of the system has the form

$$\hat{H}_{\text{eff}}^{2D} = -\frac{\hbar^2}{2m_{e_1}^*}\Delta_{\rho_1} - \frac{\hbar^2}{2m_{e_1^*}}\Delta_{\rho_2} + W_{\text{eff}}^{2D}(\rho). \tag{7.5.3}$$

Figure 7.8 shows plots of the dependence $W_{\text{eff}}(\rho)$ for case (a): vacuum – multilayer film FeSe – $SrTiO_3$ for three different values of the effective mass of an electron in *FeSe*.

For case (b) the effective potential of the electron–electron interaction is obtained from the formulas (7.2.30), (7.2.31), taking into account the symmetry of the structure: $SrTiO_3$ – monolayer film FeSe – $SrTiO_3$:

b)

$$W_{\text{eff}}(\rho) = \frac{e^2}{2\pi\varepsilon_0\rho}\left(\frac{2}{\varepsilon_1} - \frac{1}{\varepsilon_{10}}\right) - \frac{e^2}{2\varepsilon_0\underline{R}_S}\left(\frac{1}{\varepsilon_1} - \frac{1}{\varepsilon_{10}}\right)\left\{I_0\left(\frac{\rho}{\underline{R}_S}\right) - L_0\left(\frac{\rho}{\underline{R}_S}\right)\right\}, \tag{7.5.4}$$

where

$$\overline{R}_S = \left(\frac{\hbar}{2m_e^*\overline{\Omega}_S}\right)^{1/2};\ \overline{\Omega}_S = \omega_1\sqrt{\frac{\varepsilon_{10}+\varepsilon_1}{2\varepsilon_1}}.$$

Note that in the case of a symmetric $SrTiO_3$ – FeSe – $SrTiO_3$ structure, the electrons in the $\langle 2\rangle$ layer will interact with only one surface optical mode of each of the $SrTiO_3$ crystals $SrTiO_3$ ($\Omega_{S_1} = \Omega_{S_2} \equiv \overline{\Omega}_S$; $\alpha_{S_1} = \alpha_{S_2}$).

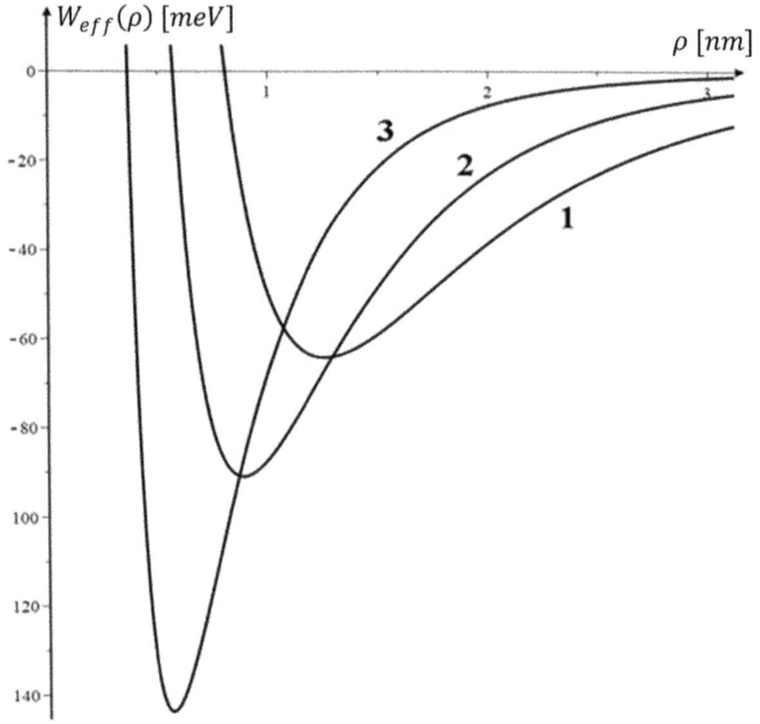

Figure 7.8. Plots of the dependence of the effective potential energy for the case of $SrTiO_3$/FeSe/vacuum. Curve 1 corresponds to $m^*=m_0$, where m_0 is the mass of the free electron, curve 2—$m^*=2m_0$, and curve 3—$m^*=5m_0$.

The plot of the dependence of $W_{\rm eff}(\rho)$ for the case of $SrTiO_3$ – FeSe – $SrTiO_3$ and various values of the effective mass of the electron is shown in figure 7.9.

In the work [44], a modification of the case (b) is discussed, when there is not one interface between $SrTiO_3$ and FeSe, but two interfaces ($SrTiO_3$ /FeSe and FeSe/$SrTiO_3$). This creates an indirect phonon attraction of electrons twice as strong as in the case of a single interface, which, in turn, will lead to an increase in T_c by about two times. Effective potential in this active case (case (b)) It can be obtained from the formula (7.5.4) by doubling the effective potential energy of the interaction of electrons.

Thus, for case (c) the effective potential is:

$$W_{\rm eff}^{2D}(\rho) = \frac{e^2}{\pi\varepsilon_0\rho}\left(\frac{2}{\varepsilon_1+1} - \frac{1}{\varepsilon_{10}+1}\right) - \frac{e^2}{\varepsilon_0 R_S}\left(\frac{1}{\varepsilon_1+1} - \frac{1}{\varepsilon_{10}+1}\right)\left\{I_0\left(\frac{\rho}{R_S}\right) - L_0\left(\frac{\rho}{R_S}\right)\right\}. \quad (7.5.5)$$

The plot of the dependence of $W_{\rm eff}(\rho)$ for case (b) and various values of the effective mass of the electron is shown in figure 7.10.

As can be seen from figures 7.8–7.10 for the limits of $\rho \ll R_S$ and $\rho \gg R_S$, the potential $W_{\rm eff}(\rho)$) of interaction between electrons has a repulsive character, but in the range of values of ρ in order R_S $W_{\rm eff}^{2D}(\rho)$ changes the behavior from repulsion to attraction.

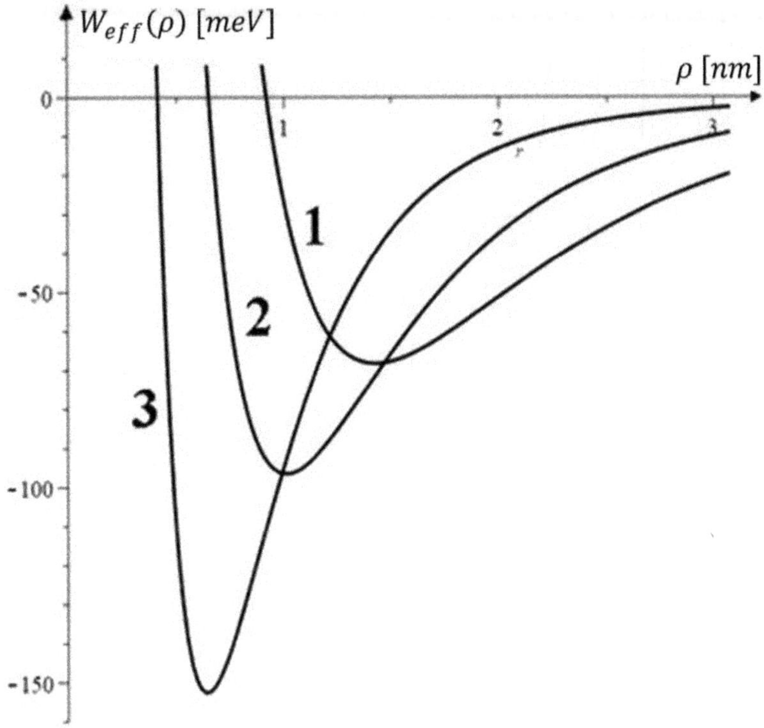

Figure 7.9. Plots of the dependence of the effective potential energy for the case of $SrTiO_3/FeSe/SrTiO_3$. Curve 1 corresponds to $m^*=m_0$, where m_0 is the mass of the free electron, curve 2—$m^*=2m_0$, and curve 3—$m^*=5m_0$.

An analysis of a similar expression for a massive crystal obtained based on the Haken method in [29]

$$W_{\text{eff}}^{3D}(r) = \frac{e^2}{\varepsilon_\infty r} - \frac{e^2}{r}\left(\frac{1}{\varepsilon_\infty} - \frac{1}{\varepsilon_0}\right)\left\{1 - e^{-\frac{r}{R_V}}\right\}, \tag{7.5.6}$$

shows that under no parameter values ε_∞, ε_0 the potential of W_{eff}^{3D}, as a function of r, does not change the nature of its behavior and is repulsive.

In the general case, when the polar crystals are different, and the layer $\langle 2 \rangle$ between them has a finite thickness, the calculation of the potential profile $W_{\text{eff}}^{2D}(\rho)$ should be performed using the formulas (7.2.30) and (7.2.31).

The tendency to weaken the repulsion of electrons in the FeSe monolayer and the appearance of attraction between them at distances of the order of their polaron radii ($\rho \sim R_S$) occurs due to interaction with the optical phonons of the $SrTiO_3$ substrate and is an important condition for the formation of bipolarons.

In the effective Hamiltonian (7.2.27), the term $W_p(\rho_{e_1}, \rho_{e_2}, d)$ (formulas (7.2.37) describes the potential energy of the interaction of electrons in a monolayer of FeSe with an additional polarization charge induced by the joint action of both electrons and located in the region of the center of mass of a two-electron system.

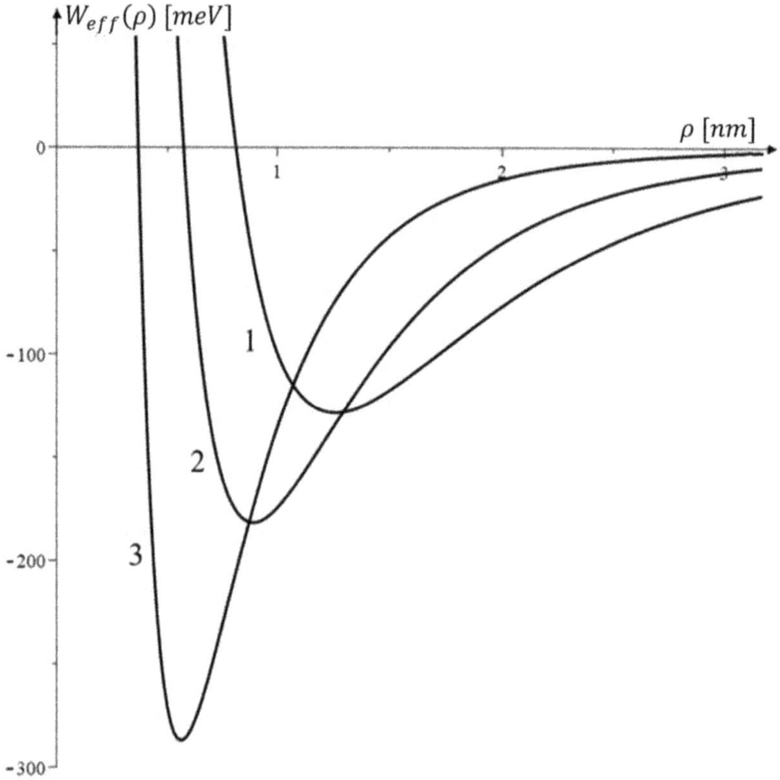

Figure 7.10. Plots of the dependence of the effective potential energy with two separate interfaces $SrTiO_3$/FeSe and FeSe/$SrTiO_3$. Curve 1 corresponds to $m^*=m_0$, where $m_0$the mass of the free electron, curve 2—$m^*=2m_0$, and curve 3—$m^*=5m_0$.

Let's estimate the density of the total surface charge for case (a), which can be calculated using the formula

$$\sigma(\rho) = 2\varepsilon_0 E_z|_{z=0} = -2\varepsilon_0 \frac{\partial V_p}{\partial z}\bigg|_{z=0} = \frac{e^2(\varepsilon_{10} - \varepsilon_1)}{2\pi(\varepsilon_1 + 1)(\varepsilon_{10} + 1)} \int_0^\infty d\eta \cdot \eta\, e_0^{-2\eta z}$$
$$\times \left\{ \frac{\lambda_1}{1 + \beta_1^2 R_S \eta^2} [J_0(\eta\,|\rho_1 - \rho|) + J_0(\eta\,|\rho_2 - \rho|)] + \frac{\lambda_2 J_0(\eta\rho)}{1 + \beta_2^2 \eta^2 \rho^2} \right\}, \tag{7.5.7a}$$

where V_p is the potential of the polarization field and z_0 is the distance to the plane in which the electrons are located (monolayer FeSe); λ_1, λ_2, β_1, β_2 are variational parameters.

Assuming $\lambda_1 = \lambda_2 = \beta_1 = \beta_2 = 1$, $z_0 = 0$ и by integrating into (7.5.7a), we obtain an explicit expression for $\sigma(\rho)$:

$$\sigma(\rho) = \frac{e^2(\varepsilon_{10} - \varepsilon_1)}{4\pi(\varepsilon_1 + 1)(\varepsilon_{10} + 1)R_S^2}$$
$$\times \left\{ 2\left[K_0\left(\frac{|\rho_1 - \rho|}{R_S}\right) - K_0\left(\frac{|\rho_2 - \rho|}{R_S}\right) \right] + \frac{\rho}{R_S} K_1\left(\frac{|\rho_1 - \rho|}{R_S}\right) \right\}; \tag{7.5.7b}$$

where $K_0(x)$, $K_1(x)$ are McDonald's functions.

Integrating the expression (7.5.7b) by giving the value of the effective electric charge e^* induced by electrons:

$$e^* = \frac{e(\varepsilon_{10} - \varepsilon_1)}{(\varepsilon_1 + 1)(\varepsilon_{10} + 1)}. \tag{7.5.7c}$$

Since $\varepsilon_{10} \gg \varepsilon_1 > 1$, then from the formula (7.5.6) for an effective charge, we obtain

$$e^* \approx e/(\varepsilon_1 + 1).$$

7.5.4 Binding energy of the bipolaron

The energy of the ground state of the bipolaron is found by the variational method.

Let's choose the test wave function of the electron pair in the form:

$$\psi(\rho) = 2(3\pi)^{-1/2}\gamma^2\rho e^{-\gamma\rho}, \tag{7.5.8}$$

where γ is the variational parameter.

The variational function of the ground state energy has the form:

$$E(\gamma, \beta, \lambda) = \langle\psi(\rho)\big|\hat{H}_{\text{eff}}(\rho, \rho_1, \rho_2, d)|\psi(\rho)\rangle. \tag{7.5.9}$$

Substituting (7.5.1), (7.5.3), (7.5.4), (7.5.8) in (7.5.9) and by integrating, we obtain an explicit expression for the variational functional of the energy of the bipolaronic state in the cases under consideration:

(a) For contact: vacuum/FeSe/$SrTiO_3$

$$\begin{aligned} E(\gamma) = {} & \frac{\hbar^2\gamma^2}{6m^*} + \frac{e^2\gamma}{3\pi\varepsilon_0}\left[\frac{1}{\varepsilon_1 + 1} - \left(\frac{1}{\varepsilon_1 + 1} - \frac{1}{\varepsilon_{10} + 1}\right)\left(4\gamma^2R_S^2 - 1\right)^{-7/2}\right. \\ & \times \left(384\gamma^6R_S^6 + 144\gamma^4R_S^4\right)\arctan\sqrt{4\gamma^2R_S^2 - 1} \\ & \left. - \left(64\gamma^6R_S^6 + 240\gamma^4R_S^4 - 8\gamma^2R_S^2 + 1\right)\sqrt{4\gamma^2R_S^2 - 1}\,\right]. \end{aligned} \tag{7.5.10a}$$

(b) For a symmetrical structure: $SrTiO_3$/FeSe/$SrTiO_3$

$$\begin{aligned} E(\gamma) = {} & \frac{\hbar^2\gamma^2}{6m^*} + \frac{e^2\gamma}{3\pi\varepsilon_0}\left[\frac{1}{\varepsilon_1} - \left(\frac{1}{\varepsilon_1} - \frac{1}{\varepsilon_{10}}\right)\left(4\gamma^2R_S^2 - 1\right)^{-7/2}\right. \\ & \times \left(384\gamma^6R_S^6 + 144\gamma^4R_S^4\right)\arctan\sqrt{4\gamma^2R_S^2 - 1} \\ & \left. - \left(64\gamma^6R_S^6 + 240\gamma^4R_S^4 - 8\gamma^2R_S^2 + 1\right)\sqrt{4\gamma^2R_S^2 - 1}\,\right]. \end{aligned} \tag{7.5.10b}$$

c) For the structure: $SrTiO_3/FeSe/SrTiO_3$ with two interfaces [45]

$$\begin{aligned} E(\gamma) = &\frac{\hbar^2\gamma^2}{6m^*} + \frac{2e^2\gamma}{3\pi\varepsilon_0}\left[\frac{1}{\varepsilon_1+1} - \left(\frac{1}{\varepsilon_1+1} - \frac{1}{\varepsilon_{10}+1}\right)\left(4\gamma^2R_S^2-1\right)^{-7/2}\right. \\ &\times\left(384\gamma^6R_S^6 + 144\gamma^4R_S^4\right)\arctan\sqrt{4\gamma^2R_S^2-1} \\ &\left. - \left(64\gamma^6R_S^6 + 240\gamma^4R_S^4 - 8\gamma^2R_S^2 + 1\right)\sqrt{4\gamma^2R_S^2-1}\right]. \end{aligned} \tag{7.5.10c}$$

The binding energy of a bipolaron is determined by the expression

$$W_b = E_b - 2E_p, \tag{7.5.11}$$

where E_b is the minimum value of the functional $E(\gamma, \beta, \lambda)$ (i.e., the energy of the bipolaronic system) and E_p is the energy of an individual polaron.

For $d \to 0$ and $\gamma^{-1} \to 0$ for E_p, we get the expression

$$E_p = -\frac{\pi}{2}\sum_{s=1,2}\alpha_s\hbar\Omega_s. \tag{7.5.12}$$

Based on the variational calculation of the binding energy W_b of the bipolaron in the FeSe layer on the $SrTiO_3$ substrate according to the formula (7.5.10a), the following was obtained:

$m_e^* = m_0$, $E_b = -10.8$ meV; $m_e^* = 2m_0$, $E_b = -17.6$ meV; $m_e^* = 5m_0$, $E_b = -31.3$ meV

In the three-layer $SrTiO_3/FeSe/SrTiO_3$ structure, the calculation was performed according to the formula (7.5.10b)

$m_e^* = m_0$, $E_b = -13.2$ meV; $m_e^* = 2m_0$, $E_b = -20, 6$ meV; $m_e^* = 5m_0$, $E_b = -35, 7$ meV.

(d) In a three-layer $SrTiO_3/FeSe/SrTiO_3$ structure with two interfaces [44, 46] (7.2.27), the calculation was performed according to the formula (7.5.10c):

$m_e^* = m_0$, $E_b = -27.3$ meV; $m_e^* = 2m_0$, $E_b = -41.2$ meV; $m_e^* = 5m_0$; $E_b = -68.8$ meV.

From the results given for the structures, monolayer FeSe film on the $SrTiO_3$ substrate and in the $SrTiO_3/FeSe/SrTiO_3$ structure, it follows that the binding energy of the bipolaron is in the range (150 – 900) K. The interaction of electrons in the FeSe conduction band with their induced polarizations and an additional positive polarization charge in the $SrTiO_3$ substrate located in the center of mass of the two-electron system (formula (7.5.7c)) plays an important role in the polarization potential well in the formation of bipolarons in a monolayer FeSe film.

The pairing of the formed polarons is facilitated by the weakening of repulsion between them, due to the interaction of electrons with an additional polarization charge, which turns into attraction, as is the case in the structures $FeSe/SrTiO_3$ and $SrTiO_3/FeSe/SrTiO_3$. As follows from expressions (7.2.30) and (7.2.31), the binding energy of the bipolaron increases with a decrease in the thickness of the FeSe film and reaches a maximum value in the limit $d \to 0$ (a monolayer FeSe film on an $SrTiO_3$ substrate), while the constant of the electron–phonon interaction α_S of the studied $FeSe/SrTiO_3$ structures $FeSe/SrTiO_3$; $SrTiO_3/FeSe/SrTiO_3$ lies in the region of intermediate and strong coupling: $\alpha \sim (1-5)$.

It should be noted that when considering the Coulomb interaction of electrons with the surface optical phonons of the substrate, the shielding effects, which lower the binding energy of the bipolaron, were not taken into account. At the same time, taking into account the exchange effects during the formation of bipolarons, as shown in [16, 18], increases their binding energy.

7.5.5 Assessment of the critical temperature in $FeSe/SrTiO_3$; $SrTiO_3/FeSe/SrTiO_3$

We will evaluate T_c based on the bipolaronic superconductivity mechanism in a fuse monolayer deposited on a polar semiconductor substrate $SrTiO_3$.

In [16], a bipolaronic model of superconductivity due to the superfluidity of a charged Bose gas was considered. Using the formula for the Bose condensation temperature under the bipolaronic superconductivity mechanism [47], we estimate T_c for case (a): a monolayer FeSe film on a polar substrate $SrTiO_3$:

$$T_c \approx 3,\, 31 \frac{\hbar^2 n_S}{m_{bp} k_0}, \tag{7.5.13}$$

where n_S is the concentration of bipolarons, $m_{bp} = 2m_p$ is the effective mass of the bipolaron, and m_p is the effective mass of the polaron.

For numerical estimates, we use the following parameter values: $m_p = (1-5)m_0$, where m_0 is the mass of the free electron and $n_S \approx (10^{13} - 10^{14})\ \text{cm}^{-2}$ [39].

As a result, for the critical temperature of the transition to the superconducting state, we obtain the interval $T_c \approx (100-300)$ K.

Considering that the binding energy of bipolarons lies in the range $E_{bp} \approx (150-900)$ K, it can be assumed that the bipolaron remains a stable quasi-particle during the formation of a superconducting Bose condensate.

Since the binding energy of the bipolaron in the case of a three-layer $SrTiO_3/FeSe/SrTiO_3$ structure (in case (b)) increases in comparison with case (a): $FeSe/SrTiO_3$ $FeSe/SrTiO_3$ by about 15%–20% and more than twice (in case (b)), the above T_c estimates will be qualitatively preserved.

The estimation of the maximum value of T_c in the model with the Einstein spectrum of optical phonons for the structures under consideration in the case of strong electron–phonon coupling according to [8] in the limit is given by the formula

$$k_0 T_c^{\max} \approx 0.13\, \hbar\Omega. \tag{7.5.14}$$

Assuming that $\hbar\Omega = \hbar\Omega_S$ is the energy of the surface optical phonon $SrTiO_3$, we obtain for the critical temperature

$$T_c^{\max} \approx 140 \text{ K},$$

which correlates with the T_c estimate made according to the formula (7.5.14).

Note that an increase in the thickness of the FeSe film leads to an exponential decrease in the binding energy of the bipolaron in comparison with the FeSe monolayer FeSe ($E_b \sim \exp(-d/R_S)$) and, accordingly, to an exponential decrease in T_c ($T_c \sim \exp(-d/R_S)$), where $R_S = (\hbar/(2m_0\omega_{SLO}))^{1/2}$ is the radius of the surface polaron. The binding energy and critical temperature T_c reach a maximum in the limit of the monolayer film FeSe ($d \to 0$).

In conclusion, it should be noted that in multilayer periodic structures, which are composite superlattices with polar semiconductor layers (such as BaO, $SrTiO_3$, etc) and metal layers (such as FeSe, CuO_2, etc), spatially extended surface phonons can play an important role in the formation of bipolarons and in the growth of T_c—new elementary excitations that have been predicted theoretically in [48] and discovered experimentally in [49–51].

7.6 Conclusion

Based on the conducted studies of bipolaronic states in inhomogeneous systems, it was found that significantly more favorable conditions for the formation of polaron pairs (bipolarons) can occur in multilayer composite structures. They can provide an independent choice of the parameters of the medium in which the electrons are located and the parameters describing the polarization properties of polar neighboring media that cause the appearance of polarons and their pairing into bipolarons.

1. The binding energy of bipolarons is in the range of $E_{bp} \approx (150 - 900)$ K (i.e., bipolarons in the studied Ginzburg sandwiches): FeSe/$SrTiO_3$ and $SrTiO_3$/FeSe/$SrTiO_3$ are stable quasi-particles and can exist in the structures under consideration at temperatures that can significantly exceed their Bose condensation temperature.
2. Calculations show that the binding energy of the bipolaron in the FeSe layer on the $SrTiO_3$ substrate strongly depends on the layer thickness ($\sim\exp(-d/R_S)$ and reaches a maximum in the limit of the monolayer film ($d \to 0$).
3. High T_c values in the studied structures are determined by the high binding energy of bipolarons due to the presence of a strongly polar semiconductor substrate $SrTiO_3$, which unequivocally confirms Ginzburg's prediction [9] about the important role of contact media in achieving high T_c.
4. In [40–42], the interaction of electrons in a monolayer FeSe film with optical phonons of the $SrTiO_3$ substrate was experimentally investigated. It has been established that the electric field associated with high-energy Fuchs-Kliewer surface phonons [52] in $SrTiO_3$penetrates the FeSe monolayer and leads to the formation of polarons in it. It was theoretically shown that the dynamic polaron-polaron interaction in the FeSe layer leads to an increase in superconductivity [53].
5. The presented theory makes it possible to simulate the system and determine the range of values of the material and geometric parameters of the layers forming multilayer structures in which T_c can be achieved in the range of room temperatures. These can be multilayer structures such as composite

superlattices, the layers of which, along with the FeSe layers, are also layers of SrO, TiO_2, BaO, and others proposed in [46].

The possibility of high-temperature bipolaronic superconductivity in structures has been established: a monolayer of FeSe deposited on a substrate of a strongly polaronic layer $SrTiO_3$ ('Ginzburg sandwiches'). The estimates of the critical temperature T_c in the studied structures and periodic systems based on them show the prospects of achieving for $T_c \sim 300$ K.

References

[1] Bogolyubov N N, Tolmachev V V and Shirkov D V 1958 *A New Approximation in the Theory of Superconductivity* (Moscow: Fizmatgiz)

[2] Bardeen J, Cooper L N and Schrieffer J R 1957 Microscopic theory of superconductivity *Phys. Rev.* **106** 162–4

Bardeen J, Cooper L N and Schrieffer J R 1957 Theory of superconductivity *Phys. Rev.* **108** 1109–75

[3] Ginzburg V L 1952 The current state of the theory of superfluidity. II. Microscopic theory *UFN* **48** 25–118

[4] Ginzburg V L 1968 The problem of high-temperature superconductivity *UFN* **95** 91–110

[5] Ginzburg V L 1970 The problem of high-temperature superconductivity. II *UFN* **101** 185–215

[6] Little W A 1964 Possibility of synthesizing an organic superconductor *Phys. Rev.* **134** A1416–25

Little W A 1965 Superconductivity at room temperature *Sci. Am.* **212** 21–7

[7] Mitsen K V and Ivanenko O M 2017 Phase diagrams of cuprates and pnictides as a key to understanding the mechanism of high-temperature superconductivity *UFN* **187** 431–41

[8] Sadovsky M V 2016 High-temperature superconductivity in FeSe monolayers *Successes Phys. Sci.* **186** 1035–57

[9] Ginzburg V L and Kirzhnits D A 1967 On the issue of high-temperature and surface superconductivity *Dokl. Acad. Sci. USSR* **176** 553–5

[10] Allender D, Bray Y and Bardeen J 1973 Model for an exciton mechanism of superconductivity *Phys. Rev.* B **7** 1020–8

[11] Phillips J C 1972 Superconductivity mechanisms and covalent instabilities *Phys. Rev. Lett.* **29** 1551–5

[12] Pekar S I 1951 *Research on the Electronic Theory of Crystals* (Moscow: Gostekhizdat)

[13] Beril S I, Starchuk A S and Fedortsov A N 2009 Bipolaron states in the quantum wires in the polar environment *J. Nanoelectron. Optoelectron.* **4** 152–8

[14] Cardenas L A, Fagot-Revurat Y, Moreau N, Kierren B and Malterre D 2009 Novel Bipolaronic ground state in K/Si(111):B *J. Surf. Sci. Nanotechnol.* **7** 259–63

[15] Pokatilov E P, Beril S I, Fomin V M and Ryabukhin G J 1992 Polaron pairing in multi-layer structures. Part 1. Bipolaron states in multi-layer structures with quantum wells *Phys. Status Solidi* B **169** 429–41

Pokatilov E P, Beril S I, Fomin V M, Riabukhin G Y and Gorjachkovskii E R 1992 Polaron pairing in multi-layer structures. Part 2. Interlayer bipolaron states in structures with quantum wells *Phys. Status Solidi* B **171** 437–45

[16] Vinetsky V L and Pashitsky E A 1974 The superfluidity of a charged bose gas and the bipolaronic mechanism of superconductivity *Ukr. Phys. J.* **20** 338

[17] Adamowski J 1989 Formation of Fröhlich bipolarons *Phys. Rev.* B **39** 3649–53

[18] Vinetsky V L and Pashitsky E A 1961 On bipolaronic states of current carriers in ion crystals *JETF* **40** 1459–68

[19] Aleksandrov A S and Ranniger J 1981 Bipolaronic superconductivity *Phys. Rev.* B **24** 1164–70

[20] Beril S I, Pokatilov E P and Ryabukhin G Y 1990 Bipolarons in structures with quantum wells *Dept. in Mold. NIINTI* No. 1196

[21] Pokatilov E P, Fomin V M and Beryl S I 1990 *Vibrational Excitations, Polarons and Excitons in Multilayer Systems and in Superlattices* (Chisinau: Stiinza) p 278

[22] Shik A Y 1992 Semiconductor structures with δ-layers *Phys. Technol. Semicond.* **26** 1161–81

[23] Beril S I and Starchuk A S 2020 Bipolaronic states in spatially separated monolayers (-layers) in multilayer structures with quantum wells *Bull. Pridnestrovian Univ.* **66** 3–11

[24] Fomin V M and Pokatilov E P 1985 Phonons and the electron–phonon interaction in multilayer systems *Phys. Status Solidi B.* **132** 69–82

[25] Fomin V M and Pokatilov E P 1985 Phonons and the electron–phonon interaction in multilayer systems *Phys. Status Solidi* B **132** 69–82

[26] Beril S I, Pokatilov E P, Fomin V M and Pogorilko G A 1985 Vanier–Mott excitons in multilayer systems *FTP* **19** 412–7

[27] Adamowski J 1985 Energy spectrum of the bound polaron *Phys. Rev.* B **32** 2588–95

[28] Bednarek S, Adamowski J and Suffczynski M 1977 Effective Hamiltonian for few-particle systems in polar semiconductors *Solid State Commun.* **21** 1–3

[29] Haken H 1956 Zur Quantumtheorie des Mehrelectronen systems im swingenden gitter *Z. Phys.* **146** 527–54

[30] Yanson I K, Bobrov N L, Rybalchenko L F, Fisun V V, Mironov O A, Chistyakov S V, Sipatov A Y and Fedorenko A I 1989 Microcontact measurements of the energy gap of superconducting superlattices based on lead chalcogenides *Lett. JETF* **49** 293–6

[31] Mironov O A, Savitsky B A and Sipatov A Y 1988 Superconductivity based on lead chalcogenides *Lett. JETF* **48** 100–2

[32] Mironov O A, Chistyakov S V, Skrylev Y I *et al* 1989 Localization of the order parameter on the dislocation grid of the mismatch of superconducting PbTe–PbS superlattices *Lett. JETF* **50** 300–3

[33] Klimin S N, Pokatilov E P and Fomin V M 1994 Bulk and interface polarons in quantum wires and dots *Phys. Status Solidi* B **184** 373–83

[34] Ginzburg V L 1964 On the question of surface superconductivity *JETF* **47** 2318

[35] Bednortz J G and Müller K A 1986 Possible high T_c superconductivity in the Ba–La–Cu–O system *Z. Phys. B: Condens. Matter* **64** 189–93

[36] Wu M K, Ashburn J R, Torng C J, Hor P H, Meng R L, Gao L, Huang Z J, Wang Y Q and Chu C W 1987 Superconductivity at 93 K in a new mixed-phase Y–Ba–Cu–O compound system at ambient pressure *Phys. Rev. Lett.* **58** 908–10

[37] Kamihara Y, Watanabe T, Hirano M and Hosono H 2008 Iron-based layered superconductor $La[O_{1-x}F_x]FeAS$ ($x = 0.05-0.12$) with $T_c = 26$ K *J. Am. Chem. Soc* **130** 3296–7

[38] Takahashi H, Igawa K, Arii K, Kamihara Y, Hirano M and Hosono H 2008 Superconductivity at 43 K in an iron-based layered compound $La_{1-x}Fe_xFeAS$ *Nature* **453** 376–378

[39] Gor'kov Lev P 2016 Peculiarities of superconductivity in the single-layer $FeSe/SrTiO_3$ interface *Phys. Rev.* B **93** 060507

[40] Zhang S *et al* 2016 Role of $SrTiO_3$ phonon penetrating into thin FeSe films in the enhancement of superconductivity *Phys. Rev.* B **94** 081116

[41] Lee J J *et al* 2014 Interfacial mode coupling as the origin of the enhancement of T_c in FeSe films on $SrTiO_3$ *Nature* **515** 245–8

[42] Zhang S *et al* 2019 Enhanced superconducting state in FeSe/$SrTiO_3$ by a dynamic interfacial polaron mechanism *Phys. Rev. Lett.* **122** 066802

[43] Beril S I, Fomin V M and Starchuk A S 2020 *Theory of Polarons, Excitons, Bipolarons and Kinetic Effects in Multilayer Structures of Various Geometries and Superlattices* (Tiraspol: Pridnestrovian University Publishing House) p 696

[44] Beril S and Starchuk A 2023 On the bipolaronic mechanism of high-temperature superconductivity in 'Ginzburg Sandwiches' FeSe–$SrTiO_3$; $SrTiO_3$–FeSe–$SrTio_3$ *Am. J. Phys. Appl.* **11** 8–20

[45] Drozdov A P, Kong P P, Minkov V S *et al* 2019 Superconductivity at 250 K in lanthanum hydride under high pressures *Nature* **569** 528–31

[46] Lee D H 2015 What makes the T_c of FeSe/$SrTiO_3$ so high? arXiv:1508.02461v1 [cond-mat. str-al]

[47] Schafroth M R 1955 Superconductivity of a charged in ideal Bose gas *Phys. Rev.* B **100.** 463–75

[48] Pokatilov E P and Beril S I 1982 Spatially extended optical modes in two-layer periodical structures *Phys. Status Solidi* B **110** k75–8
Pokatilov E P and Beril S I 1983 Electron–phonon interaction in periodic two-layer periodical structures *Phys. Status Solidi* B **118** 567–73

[49] Sood A K, Menendez J, Cardona M and Ploog K 1985 Interface vibrational modes in GaAs–AlAs superlattices *Phys. Rev. Lett.* **54** 2115–18

[50] Klein M V 1986 Phonons in semiconductor superlattices *IEEE J. Quantum Electron.* **QE-22** 1760–70

[51] Schwartz G P, Gualtieri G J, Sunder W A and Farrow L A 1987 Light scattering from confine and interface optical vibrational modes in strained-layer GaSb/AlSb superlattices *Phys. Rev.* B **36** 4868–77

[52] Fuchs R and Kliewer K L 1965 Optical modes of vibration in an ionic crystal slab *Phys. Rev.* **140** A2076–88

[53] Emin D and Hillery M S 1989 Formation of a large singlet bipolaron: application to high-temperature bipolaronic superconductivity *Phys. Rev.* B **39** 6575–93

IOP Publishing

Vibrational Excitations in Multilayer Nanostructures
Properties and manifestations
Stepan I Beril, Vladimir M Fomin and Alexander S Starchuk

Chapter 8

Kinetic effects in multilayer structures and in superlattices

8.1 Introduction

One of the tasks facing the theory is to describe experimental conditions that allow the identification of theoretically studied objects with experimentally observed ones. In the previous chapters, a theoretical description of the results of optical (spectral) studies of electronic, electron–hole, and electron–hole–phonon states at the contact of two crystals, in thin films on substrates and in multilayer structures, was presented. Optical, of course, has high accuracy and informativeness. However, it seems useful and relevant to supplement the description of the properties of electronic and electron–hole states with kinetic ones, the study of which can provide additional information. The study of various kinetic effects in thin films, multilayer structures, and superlattices has been carried out since the early sixties of the last century (see, for example, works [1]). In these and other works, it was found that specific features appear in thin semiconductor films that are absent in massive crystals, among which, first of all, the dependence of kinetic characteristics on the thickness of the crystal due to the quantization of the electronic spectrum (quantum dimensional effect) should be attributed.

In this chapter, we will focus on those kinetic effects that have not received a proper theoretical description due to the use of rough or generally incorrect approximations, in particular, the scattering of charge carriers on polar optical phonons in thin semiconductor films, inversion channels of MDS structures, and in structures with quantum wells.

8.2 Scattering of light by polar optical phonons in structures with quantum wells

Scattering of free charge carriers by polar optical phonons in structures with quantum wells has been carried out in many theoretical works [2–51]. In [2, 4], studies of the specific properties of scattering processes associated with the dimensional quantization of the energy spectrum of charge carriers in quantum wells and leading to the appearance of interference phenomena at the boundaries were carried out. At the same time, the phonon spectrum of quantum wells was assumed to be the same as in a massive crystal. In [3, 5], attempts were made to take into account the rearrangement of the spectrum of optical phonons in quantum wells. In [5], the results for the electron scattering coefficient by optical phonons were compared within the framework of several macroscopic models of electron–phonon interaction, differing in different boundary conditions, with the results corresponding to the microscopic model. It is shown that the model with zero boundary conditions for the normal components of the electric field (the nodes of the electrostatic potential created by polarization optical oscillations are located on the surface of the quantum well) has the best agreement with the microscopic model.

Diffusion in a quantum well with a step potential was considered in [6, 7], but the description of the electron–phonon interaction was given in a simplified version, without taking into account scattering on surface phonons.

A review of the results of calculations taking into account the polar scattering of electrons in two-dimensional quantum wells was published in [8, 9].

The approximation developed in [5] is based on the Hamiltonian, in which the surface and volume parts were taken from different models: continuum [6] and microscopic [10, 11], whereas in order to consistently account for them in the Hamiltonian of the electron–phonon interaction, the surface and volume parts must be obtained within the framework of one model as a result of a unitary transformation normalizing phonon modes.

This technique of obtaining normal oscillations was used in the works [12–16]; both bulk and surface vibrations were studied in the works mentioned in [9, 16, 41, 69].

The method developed in [5] is based on a model using the Hamiltonian from [10, 17]. It leads to significant discrepancies in the dependence of the surface contribution on the thickness of the well d compared with the corresponding results of the present study based on the electron–phonon interaction Hamiltonian from [12, 13]. In addition, the use of model electron–phonon interaction Hamiltonians from the works [10, 17] leads to the loss of the ability to unambiguously distinguish the influence of correct boundary conditions on the behavior of contributions depending on d for confinement phonons. Various manifestations of electron scattering effects on optical phonons for bulk crystals and superlattices are described in [18–20].

Next, it is necessary to compare with the results obtained on the basis of the approach with the correct contribution from vibrational modes.

Note that when solving the problem of scattering free charge carriers in the polar layer, the authors of [5] limited themselves to considering transitions that consider only the two lowest subbands (transitions of type $1 \rightarrow 1$, $2 \rightarrow 2$). This does not allow us to obtain the correct limit corresponding to the work [2] for the scattering coefficient at energies significantly exceeding the emission threshold.

The purpose of this section is to obtain the scattering coefficient and the relaxation rate of the momentum for scattering on polar optical phonons in quantum wells, based on the exact Hamiltonian of the electron–phonon interaction, taking into account bulk [12, 21] and surface [13, 21] optical phonons.

8.2.1 Hamiltonian and wave functions

Consider a layer of a polar crystal occupying a region of space $0 \leqslant z \leqslant d$ with dielectric permittivity ε_{20}, ε_2 and bordering on semi-infinite nonpolar media occupying a region of space $z < 0$ and $z > 0$ with dielectric permittivity ε_1, ε_3. Potential barriers at borders $z = 0$ and $z = d$ are assumed to be infinite, so the wave function of an electron in the conduction band has the form

$$\Psi(\boldsymbol{\rho}, z) = \left(\frac{2}{V}\right)^{\frac{1}{2}} U_k(\boldsymbol{\rho}, z) \exp(i\mathbf{k}_\perp \boldsymbol{\rho}) \sin(k_z z), \tag{8.2.1}$$

where $V = L_x L_y d$ is polar layer volume; $\mathbf{k}_\perp$, $k_z = (n\pi)/d$ are components of the wave vector in the plane XOY and along the axis Z accordingly; $U_k(\boldsymbol{\rho}, z)$ is the Bloch factor; and d is layer thickness.

The energy spectrum in such a system is described by the following expression:

$$E_{n,\mathbf{K}} = E_{k_\perp} + E_n = \hbar^2 k_\perp^2 + n^2 E_0, \tag{8.2.2}$$

where $E_0 = (\pi^2\hbar^2)/2m_\parallel^*$ is the energy of the lowest state of dimensional quantization; and $m_\perp^*$, $m_\parallel^*$ is effective masses in the directions perpendicular to and along the z-axis, respectively. The density of states with a fixed spin is

$$N(E_{n,\mathbf{K}}) = \frac{n \cdot m_\perp^*}{2\pi\hbar^2 d}. \tag{8.2.3}$$

An electron from the conduction band interacts with two surface (S_1, S_2) and one bulk confinement (V) by longitudinal optical modes.

The Hamiltonian of the electron–phonon interaction has the form

$$\hat{H}_{\text{e-h}} = \hat{H}^{V}_{\text{e-ph}} + \hat{H}^{S_1, S_2}_{\text{e-ph}}; \tag{8.2.4}$$

where

$$\hat{H}^{V}_{\text{e-ph}} = \sum\nolimits^{\eta, q_z}_{(q_z \neq 0)} C_{\mathbf{Q}} e^{i\eta\rho} \left[e^{iq_z z} + \frac{1}{1-C}\left(C - \frac{\cosh \eta\left(z - \frac{d}{2}\right)}{\cosh \eta \frac{d}{2}} \right) \right] \left(\hat{b}^{+}_{-\mathbf{Q}} + \hat{b}_{\mathbf{Q}} \right) \tag{8.2.5}$$

is the Hamiltonian of interaction with optical confinement phonons obtained in [12],

$$\hat{H}^{S_1}_{\text{e-ph}} = \sum\nolimits_{\eta} C^{(1)}_{\eta} \exp(i\eta\rho) \frac{\cosh\left[\eta\left(z - \frac{d}{2}\right)\right]}{\cosh \frac{\eta d}{2}} \left(\hat{b}^{+}_{-\eta, 1} + \hat{b}_{\eta, 1} \right), \tag{8.2.6}$$

$$\hat{H}^{S_2}_{\text{e-ph}} = \sum\nolimits_{\eta} C^{(2)}_{\eta} \exp(i\eta\rho) \frac{\sinh\left[\eta\left(z - \frac{d}{2}\right)\right]}{\cosh \frac{\eta d}{2}} \left(\hat{b}^{+}_{-\eta, 2} + \hat{b}_{\eta, 2} \right), \tag{8.2.7}$$

$\hat{H}^{S_1, S_2}_{\mathrm{e-ph}}$ are Hamiltonians of interaction with surface optical phonons in a layer of a polar crystal bordering nonpolar media [13, 21];

$$|C_Q|^2 = \frac{1}{L_x L_y d} \frac{4\pi\alpha_v (\hbar\omega_0)^2}{Q^2 \beta_V}; \; |C_\eta^{(i)}|^2$$

$$= \frac{1}{L_x L_y} \frac{2\pi\alpha_i (\hbar\Omega_i)^2}{\eta\beta_i \tanh \frac{\eta d}{2}}; \; \alpha_V = \frac{e^2}{4\pi\varepsilon_0\hbar}\left(\frac{1}{\varepsilon_2} - \frac{1}{\varepsilon_{20}}\right)\left(\frac{m^*}{2\hbar\omega_0}\right)^{\frac{1}{2}};$$

$$\alpha_{S_1} = \frac{e^2}{4\pi\varepsilon_0\hbar}\left(\frac{1}{\varepsilon_2 + \varepsilon \coth \frac{\eta d}{2}} - \frac{1}{\varepsilon_{20} + \varepsilon \coth \frac{\eta d}{2}}\right)\left(\frac{m^*}{2\hbar\Omega_1}\right)^{\frac{1}{2}}; \; Q^2 = \eta^2 + q_z^2;$$

$$C = \left[\frac{2}{\eta d} \tanh \frac{\eta d}{2}\right]^{\frac{1}{2}}; \; \beta_V \equiv R_V^{-1} = \left(\frac{2m^*\omega_0}{\hbar}\right); \; \beta_i = \left(\frac{2m^*\Omega_i}{\hbar}\right)^{\frac{1}{2}};$$

and $\hat{b}^+_{-\mathbf{Q}}, \hat{b}_{\mathbf{Q}}, \hat{b}^+_{-\eta_{1,2}}, \hat{b}_{-\eta_{1,2}}$ are phonon operators.

Bulk ω_0 and interface Ω_{S_i} phonon frequencies are defined by the following expressions:

$$\omega_0^2 = \left(\frac{\varepsilon_{20}}{\varepsilon_2}\right)\omega_{to}^2, \quad \Omega_1^2 = \omega_{to}^2\left(\frac{\varepsilon_{20} + \varepsilon \coth \frac{\eta d}{2}}{\varepsilon_2 + \varepsilon \coth \frac{\eta d}{2}}\right),$$
$$\Omega_2^2 = \omega_{to}^2\left(\frac{\varepsilon_{20} + \varepsilon \tanh \frac{\eta d}{2}}{\varepsilon_2 + \varepsilon \tanh \frac{\eta d}{2}}\right) \tag{8.2.8}$$

(where, for simplicity of mathematical calculations, we put $\varepsilon_1 = \varepsilon_3 = \varepsilon$).

8.2.2 Scattering coefficient and momentum relaxation rate

The scattering coefficient of an electron with a wave vector on optical phonons in the first order of perturbation theory is determined by the formula:

$$W_{\mathbf{k}} = \frac{2\pi}{\hbar} \int |\langle \mathbf{k}'|\hat{H}_{\mathrm{e-ph}}|\mathbf{k}\rangle|^2 \delta(E_{\mathbf{k}'} - E_{\mathbf{k}} \mp \hbar\omega_l)\, dN_{\mathbf{k}'}, \tag{8.2.9}$$

where $\mathbf{k}'$ is the wave vector of the scattered electron (according to which the integration is carried out); and $\hbar\omega_l$ is the optical phonon energy ($\omega_l = \Omega_{S_{1,2}}$ with $l = S_{1,2}$ and $\omega_l = \omega_0$ with $l = V$). The upper sign of the expression (8.2.9) corresponds to the emission, the lower sign corresponds to the absorption of a phonon.

The matrix element in (8.2.9) can be represented in the following form:

$$|\langle \mathbf{k}'|\hat{H}_{\mathrm{e-ph}}|\mathbf{k}\rangle|^2 = |\langle \mathbf{k}'|\hat{H}_{\mathrm{e-ph}}|\mathbf{k}\rangle|^2 + \left|\langle \mathbf{k}'|\hat{H}^{S_1, S_2}_{\mathrm{e-ph}}|\mathbf{k}\rangle\right|^2, \tag{8.2.10}$$

where

$$\left| \langle \mathbf{k}' | \hat{H}^{V}_{\text{e-ph}} | \mathbf{k} \rangle \right|^2 \frac{e^2 \hbar \omega_0}{2\varepsilon_V^* L_x L_y Q^2 d} \delta_{\mathbf{k}'_\perp, \mathbf{k}'_\perp \pm \eta} |G_V(q_z)|^2 \left\{ n(\omega_0) + \frac{1}{2} \mp \frac{1}{2} \right\}; \quad (8.2.11a)$$

$$\left| \langle \mathbf{k}' | \hat{H}^{S_1, S_2}_{\text{e-ph}} | \mathbf{k} \rangle \right|^2 = \frac{e^2 \hbar \Omega_{1, 2}}{2\varepsilon_{S_{1, 2}}^* L_x L_y \eta^2} \delta_{\mathbf{k}'_\perp, \mathbf{k}_\perp \pm \eta} |G_{S_{1, 2}}|^2 \left\{ n(\Omega_{1, 2}) + \frac{1}{2} \mp \frac{1}{2} \right\}; (8.2.11b)$$

L_x, L_y are the dimensions of the layer (in the XOY plane).

In (8.2.11a, b) the following designations have been introduced:

$$\frac{1}{\varepsilon_V^*} = \frac{1}{\varepsilon_2} - \frac{1}{\varepsilon_{20}}; \quad (8.2.12a)$$

$$\frac{1}{\varepsilon_{S_1}^*} = \frac{1}{\varepsilon_2 + \varepsilon \coth \frac{\eta d}{2}} - \frac{1}{\varepsilon_{20} + \varepsilon \coth \frac{\eta d}{2}};$$
$$\frac{1}{\varepsilon_{S_2}^*} = \frac{1}{\varepsilon_2 + \varepsilon \tanh \frac{\eta d}{2}} - \frac{1}{\varepsilon_{20} + \varepsilon \tanh \frac{\eta d}{2}}, \quad (8.2.12b)$$

where $n(\omega_l)$ is the Bose–Einstein distribution function, and the values of $G_V(q_z)$, $G_{S_{1,2}}$ are electronic parts of the corresponding matrix elements:

$$G_V(q_z) = G_R(q_z) + \frac{1}{1 - C}\left\{ C\left[\frac{\sin[(k'_z - k_z)d]}{(k'_z - k_z)d} - \frac{\sin[(k'_z + k_z)]}{(k'_z + k_z)d} \right] - G_1(\eta) \right\};$$

$$G_R(q_z) = \frac{1}{2}(G_{--} + G_{++} + G_{+-} + G_{-+}).$$

$$G_{\pm\pm} = \frac{\sin\left(\left[q_z \pm (k'_z \pm k_z)\right]\frac{d}{2}\right)}{\left[q_z \pm (k'_z \pm k_z)\right]\frac{d}{2}} \exp\left\{ i\left[q_z \pm (k'_z \pm k_z)\right]\frac{d}{2} \right\};$$

$$[G_1(\eta)]_{nm} = \frac{1}{\cosh \frac{\eta d}{2}} \left\langle \Psi_m(z) \middle| \cosh\left[\eta\left(z - \frac{d}{2}\right)\right] \middle| \Psi_n(z) \right\rangle =$$

$$= [1 + (-1)^{n+m}] \tan h \frac{\eta d}{2}\left[\frac{\eta d}{\eta^2 d^2 + (m-n)^2 \pi^2} - \frac{\eta d}{\eta^2 d^2 + (m+n)^2 \pi^2} \right];$$

$$G_{s_1}(\eta) = G_1(\eta); \; G_{s_2}(\eta) = \frac{1}{\cosh \frac{\eta d}{2}} \left\langle \Psi_m(z) \middle| sh\left[\eta\left(z - \frac{d}{2}\right)\right] \middle| \Psi_n(z) \right\rangle =$$

$$= -[1 - (-1)^{n+m}]\left\{ \frac{\eta d}{\eta^2 d^2 + (m-n)^2 \pi^2} - \frac{\eta d}{\eta^2 d^2 + (m+n)^2 \pi^2} \right\}.$$

Substituting (8.2.11) into (8.2.9), we obtain the scattering velocity between the n (initial) and m (final) subbands on bulk [7, 8, 12] and surface optical phonons, respectively:

$$W^{V}_{nm}(\mathbf{k}) = \left(\frac{e^2\omega_0 d}{8\pi\varepsilon_V^*}\right)\left\{n(\omega_0) + \frac{1}{2} \mp \frac{1}{2}\right\} \int_0^{\infty} \eta d\eta \int_0^{2\pi} d\theta\left\{F_{nm}(\eta) + \frac{1}{\eta d}\right.$$
$$\left(\frac{\delta_{m.n}C - G_1}{1 - C}\right)^2 + \frac{2}{1 - C}(\delta_{m,n}C - G_1)\left[\frac{1}{\eta^2 d^2 + (m - n)^2\pi^2}\right.$$
$$\left(1 - (-1)^{m-n}\exp(-\eta d)\right) - \tag{8.2.13}$$
$$\left.\left.\frac{1}{\eta^2 d^2 + (m + n)^2\pi^2}\left(1 - (-1)^{m+n}e^{-\eta d}\right)\right]\right\}\delta_{\mathbf{k}\prime,\mathbf{k}_\perp \pm \boldsymbol{\eta}}$$
$$\delta\left(E_{\mathbf{k}\prime} - E_{\mathbf{k}} \mp \hbar\omega_0\right);$$

$$W^{S_1, S_2}_{nm} = \sum\nolimits_{i=1,2} \int_0^{\infty} \eta d\eta \int_0^{2\pi} d\theta \left(\frac{e^2\Omega_i d}{8\pi\varepsilon^*_{S_i}}\right)$$
$$\times\left[n(\Omega)_i + \frac{1}{2} \mp \frac{1}{2}\right]\left(\frac{|G_i(\eta)|}{\eta d \tanh\frac{\eta d}{2}}\right) \quad \delta_{\mathbf{k}', \mathbf{k}_\perp \pm \boldsymbol{\eta}}\ \delta\ (E_{\mathbf{k}'} - E_{\mathbf{k}} \mp \hbar\Omega_i). \tag{8.2.14}$$

Here θ is the angle between $\boldsymbol{\eta}$, $\mathbf{k}_\perp$,

$$F_{nm}(\eta) = \left\{\frac{1 + \delta_{m,\,n}}{\eta^2 d^2 + (m - n)^2\pi^2} + \frac{1}{\eta^2 d^2 + (m + n)^2\pi^2}\right\}(1 - \xi);$$

$$\xi = \eta d(\ 1 \pm e^{-\eta d})32\ \pi^4 m^2 n^2\{\ \ [(m - n)^2\pi^2 + \eta^2 d^2]\cdot[\eta^2 d^2$$
$$+(m + n)^2\pi^2]\cdot[(1 + \delta_{m,\,n})(\eta^2 d^2 + (m + n)^2\pi^2) + (\eta^2 d^2 + (m - n)^2\pi^2)]\}^{-1};$$

$$\delta_{m,\,n} = \begin{cases}1, & m = n;\\ 0, & m \neq n.\end{cases}$$

The total scattering coefficient can be represented as

$$W^t_n(\mathbf{k}_\perp) = W^V_n(\mathbf{k}_\perp)W^{S_1}_n(\mathbf{k}_\perp) + W^{S_2}_n(\mathbf{k}_\perp), \tag{8.2.15a}$$

where

$$W^V_n\left(\mathbf{k}'_\perp\right) = \sum_{m=1}^{m\max} W^V_{nm}\left(\mathbf{k}'_\perp\right); \tag{8.2.15b}$$

$$W^{S_1, S_2}_n(\mathbf{k}_\perp) = \sum_{m=1}^{m\max} W^V_{nm}(\mathbf{k}_\perp); \tag{8.2.15c}$$

$m_{\max}$ is determined by the law of conservation of energy in accordance with the formula:

$$(m_{\max})^2 E_0 = E_n(\mathbf{k}_\perp) \pm \hbar\omega_l. \tag{8.2.16}$$

Here the upper sign corresponds to the emission, the lower sign corresponds to the absorption of phonons; $\hbar\omega_l$ is the energy of the phonon of the corresponding type.

The momentum relaxation rate plays an important role in charge transfer processes, so its calculation is of particular interest in itself. By definition, it has the following form:

$$\frac{1}{\tau(\mathbf{k})} = \frac{2\pi}{\hbar} \int \frac{|\mathbf{k}'_\perp - \mathbf{k}_\perp|}{|\mathbf{k}_\perp|} \left|\left\langle \mathbf{k}' \middle| \hat{H}_{e-ph} |\mathbf{k}\rangle\right|^2 \delta(E_{\mathbf{k}'} - E_{\mathbf{k}} \mp \hbar\omega_l) dN_{\mathbf{k}'}. \tag{8.2.17}$$

Note that the total coefficient of the momentum relaxation rate consists of bulk and surface parts. Values $[\tau^V]^{-1}_{nm}$, $[\tau^{S_1, S_2}]^{-1}_{nm}$ are expressed by the formulas (8.2.13), (8.2.14), when substituting into the integrals of the term $\pm(\eta/k)\cos\theta$.

In a member $\pm(\eta/k)\cos\theta$ the upper symbol is taken in case of absorption, and the lower one is taken in case of phonon emission.

The following parameter values were used in numerical calculations:

1. system $GaAs/Al_xGa_{1-x}As$: $x = 0.30$, m^*=0.0667 m_0, $\hbar\omega_0 = 35.2$ meV, ε_2=10.9, ε_{20}=12.5, $\varepsilon = 10.1$;
2. system CdTe/nonpolar dielectric: $m^* = 0.092$ m_0, $\hbar\omega_0$=21.7 meV, ε_2=7.13, ε_{20}=10.6, ε =2.0.

In addition, values in the range from 1 (vacuum) to 80 (limit of strong shielding induced by neighboring media) were used as values of the ε of neighboring media.

The curves in all the figures reflect the values of the scattering and relaxation rates of the momentum in units

$$W_0 = \frac{e^2}{4\pi\varepsilon_V^*\hbar}\left(\frac{2m^*\omega_0}{\hbar}\right)^{\frac{1}{2}}. \tag{8.2.18}$$

Figure 8.1 shows the dependences of the scattering coefficients W^V, $W^S{}_1$, describing scattering on confinement phonons and the even mode of surface phonons, from the electron energy in the band. In addition, the full value of the scattering coefficient is $W^{T;}$ the results of [2, 4] for scattering coefficients on polar optical phonons using a Hamiltonian with renormalized phonons are also presented. Figure 8.1(b) shows the inter-band (W^V_{inter}) and intra-band (W^V_{intra}) contributions to the scattering coefficient on bulk optical phonons.

As can be seen from figure 8.1, the behavior W^V, that is, the values of the scattering coefficient on bulk optical phonons obtained using the Hamiltonian (8.2.5), generally corresponds to the behavior of the scattering coefficient W^R, obtained on a Hamiltonian with renormalized phonons. There is a leap W^T when $E_k/\hbar\omega_0 = 1$ due to the possibility of electron scattering with phonon radiation for intra-band scattering. A number of smaller jumps are associated with smaller contributions from inter-band scattering. Corresponding jumps W^V and W^S arise when the law of conservation of energy-momentum opens scattering channels associated with subbands of dimensional

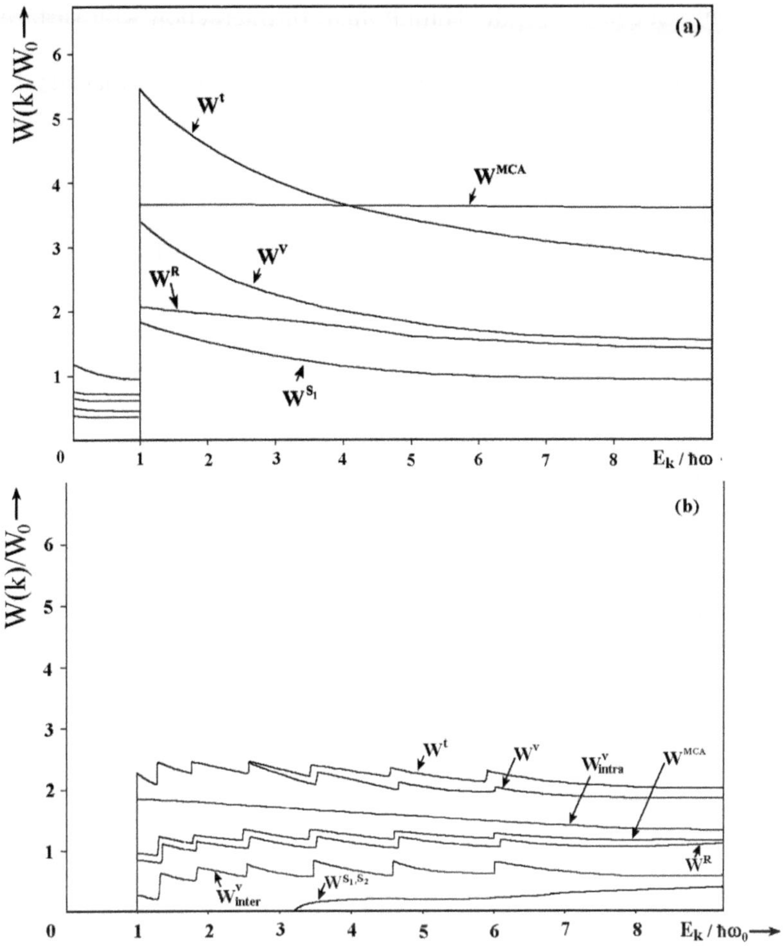

Figure 8.1. Scattering coefficient $W(k)/W_0$ depending on the electron energy in the structure GaA/vacuum: The following values are shown: W^V is the contribution from volume phonons; W^{S_1} from the even mode of surface phonons; W^{T} is total contribution: $W^T = W^V + W^{S_1} + W^{S_2}$; and W^R and W^{MCA} correspond to the calculations in [2] and according to the method of approximation of conservation of momentum, respectively. Figure 8.1(a, b) correspond to thicknesses of 2.0 and 39.0 nm.

quantization corresponding to $m > 1$. The behavior W^V detects an increase in the value of the inter-band scattering with increasing $E_k/\hbar\omega_0$ and an increase in the number of scattering channels; at the same time, with sufficiently large $E_k/\hbar\omega_0$ the value W^V, as well as W^{S_1}, decreases due to the general weakening of the electron–phonon interaction with an increase in the electron velocity. In the limit of large d, the results of this paragraph coincide with the results of [2].

Note that scattering by optical phonons is predominantly intra-band (the contribution of inter-band scattering is more than an order of magnitude less than the contribution of intra-band scattering at $d \leqslant 200$ nm). The physical reason for this phenomenon is the smooth nature of the coordinate dependence of the potential of surface phonons, which does not provide ‘mixing’ of orthogonal wave functions in matrix elements calculated on the Hamiltonian of interaction with surface optical phonons.

Figure 8.2 shows the dependences of the scattering coefficients W^V, W^{S_1}, W^R, W^{MCA}, $W^T = W^V + W^{S_1} + W^{S_2}$ from d. The scattering coefficient on the confinement of optical phonons, as indicated above, tends to zero when d values of the order of the radius of the bulk polaron are reached $R_V = (\hbar/(2m^*\omega_0))^{1/2}$ due to the weakening of the interaction of the electron with the confinement optical phonons with a decrease in d. By $d \to \infty$ W^V it reaches the volume limit. For thicknesses $d \sim 2$ nm W^R passes higher than the result of the work [3].

Behavior W^{S_1}, depending on d, reflects the behavior of the averaged magnitude of the electron–phonon interaction potentials with surface optical phonons. Note that the appearance of nonmonotonicity in behavior W^V, W^{S_1} with small d, c $d \to 0$. As can be seen from figure 8.2, when $d < 3.5$ nm surface optical phonons make a predominant contribution to the scattering coefficient.

Figure 8.3 shows the change in the values of the scattering coefficients when the dielectric permittivity of neighboring media changes. Increase ε bordering media from 1 to 80 leads, as expected, to almost complete screening of surface phonons and does not affect the confinement phonons of the polar layer in any way.

The main conclusions made for the scattering coefficients remain valid for the momentum relaxation coefficient. The difference is a faster decline in the relaxation coefficients of the momentum with increasing $E_{\mathbf{k}}$ and their smaller values, which is explained not only by positive, but also by negative contributions to the scattering coefficients.

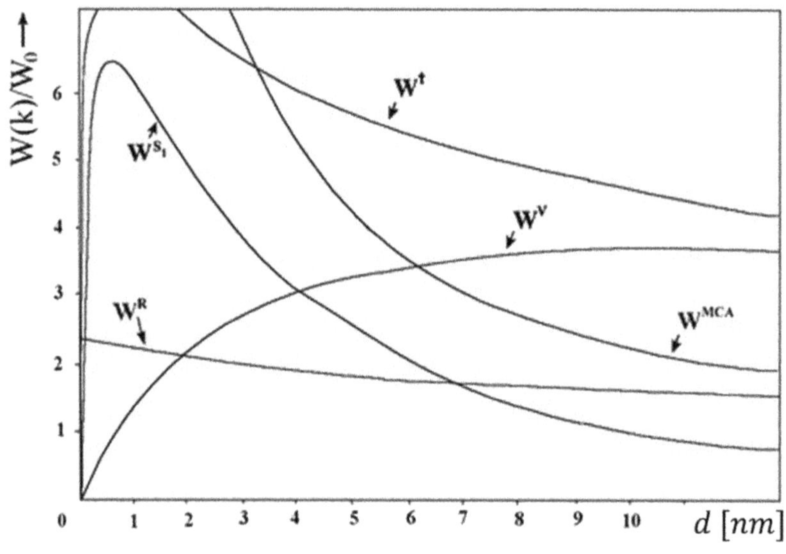

Figure 8.2. Scattering coefficient as a function of the thickness of the polar layer in the structure GaAs/vacuum. All contributions are similar to those described in the caption to figure 8.1. Electron energy $E_k = 1.2\hbar\omega_0$.

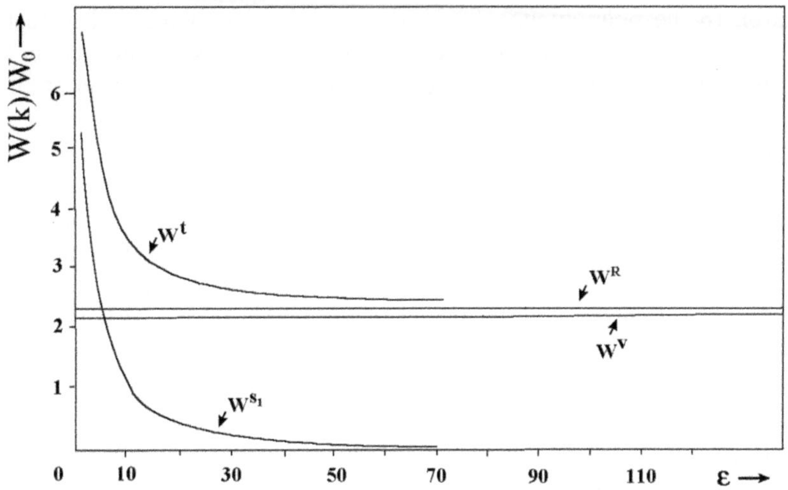

Figure 8.3. Scattering coefficient as a function of the thickness of the polar layer in the structure GaAs/AlGaAs. All contributions are similar to those described in the caption to figure 8.1. Electron energy $E_k = 1.2\hbar\omega_0$.

Thus, both the general Ridley and Riddoch approach [3] and the momentum conservation approximation [2, 5] give correct results in the region of layer thicknesses $d > 7$–$10\ R_V$. In the area of smaller thicknesses, it is fundamentally important to take into account the following features:

(1) Interaction with surface optical phonons ensures the correct behavior of the electron–phonon interaction potential at d→0. In the case of weak shielding by bordering media, scattering by optical phonons is the dominant channel of energy dissipation;
(2) Changing the selection rules for electron–phonon scattering processes with increasing electron energy in the subband;
(3) For structures with the homeopolar material of a quantum well and the polar material of bordering media, scattering occurs only on surface optical phonons;
(4) The number of contributions to scattering on surface optical phonons can be 'controlled' by a special selection of parameters of bordering media.

8.3 Mobility of charge carriers in inversion channels of the MDS structure

With a strong bending of the bands at the contact of the semiconductor with the dielectric, the potential well formed by the barrier at the interface and the electric potential in the semiconductor can be quite narrow, so that quantum mechanical effects can become significant. In most cases, a model is used to consider these effects, according to which the bands in a semiconductor are considered to be slowly changing in space and the concentration of charge carriers at a given point is a function of the local value of the energy gap between the edge of the band and the

Fermi level. In the phenomena of transport, as already mentioned in section 8.1, both bulk and surface scattering mechanisms must be taken into account. In the electric quantum limit, the potential well is so narrow that the motion of all carriers in the near-surface region in the direction of the z-axis perpendicular to the interface is quantized and is described by a wave function that vanishes only at the boundary points. A set of states with the specified dependence on z and with behavior corresponding to free charge carriers in the other two directions parallel to the interface forms a surface band. In the phenomena of carrier transport in a quasi-two-dimensional near-surface region, in many respects it can be described in the same way as carriers in volume, but with other scattering mechanisms that determine mobility [22–28]. Some of these mechanisms, such as scattering on bulk phonons, are common for both two-dimensional and three-dimensional cases, while others (e.g., scattering at the interface, surface phonons, surface roughness) are inherent only in the two-dimensional case. In this section, we will discuss scattering by surface optical phonons in inversion channels of MDS structures in which the dielectric layer is polar. Movement of charge carriers (electrons) in the inversion layer with potential energy $V(z)$ near the semiconductor–dielectric interface can be described in the approximation of the effective mass with an envelope function:

$$\psi(x, y, z) = \psi(z_e)\psi_{k_\perp}(\boldsymbol{\rho}_e) \tag{8.3.1}$$

Wave function $\psi(z_e)$ satisfies the Schrödinger equation (5.9.7), where the total potential energy $U_t(z_e)$, which moves the electron, has the form (5.9.8). In section 5.8, the results of the calculation of states in the inversion channel of MDS structures with a polar layer as a dielectric were presented, both in the general case and in the ultraquantum limit, when $U_t(z)$ can be approximated by a triangular potential.

Let's proceed to calculating the relaxation time of charge carriers when scattering on the surface optical phonons of the dielectric layer (bulk optical vibrations in the region of the inversion channel do not create fields). It is necessary to take into account that at low d the influence of the electrode becomes significant. On the one hand, the metal shields the interaction of charge carriers with surface optical phonons; on the other hand, it creates an additional attractive force that presses the electron more strongly to the surface of the dielectric, where the interaction with optical vibrations is maximal. These competing effects can lead to a nonmonotonic dependence of the binding energy on the thickness of the dielectric layer, which will manifest itself in scattering.

We calculate the mobility of charge carriers in the conducting channel of a semiconductor, taking into account the polaron effect.

1. low temperature limit ($\hbar\omega_0 > kT$).

With small T the carrier does not emit photons, since its kinetic energy is proportional to T. During absorption, the energy of the carrier increases and becomes higher than the average, which leads to the emission of a phonon immediately after its absorption. In general, the electron energy does not change and this whole push–pull process can be considered as elastic, for the description of which it is permissible to use the approximation of relaxation time, according to [29]:

$$\tau^{-1} = -\sum_{\mathbf{Q}} W(\mathbf{k}, \mathbf{Q}) \frac{Q_x}{k_{ix}} \quad . \tag{8.3.2}$$

The direction of the electric field is taken along the axis z. Calculation $W(\mathbf{k}, \mathbf{Q})$ performed similarly to the calculation in the previous paragraph, applying the 'golden rule of quantum mechanics':

$$W(\mathbf{k}, \mathbf{Q}) = \frac{2\pi}{\hbar} \left|\langle f \left| \hat{H}_{\text{e-ph}} \right| i\rangle\right|^2 \delta(E_f - E_i), \tag{8.3.3}$$

where the perturbation is the Hamiltonian averaged on the wave functions (5.4.57) and (5.2.69). The undisturbed wave functions of the system are plane waves for electrons and free oscillators for phonons:

$$\psi(Q, \rho_e) = \frac{1}{L} \exp(i\mathbf{k}_\perp \boldsymbol{\rho}) | n_Q\rangle. \tag{8.3.4}$$

Then for τ^{-1} we get

$$\tau^{-1} = \frac{m^* e^2 \omega_S n_Q}{\hbar^2 k^2} \int_{\sqrt{k^2 + k_0^2} - k}^{\sqrt{k^2 + k_0^2} + k} dQ \frac{Q^2 - k^2}{\sqrt{(2Qk_0 - Q^2 + k^2)(2k_0 Q + Q^2 - k_0^2)}} \times$$
$$\times \left(1 + \tanh^2 \frac{Qd}{2}\right) \left(1 + \frac{Q}{\beta_m}\right)^{-6} \left\{ \left[\varepsilon_3 + \varepsilon \coth \frac{Qd}{2}\right]^{-1} - \left[\varepsilon_3 + \varepsilon_{20} \coth \frac{Qd}{2}\right]^{-1} \right\}. \tag{8.3.5}$$

The expression under the sign of the square root in the denominator (8.3.5) has the structure

$$f(Q) = (Q - Q_1)(Q + Q_1)(Q_2 - Q)(Q_2 + Q).$$

Root features give a multiplier:

$$(Q - Q_1)(Q_2 - Q) = 2Q\sqrt{k_0^2 + k^2} - k_0^2 \tag{8.3.6}$$

(factor $(Q_1 + Q)(Q_2 + Q) = Q^2 + 2Q\sqrt{k_0^2 + k^2} - k_0^2$ is positively defined).

Taking into account the inequality $k_0^2 \gg k^2$, and also that the specified form is multiplied by the product $(Q - Q_1)(Q_2 - Q)$, which is proportional to k, can be up to terms proportional to k^3, we can get

$$\tau^{-1} \approx \frac{m^* e^2 \varpi_S \bar{n}_Q}{\hbar^2 k^2} \int_{\sqrt{k^2 + k_0^2} - k}^{\sqrt{k^2 + k_0^2} + k} \frac{dQ(Q - k_0)\left(1 + \tanh^2 \frac{Qd}{2}\right)}{[(Q - Q_1)(Q_2 - Q)]^{\frac{1}{2}} \left(1 + \frac{Q}{\beta_m}\right)^{-6}} \times$$
$$\times \left\{ \left[\varepsilon_3 + \varepsilon \coth \frac{Qd}{2}\right]^{-1} - \left[\varepsilon_3 + \varepsilon_{20} \coth \frac{Qd}{2}\right]^{-1} \right\}. \tag{8.3.7}$$

With small d ($Qd \ll 1$) and (8.3.7) it follows:

$$\tau^{-1} \approx \frac{\pi \bar{n} m_e^* e^2 \omega_S d\left(1 - \frac{9k}{\beta_m}\right)}{2\hbar^2 \varepsilon^*}, \tag{8.3.8}$$

which means τ^{-1} does not depend on the energy of the carrier. Independence τ^{-1} electron energy dependence also occurs when electrons are scattered by optical polarization oscillations in the volume [29]. Relationship $(\tau^{-1})^{2D}$ (formula (8.3.8)) to the corresponding expression for a massive crystal is equal to

$$\frac{(\tau^{-1})^{2D}}{(\tau^{-1})^{3D}} \sim \frac{d}{R_S}. \tag{8.3.9}$$

It follows that in the limit of thin dielectric layers ($d \ll R_S$) there is an increase in the mobility of charge carriers during scattering by optical polar phonons. Because β_m depends on the concentration of charge carriers n_e, then from (8.3.8) it follows that with growth n_e in the area of the validity of the formula (8.3.9), mobility slowly decreases:

$$\mu^{2D} \sim \left(1 - \frac{6E_i}{3eFR_S}\right)^{-1} \sim \left(1 - \frac{C_2}{n_e^{1/3}}\right).$$

In the limit of big d ($Qd \gg 1$), when the influence of the electrode becomes negligible (i.e., in the case of contact of semi-infinite polar and nonpolar crystals) we obtain:

$$\tau^{-1} \approx \frac{2\bar{n} m_e^* e \omega_S}{\hbar^2 k^2}\left(\frac{1}{\varepsilon_3 + \varepsilon_2} - \frac{1}{\varepsilon_3 + \varepsilon_{20}}\right)\frac{\beta_m^6}{\sqrt{\beta_m(\beta_m + 2k_0)}} \times \left\{1 - \frac{k^2}{4k_0}\left(\beta_m^{-1} + (\beta_m + 2k_0)^{-1}\right)(J_1 + (\beta_m + k_0)J_2)\right\}, \tag{8.3.10}$$

where

$$\beta_m^{-1} = \bar{z} = \frac{2E_i}{3eF_i},$$

and expressions for J_1 and J_2 are complex functions β_m and k_0. If F so, that $\beta_m \gg k_0$ ($\bar{z} \gg R_S$), then from expression (8.3.10) it follows:

$$\tau^{-1} \approx \frac{3\pi}{2} \exp\left\{-\frac{\hbar\omega_S}{kT}\right\} \frac{m^* e^2 \omega_S \beta_m}{\hbar^2 k_0 k}\left(\frac{1}{\varepsilon_2 + \varepsilon_3} - \frac{1}{\varepsilon_{20} + \varepsilon_3}\right), \tag{8.3.11}$$

from where it can be seen that mobility increases with increasing concentration of charge carriers. In [30], the electron mobility in the inversion n-channel on the surface was experimentally measured (100) $p - \text{Si}$, limited by scattering at the charged boundary $\text{Si} - \text{SiO}_2$. In the concentration range $n_{es} \sim 10^{12} - 10^{13}$ cm^{-2}, mobility increases linearly with growth n_S. In [31], theoretically, only the scattering mechanism on charged impurities was taken into account and a mobility value greater than the experimental one was obtained. As shown above, taking into account the additional scattering channel on SiO_2 polar optical phonons reduces mobility. At the same time, the amount of mobility depends on other parameters of the system, including the electron–phonon interaction constant, as well as on the polaron effective mass m_p^*. The small d m_p^* practically coincides with the mass band due to the strong shielding of the electron–phonon interaction by the electrode. In the case of a thick dielectric layer in the approximation of the intermediate electron–phonon interaction $m_p^* = m_e^*(1 - \pi\alpha_S/4)$, if $\bar{z} \ll R_S$.

2. In the opposite limit of high temperatures ($kT \gg \hbar\omega_S$) can be neglected k_0 under the sign of the root in the formula (8.3.5) and both in the case of absorption and emission of phonons, we obtain

$$\tau^{-1} \approx \frac{m^* e^2 \varpi_S n}{\hbar^2 k^2} \int_0^{2k} \frac{Q \quad J(Q, d)\, dQ}{\sqrt{4k^2 - Q^2}}. \tag{8.3.12}$$

For the small d and $\bar{z}$ from the expression (8.3.12) we get

$$\tau^{-1} \approx \frac{2m_e^* e^2 kT}{\hbar^3 k}\left(\frac{1}{\varepsilon_2 + \varepsilon_3} - \frac{1}{\varepsilon_{20} + \varepsilon_3}\right), \tag{8.3.13}$$

that is, as in the case of a massive crystal, $\tau \sim T^{-1}$.

The obtained results can be used to analyze experimental data from [32, 33], in which the field dependence of the mobility of charge carriers in the silicon inversion channel in the structure was measured $\text{Me} - \text{SiO}_2 - \text{Si}$. There was a shift of the nonatomic regime to the region of higher fields compared to massive silicon. Attempts to give an explanation based on the mechanisms characteristic of bulk silicon (deformation interaction, impurity scattering, etc, as well as scattering on the roughness of the boundary) did not lead to success. In [33], attention was drawn to the fact that the dielectric layer is polar and its optical vibrations create macroscopic electric fields that penetrate silicon near the contact and cause scattering of charge carriers. In silicon itself, as in a nonpolar material, there is no polar scattering. This mechanism of 'long-range' scattering leads to a significant decrease in the mobility of charge carriers at the interface. However, the model of the MDS structure proposed in [33] is simplified. The dielectric layer is assumed to be infinite, as a result, the influence of the electrode on the interaction of electrons with the optical vibrations of the dielectric was not taken into account. The analysis based on the formulas for relaxation time and mobility obtained in this paragraph shows that scattering on the surface optical phonons of the oxide layer in the considered temperature and field intervals is indeed a very effective mechanism and can serve as the main cause of the temperature and field mobility anomalies observed in experiments [32] (enhancement of the field dependence μ).

8.4 IR absorption by free charge carriers with the participation of optical phonons in structures with quantum wells

Optical transitions of free charge carriers in massive semiconductors involving polar optical phonons were considered in [34]. In the works [2–5, 35–47] theoretically and in the works [48–54] experimentally, this problem has been considered for dimensional limited polar semiconductor structures. The phonon spectrum was assumed to be the same as in a massive crystal.

In [47], a model was used in which the electron–phonon interaction was described by the Pekar–Fröhlich Hamiltonian for an infinite polar crystal, and for electrons at dimensional quantization levels, the law of conservation of momentum (the so-called conservation of momentum approximation) was assumed to be fulfilled. The analysis of the results obtained in [47] shows that in the region of small thicknesses of the quantum well, the accepted approximations lead to physically incorrect results.

The authors of works [35, 55] go beyond the approximation of conservation of momentum; however, as in [47], the phonon spectrum is assumed to be volume-like.

In this section, the IR absorption of light by free charge carriers in polar semiconductor quantum wells involving surface and bulk longitudinal optical phonons is theoretically investigated on the basis of the exact Hamiltonian of the electron–phonon interaction obtained in [13] and generalized to the case of arbitrary multilayer structures in [21, 56]. It is shown that, as in the case of Raman scattering [57–59], additional peaks due to surface phonons appear in the optical absorption spectrum.

In this section, we have adopted the effective mass approximation and Boltzmann statistics for electrons. The motion of electrons along the z-axis (directed perpendicular to the interface) is considered dimensionally quantized, and the law of dispersion for the motion of an electron in the plane of a quantum well is assumed to be parabolic.

8.4.1 Hamiltonian and wave functions

Consider a polar crystal layer occupying an area of space $0 \leqslant z \leqslant d$, with high frequency and static permittivity ε_2, ε_{20} and bordering on semi-infinite nonpolar media occupying regions of space $z < 0$ and $z > d$, with dielectric permittivity ε_1 and ε_2 accordingly. The Hamiltonian of noninteracting electrons and phonons has the form

$$\hat{H} = \frac{\hat{P}_{\parallel}^2}{2m_{\parallel}^*} + \frac{\hat{P}_{\perp}^2}{2m_{\perp}^*} + V(z) + \hat{H}_{\text{ph}}^{(V)} + \hat{H}_{\text{ph}}^{(S_1,2)}, \tag{8.4.1}$$

where $\hat{P}_{\perp}$, $P_{\parallel}$, $m_{\perp}^*$, $m_{\parallel}$ are the components of the momentum and the effective mass of the electron in the XY-plane and along the Z-axis, respectively,

$$V(z) = \begin{cases} 0, & 0 \leqslant z \leqslant d; \\ \infty, & z < 0, \ z > d; \end{cases} \tag{8.4.2}$$

is a potential barrier at the quantum well boundary;

$$\hat{H}_{\text{ph}}^{(V)} = \sum\nolimits_{\mathbf{Q}} \hbar\omega_0 \hat{a}_{\mathbf{Q}}^{+} \hat{a}_{\mathbf{Q}}; \tag{8.4.3}$$

$$H_{\text{ph}}^{(S_i)} = \sum\nolimits_{\boldsymbol{\eta}} \hbar\Omega_{S_i} \hat{b}_{S_i, \boldsymbol{\eta}}^{+} \hat{b}_{S_i, \boldsymbol{\eta}}; \ i = 1, 2 \tag{8.4.4}$$

are Hamiltonians of bulk and interface phonon modes, respectively; ω_0, Ω_{S_1}, Ω_{S_2}, $\mathbf{Q}(q_x, q_y, q_z)$, $\boldsymbol{\eta}(\eta_x, \eta_y)$ are frequencies and wave vectors of the corresponding modes; and $\hat{a}_{\mathbf{Q}}^{+}$, $\hat{a}_{\mathbf{Q}}$, $\hat{b}_{S_i, \boldsymbol{\eta}}^{+}$, $\hat{b}_{S_i, \boldsymbol{\eta}}$, $i = 1, \ 2$ are operators of the birth and annihilation of these mods.

The wave function of an electron in a quantum well has the form

$$\psi(\boldsymbol{\rho}, z) = \left(\frac{2}{V}\right)^{\frac{1}{2}} U(\boldsymbol{\rho}, z) \exp(i\mathbf{k}_{\perp}\boldsymbol{\rho}) \sin(k_z z); \tag{8.4.5}$$

where $V = L_x L_y d$ is layer volume; $\mathbf{k}_{\perp}$, $k_z = l\pi / d; \, l = 1, \ 2, \ 3, \ \ldots$ are the components of the electron wave vector along the XY-plane and along the Z-axis, respectively; $U(\boldsymbol{\rho}, z)$ is the Bloch facto; and d is the thickness of the layer.

The energy spectrum of an electron in such a system is described by the expression

$$E(\mathbf{K}) = E(\mathbf{k}_{\perp}) + E_l = \frac{\hbar^2 k_{\perp}^2}{2m_{\perp}^{*}} + l^2 E_0, \tag{8.4.6}$$

where $E_0 = \pi^2\hbar^2/(2m_{||}^{*} d^2)$ is the energy of the lowest level of dimensional quantization and $\mathbf{K} = \mathbf{k}_{\perp} + \mathbf{e}_z k_z$ is the wave vector of the electron.

The density of the states of electrons with a fixed spin is equal to

$$N[E_l(\mathbf{K})] = \frac{l m_{\perp}^{*}}{2\pi\hbar^2 d}. \tag{8.4.7}$$

An electron from the conduction band of the layer interacts with two surface (S_1, S_2) optical modes and one longitudinal confinement (V) mode [21, 56].

The Hamiltonian of the electron–phonon interaction has the form

$$\hat{H}_{\text{e-ph}} = \hat{H}_{\text{e-ph}}^{(V)} + \hat{H}_{\text{e-ph}}^{(S_1, S_2)}. \tag{8.4.8}$$

Here

$$\hat{H}_{\text{e-ph}}^{(V)} = \sum\nolimits_{(q_z \neq 0)}^{\eta > q_z} C_Q \, e^{i\mathbf{q}_{\perp}\boldsymbol{\rho}} \left[e^{i\mathbf{q}_{\perp}\boldsymbol{\rho}} + \frac{1}{1 + C}\left(C - \frac{\cosh\left[q_{\perp}\left(z - \frac{d}{2}\right)\right]}{\cosh\left(\frac{q_{\perp} d}{2}\right)} \right) \right] \left(\hat{a}_{-\mathbf{Q}}^{+} + \hat{a}_{\mathbf{Q}}\right) \tag{8.4.9}$$

is the Hamiltonian of the interaction of an electron with bulk optical phonons;

$$\hat{H}_{\text{e-ph}}^{(S_1)} = \sum_{\eta} C_{\eta}^{(1)} \exp(i\eta\rho) \frac{\cosh\left[\eta\left(z - \frac{d}{2}\right)\right]}{\cosh\left(\frac{\eta d}{2}\right)} (\hat{b}_{S_1,-\eta}^{+} + \hat{b}_{S_1,\eta}); \tag{8.4.10}$$

$$\hat{H}_{\text{e-ph}}^{(S_2)} = \sum_{\eta} C_{\eta}^{(2)} \exp(i\eta\rho) \frac{\sinh\left[\eta\left(z - \frac{d}{2}\right)\right]}{\cosh\left(\frac{\eta d}{2}\right)} (\hat{b}_{S_2,-\eta}^{+} + \hat{b}_{S_2,\eta}); \tag{8.4.11}$$

where $\hat{H}_{\text{e-ph}}^{(S_1,S_2)}$ are the Hamiltonians of electron interactions with surface optical modes in a polar crystal layer bordering nonpolar layers. The following notation is introduced in formulas (8.4.9)–(8.4.11):

$$|C_Q|^2 = \frac{1}{L_x L_y d} \cdot \frac{4\pi\alpha_V (\hbar\omega_0)^2}{Q^2 \beta_V}; \tag{8.4.12}$$

$$\alpha_V = \frac{e^2}{4\pi\varepsilon_0 \hbar} \cdot \frac{1}{\varepsilon^*} \left(\frac{m^*}{2\hbar\omega_0}\right)^{\frac{1}{2}}; \ \beta_V = R_V^{-1} = \left(\frac{2m^*\omega_0}{\hbar}\right)^{\frac{1}{2}}; \tag{8.4.13}$$

$$\left(\varepsilon_V^*\right)^{-1} = \varepsilon_2^{-1} - \varepsilon_{20}^{-1}; \ m^* = \frac{1}{2}\left(m_{\parallel}^* + m_{\perp}^*\right); \tag{8.4.14}$$

$$Q^2 = q_{\perp}{}^2 + q_z^2; \ C = \left[\left(\frac{2}{q_{\perp} d}\right) \tanh\left(\frac{q_{\perp} d}{2}\right)\right]^{\frac{1}{2}}; \tag{8.4.15}$$

$$\left| C_{\eta}^{(i)} \right|^2 = \frac{1}{L_x L_y} \cdot \frac{2\pi\alpha_{S_i} (\hbar\Omega_{S_i})^2}{\eta \beta_{S_i}}; \tag{8.4.16}$$

$$\alpha_{S_i} = \frac{e^2}{4\pi\varepsilon_0 \hbar} \cdot \frac{1}{\varepsilon_{S_i}^*} \left(\frac{m^*}{2\hbar\Omega_{S_i}}\right)^{\frac{1}{2}}; \ \beta_{S_i} = R_{S_i}^{-1} = \left(\frac{2m^*\Omega_{S_i}}{\hbar}\right)^{\frac{1}{2}}; \tag{8.4.17}$$

$$\frac{1}{\varepsilon_{S_1}^*} = \frac{1}{\varepsilon_2 + \varepsilon \cdot \tanh\left(\frac{\eta d}{2}\right)} - \frac{1}{\varepsilon_{20} + \varepsilon \cdot \tanh\left(\frac{\eta d}{2}\right)}; \tag{8.4.18}$$

$$\frac{1}{\varepsilon_{S_2}^*} = \frac{1}{\varepsilon_2 + \varepsilon \cdot \coth\left(\frac{\eta d}{2}\right)} - \frac{1}{\varepsilon_{20} + \varepsilon \cdot \coth\left(\frac{\eta d}{2}\right)}. \tag{8.4.19}$$

Frequency confinement ω_0 and superficial $\Omega_{S_i}, i = 1, 2$ phonons are found from Lidden–Sachs–Teller ratios:

$$\omega_0^2 = \frac{\varepsilon_{20}}{\varepsilon_2} \omega_{TO}^2; \tag{8.4.20}$$

$$\Omega_{S_1}^2 = \left[\frac{\varepsilon_{20} + \varepsilon \cdot \coth\left(\frac{\eta d}{2}\right)}{\varepsilon_2 + \coth\left(\frac{\eta d}{2}\right)} \right] \omega_{TO}^2; \; \Omega_{S_2}^2 = \left[\frac{\varepsilon_{20} + \varepsilon \cdot \tanh\left(\frac{\eta d}{2}\right)}{\varepsilon_2 + \tanh\left(\frac{\eta d}{2}\right)} \right] \omega_{TO}^2. \quad (8.4.21)$$

To simplify mathematical calculations, we further assume $\varepsilon_1 = \varepsilon_2 \equiv \varepsilon$ (symmetrical structure).

The Hamiltonian of the electron–phonon interaction (8.4.8)–(8.4.11) was obtained in [13] on the basis of the procedure proposed in [12], which correctly and consistently isolated normal modes in a three-layer polar structure. In addition to the obvious division into confinement and surface parts (even S_1 and odd S_2), these Hamiltonians have several properties that are important for further analysis:

firstly, unlike the Hamiltonian of the non-normalized volume spectrum, they have the correct dimensional dependence, which can be easily interpreted physically;

secondly, the constants of the electron–phonon interaction and the frequencies of surface modes (as opposed to bulk modes) contain the material parameters of the media bordering the layer in which the electrons are localized;

thirdly, as can be seen from expressions (8.4.20) and (8.4.21), the frequencies of surface vibrations depend on the thickness, as well as on the dispersion parameter, and this fact can serve as an additional characteristic feature of electron–phonon interaction processes involving phonons of the above type.

The Hamiltonian of the interaction of an electron with light has the form [60]

$$\hat{H}_{el-L}^{abs} = -\frac{e}{m^*} \left(\frac{\hbar N}{2V\pi\varepsilon\omega_2} \right)^{\frac{1}{2}} \exp(i\mathbf{q}_V\mathbf{r})(\mathbf{e} \cdot \mathbf{p}) \quad (8.4.22)$$

to absorb light and

$$\hat{H}_{el-L}^{em} = -\frac{e}{m^*} \left(\frac{\hbar(N+1)}{2V\pi\varepsilon\omega_2} \right)^{\frac{1}{2}} \exp(-i\mathbf{q}_V\mathbf{r})(\mathbf{e} \cdot \mathbf{p}) \quad (8.4.23)$$

for radiation, including spontaneous radiation.

Here N is the number of photons in the volume layer V; ω the frequency of the monochromatic light wave; $\mathbf{q}_V$ the wave vector of the light wave; $\mathbf{p}$ the quasi-momentum operator of an electron in the plane xy; and $\mathbf{e}$ is the unit polarization vector of the light wave (hereinafter assumed to be perpendicular to the axis Z).

Then the Hamiltonian of the electron–phonon-phonon interaction can be written as

$$\hat{H}_{\text{int}} = \hat{H}_{el-L}^{\text{em}} + \hat{H}_{el-L}^{\text{abs}} + \hat{H}_{\text{e-ph}}. \quad (8.4.24)$$

8.4.2 Probability of light absorption

Assuming that both the electron–phonon interaction and the interaction due to light are sufficiently small and, in addition, considering that $\omega\tau \gg 1$ (where τ is the average lifetime of an electron), we use the perturbation theory. Since in the first order of perturbation theory, the absorption of light by an electron does not occur at the specified polarization of light, we use the second order. In the second order of perturbation theory, the probability of transition to a unit of time is determined by the expression [61]

$$W_{el}(\mathbf{K}) = \frac{2\pi}{\hbar}\int\left|\sum_{\mathbf{k}'',l''}\frac{\langle \mathbf{k}', l'|\,\hat{H}_{\text{int}}\,|\,\mathbf{k}'', l''\rangle\langle \mathbf{k}'', l''\,|\,\hat{H}_{\text{int}}\,|\,\mathbf{k}, l\rangle}{E(\mathbf{k}') - E(\mathbf{K}) \mp \hbar\Omega}\right|^2 \delta\big(E(\mathbf{k}') - E(\mathbf{K}) \mp \hbar\Omega\big)\,dS', \tag{8.4.25}$$

where dS' is the number of end states in the interval $d\mathbf{k}'$ $(dS' = 2/(2\pi)^2 d\mathbf{k}')$, and the integration is performed over all the final states of the electron, l, and are the numbers of the subbands of dimensional quantization. It should be emphasized that the transitions providing light absorption described by the formula (8.4.25) are two-stage processes with the absorption (emission) of a single photon (phonon) at each stage, which can follow each other in any order [60].

Consider the possible transitions in the system and denote by $W_{l,\,l'}$ the difference between the probabilities of absorption and emission of a photon per unit of time. Then

$$W_{l,\,l'} = \frac{1}{N}\sum_{\mathbf{k}}\sum_{\mathbf{k}'}\sum_{\mathbf{q}} W\begin{pmatrix}\mathbf{k}', l'\\ \mathbf{k}, l\end{pmatrix}, \tag{8.4.26}$$

where $W\begin{pmatrix}\mathbf{k}', l'\\ \mathbf{k}, l\end{pmatrix}$ is the difference between the probabilities of transitions with absorption and emission of a photon and $\sum_{\mathbf{q}}\ldots$ means summation by wave vectors of phonons of any type (bulk, surface).

For electrons, we assume that Boltzmann statistics are applicable:

$$f_{\mathbf{k},\,l} = \exp\left(\frac{\varsigma}{T}\right)\exp\left(-\frac{E(\mathbf{K})}{T}\right) \ll 1, \tag{8.4.27}$$

where ς is the chemical potential, T *the* electronic gas temperature, and $f_{\mathbf{k},\,l}$ is electronic gas distribution function.

Summing up the probabilities of photon absorption per unit of time for transitions between different subbands of dimensional quantization, we have

$$W = \sum_{l,\,l'} W_{l,\,l'}; \tag{8.4.28}$$

where W, $W_{l,\,l'}$ can be caused by any of the phonon modes under consideration (V, S_1, S_2).

W is the total absorption probability obtained by simple summation over the phonon modes under consideration, and $W_{l,\,l'}$ describes the probability of absorption during transitions between subbands of dimensional quantization l, l' with the participation of phonons.

For bulk (confinement) phonons we have

$$W_{l,l'}^{(V)} = \frac{2T}{\varepsilon\hbar^3 d^2}\left(\frac{e^2}{4\pi\varepsilon_0}\right)^2 \frac{1}{\varepsilon_p^*}\left(\frac{\omega}{\omega_0^3}\right)\frac{\sinh\left(\frac{\hbar\omega}{2T}\right)}{\sinh\left(\frac{\hbar\omega_0}{2T}\right)}\exp\left(\frac{2\mu - (E_l + E_{l'})}{2T}\right) \times \left[H_{l,l'}^{(V)}(\omega + \omega_0) + H_{l,l'}^{(V)}(\omega - \omega_0)\right], \tag{8.4.29}$$

where

$$H_{l,l'}^{(V)}(\Omega) = H_{l,l'}^{(V,0)}(\Omega) + H_{l,l'}^{(V,1)}(\Omega) + H_{l,l'}^{(V,2)}(\Omega). \tag{8.4.30}$$

The general view of each of the members in (8.4.29) can be represented as follows:

$$H_{l,l'}^{(V,i)}(\Omega) = \frac{1}{2\pi^2}\left(\frac{\pi E_0}{T}\right)^{\frac{1}{2}} \int_{t_{\min}}^{t_{\max}} dt \cdot t^2 \exp\left(-\lambda t^2 - \frac{\mu(\Omega, l, l')}{t^2}\right) I^{(i)}(t); \; i = 1, 2, \tag{8.4.31}$$

where

$$\lambda = \frac{E_0}{4\pi^2 T};\; \mu(\Omega, l, l') = \frac{\pi^2}{4}\frac{|\hbar\Omega - (E_{l'} - E_l)|}{E_0 T}. \tag{8.4.32}$$

Expressions for $I^{(i)}(t)$ have the form

$$I^{(i,0)}(t) = \frac{1 + \delta_{l,l'}}{t^2 + \pi^2(l - l')^2} + \frac{1}{t^2 + \pi^2(l + l')^2}; \tag{8.4.33}$$

$$I^{(i,1)}(t) = \frac{-2(4\pi^2 ll')^2 t[1 - (-1)^{l+l'}\exp(-t)]}{[t^2 + \pi^2(l - l')^2]^2[t^2 + \pi^2(l + l')^2]^2}; \tag{8.4.34}$$

$$I^{(i,2)}(t) = \frac{1}{t}\left[\frac{\delta_{l,l'}C(t) - G(t)}{1 - C(t)}\right] + \frac{2}{1 - C(t)}(\delta_{l,l'}C(t) - G_1(t)) \times \left[\frac{1 - (-1)^{l-l'}\exp(-t)}{t^2 + \pi^2(l - l')^2} - \frac{1 + (-1)^{l+l'}\exp(-t)}{t^2 + \pi^2(l + l')^2}\right]. \tag{8.4.35}$$

An explicit view for $G_1(t)$ is given below.
Integration in the first term of the formula (8.4.30) gives

$$H_{l,l'}^{(V,0)}(\Omega) = \exp\left(-\frac{|\hbar\Omega - (E_l - E_{l'})|}{2T}\right) \times \left\{1 + \frac{\delta_{l,l'}}{2} - \frac{|l - l'|}{2}\sqrt{\frac{\pi E_0}{4}}\exp\left(\nu^2(\Omega, l, l')\right) \times \operatorname{erfc}\left(\nu(\Omega, l, -l')\right) + \frac{|l + l'|}{4}\sqrt{\frac{\pi E_0}{4}} \exp(\nu^2(\Omega, l, -l'))\operatorname{erfc}\left(\nu(\Omega, l, +l')\right)\right\}, \tag{8.4.36}$$

where

$$\nu^2(\Omega, l, l') = \frac{\hbar\Omega - (E_l - E_{l'})}{2\,|l + l'|\sqrt{E_0 T}} + \frac{|l + l'|\sqrt{E_0}}{2\sqrt{T}} \tag{8.4.37}$$

and is in accordance with the results obtained within the framework of the conservation of momentum approximation [47]. Expression (8.4.31) when $l' = 1,\ 2$ cannot be integrated in an analytical form. Neglecting the third term in the formula (8.4.30), we obtain the results of [35]. From a physical point of view, this is equivalent to neglecting the vanishing of the electron–phonon interaction potential at the boundaries of the quantum well.

The expression for the probability of light absorption involving surface phonon modes is similar to (8.4.29):

$$W_{l,\,l'}^{(S_i)} = \frac{2T}{\varepsilon\hbar^3 d^2}\left(\frac{e^2}{4\pi\varepsilon_0}\right)^2 \exp\left(\frac{2\varsigma - (E_l + E_{l'})}{2T}\right)\left[H_{l,\,l',\,+}^{(S_i)} + H_{l,\,l',\,-}^{(S_i)}\right];\ i = 1,\quad 2, \tag{8.4.38}$$

where

$$H_{l,ll,\pm}^{(S_i)} = \frac{1}{2\pi^2}\sqrt{\frac{\pi E_0}{T}} \times \int_{t_{\min}}^{t_{\max}} dt \cdot \frac{t^2}{\varepsilon_{S_i}^{*}(t)}\left(\frac{\omega}{\Omega_{S_i}^{3}(t)}\right)\frac{\sinh\left(\frac{\hbar\omega}{2T}\right)}{\sinh\left(\frac{\hbar\Omega_{S_i}(t)}{2T}\right)}$$
$$\exp\left(-\lambda t^2 - \frac{\mu\left(\omega - \Omega_{S_i}(t),\ l,\ l'\right)}{t^2}\right) I^{(i)}(t); \tag{8.4.39}$$

$$I^{(S_i)}(t) = \frac{|G_{S_i}(t)|}{t \cdot \tanh\left(\frac{t}{2}\right)}; \tag{8.4.40}$$

$$G_{S_1}(t) = \tanh\left(\frac{t}{2}\right)[1 - (-1)^{l-l'}]\left[\frac{1}{t^2 + \pi^2(l - l')^2} - \frac{1}{t^2 + \pi^2(l + l')^2}\right]; \tag{8.4.41}$$

$$G_{S_2}(t) = -\left[\frac{1 - (-1)^{l-l'}}{t^2 + \pi^2(l - l')^2} - \frac{1 + (-1)^{l+l'}}{t^2 + \pi^2(l + l')^2}\right]. \tag{8.4.42}$$

Light absorption coefficient K_0 can be obtained using the ratio between probability and absorption coefficient [61]:

$$K_0 = \frac{\sqrt{\varepsilon_2}}{\varepsilon} W. \tag{8.4.43}$$

8.4.3 Results and discussion

The following parameter values were used to obtain numerical results:

1. structure ($Al_xGa_{1-x}As/GaAs/Al_xGa_{1-x}As$:
   $m^* = 0.0667\,m_0$; $\hbar\omega_0 = 35.2$ meV; $x = 0.30$; $\varepsilon_2 = 10.9$; $\varepsilon_{20} = 12.5$; $\varepsilon = 10.1$, $R_V = (\hbar/(2m^*\omega_0))^{1/2} \approx 4$ nm.
2. structure 'nonpolar dielectric/CdTe/nonpolar dielectric':

$m^* = 0.092\, m_0$; $\hbar\omega_0 = 21.7$ meV; $\varepsilon_2 = 7.13$; $\varepsilon_{20} = 10.\ 6$; $\varepsilon = 7.13$, $R_V = (\hbar/(2m^*\omega_0))^{1/2} \approx 4,\ 4$ nm.

In addition, the values of dielectric permittivity for neighboring media ε were taken from the range from 1 to 80: $\varepsilon = 1$ corresponds to a vacuum, and $\varepsilon = 80$ was chosen to study the case of strong shielding of electron–phonon interaction by neighboring media.

Let's analyze the dependences of the probability of light absorption on its frequency ω, the width of the well *d*, and the dielectric constant ε neighboring media obtained as a result of numerical calculations performed for two different models: model 1 corresponds to the work [47], and model 2 corresponds to the work [35].

Figures 8.4 and 8.5 show the dependences of the probability of light absorption *W* from its frequency divided by the frequency of longitudinal bulk optical phonons (ω/ω_0) for different thicknesses *d* of the well in which the electron is localized. As can be seen from these figures, sharp peaks are observed when the frequency of light coincides with the frequencies of volume modes, and the contribution of surface modes has a maximum in the frequency range of surface vibrations. Models 1 and 2 give slightly narrower and sharper absorption curves $W^{(1)}$ and $W^{(2)}$] than obtained in these calculations.

For second-order processes with photon absorption and subsequent phonon emission, an energy deficit can be obtained from an electron. At high temperatures, when the average electron energy is significantly greater than the phonon energy,

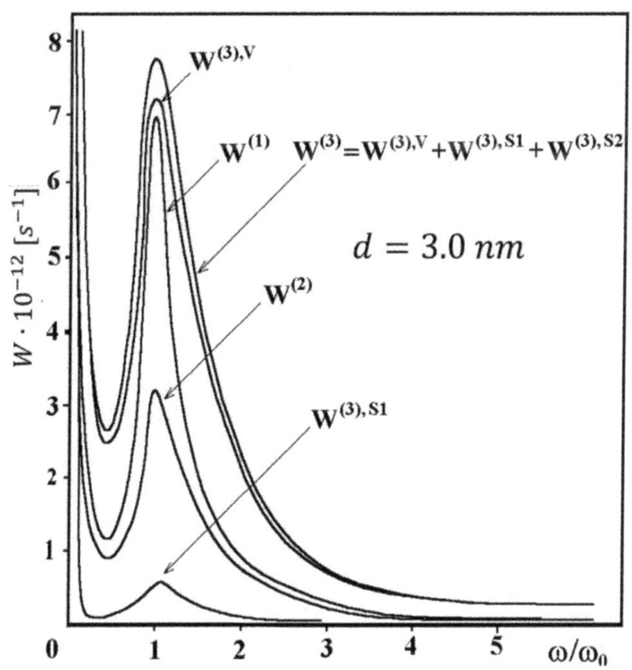

Figure 8.4. Probability of light absorption in structures GaAs − $Al_xGa_{1-x}As$, as a function of the frequency of light. Shown: (1) contribution from the participation of an even mode of surface phonons $W^{(3),\ S_1}$; (2) contribution from the participation of confinement phonons $W^{(3),\ V}$; (3) model calculation results (I) $W^{(1)}$; (4) model calculation results (II) $W^{(2)}$; and (5) $W^{(3)} = W^{(3),\ V} + W^{(3),\ S_1} + W^{(3),\ S_2}$.

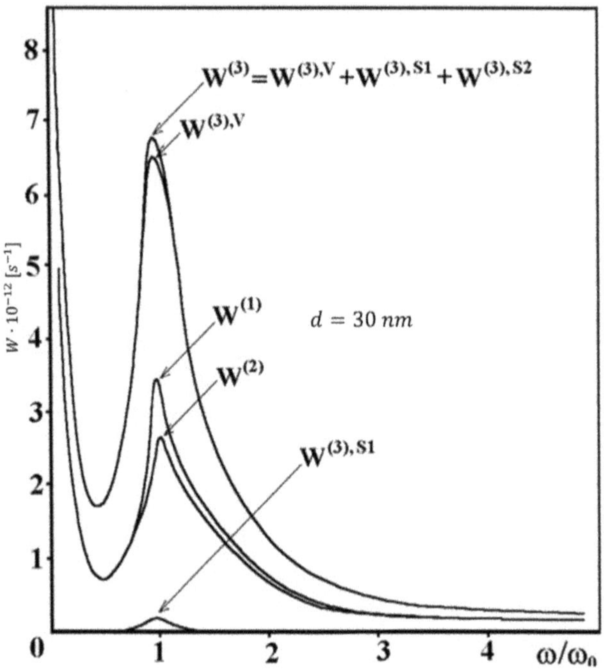

Figure 8.5. Probability of light absorption in structures GaAs – $Al_xGa_{1-x}As$, as a function of the frequency of light. Shown: (1) contribution from the participation of an even mode of surface phonons $W^{(3),\,S_1}$; (2) contribution from the participation of confinement phonons $W^{(3),\,V}$; (3) model calculation results (I) $W^{(1)}$; and (4) model calculation results (II) $W^{(2)}$; and (5) $W^{(3)} = W^{(3),\,V} + W^{(3),\,S_1} + W^{(3),\,S_2}$.

almost all electrons are able to participate in the scattering of light, and the scattering intensity increases.

Note that the values of the resonant frequency as well as the relative displacements of the resonant peaks of surface and confinement (bulk) phonons agree with the predicted formulas (8.4.29) and (8.4.38).

The drawing corresponds to the thickness $d = 3.0$ nm temperature $T = 220.0$ K.

The picture corresponds to the thickness $d = 30$ nm;temperature $T = 220.0$ K.

Figure 8.6 shows the dependences of the probability of light absorption on the thickness of the pit d for models 1, 2 and the data of this paragraph.

It is necessary to note the completely different behavior of the light absorption coefficient, depending on the type of model, in the limit of small d. So, for model 1, in the limit of small d, the value $W^{(1)}$ increases dramatically. For model 2, the probability of absorption $W^{(2)}$, calculated for the Hamiltonian of the electron–phonon interaction of Pekar-Fröhlich for a massive crystal, shows a tendency to strive for a finite value. At the same time, unlike models 1 and 2, the value obtained on the exact Hamiltonian is $W^{(3)}$, which ensures the correct behavior of the absorption probability in the limit $d \to 0$. Figure 8.6 identifies two types of contributions to the probability of absorption, which occur due to the presence of various phonon modes. Both contributions tend to zero at $d \to 0$, that's why

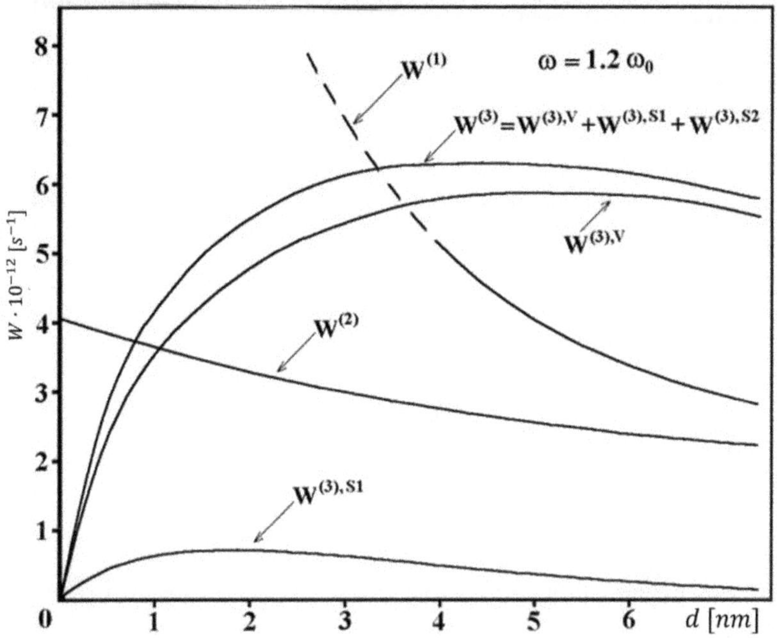

Figure 8.6. The probability of light absorption in a structure with quantum wells as a function of the thickness of the polar layer. The contributions are similar to figures 8.4 and 8.5. The magnitude of the light frequency $\omega = 1.2\omega_0$. The picture corresponds to GaAs – $Al_xGa_{1-x}As$.

$W^{(3)} \to 0$ when $d \to 0$, ($W^{(3)} = W^{(3),\,V} + W^{(3),\,S_1} + W^{(3),\,S_2}$). This behavior $W^{(3)}$ is due to the tendency of the electron–phonon interaction constant to zero in the limit $d \to 0$ as a result of the disappearance of polar material. It should be emphasized that at low d in a structure with a strong electron–phonon interaction (CdTe), the interaction with bulk phonons weakens faster than with surface phonons (figure 8.7).

In the area of large thicknesses ($d > (7 \div 10)R_V$), where R_V is the polaron radius in a bulk crystal), there is an asymptotic coincidence between the absorption probabilities involving bulk polar optical phonons, due to the similar asymptotic behavior of Hamiltonians. In addition, it should be noted that absorption involving surface polar optical modes is predominantly an intra-band process, unlike electron scattering processes involving confinement (bulk) phonons, which is due to a smooth change in the amplitude of the electron–phonon interaction for surface phonons along the z-axis.

Thus, based on the comparison of the results obtained in the framework of models 1 and 2 and the present calculation, the following conclusions can be drawn:

(1) In the area of small thicknesses ($d < (7 - 10)R_V$) the results of calculations of the probability of light absorption using the bulk Fröhlich Hamiltonian give physically incorrect results for the probability of absorption and the absorption coefficient of light.

(2) For a structure with quantum wells made of strongly polar semiconductor materials, when the thickness of the well changes, a redistribution of the efficiency of absorption processes between different modes should be expected.

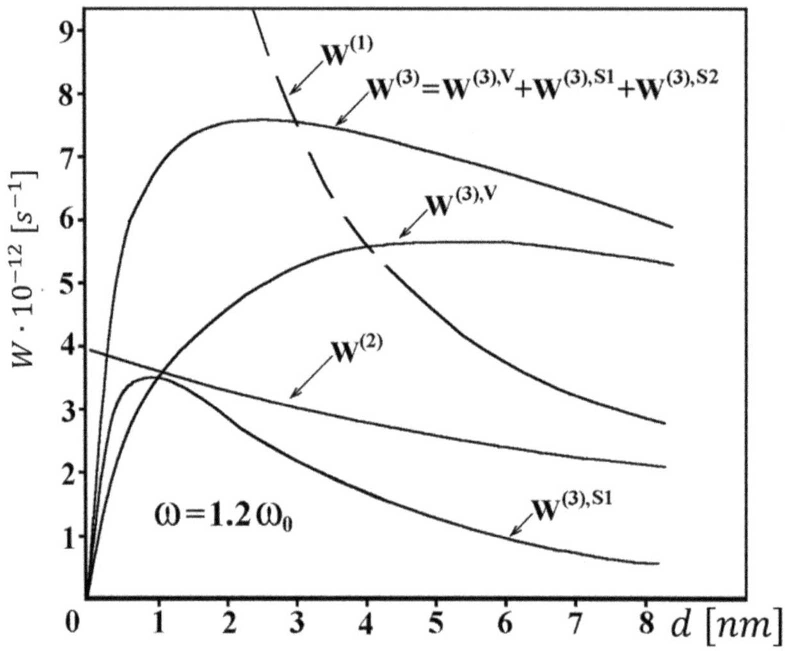

Figure 8.7. The probability of light absorption in a structure with quantum wells as a function of the thickness of the polar layer. The contributions are similar to figures 8.4 and 8.5. The magnitude of the frequency of light $\omega = 1.2\omega_0$. The picture corresponds to CdTe–nonpolar environment ($\varepsilon = 2$).

(3) From the standpoint of models 1 and 2, there is no photon–phonon resonance for structures with quantum wells made of nonpolar material bordering neighboring polar layers. At the same time, the results obtained in this paragraph indicate the possibility of such a process due to the existence of surface optical phonons.
(4) By selection of the substance of neighboring layers with large values ε_1 and ε_3 it is possible to significantly reduce the light absorption coefficient in a quantum well.

Note that for numerical comparison of the results obtained with experiments on IR absorption in thin films, a model with infinite barriers at the boundaries can be a good approximation for structures of the type CdTe/CdSe. However, for a type structure $Al_xGa_{1-x}As/GaAs$ it is necessary to take into account the final height of the barrier at the boundaries of the quantum well.

8.5 Cyclotron–phonon resonance in structures with quantum wells

Various types of resonances in massive crystals and thin semiconductor films (cyclotron–phonon, size-phonon, etc) caused by electronic transitions in the electromagnetic wave field involving optical phonons were discussed in [62–64].

In dimensionally limited crystals (films) [62], the spectrum of optical phonons is assumed to be identical to the spectrum of a massive crystal.

It is known, however, that the spectrum of optical vibrations of limited polar crystals (polar films) undergoes a radical restructuring [14, 65, 66], which should be taken into account in such studies. Otherwise, the dimensional and frequency dependence of the absorption coefficient $K(\Omega, d)$ become unfair even qualitatively.

In [62], the features of cyclotron–phonon absorption in semiconductor films in a quantizing magnetic field are theoretically investigated. The process consists of the transfer of an electron between Landau levels with simultaneous emission or adsorption of an optical phonon (cyclotron–phonon resonance). As shown in [63], when $\omega_c > \omega_0$, ω_c, ω_0—cyclotron and longitudinal optical frequencies—quantum absorption can only be associated with the transition to a higher Landau level, and the frequencies $\omega = m\omega_c \mp \omega_0$, m—an integer is resonant (the sign '−' refers to emission, '+' refers to phonon adsorption). By $\omega_c < \omega_0$ transitions with phonon emission and with similar resonant frequencies are possible (case $m = 0$ corresponds to the magnetic resonance).

To calculate the cyclotron–phonon absorption coefficient in a thin polar film, it is necessary to use a Pekar–Fröhlich Hamiltonian that takes into account the rearrangement of the phonon spectrum. Such a Hamiltonian for an arbitrary multilayer structure was obtained in [13, 21]. It summarizes the results of studies for various special cases: a polar crystal plate in vacuum [12], in nonpolar plates [67], crystal–vacuum contact [68, 69], etc.

8.5.1 Absorption coefficient

Consider a thin semiconductor film with dielectric permittivity ε_{02}, ε_2, occupying an area of space $0 \leqslant z \leqslant d$, bordering on semi-infinite identical nonpolar media $z < 0$; $z > 0$ with dielectric permittivity $\varepsilon_1 = \varepsilon_3 = \varepsilon$ (symmetric quantum well). Let the magnetic field be directed perpendicular to the film surfaces along the axis Z. We will also consider the condition fulfilled $E_g \gg E_{s.q.} = \pi^2\hbar^2 l^2/(2m_i^2 d^2)$, $i = e,\ h$—the criterion for the absence of mixing of energy bands.

We calculate the coefficient of cyclotron–phonon absorption of electromagnetic radiation by electrons whose energy spectrum has the form

$$E_{nl} = \left(n + \frac{1}{2}\right)\hbar\omega_c + E_0 \cdot l^2, \quad E_0 = \frac{\pi^2\hbar^2}{2m_e^2 d^2}, \tag{8.5.1}$$

where $n = 0, 1, 2, \ldots; l = 1, 2, 3, \ldots$.

The electron gas is assumed to be degenerate.

Following the works [62], we present the desired absorption coefficient in the form

$$K(\Omega) = \frac{\sqrt{\varepsilon(\Omega)}}{c} n_e \frac{V}{N(\Omega)}\left(1 - e^{-\frac{\hbar\Omega}{k_0 T}}\right)\mathrm{Av}_i \sum\nolimits_f 2\pi\, |\langle i\, |\widetilde{H}\, |f\rangle|^2 \delta(E_i - E_f), \tag{8.5.2}$$

where $\varepsilon(\Omega)$ is the real part of the dielectric constant (assumed to be dispersion-free); n_e is electron concentration; c is the speed of light in a vacuum; and $N(\Omega)$ is the number of photons in the initial state with frequency $\Omega = ck/\sqrt{\varepsilon(\Omega)}$. The multiplier in parentheses takes into account the forced emission of photons; $V = L_x L_y d$—normalization volume; and $E_{i,f}$—energy of the initial (i) and the final (f) conditions. The symbol Av_i means averaging over initial states with a distribution function:

$$W_{nl} = W_0 \mathrm{e}^{-\frac{\left(n+\frac{1}{2}\right)\hbar\omega_c}{kT}} \cdot \mathrm{e}^{-\frac{E_0}{kT}l^2}; \tag{8.5.3}$$

W_0 is the normalization factor:

$$W_0 = \frac{2\sinh\left(\frac{\hbar\omega_0}{2k_0T}\right)}{\sum\limits_{l=1}^{\infty}\exp\left(-\frac{E_0}{k_0T}l^2\right)}.$$

The matrix element of the transition probability in the dipole approximation has the form

$$\begin{aligned}
&\left\langle \alpha, 0_v, 0_{S_1}, 0_{S_2}, 0_k \,\middle|\, \widetilde{H} \,\middle|\, \alpha', \pm\mathbf{Q}, \pm\boldsymbol{\eta}_1, \pm\boldsymbol{\eta}_2, -\mathbf{k}\right\rangle \\
&= \sum\nolimits_{\alpha''}\left[\frac{\left\langle \alpha, 0 \,\middle|\, \hat{H}_L^V \,\middle|\, \alpha'', \pm\mathbf{Q}\right\rangle\left\langle \alpha'', 0 \,\middle|\, H_R \,\middle|\, \alpha', \pm\mathbf{k}\right\rangle}{E_\alpha - (E_{\alpha''} \pm \hbar\omega(\mathbf{Q}))}\right. \\
&\left. + \frac{\left\langle \alpha, 0 \,\middle|\, H_R \,\middle|\, \alpha'', \pm\mathbf{k}\right\rangle\left\langle \alpha'', 0 \,\middle|\, \hat{H}_L^V \,\middle|\, \alpha', \pm\mathbf{Q}\right\rangle}{E_\alpha - (E_{\alpha''} - \hbar\omega(\mathbf{k}))}\right] \\
&+ \sum\nolimits_{\alpha'', j=1,2}\left[\frac{\left\langle \alpha, 0 \,\middle|\, \hat{H}_L^{S_j} \,\middle|\, \alpha'', \pm\boldsymbol{\eta}_j\right\rangle\left\langle \alpha'', 0 \,\middle|\, H_R \,\middle|\, \alpha', \pm\mathbf{Q}\right\rangle}{E_\alpha - \left(E_{\alpha''} \pm \hbar\Omega_{S_j}\left(\boldsymbol{\eta}_j\right)\right)}\right. \\
&\left. + \frac{\left\langle \alpha, 0 \,\middle|\, H_R \,\middle|\, \alpha'', -\mathbf{k}\right\rangle\left\langle \alpha'', 0 \,\middle|\, \hat{H}_L^{S_j} \,\middle|\, \alpha', \pm\boldsymbol{\eta}_j\right\rangle}{E_\alpha - \left(E_{\alpha''} - \hbar\omega(\mathbf{k})\right)}\right].
\end{aligned} \tag{8.5.4}$$

The wave function describing the motion of an electron in a quantum well in a magnetic field directed along the axis has the form

$$|\alpha\rangle = \psi_{p_x}(x)\Phi_n(y)\psi_l(z) \equiv |p_x, \eta, l\rangle, \tag{8.5.5}$$

where

$$\psi_{p_x} = \frac{1}{\sqrt{2\pi\hbar}}\exp\left(\frac{ip_x x}{\hbar}\right) \tag{8.5.6a}$$

and describes free movement along the axis with a certain momentum p_x;

$$\Phi_n(y) = \frac{1}{\sqrt{\sqrt{\pi}a_H 2^n n!}} \exp\left[-\frac{(y-y_0)^2}{2a_H^2}\right] H_n\left(\frac{y-y_0}{a_H}\right), \tag{8.5.6b}$$

$$y_0 = \frac{cp_x}{eH}, \quad a_H = \left(\frac{\hbar}{m^*\omega_H}\right)^{1/2}, \quad \omega_H = \frac{eH}{m^*c} \tag{8.5.6c}$$

is the normalized wave function of an electron in a constant magnetic field (index $n = 0, 1, 2, \ldots$ numbers the Landau levels);

$$\psi_l(z) = \sqrt{\frac{2}{d}} \sin\left[\frac{\pi l}{d}z + \frac{\pi l}{2}\right] \tag{8.5.6d}$$

is the normalized wave function of a particle in a potential well with infinite walls; the index numbers the levels of dimensional quantization:

$$E_\alpha = \frac{p_x^2}{2m^*} + E_{n,\,l}. \tag{8.5.7}$$

The Hamiltonian of the interaction of an electron with bulk and surface optical phonons of a polar film has the form [67]

$$H_L = H_L^V + H_L^{S_1} + H_L^{S_2}, \tag{8.5.8}$$

where $H_L^{V,\,S_1,\,S_2}$ are Hamiltonians of interaction with bulk (V) and surface (S_1, S_2) optical phonons:

$$H_L^V = \sum\nolimits_{\mathbf{Q}(\eta,\,q_z)} C_\mathbf{Q} e^{i\eta\rho}\left[e^{iq_z z} + \frac{1}{1-C}\left(C - \frac{\cosh \eta\left(z - \frac{d}{2}\right)}{\cosh \frac{\eta d}{2}}\right)\right](b_{-\mathbf{Q}}^+ + b_\mathbf{Q}); \tag{8.5.9}$$

$$H_L^{S_1} = \sum\nolimits_\eta C_\eta^{(S_1)} e^{i\eta\rho} \frac{\cosh \eta\left(z - \frac{d}{2}\right)}{\cosh \frac{\eta d}{2}\left(b_{-\eta S_1}^+ + b_{\eta S_1}\right)}; \tag{8.5.10}$$

$$H_L^{S_2} = \sum\nolimits_\eta C_\eta^{(S_2)} e^{i\eta\rho} \frac{\sinh \eta\left(z - \frac{d}{2}\right)}{\cosh \frac{\eta d}{2}\left(b_{-\eta S_2}^+ + b_{\eta S_2}\right)}. \tag{8.5.11}$$

The squares of the modules of the interaction constants have the forms, respectively:

$$\left|C_Q\right|^2 = \frac{1}{V} \cdot \frac{4\pi\alpha_V(\hbar\omega_0)^2 R_V}{Q^2}, \tag{8.5.12a}$$

$$\left| C_\eta^{S_{1,\,2}} \right|^2 = \frac{1}{S} \cdot \frac{2\pi\alpha_{S_{1,2}}(\hbar\Omega_{S_{1,\,2}})^2 R_{S_{1,\,2}}}{\eta \tanh\left(\frac{\eta d}{2}\right)}, \tag{8.5.12b}$$

where

$$\omega_0 = \omega_{TO}\left(\frac{\varepsilon_0}{\varepsilon_\infty}\right)^{\frac{1}{2}};\ R_V = \left(\frac{\hbar}{2m^*\omega_0}\right)^{\frac{1}{2}};\ \alpha_V = \frac{e^2}{4\pi\varepsilon_0\hbar\Omega R_V}\left(\frac{1}{\varepsilon_\infty} - \frac{1}{\varepsilon_0}\right); \tag{8.5.12c}$$

$$\Omega_{S_1} = \omega_{TO}\left(\frac{\varepsilon_0 + \varepsilon\coth\frac{\eta d}{2}}{\varepsilon_\infty + \varepsilon\coth\frac{\eta d}{2}}\right)^{\frac{1}{2}};\ \Omega_{S_2} = \omega_{TO}\left(\frac{\varepsilon_0 + \varepsilon\tanh\frac{\eta d}{2}}{\varepsilon_\infty + \varepsilon\tanh\frac{\eta d}{2}}\right)^{\frac{1}{2}}; \tag{8.5.12d}$$

$$\alpha_{S_1} = \frac{e^2}{4\pi\varepsilon_0\hbar}\left(\frac{m^*}{2\hbar\Omega_{S_1}}\right)^{\frac{1}{2}}\left(\frac{1}{\varepsilon_\infty + \varepsilon\coth\frac{\eta d}{2}} - \frac{1}{\varepsilon_0 + \varepsilon\coth\frac{\eta d}{2}}\right); \tag{8.5.12e}$$

$$\alpha_{S_2} = \frac{e^2}{4\pi\varepsilon_0\hbar}\left(\frac{m^*}{2\hbar\Omega_{S_2}}\right)^{\frac{1}{2}}\left(\frac{1}{\varepsilon_\infty + \varepsilon\tanh\frac{\eta d}{2}} - \frac{1}{\varepsilon_0 + \varepsilon\tanh\frac{\eta d}{2}}\right); \tag{8.5.12f}$$

$$C(\eta) = \left(\frac{2}{\eta d}\tanh\frac{\eta d}{2}\right)^{\frac{1}{2}};\ R_{S_j} = \left(\frac{\hbar}{2m^*\Omega_{S_j}}\right)^{\frac{1}{2}}. \tag{8.5.12g}$$

Matrix elements of interaction with the lattice have the forms:

(a) with bulk optical phonons:

$$\begin{aligned}&\left\langle n, p_x, l, 0 \,\middle|\, \hat{H}_L^v \,\middle|\, n', p_x', l', \pm\mathbf{Q}\right\rangle = C_{\mathbf{Q}}\left[N(\mathbf{Q}) + \frac{1}{2} \pm \frac{1}{2}\right]^{1/2}\\ &\delta(p_x, p_x' \pm \hbar Q_x)M_{nn'}\left(\frac{\hbar Q_\perp}{p_H}\right)\\ &\times\left\langle l \,\middle|\, e^{iq_z z} + \frac{1}{1-C}\left(C - \frac{\cosh\eta\left(z - \frac{d}{2}\right)}{\cosh\frac{\eta d}{2}}\right) \,\middle|\, l'\right\rangle;\end{aligned} \tag{8.5.13}$$

(b) with surface optical phonons:

$$\begin{aligned}&\left\langle n, p_x, l, 0 \,\middle|\, \hat{H}_L^{S_1} \,\middle|\, n', p_x', l', \pm\eta\right\rangle = C_\eta\left[N(\eta) + \frac{1}{2} \pm \frac{1}{2}\right]^{1/2}\\ &\delta(p_x, p_x' \pm \hbar\eta_x)M_{nn'}\left(\frac{\hbar\eta}{p_H}\right)\\ &\times\left\langle l \,\middle|\, \cosh\eta\left(z - \frac{d}{2}\right) \,\middle|\, l'\right\rangle\end{aligned} \tag{8.5.14}$$

(matrix element from odd mode on the Hamiltonian $\hat{H}_L^{S_2}$ gives a negligible contribution and is not taken into account in the future).

Here:

$$M_{nn'}(\xi) = (-1)^{n-n'}\left(\frac{n'!}{n!}\right)^{1/2} \xi^{n-n'} L_{n'}^{n-n'}(\xi^2)\exp\left(-\frac{\xi^2}{2}\right); \tag{8.5.15}$$

$L_{n'}^{n-n'}(\xi^2)$ is a generalized Laguerre polynomial, $p_H = \hbar\lambda_H^{-1}$ is magnetic momentum, and $\lambda_H = (\hbar/(2m^* \ \omega_c))^{1/2}$ is magnetic length.

Next, we will consider the ultraquantum limit, in which the energy of dimensional quantization is much greater than the energy of a photon:

$$E_l \gg \hbar\Omega, \tag{8.5.16}$$

therefore, let's put $l = 1$.

In the case of a dimensional phonon resonance in a structure with a quantum well from a polar semiconductor, the matrix elements included in expressions (8.5.8) and (8.5.9) have the form

$$\left\langle 1 \left| e^{iq_z z} + \frac{1}{1-C}\left(C - \frac{\cosh \eta\left(z - \frac{d}{2}\right)}{\cosh \frac{\eta d}{2}}\right) \right| 1 \right\rangle = \frac{\sin\left(\frac{q_z d}{2}\right)}{\frac{q_z d}{2}\left[1 - \left(\frac{q_z d}{2\pi}\right)^2\right]} + \Pi\left(Q_\perp, d\right) \equiv \psi_1\left(q_z, Q_\perp, d\right). \tag{8.5.17}$$

Here:

$$\Pi(Q_\perp, d) = \frac{C}{1-C}\left(C - \frac{1}{1 + \left(\frac{Q_\perp d}{2\pi}\right)^2}\right); \tag{8.5.18a}$$

$$\left\langle 1 \left| \cosh \eta\left(z - \frac{d}{2}\right) \right| 1 \right\rangle = \frac{\sinh\left(\frac{\eta d}{2}\right)}{\left(\frac{\eta d}{2}\right)} \frac{1}{\left[1 + \left(\frac{\eta d}{2\pi}\right)^2\right]} \equiv \psi_2(\eta, d). \tag{8.5.18b}$$

The operator of the interaction of an electron with an electromagnetic wave field has the form

$$\mathbf{A} = \sum\nolimits_{\mathbf{k}}\left(\frac{2\pi c^2}{\Omega\varepsilon(\Omega)V}\right)\mathbf{e}_\mathbf{k} e^{i\mathbf{k}\mathbf{r}} a_\mathbf{k} + \text{h. c.} \tag{8.5.19}$$

Here $\mathbf{e}_\mathbf{k}$ is the unit vector of light polarization, and $b^+_{\mathbf{Q},\eta_S}$, $b_{\mathbf{Q},\eta_S}$; $a^+_\mathbf{k}$, $a_\mathbf{k}$ are the operators of the birth and annihilation of phonons and photons, respectively.

Choosing the direction $\mathbf{H}$ along the axis Z; $\mathbf{e}_\mathbf{k}$—along the axis, we get two absorption coefficients:

$$K^{\pm}(\Omega) = \sum\nolimits_{n,\,n'=0}^{\infty} K^{\pm}_{n,\,n'}(\Omega), \tag{8.5.20}$$

where

$$K^{\pm}_{n,\,n'}(\Omega) = K^{\pm}_{n,\,n',\,V}(\Omega) + K^{\pm}_{n,\,n',\,S}(\Omega). \tag{8.5.21}$$

Substituting expressions (8.5.12)–(8.5.14) into (8.5.2) and performing the necessary transformations, we obtain

$$\begin{aligned}
K^{\pm}_{n,n',V}(\Omega) = {} & A_1(\Omega)\left\{1 - \exp\left(-\frac{\hbar(\Omega)}{kT}\right)\right\} \\
& \exp\left(-\frac{E_0}{kT}\right)\frac{\alpha_V(\hbar\omega_0)^2 R_v}{\pi\hbar\omega_c} 2\sinh\left(\frac{\hbar\omega_c}{2kT}\right) \\
& \times \int_0^{\infty} Q_\perp^2 dQ_\perp \left[N_V(\omega_0) + \frac{1}{2} \pm \frac{1}{2}\right] M_{nn'}\left(\frac{\hbar Q_\perp}{p_H}\right) \\
& \delta\left[(n - n') + \frac{\Omega}{\omega_c} \mp \frac{\omega_0}{\omega_c}\right] \\
& \times \left\{\left[\frac{2(Q_\perp d)^3 + 5(2\pi)^2(Q_\perp d)}{2\left[(Q_\perp d)^2 + (2\pi)^2\right]^2} + \frac{1}{1 + \left(\frac{Q_\perp d}{2\pi}\right)^2} \cdot \frac{Q_\perp d - 1 + e^{-Q_\perp d}}{(Q_\perp d)^2}\right]\right. \\
& - \frac{1}{1 + \left(\frac{Q_\perp d}{2\pi}\right)^2} \cdot \frac{(1 - e^{-Q_\perp d})}{Q_\perp d}\left[\frac{1}{1 - C}\left(\varphi_1(Q_\perp, d) - C\right)\right] \\
& \left. + \frac{1}{2}\left[\frac{1}{1 - C}\left\{C - \varphi_1(Q_\perp, d)\right\}\right]^2\right\}.
\end{aligned} \tag{8.5.22}$$

Expression for $K^{\pm}_{n,\,n',\,S}(\Omega)$ integrates with δ-functions; as a result we get

$$K^{\pm}_{n,n\prime,S}(\Omega) = A_2(\Omega)\left\{1 - \exp\left(-\frac{\hbar(\Omega)}{kT}\right)\right\}\left(-\frac{E_0}{kT}\right)2\sinh\left(\frac{\hbar\omega_c}{2kT}\right)$$
$$\times \frac{\alpha_{S_1}(\hbar\Omega_{S_1})^2 R_{S_1}}{2\pi\tanh\left(\frac{\eta d}{2}\right)\hbar\omega_c}$$
$$\left\{\eta^2\left[N_T(\omega_0) + \frac{1}{2} \pm \frac{1}{2}\right]\left|M_{nn\prime}\left(\frac{\hbar\eta}{p_H}\right)\right|^2 |\psi_2(\eta, d)|^2\right\}\Bigg|_{\eta=\eta_0} . \tag{8.5.23}$$
$$\frac{1}{|g'(\eta_0)|},$$

where

$$g(\eta) = n - n' + \frac{\Omega}{\omega_c} \mp \frac{\Omega_{S_1}(\eta, d)}{\omega_c}; \tag{8.5.24a}$$

$$g(\eta_0) = 0; \tag{8.5.24b}$$

$$A_1(\Omega_{S_1}) = \frac{n_e\alpha_R(2\pi)^2}{4(m^*)^2\Omega_{S_1}}\left[\frac{1}{(\Omega_{S_1} + \omega_c)^2} + \frac{1}{(\Omega_{S_1} - \omega_c)^2}\right]; \tag{8.5.24c}$$

$$A_2(\Omega_{S_2}) = \frac{n_e\alpha_R(2\pi)^2}{4(m^*)^2\Omega_{S_2}}\left[\frac{1}{(\Omega_{S_2} + \omega_c)^2} + \frac{1}{(\Omega_{S_2} - \omega_c)^2}\right]; \tag{8.5.24d}$$

$\alpha_R = e^2/(\hbar c\sqrt{\varepsilon(\Omega)})$ is the coupling constant of an electron with a high-frequency field;

$$N_{TV}(\omega_0) = \left\{\exp\left(\frac{\hbar\omega_0}{kT}\right) - 1\right\}^{-1}; \; N_{TS_1}(\Omega_{S_1}) = \left\{\exp\left(\frac{\hbar\Omega_{S_1}}{kT}\right) - 1\right\}^{-1}. \tag{8.5.24e}$$

8.5.2 Calculation results

Availability of the δ-function in the expression (8.5.22) indicates that the absorption (emission) of light with the participation of bulk phonons in a quantum well, as in a massive crystal, has a resonant character and the optical spectrum will be represented by a set of separate lines with resonant frequencies determined from the expression:

$$\Omega^{\pm}_{n,\,n',\,V} = (n' - n)\omega_c \pm \omega_0. \tag{8.5.25}$$

Distinctive features of the expression for $K^{\pm}_{n,\,n',\,V}(\Omega)$ From the results obtained in [63] for a massive crystal and [62] for a structure with a quantum well are:

(1) the nature of the dimensional dependence $K^{\pm}_{n,\,n',\,V}(\Omega)$ determined by the asymptotic behavior: in the limit $d \to \infty$ from the expression (8.5.22) follows the expression of work [62] for the case $n = n' = 1$; in the limit $d \to 0$ $K^{\pm}_{n,\,n',\,V}(\Omega)$ vanishes, due to the vanishing of the function $\psi_1(q_z, Q_\perp, d)$;

(2) the position of the peaks does not depend on the thickness of the quantum well d, a is determined by the induction of a magnetic field; at the same time, the height of the peaks depends on from d: with growth d the height of the peaks also increases when $d \to \infty$ reaches a constant value. From the resonance number $m = n' - n$ the height of the peaks depends slightly.

The formula for $K^{\pm}_{n,\,n',\,S}(\Omega)$ also contains a number of features. The most significant of them are the following:

(1) the absorption is resonant, but the absorption and emission lines have a finite width determined by the dispersion of surface optical phonons, the more noticeable the smaller the width of the quantum well;

(2) the positions of the maxima of these peaks are shifted relative to the lines $\Omega^{\pm}_{n,\,n',\,V}$, defined by the formula (8.5.20);

(3) the positions of the maxima and the height of the peaks depend on d:

 (a) with decreasing d the height of the maximum increases, however, in the limit $d \to 0$ it tends to zero due to the vanishing of the electron–phonon interaction matrix element (disappearance of polar matter);

 (b) with decreasing d the maxima of the lines shift more strongly relative to the frequencies $\Omega^{\pm}_{n,\,n',\,V}$ (the latter may be significant in the analysis of experiments);

 (c) the size and position of the lines are influenced by neighboring environments.

Figures 8.4–8.7 show the patterns of peak locations for the InSb quantum well with thicknesses d=3.0; 6.0 and 10.0 nm. The vertical lines correspond to $K^{\pm}_{n,\,n',\,V}(\Omega)$ (calculated according to the formula (8.5.22)) and are located at frequencies determined by the formula (8.5.25).

By $\omega_c > \Omega$ for the position of the emission peaks, we obtain

$$\Omega = m\omega_c + \omega_0, \;\; m = 0, \;1, \;2, \;\ldots, \tag{8.5.26}$$

and absorption—in points

$$\Omega = m\omega_c\omega_0, \;\; m = 1, \;2, \;\;\ldots\,. \tag{8.5.27}$$

The transitions that make the main contribution to the intensity are transitions from the lowest Landau level. Absorption satellites have an intensity, in $\exp(\hbar\omega_0/(kT))$ times smaller.

Lines corresponding to $K^{\pm}_{n,\,n',\,S}(\Omega)$ are located from left to right of the lines $K^{\pm}_{n,\,n',\,V}(\Omega)$.

For the process of photon absorption with phonon absorption, the frequency of light must lie in the interval

$$\omega_{TO} + n'\omega_c \leqslant \Omega^+ \leqslant \omega_{LO}\left(\frac{1 + \frac{\varepsilon}{\varepsilon_0}}{1 + \frac{\varepsilon}{\varepsilon_\infty}}\right) + n'\omega_c. \tag{8.5.28}$$

In the case of photon absorption with phonon emission, the frequency of light should lie in the interval

$$n'\omega_c - \omega_{LO}\left(\frac{1 + \frac{\varepsilon}{\varepsilon_0}}{1 + \frac{\varepsilon}{\varepsilon_\infty}}\right) \leqslant \Omega^- \leqslant n'\omega_c - \omega_{TO}. \tag{8.5.29}$$

The increase, as follows from the formula (8.5.23), leads, on the one hand, to a decrease in the amplitude of the curve $K^{\pm}_{n,\,n',\,S}(\Omega)$; on the other hand, the frequency range decreases, which can be absorbed with the participation of surface optical phonons. This, in turn, leads to a decrease in the half-width of the Lorentz curves and at ε, close to ε_0, absorption will occur at frequencies close to $\Omega = n'\omega_c \pm \omega_0$.

The results shown in figures 8.8–8.10 are obtained for the following parameter values: $n_e = 10^{16}$ cm^{-3}; $T = 300$ K; $H = 1000$ Gs; ε=1; $\varepsilon_0 = 17.5$; $\varepsilon_\infty = 16$; $\omega_{TO} = 3.5 \cdot 10^{13}$ s^{-1}; $\omega_{LO} = 3.7 \cdot 10^{13}$ s^{-1};

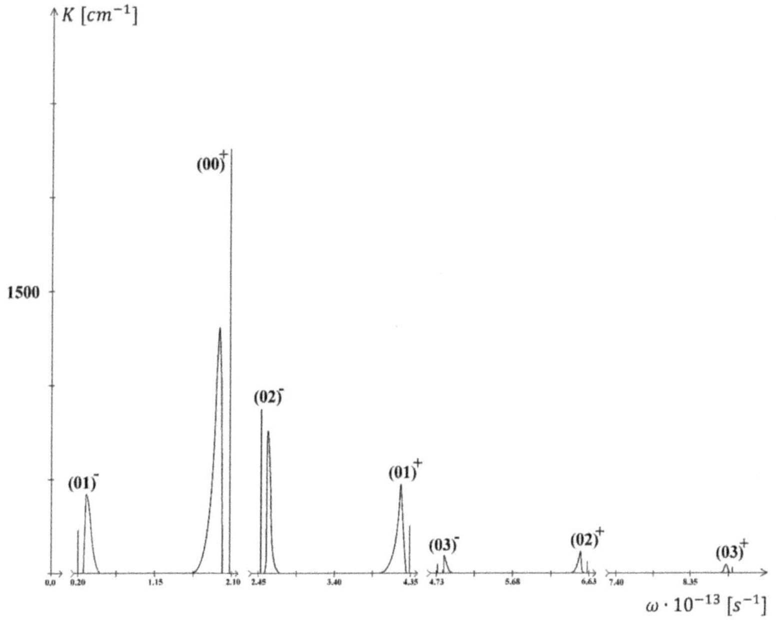

Figure 8.8. Frequency dependence of the absorption coefficient in PbTe (film thickness $d = 3\,nm$). Shown $K^{+}_{00,\,V}(\Omega)$; $K^{+}_{01,\,V}(\Omega) \cdot 10^1$; $K^{+}_{02,\,V}(\Omega) \cdot 10^1$; $K^{+}_{03,\,V}(\Omega) \cdot 10^1$; $K^{-}_{01,\,V}(\Omega) \cdot 10^1$; $K^{-}_{02,\,V}(\Omega)$; $K^{-}_{03,\,V}(\Omega) \cdot 10^1$; $K^{+}_{00,\,S}(\Omega) \cdot 10^3$; $K^{+}_{01,\,S}(\Omega) \cdot 10^4$; $K^{+}_{02,\,S}(\Omega) \cdot 10^4$; $K^{+}_{03,\,S}(\Omega) \cdot 10^4$; $K^{-}_{01,\,S}(\Omega) \cdot 10^4$; $K^{-}_{02,\,S}(\Omega) \cdot 10^3$; $K^{-}_{03,\,S}(\Omega) \cdot 10^4$.

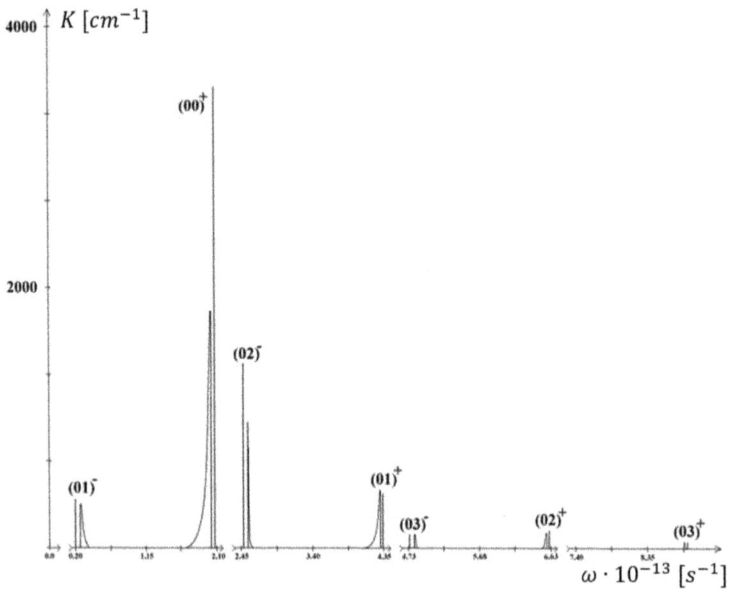

Figure 8.9. Frequency dependence of the absorption coefficient in PbTe (film thickness $d = 6$ nm). Shown $K^{+}_{00,\,V}(\Omega)$; $K^{+}_{01,\,V}(\Omega)\cdot 10^1$; $K^{+}_{02,\,V}(\Omega)\cdot 10^1$; $K^{+}_{03,\,V}(\Omega)\cdot 10^1$; $K^{-}_{01,\,V}(\Omega)\cdot 10^1$; $K^{-}_{02,\,V}(\Omega)$; $K^{-}_{03,\,V}(\Omega)\cdot 10^1$; $K^{+}_{00,\,S}(\Omega)\cdot 10^3$; $K^{+}_{01,\,S}(\Omega)\cdot 10^4$; $K^{+}_{02,\,S}(\Omega)\cdot 10^4$; $K^{+}_{03,\,S}(\Omega)\cdot 10^4$; $K^{-}_{01,\,S}(\Omega)\cdot 10^4$; $K^{-}_{02,\,S}(\Omega)\cdot 10^3$; $K^{-}_{03,\,S}(\Omega)\cdot 10^4$.

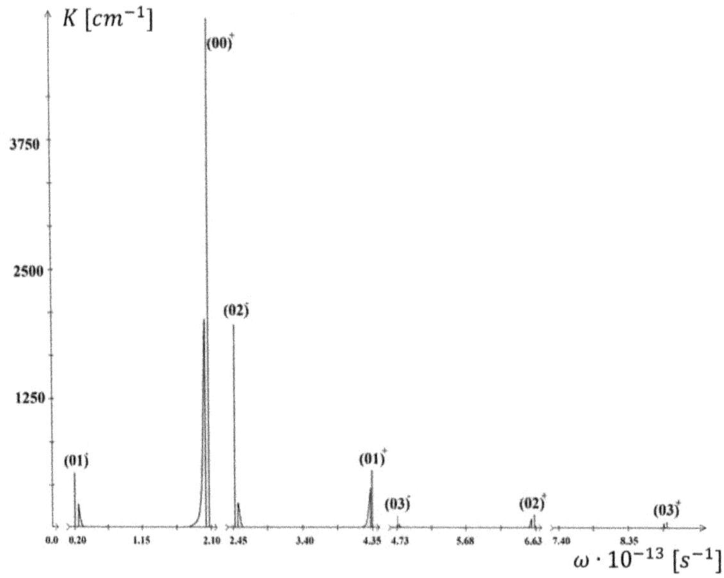

Figure 8.10. Frequency dependence of the absorption coefficient in PbTe (film thickness $d = 2$ nm). Shown $K^{+}_{00,\,V}(\Omega)$; $K^{+}_{01,\,V}(\Omega)\cdot 10^1$; $K^{+}_{02,\,V}(\Omega)\cdot 10^1$; $K^{+}_{03,\,V}(\Omega)\cdot 10^1$; $K^{-}_{01,\,V}(\Omega)\cdot 10^1$; $K^{-}_{02,\,V}(\Omega)$; $K^{-}_{03,\,V}(\Omega)\cdot 10^1$; $K^{+}_{00,\,S}(\Omega)\cdot 10^3$; $K^{+}_{01,\,S}(\Omega)\cdot 10^4$; $K^{+}_{02,\,S}(\Omega)\cdot 10^4$; $K^{+}_{03,\,S}(\Omega)\cdot 10^4$; $K^{-}_{01,\,S}(\Omega)\cdot 10^4$; $K^{-}_{02,\,S}(\Omega)\cdot 10^3$; $K^{-}_{03,\,S}(\Omega)\cdot 10^4$.

$K_{01}^{+} = 26\ \text{cm}^{-1}$; $d = 3.0$ nm; $\alpha = 0.014$;
$K_{01}^{+} = 45\ \text{cm}^{-1}$; $d = 6.0$ nm;
$K_{01}^{+} = 9\ \text{cm}^{-1}$; $d = 10.0$ nm.

In a massive crystal $K_{01}^{+} \approx 10\ \text{cm}^{-1}$ (note that in this case the lattice absorption is of the order of $10^4\ \ \text{cm}^{-1}$).

From figures 8.8 to 8.10, it can be seen that:

(1) new peaks corresponding to surface optical phonons appear in the absorption spectrum;
(2) for both bulk and surface phonons, the absorption is resonant, but the absorption peaks have a different shape: for bulk phonons, the δ-shaped peaks at frequencies $\omega = |n\omega_c \pm \omega_0|$, $n = 0, 1, 2, \ldots$, for surface Lorentz curves with maxima at $\omega = |n\omega_c \pm \Omega_S|$ and with a half-width, which is determined by the variance $\Omega_S(d)$;
(3) total absorption coefficient $K(\Omega)$ can be written in the following form: $K(\Omega) = K_V(\Omega) + K_S(\Omega)$, and as the surface part $K_S(\Omega)$, so is the bulk one $K_V(\Omega)$ have strong dimensional dependencies;
(4) neighboring environments can influence the value of $K(\Omega)$.

8.6 Quantum theory of electron emission from the 'metal–dielectric' structure in strong electric fields

The study of vacuum breakdown has shown that an important role in the initiation of vacuum discharge is played by emission processes stimulated by nonmetallic inclusions (insular or solid dielectric films, adsorbed atoms, etc) on the surface of the metal cathode [70]. It is with the presence of such inclusions that the existence of abnormally high electric field amplification coefficients is associated β, determined from the characteristics of Fowler–Nordheim [71]. The values of these coefficients can reach several hundred [72]. Cathode microarrays with a radius ratio at the apex to a height of several hundred would be easily detected on the cathode surface. Emission processes involving these nonmetallic inclusions lead to the formation of so-called cathode spots of the first type: short-lived plasma sources generated in the interelectrode gap [73, 74]. The functioning of the spots of the first type in many cases determines the initial stage of vacuum breakdown.

The issue of the correct description of the emission process from a cathode coated with nonmetallic inclusions and their role in initiating vacuum breakdown has recently become very important in connection with the development of electron–positron colliders of the TeV energy range [75]. Initiation of vacuum breakdown on the surface of the accelerating structure is the main problem limiting the rate of energy gain by particles [76]. It is possible that the presence of nonmetallic inclusions on the surface of the accelerating structure is responsible for the existence of abnormally high electric field coefficients. Emission centers with high β are potential sources of explosive electron emission, which, apparently, is the main cause of vacuum breakdown in accelerating structures [77]. The task of correctly describing emission processes involving nonmetallic inclusions is relevant for studying the initial stage of development of high-voltage discharge in gaseous media [78]. In this case, given the presence of a gaseous medium, such inclusions are inevitably present on the surface of the cathodes.

In this section, emission processes in the structure of a metal cathode–adsorbed nonmetallic film–vacuum (gas medium) are considered on the basis of a quantum mechanical description of the interaction of an electron with a fast polarization of the medium (plasma oscillations of valence electrons of a dielectric and free electrons of a metal). In this case, the tunneling electron is a quasi-particle—the Toyozawa electron polaron, having finite dimensions defined by the radius R_p [79–81]. The theory of the surface electron polaron, on the basis of which its parameters are calculated, is given in [82].

This approach allowed us to obtain an expression for the potential energy of the interaction of an electron with the induced polarization of the medium when studying the emission of electrons from a metal into a dielectric (gas medium) [21, 83–85], valid over the entire range of values of the distance x of the electron to the interface of the metal with the dielectric, including at the point $x = 0$ (the interface) at which the classical potential of the image forces diverges. In the limit $x \gg R_p$ this quantum potential passes into the electronic part of the classical potential of the image forces $U_{ie}(x) = -e^2/(4\varepsilon x)$, where ε is high-frequency dielectric permittivity of the dielectric.

8.6.1 Basic equations

The density of the emission current from the metal–dielectric structure (figure 8.11) can be expressed by the formula [86]

$$j(E, T) = e\int_{-\infty}^{\infty} N(W, T)D(W, T)dW. \tag{8.6.1}$$

Here $j(E, T)$—emission current density, e—electron charge, E—the intensity of the applied electric field, T—absolute temperature, $N(W, T)$—the number of electrons falling per unit area of the barrier in one second and having an energy close to W,

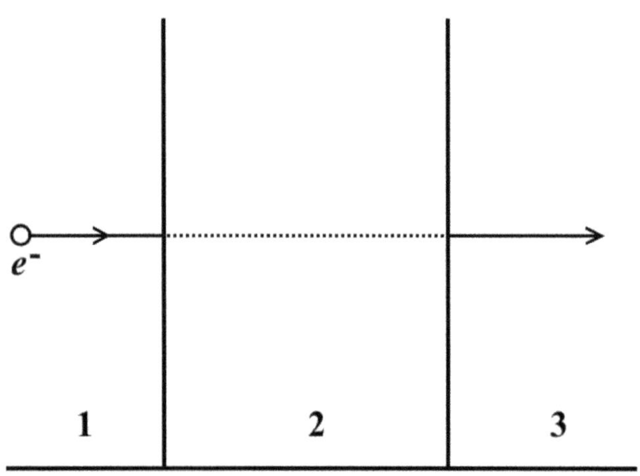

Figure 8.11. Contact metal cathode (1); nonmetallic film (2); thick d–vacuum (3).

and $D(W, E)$—transparency of the potential barrier at the contact of the cathode with the external environment.

In the Sommerfeld model, the number of electrons falling per unit area per second is

$$N(W, T) = \frac{4\pi m k_0 T}{h^3} \ln\left\{1 + \exp\left(-\frac{W - W_F}{kT}\right)\right\}, \tag{8.6.2}$$

where W_F is metal Fermi energy; k_0, h are Boltzmann and Planck constants, respectively; and m is the effective mass of the electron. The energy is counted from zero for a free electron outside the metal (figure 8.12), so the work of the output Φ_0 of the electron is equal in modulus to the Fermi energy $W_{F;}$ W is part of the energy for the movement of the electron in a direction perpendicular to the surface outside the metal:

$$W = \frac{p^2(x)}{2m} + W_t(x), \tag{8.6.3a}$$

$p(x)$ is the momentum of the electron along the normal to the surface, and $W_t(x)$ is effective potential energy of an electron (figure 8.12), having the form

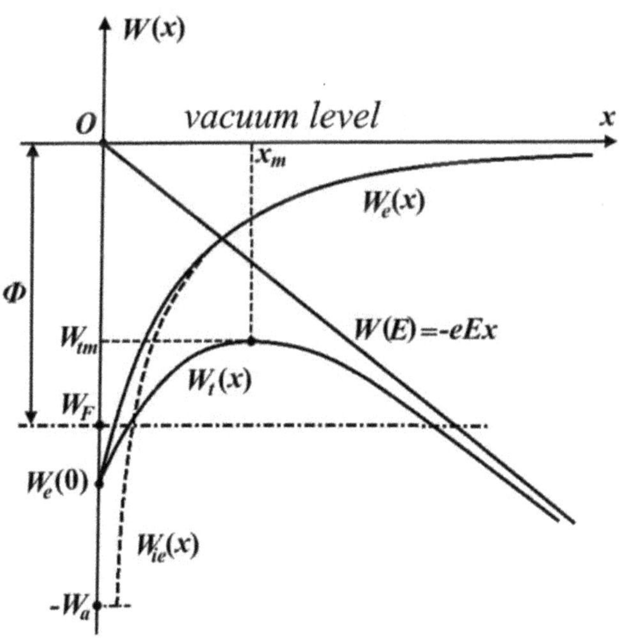

Figure 8.12. Potential energy$W(x)$ of an electron near the metal surface in the $x \geqslant 0$. Here: $W_{ie}(x)$ is classical electronic image potential; $W_e(x)$ is the quantum potential of the interaction of an electron with its induced rapid polarization; $W(E) = -eEx$ is potential energy of interaction of an electron with a field; W_a is effective potential energy of an electron inside a metal (constant value); $\Phi = W_F$ is the work of the exit (equal to Fermi energy); and W_{tm} is maximum value of $W_t(x)$; $W_e(0) = W_t(x = 0)$.

$$W_t(x) = \begin{cases} W_e(x) - eEx, & x \geqslant 0, \\ -W_a, & x < 0, \end{cases} \tag{8.6.3b}$$

where $W_e(x)$ is the quantum potential energy of the interaction of an electron with a fast polarization induced by it at the metal–dielectric contact (in the limit, it passes into the classical image potential at $x \gg R_p$, where

$$R_p = (\hbar/(2m\omega_p))^{1/2}$$

is the radius of the electron polaron and ω_p is the frequency of bulk plasma oscillations).

In [86] and subsequent works as a potential $W_e(x)$ the classical potential of the image forces is used, which has a singularity at $x = 0$ (i.e., it is incorrect both at the contact boundary itself and near it ($x \sim R_p$)).

In the works [21, 82–85] based on the polaronic theory of potential and image forces, an expression for the quantum potential of the image is obtained $W_e(x)$, describing the interaction of an electron with a fast polarization induced by it at the metal–dielectric contact, and fair in the entire range of values of the fast polarization induced by it at the metal–dielectric contact, and fair in the entire range of values $x \geqslant 0$:

$$\begin{aligned} W_e(x) = &-e^2 \int_0^\infty d\eta \;\; e^{-2\eta x}\left[1 - \frac{\eta^2}{2\left(\eta^2 + k_S^2 + k_{F_S}^2\right)}\right] \\ &\times \sum_{j=1,2} \frac{\varphi_j\left(\varepsilon_j(\eta)\right)}{\Omega_j^2(\eta)\left[1 + R_{S_j}^2\eta^2\right]} - e^2 \int_0^\infty k_\perp dk_\perp \int_0^\infty dk_x\left[1 + e^{-2k_\perp x}\right. \\ &\left. - 2e^{-k_\perp x}\cos(k_x x)\right] \frac{(\varepsilon - 1)\left[1 - \frac{k^2}{2\left(k^2 + k_S^2 + k_V^2\right)}\right]}{\varepsilon_1 k^2\left(1 + R_V^2 k^2\right)}, \end{aligned} \tag{8.6.4}$$

where

$$\begin{aligned} \varphi_1(\varepsilon_1(\eta)) = &\frac{1}{(\varepsilon_1(\eta) + 1)^2}\left[\omega_{pV}\left(\frac{(\varepsilon_1(\eta) - 1)}{\varepsilon_1(\eta)}\right)^{1/2} - \omega_p F_{12}(\Omega_1(\eta))\right]^2 \\ &\times \left[1 + F_{12}^2(\Omega_1(\eta))\right]^{-1} \cdot \left[1 + R_{S_1}^2(\eta)\eta^2\right]^{-1}; \end{aligned} \tag{8.6.5}$$

$$\begin{aligned} \varphi_2(\varepsilon_2(\eta)) = &\frac{1}{(\varepsilon_2(\eta) + 1)^2}\left[-\omega_{pV}(\eta)\left(\frac{(\varepsilon_2(\eta) - 1)}{\varepsilon_2(\eta)}\right)^{\frac{1}{2}} F_{21}(\Omega_2(\eta)) + \omega_p\right]^2 \\ &\times \left[1 + F_{21}^2(\Omega_2(\eta))\right]^{-1} \cdot \left[1 + R_{S_2}^2(\eta)\eta^2\right]^{-1}. \end{aligned} \tag{8.6.6}$$

Here $\varepsilon_j(\eta)$ is the dielectric function of a quantum dielectric, for which the expression is obtained in [85]:

$$\varepsilon_j(\eta) = 1 + \frac{\varepsilon - 1}{\left[1 + \frac{\eta^2}{\lambda^2}(\varepsilon - 1)\right]\left(1 + \frac{3\eta^2}{4\eta_F^2}\right)}; \tag{8.6.7}$$

$$\lambda^{-1} = \frac{2(\varepsilon - 1)}{\pi\sqrt{\varepsilon - 1}} R_{pS} \left\{ \sqrt{\varepsilon} + (\varepsilon - 1)\arctan\sqrt{\varepsilon} - \frac{\pi}{2}(\varepsilon - 1) \right\}; \tag{8.6.8}$$

$R_{pS_{1,2}} = (\hbar/(2m\Omega_{pS_{1,2}}))^{1/2}$ is the radius of the surface electron polaron and $\Omega_{pS_{1,2}}$ are frequencies of surface plasma oscillations;

$$k_S = 2\pi^{-1}k_F;\ k_F = (3\pi^2 N)^{1/3}.$$

Explicit expressions for quantities $F_{12}(\Omega_1(\eta))$; $F_{21}(\Omega_2(\eta))$, included in the formulas (8.6.5) and (8.6.6), are given in the monograph [21] (pp. 155 and 156).

Formulas (8.6.4)–(8.6.8) for the quantum potential $W_e(x)$ are cumbersome and not very convenient for calculations. In [84, 87, 88] it is shown that the potential energy $W_e(x)$ can be approximated with high accuracy by the expression

$$\widetilde{W}_e(x) \approx -\frac{e^2}{(4x + x_0)\varepsilon}. \tag{8.6.9}$$

Here x_0 is a parameter from the theory of polarons for which the expression is obtained:

$$x_0 = \frac{e^2}{\varepsilon W_e(0)}, \tag{8.6.10}$$

where

$$W_e(0) = -e^2 \int_0^\infty d\eta \left[1 - \frac{\eta^2}{2\left(\eta^2 + k_S^2 + k_{FS}^2\right)}\right] \times \left\{ \frac{\varphi_1(\varepsilon_1(\eta))}{\Omega_1^2(\eta)\left[1 + R_{S_1}^2\eta^2\right]} + \frac{\varphi_2(\varepsilon_2(\eta))}{\Omega_2^2(\eta)\left[1 + R_{S_2}^2\eta^2\right]} \right\}. \tag{8.6.11}$$

It is of interest to estimate the value of x_0 in comparison with the characteristic scales of the theory. For the case of 'crystal–vacuum' contact without taking into account spatial dispersion, the integral (8.6.11) is calculated exactly and for x_0 we get the expression

$$x_0 = \frac{2e^2\left(1 - \frac{R_S^2}{R_F^2}\right)}{\pi\alpha_{pS}\hbar\Omega_{pS}\left(1 + \frac{R_S^2}{R_F^2} + 2\frac{R_S^2}{R_F^2}\right)}, \tag{8.6.12}$$

where R_{pS}, α_{pS} are the radius of the surface electron polaron and the electron–plasmon interaction constant, respectively, $R_S = k_S^{-1}$, $R_F = k_F^{-1}$.

Considering that

$$\alpha_{pS} = \frac{e^2}{\hbar}\left(1 - \frac{1}{\varepsilon + 1}\right)\left(\frac{m}{2\hbar\Omega_{pS}}\right)^{1/2}; \; R_{pS} = \left(\frac{m}{2\hbar\Omega_{pS}}\right)^{1/2}, \tag{8.6.13}$$

we get

$$x_0 = \frac{4\left(1 - \frac{R_S^2}{R_F^2}\right)}{\left(1 + \frac{R_S^2}{R_F^2} + 2\frac{R_S^2}{R_F^2}\right)} R_{pS}, \tag{8.6.14}$$

i.e., $x_0 \sim R_{pS}$ and in order of magnitude coincides with the radius of the electron polaron. For typical parameter values $m = (0.1 - 1)m_0$

$$R_{pS} \sim (1 - 10) \cdot 10^{-8}\,\text{cm}, \tag{8.6.15}$$

that is, it is from one to several lattice constants.

It should be emphasized here that the approach developed in this paragraph is based on the theory of electron polarons [79–81], according to which polarons arise when interacting with plasmons of valence electrons, the characteristic frequency of which is of the order of $10^{16}\ s^{-1}$. In this regard, the polarization induced by the electron follows the motion of the electron without inertia, turning it into a quasi-particle—an electron polaron. This is the main difference from the model that takes into account the influence of electron–phonon interaction (frequency $10^{13}\ \text{s}^{-1}$) on emission characteristics [89].

Note that by some authors [90, 91] the parameter x_0. It was introduced as a 'cropping factor' in the classical image potential to eliminate the 'nonphysical divergence' on the crystal surface ($x = 0$) in it.

For the transparency of the barrier $D(E, W)$ on the contact, you can write

$$D(E, W) = \left[1 + \exp\left(-\frac{4\pi i}{\hbar}\right)\int_{x_1}^{x_2} p(x)dx\right]^{-1}, \tag{8.6.16}$$

where

$$p(x) = \sqrt{2m(W - W_t(x))}, \tag{8.6.17}$$

is coordinate-dependent electron momentum; and $W_t(x)$—the effective potential energy of the emitted electron, which, taking into account the expressions (8.6.3) and (8.6.9), can be represented as

$$W_t = -\frac{e^2}{4\varepsilon(x + x_0/4)} - eEx. \tag{8.6.18}$$

Then

$$p(x) = \sqrt{2m\left(W + \frac{e^2}{4\varepsilon\left(x + \frac{x_0}{4}\right)} + eEx\right)}. \qquad (8.6.19)$$

(In the expressions (8.6.18)–(8.6.19) and further, the CGSE system of units is used.)

The limits of integration in (8.6.16) are found from the condition $p(x) = 0$. Then according to (8.6.19)

$$W + \frac{e^2}{4\varepsilon\left(x + \frac{x_0}{4}\right)} + eEx = 0. \qquad (8.6.20)$$

To bring the integral in (8.6.16), taking into account (8.6.20), to the form considered in [86], we will replace

$$z = x + \frac{x_0}{4}. \qquad (8.6.21)$$

Then from (8.6.20) follows:

$$W\left(1 - \frac{eEx_0}{4W}\right) + \frac{e^2}{4\varepsilon z} + eEz = 0. \qquad (8.6.22)$$

Following [88], we introduce the notation:

$$\gamma = 1 - \frac{eEx_0}{4W}. \qquad (8.6.23)$$

To find the limits of integration from (8.6.20), we obtain a quadratic equation, the solution of which has the form

$$z_{1,2} = -\frac{\gamma W}{2eE} \pm \sqrt{\frac{\gamma^2 W^2}{4e^2E^2} - \frac{e}{4\varepsilon E}}. \qquad (8.6.24)$$

Introducing a new variable:

$$y_a = \frac{e}{|\gamma W|}\sqrt{\frac{eE}{\varepsilon}}. \qquad (8.6.25)$$

Then, assuming that the energy W takes only negative values, we can write

$$z_{1,2} = -\frac{\gamma W}{2eE}\left(1 \pm \sqrt{1 - y_a^2}\right). \qquad (8.6.26)$$

Taking into account the introduced notation, the integral in (8.6.16) takes the form

$$I = -\frac{2\sqrt{2m}}{\hbar} i \int_{z_1}^{z_2} \left(\gamma W + \frac{e^2}{4\varepsilon z} + eEz\right) dz. \qquad (8.6.27)$$

The resulting integral is similar to the one considered in [87]. Integration into (8.6.27) leads to the appearance of the function $v_a(y_a)$:

$$v_a(y_a) = \sqrt{\frac{1+\sqrt{1-y_a^2}}{2}} \times \left(E\left[\sqrt{\frac{2\sqrt{1-y_a^2}}{1+\sqrt{1-y_a^2}}} \right] - (1-\sqrt{1-y_a^2})K\left[\sqrt{\frac{2\sqrt{1-y_a^2}}{1+\sqrt{1-y_a^2}}} \right] \right), \tag{8.6.28}$$

where

$$K[k] = \int_0^{\pi/2} \frac{d\theta}{\sqrt{1-k^2\sin^2\theta}}, \tag{8.6.29}$$

$$E[k] = \int_0^{\pi/2} \sqrt{1-k^2\sin^2\theta}\, d\theta \tag{8.6.30}$$

are elliptic Euler integrals of the first and second kind, respectively.

Note that the function (8.6.28) differs from the Nordheim function [86] by the presence in the argument y_a polaron contribution.

Following [92], for the case $y_a \geqslant 1$ (which corresponds to the mode of autoelectronic emission) from (8.6.28) it is possible to obtain

$$v_a(y_a) = \sqrt{1+y_a} \times \left(E\left[\sqrt{\frac{1-y_a}{1+y_a}} \right] - y_a K\left[\sqrt{\frac{1-y_a}{1+y_a}} \right] \right), \tag{8.6.31}$$

Argument y_a can be defined via the y argument of the Nordheim function:

$$y_a = \frac{y}{\gamma\sqrt{\varepsilon}}. \tag{8.6.32}$$

Then from (8.6.28) for $v_a(8.5.\ y)$ (on condition $\varepsilon \approx 1$, which is performed for the nanometer range of the thickness of the adsorbed film) the expression is valid:

$$v_a(y) = \sqrt{\frac{\gamma+\sqrt{\gamma^2-y^2}}{2\gamma}} \left(E\left[\sqrt{\frac{2\sqrt{\gamma^2-y^2}}{\gamma+\sqrt{\gamma^2-y^2}}} \right] - \frac{\gamma-\sqrt{\gamma^2-y^2}}{2\gamma} K\left[\sqrt{\frac{2\sqrt{\gamma^2-y^2}}{\gamma+\sqrt{\gamma^2-y^2}}} \right] \right). \tag{8.6.33}$$

The total current through the contact is determined by an expression similar to the general formula (8.6.20) of operation [86]

$$j(E,\,T)=\frac{k_0T}{2\pi^2}\int_{-W_a}^{W_{tm}}\frac{\ln\left(1+\exp\left[-\frac{W-W_F}{kT}\right]\right)}{1+\exp\left[\frac{4}{3}\sqrt{2}E^{-\frac{1}{2}}y_a^{-\frac{3}{2}}v_a(y_a)\right]}dW$$
$$+\frac{k_0T}{2\pi^2}\int_{W_{tm}}^{\infty}\ln\left(1+\exp\left[-\frac{W-W_F}{kT}\right]\right)dW. \tag{8.6.34}$$

(According to [86], the expression (8.6.34) is written in Hartree units, i.e. j is expressed in units $m^3e^9\hbar^{-7} = 2.37\cdot10^{14}$ A cm^{-2}; E—in units $m^2e^5\hbar^{-4} = 5.14\cdot10^9$ V cm^{-1}; and W_F, kT, W, W_a, W_{tm}—in units $me^4\hbar^{-2} = 27.2$ eV.)

8.6.2 Thermoelectronic emission

At high temperatures, the emission current is caused by electrons overcoming a potential barrier with an energy higher than W_{tm}. In this case, the first integral in (8.6.34) can be neglected. Then from (8.6.34) follows:

$$j_{RS} = \frac{4\pi emk_0T}{h^3}\int_{W_{tm}}^{\infty}\ln\left(1+\exp\left[-\frac{W-W_F}{kT}\right]\right)dW. \tag{8.6.35}$$

It is obvious that the energy of the emitted electrons in the conditions under consideration is much greater than the Fermi energy. Therefore, the integrand function can be represented as

$$\ln\left(1+\exp\left[-\frac{W-W_F}{k_0T}\right]\right)\approx\exp\left[-\frac{W-W_F}{k_0T}\right]. \tag{8.6.36}$$

At the same time, from (8.6.35), taking into account (8.6.36), it follows:

$$j_{RS} = \frac{4\pi em(kT)^2}{h^3}\exp\left(-\frac{W_{tm}-W_F}{kT}\right). \tag{8.6.37}$$

Let's define the energy W_{tm}. The potential energy of an electron outside a metal cathode is determined by the expression

$$W_t = W_F + \Phi_0 - \frac{e^2}{\varepsilon(4x+x_0)} - eEx. \tag{8.6.38}$$

Here Φ_0 is the output work of the cathode material.

Energy W_t reaches a maximum at the point

$$x_m = \frac{1}{2}\sqrt{\frac{e}{E}} - \frac{x_0}{4}. \tag{8.6.39}$$

Substituting the expression (8.6.39) B (8.6.38), for W_m we get

$$W_{tm} = W_F + \Phi_0 - \frac{1}{\varepsilon}\sqrt{e^3 E} + \frac{eEx_0}{4\varepsilon}. \tag{8.6.40}$$

Formulas (8.6.37) and (8.6.40) follow the final expression for the current density of polaron thermionic emission, which is a generalized Richardson–Schottky formula:

$$j_{RS} = \frac{4\pi emk_0^2}{h^3} T^2 \exp\left[\frac{1}{k_0 T}\left(-\Phi_0 + \frac{1}{\varepsilon}\sqrt{e^3 E} - \frac{eEx_0}{4\varepsilon}\right)\right]. \tag{8.6.41}$$

As shown in [86], the Richardson–Schottky formula (and hence the formula (8.6.41)) is valid up to fields of the order of 50 MV cm^{-1}.

Let us estimate the effect of the polaron effect (i.e., the quantum nature of the image forces) on the value of the Richardson–Schottky emission current. Without taking it into account, the current density is determined by the expression following from (8.6.41):

$$j_{RS_0} = \frac{4\pi emk_0^2}{h^3} T^2 \exp\left[\frac{1}{k_0 T}\left(-\Phi + \frac{1}{\varepsilon}\sqrt{e^3 E}\right)\right]. \tag{8.6.42}$$

Then the relation j_{RS}/j_{RS0} characterizes the effect of the polaron nature of tunneling on the magnitude of the emission current:

$$\frac{j_{RS}}{j_{RS0}} = \exp\left(-\frac{eEx_0}{4\varepsilon k_0 T}\right). \tag{8.6.43}$$

Calculation results for some averaged values $\varepsilon = 3$, $x_0 = 0.3$ nm, $\Phi = 4, 5$ eV shown in figures 8.13 and 8.14. The plots in figures 8.13 and 8.14 indicate a large influence of the electron polaron effect on the emission characteristics of cathodes in strong electric fields. Moreover, in the field strength range $E \geqslant 10$ MV cm^{-1}, this effect increases with the field growth (reaching about two orders of magnitude at an electric field strength of 50 MV cm^{-1}) and decreases with increasing temperature.

Figure 8.15 shows the Richardson–Schottky current density dependences calculated using the exact formula (8.6.35) and the approximate formula (8.6.41). These dependences practically coincide in the entire domain where the Richardson–Schottky equation holds.

A decrease in the emission current with an increase in the electric field strength is associated with an increase in the effective work of the exit $\Phi(E, x_0)$. As follows from (8.6.41) and (8.6.42), the value must be used in the calculations

$$\Phi(E, x_0) = \Phi_0 + \frac{eEx_0}{4\varepsilon}. \tag{8.6.44}$$

Thus, it is shown that taking into account the electron polaron effect leads to an increase in the work of the exit over the entire range of fields and temperatures.

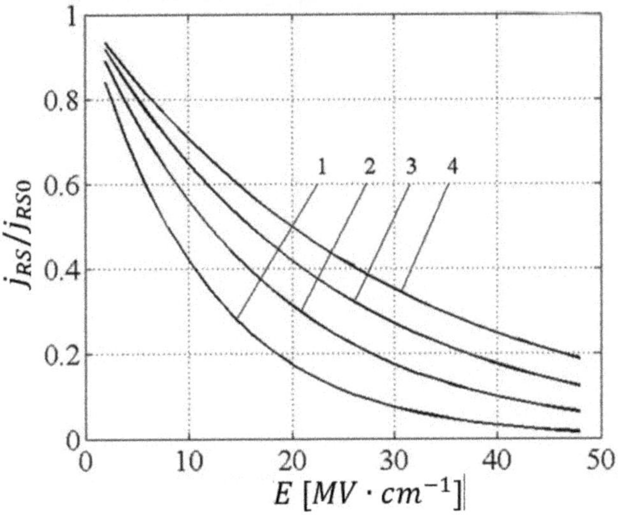

Figure 8.13. The effect of quantum image forces on the value of the emission current density: field dependence for different temperature values. 1—1000, 2—1500, 3—2000, 4—2500 K.

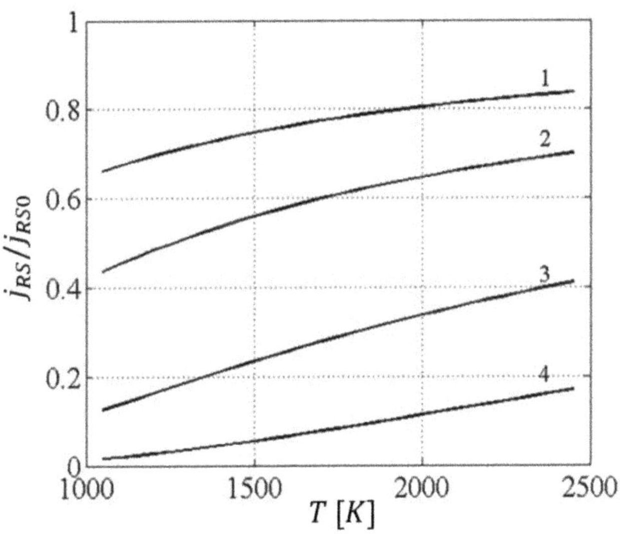

Figure 8.14. The effect of the quantum forces of the image on the value of the emission current density: temperature dependence for different values of the electric field strength: 1—5, 2—10, 3—25, 4—50 MV cm^{-1}.

In the field strength $E > 5$ MV cm^{-1}, this leads to a decrease in the emission current density by more than an order of magnitude, which is associated with the additional work of the field to move both the electron itself and the polarizing cloud following it when tunneling the electron polaron through the barrier. Formula (8.6.44) for efficient work of the exit $\Phi(E, x_0)$ confirms the results of [93, 94], in which the experimentally measured value of the contact barrier during the internal

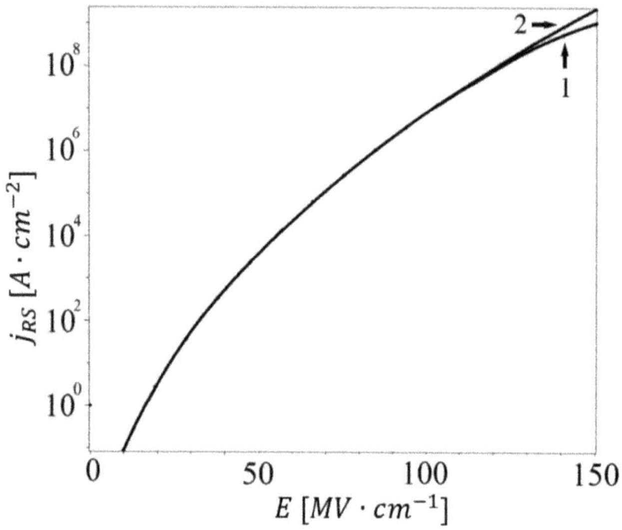

Figure 8.15. Dependence of the thermionic emission current on the electric field strength. Curve 1 is calculated according to the exact formula (8.6.35), curve 2—according to the generalized Richardson–Schottky formula (8.6.41). Parameter values: $\varepsilon = 1$, $\Phi = 4$ eV, $m = m_0$, $T = 1500$ K.

photoelectric effect in the structure $Al - SiO_2$ vacuum turned out to be greater than the actual height of the barrier, by about 0.2 eV.

8.6.3 Autoelectronic emission

Integration of expression (8.6.34) taking into account (8.6.27) and (8.6.31) leads to an equation similar to the Murphy–Good equation [86]:

$$j(E, T, x_0) = \frac{e^3E^2}{8\pi h\varepsilon^2\Phi_0 t_a^2(y_a)} \times \frac{\pi c_a kT}{\sin(\pi c_a kT)} \exp\left[-\frac{8\pi\varepsilon\sqrt{2m}\,\Phi_0^{3/2}}{3ehE}v_a(y_a)\right]. \tag{8.6.45}$$

Note that in addition to the function $v_a(y_a)$, new functions appear in the formula (8.6.45) $t_a(y_a)$ and $c_a(y_a)$ (containing polaron contributions), which can be expressed through the argument of the Nordheim function:

$$t_a(y_a) = \overline{\gamma}\sqrt{\frac{y_a}{2}}\left\{2E\left[\sqrt{\frac{y_a - y}{2y_a}}\right] - K\left[\sqrt{\frac{y_a - y}{2y_a}}\right]\right\}, \tag{8.6.46}$$

$$c_a(y_a) = \frac{4\pi\varepsilon\sqrt{2m\Phi_0}}{ehE}t_a(y_a), \tag{8.6.47}$$

where

$$\overline{\gamma} = 1 + \frac{eEx_0}{4\Phi_0}. \quad (8.6.48)$$

The temperature dependence of the autoelectronic emission current is determined by the second pre-exponential factor of the equation (8.6.45)

$$\alpha(T, E) = \frac{\pi c_a k_0 T}{\sin(\pi c_a k_0 T)}. \quad (8.6.49)$$

If $\pi c_a kT$ so little that$\alpha(T, E) \cong 1$, and the polaron effect is negligible (T. e. $x_0 = 0$), then the formula (8.6.45) comes to the Fowler–Nordheim formula:

$$j_{F-N}(E) = \frac{e^3E^2}{8\pi h\varepsilon^2\Phi_0 t^2(y)} \exp\left[-\frac{8\pi\varepsilon\sqrt{2m}\,\Phi_0{}^{3/2}}{3ehE} v\,(y)\right]. \quad (8.6.50)$$

It should be noted that formulas (8.6.45) and (8.6.50) have limitations on the values of the physical parameters included in these ratios—temperature, work of the exit, and electric field strength.

Thus, the formula (8.6.45) can be considered a generalized expression of the current density of thermoautoelectronic emission at the metal–dielectric contact coated with an adsorbed nonmetallic film, taking into account the contribution of the polaron effect and its increasing role with increasing electric field strength and temperature. It differs from the classical formula for the emission current density given in [86] by the presence of functions in the argument $v_a(y_a)$ and $t_a(y_a)$ 'polaron contribution'.

Formula (8.6.45) in the low temperature limit ($T \to 0$), under which $\alpha(T, E) = 1$, and high fields can be transformed to a simpler form if we use approximation for functions $v_a(y_a)$ and $t_a^2(y_a)$, found in [95] and applied in [96] to determine the area of the field emission barrier:

$$v_a(y_a) \approx 0,95 - 1,03y_a^2;\ t_a^2(y_a) \approx 1,1. \quad (8.6.51)$$

Substituting (8.6.51) into (8.6.45), taking into account (8.6.32) and (8.6.25), we obtain an expression for the current density of autoelectronic emission:

$$j_{\text{appr}}(E) = \frac{e^3E^2}{8,8\pi h\varepsilon^2\Phi_0} \exp\left[-\frac{8\pi\sqrt{2m}\,\Phi_0^{3/2}}{3ehE_{eff}}\right], \quad (8.6.52)$$

where E_{eff} has the form:

$$E_{\text{eff}} = \frac{E}{0,95 - \frac{1,03e^3E}{\varepsilon\left(\Phi_0 + \frac{eEx_0}{4}\right)^2}} \quad (8.6.53)$$

and includes a polaron parameter x_0.

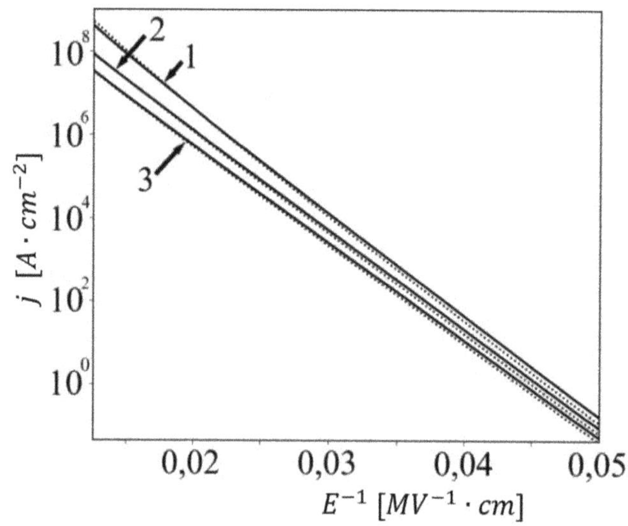

Figure 8.16. Dependences of the current density of the autoelectronic emission calculated on the basis of the formula (8.6.45) (solid lines) and the extrapolation formula (8.6.52) (dotted lines) for different values of the parameterx_0 (curves 1: $x_0 = 0$ nm, curves 2: $x_0 = 0.5$ nm, curves 3: $x_0 = 1.0$ nm), from the electric field strength E in the region of strong fields ($E = (30 - 80)$MV cm^{-1}). Values of other theory parameters: $\varepsilon = 1$, $\Phi = 4$ eV, $m = m_0$, $T = 500$ K.

Figure 8.16 shows the field dependence of the current density of the autoelectronic emission, calculated respectively by the formula (8.6.45) (solid lines) and by the extrapolation formula (8.6.52) (dotted lines) for different values of the parameter x_0 ($x_0 = 0$ nm, $x_0 = 0.5$ nm, $x_0 = 1.0$ nm), from the electric field strength in the region of strong fields ($E = (30 \div 80)$MV/cm).

The dependences of the Fowler–Nordheim current density shown in figure 8.16 in the specified field range differ by less than 3%–5%, at the same time there is a strong dependence on the parameter x_0, describing the polaron effect.

Note that in the entire range of field values in figure 8.16, when $x_0 = 0$ there is a coincidence with the results of [86], well confirmed by numerous experiments. As in the case of thermoelectronic emission, a decrease in the current density of autoelectronic emission with an increase in x_0 it is explained by an increase in the effective work of the electron output in a strong electric field due to the plasmon polaron effect.

It should be noted that the extrapolation formulas (8.6.41)—for the current density of thermoelectronic emission and (8.6.52) for the current density of autoelectronic emission, as follows from figures 8.15 and 8.16—give results that practically coincide with the results obtained by the exact formulas (8.6.35) and (8.6.45), respectively, in the interval fields from 10 to 100 MV cm^{-1}.

8.6.4 Results and discussion

The application of the polaronic theory to the consideration of physical processes at the metal–dielectric or metal–adsorbed dielectric nanofilm contact makes it possible to eliminate the singularity of the potential energy of the tunneling electron at the emitting boundary (when $x = 0$). At the fields $E > 5 \cdot 10^6$ V cm^{-1}, when the width of the potential barrier is commensurate with the radius of the electron polaron, the quantum contribution to the emission current density becomes significant. This contribution is determined by the parameter x_0, which is related to the radius of the tunneling polaron. At the same time, with an increase in the electric field strength, the influence of the polaron effect increases, which is due to an increase in the effective operation of the electron output. This effect is partially attenuated by taking into account the real value of the dielectric constant of the adsorbed nanofilm material.

As shown in [88], the parameter x_0 also appears in the formula for the current density of thermionic emission from cathodes coated with an adsorbed nonmetallic nanofilm. The plots given in [87] indicate a great influence of the polaron effect on the emission characteristics of thermocathodes under conditions of a high-voltage gas discharge. And in the field area $E \geqslant 10$ MV cm^{-1} this influence increases with the increase in the electric field strength (reaching almost two orders of magnitude at $E \sim 10^8$ V cm^{-1}) and it decreases with increasing temperature. It was also noted there that a decrease in the emission current with an increase in the electric field strength is associated with an increase in the effective operation of the output $\Phi(E, x_0)$.

8.7 Conclusion

Kinetic effects have been theoretically investigated: scattering of charge carriers on polar optical phonons in thin semiconductor films, scattering on surface optical phonons in inversion channels of MDS structures and IR absorption of light by free charge carriers, and cyclotron–phonon absorption of electromagnetic radiation in thin semiconductor films in a quantizing magnetic field. It has been established that new peaks corresponding to the frequencies of surface phonons appear in the cyclotron–phonon absorption spectrum. The absorption is resonant, but the absorption peaks have different shapes: for bulk phonons, δ-shaped peaks are observed at frequencies $\omega = |n\omega_c \pm \omega_0|$, $n = 0, 1, 2, \ldots$; for surface phonons, Lorentz curves with maxima at $\omega = |n\omega_c \pm \Omega_S|$ and with a half-width, which is determined by the variance of $\Omega_S(d)$.

References

[1] Tavger B A and Demikhovsky V Y 1964 Electron scattering by acoustic vibrations in thin semiconductor films *FTT* **6** 960–2

[2] Riddoch F A and Ridley B K 1983 On the scattering of electrons by polar optical phonons in quasi-2D quantum wells *J. Phys. C: Solid State Phys.* **16** 6971–82

[3] Ridley B K 1989 Electron scattering by confined LO polar phonons in a quantum well *Phys. Rev.* B **39** 5282–6

[4] Ridley B K 1982 The electron-phonon interaction in quasi-two-dimensional semiconductor quantum-well structures *J. Phys. C: Solid State Phys.* **15** 5899–917

[5] Rudin S and Rinecke T L 1990 Electron–LO-phonon scattering rates in semiconductor quantum wells *Phys. Rev.* B **41** 7713–7

[6] Pozhela Y and Yutsiene I 1995 Electron scattering on optical phonons in two-dimensional quantum wells with independent capture of electrons and phonons *FTP* **29** 459–68

[7] Pozela J, Juciene V, Namajunas A and Pozela K 1997 Electron-phonon scattering engineering *FTT* **31** 85–8

[8] Mirlin D N and Rodina A V 1996 Polar scattering of two-dimensional electrons in quantum wells (Review) *FTP* **38** 3201–11

[9] Bordone P and Lugli P 1994 Effect of half-space and interface phonons on the transport properties of $Al_xGa_{1-x}As/GaAs$ single heterostructures *Phys. Rev.* B **49** 8178–90

[10] Huang K and Zhu B 1988 Dielectric continuum model and Fröhlich interaction in superlattices *Phys. Rev.* B **38** 13377–86

[11] Huang K and Zhu B 1988 Long-wavelength optic vibrations in a superlattice *Phys. Rev.* B **38** 2183–6

[12] Bryksin V V and Firsov Yu A 1971 Interaction of an electron with surface phonons in an ion crystal plate *FTT* **13** 496–503

[13] Fomin V M and Pokatilov E P 1985 Phonons and the electron-phonon interaction in multilayer systems *Phys. Status Solidi* B **132** 69–82

[14] Fuchs R and Kliewer K L 1965 Optical modes of vibration in an ionic crystal slab *Phys. Rev.* **140** A2076–88

[15] Licary J J and Evrard R 1977 Electron-phonon interaction in a dielectric slab: effect of the electronic polarizability *Phys. Rev.* B **15** 2254–64

[16] Mori N and Ando T 1989 Electron–optical-phonon interaction in single and double heterostructures *Phys. Rev.* B **40** 6175–88

[17] Lassnig R Polar optical imorynterface phonons and Fröhlich interaction in double heterostructures *Phys. Rev.* B **30** 7132–7

[18] Devreese J T, De Sitter J, Johnson E J and Ngai K L 1978 New magneto-optical anomalies of impurity electrons in InSb at the two-LO-phonon region: theory and experiment *Phys. Rev.* B **17** 3207–20

[19] Shi J M, Peeters F M and Devreese J T 1995 D^- states in $GaAs/Al_xGa_{1-x}As$ superlattices in a magnetic field *Phys. Rev.* B **51** 7714–24

[20] Xiaoguang W, Peeters F M and Devreese J T 1986 Theory of the cyclotron resonance spectrum of a polaron in two dimensions *Phys. Rev.* B **34** 8800–9

[21] Pokatilov E P, Fomin V M and Beril S I 1990 *Vibrational Excitations, Polarons and Excitons in Multilayer Systems and in Superlattices* (Chisinau: Stiinza) p 288

[22] Askerov A, Pokatilov E P and Nika D L 2007 Quantitative and quantitative indicators of electron mobility, separate units with complete analytical comments *University Studio.—The Series 'Life in nature'* 249–51

[23] Fonoberov V and Balandin A A 2006 Giant enhancement of the carrier mobility in silicon nanowires with diamond coating *Nano Lett.* **6** 2442–6

[24] Nika D L, Pokatilov E P and Balandin A A 2008 Phonon-engineered mobility enhancement in the acoustically mismatched silicon/diamond transistor channels *Appl. Phys. Lett.* **93** 173111

[25] Pokatilov E P *et al* 2006 Size-quantized oscillations of the electron mobility enhancement in AlN/GaN/AlN heterostructures with InGaN nanogrooves *Appl. Ohys. Lett.* **89** 112110

[26] Pokatilov E P, Nika D L and Balandin A A 2006 Build-in filed effect on the electron mobility in AlN/GaN/AlN quantum wells *Appl. Phys. Lett.* **89** 113508

[27] Pokatilov E P, Nika D L and Balandin A A 2004 Confined electron–confined phonon scattering rates in wurzite AlN/GaN/AlN heterostructures *J. Appl. Phys.* **95** 5626–32

[28] Pokatilov E P, Nika D L and Balandin A A 2006 Electron mobility enhancement in AlN/GaN/AlN heterostructures with InGaN nanogrooves *Appl. Phys. Lett.* **89** 112110-1–113

[29] Anselm A I 1978 *Introduction to the Theory of Semiconductors* (Moscow: Nauka) p 615

[30] Herstein A, Ning T H and Fowler A B 1976 Electron scattering in silicon inversion layers by oxide and surface roughness *Surf. Sci.* **58** 178–87

[31] Stern F 1978 Two subband screening and transport in (001) silicon inversion layers *Surf. Sci.* **73** 1977–2006

[32] Hess K 1978 Review of experimental aspects of hot electron transport in MOS-structure *Solid-State Electron* **21** 123–6

[33] Hess K and Vogl P 1979 Remote polar phonon scattering in silicon inversion layers *Solid State Commun.* **30** 807–9

[34] Hai G Q, Peeters F M and Devreese J T 1993 Polaron-cyclotron-resonance spectrum resulting from interface-and slab-phonon modes in a GaAs/AlAs quantum well *Phys. Rev.* B **47** 10358–74

[35] Gurevich V L, Parshin D A and Stengel K E 1988 Absorption of light by free carriers with the participation of optical phonons in quasi–two–dimensional systems *FTT* **30** 1466–75

[36] Osipov V V, Selyakov A Y and Foygel M 1998 In interband absorption of long-wave radiation in Delta–doped superlattices based on single–crystal wide–band semiconductors *FTP.* **32** 221–6

[37] Adamska H and Spector H N 1984 Free carrier absorption in quantum well structures for polar optical phonon scattering *J. Appl. Phys.* **56** 1123–7

[38] Babiker M, Ghosal A and Ridley B K 1989 Intrasubband transitions and well capture via confined, guided and interface L0 phonons in superlattices *Superlattices Microstruct.* **5** 133–6

[39] Comas F, Trallero Giner C and Leon H 1986 Quantum theory of free-carrier absorption in quasi-two-dimensional semiconducting structures. Degenerate carrier case *Phys. Status Solidi* B **138** 219–27

[40] Constantinou N C and Ridley B K 1990 Electron energy relaxation via LO-phonon emission in free standing GaAs wafers *J. Phys.: Condens. Matter* **2** 7465–71

[41] Hassan H H and Spector H N 1986 Optical absorption in semiconducting quantum-well structures: indirect interband transitions *Phys. Rev.* B **33** 5456–60

[42] Haupt R and Wender L 1991 Electron-phonon interaction and electron scattering by modified confined LO phonons in semiconductor quantum wells *Phys. Rev.* B **44** 1850–60

[43] Hess K 1979 Impurity and phonon scattering in layered structures *Appl. Phys. Lett.* **35** 484–90

[44] Price P J Polar-optical-mode scattering for an ideal quantum-well heterostructure *Phys. Rev.* B **30** 2234–5

[45] Sinha C and Mikhopadkyay S 1997 Scattering of a polaron in the presence of a laser field *J. Phys. Condens. Matter* **9** 9597–600

[46] Spector H N 1983 Free-carrier absorption in quasi-two-dimensional semiconducting structures *Phys. Rev.* B **28** 971–6

[47] Trallero Giner C and Anton M 1986 Quantum theory of free-carrier absorption in quasi-two-dimensional semiconducting structures *Phys. Status Solidi* B **133** 563–72
[48] Lambin P, Vigneron J P, Lucas A A, Thiry P A, Liehr M, Pireaux J J, Caudano R and Kuech T J 1986 Observation of long-wavelength interface phonons in a GaAs/AlGaAs superlattice *Phys. Rev. Lett.* **56** 1842–5
[49] Lassnig R 1984 Optical interface phonons and Fröhlich interaction in double heterostructures *Phys. Rev.* B **30** 7132–7
[50] Maciel A C, Campelo Cruz L C and Ryan J F 1987 Resonant Raman scattering from confined phonons and interface phonons in a GaAs/GaAlAs superlattice *J. Phys. C: Solid State Phys.* **20** 3041–6
[51] Schwartz G P, Gualtieri G J, Sunder W A and Farrow L A 1987 Light scattering from quantum confined and interface optical vibrational modes in strained-layer GaSb/AlSb superlattices *Phys. Rev.* B **36** 4868–77
[52] Seilmeier A, Hübner H-J, Abstreiter G, Weimann G and Schlapp W Intersubband relaxation in GaAs − $Al_xGa_{1-x}As$ quantum well structures observed directly by an infrared bleaching technique *Phys. Rev. Lett.* **59** 1345–48
[53] Suh E-K, Bartholomev D U, Ramdas A K, Rodrigues S *et al* Raman scattering from superlattices of diluted magnetic semiconductors *Phys. Rev.* B **36** 4316–31
[54] Tatham M C, Ryan J F and Foxon C T Time-resolved Raman measurements of intersubband relaxation in GaAs quantum wells *Phys. Rev. Lett.* **63** 1637–40
[55] Lutsky V N, Korneev D N and Elinson A Y 1966 Observation of quantum dimensional effects in bismuth films by tunneling spectroscopy *Lett. JETF* **4** 267–70
[56] Pokatilov E P and Beril S I 1983 Electron-phonon interaction in periodic two-layer structures *Phys. Status Solidi* B **118** 567–73
[57] Cardona M 1989 Folded, confined, interface, surface, and slab vibrational modes in semiconductor superlattices *Superlattices Microstruct.* **5** 27–42
[58] Klein M V 1986 Phonons in semiconductor superlattices *IEEE J. Quantum Electron.* **QE-22** 1760–70
[59] Sood A K, Menendez J, Cardona M and Ploog K 1985 Interface vibrational modes in GaAs–AlAs superlattices *Phys. Rev. Lett.* **54** 2115–8
[60] Schiff L 1959 Quantum mechanics (Moscow: Publishing House of Foreign Literature) p 473
[61] Ridley B 1986 *Quantum Processes in Semiconductors* (Moscow: Mir) p 304
[62] Bass F G and Bakanas R K 1976 Features of cyclotron-phonon absorption in semiconductor films *FTT.* **18** 2672
[63] Bass F G and Levinson I B 1965 Cyclotron-phonon resonance in semiconductors *JETF* **49** 914–24
[64] Bass F G and Matulis A Y 1970 Dimensional phonon jumps of light absorption in semiconductor films *FTT* **12** 2039–41
[65] Bryksin V V, Mirlin D N and Firsov Y A 1973 Surface optical phonons in ionic crystals *UFN* **113** 30–63
[66] Fuchs R, Kliewer K L and Pardey W J 1966 Optical properties of an ionic crystal slab *Phys. Rev.* **150** 589–96
[67] Beril S I, Pokatilov E P and Cheban I S 1983 Wannier–Mott excitons in a plate of a polar crystal of finite thickness *FTT* **25** 2661
[68] Evans E and Mills D L 1973 Interaction of slow electron with the surface of model dielectric: theory of surface polarons *Phis. Rev.* B **8** 4004–18
[69] Sac J 1972 Theory of surface polaron *Phis. Rev.* B **6** 3981–6

[70] Mesyats G A and Proskurovsky D I 1989 *Pulsed Electrical Discharge in Vacuum* (Berlin: Springer)
[71] Latham R V 1995 *High Voltage Vacuum Insulation: Basic Concepts and Technological Practice* (Amsterdam: Elsevier)
[72] Cox B M and Williams W T 1977 Field-emission sites on unpolished stainless steel *J. Phys. D: Appl. Phys.* **10** L5–9
[73] Anders A 2008 *Cathodic Arcs: From Fractal Spots to Energetic Condensation* (New York: Springer)
[74] Mesyats G A 2000 *Cathode Phenomena in a Vacuum Discharge: The Breakdown, the Spark, and the Arc* (Moscow: Nauka)
[75] *A 3 TeV e+e− Linear Collider Based on CLIC Technology*, ed. G. Guignard, CERN Report No. CERN 2000-008, 2000
[76] Wuensch W *Advances in the Understanding of the Physical Processes of Vacuum Breakdown* CERN-OPEN-2014-028. CLIC-Note-1025, CERN, Geneva, May 2013
[77] Barengolts S A, Mesyats V G, Oreshkin V I, Oreshkin E V, Khishchenko K V, Ulimanov I V and Tsventoukh M M 2018 Mechanism of vacuum breakdown in radio-frequency accelerating structures *Phys. Rev. Accel. Beams* **21** 061004
[78] Korolev Y D and Mesyats G A 1998 *Physics of Pulsed Breakdown in Gases* (URO Press)
[79] Hermanson J 1974 The Self-Energy Problem in Quantum Dielectrics *Elementary Excitations in Solids, Molecules, and Atom, Part B* (Berlin: Springer) 199–211
[80] Hermanson V 1972 Simple model of electronic correlation in insulators *Phys. Rev.* **6** 2427–32
[81] Toyozawa Y 1954 Theory of the electronic polaron and ionization of a trapped electron by an exciton *Prog. Theor. Phys.* **12** 421–36
[82] Beril S I and Pokatilov E P 1978 Surface states in a quantum dielectric *HTP* **12** 2030–3
[83] Pokatilov E P, Beril S I and Fomin V M 1988 Potentials and image forces of the electron polaron model *Surface* **5** 5–12
[84] Beril S I, Pokatilov E P, Goryachkovskii E P and Semenovskaya N N 1993 Polaron theory of the image potential considering the spatial dispersion *Phys. Status Solidi B.* **176** 347–53
[85] Pokatilov E P, Beril S I and Fomin V M 1988 Image potentials and image forces in the polaron theory *Phys. Status Solidi* **B 147** 163–72
[86] Murphy E I and Good A H 1956 Thermionic emission, field emission, and the transition region *Phys. Rev.* **102** 1464–73
[87] Barengolts Y A and Beril S I 2016 Polaron theory of the emission current in a cathode-adsorbed nanofilm system at the initial stage of a high-voltage gas discharge *IFMBE Proc.: 3rd Int. Conf. on Nanotechnologies and Biomedical Engineering* 55 230–3
[88] Barengolts Y A and Beril S I 2014 The effect of molecules adsorbed to the cathode surface on the characteristics of a high-voltage gas discharge *IEEE Trans. Plasma Sci.* **42** 3109–12
[89] Reich K V and Eidelman E D 2009 Effect of electron-phonon interaction on field emission from carbon nanostructures *Europhys. Lett.* **85** 47007
[90] Shikin V B and Monarkha Y P 1975 Surface charges in helium *FTT* **1** 957–83
[91] Edelman V S 1980 Levitating electrons *UFN.* **130** 675–704
[92] Modinos A 1984 *Field, Thermionic, and Secondary Electron Emission Spectroscopy* (New York: Plenum) p 320
[93] Harstein A and Weinberg Z A 1978 On the nature of the image force in quantum mechanics with application to photon assisted tunnelling and photoemission *J. Phys. C: Solid State Phys.* **11** L469–73

[94] Harstein A and Weinberg Z A 1979 Unified theory of internal photoemission and photon-assisted tunneling *Phys. Rev.* B **20** 1335–8
[95] Shrednik V N 1974 Theory of autoelectronic emission of metals *Non-Heated Cathodes* (Moscow: Soviet Radio) pp 166–9
[96] Popov E O, Kolosko A G, Chumak M A and Filippov S V 2019 Ten ways to determine the area of field emission *ZhTF* **89** 1615–25

Chapter 9

Raman light scattering in multilayer systems and superlattices

9.1 Introduction

In chapter 9, the electrodynamic theory of Raman light scattering in multilayer structures with an arbitrary number of layers and, in particular, in a superlattice is developed based on the modulation transmission method discussed in chapter 1, in which the theory of potential in multilayer systems of planar geometry was constructed.

9.2 Wave equations taking into account Raman scattering of light

In previous chapters, it was shown that there is a radical restructuring of the energy spectrum of phonons and other types of vibrational excitations in multilayer structures compared with the spectra of massive crystals of the same materials that make up individual layers. In particular, 'captive' volumetric and spatially extended surface optical phonons arise in superlattices. Their existence is manifested in the Raman spectra of light, experimental studies that provide valuable information about the properties of vibrational excitations inherent in superlattices.

In the paper [1] the spectral peaks of Raman scattering of light for superlattices GaAs/AlAs, observed in the frequency shift regions of the order of 280 and 380 cm^{-1}, for the first time were interpreted as a manifestation of spatially extended surface optical phonons. As a result of systematic studies of the first- and second-order Raman scattering spectra for superlattices GaAs/AlAs and $GaAs/GaAs_{1-x}P_x$ [1–6] clear peaks of light scattering (up to high quantization orders) were detected by 'crushed' longitudinal acoustic phonons and optical ones as 'trapped' in layers GaAs and AlAs, so with spatially extended surface phonons. For superlattices GaSb/AlSb in the frequency shift region of the order of 230 and 335 cm^{-1}, Raman scattering peaks associated with spatially extended surface phonons were noted [7].

doi:10.1088/978-0-7503-6164-4ch9

Here, the electrodynamic theory of Raman scattering of light in multilayer structures with an arbitrary number of layers and, in particular, in a superlattice is developed. For this purpose, the modulation transmission method formulated in section 1.3 is used. Consider a multilayer structure with an arbitrary number of layers. Maxwell's equations in each of the layers ($n = I - 1, \ldots, K + 1$) contain magnetic induction

$$\mathbf{B}_n(\mathbf{r}, t) = \mu_0 \int_{-\infty}^{t} \overleftrightarrow{\mu}_n(t - t')\mathbf{H}_n(\mathbf{r}, t')dt' \tag{9.2.1}$$

and electric displacement

$$\mathbf{D}_n(\mathbf{r}, t) = \varepsilon_0 \int_{-\infty}^{t} \overleftrightarrow{\varepsilon}_n(t - t')\mathbf{E}_n(\mathbf{r}, t')dt' + \mathbf{P}_n(\mathbf{r}, t). \tag{9.2.2}$$

The second term describing the Raman scattering of light in that layer will be specified further.

Let's perform a Fourier transform of the time dependence of vector fields by formulas of the form

$$\mathbf{E}_n(\mathbf{r}, \omega) = \frac{1}{2\pi} \int_{-\infty}^{+\infty} e^{i\omega t}\mathbf{E}_n(\mathbf{r}, t)dt. \tag{9.2.3a}$$

Since from the realness of the vector field $\mathbf{E}_n(\mathbf{r}, t)$ it follows that

$$\mathbf{E}(\mathbf{r}, -\omega) = \mathbf{E}_n^*(\mathbf{r}, \omega), \tag{9.2.3b}$$

and it is enough to find Fourier images for non-negative ω:

$$\mathbf{E}_n(\mathbf{r}, t) = \int_0^{\infty} [e^{-i\omega t}\mathbf{E}_n(\mathbf{r}, \omega) + e^{i\omega t}\mathbf{E}_n^*(\mathbf{r}, \omega)\Big]d\omega. \tag{9.2.3c}$$

Introducing the dielectric function in accordance with (123.4) [8]

$$\overleftrightarrow{\varepsilon}_n(\omega) = \int_0^{\infty} \overleftrightarrow{\varepsilon}_n(t)e^{i\omega t}dt \tag{9.2.4}$$

and similarly to $\overleftrightarrow{\mu}_n(\omega)$, transform the material equations (9.2.1) and (9.2.2):

$$\mathbf{B}_n(\mathbf{r}, \omega) = \mu_0 \overleftrightarrow{\mu}_n(\omega)\mathbf{H}_n(\mathbf{r}, \omega); \tag{9.2.5}$$

$$\mathbf{D}_n(\mathbf{r}, \omega) = \varepsilon_0 \overleftrightarrow{\varepsilon}_n(\omega)\mathbf{E}_n(\mathbf{r}, \omega) + \mathbf{P}_n(\mathbf{r}, \omega). \tag{9.2.6}$$

In the framework of linear field quantum mechanical calculation (e.g., section 19 in [9]) for the polarizability of the transition is obtained by an expression of the form

$$\mathbf{P}_n(\mathbf{r}, t) = \varepsilon_0 \int_0^{\infty} \Big[e^{-i\omega' t}\overleftrightarrow{\chi}_n(\mathbf{r}, t; \omega')\mathbf{E}_n(\mathbf{r}, \omega') + e^{i\omega' t}\overleftrightarrow{\chi}_n^{*}(\mathbf{r}, t; \omega')\mathbf{E}_n^*(\mathbf{r}, \omega')\Big]d\omega'. \tag{9.2.7}$$

where the dielectric permittivity tensor of the transition $\overleftrightarrow{\chi}_n(\mathbf{r}, t;\omega')$ is modulated by elementary excitations in the nth layer. We define the Fourier image of this tensor by a formula similar to (9.2.3a):

$$\overleftrightarrow{\chi}_n(\mathbf{r}, \omega; \omega') = \frac{1}{2\pi}\int_{-\infty}^{\infty} e^{i\omega t}\overleftrightarrow{\chi}_n(\mathbf{r}, t; \omega')dt. \tag{9.2.8}$$

Then from the structure of the Fourier image of the right part of (9.2.7)

$$\begin{aligned}\mathbf{P}_n(\mathbf{r}, \omega) = \varepsilon_0 \int_0^{\infty}\Big[&\overleftrightarrow{\chi}_n(\mathbf{r}, \omega - \omega'; \omega')\mathbf{E}_n(\mathbf{r}, \omega')+ \\ &+ \overleftrightarrow{\chi}_n^{*}(\mathbf{r}, -\omega - \omega'; \omega')\mathbf{E}_n^{*}(\mathbf{r}, \omega')\Big]d\omega'\end{aligned} \tag{9.2.9}$$

the manifestation of combinational frequencies in the transition polarization spectrum is obvious.

As a result the described Fourier transform from Maxwell's equations follows:

—the relationship between the strengths of magnetic and electric fields

$$\mathbf{H}_n(\mathbf{r}, \omega) = -i[\omega\mu_0\overleftrightarrow{\mu}_n(\omega)]^{-1}\mathrm{rot}\,\mathbf{E}_n(\mathbf{r}, \omega); \tag{9.2.10}$$

—the wave equation

$$\mathrm{rot}\Big[\overleftrightarrow{\mu}_n^{-1}(\omega)\mathrm{rot}\,\mathbf{E}_n(\mathbf{r}, \omega)\Big] = \omega^2\Big[c^{-2}\overleftrightarrow{\varepsilon}_n(\omega)\mathbf{E}_n(\mathbf{r}, \omega) + \mu_0\mathbf{P}_n(\mathbf{r}, \omega)\Big]; \tag{9.2.11}$$

—Gauss theorems in differential form in the absence of third-party charges

$$\mathrm{div}\,[\varepsilon_0\overleftrightarrow{\varepsilon}_n(\omega)\mathbf{E}_n(\mathbf{r}, \omega) + \mathbf{P}_n(\mathbf{r}, \omega)] = 0; \tag{9.2.12}$$

$$\mathrm{div}\,[\overleftrightarrow{\mu}_n(\omega)\mathbf{H}(\mathbf{r}, \omega)] = 0, \tag{9.2.13}$$

the second of which is satisfied automatically due to (9.2.10). The obtained equations are the basis for describing the Raman scattering of light with arbitrary anisotropy of the dielectric and magnetic properties of the layers. In the case of scalar functions $\varepsilon_n(\omega)$ and $\mu_n(\omega)$ equations (9.2.11) and (9.2.12) take the form

$$(\mathrm{grad\,div} - \Delta)\mathbf{E}_n(\mathbf{r}, \omega) = \omega^2[c^{-2}\varepsilon_n(\omega)\mu_n(\omega)\mathbf{E}_n(\mathbf{r}, \omega) + \mu_0\mu_n(\omega)\mathbf{P}_n(\vec{r}, \omega)]; \tag{9.2.14}$$

$$\mathrm{div}\,\mathbf{E}_n(\mathbf{r}, \omega) = -[\varepsilon_0\varepsilon_n(\omega)]^{-1}\mathrm{div}\,\mathbf{P}_n(\mathbf{r}, \omega). \tag{9.2.15}$$

Substituting (9.2.15) into (9.2.14) allows us to write the wave equation in a compact formula:

$$\Delta\mathbf{E}_n(\mathbf{r}, \omega) = -\omega^2c^{-2}\varepsilon_n(\omega)\mu_n(\omega)\mathbf{E}_n(\mathbf{r}, \omega) - \mathbf{J}_n(\mathbf{r}, \omega), \tag{9.2.16a}$$

where the function

$$\mathbf{J}_n(\mathbf{r}, \omega) = \omega^2\mu_0\mu_n(\omega)\mathbf{P}_n(\mathbf{r}, \omega) + [\varepsilon_0\varepsilon_n(\omega)]^{-1}\mathrm{grad\,div}\,\mathbf{P}_n(\mathbf{r}, \omega) \tag{9.2.16b}$$

It is determined by the polarization of the transition (9.9).

Next, we perform the Fourier transform of vector fields with respect to variables $\boldsymbol{\rho} = (x, y)$ in a plane perpendicular to the stratification axis:

$$\mathbf{E}_n(\boldsymbol{\eta}, z, \omega) = \frac{1}{(2\pi)^2} \int e^{-i\boldsymbol{\eta}\boldsymbol{\rho}} \mathbf{E}_n(\boldsymbol{\rho}, z, \omega) d^2\rho. \qquad (9.2.17)$$

As a result, the wave equation (9.2.16a) is reduced to a second-order differential equation:

$$[\nabla_z^2 + k_n^{\|2}(\omega, \eta)]\mathbf{E}_n(\boldsymbol{\eta}, z, a) = -\mathbf{J}_n(\boldsymbol{\eta}, z, \omega), \qquad (9.2.18a)$$

where the designation is entered

$$k_n^{\|2}(\omega, \eta) = \omega^2 c^{-2} \mu_n(\omega) \varepsilon_n(\omega) - \eta^2, \qquad (9.2.18b)$$

with the source

$$\begin{aligned} \mathbf{J}_n(\boldsymbol{\eta}, z, \omega) &= \omega^2 \mu_0 \mu_n(\omega) \mathbf{P}_n(\boldsymbol{\eta}, z, \omega) \\ &- \left[\varepsilon_0 \varepsilon_n(\omega)\right]^{-1} (\boldsymbol{\eta} - i\mathbf{e}_3 \nabla_z) \left[(\boldsymbol{\eta} - i\mathbf{e}_3 \nabla_z) \mathbf{P}_n(\boldsymbol{\eta}, z, \omega)\right] \end{aligned} \qquad (9.2.18c)$$

Equation (9.2.15) is transformed in the same way:

$$(\boldsymbol{\eta} - i\mathbf{e}_3\nabla_z)\mathbf{E}_n(\boldsymbol{\eta}, z, \omega) = -[\varepsilon_0 \varepsilon_n(\omega)]^{-1} (\boldsymbol{\eta} - i\mathbf{e}_3\nabla_z)\mathbf{P}_n(\boldsymbol{\eta}, z, \omega). \qquad (9.2.19)$$

The Fourier image of the polarization of the transition (9.2.19) is reduced to the expression

$$\begin{aligned} \mathbf{P}_n(\boldsymbol{\eta}, z, \omega) = \varepsilon_0 \int_0^\infty d\omega' \int d^2\eta' [\overleftrightarrow{\chi}_n(\boldsymbol{\eta} - \boldsymbol{\eta}', z, \omega - \omega'; \omega')\mathbf{E}_n(\boldsymbol{\eta}', z, \omega') \\ + \overleftrightarrow{\chi}_n^*(-\boldsymbol{\eta} - \boldsymbol{\eta}', z, -\omega - \omega'; \omega')\mathbf{E}_n^*(\boldsymbol{\eta}', z, \omega'). \end{aligned} \qquad (9.2.20)$$

Let us consider, as is usually done in the theory of scattering in solids (see (2.6) in [10] or (2.26) in [11]), the modulation of the dielectric permittivity tensor of the transition by vibrational excitations of various branches of j with a certain value of the wave vector $\mathbf{q}_j$:

$$\overleftrightarrow{\chi}_n(\mathbf{r}, t; \omega') = \sum_\nu \left[\overleftrightarrow{\chi}_n^{(+)}(\nu, t; \omega')e^{i\mathbf{q}_j\mathbf{r}} + \overleftrightarrow{\chi}_n^{(-)}(\nu, t; \omega')e^{-i\mathbf{q}_j\mathbf{r}}\right], \qquad (9.2.21)$$

where under the summation sign for all mods $\nu \equiv (\mathbf{q}_j, j)$ the first term corresponds to transitions with absorption, and the second one corresponds to the emission of the corresponding quantum. The Fourier image of the dielectric permittivity tensor of the transition (9.2.21) has the form

$$\begin{aligned} \overleftrightarrow{\chi}_n(\boldsymbol{\eta}, z, \omega; \omega') = \sum_\nu \Big[&\overleftrightarrow{\chi}_n^{(+)}(\nu, \omega; \omega')e^{iq_j^{\|}z}\delta(\boldsymbol{\eta} - \mathbf{q}_j^{\perp}) \\ &+ \overleftrightarrow{\chi}_n^{(-)}(\nu, \omega; \omega')e^{iq_j^{\|}z}\delta(\boldsymbol{\eta} + \mathbf{q}_j^{\perp})\Big], \end{aligned} \qquad (9.2.22)$$

in which the notation is used $\mathbf{q}_j = \mathbf{q}_j^{\perp} + q_j^{\|}\mathbf{e}_3$, $\mathbf{q}_j^{\perp} \perp \mathbf{e}_3$. Substituting this expression into the first part (9.2.20) gives

$$\begin{aligned}\mathbf{P}_n(\boldsymbol{\eta}; z, \omega) = \sum_{\nu}\int_0^{\infty} d\omega' \int d^2\eta' \Big\{ &\mathbf{P}_n^{(+)}(\nu; z; \omega; \omega', \boldsymbol{\eta}')\delta\left(\boldsymbol{\eta} - \boldsymbol{\eta}' - \mathbf{q}_j^{\perp}\right)\\ &+ \mathbf{P}_n^{(-)}(\nu, z; \omega; \omega'; \boldsymbol{\eta}')\delta\left(\boldsymbol{\eta} - \boldsymbol{\eta}' + \mathbf{q}_j^{\perp}\right)\\ &+ \mathbf{P}_n^{(+)*}(\nu, z; -\omega; \omega'; \boldsymbol{\eta}')\delta\left(\boldsymbol{\eta} + \boldsymbol{\eta}' + \mathbf{q}_j^{\perp}\right)\\ &+ \mathbf{P}_n^{(-)*}(\nu, z; -\omega; \omega'; \boldsymbol{\eta}')\delta\left(\boldsymbol{\eta} + \boldsymbol{\eta}' - \mathbf{q}_j^{\perp}\right)\Big\},\end{aligned} \tag{9.2.23}$$

where the functions are introduced:

$$\mathbf{P}_n^{(\pm)}(\nu; z; \omega; \omega'; \boldsymbol{\eta}') = \varepsilon_0 \overleftrightarrow{\chi}_n^{(\pm)}(v; \omega - \omega'; \omega')e^{\pm i q_j^{\|} z}\mathbf{E}_n(\boldsymbol{\eta}'; z; \omega'). \tag{9.2.24}$$

Thus, the form of the dielectric permittivity tensor of the transition (9.2.21) directly follows the structure (9.2.23) of the Fourier image of the corresponding polarization, and hence the describing Raman scattering of the light source (9.2.18c) in the wave equation (9.2.18a), which turns out to be integro-differential with respect to $\mathbf{E}_n(\boldsymbol{\eta}, z, \omega)$.

9.3 Solving wave equations

In typical experiments on Raman scattering of light (see, e.g., [2]), the exciting laser electromagnetic wave is monochromatic. Let it be characterized by frequency ω_L and the wave vector $\boldsymbol{\eta}_L$ in the plane XOY:

$$\mathbf{E}_{nL}(\mathbf{r}, t) = \mathrm{Re}[\mathbf{E}_{nL}(z)e^{i\boldsymbol{\eta}_L\boldsymbol{\rho} - i\omega_L t}]. \tag{9.3.1a}$$

The Fourier image of the field of this wave in accordance with (9.2.3a) and (9.2.17) has the structure

$$\begin{aligned}\mathbf{E}_{nL}(\boldsymbol{\eta}, z, \omega) = \frac{1}{2}\Big[&\mathbf{E}_{nL}(z)\delta(\boldsymbol{\eta} - \boldsymbol{\eta}_L)\delta(\omega - \omega_L)\\ &+ \mathbf{E}_{nL}^{*}(z)\delta\left(\boldsymbol{\eta} + \boldsymbol{\eta}_L\right)\delta\left(\omega + \omega_L\right)\Big].\end{aligned} \tag{9.3.1b}$$

Next, we will consider the components of the dielectric permittivity tensor of the transition (9.2.21) as quantities containing a small parameter, which is determined by the mechanism of Raman scattering of light. This may be, for example, the amplitude of the interaction of a subsystem of band charge carriers (electrons, holes, polarons, or excitons) with a subsystem of vibrational excitations (phonons, polaritons). The specific magnitude of such an amplitude for multilayer systems is found in section 3.7. From (9.2.24) it is clear that the polarization (9.2.23) for this parameter has an order not lower than the first. Therefore, limiting ourselves to zero-order quantities, we find from (9.2.18a) that the amplitude of the electric field intensity of the exciting wave satisfies the homogeneous equation

$$(\nabla_z^2 + k_{nL}^{\parallel 2})\mathbf{E}_{nL}(z) = 0, \tag{9.3.2}$$

where $k_{nL}^{\parallel 2} = k_n^{\parallel 2}(\omega_L, \eta_L)$. At the same time, the transversity of this field follows from (9.2.19):

$$(\boldsymbol{\eta}_L - i\mathbf{e}_3 \nabla_z)\mathbf{E}_{nL}(z) = 0. \tag{9.3.3}$$

The physical meaning of the approximation made is that the effects of wave generation with combinational frequencies are neglected $\omega \neq \omega_L$ on the propagation of an exciting wave with ω_L.

The solution of the wave equation (9.3.2) for different polarization of the exciting wave is found in chapter 1 (see formulas (3.4.1) and (3.2.1)). It is convenient to introduce dimensionless amplitudes of the electric field in a multilayer system (conversion coefficients):

$$E_L^{\perp}(z) = \widetilde{E}_L^{\perp}(z)/E_{iL}^{\perp}. \tag{9.3.4}$$

To write down the boundary values of these coefficients explicitly, we successively substitute (3.4.3) and (3.4.8c) in (3.2.1):

$$E_L^{\perp}(z_n)\frac{\mu_{I-1}}{\mathcal{R}_k} = [D_{n,\,I-1}\mathcal{R}_K + D_{nK}D_{K,\,I-1}e^{i\lambda_{K+1}}\mu_{K+1}]E_L^{\perp}(z_{I-2}). \tag{9.3.5}$$

The modulation transfer matrices are used here $D_{nk}(I-1|K+1)$; moreover, the frequency and wave vector are ω_L and $\boldsymbol{\eta}_L$. Then based on (3.4.8a) we find the relationship between the amplitudes of the electric field at the outer boundary $E_L^{\perp}(z_{I-2})$ and in the falling wave $E_{iL}^{\perp}$:

$$E_L^{\perp}(z_{I-2}) = -2i\sin\lambda_{I-1}\mathcal{R}_k[\ell^+]^{-1}E_{iL}^{\perp}. \tag{9.3.6}$$

Substituting it into (9.3.5), taking into account the definition of (3.5.6b), we obtain the conversion coefficients:

$$\widetilde{E}_n^{\perp}(z_n) = -\frac{2i\overline{\mu}_{I-1}}{\ell^+}[D_{n,\,I-1}\mathcal{R}_K + D_{nK}D_{K,\,I-1}e^{i\lambda_{K+1}}\mu_{K+1}]. \tag{9.3.7}$$

In particular, for a semibounded superlattice on the basis of (4.3.2a) and (4.3.2b) using the notation (3.5.12), we establish that these transformation coefficients

$$\widetilde{E}_L^{\perp}(z_{I+2J-1}) = \left(\widetilde{R}_c^- - \widetilde{R}_c^+\right)(\widetilde{R}_c^-)^{-1}\alpha_L^{-J}; \tag{9.3.8a}$$

$$\widetilde{E}_L^{\perp}(z_{I+2J}) = \left(\widetilde{R}_c^- - \widetilde{R}_c^+\right)(\widetilde{R}_c^-)^{-1}\left[\mu_a + \alpha_L^{-1}\mu_b\right]\nu^{-1}\alpha_L^{-J}(|\,\alpha_L\,| > 1) \tag{9.3.8b}$$

are power functions of the index of the period J. The absorption of the exciting electromagnetic wave in the inner layers is taken into account by the imaginary part $\varepsilon_n''(\omega)$ of the dielectric functions.

The intensity of the incident exciting wave in an external nonabsorbing medium ($n = I - 1$) in the case of s-polarization is determined by the amplitude of the electric field:

$$I_{iL}^{(s)} = \frac{1}{2}\sqrt{\frac{\varepsilon_0\varepsilon_{I-1}(\omega_L)}{\mu_0\mu_{I-1}(\omega_L)}} \mid E_{iL}^{\perp} \mid^2, \tag{9.3.9}$$

and in the case of p-polarization—by the amplitude of the magnetic field:

$$H_{iL}^{\perp} = -\omega_L\varepsilon_0\varepsilon_{I-1}(\omega_L)\left(k_{I-1,\,L}^{\parallel}\right)^{-1}E_{iL}^{\perp}; \tag{9.3.10a}$$

$$I_{iL}^{(p)} = \frac{1}{2}\sqrt{\frac{\mu_0\mu_{I-1}(\omega_L)}{\varepsilon_0\varepsilon_{I-1}(\omega_L)}} \mid H_{iL}^{\perp} \mid^2, \tag{9.3.10b}$$

It should be noted that in a number of experiments on the observation of Raman scattering of light [2, 5], the illumination of the superlattice by exciting p-polarized laser light was carried out at the Brewster angle, at which the minimum reflection coefficient is achieved. The results of the calculation of this coefficient according to the formulas (3.12) and (3.5.13) in the case of superlattice illumination GaAs/AlAs p-polarized laser light with wavelength $\lambda_L = 514.5$ nm are shown in figure 9.1 depending on the angle of incidence. It follows from the graph that Brewster's angle $\theta_{I-1,\,B} = 76°00'$. For comparison, we give the values of the Brewster angle for reflection from GaAs: $\theta_{I-1,\,B} = 76°30'$ and AlAs: $\theta_{I-1,\,B} = 73°10'$. It corresponds to the optimal reflection coefficient $R_L = 2.254 \cdot 10^{-3}$ and the decrement of decreasing $|\alpha_L| = 1.041$. If due to the smallness of the minimum coefficients R_L neglect reflection ($|r| \to 0$), then in formulas (9.3.8a), (9.3.8b) the multiplier is simplified:

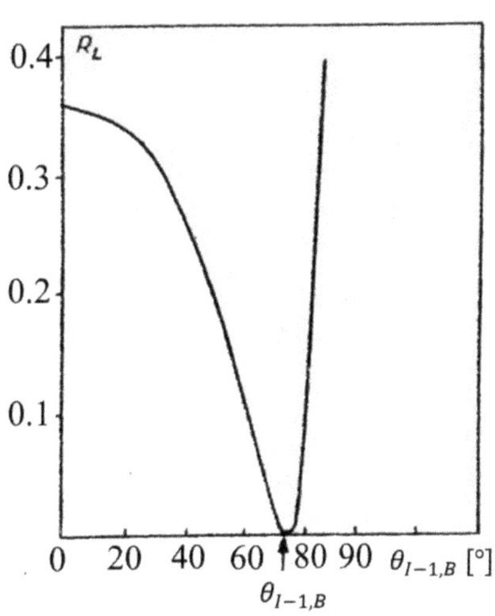

Figure 9.1. Dependence of the reflection coefficient of exciting p-polarized light on the angle of incidence for a superlattice GaAs/AlAs: GaAs (a) – $l_a = 5.44$ nm; $\varepsilon_a(\omega_L)$—based on the results of the Kramers-Kronig analysis of reflection spectra [13]; AlAs (b) – $l_b = 1.67$ нм; $\varepsilon_b(\omega)$—according to the interpolation formula [14].

$$\widetilde{E}_L^{\perp}(z_{I+2J-1}) \approx \alpha_L^{-J}; \tag{9.3.11a}$$

$$\widetilde{E}_L^{\perp}(z_{I+2J}) \approx \left[\mu_a + \alpha_L^{-1}\mu_b\right]\nu^{-1}\alpha_L^{-J}. \tag{9.3.11b}$$

To describe the generation of electromagnetic waves with combination frequencies $\omega \neq \omega_L$ in the first order, according to the small parameter included in the components of the dielectric permittivity tensor of the transition, it is sufficient in the right part (9.2.24) to substitute the electric field strength of the exciting wave (9.3.1b):

$$\begin{aligned} \mathbf{P}_n(\boldsymbol{\eta}, z, \omega) = \frac{1}{2}\sum_{\nu}\Big\{&\mathbf{P}_{n\nu}^{(+)}(z, \omega)\delta\left(\boldsymbol{\eta} - \boldsymbol{\eta}_L - \mathbf{q}_j^{\perp}\right) \\ &+ \mathbf{P}_{n\nu}^{(-)}(z, \omega)\delta\left(\boldsymbol{\eta} - \boldsymbol{\eta}_L + \mathbf{q}_j^{\perp}\right) \\ &+ \mathbf{P}_{n\nu}^{(+)*}(z, -\omega)\delta\left(\boldsymbol{\eta} + \boldsymbol{\eta}_L + \mathbf{q}_j^{\perp}\right) \\ &+ \mathbf{P}_{n\nu}^{(-)*}(z, -\omega)\delta\left(\boldsymbol{\eta} + \boldsymbol{\eta}_L - \mathbf{q}_j^{\perp}\right)\Big\}, \end{aligned} \tag{9.3.12}$$

where the designations are introduced:

$$\mathbf{P}_{n\nu}^{(\pm)}(z, \omega) = \varepsilon_0 \overleftrightarrow{\chi}_n^{(\pm)}(\nu, \omega - \omega_L; \omega_L)e^{\pm iq_j^{\parallel}z}\mathbf{E}_{nL}. \tag{9.3.13}$$

In this case (9.2.18a) it is not a homogeneous differential equation in which the source (9.2.18c) has a frequency structure (9.3.12). Consequently, the electromagnetic field has the same structure. Generation and propagation of electromagnetic waves characterized by two-dimensional wave vectors

$$\boldsymbol{\eta}_{\nu}^{(\pm)} = \boldsymbol{\eta}_L \pm \mathbf{q}_j^{\perp}, \tag{9.3.14}$$

are described by the equation following from (9.2.18a)

$$\left[\nabla_z^2 + k_n^{(\pm)\parallel 2}\right]\mathbf{E}_{n\nu}^{(\pm)}(z, \omega) = -\mathbf{J}_{n\nu}^{(\pm)}(z, \omega), \tag{9.3.15a}$$

where $k_n^{(\pm)\parallel 2} \equiv k_n^{\parallel 2}(\omega, \eta_{\nu}^{(\pm)})$. In accordance with (9.2.18c) (9.3.13), the source is given the form

$$\begin{aligned} \mathbf{J}_{n\nu}^{(\pm)}(z, \omega) = e^{\pm iq_j^{\parallel}z}\Big\{&\omega^2c^2\mu_n(\omega)\mathbf{e}_{\alpha} - \varepsilon_n^{-1}(\omega)\left[\boldsymbol{\eta}_{\nu}^{(\pm)} + \mathbf{e}_3\left(\pm q_j^{\parallel} - i\nabla_z\right)\right]\Big\} \\ &\cdot\left[\eta_{\nu}^{(\pm)\alpha} + \delta^{\alpha 3}\left(\pm iq_j^{\parallel} - i\nabla_z\right)\right]\Big\}\chi_n^{(\pm)\alpha\beta}(\nu; \omega - \omega_L; \omega_L)E_{nL}^{\beta}(z), \end{aligned} \tag{9.3.15b}$$

where tensor analysis notation is used.

Similarly, equation (9.3.19) follows

$$[\boldsymbol{\eta}^{(+)} - i\mathbf{e}_3\nabla_z]\mathbf{E}_{n\nu}^{(\pm)}(z, \omega) = \\ = -e^{\pm iq_j^{\|}z}\varepsilon_n^{-1}(\omega)\Big[\eta_\nu^{(\pm)\alpha} + \delta^{\alpha 3}\Big(\pm q_j^{\|} - i\nabla_z\Big)\Big]\chi_n^{(\pm)\alpha\beta}(\nu;\omega - \omega_L;\omega_L)E_{nL}^{\beta}(z), \tag{9.3.16}$$

from where after a single differentiation by z and substitution $\nabla_z^2\ E_{n\nu}^{(\pm)\|}(z, \omega)$ from (9.3.15a) with the source $J_{n\nu}^{(\pm)\|}(z, \omega)$ (9.3.15b) we find the component of the electric field along the stratification axis:

$$\mathbf{E}_{n\nu}^{(\pm)}(z, \omega) = \frac{1}{k_{n\nu}^{(\pm)\|2}}\{i\ \nabla_z\ \boldsymbol{\eta}_\nu^{(\pm)}E_{n\nu}^{(\pm)\perp}(z, \omega) - \\ - e^{\pm iq_j^{\|}z}\omega^2 c^{-2}\mu_n(\omega)\delta^{\alpha 3}\chi_n^{(\pm)\alpha\beta}(\nu;\omega - \omega_L;\omega_L)E_{nL}^{\beta}(z)\}. \tag{9.3.17}$$

In this regard, comprehensive information about the electromagnetic wave is contained in the component of the electric field across the stratification axis.

For the specified component from (9.3.15a), the equation follows directly:

$$\Big[\nabla_z^2 + k_n^{(\pm)\|2}\Big]\mathbf{E}_{n\nu}^{(\pm)\perp}(z, \omega) = -\mathbf{J}_{n\nu}^{(\pm)\perp}(z, \omega). \tag{9.3.18}$$

The solution of the obtained inhomogeneous differential equation satisfying the boundary conditions

$$\mathbf{E}_{n\nu}^{(\pm)\perp}(z, \omega)\big|_{z=z_{n-1}} = \mathbf{E}_\nu^{(\pm)\perp}(z_{n-1}, \omega);\ \mathbf{E}_{n\nu}^{(\pm)\perp}(z, \omega)\big|_{z=z_n} = \mathbf{E}_\nu^{(\pm)\perp}(z_n, \omega),$$

has the form ($z_{n-1} \leqslant z \leqslant z_n$)

$$\mathbf{E}_{n\nu}^{(\pm)\perp}(z, \omega) = \Big[\sin\lambda_{n\nu}^{(\pm)}\Big]^{-1}\Big\{\mathbf{E}_\nu^{(\pm)\perp}(z_n, \omega)\sin\Big[k_{n\nu}^{(\pm)\|}(z - z_{n-1})\Big] \\ + \mathbf{E}_\nu^{(\pm)\perp}(z_{n-1}, \omega)\sin\Big[k_{n\nu}^{(\pm)\|}(z_n - z)\Big]\Big\} + \boldsymbol{\theta}_{n\nu}^{(\pm)\perp}(z, \omega). \tag{9.3.19}$$

Here $\lambda_{n\nu}^{(\pm)} \equiv l_n k_{n\nu}^{(\pm)\|}$ and the integral transformation of the source is introduced:

$$\boldsymbol{\theta}_{n\nu}^{(\pm)\perp}(z, \omega) = \boldsymbol{\Phi}\Big[\mathbf{J}_{n\nu}^{(\pm)\perp}(z', \omega)\Big], \tag{9.3.20a}$$

defined by an explicit expression

$$\boldsymbol{\Phi}\Big[\mathbf{J}_{n\nu}^{(\pm)\perp}(z, \omega)\Big] = \Big[k_{n-1}^{(\pm)\|}\sin\lambda_{n-1}^{(\pm)}\Big]^{-1}\int_{z_{n-1}}^{z_n}\mathbf{J}_{n-1}^{(\pm)\perp}(z', \omega) \\ \Big\{\sin\Big[k_{n\nu}^{(\pm)\|}(z_n - z')\Big]\sin\Big[k_{n\nu}^{(\pm)\|}(z - z_{n-1})\Big]\theta(z' - z) \\ + \sin\Big[k_{n\nu}^{(\pm)\|}(z' - z_{n-1})\Big]\sin\Big[k_{n\nu}^{(\pm)\|}(z_n - z)\Big] \\ \theta(z - z')\Big]\Big\}dz. \tag{9.3.20b}$$

With an arbitrary direction of propagation of the scattered wave other than normal ($\boldsymbol{\eta}_\nu^{(\pm)} \neq 0$), the source is conveniently decomposed into orts *s*- and *p*-polarization:

$$\boldsymbol{\sigma}_\nu^{(\pm)} = \frac{\left[\boldsymbol{\eta}_\nu^{(\pm)}, \mathbf{e}_3\right]}{\eta_\nu^{(\pm)}}; \boldsymbol{\pi}_\nu^{(\pm)} = \frac{\boldsymbol{\eta}_\nu^{(\pm)}}{\eta_\nu^{(\pm)}}. \tag{9.3.21a}$$

Considering these definitions as equations with respect to the orts of the Cartesian coordinate system, we find for $\alpha = 1,\ 2$:

$$\mathbf{e}_\alpha = \frac{1}{\eta_\nu^{(\pm)}}\left[\eta_\nu^{(\pm)\alpha}\boldsymbol{\pi}_\nu^{(\pm)} + \varepsilon^{\alpha\gamma3}\eta_\nu^{(\pm)\gamma}\boldsymbol{\sigma}_\nu^{(\pm)}\right]. \tag{9.3.21b}$$

As a result, the desired decomposition follows from (9.3.15b):

$$\mathbf{J}_{n\nu}^{(\pm)\perp}(z, \omega) = J_{n\nu}^{(\pm)(s)}(z, \omega)\boldsymbol{\sigma}_\nu^{(\pm)} + J_{n\nu}^{(\pm)(p)}(z, \omega)\boldsymbol{\pi}_\nu^{(\pm)} \tag{9.3.22a}$$

with the components

$$\left\| \begin{matrix} J_{n\nu}^{(\pm)(s)}(z, \omega) \\ J_{n\nu}^{(\pm)(p)}(z, \omega) \end{matrix} \right\| = \frac{e^{\pm iq_j^\| z}}{\eta_\nu^{(\pm)}} \left\| \begin{matrix} \omega^2 c^{-2}\mu_n(\omega)\varepsilon^{\alpha\gamma3}\eta_\nu^{(\pm)\gamma} \\ \varepsilon_n^{-1}(\omega)\left[\eta_\nu^{(\pm)\alpha}k_{n\nu}^{(\pm)\|2} - \eta_\nu^{(\pm)2}\delta^{\alpha3}\left(\pm q_j^\| - i\nabla_z\right)\right] \end{matrix} \right\| \tag{9.3.22b}$$
$$\times\, \chi_n^{(\pm)\alpha\beta}(\nu; \omega - \omega_L; \omega_L)E_{nL}^\beta(z).$$

Due to the linearity of the integral transformation (9.3.20b) for the function (9.3.20a), we find a decomposition similar to (9.3.22a):

$$\boldsymbol{\theta}_{n\nu}^{(\pm)\perp}(z, \omega) = \theta_{n\nu}^{(\pm)(s)}(z, \omega)\boldsymbol{\sigma}_\nu^{(\pm)} + \theta_{n\nu}^{(\pm)(p)}(z, \omega)\boldsymbol{\pi}_\nu^{(\pm)}; \tag{9.3.23a}$$

$$\theta_{n\nu}^{(\pm)(s,\, p)}(z, \omega) = \Phi\left[J_{n\nu}^{(\pm)(s,\, p)}(z', \omega)\right]. \tag{9.3.23b}$$

In the case of *s*- polarization of the exciting electromagnetic wave

$$\begin{cases} E_{nL}^\beta(z) = E_{nL}^\perp(z)\sigma_L^\beta \equiv E_{nL}^\perp(z)\eta_L^{-1}[\boldsymbol{\eta}_L, \mathbf{e}_3], \ \beta = 1, 2; \\ E_{nL}^3(z) \equiv E_{nL}^\|(z) = 0, \end{cases}$$

performing an integral transformation (9.3.20b) above the right part (9.3.22b), we find

$$\left\| \begin{matrix} \theta_{n\nu}^{(\pm)(s)}(z, \omega) \\ \theta_{n\nu}^{(\pm)(p)}(z, \omega) \end{matrix} \right\|$$
$$= \frac{1}{\eta_L^{(\pm)}\eta_L} \left\| \begin{matrix} \omega^2 c^{-2}\mu_n(\omega)\varepsilon^{\alpha\gamma3}\eta_\nu^{(\pm)\gamma}\mathcal{F}_1^{(\pm)}(z) \\ \varepsilon_n^{-1}(\omega)\left[\eta_\nu^{(\pm)\alpha}k_{n\nu}^{(\pm)\|2}\mathcal{F}_1^{(\pm)}(z) - \eta_\nu^{(\pm)2}\delta^{\alpha3}\left(\pm q_j^\|\mathcal{F}_1^{(\pm)}(z) - i\mathcal{F}_2^{(\pm)}(z)\right)\right] \end{matrix} \right\| \tag{9.3.24}$$
$$\times\, \chi_n^{(\pm)\alpha\beta}(\nu, \omega - \omega_L; \omega_L)[\boldsymbol{\eta}_L, \mathbf{e}_3]^\beta.$$

Here the functions are introduced:

$$\mathcal{F}_1^{(\pm)}(z) = \Phi\left[e^{\pm iq_j^{\|}z'} E_{nL}^{\perp}(z')\right]; \tag{9.3.25a}$$

$$\mathcal{F}_2^{(\pm)}(z) = \Phi\left[e^{\pm iq_j^{\|}z'}\ \nabla_{z'}\ E_{nL}^{\perp}(z')\right], \tag{9.3.25b}$$

which are taking into account the explicit dependency $E_{nL}^{\perp}$ (3.4.1c) finally presented in a single form:

$$\begin{aligned} \mathcal{F}_i^{(\pm)}(z) = &\left[(-k_{n\nu}^{(\pm)\|2} + k_{nL}^{\|2} + q_j^{\|2})^2 - 4k_{nL}^{\|2}q_j^{n2}\right]^{-1} \\ &\times \left\{f_j^{(\pm)}(z) - \left[\sin\lambda_{n\nu}^{(\pm)}\right]^{-1}\left[f_i^{(\pm)}(z_n)\sin k_{n\nu}^{(\pm)\|}(z - z_{n-1}) + \right.\right. \\ &\left.\left.+ f_i^{(\pm)}(z_{n-1})\right]\sin k_{n\nu}^{(\pm)\|}(z_n - z)\right]\right\}, \end{aligned} \tag{9.3.26a}$$

where

$$f_1^{(\pm)}(z) = e^{\pm iq_j^{\|}z}\left[(-k_{n\nu}^{(\pm)\|2} + k_{nL}^{\|2} + q_j^{\|2})E_{nL}^{\perp}(z) \pm 2iq_j^{\|}\nabla_z E_{nL}^{\perp}(z)\right]; \tag{9.3.26b}$$

$$f_2^{(\pm)}(z) = e^{\pm iq_j^{\|}z}\left[(-k_{n\nu}^{(\pm)\|2} + k_{nL}^{\|2} + q_j^{\|2})\ \nabla_z\ E_{nL}^{\perp}(z) \mp 2ik_{nL}^{\|2}q_j^{\|}E_{nL}^{\perp}(z)\right]. \tag{9.3.26c}$$

In the case of p-polarization of the exciting electromagnetic wave

$$\begin{cases} E_{nL}^{\beta}(z) = E_{nL}^{\perp}(z)\pi_L^{\beta} \equiv E_{nL}^{\perp}(z)\dfrac{\eta_L^{\beta}}{\eta_L}, \quad \beta = 1, 2; \\ E_{nL}^{3}(z) \equiv E_{nL}^{\|}(z) = \dfrac{i\eta_L}{k_{nL}^{\|2}}\nabla_z E_{nL}^{\perp}(z) \end{cases}$$

the components (9.3.23b) are described by the equation

$$\begin{aligned} &\left\| \begin{matrix} \theta_{n\nu}^{(\pm)(S)}(z, \omega) \\ \theta_{n\nu}^{(\pm)(p)}(z, \omega) \end{matrix} \right\| = \frac{1}{\eta_\nu^{(\pm)}\eta_L} \\ &\times \left\| \begin{matrix} \omega^2c^{-2}\mu_n(\omega)\varepsilon^{\alpha\gamma3}\eta_\nu^{(\pm)\gamma}F_1^{(\pm)\alpha}(z) \\ \varepsilon_n^{-1}(\omega)\left[\eta_\nu^{(\pm)\alpha}k_{n\nu}^{(\pm)\|2}F_1^{(\pm)\alpha}(z) - \eta_\nu^{(\pm)2}\delta^{\alpha3}(\pm q_j^{\|}F_1^{(\pm)\alpha}(z) - iF_2^{(\pm)\alpha}(z))\right] \end{matrix} \right\|, \end{aligned} \tag{9.3.27a}$$

where

$$\begin{aligned} F_1^{(\pm)\alpha}(z) = &\chi_n^{(\pm)\alpha\beta}(\nu, \omega - \omega_L, \omega_L)\eta_L^{\beta}\ \mathcal{F}_1^{(\pm)}(z) \\ &+ +\chi_n^{(\pm)\alpha3}(\nu, \omega - \omega_L; \omega_L)\frac{i\eta_L^2}{k_{nL}^{\|2}}\mathcal{F}_2^{(\pm)}(z); \end{aligned} \tag{9.3.27b}$$

$$F_2^{(\pm)\alpha}(z) = \chi_n^{(\pm)\alpha\beta}(\nu, \omega - \omega_L; \omega_L)\eta_L^\beta \mathcal{F}_2^{(\pm)}(z) - \\ - \chi_n^{(\pm)\alpha 3}(\nu, \omega - \omega_L; \omega_L) i\eta_L^2 \mathcal{F}_1^{(\pm)}(z). \tag{9.3.27c}$$

With a normal distribution of the scattered wave ($\boldsymbol{\eta}_L^{(\pm)} = 0$) the source (9.3.15b) is represented as an ort decomposition of the Cartesian coordinate system:

$$\mathbf{J}_{n\nu}^{(\pm)\perp}(z, \omega) = \sum_{\alpha=1}^{2} J_{n\nu}^{(\pm)\alpha}(z, \omega)\mathbf{e}_\alpha, \tag{9.3.28a}$$

with the components

$$J_{n\nu}^{(\pm)\alpha}(z, \omega) = \mathrm{e}^{\pm iq_j^{\|}z}\omega^2 c^{-2}\mu_n(\omega)\chi_n^{(\pm)\alpha\beta}(\nu, \omega - \omega_L; \omega_L)E_{nL}^\beta(z). \tag{9.3.28b}$$

In a similar decomposition for the function (9.3.20a)

$$\boldsymbol{\theta}_{n\nu}^{(\pm)}(z, \omega) = \sum_{\alpha=1}^{2} \theta_{n\nu}^{(\pm)\alpha}(z, \omega)\mathbf{e}_\alpha \tag{9.3.29a}$$

the components are defined by the transformation (9.3.20b):

$$\theta_{n\nu}^{(\pm)\alpha}(z, \omega) = \Phi\left[J_{n\nu}^{(\pm)\alpha}(z, \omega)\right]. \tag{9.3.29b}$$

This transformation in the case of *s*-polarization of the exciting electromagnetic wave leads to

$$\theta_{n\nu}^{(\pm)\alpha}(z, \omega) = \omega^2 c^{-2}\mu_n(\omega)\chi_n^{(\pm)\alpha\beta}(\nu, \omega - \omega_L; \omega_L)[\boldsymbol{\eta}_L, \mathbf{e}_3]^\beta \eta_L^{-1}\mathcal{F}_1^{(\pm)}(z), \tag{9.3.30}$$

and in the case of *p*-polarization of the exciting electromagnetic wave to

$$\theta_{n\nu}^{(\pm)\alpha}(z, \omega) = \omega^2 c^{-2}\mu_n(\omega)\eta_L^{-1}F_1^{(\pm)\alpha}(z), \tag{9.3.31}$$

where the functions (9.3.26a), (9.3.26b), and (9.3.27b) are used. Thus, the solution of the wave equations has been completed taking into account the Raman scattering of light with an arbitrary structure of the dielectric permittivity tensor of the transition $\chi_n^{(\pm)\alpha\beta}(\nu, \omega - \omega_L; \omega_L)$.

In the case of a superlattice GaAs/AlAs in accordance with the selection rules for Raman scattering of light (table 9.2 of the work [12]) corresponding to active irreducible representations A_1 and B_2 and crystal class D_{2d} the tensors of the dielectric permittivity of the transition have the structure, respectively:

$$\begin{Vmatrix} a & 0 & 0 \\ 0 & a & 0 \\ 0 & 0 & b \end{Vmatrix} \text{ and } \begin{Vmatrix} 0 & d & 0 \\ d & 0 & 0 \\ 0 & 0 & 0 \end{Vmatrix},$$

and the total tensor:

$$\overleftrightarrow{\chi}_n = \begin{Vmatrix} \chi_n^{11} & \chi_n^{12} & 0 \\ \chi_n^{12} & \chi_n^{11} & 0 \\ 0 & 0 & \chi_n^{33} \end{Vmatrix}. \tag{9.3.32}$$

From this place arguments $(\nu, \omega - \omega_L; \omega_L)$ go down. This makes it possible to concretize the decomposition components found above (9.3.23a) and (9.3.29a).

For an arbitrary direction of propagation of the scattered wave other than normal, for the s-polarization of the exciting wave from (9.3.24) follows:

$$\begin{Vmatrix} \theta_{n\nu}^{(\pm)(s)}(z, \omega) \\ \theta_{n\nu}^{(\pm)(p)}(z, \omega) \end{Vmatrix} = \frac{1}{\eta_L^{(\pm)}\eta_L} \begin{Vmatrix} \frac{\omega^2}{c^2}\mu_n(\omega)\left[\chi_n^{(\pm)11}(\boldsymbol{\eta}_L^{(\pm)}\boldsymbol{\eta}_L) - \chi_n^{(\pm)12}(\eta_\nu^{(\pm)x}\eta_L^y + \eta_v^{(\pm)y}\eta_L^x)\right] \\ \frac{k_{n\nu}^{(\pm)\|2}}{\varepsilon_n(\omega)}\left\{\chi_n^{(\pm)11}\left[\boldsymbol{\eta}_\nu^{(\pm)}, \boldsymbol{\eta}_L\right]^z - \chi_n^{(\pm)12}(\eta_\nu^{(\pm)x}\eta_L^x - \eta_\nu^{(\pm)y}\eta_L^y)\right\} \end{Vmatrix} \mathcal{F}_1^{(\pm)}(z). \tag{9.3.33}$$

$$\begin{Vmatrix} \theta_{n\nu}^{(\pm)(s)}(z, \omega) \\ \theta_{n\nu}^{(\pm)(p)}(z, \omega) \end{Vmatrix} = (\eta_\nu^{(\pm)}\eta_L)^{-1}.$$

$$\begin{Vmatrix} \frac{\omega^2}{c^2}\mu_n(\omega)\left\{\chi_n^{(\pm)11}\left[\boldsymbol{\eta}_L, \boldsymbol{\eta}_\nu^{(\pm)}\right]^z - \chi_n^{(\pm)12}(\eta_\nu^{(\pm)x}\eta_L^x - \eta_\nu^{(\pm)y}\eta_L^{(\pm)y})\right\}\mathcal{F}_1^{(\pm)}(z) \\ \frac{1}{\varepsilon_n(\omega)}\left\{k_{n\nu}^{(\pm)\|2}\left[\chi_n^{(\pm)11}(\boldsymbol{\eta}_\nu^{(\pm)}, \boldsymbol{\eta}_L) + \chi_n^{(\pm)12}(\eta_\nu^{(\pm)x}\eta_L^y + \eta_\nu^{(\pm)y}\eta_L^x)\right]\right. \\ \left.\mathcal{F}_1^{(\pm)}(z) + \eta_\nu^{(\pm)2}\eta_L^2\chi_n^{(\pm)33}\mathcal{F}_3^{(\pm)}(z)\right. \end{Vmatrix}, \tag{9.3.34}$$

Similarly, substituting (9.3.32) into (9.3.27) gives for the p-polarization of the exciting wave where in the notation (9.3.26a)

$$f_3^{(\pm)}(z) = e^{\pm iq_j^{\|}z} \left[(-k_{n\nu}^{(\pm)\|2} + k_{nL}^{\|2} + q_j^{\|2})E_{nL}^{\perp}(z) \mp \frac{iq_j^{\|}}{k_{nL}^{\|2}}(-k_{n\nu}^{(\pm)\|2} - k_{nL}^{\|2} + q_j^{\|2})\nabla_z E_{nL}^{\perp}(z)\right]. \tag{9.3.35}$$

As a result of using the tensor (9.3.32) with normal propagation of the scattered wave for the s-polarization of the exciting wave from (9.3.30) follows:

$$\left\| \begin{array}{c} \theta_{n\nu}^{(\pm)x}(z,\omega) \\ \theta_{n\nu}^{(\pm)y}(z,\omega) \end{array} \right\| = \frac{\omega^2}{c^2}\frac{\mu_n(\omega)}{\eta_L} \left\| \begin{array}{c} \chi_n^{(\pm)11}\eta_L^y - \chi_n^{(\pm)12}\eta_L^x \\ -\chi_n^{(\pm)11}\eta_L^x + \chi_n^{(\pm)12}\eta_L^y \end{array} \right\| \mathcal{F}_1^{(\pm)}(z), \tag{9.3.36}$$

and for the p-polarization of the exciting wave (9.3.31):

$$\left\| \begin{array}{c} \theta_{n\nu}^{(\pm)x}(z,\omega) \\ \theta_{n\nu}^{(\pm)y}(z,\omega) \end{array} \right\| = \frac{\omega^2}{c^2}\frac{\mu_n(\omega)}{\eta_L} \left\| \begin{array}{c} \chi_n^{(\pm)11}\eta_L^x + \chi_n^{(\pm)12}\eta_L^y \\ \chi_n^{(\pm)11}\eta_L^y + \chi_n^{(\pm)12}\eta_L^x \end{array} \right\| \mathcal{F}_1^{(\pm)}(z). \tag{9.3.37}$$

It follows from the obtained formulas that when Raman scattering of light is observed, polarized separation of contributions of vibrational excitations of different symmetry is possible. So, in the case of Raman scattering of light in geometry 'backwards' in the plane of incidence of the exciting wave (in particular, in the direction of reflection when illuminating a multilayer structure with p-polarization light at Brewster angle), when for certainty $\boldsymbol{\eta}_L = \eta_L^x \mathbf{e}_1$ and $\boldsymbol{\eta}_\nu^{(\pm)} = \eta_\nu^{(\pm)x}\mathbf{e}_1$, from (9.3.33) and (9.3.34) it directly follows that the generation of waves with combination frequencies having the same type of polarization as the exciting wave (polarized scattering) is due to A_1-oscillation, whereas the generation of waves with combination frequencies having a different type of polarization than the exciting wave (depolarized scattering)—B_2-fluctuations. In the case of normal propagation of a wave scattered in geometry 'backwards', when $\boldsymbol{\eta}_\nu^{(\pm)}=0$ and for certainty again $\boldsymbol{\eta}_L = \eta_L^x \mathbf{e}_1$, from (9.3.36) and (9.3.37) it can be seen that the generation of waves with combinational frequencies polarized parallel to the exciting wave (polarized scattering) is due to A_1-oscillation, while the generation of waves with combination frequencies, polarized perpendicular to the exciting wave (depolarized scattering), B_2-fluctuations. This conclusion was formulated in [5] for the case when the exciting light falls normally to the surface of the scattering multilayer structure.

9.4 Determination of boundary amplitudes and intensity of waves with combinatorial frequencies

To complete the task, it is necessary to find the amplitudes of the electric field of waves with combinational frequencies on the interface surfaces of the layers included in the overall result (9.3.19). Let us first consider the Raman scattering of light in an arbitrary direction other than normal.

In the case of s-polarization of the scattered wave, the components of the electric field are

$$\mathbf{E}_{n\nu}^{(\pm)\perp}(z,\omega) = \boldsymbol{\sigma}_\nu^{(\pm)} E_{n\nu}^{(\pm)\perp}(z,\omega);\ \mathbf{E}_{n\nu}^{(\pm)\parallel}(z,\omega) = 0, \tag{9.4.1}$$

and the tangential component of the magnetic field in accordance with (4.18b) is

$$\mathbf{H}_{n\nu}^{(\pm)\perp}(z,\omega) = \boldsymbol{\pi}_\nu^{(\pm)}[i\omega\,\mu_0\mu_n(\omega)]^{-1}\,\nabla_z\, E_{n\nu}^{(\pm)\perp}(z,\omega). \tag{9.4.2}$$

Boundary conditions of continuity of the tangential components of the magnetic field (9.4.2) on the shockless surfaces of the sections of the layers z_n using (9.3.19) and notation

$$\mu_{n\nu} = \overline{\mu}_{n\nu}\left[\sin \lambda_{n\nu}^{(\pm)}\right]^{-1};\ \nu_{n\nu} = \mu_{n\nu} \cos \lambda_{n\nu}^{(\pm)} + \mu_{n+1,\,\nu} \cos \lambda_{n+1,\,\nu}^{(\pm)}; \tag{9.4.3}$$

$$\overline{\mu}_{n\nu} = \mu_{n\nu}^{(s)} \equiv k_{n\nu}^{(\pm)\|}[\mu_n(\omega)]^{-1} \tag{9.4.4}$$

take the form ($n = I - 1, \ldots, K$)

$$\begin{aligned} &-\mu_{n\nu} E_{\nu}^{(\pm)\perp}(z_{n-1}, \omega) + \nu_{n\nu} E_{\nu}^{(\pm)\perp}(z_n, \omega) \\ &-\mu_{n+1,\nu} E_{\nu}^{(\pm)\perp}(z_{n+1}, \omega) = \Sigma_{n\nu}^{(\pm)}(\omega), \end{aligned} \tag{9.4.5a}$$

where the first part

$$\Sigma_{n\nu}^{(\pm)}(\omega) = \Sigma_{n-1,\,\nu}^{'(\pm)}(z_n, \omega) - \Sigma_{n\nu}^{'(\pm)}(z_n, \omega) \tag{9.4.5b}$$

is expressed through the function

$$\Sigma_{n\nu}^{'(\pm)}(z, \omega) = \Sigma_{n\nu}^{'(\pm)(s)}(z, \omega) \equiv \mu_n^{-1}(\omega)\ \nabla_z\ \theta_{n\nu}^{(\pm)(s)}(z, \omega) \tag{9.4.6}$$

In the case of p-polarization of the scattered wave, the components of the electric field, taking into account (9.3.17), are

$$\mathbf{E}_{n\nu}^{(\pm)\perp}(z, \omega) = \boldsymbol{\pi}_{\nu}^{(\pm)} E_{n\nu}^{(\pm)\perp}(z, \omega); \tag{9.4.7a}$$

$$\begin{aligned} &E_{n\nu}^{(\pm)\|}(z, \omega) = \left[k_{n\nu}^{(\pm)\|}\right]^{-2} \\ &\left\{i\eta_{\nu}^{(\pm)}\ \nabla_z\ E_{n\nu}^{(\pm)\perp}(z, \omega) - \mathrm{e}^{\pm iq_j z}\omega^2 c^{-2}\mu_n(\omega)\chi_n^{(\pm)3\beta} E_{nL}^{\beta}(z)\right\}, \end{aligned} \tag{9.4.7b}$$

and the tangential component of the magnetic field in accordance with (1.9.9)

$$\mathbf{H}_{n\nu}^{(\pm)}(z) = \boldsymbol{\sigma}_{\nu}^{(\pm)} \frac{i\omega\ \varepsilon_0}{k_{n\nu}^{(\pm)\|2}}\{\varepsilon_n(\omega)\ \nabla_z\ E_{n\nu}^{(\pm)\perp}(z, \omega) + i\mathrm{e}^{\pm iq_j z}\eta_{\nu}^{(\pm)}\chi_n^{(\pm)3\beta} E_{nL}^{\beta}(z)\Big\}. \tag{9.4.8}$$

The above boundary conditions for the continuity of the tangential components of the magnetic field using (9.3.19), (9.4.3) and the notation

$$\overline{\mu}_{n\nu} = \overline{\mu}_{n\nu}^{(p)} \equiv \varepsilon_n(\omega)\left[k_{n\nu}^{(\pm)\|}\right]^{-1} \tag{9.4.9}$$

are reduced to the previous form (9.4.5a), where now in (9.4.5b)

$$\begin{aligned} &\Sigma_{n\nu}^{'(\pm)}(z, \omega) = \Sigma_{n\nu}^{'(\pm)(p)}(z, \omega) \\ &\equiv \left[k_{n\nu}^{(\pm)\|}\right]^{-2}\{\varepsilon_n(\omega)\ \nabla_z\ \theta_{n\nu}^{(\pm)(p)}(z, \omega) + i\mathrm{e}^{\pm iq_j z}\eta_{\nu}^{(\pm)}\chi_n^{(\pm)3\beta} E_{nL}^{\beta}(z_n)\}. \end{aligned} \tag{9.4.10}$$

Finally, with normal propagation of the scattered wave

$$\mathbf{E}_{n\nu}^{(\pm)\perp}(z, \omega) = \sum_{a=1,2} E_{n\nu}^{(\pm)\perp\alpha}(z, \omega)\mathbf{e}_a; \tag{9.4.11}$$

$$\mathbf{H}_{n\nu}^{(\pm)\perp}(z, \omega) = [i\omega\ \mu_0\mu_n(\omega)]^{-1} \sum_{\alpha,\beta=1,2} \varepsilon^{\alpha\beta 3}\ \nabla_z\ E_{n\nu}^{(\pm)\perp\alpha}(z, \omega)\mathbf{e}_\beta. \tag{9.4.12}$$

The abovementioned boundary conditions for the continuity of the tangential components of the magnetic field using the following notation from (9.4.4) or (9.4.9)

$$\overline{\mu}_{n\nu} \equiv \frac{c}{\omega}\sqrt{\frac{\varepsilon_n(\omega)}{\mu_n(\omega)}} \tag{9.4.13}$$

again take the form (9.4.5a) for each of the Cartesian components $E_{n\nu}^{(\pm)\alpha}(z, \omega)$, where in (9.4.5b)

$$\Sigma_{n\nu}^{'(\pm)}(z, \omega) = \Sigma_{n\nu}^{'(\pm)\alpha}(z, \omega) \equiv [\mu_n(\omega)]^{-1}\ \nabla_z\ \theta_{n\nu}^{(\pm)(\alpha)}(z, \omega). \tag{9.4.14}$$

The solution of an inhomogeneous system of algebraic equations (9.4.5a) is obtained by the formulas derived earlier in the theory of potential for multilayer structures (1.3.7a) and (1.3.7b), in the following form:

$$\begin{aligned} E_{n\nu}^{(\pm)\|}(z_n, \omega) &= D_{n,\,I-1,\,\nu}\mu_{I-1,\,\nu}E_\nu^{(\pm)\perp}(z_{I-2}, \omega) \\ &+ D_{nk\nu}\ \mu_{K+1,\,\nu}E_\nu^{(\pm)\perp}(z_{K+1}, \omega) + \Phi_{n\nu}^{(\pm)}(\omega); \end{aligned} \tag{9.4.15}$$

$$\Phi_{n\nu}^{(\pm)}(\omega) = \sum_{k=I-1}^{K} D_{nk\nu}\ \Sigma_{n\nu}^{(\pm)}(\omega), \tag{9.4.16}$$

where the modulation transfer matrices $D_{nk\nu} = D_{nk\nu}\ (I-1⍰K+1)$ are calculated with the above coefficients (9.4.4), (9.4.9), or (9.4.13) for an electromagnetic wave with a frequency of ω and a two-dimensional wave vector $\boldsymbol{\eta}_\nu^{(\pm)}$.

In a typical experimental situation, Raman scattering of light does not occur in the outer layers: $\chi_n^{(\pm)}(\nu, \omega-\omega_L; \omega_L) = 0$ for $n = I-1, K+1$. Comparison of the solution (9.3.19) in these layers with the type of fields corresponding to the propagation of waves with combinational frequencies 'backwards' from the scattering layers

$$E_{I-1,\,\nu}^{(\pm)\perp}(z, \omega) = e^{-ik_{I-1,\nu}^{(\pm)\|}(z-z_{I-1})}E_{b\nu}^{(\pm)\perp}(\omega) \tag{9.4.17a}$$

and 'forward' from the scattering layers

$$E_{K+1,\,\nu}^{(\pm)\perp}(z, \omega) = e^{-ik_{K+1,\nu}^{(\pm)\|}(z-z_K)}E_{f\nu}^{(\pm)\perp}(\omega), \tag{9.4.17b}$$

allows you to immediately express the corresponding amplitudes through the boundary values of the amplitudes on the outer surfaces of the system:

$$E_{b\nu}^{(\pm)\perp}(\omega) = e^{-i\lambda_{I-1,\nu}^{(\pm)}} E_{\nu}^{(\pm)\perp}(z_{I-2}, \omega); \tag{9.4.18a}$$

$$E_{f\nu}^{(\pm)\perp}(\omega) = e^{-i\lambda_{K+1,\nu}^{(\pm)}} E_{\nu}^{(\pm)\perp}(z_{K+1}, \omega). \tag{9.4.18b}$$

It is obvious that in the outer layers there are no waves with combinational frequencies that would propagate towards the scattering layers. Hence follows:

$$\begin{cases} E_{\nu}^{(\pm)\perp}(z_{I-2}, \omega) = e^{i\lambda_{I-1,\nu}^{(\pm)}} E_{\nu}^{(\pm)\perp}(z_{I-1}, \omega); \\ E_{\nu}^{(\pm)\perp}(z_{K+1}, \omega) = e^{i\lambda_{K+1,\nu}^{(\pm)}} E_{\nu}^{(\pm)\perp}(z_{K}, \omega). \end{cases} \tag{9.4.19ab}$$

Substituting into these relations the amplitudes of the field on the inner surfaces of the section from (9.4.15) and solving the resulting system of inhomogeneous algebraic equations, we find the amplitudes at the outer boundaries of the system, and then the amplitudes of scattered electromagnetic waves (9.4.18a) and (9.4.18b):

$$E_{b\nu}^{(\pm)\perp}(\omega) = \frac{1}{\ell_{\nu}^{+}}\left[\mathcal{R}_{K\nu}\, \Phi_{I-1,\,\nu}^{(\pm)}(\omega) + e^{i\lambda_{K+1,\nu}^{(\pm)}} D_{I-1,\,\nu}\mu_{K+1,\,\nu}\Phi_{K\nu}^{(\pm)}(\omega)\right]; \tag{9.4.20a}$$

$$E_{f\nu}^{(\pm)\perp}(\omega) = \frac{1}{\ell_{\nu}^{+}}\left[\mathcal{R}_{I-1,\,\nu}^{+}\Phi_{K\nu}^{(\pm)}(\omega) + e^{i\lambda_{I-1,\nu}^{(\pm)}} D_{K,\,I-1,\,\nu}\mu_{I-1,\,\nu}\Phi_{I-1,\,\nu}^{(\pm)}(\omega)\right], \tag{9.4.20b}$$

where the notation (3.4.9a)–(3.4.9c) is used, and the determinant ℓ_{ν}^{+} assumed to be nonzero. This completes the calculation of the electric fields of scattered waves with combinational frequencies, since the modulation transmission matrices for various multilayer structures were found explicitly earlier (see chapter 1). The results generalize the theory of Raman scattering of light proposed in [10] on the basis of the idea of modulation by phonons of the dielectric permittivity tensor for a system of a particular type consisting of two semi-infinite media–vacuum and metal.

The Poynting vector for the resulting scattered wave has the form

$$\mathbf{S}_n(\mathbf{r}, t) = \langle \mathbf{E}_n(\mathbf{r}, t), \mathbf{H}_n(\mathbf{r}, t) \rangle, \tag{9.4.21a}$$

where according to the structure of the right part (9.3.12)

$$\mathbf{E}_n(\mathbf{r}, t) = \mathrm{Re}\int_{-\infty}^{\infty} d\omega\, e^{-i\omega t}\sum_{\nu}\left[\mathbf{E}_{n\nu}^{(\pm)}(z, \omega)e^{i\eta_{\nu}^{(+)}\rho} + \mathbf{E}_{n\nu}^{(-)}(z, \omega)e^{i\eta_{\nu}^{(-)}\rho}\right]; \tag{9.4.21b}$$

and angle brackets denote averaging over a statistical ensemble of vibrational excitations (compare with section 20 of [9]). Further, we take into account that according to the quantum mechanical derivation of the dielectric permittivity tensor of the transition (9.2.21), the components of the electric (and similarly magnetic) field strength found above are represented as

$$\mathbf{E}_{n\nu}^{(+)}(z, \omega) = \boldsymbol{\mathcal{E}}_{n\nu}^{(+)}(z, \omega)\langle N_{\nu} - 1\, |\hat{b}_{\nu}(\omega - \omega_L)| N_{\nu}\rangle; \tag{9.4.22a}$$

$$\mathbf{E}_{n\nu}^{(-)}(z,\omega) = \boldsymbol{\mathcal{E}}_{n\nu}^{(-)}(z,\omega)\left\langle N_\nu + 1 \,\middle|\, \hat{b}_\nu^+(\omega - \omega_L) \middle|\, N_\nu \right\rangle \tag{9.4.22b}$$

for anti-Stokes and Stokes processes, respectively.

Taking into account that the various modes of vibrational excitations are uncorrelated, from (9.4.21a) we find

$$\begin{aligned} \mathbf{S}_n(z,t) = \frac{1}{2}\mathrm{Re}\int_{-\infty}^{\infty} d\omega_1 \int_{-\infty}^{\infty} d\omega_2 e^{-i(\omega_2-\omega_1)t} \sum_v \{[\boldsymbol{\mathcal{E}}_{n\nu}^{(+)}(z,\omega_1), \boldsymbol{\mathcal{H}}_{n\nu}^{(+)*}(z,\omega_2)] \times \\ \left\langle \hat{b}_\nu^+(-\omega_2 + \omega_L)\hat{b}_\nu(\omega_1 - \omega_2) \right\rangle \\ + \left[\boldsymbol{\mathcal{E}}_{n\nu}^{(-)}(z,\omega_1), \boldsymbol{\mathcal{H}}_{n\nu}^{(-)*}(z,\omega_2)\right]\left\langle \hat{b}_\nu(-\omega_2 + \omega_L)\hat{b}_\nu^+(\omega_1 - \omega_L) \right\rangle \Big\}. \end{aligned} \tag{9.4.23}$$

For stationary processes, the correlation functions included in (9.4.23) contain a delta function:

$$\left\langle \hat{b}_\nu^+(-\omega_2)\hat{b}_\nu(\omega_1) \right\rangle = \delta(\omega_1 - \omega_2)L_\nu^{(+)}(\omega_1); \tag{9.4.24a}$$

$$\left\langle \hat{b}_\nu(-\omega_2)\hat{b}_\nu^+(\omega_1) \right\rangle = \delta(\omega_1 - \omega_2)L_\nu^{(-)}(\omega_2), \tag{9.4.24b}$$

where spectral densities are introduced:

$$L_\nu^{(+)}(\omega) = (2\pi)^{-1}\int_{-\infty}^{\infty} d\tau\, e^{i\omega t}\left\langle \hat{b}_\nu^+\hat{b}_\nu(\tau) \right\rangle; \tag{9.4.25a}$$

$$L_\nu^{(-)}(\omega) = (2\pi)^{-1}\int_{-\infty}^{\infty} d\tau\, e^{i\omega t}\left\langle \hat{b}_\nu\hat{b}_\nu^+(\tau) \right\rangle. \tag{9.4.25b}$$

This makes it possible to perform one of the frequency integrations in (9.4.23):

$$\begin{aligned} \mathbf{S}_n(z) = \frac{1}{2}\mathrm{Re}\int_{-\infty}^{\infty} d\omega \sum_\nu \Big\{\left[\boldsymbol{\mathcal{E}}_{n\nu}^{(\pm)}(z,\omega), \boldsymbol{\mathcal{H}}_{n\nu}^{(+)*}(z,\omega)\right] \\ \times L_\nu^{(+)}(\omega - \omega_L) + [\boldsymbol{\mathcal{E}}_{n\nu}^{(-)}(z,\omega), \boldsymbol{\mathcal{H}}_{n\nu}^{(-)*}(-z,\omega)]L_\nu^{(-)}(\omega - \omega_L)\}. \end{aligned} \tag{9.4.26}$$

For damped vibrational excitations described by a complex frequency $\Omega_\nu = \omega_\nu - i\Gamma_\nu$, spectral densities in (9.4.26) take the form of Lorentzians:

$$L_\nu^{(\pm)}(\omega - \omega_L) = \left(\overline{N}_\nu + \frac{1}{2} \mp \frac{1}{2}\right)\frac{1}{\pi}\frac{\Gamma_v}{(\omega - \omega_L \mp \omega_\nu)^2 + \Gamma_\nu^2}, \tag{9.4.27a}$$

which have maxima at $\omega = \omega_L + \omega_\nu$ for anti-Stokes and when $\omega = \omega_L - \omega_\nu$—for Stokes processes.

As can be seen from (9.4.26), contributions from processes involving different modes of vibrational excitations are additive, as are contributions from anti-Stokes and Stokes processes involving vibrational excitations of the same mode:

$$\mathbf{S}_n(z) = \sum_{\nu}\left[\mathbf{S}_{n\nu}^{(+)}(z) + \mathbf{S}_{n\nu}^{(-)}(z)\right], \tag{9.4.28a}$$

where the partial vectors have a spectral distribution density:

$$\mathbf{S}_{n\nu}^{(\pm)}(z) = \int_{-\infty}^{\infty} d\omega\, \mathbf{S}_{n\nu}^{(\pm)}(z, \omega) \tag{9.4.28b}$$

$$\mathbf{S}_{n\nu}^{(\pm)}(z, \omega) = \frac{1}{2}\mathrm{Re}\left[\boldsymbol{\mathcal{E}}_{n\nu}^{(\pm)}(z, \omega), \boldsymbol{\mathcal{H}}_{n\nu}^{(\pm)*}(z, \omega)\right]L_{\nu}^{(\pm)}(\omega - \omega_L). \tag{9.4.28c}$$

Hence, the spectral density of the intensity of an electromagnetic wave scattered in a certain direction, characterized by a given two-dimensional wave vector, follows directly $\boldsymbol{\eta}_{\nu}^{(\pm)} \equiv \boldsymbol{\eta}$:

$$I_{\nu}^{(\pm)}(\boldsymbol{\eta}, z, \omega) = \sum_{\bar{\nu}}\left|\, \frac{1}{2}\mathrm{Re}\,\overline{\left[\boldsymbol{\mathcal{E}}_{n\nu}^{(\pm)}(z, \omega), \boldsymbol{\mathcal{H}}_{n\nu}^{(\pm)*}(z, \omega)\right]}L_{\nu}^{(\pm)}(\omega - \omega_L)\right|. \tag{9.4.29}$$

Here $\bar{\nu} = (q_j^{\|}, j)$ is a set of multidimensional indices of oscillation branches j and a set of wave vector components along the stratification axis $q_j^{\|}$, joint with the assignment of the components of the wave vector across the stratification axis $\mathbf{q}_j^{\perp} \equiv \pm\boldsymbol{\eta}_{\nu}^{(\pm)} \mp \boldsymbol{\eta}_L$, and the horizontal line indicates averaging over the phases of the specified wave that occur during generation processes in different layers of the system under consideration. In the 'backward' scattering configuration, the spectral density of the intensity of the scattered wave in an external nonabsorbing medium ($n = I - 1$) at s-polarization, it is determined similarly to (9.4.9) by the amplitude of the electric field (9.4.20a):

$$I_b^{(\pm)(s)}(\boldsymbol{\eta}, \omega) = \frac{1}{2}\sqrt{\frac{\varepsilon_0\varepsilon_{I-1}(\omega)}{\mu_0\mu_{I-1}(\omega)}}\sum_{\bar{\nu}}\overline{\left|\,\boldsymbol{\mathcal{E}}_{b\nu}^{(\pm)\perp}(\omega)\right|^2}\, L_{\nu}^{(\pm)}(\omega - \omega_L), \tag{9.4.30}$$

and in the case of p-polarization, similarly (9.3.10a), (9.3.10b)—by the amplitude of the magnetic field:

$$H_{b\nu}^{(\pm)\perp}(\omega) = \omega\varepsilon_0\varepsilon_{I-1}(\omega)\left[k_{I-1,\,\nu}^{(\pm)\|}\right]^{-2}E_{b\nu}^{(\pm)\perp}(\omega); \tag{9.4.31a}$$

$$I_b^{(\pm)(p)}(\boldsymbol{\eta}, \omega) = \frac{1}{2}\sqrt{\frac{\mu_0\mu_{I-1}(\omega)}{\varepsilon_0\varepsilon_{I-1}(\omega)}}\sum_{\bar{\nu}}\overline{\left|\,\boldsymbol{\mathcal{H}}_{b\nu}^{(\pm)\perp}(\omega)\right|^2}\, L_{\nu}^{(\pm)}(\omega - \omega_L), \tag{9.4.31b}$$

where the notation of the form (9.4.22a), (9.4.22b) is used. Using (9.4.31a), it is also possible in this case to express the spectral intensity density through the amplitude of the electric field of the scattered wave:

$$I_b^{(\pm)(p)}(\boldsymbol{\eta}, \omega) = \frac{1}{2}\sqrt{\frac{\mu_0\mu_{I-1}(\omega)}{\varepsilon_0\varepsilon_{I-1}(\omega)}}\left|\frac{\omega\varepsilon_0\varepsilon_{I-1}(\omega)}{k_{I-1,\,\nu}^{(\pm)\|}}\right|^2\sum_{\bar{\nu}}\overline{\left|\,\mathcal{E}_{b\nu}^{(\pm)\perp}(\omega)\right|^2}\, L_{\nu}(\omega - \omega_L). \tag{9.4.31c}$$

In the case of normal propagation of a scattered wave, the expressions (9.4.30) and (9.4.31v) are equivalent.

Taking into account the dependencies (9.3.8a) and (9.3.8b) it is convenient to introduce a dimensionless function:

$$F^{(\pm)}(\pm(\eta - \eta_L), \omega) = \sum_{\overline{\nu}} |\overline{\tilde{\mathcal{E}}_{b\nu}^{(\pm)\perp}(\omega)|^2} L_\nu^{(\pm)}(\omega - \omega_L), \tag{9.4.32}$$

by means of which the spectral intensity densities (9.4.30) and (9.4.31c) are expressed as

$$I_b^{(\pm)(s)}(\eta, \omega) = \frac{1}{2}\sqrt{\frac{\varepsilon_0 \varepsilon_{I-1}(\omega)}{\mu_0 \mu_{I-1}(\omega)}} |E_{iL}|^2 F^{(\pm)}(\pm(\eta - \eta_L), \omega); \tag{9.4.33a}$$

$$I_b^{(\pm)(p)}(\eta, \omega) = \frac{1}{2}\sqrt{\frac{\mu_0 \mu_{I-1}(\omega)}{\varepsilon_0 \varepsilon_{I-1}(\omega)}} \left| \frac{\omega \varepsilon_0 \varepsilon_{I-1}(\omega)}{k_{I-1,\,\nu}^{(\pm)\|}} \right|^2 |E_{iL}|^2 F^{(\pm)}(\pm(\eta - \eta_L), \omega). \tag{9.4.33b}$$

If the external medium is a vacuum, and the scattered light is *s*-polarized or propagates normally to the surface, then it follows from (9.4.33a), (9.4.33b) that (9.4.32) is a form function of the spectral intensity density. The spectral density of the intensity of the wave scattered 'forward' is determined by completely similar formulas through the amplitude (9.4.20b). The analysis of the angular distribution of the intensity of the scattering of waves based on (9.4.28a) gives a differential effective cross-section or other observable characteristics of Raman scattering of light in multilayer structures.

9.5 Raman scattering of light in superlattices

Considering Raman scattering of light in a superlattice, we note that the solution (9.4.20a) for a semibounded periodic structure is simplified:

$$E_{b\nu}^{(\pm)\perp}(\omega) = \left[\mathcal{R}_{I-1,\,\nu}^{+}\right]^{-1} \Phi_{I-1,\,\nu}^{(\pm)\perp}(\omega), \tag{9.5.1a}$$

where in accordance with (9.4.16) and (9.4.5b)

$$\Phi_{I-1,\,\nu}^{(\pm)}(\omega) = \sum_{k=I}^{\infty}\left[D_{I-1,\,k-1,\,\nu}\Sigma_{k\nu}^{'(\pm)}(z_{k-1}) - D_{I-1,\,k\nu}\,\Sigma_{k\nu}^{'(\pm)}(z_k)\right] \tag{9.5.1b}$$

after index conversion $k \longrightarrow k - 1$ in the first amount.

Substitute in (9.5.1a), (9.5.1b) explicit expressions of the elements of the modulation transfer matrix, using the notation (4.10.4c) with (4.13.9a) and (4.13.9b) for α_ν ($|\alpha_\nu| > 1$) and (4.13.12) for $\hat{R}_{c\nu}^{-}$:

$$E_{b\nu}^{(\pm)\perp}(\omega) = \sum_{L=0}^{\infty}\sum_{v=a,b} E_{b\nu,\,vL}^{(\pm)\perp}(\omega), \tag{9.5.2}$$

where the terms

$$\left\| \begin{matrix} E^{(\pm)\perp}_{b\nu,\, aL}(\omega) \\ E_{b\nu,\, bL}(\omega) \end{matrix} \right\| = \alpha_\nu^{-L}(\widetilde{R}^{-}_{c\nu}) \times \left\| \begin{matrix} \Sigma'^{(\pm)}_{I+2L,\, \nu}(z_{I+2L-1},\, \omega) - \left(\mu_{a\nu} + \alpha_\nu^{-1}\mu_{b\nu}\right)\nu_\nu^{-1}\Sigma'^{(\pm)}_{I+2L,\, \nu}(z_{I+2L},\, \omega) \\ \left(\mu_{a\nu} + \alpha_\nu^{-1}\mu_{b\nu}\right)\nu_\nu^{-1}\Sigma'^{(\pm)}_{I+2L+1,\, \nu}(z_{I+2L},\, \omega) - \alpha_\nu^{-1}\Sigma'^{(\pm)}_{I+2L+1,\, \nu}(z_{I+2L+1},\, \omega) \end{matrix} \right\| \tag{9.5.3}$$

describe the generation of electromagnetic waves with a frequency of ω in layers of the corresponding type a, b from Lth period. Multiplier α_ν^{-L} takes into account the attenuation of the amplitudes of these waves as they propagate to the outer boundary of the superlattice.

If we assume that the explicit dependence of the Hamiltonian of the electron-vibrational interaction on the z-coordinate along the stratification axis enters through matrix elements into the tensors of the dielectric permittivity of the transition $\overleftrightarrow{\chi}^{(\pm)}_n(\nu,\, t;\omega')$, then the exponents in (9.2.21) should be put $q_j^{\|} = 0$. Further, we note that Raman scattering of light can be caused by the interaction of an electronic subsystem with various types of phonons excited in periodic structures (see classification in section 4.9). When Raman scattering of light is caused by the interaction of an electronic subsystem with 'captive' volumetric optical phonons, then the multi-dimensional index of the oscillation branch is $j = (BO,\, P,\, v,\, L,\, m)$, where $P = L,\, T$ is the type of the polarization, $v = a,\, b$ the type of the layer, L the number of the period, $m = 1,\ 2,\ 3,\ \ldots$ the the order of dimensional quantization, $q_j^{\|} = \pi m/l_v$, and the tensors of the dielectric permittivity of the transition are periodic:

$$\overleftrightarrow{\chi}^{(\pm)}_{I+2,\, L(+1)}(\nu',\, t;\omega') = \overleftrightarrow{\chi}^{\pm}_{a(b)}(\nu',\, t;\omega')\delta_{a(b),\, v'}\delta_{L,\, L'}. \tag{9.5.4}$$

When Raman scattering of light is caused by the interaction of an electronic subsystem with spatially extended optical phonons, then in the index $j = (SO,\, s,\, \kappa)$, where $s = I,\, \ldots,\ N_0$ is the oscillation branch number, κ is the the translational wave vector and, as follows from the form of the electron–phonon interaction Hamiltonian (3.7.19b), the dependence of the dielectric permittivity tensors of the transition on the layer number reduces to a phase multiplier:

$$\overleftrightarrow{\chi}^{(\pm)}_{I+2L+(1)}(\nu,\, t;\omega') = \mathrm{e}^{\pm i\kappa\, \mathcal{L}L}\overleftrightarrow{\chi}^{\pm}_{a(b)}(\nu,\, t;\omega'), \tag{9.5.5}$$

where $\mathcal{L}$ is the period of the superlattice.

When Raman scattering of light occurs due to the interaction of the electronic subsystem with surface spatially decreasing optical phonons, $j = (DO,\, s)$, where $s = 1,\ 2$ is the number of the oscillation branch, then by virtue of (4.10.13a) and (4.10.13b) the dependence of the dielectric permittivity tensors of the transition on the layer number has a power-law character ($|\alpha| > 1$):

$$\overleftrightarrow{\chi}^{(\pm)}_{I+2,\, L(+1)}(\nu,\, t;\omega') = \alpha^{-L}\overleftrightarrow{\chi}^{\pm}_{a(b)}(\nu,\, t;\omega'). \tag{9.5.6}$$

Similar transformational properties of transition dielectric permittivity tensors can also be used in the interaction of an electronic subsystem with acoustic phonons in periodic structures. Based on these properties and taking into account the power dependence (9.3.11a), (9.3.11b) of the amplitudes of the electric field of the exciting wave on the period number in (9.3.26b), (9.3.26c), (9.3.31), and (9.4.10), we find from (9.5.3):

$$\left\| \begin{matrix} E^{(\pm)\perp}_{b\nu,\, aL}(\omega) \\ E_{b\nu,\, bL}(\omega) \end{matrix} \right\| = \frac{\alpha_\nu^{-L}\alpha_L^{-L}}{\widetilde{R}^{-}_{c\nu}}\ell^L$$
$$\times \left\| \begin{matrix} \Sigma'^{(\pm)}_{a,\,\nu}(z_{I-1},\,\omega) - \left(\mu_{a\nu} + \alpha_\nu^{-1}\mu_{b\nu}\right)\nu_\nu^{-1}\Sigma'^{(\pm)}_{a,\,\nu}(z_I,\,\omega) \\ \left(\mu_{a\nu} + \alpha_\nu^{-1}\mu_{b\nu}\right)\nu_\nu^{-1}\Sigma'^{(\pm)}_{b,\,\nu}(z_I,\,\omega) - \alpha_\nu^{-1}\Sigma'^{(\pm)}_{b,\,\nu}(z_{I-1},\,\omega) \end{matrix} \right\|, \quad (9.5.7a)$$

where the functions $\Sigma^{I(\pm)}_{v\,\nu}(z,\,\omega)$ are determined by the formulas (9.4.6), (9.4.10), or (9.4.14) with the corresponding layer-type tensor of the dielectric permittivity of the transition $\overleftrightarrow{\chi}^{(\pm)}_{v}(\nu,\,\omega - \omega_L;\omega_L)$. The coefficient ℓ depending on the type of phonons has the following values:

$$\ell_{BO} = 1;\ \ell_{SO} = e^{\pm i\kappa\,\mathcal{L}};\ \ell_{DO} = \alpha^{-1}. \quad (9.5.7b)$$

Taking into account that the waves with combinational frequencies generated in different layers are mutually incoherent due to the lack of correlation of the phases of the wave functions of the electronic subsystem, we obtain the expression entering the spectral planes of intensity (9.4.30) and (9.4.31c):

$$\overline{\left|\mathcal{E}^{(\pm)\perp}_{b\nu}(\omega)\right|^2} = \sum_{L'=0}^{\infty}\sum_{v'=a,b}\left|\mathcal{E}^{(\pm)\perp}_{b\nu,\,v'L'}(\omega)\right|^2, \quad (9.5.8)$$

where the notation of the form (9.4.22a), (9.4.22b) is used.

Since the dependence of the amplitudes of the electric field of the scattered wave (9.5.7a) on the period number is fully revealed, it is possible to perform summation by this number in (9.5.8), which in the case of 'captive' volumetric phonons gives

$$\overline{\left|\mathcal{E}^{(\pm)\perp}_{b\nu}(\omega)\right|^2} = \left|\mathcal{E}^{(\pm)\perp}_{b\nu,\,vL}(\omega)\right|^2, \quad (9.5.9a)$$

and in the case of spatially extended or surface spatially decreasing:

$$\left|\mathcal{E}^{(\pm)\perp}_{b\nu}(\omega)\right|^2 = \left[1 - \left|\alpha_\nu^{-1}\alpha_L^{-1}\ell\right|^2\right]^{-1}\sum_{v=a,b}\left|\mathcal{E}^{(\pm)\perp}_{b\nu,\,v\,0}(\omega)\right|^2. \quad (9.5.9b)$$

Finally, in accordance with the definition (9.4.32), we complete the calculation of the form function of the spectral plane of intensity:

$$F^{(\pm)\perp}(\pm(\eta - \eta_L),\,\omega) = \sum_{j}\left[1 - \left|\alpha_\nu^{-1}\alpha_L^{-1}\ell\right|^2\right]^{-1}\sum_{v=a,b}\left|\tilde{\mathcal{E}}^{(\pm)\perp}_{b\nu,\,v\,0}(\omega)\right|^2 L^{(\pm)\perp}_{\nu}(\omega - \omega_L), \quad (9.5.10)$$

Table 9.1. Modulation transfer functions from the dielectric permittivity tensor of the transition to the spectral density of the Stokes component of scattered light.

Raman scattering geometry	Polarization	v	$\omega - \omega_L = 250\ \text{cm}^{-1}$	$\omega_L - \omega = 300\ \text{cm}^{-1}$
In the direction of reflection of the exciting radiation	p	a	$4.523 \cdot 10^{-4}$	$4.532 \cdot 10^{-4}$
		b	$4.609 \cdot 10^{-5}$	$4.603 \cdot 10^{-5}$
	s	a	$1.652 \cdot 10^{-3}$	$1.658 \cdot 10^{-3}$
		b	$1.441 \cdot 10^{-4}$	$1.440 \cdot 10^{-4}$
Normally to the surface	—	a	$1.140 \cdot 10^{-3}$	$1.142 \cdot 10^{-3}$
		b	$1.033 \cdot 10^{-4}$	$1.036 \cdot 10^{-3}$

In the calculation, the same values of the nonzero components of the dielectric permittivity tensor of the transition were taken.

where for 'captive' volumetric optical phonons$\bar{j} = (BO, P, m)$, and for other types of optical phonons $\bar{j} \equiv j$.

Table 9.1 shows the results of numerical calculation of the modulation transfer functions from the dielectric permittivity tensor of the transition to the spectral density

$$\frac{1}{1 - |\alpha_\nu \alpha_L|^{-2}} \frac{\left|\tilde{\varepsilon}^{(-)\perp}_{b\nu,\, v\, 0}(\omega)\right|^2}{\left|\chi^{(-)}_{ik}(\nu; \omega - \omega_L; \omega_L)\right|^2}$$

of Stokes components of scattered light for a superlattice GaAs/AlAs with the parameters given in figure 9.2, when illuminated by exciting s-polarized radiation with $\lambda = 514.5$ nm at Brewster's angle. It follows from the table 9.1, firstly, that the contribution of GaAs layers to Raman scattering of light is about an order of magnitude higher than that of AlAs layers, with all the specified geometries of this effect and, secondly, that the modulation transfer functions are very weakly dispersed in the range of $250\ \text{cm}^{-1} < \omega_L - \omega < 300\ \text{cm}^{-1}$. If we assume that the components of the dielectric permittivity tensor of the transition in the actual summation domain in (9.5.10) change negligibly little (cf [10]), then the form function of the spectral density of the scattered light intensity is proportional to

$$\sum_{\bar{j}} L_\nu^{(\pm)}(\omega - \omega_L).$$

The found structure of the form function (9.5.10) allows us to draw the following conclusions regarding the spectrum of Raman scattering of light in superlattices:

1. Due to the interaction of the electronic subsystem with 'captive' volumetric optical phonons, a series of peaks with maxima at $\omega_{vm} = \omega_v(\mathbf{q}^\perp, \pi m/l_v)$; $v = a, b$; $m = 1, 2, \ldots$, where $\omega_v\,(\mathbf{q})$ is the law of spatial dispersion of volumetric phonons in a layer of type$v$. Functional dependence of the electron–phonon interaction Hamiltonian on the coordinate $[z - (z_{n-1} + z_n)/2]$ in the n-th the layer is even if m is even, and odd if m

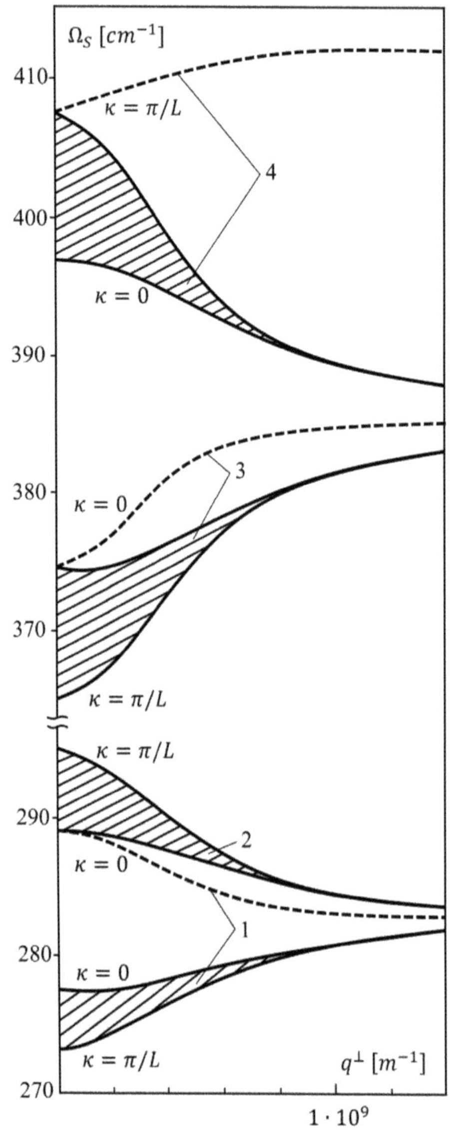

Figure 9.2. The laws of dispersion of the real part of the frequencies of spatially extended (solid lines) and surface spatially decreasing (dashed lines) phonons in a superlattice GaAs/AlAs. For GaAs (a) $l_a = 5.44$ nm; $\bar{\varepsilon}_a = 10.9$; $\omega_{La} = 295$ cm^{-1}; $\omega_{Ta} = 273$ cm^{-1}; $\gamma_a = 0.01$ [15]; AlAs (b) $l_b = 1.67$ nm; $\bar{\varepsilon}_b = 8.5$; $\omega_{Lb} = 407$ cm^{-1}; $\omega_{Tb} = 364$ cm^{-1}; $\gamma_b = 0.01$ [15].

is odd. Consequently, the vibrational excitations at even and odd m are related respectively to A_1- and B_2-types. In accordance with the rules of selection of the components of the dielectric permittivity tensor of the transition for the superlattice GaAs/AlAs, formulated in section 9.3, peaks with even m appear in polarized, and peaks with odd m in depolarized scattering [5]. The width of the peaks is determined by the spectral density of the correlation functions (9.4.25a), (9.4.25b).

2. Due to the interaction of the electronic subsystem with spatially extended optical phonons, a contribution to the form function arises (9.5.10):

$$F_{SO}^{(\pm)\perp}(\pm(\eta - \eta_L), \omega) = = \sum_{s=1}^{4} \frac{\mathcal{L}}{\pi} \int_0^{\pi/\mathcal{L}} d\kappa \, [1 - |\alpha_\nu \alpha_L|^{-2}]^{-1} \sum_{\nu=a,b} \left| \tilde{\mathcal{E}}_{b\nu,\, v\, 0}^{(\pm)\perp}(\omega) \right|^2 L_\nu^{(\pm)\perp}(\omega - \omega_L), \tag{9.5.11}$$

where the transition from the translational wave vector summed by possible values to integration is carried out. In figure 9.1, the laws of dispersion of spatially extended phonons in a superlattice are represented by solid lines GaAs/AlAs, calculated numerically by the dispersion equation (4.10.7) for limit values $\kappa = 0$ and $\kappa = \pi/\mathcal{L}$ using complex dielectric functions

$$\varepsilon_v(\omega) = \bar{\varepsilon}_v \frac{\omega_{Lv}^2 - \omega^2 - i\omega\, \omega_{Tv}\gamma_v}{\omega_{Tv}^2 - \omega^2 - i\omega\, \omega_{Tv}\gamma_v}, \tag{9.5.12}$$

in which attenuation is taken into account. In this regard, the roots of the specified dispersion equation are complex numbers:

$$\Omega_s(\kappa, q^\perp) = \omega_s(\kappa, q^\perp) - i\Gamma_s(\varkappa, q^\perp). \tag{9.5.13}$$

The four phonon zones are numbered in ascending order of the real part of the frequency (9.5.13). The second branch of the surface spatially decreasing phonons is suppressed when $q^\perp = 4, 24 \cdot 10^7\ \mathrm{m}^{-1}$ due to violation of the decreasing condition $|\alpha| > 1$. Very weakly dispersing imaginary part of frequencies (в cm^{-1}) with $q^\perp = 0$ compose:

– for spatially extended phonons

$$\Gamma_1 = 1.371;\ \Gamma_2 = 1.374;\ \Gamma_3 = 1.811;\ \Gamma_4 = 1.814;$$

– for surface spatially decreasing phonons

$$\Gamma_1 = 1.374;\ \Gamma_2 = 1.366;\ \Gamma_3 = 1.811;\ \Gamma_4 = 1.820.$$

It can be seen from the graphs that taking into account the attenuation removes the degeneration of the lower (GaAs-type) and upper (AlAs-type) pairs of branches that existed without such accounting (see figure 9.2). Note also that the limiting splitting of branches in these pairs as well as the imaginary part of the frequency (9.5.13) in the region $\gamma_v \lesssim 0, 1$ are directly proportional to the attenuation factors γ_v from (9.5.12).

Next, in (9.5.11), we move from integration over the translational wave vector to integration over the corresponding frequency interval:

$$F_{SO}^{(\pm)\perp}\left(\pm(\eta-\eta_L),\omega\right)=\sum_{s=1}^{4}\int_{\omega_{s\,\min}(q^\perp)}^{\omega_{s\,\max}(q^\perp)}d\omega_s\left\{D_s\left[1-\left|\alpha_v\,\alpha_L\right|^2\right]^{-1}\right.$$
$$\left.\times\sum_{v=a,b}\left|\tilde{\mathcal{E}}_{bv,v\,0}^{(\pm)\perp}(\omega)\right|^2 L_v^{(\pm)}(\omega-\omega_L)\right\}\Bigg|_{\kappa=\kappa(\omega_S)},\tag{9.5.14a}$$

where we use explicit dependency$\kappa=\kappa(\omega_S)$ for fixed $q^\perp$. The limits of integration are the essence

$$\omega_{1,\,3\,\min}(q^\perp)\equiv\omega_{1,\,3}(\pi/\mathcal{L},\,q^\perp);\ \omega_{1,\,3\,\max}(q^\perp)\equiv\omega_{1,\,3}(0,\,q^\perp);\tag{9.5.14b}$$

$$\omega_{2,\,4\,\min}(q^\perp)\equiv\omega_{2,\,4}(0,\,q^\perp);\ \omega_{2,\,4\,\max}(q^\perp)\equiv\omega_{2,\,4}(\pi/\mathcal{L},\,q^\perp).$$

and the density of states of spatially extended phonons is introduced

$$D_s(\varkappa)=\frac{\mathcal{L}}{\pi}\left|\frac{d\kappa}{d\omega_s(\kappa,\,q^\perp)}\right|\tag{9.5.14c}$$

for $\omega_{s\,\min}(q^\perp)\leqslant\omega_s(\kappa,\,q^\perp)\leqslant\omega_{s\,\max}(q^\perp)$.

In the conditions of approximate constancy of the spectral density (9.5.9b) discussed above, taking this function out from under the sign of the integral in (9.5.14a), we obtain that the form function

$$F_{SO}^{(\pm)\perp}(\pm(\eta-\eta_L),\,\omega)\approx\sum_{s=1}^{4}[1-|\alpha_v\,\alpha_L|^{-2}]^{-1}\sum_{v=a,b}\left|\tilde{\mathcal{E}}_{bv,\,v\,0}^{(\pm)\perp}(\omega)\right|^2\mathcal{F}_s^{(\pm)}(\mathbf{q}_\perp,\,\omega)\tag{9.5.15a}$$

is determined both by the spectral density of correlation functions (9.4.25a) and (9.4.25b), and by the density of states of spatially extended phonons:

$$\mathcal{F}_s^{(\pm)}(q^\perp,\,\omega)=\int_{-\infty}^{\infty}d\omega_s\left\{\overline{D}_s(\kappa)L_v^{(\pm)}(\omega-\omega_L)\right\}\Big|_{\kappa=\kappa(\omega_s)},\tag{9.5.15b}$$

where

$$\overline{D}_s(\kappa)|_{\kappa=\kappa(\omega_s)}=\begin{cases}D_s(\kappa)|_{\kappa=\kappa(\omega_s)},\ \omega_s\in(\omega_{s\,\min}(q^\perp),\,\omega_{s\,\max}(q^\perp))\\ 0,\ \omega_s\notin(\omega_{s\,\min}(q^\perp),\,\omega_{s\,\max}(q^\perp))\end{cases}\tag{9.5.15c}$$

For initial sections $q^\perp\to 0$ dispersion curves in disregard of attenuation from equation (3.2.7) at $\kappa\neq 0$ we find

$$\omega_1(\kappa,\,q^\perp)=\omega_{Ta}+\frac{A_1(q^\perp)}{\sin^2\frac{\kappa\mathcal{L}}{2}};\tag{9.5.16a}$$

$$A_1(q^\perp)=\varepsilon_a[8\varepsilon_b(\omega_{Ta})\omega_{Ta}]^{-1}\left(\omega_{La}^2-\omega_{Ta}^2\right)q^{\perp 2}l_a l_b;\tag{9.5.16b}$$

$$\omega_2(\kappa, q^\perp) = \omega_{La} - \frac{A_2(q^\perp)}{\sin^2 \frac{\kappa\mathcal{L}}{2}}; \tag{9.5.16c}$$

$$A_2(q^\perp) = \varepsilon_b(\omega_{La})[8\varepsilon_a\omega_{La}]^{-1}\left(\omega_{La}^2 - \omega_{Ta}^2\right)q^{\perp 2}l_a l_b; \tag{9.5.16d}$$

the dispersion curves have a similar appearance for the upper ones (AlAs-type branches) $s = 3,\ 4$. Hence, for the density of states (9.5.14b), the expression follows:

$$D_s(\kappa) = [\pi A_s(q^\perp)]^{-1} \cos^{-1}\frac{\kappa\mathcal{L}}{2} \sin^3\frac{\kappa\mathcal{L}}{2}. \tag{9.5.17}$$

Explicit dependencies directly follow from formulas (9.5.16a) and (9.4.26c):

$$\kappa(\omega_1) = \frac{2}{\mathcal{L}}\arcsin\frac{A_1(q^\perp)}{\omega_1 - \omega_{Ta}}; \tag{9.5.18a}$$

$$\kappa(\omega_2) = \frac{2}{\mathcal{L}}\arcsin\frac{A_2(q^\perp)}{\omega_{La} - \omega_2}, \tag{9.5.18b}$$

which allow us to find the densities of states (9.5.17) in the final form:

$$D_1(\kappa)|_{\kappa=\kappa(\omega_1)} = \frac{\sqrt{A_1(q^\perp)}}{\pi(\omega_1 - \omega_{Ta})\sqrt{\omega_1 - \omega_{1\,\min}(q^\perp)}}; \tag{9.5.19a}$$

$$D_2(\varkappa)|_{\varkappa=\varkappa(\omega_2)} = \frac{\sqrt{A_2(q^\perp)}}{\pi(\omega_{La} - \omega_2)\sqrt{\omega_{2\,\max}(q^\perp) - \omega_2}}. \tag{9.5.19b}$$

Integrating the expressions found, we find the fraction of the number of vibrational modes in the interval from $\omega_s(\pi/\mathcal{L}, q^\perp)$ to ω_s:

$$\begin{aligned}\xi(\omega_s) &\equiv \left|\int_{\omega_s\left(\frac{\pi}{\mathcal{L}},\ q^\perp\right)}^{\omega_s} D_s(\kappa(\omega_s))d\omega_s\right| \left\{\int_{\omega_{s\,\min}(q^\perp)}^{\omega_{s\,\max}(q^\perp)} D_s(\kappa(\omega_s))d\omega_s\right\}^{-1} \\ &= \arctan\left[A_s^{-1}(q_\perp)\left|\ \omega_s - \omega_s\left(\frac{\pi}{\mathcal{L}},\ q^\perp\right)\right|\right]^{-\frac{1}{2}} \\ &\times \arctan\left[A_s^{-1}(q_\perp)|\omega_{2\,\max}(q^\perp) - \omega_{s\,\min}(q^\perp)|\right]^{-1/2},\end{aligned} \tag{9.5.20a}$$

and then—the length of the frequency interval $\Delta\omega_s(\xi)$, for which this fraction is a given number $\xi < 1$:

$$\begin{aligned}\Delta\omega_s(\xi) &\equiv |\omega_s(\xi) - \omega_s(\pi/\mathcal{L}, q^\perp)| = \\ &= A_s(q^\perp)\,\mathrm{tg}^2\left\{\xi \arctan\left[A_s^{-1}(q^\perp)|\omega_{s\,\max}(q^\perp) - \omega_{s\,\min}(q^\perp)|\right]^{-1/2}\right\}.\end{aligned} \tag{9.5.20b}$$

With $q^{\perp} \to 0$ the right part turns into $A_s(q^{\perp})\,\mathrm{tg}^2(\pi\xi/2)$, so half of the total number of vibrational modes falls on the interval $\Delta\omega_s(1/2) = A_s(q^{\perp})$.

In the limit $\Gamma_\nu \ll A_s(q^{\perp})$ the spectral density of the correlation functions (9.4.25a) and (9.4.25b) is reduced to δ-function. Consequently, the shape of the peaks of Raman scattering of light, as follows from (9.5.15), basically repeats the shape of the density of states at the corresponding frequency:

$$\mathcal{F}_s^{(\pm)}(\mathbf{q}^{\perp}, \omega) \approx \overline{D}_s(\kappa)|_{\kappa=\kappa(\pm(\omega-\omega_L))}. \tag{9.5.21}$$

In the opposite limit $\Gamma_v \gg A_s(q^{\perp})$, when the spectral density of correlation functions (9.4.25a), (9.4.25b) changes little in the interval $(\omega_{s\,\min}(q^{\perp}),\ \omega_{s\,\max}(q^{\perp}))$, it determines mainly the shape of the peaks of Raman scattering of light:

$$\mathcal{F}_s^{(\pm)}(\mathbf{q}^{\perp}, \omega) \approx L_\nu^{(\pm)}(\omega - \omega_L)|_{\kappa=\kappa\left(\omega_s\left(\frac{\pi}{L}, q^{\perp}\right)\right)} \int_{\omega_{s\,\min}(q^{\perp})}^{\omega_{s\,\max}(q^{\perp})} d\omega_s D_s(\kappa)|_{\kappa=\kappa(\omega_s)}. \tag{9.5.22}$$

In the general case of an arbitrary relation between Γ_v and $A_s(q^{\perp})$ the shape of the Raman scattering peaks depends on both factors included in (9.5.15). Note also that, since when $s = 1,\ 4$ and $s = 2,\ 3$ the functional dependence of the electron–phonon interaction Hamiltonian on the z-coordinate, measured from the middle of the layer, is predominantly even and odd (for any value of k), insofar as, in accordance with the rules for selecting the components of the dielectric permittivity tensor of the transition specified in section 9.3, peaks with $s = 1,\ 4$ are manifested in the polarized, and peaks with $s = 2,\ 3$ in depolarized scattering. For lower branches (GaAs-type) such peaks of Raman scattering of light were observed in [1, 2], and the theoretical results described above are in satisfactory agreement with experimental data.

3. The contribution to the form function as a result of the interaction of the electronic subsystem with surface attenuated optical phonons is determined in accordance with the formula (9.5.10) by the spectral density of correlation functions. In figure 9.2, the laws of dispersion of surface-damped phonons in a superlattice are represented by dashed lines GaAs/AlAs, calculated numerically by the dispersion equation (4.10.18) using complex dielectric functions (9.5.12). The imaginary part of the frequencies of these phonons takes values close to those available in the case of spatially extended phonons. Since the dispersion curves of the real part of the frequencies $\omega_s(q^{\perp})$, and with them, the positions of the peaks of Raman scattering of light pass differently than in the previous case, then according to the spectra of Raman scattering of light obtained in a fairly wide range of values $|\eta - \eta_L|$, it is possible to identify the type of superlattice phonons responsible for these peaks, spatially extended or surface spatially decreasing. The manifestation of the latter, as follows from (9.5.7b) and (9.5.10), significantly weakens as the decrement of decrease increases α.

9.6 Conclusion

A wide experimental study of Raman scattering spectra, including their dependence on the geometric characteristics of scattering and the type of polarization of scattered radiation, makes it possible to diagnose vibrational disturbances and evaluate the parameters of interaction with them of the electronic subsystem in multilayer structures and in superlattices.

9.7 Summary

The monograph generalizes the results of the authors' original research in the field of the electron-vibrational processes and states of the electron–hole–phonon systems in multilayer planar, cylindrical, and spherical structures and planar superlattices. Much attention has been paid to the description of the polarization vibrational states of the lattice and the electron plasma and the consideration of the electron-vibrational systems in quantum wells of various geometries. Optical and transport effects include scattering of free charge carriers by polar bulk and surface optical phonons, infrared absorption of light by free electrons assisted by optical phonons in planar multilayer structures with quantum wells, quantum theory of electron emission from the metal–dielectric structures in strong electric fields, as well as the cyclotron–phonon resonance in structures with quantum wells due to the renormalization of the phonon spectrum of multilayer systems.

The normal polarization states calculated without and with the retardation (polaritons and phonons, respectively) are classified in the model of the spatial distribution of electric and magnetic field strengths, as well as the field changes throughout the structure (surface spatially decreasing and surface spatially extended states). The theory of the potential due to the surface and bulk charges and all types of polarization (surface and bulk, inertial and inertialess) in multilayer structures of various geometries and superlattices was developed. Derivation of the Pekar–Fröhlich Hamiltonian was performed for an arbitrary planar multilayer structure with quantum wells, for multilayer structures of cylindrical and spherical geometries, and for a planar superlattice containing layers with polar optical phonons.

Based on the developed approximation, the potential energy of the electron self-action in planar, cylindrical, and spherical multilayer structures and planar superlattices has been derived, from which the potential energy of the image forces follows in the quasi-classical limit. The contributions from the self-action effect to the potential relief of the band edges for various multilayer structures were analytically calculated. The polaronic states at the contact of polar and nonpolar crystals, in a polar film, at the contact of two polar crystals, in a superlattice, and a quantum wire in a dielectric matrix have been studied. The contributions to the renormalization of the energy and the effective mass of the electron due to the polaron effect and the self-action were calculated and analyzed.

Exciton, biexciton, and impurity states in homeopolar multilayer structures with quantum wells were investigated. The change in the Coulomb electron–hole interaction and the effects of self-action have been taken into account. The theory of exciton states in thin polar films and quantum wires in a polar dielectric matrix was developed. Exciton size effects were divided into two types: band effects (the edges of the conduction band and valence band are shifted due to the electron and hole size quantization, the polaron effect, and the self-action for electrons and holes) and intraexciton effects (conversion of the electron–hole interaction, shielding by bulk and surface optical vibrations). It is concluded that none of the contributions from those effects can be neglected. The general theory was used to calculate the experimentally measurable energy of the photon generating the exciton, as well as the binding energy of the exciton. The calculations were carried out using the parameters of polar crystals PbI_2 and CdTe (in the latter case, the degeneracy of the hole band and the nonparabolicity of the electron band are taken into account); the results were in good agreement with the available experimental data. The nature of the bipolaronic states with high binding energy was investigated and established for multilayer systems with alternating quantum layers of polar and nonpolar semiconductors (semimetals) and in quantum wires in polar dielectric media (the Little superconductivity mechanism). The mechanism of the experimentally observed high-temperature superconductivity (HTSC) in FeSe thin films on $SrTiO_3$ substrates was theoretically investigated. Using the theory of large-radius bipolaronic states, developed on the basis of the exact Hamiltonian electron–phonon interaction for arbitrary multilayer structures, the bipolaronic mechanism of Cooper pairing of polarons in monolayers of FeSe on $SrTiO_3$ substrates and in three-layer structures of $SrTiO_3$–FeSe–$SrTiO_3$ (typical 'Ginzburg sandwiches') were studied. The approach proposed by Ginzburg to enhance the electron–phonon interaction and to achieve HTS by separating the regions of electron arrangement (forming Cooper pairs or bipolarons) from the exciton excitation regions (inducing the inertial polarization) makes it possible to implement criteria for the formation of bipolaronic states in multilayer structures with high binding energy, due to the possibility of choosing optimal geometric and material parameters (layer thicknesses, dielectric permittivity, optical frequencies, effective masses of charge carriers). The binding energy of bipolarons in these structures was in the range of values at which bipolarons remain stable and can exist at temperatures significantly higher than their Bose–Einstein-Condensation temperature. A generalization of the theory of thermo-autoelectronic emission in MDS (Metal–Dielectric–Semiconductor) structures was carried out in the case of the manifestation of the quantum nature of image forces in high and ultrahigh electric fields, in which the electronic polaron effect was shown to be important. The theory of Raman scattering in planar multilayer structures and semi-infinite superlattices was presented. The scattering involving surface spatially decreasing and surface spatially extended vibrational excitations specific to planar superlattices was discussed. On this basis, experimental data on Raman scattering spectra were interpreted.

The theoretical apparatus represented in this monograph can be used to study various electronic and optoelectronic processes in advanced planar and curvilinear multilayer structures with quantum wells and in superlattices in the presence of external fields.

References

[1] Sood A K, Menendez J, Cardona M and Ploog A K 1985 Interface vibrational modes in GaAs–AlGaAs superlattices *Phys. Rev. Lett.* **54** 2115–8

[2] Bayramov B H 1987 Inelastic light scattering in superlattices with GaAs/AlAs and GaAs/ $GaAl_{1-x}P_x$ quantum-dimensional heterostructures Preprint No. 1192. L *LFTI* 32

[3] Ratnikov P V and Silin A P 2018 Two-dimensional graphene electronics: current state and prospects *UFN* **188** 1249–87

[4] Pokatilov E P, Nika D L and Balandin A A 2004 Confined electron-confined phonon scattering rates in wurtzite AlN/GaN/AlN heterostructures *J. Appl. Phys.* **95** 5626–32

[5] Sood A K, Menendez J, Cardona M and Ploog K 1985 Resonance Raman scattering by confined LO and TO phonons in GaAs – AlAs superlattices *Phys. Rev. Lett.* **54** 2111–4

[6] Sood A K, Menendez J, Cardona M and Ploog K 1985 Second-order Raman scattering by confined optical phonons and interface vibrational modes in GaAs–AlAs superlattices *Phys. Rev.* B **32** 1412–4

[7] Schwartz G P, Gualtieri G J, Sunder W A and Farrow L A 1987 Light scattering from quantum confine and interface optical vibrational modes in strained-layer GaSb/AlSb superlattices *Phys. Rev.* B **36** 4868–77

[8] Landau L D and Lifshits E M 1976 *Statistical Physics. Part 1* (Moscow: Nauka) 584

[9] Born M I and Kun H 1958 *Dynamic Theory of Crystal Lattices* (Moscow: IL) 488

[10] Mills D L, Maradudin A A and Burstein E 1970 Theory of the Raman effect in metals *Ann. Phys.* **56** 504–55

[11] 1979 *Light Scattering in Solids* ed M Cardona (Moscow: Mir) p 392

[12] Poulet A and Mathieu J-P 1974 *Vibrational Spectra and Symmetry of Crystals* (Moscow: Mir) p 440

[13] Sobolev V V 1979 *Optical Fundamental Spectra of Compounds of Group AIII VV* (Chisinau: Stiinza) p 288

[14] Pern R E and Onton A 1971 Refractive index of AIAs *J. Appl. Phys.* **42** 3499–500

[15] Willardson R and Bira A (ed) 1970 *Optical Properties of Semiconductors. Semiconductor Compounds of Type AIIIBV* (Moscow: Mir) p 488

www.ingramcontent.com/pod-product-compliance
Ingram Content Group UK Ltd.
Pitfield, Milton Keynes, MK11 3LW, UK
UKHW051910100926
8556IPUK00005B/826